GEOTECHNICAL SPECIAL PUBLICATION NO. 180

FROM RESEARCH TO PRACTICE IN GEOTECHNICAL ENGINEERING

SPONSORED BY
The Geo-Institute of the American Society of Civil Engineers

EDITED BY
James E. Laier
David K. Crapps
Mohamad H. Hussein

Published by the American Society of Civil Engineers

Library of Congress Cataloging-in-Publication Data

From research to practice in geotechnical engineering / sponsored by the Geo-Institute of the American Society of Civil Engineers ; edited by James E. Laier, David K. Crapps, Mohamad H. Hussein.

p. cm.

Includes bibliographical references and index.

ISBN-13: 978-0-7844-0962-6

ISBN-10: 0-7844-0962-5

1. Soil mechanics. 2. Rock mechanics. I. Schmertmann, John H. II. Laier, James E. III. Crapps, David K. IV. Hussein, Mohamad H. V. American Society of Civil Engineers. Geo-Insitute

TA710.F73 2008

624.1'51--dc22 2008002410

American Society of Civil Engineers
1801 Alexander Bell Drive
Reston, Virginia, 20191-4400

www.pubs.asce.org

ISBN 0-7844-0962-5
ISBN 13 978-0-7844-0962-6
Manufactured in the United States of America.

Geotechnical Special Publications

1 *Terzaghi Lectures*
2 *Geotechnical Aspects of Stiff and Hard Clays*
3 *Landslide Dams: Processes, Risk, and Mitigation*
7 *Timber Bulkheads*
9 *Foundations & Excavations in Decomposed Rock of the Piedmont Province*
11 *Dynamic Response of Pile Foundations–Experiment, Analysis and Observation*
14 *Geotechnical Aspects of Karst Terrains*
15 *Measured Performance Shallow Foundations*
16 *Special Topics in Foundations*
17 *Soil Properties Evaluation from Centrifugal Models*
18 *Geosynthetics for Soil Improvement*
19 *Mine Induced Subsidence: Effects on Engineered Structures*
21 *Hydraulic Fill Structures*
22 *Foundation Engineering*
23 *Predicted and Observed Axial Behavior of Piles*
24 *Resilient Moduli of Soils: Laboratory Conditions*
25 *Design and Performance of Earth Retaining Structures*
27 *Geotechnical Engineering Congress*
28 *Detection of and Construction at the Soil/Rock Interface*
29 *Recent Advances in Instrumentation, Data Acquisition and Testing in Soil Dynamics*
32 *Embankment of Dams–James L. Sherard Contributions*
33 *Excavation and Support for the Urban Infrastructure*
34 *Piles Under Dynamic Loads*
35 *Geotechnical Practice in Dam Rehabilitation*
37 *Advances in Site Characterization: Data Acquisition, Data Management and Data Interpretation*
39 *Unsaturated Soils*
40 *Vertical and Horizontal Deformations of Foundations and Embankments*
41 *Predicted and Measured Behavior of Five Spread Footings on Sand*
42 *Serviceability of Earth Retaining Structures*
43 *Fracture Mechanics Applied to Geotechnical Engineering*
44 *Ground Failures Under Seismic Conditions*
45 *In Situ Deep Soil Improvement*
46 *Geoenvironment 2000*
47 *Geo-Environmental Issues Facing the Americas*
48 *Soil Suction Applications in Geotechnical Engineering*
49 *Soil Improvement for Earthquake Hazard Mitigation*
50 *Foundation Upgrading and Repair for Infrastructure Improvement*
51 *Performance of Deep Foundations Under Seismic Loading*
52 *Landslides Under Static and Dynamic Conditions–Analysis, Monitoring, and Mitigation*
53 *Landfill Closures–Environmental Protection and Land Recovery*
54 *Earthquake Design and Performance of Solid Waste Landfills*
55 *Earthquake-Induced Movements and Seismic Remediation of Existing Foundations and Abutments*
56 *Static and Dynamic Properties of Gravelly Soils*
57 *Verification of Geotechnical Grouting*
58 *Uncertainty in the Geologic Environment*
59 *Engineered Contaminated Soils and Interaction of Soil Geomembranes*
60 *Analysis and Design of Retaining Structures Against Earthquakes*
61 *Measuring and Modeling Time Dependent Soil Behavior*
62 *Case Histories of Geophysics Applied to Civil Engineering and Public Policy*
63 *Design with Residual Materials: Geotechnical and Construction Considerations*
64 *Observation and Modeling in Numerical Analysis and Model Tests in Dynamic Soil-Structure Interaction Problems*
65 *Dredging and Management of Dredged Material*
66 *Grouting: Compaction, Remediation and Testing*
67 *Spatial Analysis in Soil Dynamics and Earthquake Engineering*
68 *Unsaturated Soil Engineering Practice*
69 *Ground Improvement, Ground Reinforcement, Ground Treatment: Developments 1987-1997*
70 *Seismic Analysis and Design for Soil-Pile-Structure Interactions*
71 *In Situ Remediation of the Geoenvironment*
72 *Degradation of Natural Building Stone*
73 *Innovative Design and Construction for Foundations and Substructures Subject to Freezing and Frost*

74 *Guidelines of Engineering Practice for Braced and Tied-Back Excavations*
75 *Geotechnical Earthquake Engineering and Soil Dynamics III*
76 *Geosynthetics in Foundation Reinforcement and Erosion Control Systems*
77 *Stability of Natural Slopes in the Coastal Plain*
78 *Filtration and Drainage in Geotechnical/Geoenvironmental Engineering*
79 *Recycled Materials in Geotechnical Applications*
80 *Grouts and Grouting: A Potpourri of Projects*
81 *Soil Improvement for Big Digs*
82 *Risk-Based Corrective Action and Brownfields Restorations*
83 *Design and Construction of Earth Retaining Systems*
84 *Effects of Construction on Structures*
85 *Application of Geotechnical Principles in Pavement Engineering*
86 *Big Digs Around the World*
87 *Jacked Tunnel Design and Construction*
88 *Analysis, Design, Construction, and Testing of Deep Foundations*
89 *Recent Advances in the Characterization of Transportation Geo-Materials*
90 *Geo-Engineering for Underground Facilities*
91 *Special Geotechnical Testing: Central Artery/Tunnel Project in Boston, Massachusetts*
94 *Performance Confirmation of Constructed Geotechnical Facilities*
95 *Soil-Cement and Other Construction Practices in Geotechnical Engineering*
96 *Numerical Methods in Geotechnical Engineering: Recent Developments*
97 *Innovations and Applications in Geotechnical Site Characterization*
98 *Pavement Subgrade, Unbound Materials, and Nondestructive Testing*
99 *Advances in Unsaturated Geotechnics*
100 *New Technological and Design Developments in Deep Foundations*
101 *Slope Stability 2000*
102 *Trends in Rock Mechanics*
103 *Advances in Transportation and Geoenvironmental Systems Using Geosynthetics*
104 *Advances in Grouting and Ground Modification*
105 *Environmental Geotechnics*
106 *Geotechnical Measurements: Lab & Field*
107 *Soil Dynamics and Liquefaction 2000*
108 *Use of Geophysical Methods in Construction*
109 *Educational Issues in Geotechnical Engineering*
110 *Computer Simulation of Earthquake Effects*
111 *Judgment and Innovation: The Heritage and Future of the Geotechnical Engineering Profession*
112 *Soft Ground Technology*
114 *Soils Magic*
115 *Expansive Clay Soils and Vegetative Influence on Shallow Foundations*
116 *Deep Foundations 2002: An International Perspective on Theory, Design, Construction, and Performance*
117 *Discrete Element Methods: Numerical Modeling of Discontinua*
118 *A History of Progress: Selected U.S. Papers in Geotechnical Engineering*
119 *Soil Behavior and Soft Ground Construction*
120 *Grouting and Ground Treatment*
121 *Probabilistic Site Characterization at the National Geotechnical Experimentation Sites*
122 *Sinkholes and the Engineering and Environmental Impacts of Karst*
123 *Recent Advances in Materials Characterization and Modeling of Pavement Systems*
124 *GeoSupport 2004: Drilled Shafts, Micropiling, Deep Mixing, Remedial and Specialty Foundation Systems*
125 *Current Practices and Future Trends in Deep Foundations*
126 *Geotechnical Engineering for Transportation Projects*
127 *Recycled Materials in Geotechnics*
128 *Soil Constitutive Models: Evaluation, Selection, and Calibration*
129 *Advances in Designing and Testing Deep Foundations*
130 *Advances in Pavement Engineering*
131 *Contemporary Issues in Foundation Engineering*
132 *Advances in Deep Foundations: In Memory of Michael W. O'Neill*
133 *Earthquake Engineering and Soil Dynamics*
134 *Soil Dynamics Symposium in Honor of Professor Richard D. Woods*
135 *Erosion of Soils and Scour of Foundations*
136 *Innovations in Grouting and Soil Improvement*

137 *Legal and Liability Issues in Geotechnical Engineering*
138 *Site Characterization and Modeling*
139 *Calibration of Constitutive Models*
140 *Slopes and Retaining Structures under Seismic and Static Conditions*
141 *International Perspectives on Soil Reinforcement Applications*
142 *Waste Containment and Remediation*
143 *Geomechanics: Testing, Modeling, and Simulation*
144 *Sinkholes and the Engineering and Environmental Impacts of Karst*
145 *Seismic Performance and Simulation of Pile Foundations in Liquefied and Laterally Spreading Ground*
146 *Asphalt Concrete: Simulation, Modeling and Experimental Characterization*
147 *Unsaturated Soils 2006*
148 *Advances in Unsaturated Soil, Seepage, and Environmental Geotechnics*
149 *Site and Geomaterial Characterization*
150 *Soil and Rock Behavior and Modeling*
151 *Advances in Earth Structures: Research to Practice*
152 *Ground Modification and Seismic Mitigation*
153 *Foundation Analysis and Design: Innovative Methods*
154 *Pavement Mechanics and Performance*
155 *Underground Construction and Ground Movement*
156 *Geomechanics II: Testing, Modeling, and Simulation*
157 *Computer Applications in Geotechnical Engineering*
158 *Contemporary Issues in Deep Foundations*
159 *Case Studies in Earth Retaining Structures*
160 *Dynamic Response and Soil Properties*
161 *Embankments, Dams, and Slopes: Lessons from the New Orleans Levee Failures and Other Issues*
162 *Problematic Soils and Rocks and In Situ Characterization*
163 *Geoenvironmental Engineering*
164 *Innovative Applications of Geophysics in Civil Engineering*
165 *Geosynthetics in Reinforcement and Hydraulic Applications*
166 *Educational Activities in Geotechnical Engineering*
167 *Geotechnics of Soil Erosion*
168 *Grouting for Ground Improvement: Innovative Concepts and Applications*
169 *Soil and Material Inputs for Mechanistic-Empirical Pavement Design*
170 *Probabilistic Applications in Geotechnical Engineering*
171 *Advances in Shallow Foundations*
172 *Soil Improvement*
173 *Advances in Measurement and Modeling of Soil Behavior*
174 *Designing Our Underground Space*
175 *Field Measurements in Geomechanics 2007*
176 *Analysis of Asphalt Pavement Materials and Systems: Emerging Methods*
177 *GeoCongress 2008: Geotechnics of Waste Management and Remediation*
178 *GeoCongress 2008: Geosustainability and Geohazard Mitigation*
179 *GeoCongress 2008: Characterization, Monitoring, and Modeling of GeoSystems*

Preface

This Geotechnical Special Publication (GSP), sponsored by the American Society of Civil Engineers' Geo-Institute, honors Professor John H. Schmertmann, Ph.D., P.E., for his distinguished career as a teacher, a research engineer and a practitioner. Dr. Schmertmann, Professor Emeritus at the University of Florida and a Professional Engineer, has a distinguished record in Research and Practice receiving numerous awards and accolades from his peers, some of which are noted in this Volume. He has contributed much to the Geotechnical Engineering Profession through his publications, reports and lectures.

Dr. Schmertmann is proud of the accomplishments and contributions made by many of his former students. The Volume lists his Senior Honors, Masters and Ph.D. Degree students as well as those receiving Post Doctoral fellowships to work with him. He taught his students and those working with him in research and practice superb analytical skills as well as special problem solving techniques which they appreciate and continue to use. Dr Schmertmann is an advocate of the Method of Multiple Working Hypotheses, the Observational Method, and E-prime English. He has successfully adapted these methods to his geotechnical engineering work; but always gives ample credit to others for their origination. The first technique allows the engineer to organize sometimes conflicting information, to make and test hypotheses and to arrive at the most likely correct conclusion. Observations confirm right choices; but, sometimes more importantly, allow the engineer to avoid or minimize the effects of surprises when all does not go as planned. E-prime English eliminates the passive voice resulting in a clear association between the subject and verb which improves the written communication skills of engineers.

The 28 papers contained in this GSP reflect the theme of the Volume, "From Research to Practice". Most of the papers were submitted by invited authors and a few were accepted following a general call for abstracts. The volume also contains 16 of Dr. Schmertmann's "classic" papers considered by the editors as representative of his contributions to the geotechnical engineering literature. The majority of the included papers were presented as part of a special "Symposium in Honor of John Schmertmann" at GeoCongress 2008: The Challenge of Sustainability in the Geoenvironment, the Annual Congress of the Geo-Institute of ASCE, held in New Orleans, LA, March 9-12, 2008. The GSP was unveiled during the symposium and most of the prestigious authors were there to present their papers personally.

The editors thank the authors for their contributions to this volume. All papers are eligible for discussion in the *Journal of Geotechnical and Geoenvironmental Engineering* and are eligible for ASCE/G-I awards.

Each of the papers accepted for publication in this volume was independently and expertly peer reviewed by at least two reviewers in accordance with ASCE-G-I publishing standards and guidelines. The editors appreciate the thoughtful reviews and constructive suggestions made to improve the papers herein. Some of the reviews were made under tight deadlines and we appreciate the assistance of those willing to work with us to meet our schedule. The following are the names of the reviewers who assisted with this important task.

Brian Anderson
C.E. Burke
C. Basnett
Robert Bea
Jean-Louis Briaud
Paul Bullock
Dan Brown
William Camp
Ray Castilli
David K. Crapps
V. Diyaijee
R. A. Failmetzger
Bengt Fellenius
John Garlanger
Sarah Gassman
George Goble
Jack Hayes
Mohamad Hussein
V. Komurka
N. Massoudi
Mike Holloway
Mike McVay
James Mitchell
Gary Kuhns
James Laier
Mike Lewis
Garland Likins
Franke Rausche
Lymon Reese
David Rempe
Ken Stokoe
John Schmertmann
Julian Seidel
M. Suleiman
R. Thompson
Frank Townsend
Anwar Wissa

The Deep Foundations Committee of the Geo-Institute originated the idea and deserves major credit for setting up this volume to honor Dr. John H. Schmertmann for a distinguished career in geotechnical engineering during which he has and continues to bring research to practice in our field.

The editors have been careful in compiling the lists of students, post doctoral fellows working with Dr. Schmertmann and the reviewers that have contributed to the quality of the papers in this volume. We appreciate the contributions made by all and we apologize for any names inadvertently omitted from the lists included herein.

The Editors

David K. Crapps
Principal Engineer
David K. Crapps, Inc.
Gainesville, Florida
dcrapps@attglobal.net

James E. Laier
Chief Executive Officer
Southern Earth Sciences, Inc.
Mobile, Alabama
jlaier@soearth.com

Mohamad H. Hussein
Vice President
GRL Engineers, Inc
Orlando, Florida
MHGRL@pile.com

David K. Crapps John H. Schmertmann James E. Laier Mohamad H. Hussein

Biography

John H. Schmertmann

Civil engineer, and Professor Emeritus from the University of Florida Dept. of Civil and Coastal Engineering, Dr. John H. Schmertmann began life in New York City on 2 December 1928 with German immigrant parents Margaret A. Carstens and Johannes C. Schmertmann. He moved to Gainesville FL in 1956 with his new bride, the former Pauline A. Grange, with whom he had four children: Carl (1959), Gary (1960), Neil (1963) and Joy (1969).

He received his undergraduate degree from M.I.T. in 1950 and later did graduate work at Northwestern University from which he obtained a Ph.D. in 1962, all in civil engineering. He continued his education, with his family along, during sabbaticals in a 1962-3 postdoctoral fellowship at the Norwegian Geotechnical Institute in Oslo and as a visiting scientist in 1971-2 at the Canadian National Research Council in Ottawa. After working for Mueser-Rutledge in NYC from 1951-54 he served in the U.S. Army Corps of Engineers from 1954-56, and then joined the UF faculty in 1956 as a Assistant Professor of Civil Engineering, became a full Professor in 1965 in charge of the geotechnical section and retired from the faculty in 1978.

In 1978 Dr. Schmertmann and Dr. David K. Crapps started the Gainesville consulting engineering firm of Schmertmann and Crapps Inc., retiring in 1997, and then formed his own company, John H. Schmertmann, Inc. In 1991 he co-founded Loadtest, Inc., an international engineering company with headquarters in Gainesville. He remains active in both companies.

Some of the better known consulting projects in which he played a significant role include the geotechnical investigation and design of the Sunshine Skyway Bridge replacement near Tampa FL, moving the Cape Hatteras lighthouse, and work on the crawlerways and launchpads for NASA at Cape Canaveral. He has lived in Gainesville FL for 52 years and locally did laboratory and field research, including via consulting, to call attention to the shrink-swell clay foundation problems in the area and develop practical solutions.

Dr. Schmertmann considers himself as a generalist in geotechnical engineering. He made many and varied contributions to geotechnical engineering through his research, many publications, well over a hundred lectures in the USA and throughout the world, and consulting work. These include significant theoretical and practical contributions to the specialty areas of consolidation testing, settlement analysis, seepage and piping, fundamental shear strength in clays, shrink/swell clay behavior, insitu testing, slope stability, and soil ageing.

He developed design methods used throughout the world, with his strain factor settlement calculation method perhaps the most widely known. His extensive work with insitu testing includes a leading role in understanding the statics and dynamics of the Standard Penetration Test, in introducing the Dutch cone penetration and Marchetti dilatometer tests into the USA, and in developing the associated ASTM standards. In recognition of his eminence with insitu testing, Dr. Schmertmann became the general chairman of the First International Symposium on Penetration Testing (ISOPT-1), at Orlando, Florida, in 1988.

His 1989 25th Terzaghi Lecture introduced new soil ageing concepts. More recently, Dr. Schmertmann has helped pioneer the introduction of the Osterberg cell method for the unlimited high capacity static testing of deep foundations. He also introduced new design methods for calculating piping through embankment sands and the transient seepage and stability effects of rainfall infiltration into slopes.

In recognition of his achievements he received various honors and awards from the American Society of Civil Engineers (ASCE): beginning with the Collingwood Prize in 1956 and currently ending with this volume. It includes a sampling of his publications in the 52-year interval. With his election in 1984 to the National Academy of Engineering, he achieved one of the highest honors in engineering.

He lived through the transition from the slide rule to the computer. But geotechnical engineering involves geology and the surprises inherent in dealing with Mother Nature. He often admonished his computer-oriented students and colleagues with "Don't forget to test the soils!" If he has a specialty he would say "testing". He believes a good test has more value than many theories, and many tests even more... He also advocates "simplicity" when appropriate, as illustrated in his paper in this volume.

A review of his List of Publications will show that Dr. Schmertmann has remained very active in sharing his varied geotechnical insights and experience with other engineers. In 2007 he submitted two papers to the ASCE on the subject of "The Shear Strength of Clays" – the culmination of extensive and original laboratory research done long ago and to which he has given much thought over many years. When it comes to technical papers, some have classified him as a 'lumper' vs. a 'splitter'. He accumulates research and experience and, typically 10 or more years later, eventually lumps it together in a paper. In this way he continues to influence his profession.

Many colleagues and other engineers have contributed to Dr. Schmertmann's professional growth. Here he especially recognizes those mentors who saw his potential and helped greatly to influence his thinking and enhance his early career. In alphabetical order: Drs.Laurits Bjerrum, T. William Lambe, Jorj O. Osterberg, Frank E. Richart Jr. and Philip C. Rutledge. Since then, over the last approximately 30 years, he has shared many experiences and learned from many engineering discussions with Dr. David K. Crapps – his former Ph.D. student at the UF and then his partner in Schmertmann & Crapps, Inc. Similarly, for approximately the last 20 years he has shared and learned from John ('Jack') A. Hayes, the President of Loadtest Inc.

Ending on a personal note: He and his wife Pauline like to travel and camp. They have visited more than 40 countries together and camped all over the USA and Canada, and in Europe – in addition to his professional travels. Dr. Schmertmann has had a lifetime interest in fishing and Zeppelins. He saw the Zeppelin "Hindenburg" in 1936 and arranged to fly in a modern Zeppelin over Munich in 2005. Physically active, he used a bicycle to commute to the UF, still uses one to commute to his current offices, and renewed his interest in golf after his 1997 'retirement' from Schmertmann and Crapps, Inc.

List of Publications and Lecture/Seminar Locations

(Compiled through August 2007)

Papers in Technical Journals and Magazines (Subject to Formal Peer Review)

Schmertmann, J.H. (1953), "Estimating the True Consolidation Behavior of Clay from Laboratory Test Results", ASCE, *Proceedings* , Separate No. 311, 26 pp.

*Schmertmann, J.H. (1955), "The Undisturbed Consolidation Behavior of Clay", ASCE, *Transactions*, Vol. 120, pp. 1201-1233.

Schmertmann, J.H. and Hall, J.R. Jr. (1961), "Cohesion after Non-Hydrostatic Consolidation", ASCE, *Journal of the Soil Mechanics and Foundations Division*, Paper 2881, pp. 39-60.

Schmertmann, J.H. (1962), "Comparisons of One and Two Specimen CFS-Tests", ASCE, *Journal of the Soil Mechanics and Foundations Division*, pp. 169-205.

Schmertmann, J.H. (1963), "Generalizing and Measuring the Hvorslev Effective Components of Shear Resistance", *ASTM STP 361*, pp. 147-157.

Schmertmann, J.H. (1966), "Penetrometers – A Choice for Soil Exploration", *The Florida Architect*, pp. 4-6.

Schmertmann, J.H. (1967), "Static Penetrometers for Soil Exploration", *Civil Engineering*, pp. 71-73.

Schmertmann, J.H. (1969), "Swell Sensitivity", Technical Note in *Geotechnique*, pp. 530-533.

**Schmertmann, J.H. (1970), "Static Cone to Compute Static Settlement over Sand", ASCE, *Journal of the Soil Mechanics and Foundations Division*, pp. 101-1043. Reprinted in ASCE GSP 118, Vol. 2 (2003), *A History of Progress – Selected U.S. Papers in Geotechnical Engineering*, W.A. Marr Ed., pp. 1372-1404.

Schmertmann, J.H. (1970), "The Screw-Plate Load Test", *Special Procedures for Testing Soil and Rock for Engineering Purposes, ASTM, STP 479*, pp. 81-85.

Schmertmann, J.H. (1970), "Deep Static Cone Penetration Test", *Special Procedures for Testing Soil and Rock for Engineering Purposes, ASTM, STP 479*, pp. 71-77.

Schmertmann, J.H. (1974), "A Dewatering Case History in Florida", ASCE, *Journal Construction Division*, pp. 377-393.

Hartmann, J.P. and Schmertmann, J.H. (1975), "FEM Study of Elastic Phase of Pressuremeter Test", ASCE, *Specialty Conference on In-Situ Testing*, Vol. 1, pp. 190-207.

Laier, J.E., Schmertmann, J.H. and Schaub, J.H. (1975), "Effect of Finite Pressuremeter Length in Dry Sand", *Ibid.*, Vol. 1, pp. 241-259.

***Schmertmann, J.H. (1975), "Measurement of Insitu Shear Strength", ASCE, *Proceedings of a Conference on insitu measurement of soil properties*,

Vol. II, State-of-the-Art Report for Session III, pp. 57-138. Discussions and closure to p. 179.

Schmertmann, J.H., Smith, T.V. and Ho, R. (1978), "Example of an Energy Calibration Report on a Standard Penetration Test (ASTM Standard Dl586-67) Drill Rig", Technical Note, *ASTM Geotechnical Testing Journal*, Vol. 1, No. 1, pp. 57-61.

Schmertmann, J.H., Hartman, J.P. and Brown, P.R. (1978), "Improved Strain Influence Factor Diagrams", Technical Note, *ASCE Journal of the GED*, GT8, pp. 1131-1135.

Schmertmann, J.H. (1978), "Use the SPT to Measure Dynamic Soil Properties? - Yes, But...!", ASTM STP 654, pp. 341-355.

Schmertmann, J.H. (1979), "Statics of SPT", *ASCE Journal of the GED*, GT5, pp. 655-670.

****Schmertmann, J.H. and Palacios, A. (1979), "Energy Dynamics of SPT", *ASCE Journal of the GED*, GT8, pp. 909-926.

Schmertmann, J.H. and Crapps, D.K. (1980), "Slope Effect on House Shrink-Swell Movements", *ASCE Journal of the GED*, GT12, pp. 1327-1344.

Schmertmann, J.H. (1981), "A General Time-Related Soil Friction Increase Phenomenon", *ASTM STP 740*, pp. 456-484.

Schmertmann, J.H. (1983), "A Simple Question About Consolidation", *ASCE Journal of Geotechnical Engineering*, Technical Note, Vol. 109, No. 1, pp. 119-122. Discussions and Closure in Vol. 110, No. 5, May 1984, pp. 665-673.

Handy, R.L., Schmertmann, J.H., and Lutenegger, A.J. (1985), "Borehole Shear Tests in a Shallow Marine Environment", *Strength Testing of Marine Sediments: Laboratory and In-Situ Measurements, ASTM STP 833*, R. C. Chaney and K. R. Demars, Eds., American Society for Testing and Materials, Philadelphia, pp. 140-153.

Sonnenfeld, S.L., Schmertmann, J.H. and Williams, R.C. (1985), "A Bridge Site Investigation Using SPTS, MPMTS, and DMTS from Barges", *Strength Testing of Marine Sediments: Laboratory and In-Situ Measurements, ASTM STP 883*, R. C. Chaney and K. R. Demars, Eds., American Society for Testing and Materials, Philadelphia, pp. 515-535.

Schmertmann, J.H. (1986), "Dilatometer to Compute Foundation Settlement", *Proc. ASCE Specialty Conf. on Use of In Situ Tests For Geotechnical Engineering*, Va. Tech, Blacksburg, VA., *Geotechnical Spec. Publ. No. 6*, pp. 303-321.

Schmertmann, J.H. (1986), Baker, W., Gupta, R. and Kessler, K., "CPT/DMT QC of Ground Modification at a Power Plant", Va. Tech, Blacksburg, Va., ASCE, *Geotechnical Spec. Publ. No. 6*, pp. 985-1001.

Schmertmann, J.H. (1988), "Dilatometers Settle In", *Civil Engineering*, pp. 68-70.

Schmertmann, J.H. (1989), "Density Tests Above Zero Air Voids Line", *ASCE Journal of Geotechnical Engineering*, Vol. 115, No. 7, pp. 1003-1018. Closure: Vol. 117, No. 9, p. 1450.

Schmertmann, J.H. (1991), "The Mechanical Aging of Soils", The 1989, 25th Terzaghi Lecture, *ASCE Journal of the Geotechnical Engineering Division*, Vol. 117, No. 9, pp. 1285-1330. Two *discussions* and *closure* in Vol. 118, No. 12, Dec 92, pp. 2009-2014.

Schmertmann, J.H. and Henry, J. F. (1992), "A Design Theory for Compaction Grouting", *ASCE Proc. Grouting, Soil Improvement and Geosynthetics - 92*. pp. 215-228.

Schmertmann, J.H. (1993), "Conical Test Load Measuring Compressibility", Technical Note, *ASCE Journal of the Geotechnical Engineering Division*, Vol. 119, No. 5, pp. 965-971.

Schmertmann, J.H. (2005), "Stress Diffusion Experiment in Sand", *ASCE JGGE*, V.131, N. 1, pp. 1-10.

Bullock, P.J. and Schmertmann, J.H. (2005), "Side Shear Setup I: Test Piles Driven in Florida", *ASCE JGGE*, V.131, N. 3, pp. 292-300.

Bullock, P.J. and Schmertmann, J.H. (2005), "Side Shear Setup II: Results from Florida Test Piles", *ASCE JGGE*, V.131, N. 3, pp. 301-310.

Schmertmann, J.H. (2006), "Estimating Slope Stability Reduction Due to Rain Infiltration Mounding", *ASCE JGGE*, V.132, N.9, pp. 1219-1228.

Schmertmann, J.H. (2008), "The Components of Clay Shear, Part I", submitted for publication in *ASCE JGGE*, approx. 10 pp.

Schmertmann, J.H. (2008), "The Components of Clay Shear, Part II", submitted for publication in *ASCE JGGE*, approx. 11 pp.

* Paper won 1956 ASCE national COLLINGWOOD PRIZE

** Paper won 1971 ASCE national NORMAN MEDAL (highest society award)

*** Paper won 1977 ASCE national STATE OF THE ART OF CIVIL ENGINEERING AWARD

**** Paper won 1981 ASCE divisional (Geot. Engineering) MIDDLEBROOKS AWARD

Discussions (Subject to Peer Review)

Schmertmann, J.H. (1956), "Effect of Sample Disturbance on the Strength of a Clay", by Max L. Calhoon. A discussion in *ASCE Transactions*, Vol. 121, pp. 940-950.

Schmertmann, J.H. (1961), "The Influence of Strain on the Shearing Resistance of Sensitive Clay", by Carl. B. Crawford. A discussion in *ASTM Proceedings*, Vol. 61, pp. 1271-1273.

Schmertmann, J.H. (1963), "Friction and Cohesion of Saturated Clays", by T. H. Wu. A discussion in *ASCE Journal of the Soil Mechanics and Foundations Division*, pp. 260-268.

Schmertmann, J.H. (1963), "Direct Shear Test for Effective Stress Parameters", by H. H. O'Neill. A discussion in *ASCE Journal of the Soil Mechanics and Foundations Division*, pp. 158-161.

Schmertmann, J.H. (1964), "Stability of Cuts in Soft Clays", by T. C. Kenney. A discussion in *ASCE Journal of the Soil Mechanics and Foundations Division*, pp. 183-190.

Schmertmann, J.H. (1965), "Interpretation of the Consolidation Test", by C. W. Crawford. A discussion in *ASCE Journal of the Soils Mechanics and Foundations Division*, pp. 131-135.

Schmertmann, J.H. (1965), "In-Situ Strength Characteristics of Soft Clays", by I. Noorany and H. B. Seed. A discussion in *ASCE Journal of the Soils Mechanics and Foundations Division*, pp. 123-127.

Schmertmann, J.H. (1966), "Stability of Slopes in Anisotropic Soils", by K. Y. Lo. A discussion in *ASCE Journal of the Soil Mechanics and Foundations Division*, pp. 203-207.

Schmertmann, J.H. (1966), "Standard Penetration Tests: Its Uses and Abuses", by F. A. Fletcher. A discussion in *ASCE Journal of the Soil Mechanics and Foundations Division*, pp. 130-133.

Schmertmann, J.H. (1969), Discussion of: "Deepsounding Test Results and The Settlement of Spread Footings on Normally Consolidated Sands", by. D. Thomas, *Geotechnique*; pp. 316-317.

Schmertmann, J.H. (1969), Discussion of V. de Mello's state-of-the-art paper, Proc. 7th ICSM&FE, Mexico City, Vol. 3, pp. 250-1.

Schmertmann, J.H. (1970), Discussion of: "Sand Densification by Heavy Vibratory Compactor", ASCE paper 6656, *ASCE Journal of the Soil Mechanics and Foundations Division*; pp. 363-365.

Schmertmann, J.H. (1970), Discussion of: "Tests on Instrumented Piles, Ogeechee River Site", A. Vesic, ASCE paper 7170, *ASCE Journal of the Soil Mechanics and Foundations Division*; pp. 1483-1484.

Schmertmann, J.H. (1971), "The Importance of Side Friction and Lateral Stress to the SPT-N Value", Discussion to Session I, *Proceedings IV PanAmerican Congress on Soil Mechanics and Foundation Engineering*, Puerto Rico, Vol. III, pp. 90-98.

Schmertmann, J.H. (1972), Discussion of: "Simplified Procedure for Evaluating Soil Liquefaction Potential", Seed and Idriss, *ASCE Journal of the Soil Mechanics and Foundations Division*, Vol. 97, SM9; pp. 430-433.

Schmertmann, J.H. (1973), Discussion of Task Committee's report for Foundation Design Manual, *ASCE Journal of the Soil Mechanics and Foundations Division*, pp. 477-8.

Schmertmann, J.H. (1974), "CPT to Help Understand SPT", *Discussion to ESOPT I*, Stockholm, Vol. 2.1, p. 146.

Schmertmann, J.H. (1974), Discussion of "Pore Pressures that Produce Noncon-Servative q_c DATA", *Ibid.*, pp. 146-150.

Schmertmann, J.H. (1977), and Morgenstern, N.R., Discussion of Main Session I State-of-the-Art Report "Stress-Deformation and Strength Characteristics" by C.C. Ladd, *IX ICSMFE*, Tokyo, Vol. 3, pp. 356-358.

Schmertmann, J.H. (1979), Discussion of General Report by Sowers, G. F., *Proceedings International Symposium on Soil Mechanics*, Oaxaca, Mexico, Vol. 1, pp. 25-26.

Schmertmann, J.H. and Woods, R.D. (1980), Discussion of: "Anisotropic Shear Modulus Due to Stress Anisotropy", by S.K. Roesler in Jul 79 *ASCE Journal GED*; GT8, pp. 958-960.

Schmertmann, J.H. (1981), Discussion of: "In Situ Test by Flat Dilatometer" by S. Marchetti in *ASCE Journal GED*, Mar 80; pp. 831-832.

Schmertmann, J.H. (1984), Discussions of: "Undrained Strength Characteristics of Normally Consolidated Aged Clay" by H. Hanzawa, and "Increase in Undrained

Shear Strength Due to Secondary Compression" by K. Yasuhara and S. Ue; *Soils and Foundations* (Japan), Sep 83; pp. 158-160.

Schmertmann, J.H. (1984), Discussion of: "Three Case Studies for Short Term Stability of Soft Clay Deposits" by H. Hanzawa, *Soils and Foundations* (Japan), Jun 83; pp. 160-161.

Schmertmann, J.H. (1984), Discussion of: "Reproducible SPT Hammer Force with an Automatic Free Fall SPT Hammer System" by C. O. Riggs et. al., *ASTM Geotechnical Testing Journal*, Vol. 7, No. 7, pp. 167-8.

Schmertmann, J.H. (1984), Discussion of: "Time Effects on the Stress-Strain Behavior of Natural Soft Clays" by J. Graham et. al., *GEOTECHNIQUE*, Vol. XXXIV, No. 3, pp. 443-445.

Schmertmann, J.H. (1985), Discussion of: "Stress Ratio Effects on Shear Modulus of Dry Sands", by P. Yu and F. E. Richart, Jr., *ASCE Journal of Geotechnical Engineering*, March, 1984; pp. 1153-4.

Schmertmann, J.H. (1987), Discussion of: "Time-Dependent Strength Gain in Freshly Deposited or Densified Sand" by J. Mitchell and Z. V. Solymar (Nov. 84), *ASCE Journal GED*, Vol. 113, No. 2, pp. 173-175.

Schmertmann, J.H. (1987), Discussion of: "Large-Scale Cyclic Shear Bin to Evaluation Methods for Mitigating Liquefaction Hazard", *Soils and Foundations*, Vol. 27, No. 1, pp. 104-5.

Schmertmann, J.H. (1988), Discussion of: "Undrained Shear Behavior of Clays Undergoing Long-Term Anisotropic Consolidaiton", *Soils and Foundations* (Japan), Vol. 28, No. 1, p. 134.

Schmertmann, J.H. (1989), Discussion of Landva, A. O. et. al., "Performance of a Raft Foundation Supporting a Multistorey Structure", *Canadian Geotechnical Journal*, 26: 185-186.

Schmertmann, J.H. (1991), Discussion of: "Settlement of Shallow Foundations on Granular Soils", by Leonards, G.A. and Frost, J.D., *ASCE Journal GED*, V117, N1, pp. 179-181.

Schmertmann, J.H. (2000), Discussion of "Triaxial Behavior of Kaolinite in Different Pore Fluids", by A. Anandarajah and D. Zhao, *ASCE JGGE* V.127, N.5, p. 463.

Schmertmann, J.H. (2000), Discussion of: "The Development of Shear Strength for Sedimentary Soft Clay with Respect to Aging Effect", by Y. Z. Tang and T Tsuchida. Discussion in *Soils and Foundations*.

Schmertmann, J.H. (2002), Discussion of Foster, M., Fell, R., and Spannagle, M. of "A Method for Assessing the Relative Likelihood of Failure of Embankment Dams by Piping", *Can. Geotech. J*, Vol. 39, pp. 495-6.

Schmertmann, J.H. (2004), Discussion of "Effect of Live Load Surcharge on Retaining Walls and Abutments", BY J. S. Kim and R. M. Barker, *ASCE JGGE*, V130(4), 446-447.

Schmertmann, J.H. (2004), Discussion of "Time for Development of Internal Erosion and Piping in Embankment Dams", by Fell et. al., *ASCE JGGE*, V130, N9, p. 980.

Schmertmann, J.H. (2006), Discussion of "Ultimate Lateral Resistance to Piles in Cohesionless Soils" by L. Zhang and F. Silva, *ASCE JGGE*, V132, N8, 1108-1109.

Schmertmann, J.H. (2007), Discussion of: “Energy Efficiency for Standard Penetration Tests”, by E. Odebraecht et. al., *ASCE JGGE*, V133, N4, 487-488.

Monaco, P. and Schmertmann, J.H. (2007), Discussion of: “Accounting for Soil Aging When Assessing Liquefaction Potential”, by Leon, E., Gassman, S., and Talwani, P., *ASCE JGGE*, September, v.133, N.9, pp. 1177-1179.

Published Reports, Manuals, and Papers in Conference Proceedings

Schmertmann, J.H. (1958), "A Thermally Controlled Soil Freezing Cabinet", U. S. Army Corps of Engineers; Snow, Ice and Permafrost Research Establishment *Technical Report No. 50*, 35 pp.

Richart, F.E. Jr., and Schmertmann, J.H. (1958), "The Effect of Seepage on the Stability of Sea Walls", *Proceedings Sixth International Conference on Coastal Engineering*, pp. 794-817.

Schmertmann, J.H. and Osterberg, J.O. (1960), "An Experimental Study of the Development of Cohesion and Friction with Axial Strain in Saturated Cohesive Soils", *Proceedings ASCE Research Conference on Shear Strength of Cohesive Soils*, Boulder, Colorado, pp. 643-694.

Schmertmann, J.H. and Taylor, R.S. (1965), "Quantitative Data From A Patterned Ground Site Over Permafrost", U. S. Army Cold Regions Research and Engineering Laboratory, *Research Report 96*, 79 pp.

Schmertmann, J.H. (1967), "Guidelines for Use in the Soils Investigation and Design of Foundations for Bridge Structures in the State of Florida", *Florida State Road Department Research Bulletin 121*, 87 pp. plus appendices.

Schmertmann, J.H. (1969), "Dutch Friction-Cone Penetrometer Exploration of Research Area at Field 5, Eglin AFB, Florida", *Contract Report S-69-4*, U. S. Army Corps of Engineers, Waterways Experiment Station, Vicksburg, Miss., 50 pp.

Schmertmann, J.H. (1971), "Special Techniques for Soil Investigation", *18th Annual SM & FE Conference*, University of Minnesota, held March 26, 1970, Proceedings, 37 pp.

Schmertmann, J.H. (1972), "Effects of In-Situ Lateral Stress on Friction-Cone Penetrometer Data in Sands", *Fugro Sondeer Symposium* (Holland), pp. 37-39.

Schmertmann, J.H. (1973), "Settlements of Structures on Granular Soils – Estimating Procedures and Case Histories", ASCE NYC Met. Section Seminar on Settlement of Structures, *Proceedings*, Part II, 26 pp.

Schmertmann, J.H. (1974), "Penetration Testing in the USA", State-of-the-Art Report, *Proceedings European Symposium on Penetration Testing* (ESOPT), Stockholm, Vol. 1, pp. 217-218.

Schmertmann, J.H. (1974), "Penetration Pore Pressure Effects on Quasi-Static Cone Bearing, q_c, *Ibid.*, Vol. 2.2, pp. 345-351.

Schmertmann, J.H. (1976), "In Situ Determinations of Soil Parameters", contribution to lecture notes for ASCE, National Capital Section, *1976 Seminar on Field and Laboratory Determinations of Soil Parameters for Geotechnical Design*, 86 pp.

Schmertmann, J.H. (1978), "Study of Feasibility of Using WISSA-Type Piezometer Probe to Identify Liquefaction Potential of Saturated Fine Sands", U.S. Army Engineers Waterways Experiment Station, *Technical Report S-78-2*, 73 pp.

Schmertmann, J.H. (1978), "The Statics and Dynamics of the Standard Penetration Test", *Proceedings FHWA Symposium on Site Exploration in Soft Ground using In-situ Techniques*, FHWA TS-80-202, pp. 145-205.

Schmertmann, J.H. (1978), "Effect of Shear Stress on Bulk Modulus of Sand", U.S. Army Engineers Waterways Experiment Station, *Technical Report S-78-16*, 45 pp. + 3 Append.

Schmertmann, J.H. (1979), "Reliability of the SPT as the Sole Testing Method in South Florida", *Proceedings Geotechnical Engineering in South Florida for non-geotechnical engineers*, Workshop by South Florida Section ASCE and Florida International Univ., April, Ft. Lauderdale, 18 pp.

Schmertmann, J.H. (1982), "A Method for Determining the Friction Angle in Sands from the Marchetti Dilatometer Test (DMT)", *Proceedings of the Second European Symposium on Penetration Testing* (ESOPT II), Amsterdam, Vol. 2, pp. 853-861.

Schmertmann, J.H. (1983), "Past, Present and Future of the DMT", *Proceedings of the First International Dilatometer Conference*, Edmonton, sponsored by Mobile Augers Research, Ltd., pp. 1-6.

Schmertmann, J.H. and Crapps, D. K. (1983), "Use of Insitu Penetration Tests to Aid Pile Design and Installation", *Proceedings of the '83 GEOPILE Conference*, sponsored by the Associated Pile and Fitting Corp, pp. 27-48.

Schmertmann, J.H. (1985), "Introduction to the Marchetti Dilatometer Test", *Proc. Conf. on Innovation in Geotechnical Engineering*, sponsored by Central PA Section A.S.C.E. and PennDOT, Harrisburg, PA, April 18-19, 11 pp.

Schmertmann, J.H. (1985), "Measure and Use the Insitu Lateral Stress", *The Practice of Foundation Engineering*, a volume honoring Jorj O. Osterberg and published by the Dept. of Civil Engineering, Northwestern University, pp. 189-213.

Schmertmann, J.H. (1986), "Some 1985-86 Developments in Dilatometer Testing and Analyses", *Proc. Conf. on Geotechnical Engineering Practice*, sponsored by Central PA Section A.S.C.E. and PennDOT, Harrisburg, PA, April 17-18, 20 pp.

Schmertmann, J.H. (1987), "Pile Load Distribution Determined from Tip Telltales", Vol. 13.10, XIII C.G.T., Univ. of Torino (Italy), 10-12 Nov., 33 pp. (also published in 1988 Chicago Geotechnical-Structural Lecture Series, Illinois Section ASCE & Ill. Inst. of Technology. (also published in *Inzynieria Morska*, 1988, No. 4, pp. 174-6 (in Polish)).

Schmertmann, J.H. (1988), "The Coefficient of Consolidation Obtained from p—2 Dissipation in the DMT", *Proc. Conf. Developments in Geotechnical Engineering*, Central PA Section ASCE and PennDOT, 7-8 Sep., 18 pp. (also in GPE, Inc. DMT DIGEST #10, Mar 88).

Schmertmann, J. H. (1993), "The Bottom-Up, Osterberg Cell Method for Static Testing of Shafts and Piles", *Proc. Progress in Geotechnical Engineering Practice*, Hershey PA, 12-14 Apr 93, Central PA ASCE, PennDOT, 7 pp.

Schmertmann, J.H. (1993), "Update on the Mechanical Aging of Soils (25th Terzaghi Lecture)". *Symposium*, titled "Sobre Envejecimiento de Suelos", Sponsored by The Mexican Society of Soil Mechanics, Mexico City, 26 Aug 93. 30 pp.

Schmertmann, J. H. (1994), "A Conical Test Load (CTL) Load Test". *Proc. Soil-Structure Interaction and Environmental Issues*, Hershey PA, 12-14 Sep, Central PA ASCE Section, PennDOT, 19 pp.

Schmertmann, J.H. and Crapps, D. K. (1994), "Past, Present and Future in Deep Foundations, with Florida Emphasis", *Proc. Int. Conf. on Design and Construction of Deep Foundations*, U.S. Federal Highway Administration (FHWA), Keynote paper, Vol. 1, pp. 188-208.

Schmertmann, J.H. (1995), "CPT'63(+) - A History in the USA from a University of Florida Viewpoint", *Proc. CPT'95*, Swedish Geotechnical Society, Linkoping, Sweden, Vol 3, pp. 269-275

Schmertmann, J.H. and Hayes, J. A. (1997), "The Osterberg Cell and Bored Pile Testing - A Symbiosis", *Proceedings, The Third International Geotechnical Engineering Conference*, Cairo University, Egypt, 5-8 Jan 1997, pp. 139-167.

Schmertmann, J.H. and Hayes, J. A. (1997), "Observations from Osterberg Cell Tests of Bored Piles", *Fulcrum Newsletter*, Deep Foundations Institute, pp. 11-14.

Schmertmann, J.H. and Hayes, J. A. (1997), "Observations from Osterberg Cell Load Tests of Bored Piles", *Proceedings*, Central PA Section ASCE & Penn DOT Conf. On EXCELLENCE IN ENGINEERING, 23-24 Oct 97, Harrisburg, PA. 4 pp.

Schmertmann, J.H., Hayes, J. A., Molnit, Thomas and Osterberg, J. O. (1998), "O-Cell Testing Case Histories Demonstrate the Importance of Bored Pile (Drilled Shaft) Construction Technique", *Proc. 4th International Conf. On Case Histories in Geotechnical Engineering*, St. Louis, MO, 9-12 March, pp. 1103-1115. ("Published" in CD form by Univ. of Missouri-Rolla).

Schmertmann, J.H. (2000), "The No-Filter Factor of Safety Against Piping Through Sands", *ASCE GeoInstitute Geotechnical Special Publication No. 111 – Judgement and Innovation*, Ed. By F. Silva and E. Kavazanjian, pp. 65-132.

Crapps, D.K. and Schmertmann, J.H. (2002), "Compression Top Load Reaching Shaft Bottom – Theory vs. Tests", *Proc.* Deep Foundations 2002, ASCE Geotechnical Special Publication No. 116, V. 2, pp. 1533-1549.

Schmertmann, J.H. (2008), "Some Simple Methods I Used in Research and Practice", submitted for publication in *ASCE GSP 180*, approx. 15 pp.

Special Publications

Schmertmann, J.H. (1950), "The Differential Thermal Analyzer as a Routine Test for the Mineral Determination of Foundation Soils", Report to the M.I.T. Soil Stabilization Project, 99 pp.

Schmertmann, J.H. (1962), "Electronic Computer Calculations for the Separation of Effective Shear Resistance Components from an IDS-Test", Norwegian Geotechnical Institute (NGI) Internal Report F.202.8, 17 pp.

Schmertmann, J.H. (1963), "Reconsolidation Study of Maglerud Quick Clay", NGI Internal Report F. 234, 42 pp.

Schmertmann, J.H. (1963), "Laboratory and Computer Study of a Series of Direct Shear Tests on Sand", NGI Internal Report F.202.9, 60 pp.

Schmertmann, J.H. (1972), *Series of Internal Reports to Division of Building Research*, Canadian National Research Council: (Project G-1) No. 1 - "Static Cone Penetration Tests" No. 2 – "Hydraulic Fracturing to Determine K'_o and Tensile Strength" No. 3 - "Pressuremeter Tests in Leda Clay"

Nottingham, L.C. and Schmertmann, J.H. (1975), "An Investigation of Pile Capacity Design Procedure", University of Florida, College of Engineering, EIES, Dept. of Civil Engineering, *Final Report D629 to the Florida DOT*, 159 pp. plus appendix.

Schmertmann, J.H., Crapps, D.K. and Marquardt, John (1976), "A National Survey of State Department of Transportation Bridge Foundation Practice", University of Florida, College of Engineering, EIES, Department of Civil Engineering, *Supplemental Report D657*, to the Florida DOT, 80 pp.

Schmertmann, J.H. (1976), "The Shear Behavior of Soil with Constant Structure", *Laurits Bjerrum Memorial Volume*, Norwegian Geotechnical Institute, pp. 65-98.

Schmertmann, J.H. and Crapps, D. K. (1977), "Wave Equation Study and Recommended Specifications for the Impact Driving of Prestressed Concrete Piles", University of Florida, College of Engineering, EIES, Department of Civil Engineering, *Final Report D657*, to the Florida DOT, 214 pp.

Schmertmann, J.H. and Smith, T.V. (1977), "A Summary of SPT Energy Calibration Services Performed for the Florida DOT Under Service Contract 99700-7150-010", University of Florida, College of Engineering, Department of Civil Engineering, EIES, *Final Research Report 245-D73*, 21 pp. plus appendices.

Schmertmann, J.H. and Bertin, Y. (1978), "A Centrifuge Model Study to Help Determine the Factor of Safety Against Sinkhole Collapse", University of Florida, College of Engineering, EIES, Dept. of Civil Engineering, *Final Report Research Project 245-D59*, DOT Project No.: 9970-7173-010, 122 pp.

Schmertmann, J.H. (1978), "Guidelines for the Cone Penetration Test Performance and Design", U.S. Department of Transportation, FHWA-TS-78-209, 145 pp.

Schmertmann, J.H. (1982), "Investigation of the Use of Cyclic, Tortional, Screwplate Test to Determine the Insitu Liquefaction Potential of Cohesionless Soils", Final Report to the Nuclear Regulatory Commission, Contract No. NRC-04-77-171, by the University of Florida, Department of Civil Engineering, 2 volumes.

Townsend, F.C., Schmertmann, J.H., Logan, T.J., Pietrus, T.J. and Wong, Y.W. (1982), "An Analytical and Experimental Investigation of a Quantitative Theory for Piping in Sand", Final Research Report for Florida Power and Light Company, by the Department of Civil Engineering, EIES, College of Engineering, University of Florida, 100 pp.

Schmertmann, J.H. (1983-91), *DMT DIGESTS*:, distributed by GPE Inc., Gainesville, FL., Schmertmann, J.H., Editor. #1 - Apr 83, #2 - Jul 83, #3 - Feb 84, #4 - Jun 84, #5 - Feb 85, #6 - Jul 85, #7 - Mar 86, #8 - Aug 86, #9 - May 87, #10 - May 88, #11 - Apr 89, #12 – Dec 91.

Schmertmann, J.H. (1984), "The New In-Situ Marchetti Dilatometer Test", *Geotechnical News*, Vol. 2, No. 3, pp. 34-35.

Schmertmann, J.H. (1988), "Guidelines for Using the CPT, CPTU and Marchetti DMT for Geotechnical Design", Report, FHWA-PA-87-024+84-24, 4 Vols. prepared for Penn. Dept. of Transportation.

Vol. I:Smmmary (78 pp. incl. CPT & DMT bibliographies) NITS A. No. PB88-211636AS ($14.95).

Vol. II:Using CPT and CPTU Data (235 pp., by Robertson & Campanella), NITS A. No. PB88-211644AS ($32.95).

Vol. III:DMT Test Methods and Data Reduction (183 pp.) NITS A. No. PB88-211651AS ($19.95).

Vol. IV:DMT Design Methods and Examples (135 pp.) NITS A. No. PB88-211669AS ($19.95).

Schmertmann, J.H. (1995), "Use and Misuse of the Standard Penetration Test in Design and Construction", 5th Mueser Rutledge Lecture, New York City, 24 Oct 95, 5 p. handout

Schmertmann, J.H. (1995), "Soil Arching and the Spanning of Voids",16 pp., Compilation of lectures presented at the Kungl Tekniska Hogskolan (KTH), given 2 Oct 95, published by KTH. Also in the PennDOT-ASCE 1999 Conf. Proc., Hershey, PA, and as the First Distinguished Lecture in Civil Engineering, 1998, Univ. of Mass., Amherst, MA.

Bullock, P.J. and Schmertmann, J.H. (2003), "Determining the Effect of Stage Testing on the Dimensionless Pile Side Shear Setup Factor", University of Florida, Dept. of Civil and Coastal Engineering, Final Report for FLDOT, Contract No. BC-354, RPWO No. 27, 93 pp.

List of Organizations Where Dr. John H. Schmertmann Presented Lectures and Seminars

(More than once at some, compiled through August 2007)

UNIVERSITIES

Alberta
Belo Horizonte (Brazil)
British Columbia
Cairo (Egypt)
Calgary
Chalmers (Sweden)
Cornell
Florida
Florida Atlantic
Georgia Tech
India Instit. Tech. (New Delhi)
Kentucky
Laval
Louisiana State
Manitoba
Massachusetts (Amherst)
Megill
Michigan
Minnesota
Missouri, Rolla
New Brunswick (Canada)
North Carolina State
Northwestern
Nova Scotia Tech. College
Norwegian Technical
Ottawa (Canada)
Purdue
Saskatchewan
Stanford
Stockholm Technical
Surrey (England)
Texas (Austin)
Torino (Italy)
Toronto
Tufts

RESEARCH ORGANIZATIONS

Canadian National Research Council, DBR
NASA, Cape Canaveral
Norwegian Geotechnical Institute
Swedish Geotechnical Institute
W.E.S., Vicksburg, Mississippi

ENGINEERING CONSULTANTS

Dames & Moore, Los Angeles,
Froehling & Robertson, Charlotte NC
Fugro California, Inc.
(now Earth Technology
Corp., Long Beach CA)
Fugro-Cesco (Holland)
Law Engineering Company, Atlanta
McClelland Engineers, Houston
Soil Exploration Company, Minnesota
Woodward-Group, California
Statoil, Stavanger, Norway
URS, Tampa FL
Haywood Baker, Tampa FL

PROFESSIONAL SOCIETIES

ASCE:

Ames, IA
Atlanta, GA
Baltimore, MD
Boston, MA
Chicago, IL
Cocoa Beach, FL
Detroit, MI
Edmonton, AB
Ft. Lauderdale, FL
Gainesville, FL
Harrisburg, PA
Hershey, PA
Houston, TX
Jacksonville, FL
Lake City, FL
Lakeland, FL
Miami, FL
Montreal, PQ
New Orleans, LA
New York City, NY
Ocala, FL
Omaha, NE
Phoenix, AZ
St. Louis, MO
Tallahassee, FL
Tampa, FL
Washington, D.C

ASTM:

New Orleans, LA
San Francisco, CA

Florida Engineering Society

Daytona, FL
Jacksonville, FL
San Juan, PR

MISCELLANEOUS:

Amsterdam (Netherlands)
Belo Horizonte (Brazil)
Boston (BSCE)
Charleston, SC
Fredericton, NB
Kuala Lumpur (Malaysia)
Linkoping (Sweden)
Mexico City, Mexico
Milano (Italy)
New Delhi, India
Orlando, FL
Oslo, Norway
Santa Barbara, CA
Singapore
Stockholm, Sweden
Oaxaca (Mexico)

PROFESSOR SCHMERTMANN'S UNIVERSITY OF FLORIDA STUDENTS

Post Doctoral Fellows

Al-Awkati, Zaid (1974)
Arnold, Maurice (1977)
Campanella, Richard G. (1977)
Holden, James C. (1971)
Senneset, Kaare (1978)
Silvano, Marchetti (1980)
Tumay, Mehmet T. (1975)
Woods, Richard (1976)

Ph.Ds, with Dissertations

Bullock, Paul J. (1999)
"Pile Friction Freeze: A Field and Laboratory Study" 552 pp.

Crapps, David K. (1977)
"Wave Equation Study and Recommended Specifications for the Impact Driving of Prestressed Concrete Piles" 394 pp.

Hartman, John P. (1974)
"Finite Element Parametric Study of Vertical Strain Influence Factors and the Pressuremeter Test to Estimate the Settlement of Footings on Sand" 407 pp.

Heller, Lyman W. (1971)
"The Particle Motion Field Generated by the Torsional Vibration of a Circular Footing on Sand" 270 pp.

Ho, Kum-Hung (1971)
"Theoretical and Experimental Relationships between Stress Dilatency and IDS Strength Components" 230 pp.

Laier, James E. (1973)
"Effects of Pressuremeter Probe Length/Diameter Ratio and Borehole Disturbance on Pressuremeter Test Results in Dry Sand" 204 pp.

Nottingham, Larry C. (1975)
"Use of Quasi-Static Friction Cone Penetrometer Data to Predict Load Capacity of Displacement Piles" 553 pp.

Palacios, Alejandro (1977)
"The Theory and Measurement of Energy Transfer during Standard Penetration Test Sampling"

Masters

Bea, Robert G.
Bertin, Yves
Brown, P. Ralph
Bunnel, Daniel B.
Caillemer, Baudouin N.
Cook, Walter K.
Didonato, Kenneth M.
Driggers, Jamie F.
Dunkelberger, Douglas S.
Edmonds, Gregory P.
Freed, David L.
Gallet, Alain J.
Gould, Jon H.
Hall, John R. Jr.
Hancock, Scott
Hanskat, Charles F.
Harman, Don E.
Jackson, Billy E.
Johnson, Michael A.
Kehoe, Steven P.
Lhuer, Jean-Michel
Lingo, Edward W.
Logan, Thomas J.
MacFarlane, James W.
Malerk, Thomas
Margas, Hristo
Marquardt, John
Marshall, A. Franklin Jr.
Nicholson, Glenn A.
Oenbrink, Stephen J.
Piton, Henri
Prochaska, Billy R.
Reese, Jules D.
Smith, Terrill V.
Strømann, Helle
Topshøj, Anne-Grethe
Varner, Michael
Villafane, German A.
Weber, Richard P.

Senior Honors

Burke, A.
Karkowski, R.
Marshall, Peter W.

Contents

Selected papers authored by John H. Schmertmann

The Undisturbed Consolidation Behavior of Clay 1
John H. Schmertmann
From *ASCE Transactions*, 1955, 120, 1201-1233. Reproduced with permission.

Generalizing and Measuring the Hvorslev Effective Components of Shear Resistance 34
John H. Schmertmann
From *Laboratory Shear Testing of Soils,* ASTM Special Technical Publication No. 361, 1969, 147-157. Reproduced with permission.

Swell Sensitivity 45
J. H. Schmertmann
From *Géotechnique*, December, 1969, 530-533. Reproduced with permission.

Static Cone to Compute Static Settlement over Sand 49
John H. Schmertmann
From *J. Soil Mechanics and Foundations Div, 1970,*. 96(3), 1011-1043. Reproduced with permission.

Discussion of *The Standard Penetration Test* 82
J. H. Schmertmann
From *Proc. of the Fourth Panamerican Conference on Soil Mechanics and Foundation Engineering,* ASCE, San Juan, Puerto Rico,1971. 90-99. Reproduced with permission.

Effects of In Situ Lateral Stress on Friction—Cone Penetrometer Data 92
John H. Schmertmann
From *Fugro Sondeer Symposium*, Holland, 1972, 37-39. Reproduced with permission.

Dewatering Case History in Florida 95
John H. Schmertmann
From *Journal of the Construction Division,* 1974, 100(3), 377-393. Reproduced with permission.

Slope Effect on House Shrink-Swell Movements 112
John H. Schmertmann and David K. Crapps
From *Journal of the Geotechnical Engineering Division,* 1980, 106(12), 1327-1344. Reproduced with permission.

Dilatometers Settle In 130
John H. Schmertmann
From *Civil Engineering*, 1988, March, 68-70. Reproduced with permission.

Density Tests above Zero Air Voids Line 133
John H. Schmertmann
From *J. of Geotechnical Engineering,* 1989, 115(7), 1003-1018. Reproduced with permission.

A Design Theory for Compaction Grouting 149
John H. Schmertmann and James F. Henry
From *Grouting, Soil Improvement, and Geosynthetics* (Geotechnical Special Publication No. 30), 1992, 215-228. Reproduced with permission.

Conical Test Load Measuring Compressibility 163
John H. Schmertmann
From *J. of Geotechnical Engineering,* 1993, 119(5) 965-971. Reproduced with permission.

CPT'63(+)—A History in the USA from a University of Florida Viewpoint 170
John H. Schmertmann
From *Proc. CPT'95,* Swedish Geotechnical Society, Linkoping, Sweden, 1995, 269-275. Reproduced with permission.

Observations from Osterberg Cell Tests of Bored Piles 177
John H. Schmertmann and John A. Hayes
From *Fulcrum Newsletter*, Deep Foundations Institute, Winter '96-'97, 11-14. Reproduced with permission.

Discussion of "Time for Development of Internal Erosion and Piping in Embankment Dams" by Robin Fell, Chi Fai Wan, John Cyganiewicz, and Mark Foster 181
John H. Schmertmann
From *J. of Geotechnical and Geoenvironmental Engineering,* 2004, 130(9), 980-981. Reproduced with permission.

Estimating Slope Stability Reduction due to Rain Infiltration Mounding 183
John H. Schmertmann
From *J. of Geotechnical and Geoenvironmental Engineering,* 2006, 132(9), 1219-1228. Reproduced with permission.

Invited Papers

Schmertmann's Swell Sensitivity—Revisited 193
Alan J. Lutenegger

Physical Modeling of Raft on Settlement Reducing Piles 206
Vincenzo Fioravante, Daniela Giretti, and Michele Jamiolkowski

Simple Shear Testing and Behavior of Norwegian Quick Clays—Revisited 230
Arvid Landva

"Liquepaction": Hydraulic Compaction of Soil near Rammed Piers 251
Richard L. Handy

From Theory to Implementation of a CPT-Based Probabilistic and Fuzzy Soil Classification 259
Mehmet T. Tumay, Murad Y. Abu-Farsakh, and Zhongjie Zhang

Experimental and Analytical Investigation of Geogrid MSE Walls 277
D. V. Reddy and Fernando Navarrete

In Situ Tests by Seismic Dilatometer (SDMT) 292
Silvano Marchetti, Paola Monaco, Gianfranco Totani, and Diego Marchetti

Analysis of Post-Installation Dynamic Load Test Data for Capacity Evaluation of Deep Foundations 312
Frank Rausche, Garland Likins, and Mohamad H. Hussein

The Role of Ram Momentum in Pile Driving—Part IV: Applications 331
David K. Crapps

Design of Foundations for a Wind Turbine Employing Modern Principles 351
Lymon C. Reese and Shin-Tower Wang

Axial Pile Capacity of Large Diameter Cylinder Piles 366
Peter Lai, Michael McVay, David Bloomquist, and Dhuruva Badri

Effective Stress Analysis and Set-up for Shaft Capacity of Piles in Clay 384
Bengt H. Fellenius

Wave Forces on Bridge Decks during Hurricanes and Impact on the Foundation 407
Jean-Louis Briaud and Hamn-Ching Chen

Vibration and Shear Wave Velocity Measurements at the I-10 Blackwater River Bridge 417
Richard D. Woods and David K. Crapps

Mitigation of Liquefaction Potential of Silty Sands 433
James K. Mitchell

Loading Tests: Interpretation and Prediction of their Results 452
Luciano Décourt

The Easy Button for Driven Pile Setup: Dynamic Testing 471
Paul J. Bullock

Downdrag in Pile Design: The Positive Aspects of Negative Skin Friction 489
Gary L. Kuhns

Anatomy of Foundation Performance Involving Three Grain Silos Systematically Loaded to Impending Failures 507
James E. Laier, Gary D. E. Cowles, and Mark E. White

Seepage through Rockfill Dams in Narrow Valleys 522
Ali Soleimanbeigi and Fardin Jafarzadeh

Site Characterization Philosophy and Liquefaction Evaluation of Aged Sands—A Savannah River Site and Bechtel Perspective 540
Michael R. Lewis, Ignacio Arango, and Michael D. McHood

Design and Construction of Earthen Reservoirs in Florida (1967-2007) 559
John E. Garlanger and Thomas J. Leto

Failure of the New Orleans 17th Street Canal Levee and Floodwall during Hurricane Katrina **574**
Robert G. Bea

Geotechnical Issues for Closure of Ammonia Soda Ash Waste Ponds **594**
Gary R. Schmertmann, Robert C. Bachus, and Keith D. Roberts

Slope Failure at a Major Cut for a Motorway in Austria—Its Cause and Its Stabilisation **616**
Martin Fuchsberger

Benefits of In-House Collaboration of Geotechnical and Structural Engineering Services **626**
David L. Freed

The Quest for Quality in Deep Foundations **636**
John A. (Jack) Hayes

Some Simple Methods I Used in Research and Practice **653**
John H. Schmertmann

Indexes

Subject Index **669**

Author Index **671**

AMERICAN SOCIETY OF CIVIL ENGINEERS

Founded November 5, 1852

TRANSACTIONS

Paper No. 2775

THE UNDISTURBED CONSOLIDATION BEHAVIOR OF CLAY

By John H. Schmertmann,[1] J. M. ASCE

With Discussion by Messrs. Robert L. Crisp, Jr., and John H. Schmertmann

Synopsis

Quantitative methods for correcting the results of the laboratory consolidation test for the effects of sample disturbance are presented. These methods are developed by attempting to answer several questions: What is the basic laboratory behavior and field behavior of a consolidating clay? How is this behavior modified by sample disturbance? What is the pattern of such modifications? These patterns are used as a basis for estimating, by extrapolation, the completely undisturbed conditions. Finally, an example of the results obtained by the use of the suggested methods is presented.

The techniques suggested were developed primarily from observations made during the performance of routine tests and analyses. Examples are presented of the various ways in which sample disturbance can occur and the manner in which this disturbance can alter the results of the consolidation test. Some observations are made indicating the pattern of this disturbance with respect to the steepness of the virgin slope, the lowering of equilibrium void ratios, and the rebound slopes.

It is emphasized that the rebound slope is important in interpreting the results of consolidation tests. Results of laboratory tests and a settlement analysis are presented which indicate that the rebound slope may be an important guide in the estimation of recompression settlements for many clay formations. As shown by an example, recompression settlement may be of significant magnitude; it is therefore worthy of careful study.

Alternate methods are introduced for estimating the preconsolidation stress. Because of the manner in which the position of the virgin curve is changed by sample disturbance, the use of a graphical method for determining a minimum

Note.—Published, essentially as printed here, in October, 1953, as *Proceedings-Separate No. 311*. Positions and titles given are those in effect when the paper or discussion was received for publication.

[1] United States Army, Snow, Ice, and Permafrost Research Est., Wilmette, Ill.; formerly Junior Soils Engr., Moran, Proctor, Mueser & Rutledge, New York, N. Y.

possible preconsolidation stress is suggested. Several examples are presented to show how water-content data can be useful in determining preconsolidation profiles in clay layers having uniform consolidation characteristics. A "void-ratio-reduction-pattern" technique is introduced as another method of determining preconsolidation values.

Sample disturbance in the laboratory is shown to be a contributing cause of too-slow predictions of the rate of settlement of sensitive clays. A time-settlement analysis is presented in which the results of laboratory consolidation tests were corrected for the effects of sample disturbance.

Because of limited data the application of the methods presented is restricted to saturated natural clays of sedimentary or glacial-sedimentary origin.

INTRODUCTION

General Purpose.—The laboratory consolidation test is the basis for computing the settlement of structures overlying compressible clay deposits. This test requires expensive equipment, expert supervision, and considerable time. Each test is therefore a definite technical and financial investment.

Unfortunately, the structure of every test sample is partly altered as a result of the stress changes that occur when the sample is removed from the ground and the distortions that are caused by the sampling operation and subsequent laboratory handling. A certain part of this disturbance has proved unavoidable.[2] Such sample disturbance distorts the results obtained from the laboratory consolidation test. It is therefore technically necessary and economically desirable to be able to correct the results of the laboratory test for the effects of sample disturbance.

It is intended that quantitative methods be presented herein whereby the results of a laboratory consolidation test, modified by the inevitable sample disturbance, can be corrected for the effects of this disturbance and thus be used to better predict the true behavior of the clay in the field.

Scope.—The correction of the results of laboratory consolidation tests is investigated as follows:

1. Examination of the basic behavior of a consolidating clay.
2. Observations of the manner in which this behavior is modified by sample disturbance.
3. Investigation of the pattern of such modifications.
4. The use of such patterns as a basis for estimating, by extrapolation, the completely undisturbed conditions.

There are also included herein investigations of the preconsolidation stress, of recompression settlements and rebound heave, and of the effects of sample disturbance on the laboratory time curves. An example of the results obtained by the use of the suggested methods is presented.

The data and methods introduced are applicable only to natural cohesive soils. The laboratory testing and job experience are limited to saturated or-

[2] "Subsurface Exploration and Sampling of Soils for Civil Engineering Purposes," by M. J. Hvorslev, Waterways Experiment Station, Vicksburg, Miss., November, 1949, p. 189.

ganic and inorganic sedimentary clay deposits and to clays deposited as a result of glacial action; therefore, partly saturated and residual clays are excluded from this work.

TERMINOLOGY, EQUIPMENT, AND TECHNIQUES

Terminology.—The term "consolidation test" refers to a fixed ring or floating ring, standard, one-dimensional laboratory test. The term "pressure-void ratio diagram" refers to the semilog method of indicating the results of a laboratory consolidation test.

Equipment.—The consolidometer equipment,[3] both floating ring and fixed ring, utilized dead-weight loading at a lever ratio of 10 to 1. The consolidometer ring was 1.00 in. high, with an inside diameter of 2.50 in. All sample compression was measured with dials reading directly to 0.0001 in.

All undisturbed samples were obtained with a thin-walled, fixed-piston sampler. The sampler has a length of 30 in., an inside diameter of 3.0 in., with a wall of No. 16 to No. 20 gage metal. Whenever possible, the sampler was pressed into the soil with a rapid, continuous push.

Techniques.—The consolidation tests were performed using the standard geometric progression of load increments. A continuous time-compression plot (semilog) was kept for each load increment; the subsequent increment was not applied until the increment was determined to be at the stage of secondary compression. The final pressure-void ratio diagrams were computed by adjusting all load-increment compressions for a standard laboratory time interval—that time being one at which all increments were in secondary compression. This time was generally 500 min, 1,000 min, or 1,440 min. Any swelling noted during a load increment was immediately arrested by the application of the next load increment.

A careful determination of the initial height (to 0.0001 in.) and weight (to 0.01 gram) of the test sample was made prior to each consolidation test. A specific-gravity test, using a pycnometer, was also made. This information, in addition to data concerning the final water content of the test sample, permitted an accurate computation of the void ratio and the degree of saturation at the beginning and end of each test. It was therefore possible to make the important cross-check between computed and measured over-all compressions and thus assure that there were no major inaccuracies in the computation of the best results.

EXISTING INFORMATION

The existing (1954) information on the subject of the effects of sample disturbance is largely qualitative in scope. This investigation can be considered to be an attempt to develop this qualitative information and the additional information presented herein into a systematic, quantitative procedure for correcting results of laboratory consolidation tests for the effects of sample disturbance. Perhaps the most comprehensive development of some of the qualitative concepts considered herein has been made[4] by P. C. Rutledge, M. ASCE. A

[3] "Recent Developments in Soil Testing Apparatus," by P. C. Rutledge, *Journal*, Boston Soc. of Civ. Engrs., October, 1935.

[4] "Relation of Undisturbed Sampling to Laboratory Testing," by P. C. Rutledge, *Transactions*, ASCE, Vol. 109, 1944, p. 1155.

thesis by T. Van Zelst, J.M. ASCE, completed in 1948 at Northwestern University, at Evanston, Ill., included pertinent data and analyses of special significance; Mr. Van Zelst's information is utilized subsequently. Mr. Van Zelst also summarized the available information (as of 1948) on factors that affect the results of laboratory consolidation tests.

Basic Consolidation Behavior

Laboratory.—The basic laboratory behavior of a clay consolidating under the influence of stresses higher than those previously experienced by the sample

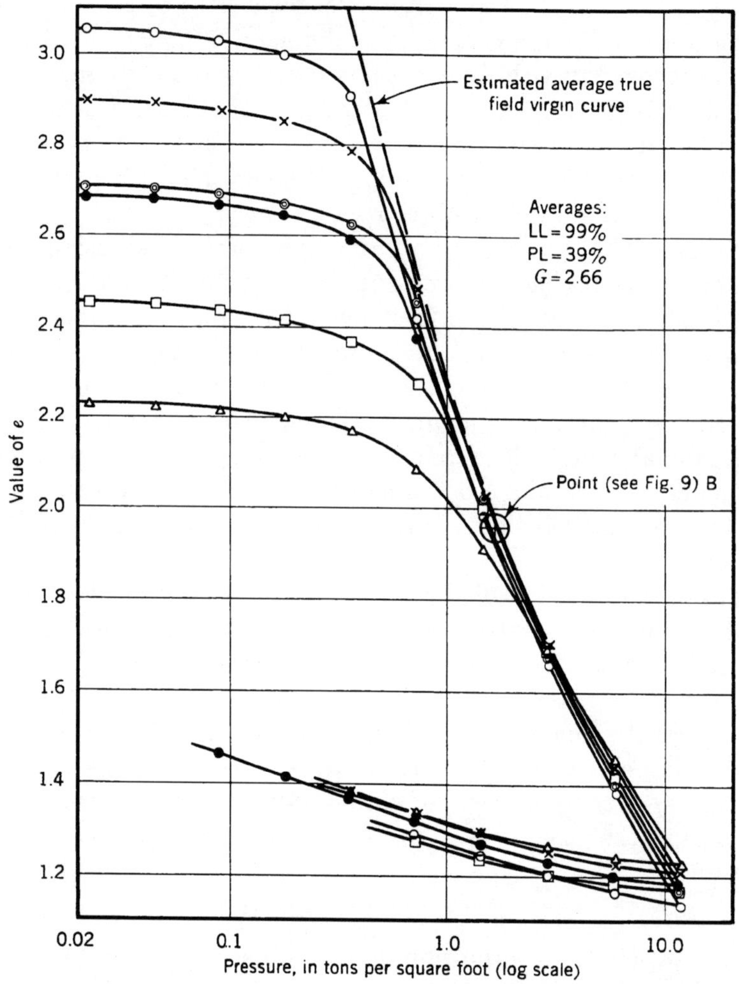

Fig. 1.—Example of Unique Position of Laboratory Virgin Slope

is such that its void ratio, e (or water content, w), steadily reduces with increasing intergranular soil pressures. Except in the case of significant long-time secondary consolidation, there is a new equilibrium void ratio for each new effective stress. When several of these points are plotted on a semilog plot, the resulting steep line, with flat curvature, is termed the "virgin consolidation slope."

In order to demonstrate that, for a given soil, the position of the virgin slope is a unique property of that soil, laboratory test data are presented in Fig. 1.

UNDISTURBED CONSOLIDATION

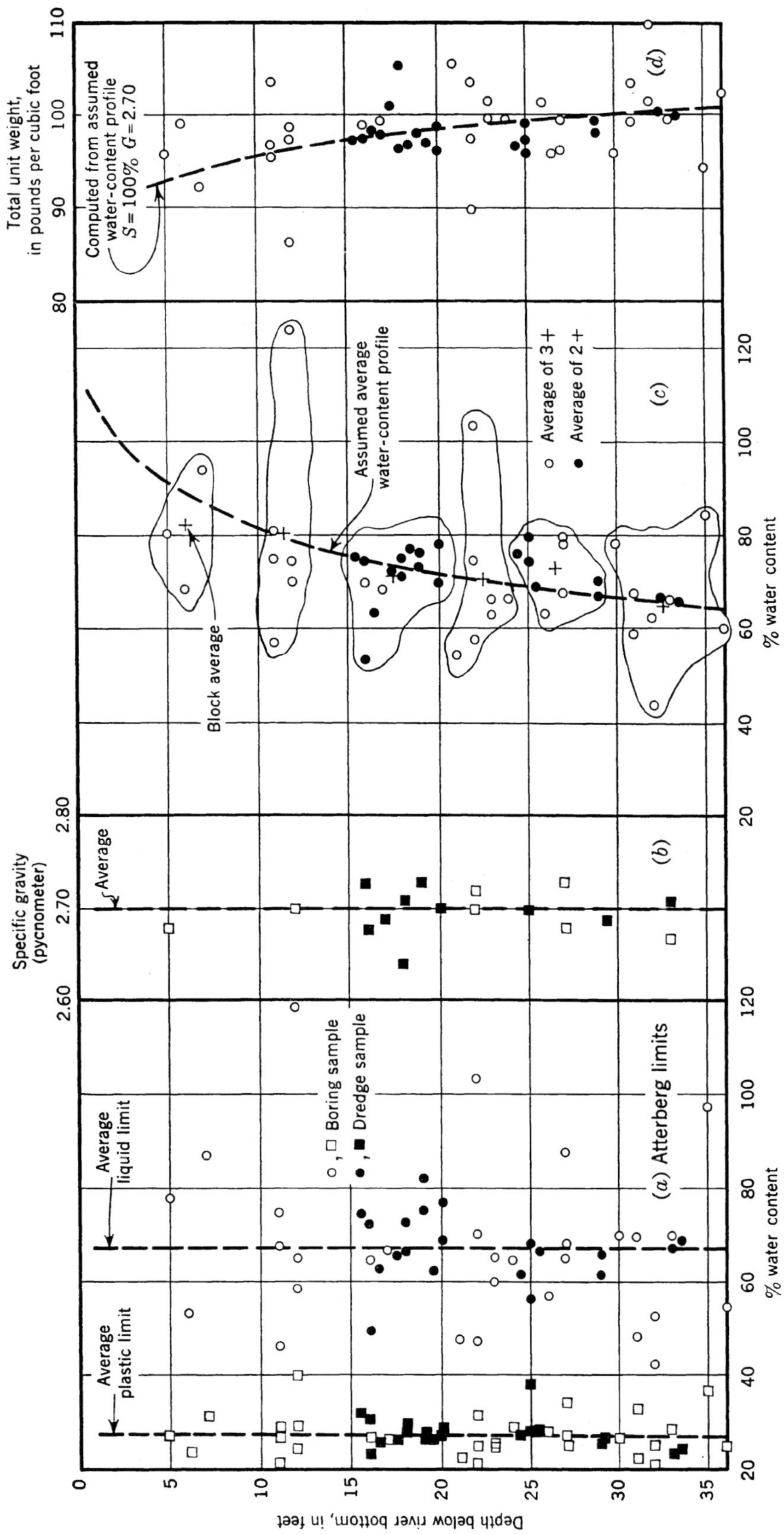

FIG. 2.—LABORATORY TEST DATA FROM RIVER-DEPOSITED, ORGANIC, SILTY CLAY

UNDISTURBED CONSOLIDATION

The six consolidation tests were performed on six separate samples obtained from a single boring and distributed throughout the uniform lower half of a 60-ft New York post-glacial marine river deposit of organic silty clay. This clay had an average liquid limit (LL) of 99%, a plastic limit (PL) of 39%, and a specific gravity (G) of 2.66. The range of from 3.08 to 2.24 in the initial void ratio, e_0, is the result of varying degrees of consolidation that had occurred under the load of a recent fill at the site. Despite the unavoidable minor

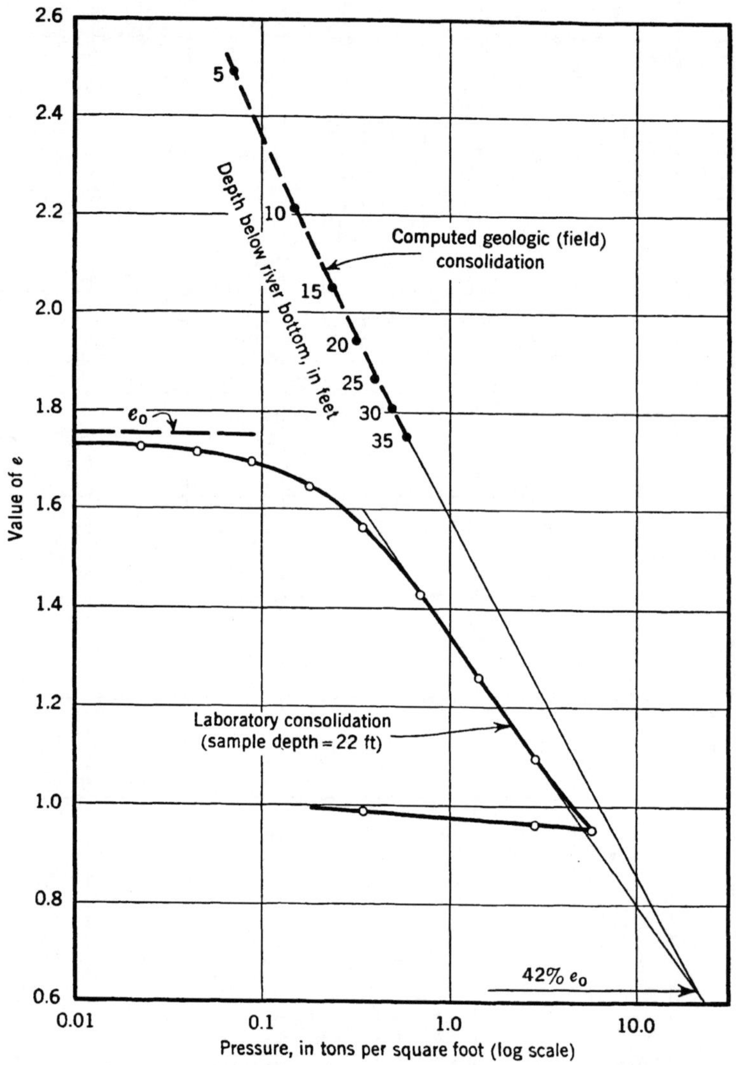

FIG. 3.—COMPARISON OF LABORATORY AND FIELD CONSOLIDATION

sample differences, it can be seen that the position of the virgin slope is basically independent of the interrelated properties of e_0 and the preconsolidation stress, P_c.

Laboratory results similar to those noted in Fig. 1 have been obtained from other uniform soil formations—for example, the consolidation-test data on Mexico City clay.[4] It therefore appears likely that a unique virgin slope is a laboratory property of all saturated, natural clay soils.

UNDISTURBED CONSOLIDATION

Field.—In order to investigate effectively the basic behavior of a clay as it consolidates in the field it is necessary that only proven uniform clay deposits having known stress histories be considered. Such a clay deposit is used in the subsequent investigation.

In Fig. 2 there are presented the results of laboratory tests on soil specimens from a bed of marine, river-deposited, organic, silty clay. This river-bed formation is finely layered with organic silty clay and silty fine sand. The sand layers range in thickness from a thin sifting of sand to a maximum of $\frac{1}{8}$ in. The thin sand layers are a minor part of the entire formation. The laboratory

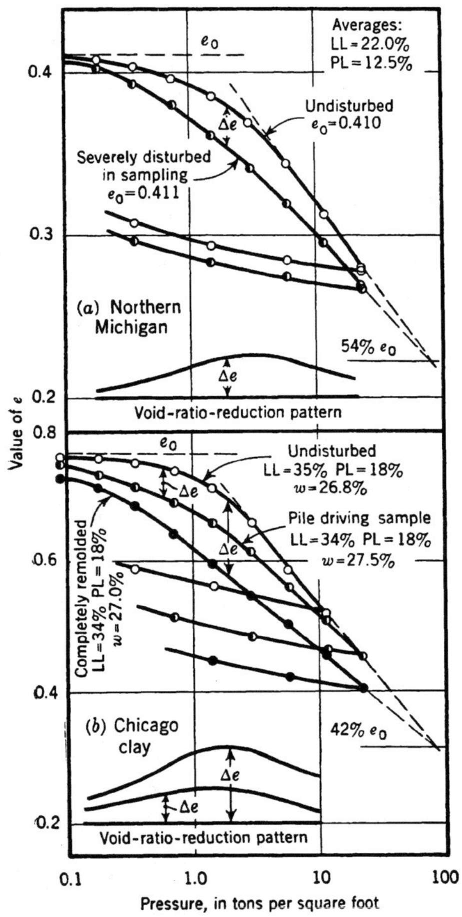

FIG. 4.—COMPARISONS OF UNDISTURBED AND DISTURBED CONSOLIDATION TESTS ON GLACIAL SOILS

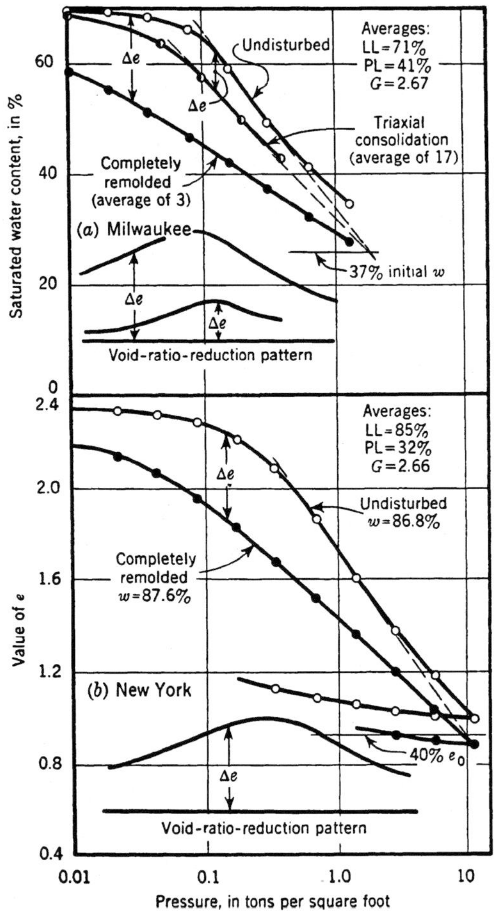

FIG. 5.—COMPARISONS OF UNDISTURBED AND DISTURBED CONSOLIDATION TESTS ON ORGANIC SOILS

samples were taken from twelve borings and five dredge positions across a 4,000-ft river width. To permit correlation with overburden pressure, all data are plotted in terms of depth below river bottom.

The results of the laboratory Atterberg limit tests and specific-gravity tests (Figs 2(*a*) and 2(*b*)) are presented to show the over-all uniformity of the river deposit. Fig. 2(*c*) is a plot of the natural water-content data in which each point represents the average of at least two or three determinations of water content. In Fig. 2(*c*) there is shown the average water-content profile determined on the basis of the most logical curve through block averages of the detailed data. The upper part of the assumed water-content profile is sharply curved because the upper 5 ft of the river formation was reported to be in a "semifluid" condition and could not be sampled. It is probable that this semifluid state results from disturbance of the very soft clay which has a water content much higher than the liquid limit. In Fig. 2(*d*) there are shown the total unit weights determined by weight and volume measurements made on the entire sample before it was removed from the sampler tube, together with a unit-weight profile computed from the assumed water-content profile. The computed profile is in agreement with the observed data. Therefore, the data presented in Fig. 2 indicate an essentially uniform soil formation which exhibits an average decrease in water content with depth. The numerous thin sand seams are believed to be at least partly responsible for the scatter of the data in Fig. 2.

Stress History.—This river deposit was under an average of 35 ft of water at the time of sampling. It is believed that its surface has never been above water. Thus, there was no preconsolidation by surface drying, which is often found in otherwise normally consolidated clay deposits. This conclusion, together with the belief that the overburden never significantly exceeded that presently existing, is supported by the observation of the very soft consistency of the upper 5 ft of the deposit. The many thin, fine-sand layers noted throughout this formation make it seem probable that rapid drainage followed deposition, with the result that the soil is normally consolidated.

If normal consolidation is assumed, it is possible to compute the pressure-void ratio plot for the field consolidation of this soil formation. Overburden stresses can be computed directly from the computed profile of average unit weights, and the void ratio can be computed from the average water-content profile. Such computations have been made for 5-ft increments of depth below river bottom; a semilog plot of the results is shown in Fig. 3.

Comparison Between Laboratory and Field.—Also shown in Fig. 3 is a laboratory consolidation test on a sample taken at a depth of 22 ft. This test sample was the less disturbed of the two consolidation-test samples obtained from the site. It is important to note that the computed geologic (or field) consolidation curve has the typical flat curvature observed in laboratory virgin slopes (Fig. 1). The computed field virgin slope also agrees with the observed virgin slope for a laboratory consolidation test on a partly disturbed sample of the same soil.

The preceding analysis and several others, which include Detroit (Mich.) and Chicago (Ill.) clay formations—but which required more indirect reasoning—have consistently led the writer to the conclusion that many natural clays consolidate in the field along virgin slopes similar, if not identical, to those obtained from a completely undisturbed laboratory consolidation test.

EFFECTS OF SAMPLE DISTURBANCE

Previous Observations.—Mr. Rutledge[5] has summarized the general effects of sample disturbance on the laboratory consolidation test, as follows:

[5] "Relation of Undisturbed Sampling to Laboratory Testing," by P. C. Rutledge, *Transactions*, ASCE, Vol. 109, 1944, p. 1164.

UNDISTURBED CONSOLIDATION

"(1) It decreases the void ratio at which the soil will carry any given vertical stress;

"(2) It obscures the previous stress history of the soil and its preconsolidation load; and

"(3) The straight-line portion of the remolded compression curve is displaced downward from the laboratory virgin compression curve, and its slope, or the rate of decrease in void ratio with increasing stress, is less.

"These results show definitely the effects of sample disturbance over the range from the best undisturbed samples which have been obtained for laboratory test to complete remolding. Any intermediate degree of disturbance must result in a compression curve which falls between the curves that would be obtained for the same soil in these two known limiting conditions."

In Figs. 4 and 5 there are presented data from two glacial soils and two organic soils showing the typical results of a consolidation test; the tests were made on the most undisturbed samples obtainable and on other completely and partly disturbed samples of the same soil. Such disturbance can occur in a variety of ways: In Fig. 4(*b*) is presented an example of in situ disturbance (pile driving). The sample was taken approximately 400 days after driving and was partly remolded. An example of excessive disturbance in removing the sample from the ground (poor sampling technique) is presented in Fig. 4(*a*). The triaxial test results presented in Fig. 5(*a*) for Milwaukee (Wis.) represent laboratory disturbance by sample manipulation; these triaxial samples required more trimming surface per unit volume than did the one-dimensional consolidation test samples. The Milwaukee results are from the work of **M.** L. Calhoon,[6] J. M. ASCE. Finally, in Fig. 4(*b*) and Figs. 5(*a*) and 5(*b*) are examples of deliberate laboratory remolding.

In Fig. 6 there are shown pressure-void ratio curves obtained by Mr. Van Zelst for a glacial lake clay from Crookston, Minn. These curves summarize the results of work concerning the effects of varying degrees of sample disturbance on the results of a consolidation test. Various degrees of disturbance were obtained by using a constant-diameter ring and varying the height of the sample, thereby varying the percentage of the total volume that was disturbed by sample trimming.

The data in Figs. 4, 5, and 6 are additional evidence of the validity of Mr. Rutledge's conclusions.

Additional Observations.—The data in Figs. 4, 5, and 6, however, permit some additional observations.

Initial Virgin Slope.—Straight-line extensions of the initial virgin slopes of consolidation tests on samples disturbed in varying degrees intersect within a narrow range of void ratios. The term, initial virgin slope, is herein defined as the slope of the straight line drawn through the first two laboratory test points which fall on the virgin part of the pressure-void ratio diagram.

In Figs. 3 and 6 it can be seen that the intersection of the field initial virgin slope with the initial virgin slopes of laboratory consolidation tests ranges from 40% to 46% of the initial void ratio of the sample. In Figs. 4, 5, and 14 (to be presented subsequently) the intersection range for undisturbed and more-

[6] "The Effect of Remolding on the Strength of a Saturated Clay," by M. L. Calhoon, thesis presented to Northwestern University, Evanston, Ill., in 1948 in partial fulfilment of the requirements for the degree of Master of Science.

disturbed samples is between 37% and 60% of e_0. If the unusually dense silty clay shown in Fig. 4(*a*) and the sample disturbed by pile driving 400 days prior to sampling—thereby permitting possible thixotropic action (Fig. 4(*b*))—are excepted, the intersection range becomes from 37% to 46% of e_0.

On the limited basis of these and similar data, it is believed that an initial virgin-slope intersection point at 42% of e_0 would be a good estimate for most clays.

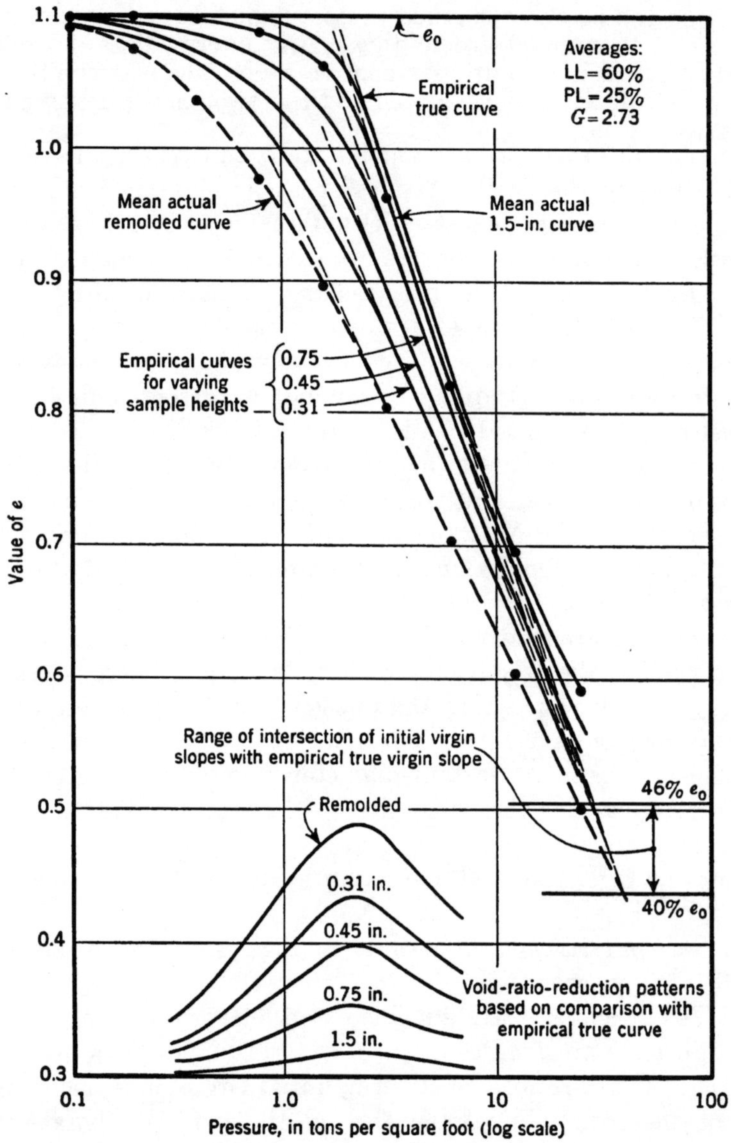

FIG. 6.—EFFECT OF SAMPLE DISTURBANCE ON THE PRESSURE-VOID RATIO DIAGRAM

Void-Ratio-Reduction Pattern.—For each set of comparative consolidation tests there has been plotted a void-ratio-reduction pattern. This term is used herein to describe the curve that shows the pattern of the reduction in void ratio of the disturbed sample test when it is compared to the undisturbed sample test. With respect to the least-disturbed consolidation-test curve, the void-ratio-reduction curves for more-disturbed samples increase symmetrically

UNDISTURBED CONSOLIDATION

from a small value at low stresses to a maximum value at or near the preconsolidation stress—and then decrease again along the virgin slope. The data in Fig. 6 indicate that the same pattern may exist between the clay in the field and the laboratory sample.

Rebound Slope.—Finally, disturbance does not appear to alter significantly the rebound slope. Examination of the data from Mr. Van Zelst's thesis and other test data supports this preliminary observation.

Geologic Rebound of a Preconsolidated Clay

Laboratory investigations have shown that it is common to find clays that have been subjected—in their geologic history—to pressures that are greater

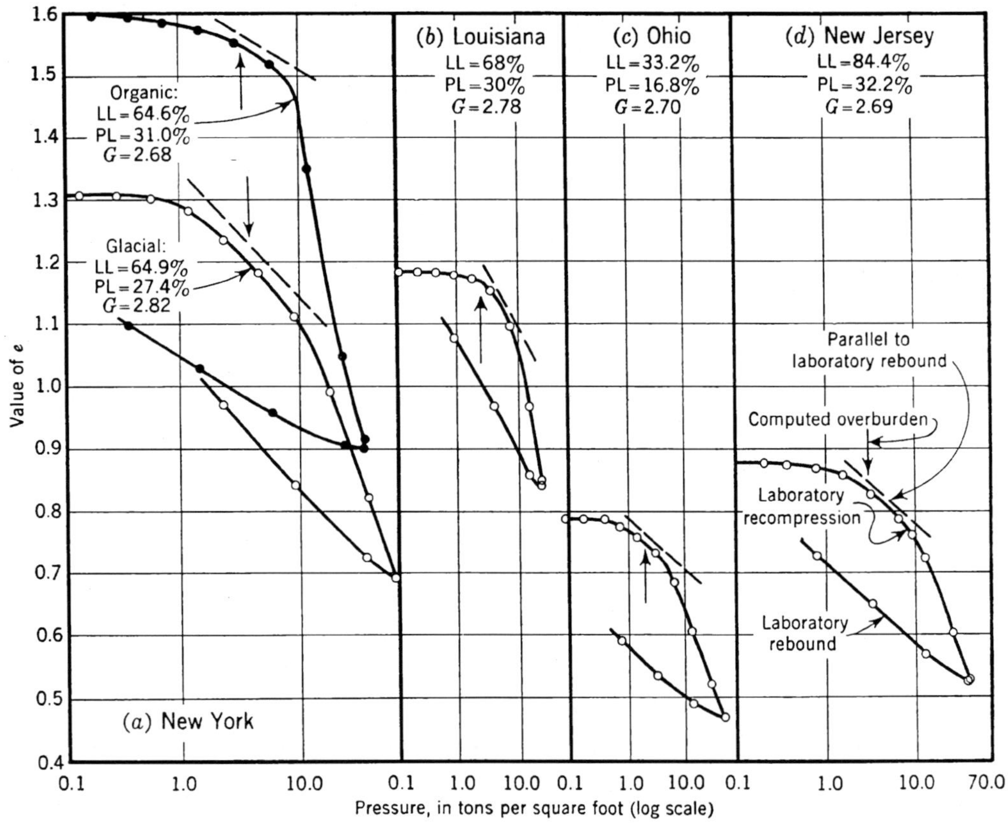

Fig. 7.—Comparisons of Laboratory Recompression and Rebound Slopes

than those presently existing in the form of overburden; such clay is said to be preconsolidated. During the process of stress release from the preconsolidation stress to that stress now existing, the soil must have swelled or rebounded. One must estimate the magnitude of this rebound when making a complete settlement analysis.

In order to investigate this problem one must imagine a laboratory consolidation test on a highly preconsolidated clay. What would be the shape of the recompression part of the pressure-void ratio diagram? Long rebounds and subsequent recompressions of laboratory samples reveal a shallow hysteresis loop in which the recompression is approximately parallel to and at a slightly higher void ratio than the rebound. If a similar rebound occurred in the field, one would therefore also expect the field recompression curve to be approximately paralled to the field rebound curve, with a shallow loop between the curves.

In Fig. 7 there are shown laboratory consolidation-test results on a variety of highly preconsolidated clays. These results are typical of those of many other tests on such clays. Any swelling under light loads was immediately arrested by the application of the next load increment. The three-slope character of such tests is shown by the approximately constant void-ratio line during the loading interval when swelling was restrained, then by the distinct recompression slope, and finally by the steep virgin slope. For each test there is indicated the computed existing overburden stress, and a dashed line has been drawn parallel to the laboratory rebound slope to facilitate comparison with the recompression slope. The recompression curves are approximately parallel to the laboratory rebound slopes, thus indicating the effect of a field rebound whose slope approximates the laboratory rebound slope.

Previously, evidence was presented which indicated that the laboratory virgin slope is a reasonable indication of virgin-consolidation behavior in the field. Thus, there is reason to believe that the laboratory recompression slope, which precedes the virgin slope, should also be representative of field behavior. Because the field recompression curve should be parallel to the field rebound curve and laboratory recompression curves are observed to be parallel to laboratory rebound slopes, the clay behaves as if its field rebound occurred at a slope approximately parallel to the observed laboratory rebound which concludes a consolidation test.

As previously shown, rebound slopes do not appear to be significantly altered by sample disturbance. Therefore, it should be possible to estimate recompression settlements that are more or less independent of the usual disturbance of test samples. This problem will be investigated more completely subsequently.

A number of recompression-type settlement analyses have been performed based on the assumption that field and laboratory rebound slopes are parallel. The most successful of these analyses is discussed in detail subsequently. The results of this analysis, together with actual field observations, are presented in Table 2 (to be presented subsequently). To date (1954), the settlement data available for checking the previously cited computations have agreed satisfactorily, thus supporting the approximate parallel-relationship hypothesis for the clays at the sites analyzed.

THE PRECONSOLIDATION STRESS

Determination of the preconsolidation stress is very important in estimating the true consolidation behavior of the clay in the field. Unfortunately, distortion of the curvature characteristics of the laboratory consolidation sample, from which the previous stress history of the soil is inferred, is one of the most serious results of sample disturbance. The graphical construction suggested[7] by A. Casagrande, M. ASCE, is the method most commonly used for estimating

[7] "Determination of the Preconsolidation Load and Its Practical Significance," by A. Casagrande, *Proceedings*, International Conference on Soil Mechanics and Foundation Eng., School of Eng., Harvard Univ., Cambridge, Mass., Vol. III, 1936.

UNDISTURBED CONSOLIDATION

the value of the preconsolidation stress. In this technique varying degrees of sample disturbance are not completely accounted for. As can be seen by using this construction on the data presented in Figs. 4, 5, 6, and 16, increasing disturbance makes the choice of the minimum radius of curvature quite difficult and generally results in lower determinations of preconsolidation stress.

There are presented subsequently several alternate techniques which may also be useful in determining preconsolidation values.

Minimum Possible Preconsolidation Stress.—In Fig. 8 there is illustrated a method that can be used to determine a minimum possible preconsolidation stress. This method was originally suggested by Mr. Rutledge and has been

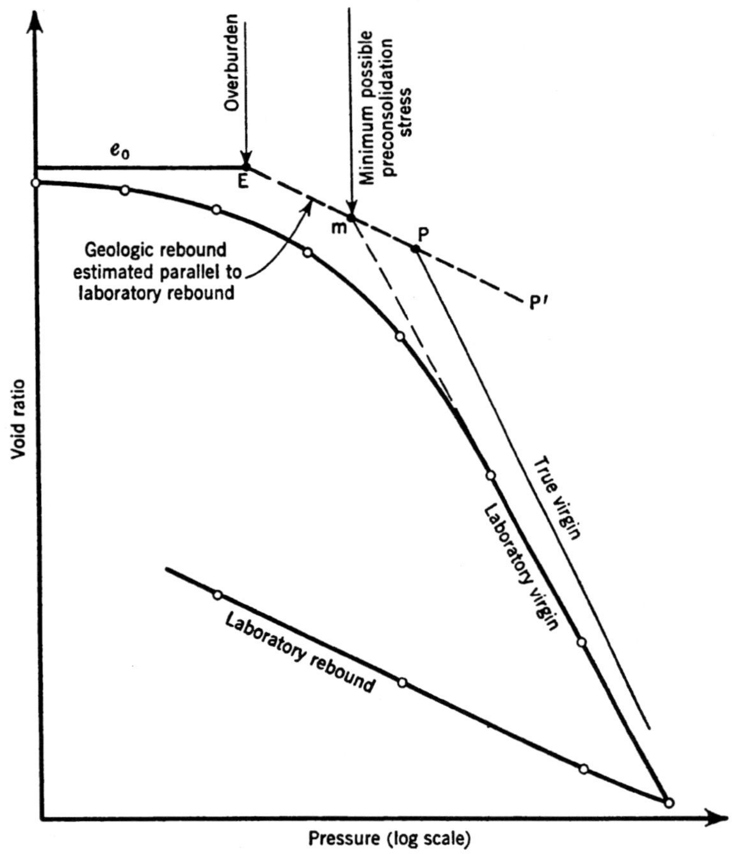

FIG. 8.—DETERMINATION OF MINIMUM POSSIBLE PRECONSOLIDATION STRESS

slightly modified herein. In Fig. 8, point E represents the presently existing field overburden and void ratio. Line E–P′ represents the geologic rebound constructed parallel to the laboratory rebound through point E. The laboratory initial virgin slope is extended until it intersects either the initial void-ratio line or geologic rebound line at point m. From the knowledge of the effects of sample disturbance, one knows that the true virgin slope must lie to the right of the laboratory slope. Therefore point m must always be at a lower stress than the true preconsolidation value determined at point P.

If it can be stated with a reasonable degree of certainty that the preconsolidation stress cannot be below a certain value, one has established a lower limit that may have engineering significance. If this lower limit is higher than

UNDISTURBED CONSOLIDATION

the overburden pressure known to be effective at the sample elevation, one may be reasonably certain that some, if not all, of the consolidation settlement will be recompression. However, if the lower limit is consistently below the computed overburden stress, the analyst then has reason for believing the soil to be normally consolidated and—on rare occasions—under-consolidated.

Water-Content Data.—The water content of a cohesive soil is independent of sample disturbance. Furthermore, it is one of the simplest properties to determine in the laboratory. Information about water content is therefore potentially useful in estimating the undisturbed properties of natural clays.

Although it is rare to find an entire clay layer that is uniform, it is not uncommon to find clay formations with a part of their thickness exhibiting uniform consolidation characteristics. Uniformity can be demonstrated either directly by consolidation tests or indirectly by classification tests. If such a condition can be found, the water-content profile through this part of the formation can be utilized to estimate the true preconsolidation values.

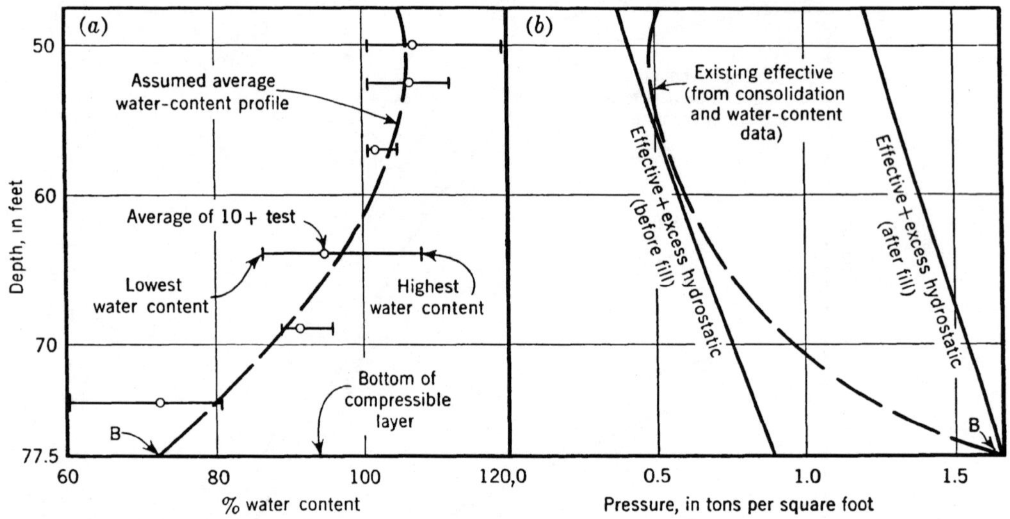

FIG. 9.—EXAMPLE OF DETERMINATION OF PRECONSOLIDATION FROM NATURAL WATER-CONTENT PROFILE

An example of such an estimate is the data presented in Figs. 2 and 3, in which a study of the water-content profile indicated normal consolidation. For another example, Fig. 9 is presented. In this case the lower half of a 60-ft, marine, organic clay deposit was demonstrated to have uniform consolidation properties by the tests presented in Fig. 1. In Fig. 9(*a*) are contained the water-content data and assumed profile for this part of the formation. In this analysis it was possible to establish a direct relationship between water content and effective stress. This correlation was possible because this was a case of double-drainage consolidation under a recent fill. The effective stress at the bottom of the compressible layer was therefore the total stress at that elevation; it could be computed readily. Points B in Fig. 9 indicate the water content and effective stress at the bottom of the layer.

The effective stress-void ratio computation for point B is shown by the circled cross in Fig. 1. This cross represents a point on the true average virgin curve of this soil. The heavy dashed line in Fig. 1, which passes through point

UNDISTURBED CONSOLIDATION

B, is an estimate of the average true field virgin curve over the void-ratio range shown. By converting the water-content profile to void ratio and using the estimated true virgin curve, it is possible to compute the effective-stress profile at the time of sampling.

The results of this computation are shown in Fig. 9(*b*) in relation to the total of effective stresses and excess hydrostatic stresses existing before and

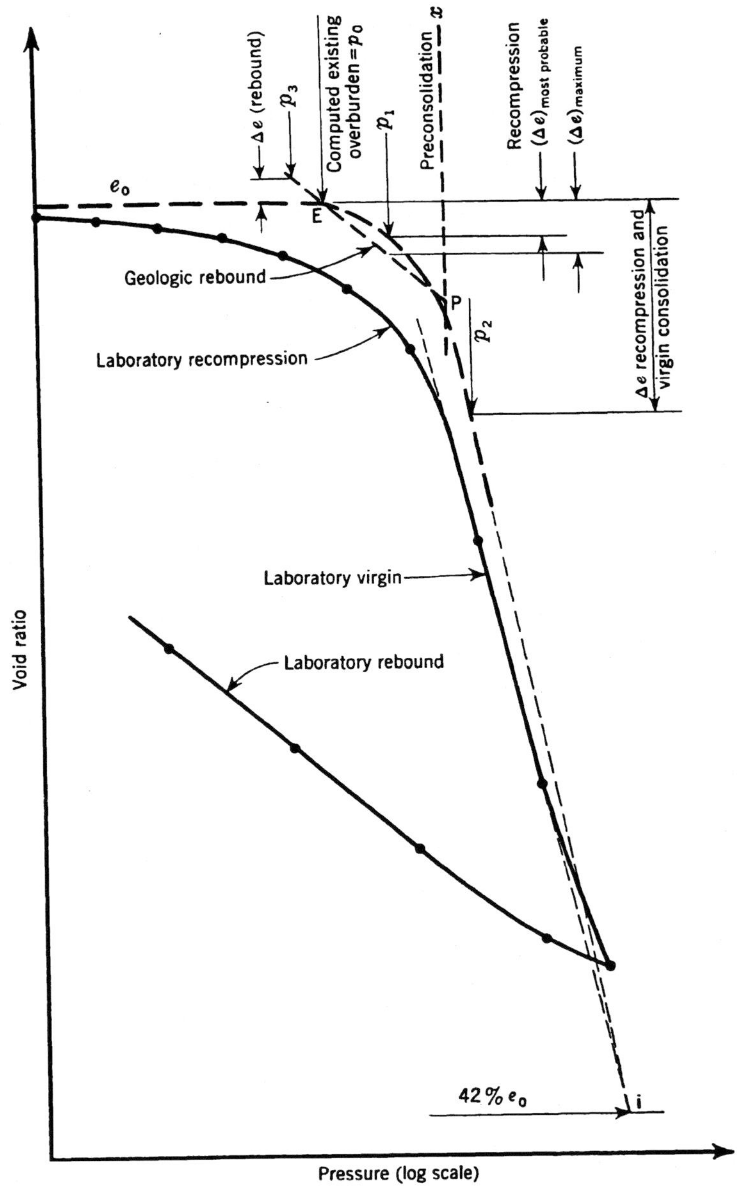

Fig. 10.—Reconstruction of True One-Dimensional Consolidation Behavior

after the fill was placed. After the effective-stress distribution was determined for the lower half of the formation, it was relatively easy to estimate the distribution for the upper half. Settlement computations including this complete distribution gave a close agreement with actual observations following the placing of the fill; normal consolidation at the time the fill was placed was assumed.

UNDISTURBED CONSOLIDATION

In cases in which no direct relationship between water content and effective stress can be determined, trial-and-error procedures must be used until the water-content profile checks with an assumed preconsolidation profile.

A method that may permit the successful application of water-content data to the determination of the curvature of the preconsolidation profile involves use of the water content-plasticity ratio.[8]

Void-Ratio-Reduction Patterns.—The best method of determining the preconsolidation stress directly from an individual consolidation test is based on a thorough understanding of the nature of the distortion of the pressure-void ratio diagram caused by sample disturbance. The void-ratio-reduction patterns presented in Figs. 4, 5, and 6 can logically be used. It is suggested that these patterns be utilized by choosing various possible preconsolidation values, reconstructing the estimated true consolidation test for each chosen value, and picking the highest value that gives a symmetrical void-ratio-reduction pattern. The results obtained by the writer are encouraging but inconclusive (as of 1954).

RECONSTRUCTION AND USE OF THE ESTIMATED TRUE PRESSURE-VOID RATIO DIAGRAM

Reconstruction.—In order to reconstruct the estimated true one-dimensional void ratio-consolidation pressure diagram, all the preceding material must be utilized. Beginning with the laboratory consolidation test, one should use Fig. 10 in the following step-by-step procedure:

1. Point E is plotted, which represents the effective stress and the void ratio existing in the field at the sample location. The effective stress is usually the computed overburden pressure. If the clay is not permitted to swell during storage, the field void ratio is the computed saturated initial void ratio, e_0, of the consolidation test sample.

2. Because point E represents the initial conditions for field compression, a perfect stress transfer from field to sampler to consolidometer would result in a test beginning with a constant e_0-line to point E.

3. The value of P_c is then determined by the most suitable method. For example—the use of water-content data is suggested for a uniform clay formation, and the Casagrande graphical construction[6] for a test with a steep virgin slope and a clearly defined point of minimum curvature. A constant stress line, x–x, is constructed at the chosen preconsolidation stress value.

4. The geologic rebound is constructed parallel to the laboratory rebound, goes through point E and intersects line x–x at the preconsolidation point P.

5. The initial laboratory virgin slope is extended as a straight line until it intersects the 42% e_0-line at point i. The line i–P is then constructed; this line represents the position and slope of the upper part of the estimated true virgin curve.

6. A recompression curve is estimated for the geologic rebound line E–P. This curve probably merges with the estimated true initial virgin slope in such a manner that the graphical construction[6] at the point of sharpest curvature will yield the preconsolidation point P.

---

[8] "Application of Soil Mechanics in Designing Building Foundations," by A. Casagrande and R. E. Fadum, *Transactions*, ASCE, Vol. 109, 1944, p. 383.

UNDISTURBED CONSOLIDATION

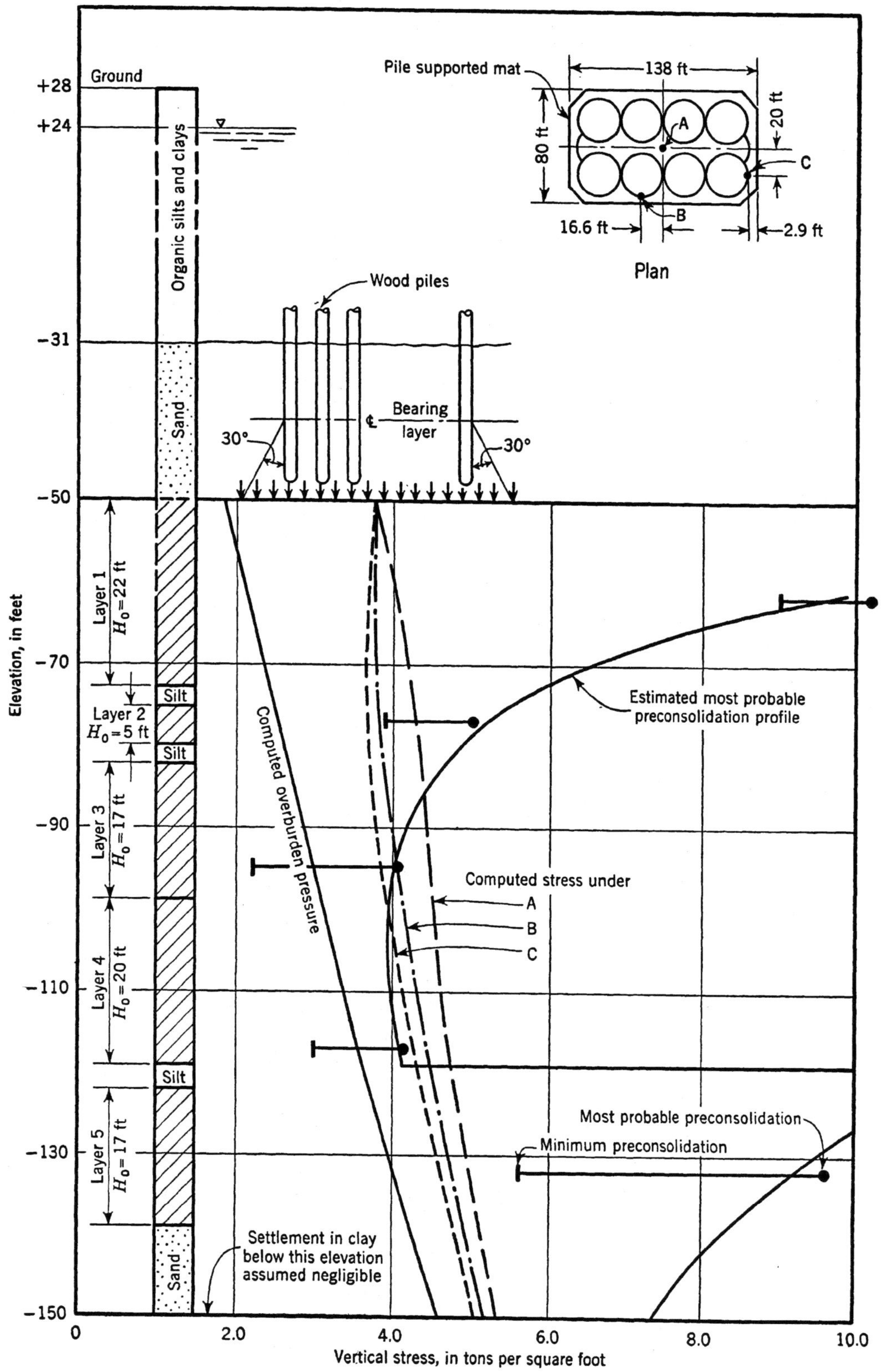

FIG. 11.—STRESS PROFILES UNDER SETTLEMENT POINTS

UNDISTURBED CONSOLIDATION

The heavy dashed line in Fig. 10 indicates the final reconstruction of the estimated true field pressure-void ratio diagram.

If it is determined that the clay is not overconsolidated, the construction of the geologic rebound and estimation of a recompression curve are unnecessary. The effective stress is either the overburden or some lower value on the initial-void-ratio line. The estimated true initial virgin slope is determined in the manner outlined in step 5.

Use.—The reconstructed true pressure-void ratio diagram can be used directly for settlement computations. As shown in Fig. 10, a stress increase of from p_0 to p_1 results in a void-ratio reduction along the recompression part of the consolidation test. Because the amount of loop between the geologic rebound and the recompression curve is presently (1954) a matter of judgment, two changes in the recompression void ratio have been indicated. The most probable change is based on the estimated recompression curve. For the maximum void-ratio change no loop is assumed, and therefore the recompression is back along the geologic rebound. All laboratory test data indicate the existence of some loop.

If the stress increase is from p_0 to p_2, both recompression and virgin consolidation occur. The reduction in void ratio for this load increase is indicated

TABLE 1.—VOID RATIOS AND PRESSURES IN FIG. 11

Item	Layer				
	1	2	3	4	5
Value of e_1 (in situ void ratio)	1.11	1.19	1.18	1.06	0.852
Value of p_1 (overburden stress), in tons per square foot	8.5	5.1	4.2	4.0	9.0
Value of p_c (preconsolidation stress), in tons per square foot	2.15	2.56	2.89	3.34	3.97

in Fig. 10 and is independent of any assumed recompression curve. Should there be a net decrease in load from p_0 to p_3, the soil would continue to rebound along the established geologic rebound slope and yield the void-ratio increase shown in Fig. 10. Examples of this occurrence can be seen in the consolidation-test results presented in Fig. 7. The glacial lake sample (Fig. 7(*a*)) and the Ohio clay (Fig. 7(*d*)) were both varved. Therefore, the clay in the sampler was able to swell during storage by drawing water from the silt and sand par stof the sample. The relationship between the computed overburden stress and the lower stress at the beginning of the recompression slope indicates that swell occurred. This swell was along the established rebound slope.

A SETTLEMENT ANALYSIS

General Approach.—In order to give practical significance to the previously developed theoretical concepts, an interesting settlement analysis will be reviewed which permits the use of some of these concepts.

It was proposed that additional storage facilities be constructed adjacent to a rigid eight-bin silo structure that had been built in 1926. In order to determine the accuracy with which one could compute the settlement of the new structure, settlement analyses were performed for three locations on the original structure for which load and settlement records had been kept. The uniform soil conditions of the entire site permitted this procedure. The most significant phases of this settlement analysis are as follows:

After a series of preliminary borings had established the uniformity of the subsoil conditions, a single undisturbed-sample boring was made in the area of the proposed structure. A schematic log for this boring is indicated in Fig. 11. Average void-ratio and pressure data for each of the five layers are presented

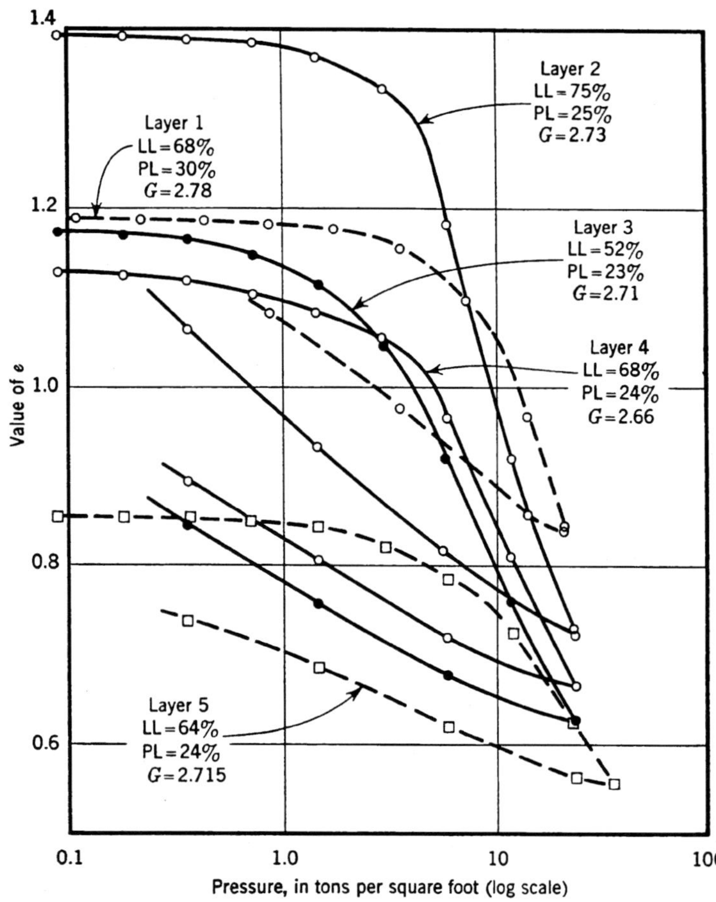

FIG. 12.—LABORATORY CONSOLIDATION-TEST DATA USED IN SETTLEMENT ANALYSIS

in Table 1. In this case the compressible layer causing settlements (shown crosshatched) was a medium to stiff, inorganic, river-deposited, gray-green Pleistocene clay extending from El. −50 to El. −139. This layer contained occasional thin silt lenses and pockets and three silt layers which are indicated on the profile. Numerous water-content tests were performed on undisturbed samples from this compressible layer. On the basis of these tests plus the boring-log information, five settlement layers for computation were established. A consolidation test was performed on an undisturbed sample from each layer in order to determine the consolidation characteristics of each layer and the overall preconsolidation profile. The results of the laboratory consolidation tests

UNDISTURBED CONSOLIDATION

are presented in Fig. 12. The elevations of the consolidation-test samples, the minimum and the most probable preconsolidation estimates, and the estimated most probable preconsolidation profile are shown in Fig. 11.

Analysis Techniques.—For this analysis it was decided to use the previously cited void-ratio-reduction-pattern method of estimating the most probable preconsolidation values. An example of this method is shown in Fig. 13. In Fig. 13 various possible preconsolidation values, P_c, were tried, and an estimated true curve was reconstructed for each trial. The resulting void-ratio-

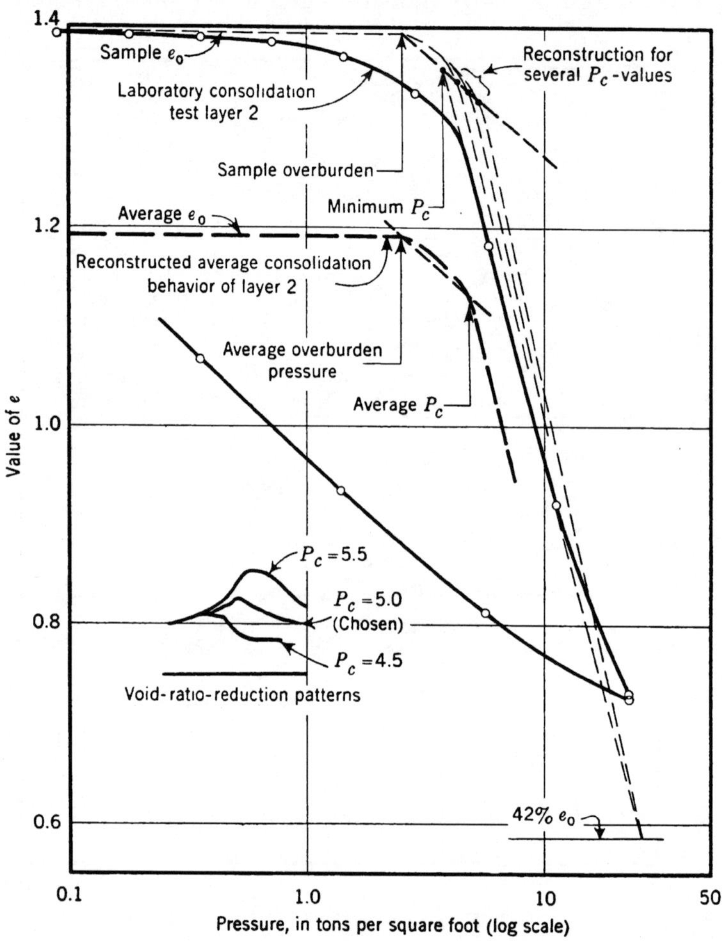

FIG. 13.—TYPICAL UTILIZATION OF LABORATORY PRESSURE-VOID RATIO DIAGRAM

reduction patterns for three trials are indicated, as well as the pattern that was chosen.

In Fig. 11 there is shown the plan of the foundation mat of the silo structure; included are the locations of the three points for which settlement analyses were performed. The recorded average live load plus computed dead load—which were uniformly distributed over the area of the foundation mat—was 2.23 tons per sq ft. This load was transmitted through wood piles into a 19-ft bearing layer of sand, the top of which was at El. −31. The piles almost penetrate this sand layer. The assumption made to account for the load-distributing action of this bearing layer is shown in Fig. 11. By use of this assumption, the

UNDISTURBED CONSOLIDATION

average stress increase at the top of the clay formation was determined as 1.90 tons per sq ft. The Boussinesq elastic stress-distribution charts for rectangular areas were then used to estimate the stress increase under the three settlement points.[9] These stress increases exist in addition to the computed existing overburden pressure and are so plotted in Fig. 11.

The average saturated void ratio for each computational layer was determined from the natural water-content data. The average initial and preconsolidation stresses were determined from the overburden and preconsolidation stress profiles. The values are indicated on the log in Table 1. These values, together with the reconstructed virgin-slope and rebound-slope magnitudes from the appropriate consolidation test, were then used to reconstruct the average true one-dimensional consolidation pressure-void ratio diagram for the entire layer. An example of this reconstruction for layer 2 is shown in Fig. 13.

There have now been reconstructed five average true consolidation tests, each one representative of a certain part of the entire compressible layer. The next step was to indicate the appropriate average total overburden plus structural stress under settlement points A, B, and C on each of the average true pressure-void ratio diagrams. As was shown in Fig. 10, the void-ratio reductions can be determined directly from the reconstructed pressure-void ratio diagrams. The void-ratio reduction, Δe, can then be converted into settlement by

TABLE 2.—Settlement[a] of Structure in Fig. 11

Point	Computed (Flexible Structure)						Observed (rigid structure)
	Layer[b]					Total	
	(1)	(2)	(3)	(4)	(5)		
A	5.0-R	1.1-R	3.9-RV	4.7-RV	0.8-R	15.5	12.3
B	4.4-R	0.8-R	1.9-RV	3.3-RV	0.5-R	10.9	11.3
C	4.3-R	0.7-R	1.4-R	1.8-RV	0.4-R	8.6	9.9
Average						11.7	11.2

[a] In inches. [b] R denotes recompression, RV denotes recompression, then virgin.

$$\Delta H = H_0 \left(\frac{\Delta e}{1 + e_0} \right) \qquad (1)$$

in which ΔH is the change in height or settlement, and H_0 represents initial height of the compressible layer.

Comparison Between Computed and Observed Settlement.—In Table 2 there are shown (1) the results of the settlement computations for each layer, (2) whether the settlements are recompression or both recompression and virgin, and (3) the observed settlement of the rigid structure. Because the time-settlement records indicate that settlements are complete, all computations are for ultimate settlement.

[9] "Fundamentals of Soil Mechanics," by D. W. Taylor, John Wiley & Sons, Inc., New York, N. Y., 1948, p. 255.

UNDISTURBED CONSOLIDATION

The observed settlements presented for points B and C are the average of observations made at the four symmetrical locations for each, as noted in Fig. 11. The observed settlement at point A was obtained by adding the measured dishing of the mat to the settlements measured at the edges of the silos.

The correlation between computed settlements and observed settlements indicated in Table 2 is an example of the successful application of the theoretical concepts that have been developed. It is particularly significant that most of the computed settlement was recompression settlement. Hence, this is an example of the successful computation of recompression settlements by a rational method. As shown by this example, such settlements can be of a significant magnitude.

Laboratory Time-Consolidation Data

Permeability and Sample Disturbance.—One often hears the comment that predicted rates of settlement, based on direct extrapolations from the laboratory time-consolidation curves, are slower than those actually observed in the field. It is usually suggested that this is particularly true for organic soils. The extrapolation is usually based on the height of the test sample and the height of the compressible layer by use of the conventional square relationship.[10] Any discrepancy between prediction and field observation is generally attributed to the existence of unknown drainage layers or two-dimensional or three-dimensional drainage rather than the one-dimensional drainage of the laboratory consolidation test.

These explanations are probably correct in many cases. However, the writer believes that in some instances the discrepancy can be accounted for by the effects of sample disturbance on the laboratory time curves. This opinion is based on the following:

1. Sample permeability should be related to the void structure of the sample. Disturbance may alter any natural interconnected void structure so as to break down this structure and thus reduce permeability.[11]

Fine-grained organic soils generally have a sensitive structure and thus may readily have their permeabilities reduced by sample disturbance. Those who have had experience in classifying organic "silt" will probably agree with this statement. These samples, when in an undisturbed state, break with a vertical face, have a fine-grit surface texture, and often have a surface water sheen—all characteristics of the silt. However, when these samples are remolded to test their plasticity, they lose their original characteristics and behave like organic clays. Thus, these samples often present a field-classification difficulty and—incidentally—provide a striking example of how permeability can be altered by sample disturbance.

2. Several time-settlement predictions, using the H^2-relationship, involving compressible organic soils have come to the attention of the writer. These predictions compared favorably with field observations. The best example is at the site from which the data in Figs. 1 and 9 were obtained. An exten-

[10] "Theoretical Soil Mechanics," by K. Terzaghi, John Wiley & Sons, Inc., New York, N. Y., 1943, p. 282.

[11] "The Structure of Inorganic Soil," by T. William Lambe, *Proceedings-Separate No. 315*, ASCE, October, 1953.

sive surface fill, plus the test-proven fact that horizontal permeability and vertical permeability were equal, assured the accuracy of the theoretical drainage pattern. The successful time-settlement computations obtained on this project are believed to be in part the result of using time curves from consolidation-test samples which showed a minimum of sample disturbance.

3. Sensitive organic soils sometimes do not yield suitably undisturbed consolidation-test data. In one such case the writer successfully attempted to correct the time-curve data for the effects of sample disturbance. Some of the details of this investigation are presented subsequently.

Example of Correction for Sample Disturbance.—A 5,500,000-gal oil tank, 140 ft in diameter, was placed on approximately 18 ft of new granular fill over-

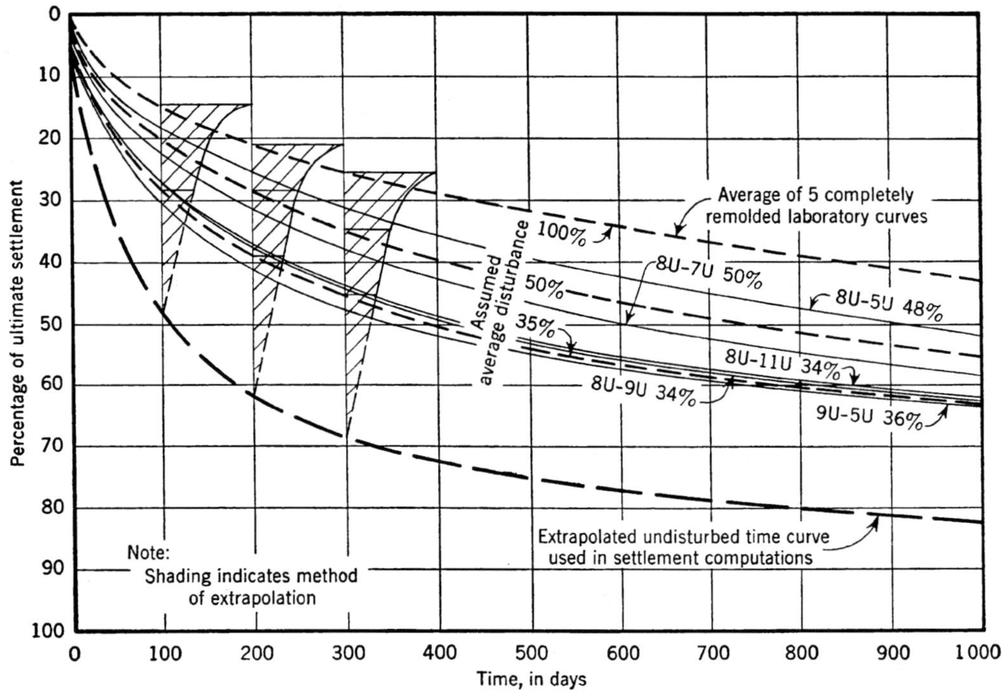

FIG. 14.—EXTRAPOLATION OF UNDISTURBED TIME-SETTLEMENT CURVE FOR 22-FT LAYER FROM DISTURBED LABORATORY CONSOLIDATION TESTS

lying a 22-ft layer of compressible, river-deposited, marine, organic silty clay. Settlements became so serious that remedial measures were required. The nature of these remedies depended on the reason for the settlement—whether entirely because of consolidation, or partly because of a bearing-capacity failure. A careful time-settlement analysis was therefore performed to determine whether consolidation alone would account for the recorded time-settlement behavior of the structure.

Fortunately, at this site the drainage pattern was easily determined. Six borings around the circumference of the tank established that no intermediate drainage layers existed within the 22-ft compressible stratum. The borings also established the existence of a silty fine-sand drainage layer under the compressible stratum. The newly placed fill was known to be coarse and clean enough to permit drainage to the surface. The heavy fill blanket and large-diameter tank located over a relatively thin compressible formation assured that the drainage was essentially one-dimensional and of the double-drainage type.

Undisturbed samples obtained at this site proved to be much more disturbed than those from which the data in Fig. 1 were obtained. This finding, together with the observation of a systematic decrease in permeability with increased

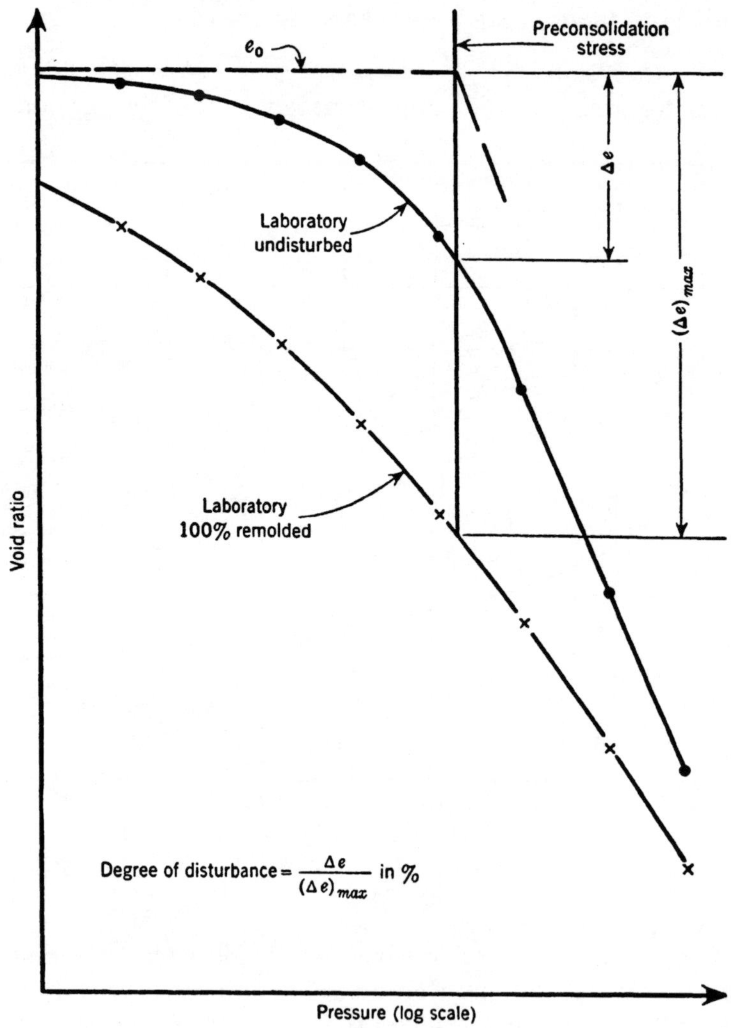

FIG. 15.—METHOD OF DETERMINING DEGREE OF DISTURBANCE

disturbance, revealed the necessity of attempting to correct the consolidation-test data for the effects of sample disturbance.

Five undisturbed-sample tests and two completely remolded-sample consolidation tests were performed on samples from this site. Typical virgin slope-time curves from these tests, adjusted by the conventional square relationship for a 22-ft layer, are plotted with light solid lines in Fig. 14. Each test curve was assigned a degree of disturbance, as shown in Fig. 15, by a method utilizing consolidation-test results. Three groups of time curves were then assigned

UNDISTURBED CONSOLIDATION

average disturbance values, and the completely undisturbed-sample time curve shown in Fig. 14 was extrapolated by the method shown in Fig. 15.

In Fig. 16 there are shown the average undisturbed and completely remolded pressure-void ratio diagrams. By utilizing the extrapolated time curve from Fig. 14 and assuming normal consolidation to exist at the time the fill was placed, an average existing effective stress was computed at a specific point under the tank. Based on water-content data, the average existing void ratio at this point was also computed. The result of these computations was the stress-void ratio point indicated by the cross in Fig. 16. This point was assumed to be a point on the field virgin slope. An undisturbed field virgin slope was constructed through this point in Fig. 16.

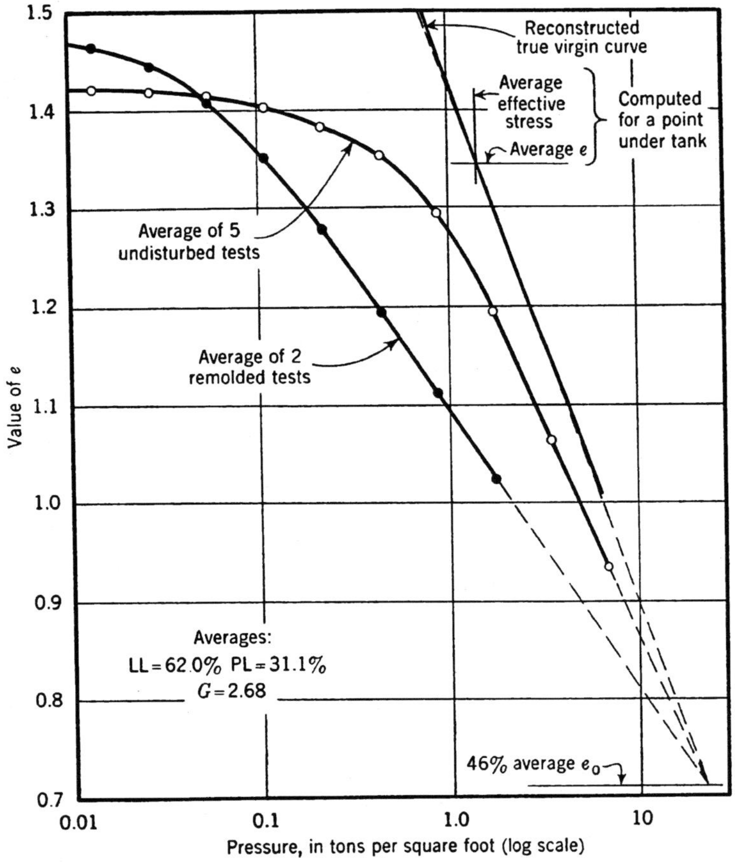

FIG. 16.—DETERMINATION OF TRUE VIRGIN CURVE

By utilizing the undisturbed time-settlement curve in Fig. 14 and the undisturbed virgin curve in Fig. 16, time-settlement computations were made for three points under the tank where settlements have been recorded. In Fig. 17 there are indicated the comparisons of observed and computed time-settlement curves for the three points analyzed. The conclusion, checked by stability analysis and subsequent structure performance, was that the observed settlements at the site were caused entirely by consolidation of the 22-ft organic silty clay stratum.

UNDISTURBED CONSOLIDATION

The favorable time-settlement comparisons indicated in Fig. 17 show that predictions based on uncorrected laboratory test results would have been badly misleading. At this site this error would have resulted in rate predictions that would have been much too low.

CONCLUSIONS

1. Evidence has been presented which indicates that at least some natural clays behave in the field in a manner similar to that predicted by laboratory virgin-slope and rebound-slope data from a consolidation test.

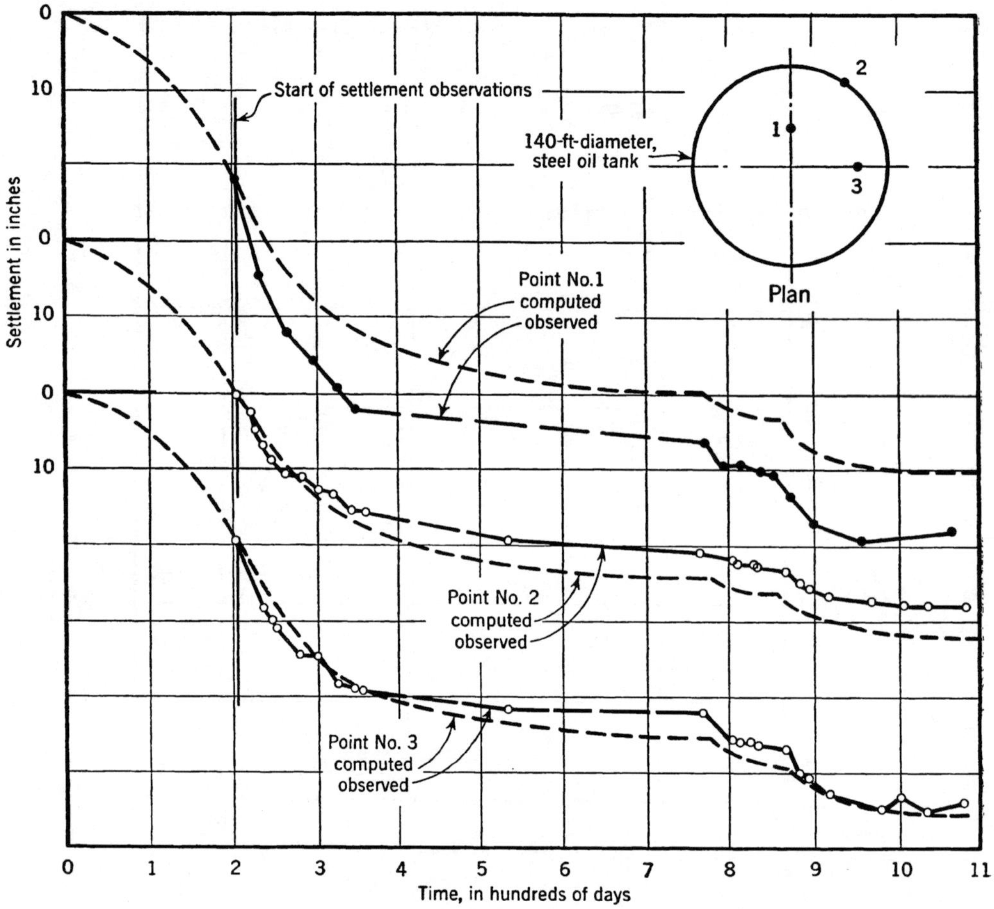

FIG. 17.—COMPARISONS OF OBSERVED AND COMPUTED TIME-SETTLEMENT CURVES

2. Alternate methods are available for the important estimate of the preconsolidation profile.

3. It is possible to perform a systematic reconstruction to estimate the true pressure-void ratio diagram from the laboratory consolidation test of an always partly disturbed sample.

4. Construction of the estimated, true pressure-void ratio diagram permits the simple computation of recompression, as well as virgin void-ratio changes.

5. It may sometimes be necessary to correct the laboratory time-curve data for the effects of sample disturbance.

UNDISTURBED CONSOLIDATION

6. In order to minimize any errors involved in the suggested methods of extrapolation, it is desirable to obtain samples as little disturbed as possible for laboratory use.

ACKNOWLEDGMENTS

The files of Moran, Proctor, Mueser & Rutledge, Consulting Engineers of New York, N. Y., were the source of most of the data contained herein. Their soil mechanics laboratory staff, under the supervision of James D. Parsons, M. ASCE, is responsible for the design of the undisturbed sampler and the quality of the laboratory data used as illustrations in this paper.

Appreciation is expressed to Mr. Rutledge, Mr. Parsons, and J. O. Osterberg, A.M. ASCE, professor of civil engineering at Northwestern University, for their suggestions and encouragement.

DISCUSSION

ROBERT L. CRISP, JR.,[12] J.M. ASCE.—A valuable contribution to the field of soil mechanics has been made by Mr. Schmertmann. The effects of previous geological history and sample disturbance have been recognized as primary factors in selecting representative compression curves from laboratory consolidation tests.

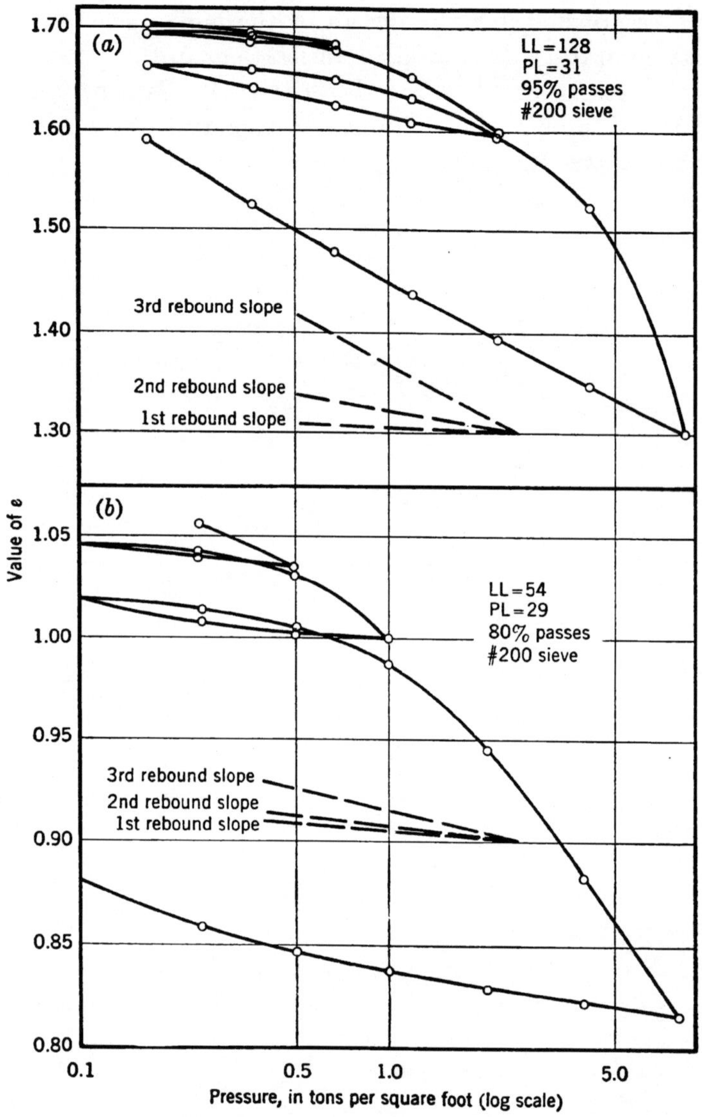

FIG. 18.—RELATIONSHIP BETWEEN VOID RATIO AND PRESSURE

With reference to "Geologic Rebound of a Preconsolidated Clay" and "Reconstruction and Use of the Estimated True Pressure-Void Ratio Dia-

[12] Engr., Soils Mechanics, South Atlantic Div. Lab., Corps of Engrs., U. S. Dept. of the Army, Marietta, Ga.

gram," it has been the experience of the writer that the slope of the rebound curve of clay soils is a function of the inherent structure of the soil and the consolidation pressure from which rebound begins. Considerable variation in the slope of rebound curves has been noted for samples subjected to repeated cycles of loading and unloading. In general, the slope of the rebound curve tends to increase with an increase in applied stress. Therefore, the rebound curve will have innumerable slopes, each a function of the maximum applied laboratory stress. In Fig. 18 there are shown plots of laboratory test data for samples subjected to cyclic loading. The variation in slope is quite evident in the graphical comparison in each pressure-void ratio diagram.

It is believed that an arbitrary selection of the rebound curve, obtained by reducing the applied stress from the maximum, to approximate geologic rebound can lead to erroneous corrections in locating point P in Fig. 10. The location of this point is critical with respect to settlement computations because the slope of the virgin curve is dependent on the location of point P. In addition, the slope of the recompression curve approximates the slope of the rebound curve only when the rebound curve is very flat. This situation is generally encountered when the applied load is less than the preconsolidation pressure. Therefore, extreme care should be used when reconstructing the recompression part of the true curve from laboratory rebound curves.

JOHN H. SCHMERTMANN,[13] J. M. ASCE.—The validity of the parallel relationship between laboratory recompression and rebound slopes has been questioned by Mr. Crisp. It is agreed that the rebound slope varies with the physical and mineralogical structure of the clay and the void ratio (which is a function of the consolidation pressure) from which the rebound begins. However, an attempt will be made herein to show that this variation is somewhat predictable and that recompression-slope magnitudes (all slopes being measured by Δe over one pressure cycle on a conventional semilog *p-e* plot) can be estimated without the extreme care that Mr. Crisp believes to be necessary.

Some preliminary work indicates that for many clays the change in the magnitude of the rebound slope with decreasing void ratio may be primarily a function of the rebound slope C_r, and the void ratio at the beginning of the rebound, e_r. This relationship is tentatively expressed as

$$\log \frac{C_{r1}}{C_{r2}} = 2.5 \log \frac{e_{r2} + 1}{e_{r1} + 1} \qquad (2)$$

The factor 2.5 is an average value which subsequent work may show to vary predictably with clay structure—perhaps as expressed by Atterberg limits. Laboratory rebound and recompression cycles show that the recompression slope, C_x, can be approximated by 1.15 C_r. Under these conditions it is believed that the best method of plotting available data of comparisons between laboratory recompression and rebound slopes would be a logarithmic plot of $(e_r + 1)/(e_x + 1)$ versus C_x/C_r—in which e_x is the void ratio at the beginning of the rebound that precedes the recompression (the slope of which is C_x) and in which

[13] U. S. Army, Snow, Ice, and Permafrost Research Establishment, Wilmette, Ill.; formerly Junior Soils Engr., Moran, Proctor, Mueser & Rutledge, New York, N. Y.

SCHMERTMANN ON UNDISTURBED CONSOLIDATION

e_r and C_r refer to the final rebound of the laboratory consolidation test. Such a plot is presented in Fig. 19.

The data in Fig. 19 were obtained from fifty-nine individual consolidation tests on preconsolidated clays showing distinct recompression slopes. Fig. 19 includes only comparisons from samples that were slightly disturbed. As shown by the legend, the type of the symbol denotes geologic origin, and the size of the symbol denotes a qualitative estimate of sample disturbance. These data include forty-nine tests on clays taken from sixteen sites in nine states of the United States and ten tests from five sites in four foreign countries. In nine of the tests an intermediate virgin rebound-reload cycle was performed, and therefore these tests yielded more than one comparison point.

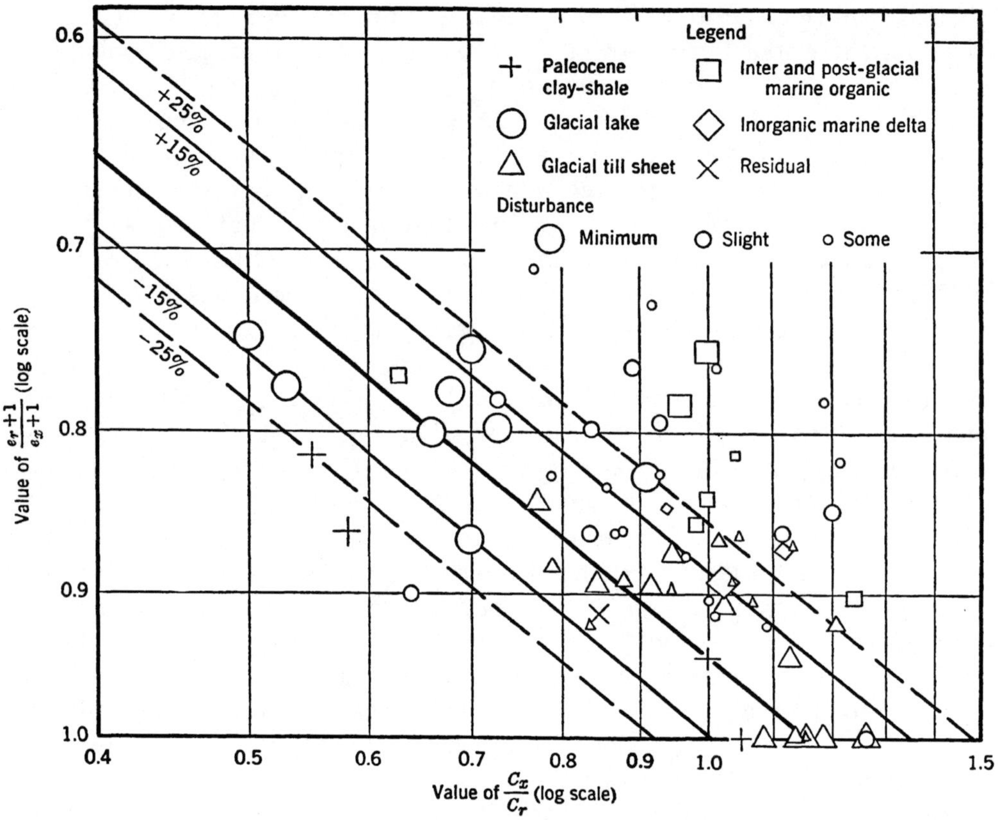

FIG. 19.—VARIATION OF REBOUND SLOPE WITH VOID RATIO

The scatter of the data in Fig. 19 was expected. The magnitude of the recompression slope, C_x, is significantly increased by sample disturbance. As disturbance is known to be variable one should expect scatter. In addition, parameters other than void ratio may be important.

By using a system of weighing individual results according to sample disturbance—minimum, 5; slight, 3; some, 1—the following conclusions can be drawn from the data presented:

The heavy solid line in Fig. 19, the slope of which independently supports the 2.5 factor in Eq. 2, indicates the most probable variation of the recompression slope with void ratio. The light solid lines indicate the ± 15% zone of error about the heavier line. Based on the data presented, one may have 65% confidence that, if one uses the relationship expressed by the heavy line, the error in C_x will not be greater than 15%. The data points within the dashed lines indicate that one may have 80% confidence that the error will not be greater than 25%. The latter error is within the practical accuracy of settlement analyses. The maximum error is approximately 45%. It is therefore believed that the heavy line in Fig. 19 represents a usable empirical relationship.

When using Fig. 19 one must remember that e_x represents the void ratio from which the field rebound began. Unlike e_r, e_x is not definitely known and

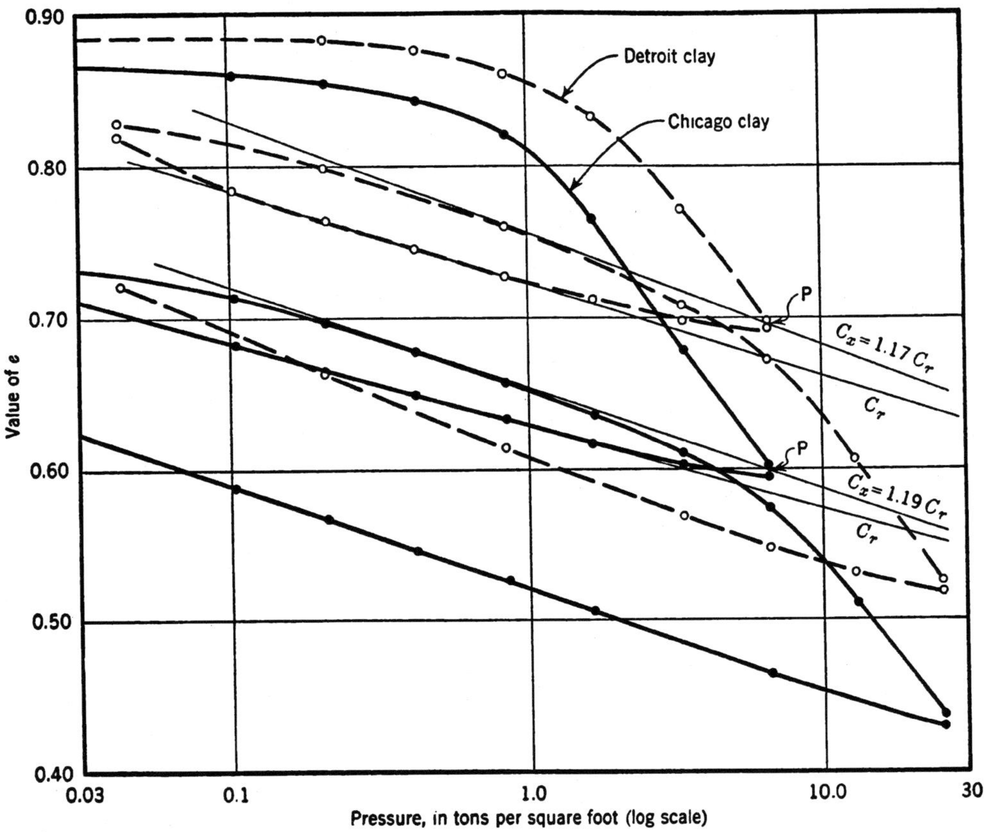

FIG. 20.—TWO COMPARISONS OF REBOUND AND SUBSEQUENT RECOMPRESSION

must be estimated from the pressure-void ratio diagram. As examples, the writer chose the following e_x-values for the tests presented in Fig. 7: (*a*) 1.53 and 1.15, (*b*) 1.06, (*c*) 0.73, and (*d*) 0.77. These e_x-values were easy to estimate because the tests showed a minimum of sample disturbance and the recompression slopes were distinct. In more difficult cases a short, graphical, successive approximation procedure, working back from the in situ pressure (lower pressure if swelling occurred) and the void ratio, will be helpful.

It is interesting to note in Fig. 19 that, for the nine tests presented, a cyclic rebound and recompression procedure (in which $e_x = e_r$) yields an average (C_x/C_r)-value of 1.15, with a range of ± 0.10. Thus, a laboratory rebound

SCHMERTMANN ON UNDISTURBED CONSOLIDATION

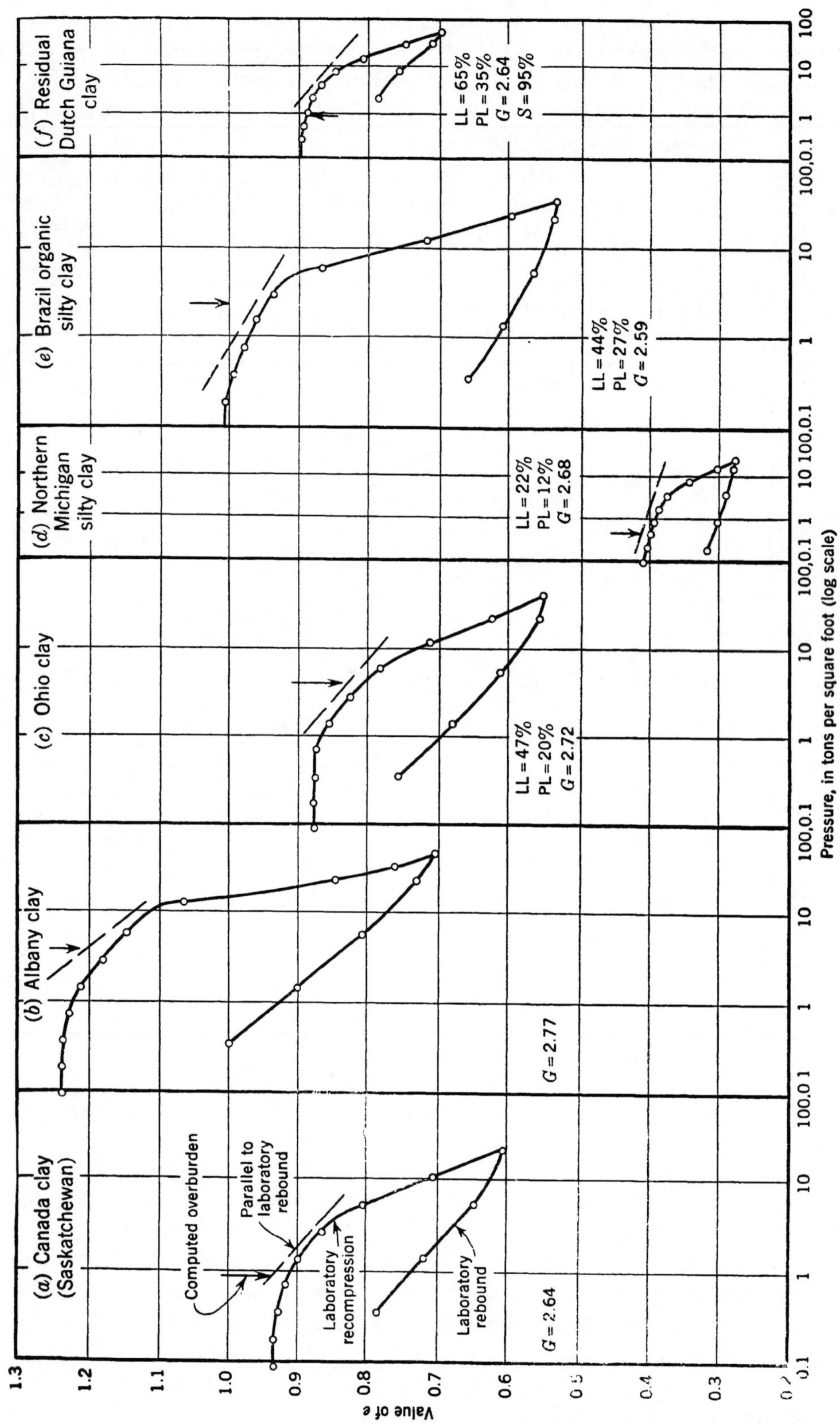

FIG. 21.—SUPPLEMENTARY COMPARISONS OF LABORATORY RECOMPRESSION AND REBOUND SLOPES

slope is approximately 87% of the average slope of the subsequent recompression. The reason for the steeper recompression slope can be seen by comparing the values of C_x and C_r shown in Fig. 20. These laboratory test results were obtained at Northwestern University from two carefully cut, test-pit samples.

The information in Figs. 19 and 20 makes it possible to recommend the following changes in two of the steps listed under the heading, "Reconstruction and Use of the Estimated True Pressure-Void Ratio Diagram":

4. Construct the geologic rebound by estimating the e_x-value and determining the ratio $(e_r + 1)/(e_x + 1)$. The value of C_x/C_r can then be taken as $0.87\ C_x$.

This geologic rebound is drawn through point E (or the lower stress on the e_0-line if the sample swelled) and intersects line x-x at the preconsolidation point P.

6. Reference to Fig. 20 shows that the C_x-lines, drawn tangent to the average recompression slope, pass closely through the preconsolidation points P. This phenomenon was noted for most of the other cycled tests. Thus, a line with a slope of C_x (determined in step 4) constructed through point P may be useful as an upper-tangent guide line for the reconstruction of the field recompression curve.

If there is no laboratory-test evidence to the contrary, these procedures are recommended as the best empirical relationships presently (1955) available for reconstructing the recompression part of field consolidation behavior.

Mr. Crisp believes that the recompression curve approximates the slope of the rebound curve only in instances where the rebound curve is very flat. Although the rebounds from the test results plotted in Fig. 7 are quite steep, the distinct recompression slopes closely approximate them. Fig. 21 is presented to supplement the data in Fig. 7. According to the data presented in Fig. 18, Mr. Crisp's opinion may be based on a comparison of the final rebound, which begins from the virgin slope, with rebounds beginning at the recompression slope. Because virgin consolidation is believed to be associated with a much more severe change of the clay structure, the two rebound phenomena may not be directly comparable. In this connection it is interesting to note that the upper test in Fig. 18, which Mr. Crisp privately reported to have overburden and preconsolidation pressures of 2.5 tons per sq ft and 6 tons per sq ft, has an over-all recompression slope (upper void-ratio points) that is somewhat steeper than the final rebound slope.

As can be seen in Fig. 10, the reconstructed slope of the upper part of the virgin curve, drawn between points P and i, is not sensitive to the position of point P. A large error in estimating the field rebound slope has a relatively slight influence on the reconstructed virgin slope.

Mr. Crisp is thanked for his thoughtful criticisms and for emphasizing the need to present a more comprehensive review of the laboratory data on comparisons between rebound and recompression slopes.

GENERALIZING AND MEASURING THE HVORSLEV EFFECTIVE COMPONENTS OF SHEAR RESISTANCE

By John H. Schmertmann[1]

Synopsis

The Hvorslev effective components are reviewed, and it is argued that the usual assumption of equal cohesion at equal void ratio can be in error because void ratio is not a sufficient description of soil structure. To avoid this error, and to permit investigating the components as functions of strain, it is necessary to redefine and generalize them. This is done. An experimental procedure, called the IDS test, is described which permits separation of the generalized components in a direct manner from a single specimen. The test is versatile as well as practical. The few available comparisons between the Hvorslev and generalized components are presented and are favorable.

Hvorslev Components

Hvorslev (**1,2**)[2] presented and later reviewed his work in determining shear strength[3] components, which he called effective cohesion, c_e, and effective friction, φ_e. He formulated the results of his experiments with two remolded clays, and the resulting equation for each had the simple Coulomb form, thus defining c_e and φ_e. Because these components were based on strength behavior at different effective stress but equal void ratio on the failure plane at failure, and c_e at a given void ratio appeared independent of whether in the virgin, rebound, or recompression phase of stress history, they have also become commonly known as the "fundamental" or "true" components of soil strength.

The concept of these components has endured with few corrections or additions since the work of Hvorslev. Bjerrum (**3**, p. 80) Bjerrum and Wu (**4**), and Jakobson (**5**) have indicated that for some undisturbed clays, or clays remolded at water contents below the liquid limit, there exists an "origin cohesion" to be added to the cohesion which is a function of only void ratio. Gibson (**6**) showed that the components determined from drained tests must be corrected for volume-change effects, if any.

Becuase the determination of c_e and φ_e is a laborious procedure often requiring unobtainable duplicate specimens, it has been done in research for only very few undisturbed soils, and perhaps never for purely practical work. Yet the effective components have had great practical

[1] Associate professor of civil engineering, University of Florida, Gainesville, Fla.

[2] The boldface numbers in parentheses refer to the list of references appended to this paper.

[3] The term shear "resistance" refers to the shear stress mobilized by the soil at any given condition, while shear "strength" refers only to the shear resistance at a defined failure condition.

LABORATORY SHEAR TESTING OF SOILS

significance. Their mere existence has been the rallying-point for the continued use of the component philosophy of soil behavior. Their use permitted the mathematical studies which are the theoretical consolidation and consideration of the comparisons between different types of tests for undrained strength, and Jakobson's(5) further extension to include origin cohesion and octahedral normal stress.

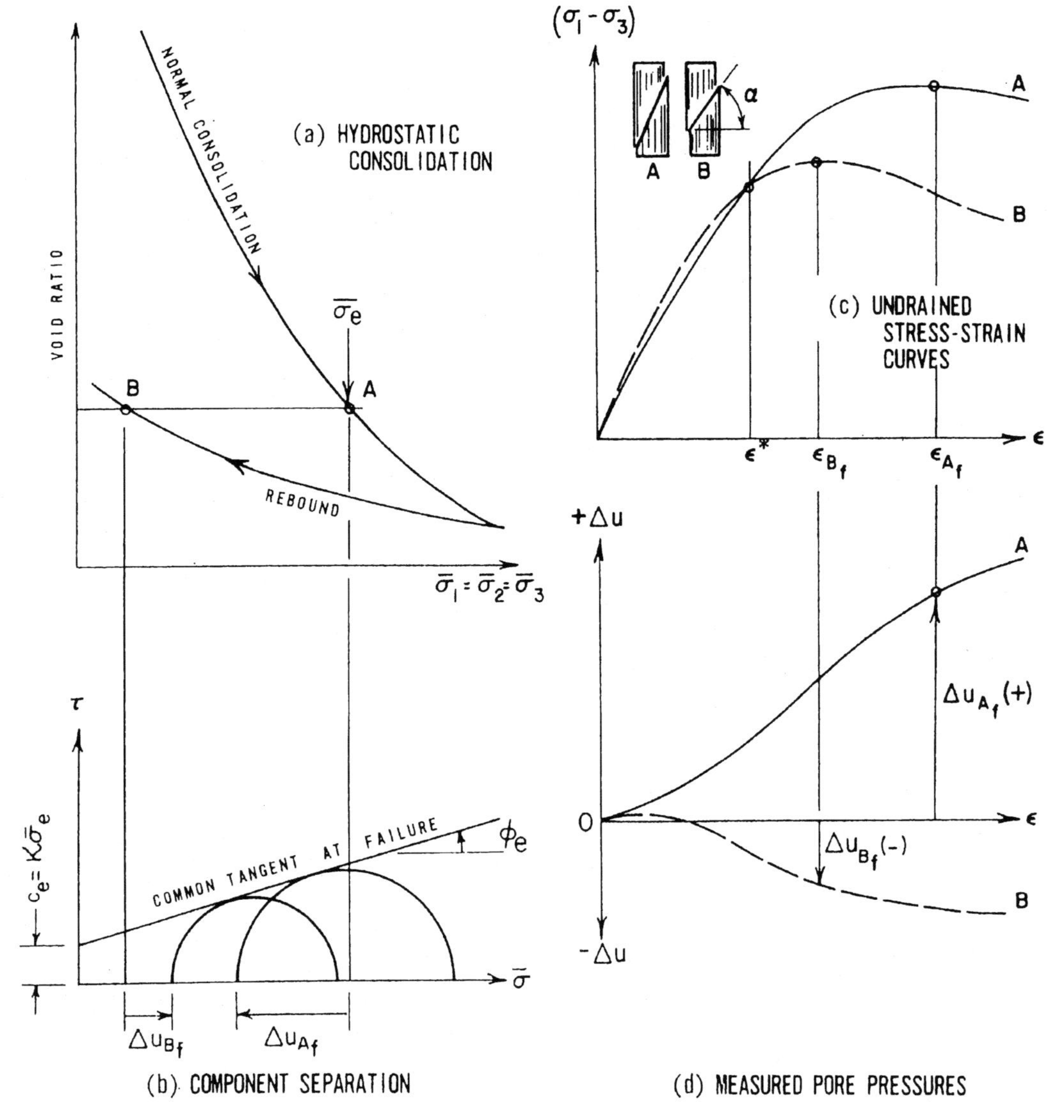

FIG. 1—Hydrostatically Consolidated, Undrained Triaxial Tests (Fictional) to Determine Effective Components and Illustrate Their Physical Meaning.

bases for some common empirical methods of applying soil strength analysis in practice. Some examples are Skempton's (7) analysis of undrained strength and the $\varphi = 0$ analysis, Hansen and Gibson's (8) extension to the case of anisotropic Bjerrum (3, p. 79) and Skempton and Bishop (9) used the effective components to show the relationship between drained and undrained strength envelopes and between water content and undrained strength.

SCHMERTMANN ON COMPONENTS OF SHEAR RESISTANCE

VOID RATIO AND STRUCTURE

Intended Physical Meaning of Effective Strength Components:

The well-known experimental method suggested by Terzaghi (10) to determine c_e and φ_e, which also is generally considered to illustrate excellently the physical meaning of the effective components, is reviewed in Fig. 1(*a*) and (*b*). This method and its relationship to four other methods is also discussed by Bjerrum (3). Following Hvorslev (1), the assumption is made that because tests *A* and *B* have the same void ratio at failure they have the same c_e, and all strength differences are due to $(\Delta\bar{\sigma}) \tan \varphi_e$. Most investigators would agree that effective cohesion, because it is believed to be "fundamental," is some function of the noneffective stresses or bonds between particles and of the details of the extremely complex geometry of particle spacing and contacts. We can say that c_e is a function of the catch-all term "soil structure" or just "structure." Methods-of-testing will be considered a constant here, and possible dilatancy corrections, as suggested by Gibson, are also not considered for simplicity.

When assuming equal cohesion at equal void ratio it is therefore also assumed that equal void ratio means equal structure. Of course, Hvorslev realized this point. For example, he wrote: "Furthermore, it must also be assumed that there is no significant difference in the geometric structure of the test specimens in a given series at the time of failure" (2), p. 206.

The meaning intended for these components is then:

> These are the two components of a soil's shear strength in which one, effective friction, is completely and linearly responsible for the difference in shear strength between two specimens with the same structure but with different effective stress on the failure plane. The other, effective cohesion, is the strengh remaining after extrapolating to zero effective stress.

But this intended meaning may not be clearly achieved by an experimental determination using the Terzaghi method or a variation of it. Some of the following discussion refers to Fig. 1, in which is presented imaginary data from a simple 2-test series to determine the effective components by the Terzaghi method. For illustration, the author has imagined test data which involve and perhaps exaggerate several difficulties in evaluating the significance or accuracy of the components thus separated.

Two Factors Influencing Structure:

Void ratio is of course an important gross measure of soil structure. However, it tells us nothing about the important details of structure. For example, an undisturbed and a remolded sensitive clay can have the same void ratio, but structure is certainly very different. Even assuming void ratio is necessary to describe structure, the question is whether it is sufficient. The author argues that in general it is not, but in the special circumstances originally encountered by Hvorslev it could be sufficient.

Overconsolidation—Fig. 1(*c*) and (*d*) illustrate stress-strain and pore pressure-strain curves that might be expected from two consolidated-undrained triaxial tests such as *A* and *B*. The initial modulus is a measure of the "stiffness" of the soil structure. It could well be stiffer and therefore obviously different for test *B* in spite of lower effective stress. Pore pressure is a measure of the tendency of a saturated structure to reduce volume with strain. The structure of *B* wants to expand while that of *A* wants to contract. During consolidation it is necessary to imagine the occurrence of micro-shear failure, or slippages, at many contact points throughout the mass. These slippages permit the platey particles to assume a more stacked (dispersed) arrangement which is a necessary corequisite to the void ratio reduction. Hvorslev already noted this effect. It is impossible that this many slippages are all reversible with rebound swell when the consolidation stresses are reduced in test B. The consolidation-rebound modified structure at B must be different, perhaps very much so, from the simple consolidation modified structure at A.

Consider now the practical case of an undisturbed soil with always-possible significant brittle cementing, or other bonding, between its particles. It then would have a true tensile strength in terms of effective stresses. What remains of this strength at failure must be part of "origin cohesion" and be included in c_e. It also seems likely that the strains of overconsolidation will irreversibly destroy some of these bonds that would otherwise not be destroyed, and then the total c_e at B will be less than at A (Fig. 1). Thus the assumption of equal c_e at equal void ratio would be in error and lead to values of c_e that are too low and of φ_e that are too high. Mencl (11) has made the same observation. It is known that determinations of φ_e for some undisturbed clays yield greater values than after remolding—which may, in part, be due to the above effect (12,13). Another possibility is discussed below.

Also implicit in the component determination illustrated by Fig. 1(b) is the assumption that both specimens fail on the plane of common tangency, which has the same inclination ($\alpha = 45 + \varphi_e/2$) in both specimens. But, as illustrated in Fig. 1(c), specimen A may fail along a plane with a different inclination than B. If ($\alpha_A \neq 45 + \varphi_e/2 = \alpha_B$), then the use of the common tangent gives too-high values of φ_e. On the other hand, if ($\alpha_A = 45 + \varphi_e/2 \neq \alpha_B$), then one obtains too-low values of φ_e. It seems likely that the less-disturbed specimen, A, will retain more of any natural brittle and/or anisotropic structure and thus $|\alpha_A - (45 + \varphi_e/2)| > |\alpha_B - (45 + \varphi_e/2)|$, and φ_e will be too high.

Shear Strain[4]—One has only to observe the development of a failure plane in a compression test to realize that shear strain can produce a structural change in the soil. The slickensides observed in some natural clays are evidence of the permanence of such structural changes. Structural distortion accompanying shear strain causes many local and irreversible slippages between particles as the structure struggles to mobilize its internal resistance to the forces producing the distortion. Recent investigations indicate that shear strain causes the particles to assume a more parallel, or dispersed, structure with increasing strain (14,15,16).

Referring to Fig. 1(c), suppose one tried to determine the effective components at different values of strain. At strain below ϵ^* this would lead to the unplausible result of negative φ_e. At greater strain it might lead to negative c_e. Obviously, this is not a correct procedure. Yet to consider together in Fig. 1(b) the two failures, each at different strain, is seemingly correct—at least in Hvorslev's investigation. Why? It can only be so if the initial differences in structure due to the overconsolidation were erased by the structural changes of strain, so that at failure at the same void ratio the structures were essentially identical. Under what conditions might this occur?

Hvorslev's Experiments — Hvorslev

[4] The writer does not differentiate between recoverable and nonrecoverable deformations or between elastic and plastic strains. The term "strain" as used here means total deformation divided by an appropriate undeformed dimension.

SCHMERTMANN ON COMPONENTS OF SHEAR RESISTANCE

started with soils remolded at the liquid limit, so that brittle bonding or a highly flocculated initial structure were not possible. His use of direct and torsion shear machines tended to disperse the particles during consolidation in the same direction as the subsequent shear strains would tend to do and thus make a high degree of dispersion at failure all the more likely. Translating Hvorslev's failure strain from direct and torsion machines into the now more familiar compressive strain, and ignoring the possibility of greater strain in failure planes, the author obtained the following (17, p. 261):

AXIAL STRAIN AT FAILURE, PER CENT.

Soil	Normally Consoli-dated	Overcon-solidation ratio = 5
Vienna clay............	40	4
Little Belt clay.........	26	9

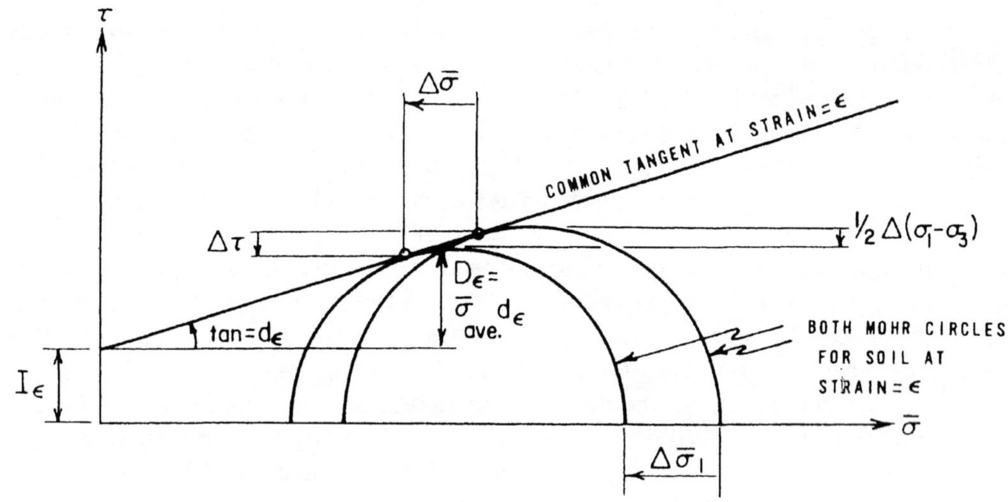

$$d_\epsilon = \tan\left\{\sin^{-1}\left[\frac{\Delta(\sigma_1-\sigma_3)}{2\Delta\bar{\sigma}-\Delta(\sigma_1-\sigma_3)}\right]\right\} \quad \ldots\ldots\ldots\ldots (3)$$

$$I_\epsilon = \frac{(\sigma_1-\sigma_3)-[2\bar{\sigma}_1-(\sigma_1-\sigma_3)]\sin(\tan^{-1}d_\epsilon)}{2\cos(\tan^{-1}d_\epsilon)} \quad \ldots\ldots\ldots\ldots (4)$$

NOTE: $(\sigma_1-\sigma_3)$ AND $\bar{\sigma}_1$ FROM SAME, BUT EITHER, CIRCLE.

FIG. 2—Separation of Generalized Components, Illustrated for Plane of Envelope Tangency in a Triaxial Test.

It seems reasonable to conclude that failure strain was large in these tests, making it still more likely that pre-shear differences in structure were erased by the shear strain to failure.

Can we rely on this structural sequence to occur with other soils, such as undisturbed soils with a highly developed initial flocculated and/or anisotropic structure, or brittle soils which do not fail at

LABORATORY SHEAR TESTING OF SOILS

high strain, or in other tests such as the triaxial where consolidation and shear dispersion directions are less likely to coincide? The author thinks not. In a recent paper Olson (18) indicates he may have similar thoughts. Even if we could rely on this for the failure condition, we definitely should not for the important strains less than failure. Any consideration of other strains must be preceded by more generalized definitions for the effective components.

basis, more generalized definitions for the effective components can now be written:

D_ϵ = *Dependent Component*, component of shear resistance mobilized on any plane and at any strain ϵ, which is dependent on effective stress on that plane according to the equation:

$$D_\epsilon = \left[\bar{\sigma}\left(\frac{\Delta\tau}{\Delta\bar{\sigma}}\right)\right]_{\Delta\bar{\sigma}\to 0} = \left[\sigma\left(\frac{d\tau}{d\bar{\sigma}}\right)\right] = \bar{\sigma}d_\epsilon \ldots (1)$$

when, at ϵ, the shear resistance on that

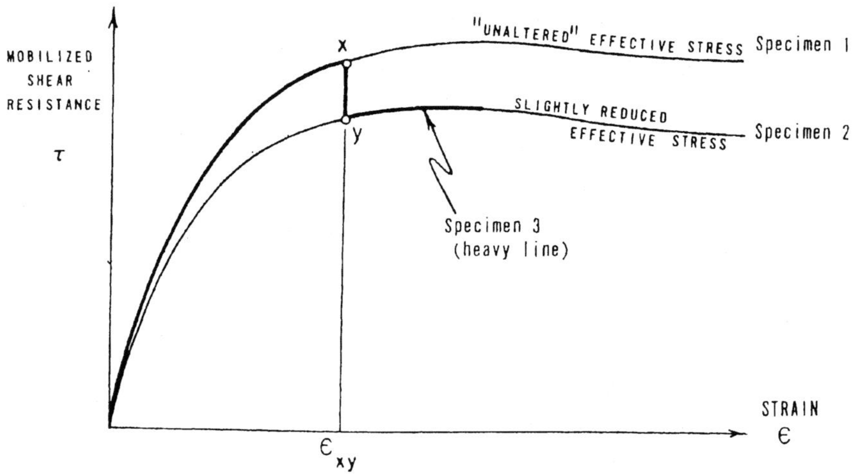

FIG. 3—Curve Hopping with One Specimen.

GENERALIZATION

Definitions:

Because structure is a function of stress history and strain, the components, or at least c_e, must also be. Furthermore, there is no reason to believe that c_e or φ_e (or both) are maximum values or that they both reach a maximum at the same strain, or that they are not indepenent during strain. Strain must be considered an independent variable. Then the concept of a failure is no longer of special significance, nor is the need to restrict the component separation to some future failure plane. Complete generalization requires that any plane can be considered. Using the intended meaning of the effective strength components as a plane changes by $\Delta\tau$ due to $\Delta\bar{\sigma}$. Because there must be no change in soil structure, $\Delta\sigma$ must approach zero.[5]

I_ϵ = *Independent Component*, remaining component defined by:

$$I_\epsilon = \tau_\epsilon - D_\epsilon \ldots\ldots\ldots\ldots (2)$$

when τ_ϵ is the total shear resistance mobilized at ϵ on the plane considered.

In the incremental form necessary in experiments, the definitions are illustrated by Fig. 2 for the plane of envelope tangency (note similarity to Fig. 1(*b*)). The equations for this case are also given.

The author has taken the liberty of

[5] The certain, but perhaps minor, additional influence on $\Delta\tau$ of normal stress changes on the planes orthogonal to the plane considered is now ignored for simplicity. Therefore, this generalization must be considered incomplete.

SCHMERTMANN ON COMPONENTS OF SHEAR RESISTANCE

changing the symbols for these generalized effective components. In previous publications he merely changed c_e to c_ϵ, and φ_e to φ_ϵ to emphasize they were functions of strain, ϵ. However, the physical meaning of these components may prove to differ considerably from the usual concepts of soil cohesion and friction. Therefore, continuing the philosophy of generalizing, they are merely called the components that are "dependent" and "independent" of effective stress, or D_ϵ and I_ϵ, respectively.

Experimental Approach:

The objective when measuring D_ϵ and I_ϵ is to obtain two effective stress conditions with the soil at the identically same structure and to do this at any strain. This seems impossible, and it is. Structure changes with effective stress and cannot be exactly the same, but perhaps it can be sufficiently so for a good approximation of the ideal. The measuring problem becomes one of achieving a relative maximum of effective stress change for a minimum of the inevitable structural changes.

Imagine that we can obtain two shear test specimens that are genuinely identical at the start of a shear test (zero strain in Fig. 3). We shear specimen 1 by some test method and obtain the curve shown. Then just before shearing specimen 2 we impose a slight change in the starting effective stresses. We then shear with the same test method and obtain the other stress-strain curve at all points of which effective stress is slightly different from 1. At any strain chosen we have two mobilized shear resistance values and two effective conditions and can therefore separate D_ϵ and I_ϵ, as shown in Fig. 2 for the plane of envelope tangency.

How great must the slight change be? The answer is the minimum necessary to separate the two circles in Fig. 2, so that a tangent can be determined with sufficient accuracy.

Curve-Hopping and Use of One Specimen:

The great practical difficulty in a component measurement method such as that just suggested (Fig. 3) is that it is often impossible to obtain two or more specimens that are sufficiently identical. Consider now a different technique with specimen 3. The small change in the initial effective stress condition of specimen 2 could also have been imposed at any point during strain (assuming it could be imposed without change of strain) such as x at ϵ_{xy}. If the subsequent shear resistance of specimen 3 at ϵ_{xy} is the same as that for specimen 2 (point y), then only specimen 3 is needed for component separation at ϵ_{xy}.

The only possible difference in structure between points x and y is due to the imposed decrease in effective stress which in turn results in a decrease in the shear resistance at ϵ_{xy}. The author's experience is that an adequate "hop" between curves results in a void ratio increase Δe of less than 1 part per 100. It has been a pleasant surprise to discover that it is possible to make several such hops during a strain-controlled test, and to make them in both the x-y and y-x directions, without greatly changing the subsequent position of the stress-strain curves (19). This curve-hopping technique is then at least approximately reversible with continuing strain. Either very little structural change occurs with each hop or the shear resistance effects of any change are almost recoverable with continued strain.

It is the author's opinion that the 1-specimen, curve-hopping separation of the generalized effective components results in the minimum structural difference between two conditions of effective stress that is presently possible with common soil mechanics techniques. It is at least an approximation of the minimum necessary.

LABORATORY SHEAR TESTING OF SOILS

The IDS Test:

The objective of this test is to separate the *I* and *D* effective components at any strain—hence the abbreviation IDS (CFS was used in previous publications because of previous cohesion and friction terminology). Curve-hopping can be used to the extent required. Two examples are given in Fig. 4. If the components are only desired for the after-failure shear resistance condition, then a hopping sequence such as *a-b-c-d-e-f* in Fig. 4(*a*), with interpolation between *b-e*, is sufficient for a separation at ϵ_{af}. If a continuous curve of the variation of *D* and *I* with strain is desired, then many hops may be required, as shown by the solid line in Fig. 4(*b*). After interpolation, again indicated by dashed lines, the component separation can be made at strains such as ϵ_1 to ϵ_5 and a curve fitted to illustrate their continuous variation with strain.

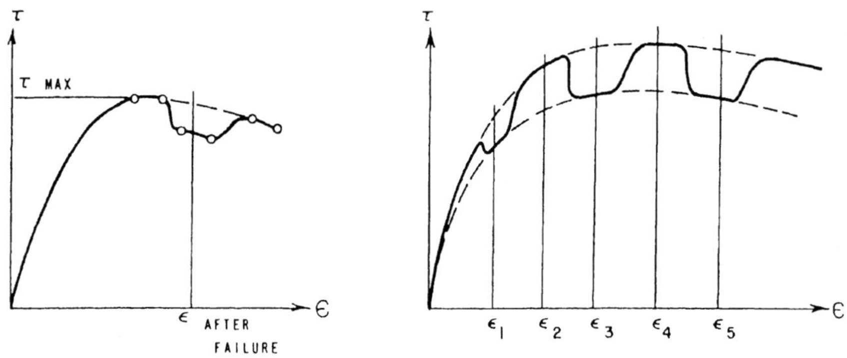

(a) IDS TEST AFTER CONVENTIONAL FAILURE

(b) COMPLETE IDS TEST

FIG. 4—Examples of After-Failure and Complete One-Specimen IDS Tests.

Variations in IDS Test—The concept of this test has proven to be applicable in many variations. It is the imposed change in effective stress that triggers the curve-hop. This change can be performed in any convenient manner as for example, two levels of $\bar{\sigma}_1$ wherein pore pressure is suitably controlled, or two levels of pore pressure or confining stress in drained tests, or two levels of constant volume in undrained tests with pore pressure measured. Wu et al used the test, in multi-specimen form, to study relaxation and creep. Bea (20), and Schmertmann and Hall (21) also studied time phenomena with 1-specimen tests using a short extrapolation to obtain pre-test conditions.

Equipment Requirements and Technique—To separate effective stress components on a particular plane at a particular strain it is only necessary to know mobilized shear resistance for two known effective stress conditions on that plane at that strain. The design of the equipment does not matter provided it can give this information. Presently, the triaxial is most suitable because of the relative ease of effective stress measurement. Any plane can be chosen. The direct shear apparatus is also suitable if the analysis is restricted to drained tests and the horizontal plane, as done by Landva (22). Theoretically even the undrained field vane test would be suitable to determine the average *d* component for the failure cylinder if somehow one could impose a controlled effective stress increment. The measurement precision required with any equipment can be estimated by review of the example given by the author in Table 5 of Ref. (19).

SCHMERTMANN ON COMPONENTS OF SHEAR RESISTANCE

The most difficult part of the 1-specimen technique is to ensure that the specimen has come to stress equilibrium after the imposition of an effective stress hop. The difficulty increases with decreasing permeability and increase in swelling tendency. The experimental techniques for the IDS test are intended to overcome this problem.

fixed to a specified strain nor are they necessarily maximum values. As shown in Fig. 1(c), it is likely that the NC and OC specimens fail at a different strain. Then to what strain should c_e and $\tan \varphi_e$ be assigned for comparison with I and d? At conventional failure the sum of both components is a maximum, but it is very likely that both are not at their

TABLE 1—AVAILABLE COMPARISONS OF $\tan \varphi_e$ AND $d_{max} = \max d_\epsilon$.

Soil	$\tan \varphi_e$	d_{max}	Failure Criterion	Notes	Source
Remolded kaolinite	0.51		drained test $(\sigma_1 - \sigma_3)$ and $\bar{\sigma}_1/\bar{\sigma}_3$	constant volume direct shear tests	O'Neil (**25**)
	0.44 to 0.52		$\bar{\sigma}_1/\bar{\sigma}_3$	undrained triaxial	Schmertmann (**17**)
		0.51	$\bar{\sigma}_1/\bar{\sigma}_3$	constant volume IDS after failure	
		0.49	$\bar{\sigma}_1/\bar{\sigma}_3$	constant $\bar{\sigma}_1$ IDS	
Glacial Lake clay, PI = 32 per cent					
Undisturbed	0.19		$(\sigma_1 - \sigma_3)$	undrained triaxial	Wu et al (**15**)
		0.18	$(\sigma_1 - \sigma_3)$	multi-specimen, constant volume IDS	
Remolded	0.42		$(\sigma_1 - \sigma_3)$ at 10 per cent axial strain	undrained triaxial	
		0.41		multi-specimen constant volume IDS	
Undisturbed Leda clay		0.25	$(\sigma_1 - \sigma_3)$ and $\bar{\sigma}_1/\bar{\sigma}_3$	constant $\bar{\sigma}_1$ IDS	Schmertmann (**19**)
	0.31		Indirectly inferred before failure		Crawford (**26**)

COMPARISON BETWEEN HVORSLEV AND GENERALIZED COMPONENTS

In this paper the author attempts to show that the effective shear resistance components, D_ϵ and I_ϵ, can logically be considered a generalization of the effective strength components originally proposed by Hvorslev. If true, then at failure one should expect to be able to demonstrate that $I = c_e$ and $d = \tan \varphi_e$. But the comparison is more difficult because the c_e and φ_e components are not maximum values, so one cannot compare with both I_{max} and d_{max}. These comparison difficulties are illustrated by the author (**17**). A further complication is that Hvorslev used drained tests, and his work does not specify whether to use $(\sigma_1 - \sigma_3)$ or $(\bar{\sigma}_1/\bar{\sigma}_3)$ failure in undrained tests. It has become common to use $(\sigma_1 - \sigma_3)_{max}$ but the structural condition of the soil is probably more unique at $(\bar{\sigma}_1/\bar{\sigma}_3)_{max}$ (**23,18,24**).

Table 1 lists the presently available

comparisons of tan φ_e and d_{max}. This comparison was chosen as most likely to show agreement. IDS tests indicate it is very likely that I is past its maximum and decreasing at the strain of maximum shear resistance, which is also the strain of $(I + D)_{max}$. Therefore if $\bar{\sigma}$ is not changing too rapidly with strain, this is also approximately the strain of d_{max}. Maximum shear resistance is also the point for the tan φ_e determination and therefore should usually be in approximate agreement with d_{max}. The data presented generally support this argument. Unfortunately the list is brief and more comparisons are certainly needed.

Wu (27) offers some additional, though indirect, evidence. Using data from four remolded clays found in the literature, he plotted the variation of d_{max} with plasticity index and found agreement with a published tan φ_e versus PI curve averaged from many other remolded clays. Perhaps significantly, a similar comparison using undisturbed clays showed poor agreement, with d_{max} much less than tan φ_e. As discussed here, the determination of φ_e for undisturbed clays can involve serious errors with resulting too-high φ_e values.

CONCLUSION

The author recommends that the D_ϵ and I_ϵ shear resistance components be considered as possible generalizations of the Hvorslev effective strength components. If true, their determination is greatly simplified by the various forms of the IDS test. However, regardless of the merit of this argument, it is necessary for investigators to keep in mind the possibility that equal void ratio may not mean equal soil structure (cohesion) when determining the effective strength components for the conditions in their tests.

REFERENCES

(1) M. J. Hvorslev, "Ueber die Festigkeitzeigenschaften gestörter bindiger Böden," Thesis, 159 pages. Published by Danmarks Naturvidenskabelige Samfund, Ingenirvidenskabelige Skrifter, Series A, Nr. 45 Copenhagen, 1937.

(2) M. J. Hvorslev, "Physical Components of the Shear Strength of Saturated Clays," American Society of Civil Engineers Research Conference on Shear Strength of Cohesive Soils, 1960, p. 169.

(3) L. Bjerrum, "Theoretical and Experimental Investigations on the Shear Strength of Soils," Norwegian Geotechnical Institute, Publication No. 5., 1954.

(4) L. Bjerrum and T. H. Wu, "Fundamental Shear Strength Properties of Lilla Edet Clay," *Géotechnique*, Vol. 10, No. 3, 1960.

(5) B. Jakobson, "Influence of Sample Type and Testing Method on Shear Strength of Clay Samples," *Proceedings*, No. 8, Royal Swedish Geotechnical Institute, Stockholm.

(6) R. E. Gibson, "Experimental Determination of the True Cohesion and True Friction in Clays," *Proceedings*, 3rd International Conference on Soil Mechanics and Foundation Engineering, Vol. 1, 1953, p. 126.

(7) A. W. Skempton, "The Effective Stresses in Saturated Clays Strained at Constant Volume," *Proceedings*, International Congress of Applied Mechanics, Vol. II, 1948, p. 378.

(8) J. B. Hansen and R. E. Gibson, "Undrained Shear Strengths of Anisotropically Consolidated Clays," *Géotechnique*, Vol. 1, No. 3, 1948.

(9) A. W. Skempton and A. W. Bishop, "Soils," in *Building Materials*, by M. Reiner, chapter X, 1954, p. 470.

(10) K. Terzaghi, "Die Coulombsche Gleichung für den Scherwiderstand bindiger Boden," *Bautechnik*, Vol. 16, 1938, pp. 343–346. Translated in *From Theory to Practice in Soil Mechanics*, by L. Bjerrum.

(11) V. Mencl, Discussion in American Society of Civil Engineers *Journal Soil Mechanics and Foundations Division*, Feb. 1963, p. 259.

(12) L. Bjerrum and N. E. Simons, "Comparison of Shear Strength Characteristics of Normally Consolidated Clays," American Society of Civil Engineers Research Con-

ference on Shear Strength of Cohesive Soils, 1960, p. 720.

(13) K. Y. Lo, "Shear Strength Properties of a Sample of Volcanic Material of the Valley of Mexico," *Geotechnique*, Vol. XII, No. 4, Dec. 1962, p. 314.

(14) H. B. Seed and C. K. Chan, "Structure and Strength Characteristics of Compacted Clays," American Society of Civil Engineers, *Journal Soil Mechanics and Foundations Division*, Vol. 85, No. SM5, 1959, pp, 87–128.

(15) T. H. Wu, A. G. Douglas and R. D. Goughnour, "Friction and Cohesion of Saturated Clays," American Society of Civil Engineers, *Journal Soil Mechanics and Foundations Division*, June, 1962.

(16) R. Pusch, "Clay Particles—Their Size, Shape and Arrangement in Relation to Some Important Physical Properties of Clays," Handlingar N. 40, *Transactions*, The Nat. Swedish Council for Bldg. Res., Stockholm, 1962.

(17) J. H. Schmertmann, Discussion, in American Society of Civil Engineers, *Journal Soil Mechanics and Foundations Division*, Feb. 1963, p. 263.

(18) R. E. Olson, "Shear Strength Properties of Sodium Illite," American Society of Civil Engineers, *Journal Soil Mechanics and Foundations Division*, Feb. 1963, p. 206.

(19) J. H. Schmertmann, "Comparisons of One and Two-Specimen CFS Tests," American Society of Civil Engineers, *Journal Soil Mechanics and Foundations Division*, Dec. 1962.

(20) R. G. Bea, "An Experimental Study o Cohesion and Friction During Creep in Saturated Clay," Masters Thesis, University of Florida, June 1960. Partly published in discussion, American Society of Civil Engineers, *Journal Soil Mechanics and Foundations Division*, Feb. 1963, p. 268.

(21) J. H. Schmertmann, and J. R. Hall, Jr., "Cohesion After Non-Hydrostatic Consolidation," American Society of Civil Engineers, *Journal Soil Mechanics and Foundations Division*, Aug. 1961.

(22) A. Landva, "En eksperimentell undersøkelse av skjaerfastheten i normalkonsoliderte leirer," Norwegian Geotechnical Institute, Internal Report F. 175, 1962.

(23) L. Bjerrum, "The Effective Shear Strength Parameters of Sensitive Clays," *Proceedings*, Fifth International Conference on Soil Mechanics and Foundation Engineering, Vol. I, p. 27; and Norwegian Geotechnical Institute, Publ. No. 45, 1961.

(24) N. E. Simons, "The Effect of Overconsolidation on the Shear Strength Characteristics of an Undisturbed Oslo Clay," American Society of Civil Engineers, *Proceedings*, Research Conference on Shear Strength of Cohesive Soils, 1960, p. 761.

(25) H. M. O'Neil, "Direct Shear Test for Effective Stress Parameters," American Society of Civil Engineers, *Journal Soil Mechanics and Foundations Division*, paper 3232, 1962.

(26) C. W. Crawford, "The Influence of Strain on Shearing Resistance of Sensitive Clay," *Proceedings*, Am. Soc. Testing Mats., Vol. 61, 1961.

(27) T. H. Wu, Discussion, Am. Soc. Civil Engrs. *Journal Soil Mechanics and Foundations Division*, July, 1963, p. 230.

TECHNICAL NOTE

SWELL SENSITIVITY

J. H. Schmertmann, Ph.D., P.E., M.A.S.C.E.*

It appears that some, but not all, clays significantly increase their swell behaviour when subjected to the repeated, large shear strains of mechanical remoulding. In the absence of a descriptive term for this phenomenon, the Author calls it 'swell sensitivity' and expresses it quantitatively[1] by the ratio of swelling index remoulded/swelling index undisturbed, or C_{s_R}/C_{s_U}. Knowledge of swell sensitivity proved important at one foundation site. Many such sites probably exist. Although engineers have probably observed swell sensitivity many times, it appears to have received little attention in the literature. This Note reviews the phenomenon, presents an example and discusses the type of clay which could have an important swell sensitivity.

According to the current state of the art, clays with high swell potential can be identified by their high activity and/or high liquid limits and/or low shrinkage limit (Seed *et al.*, 1962; Holtz and Gibbs, 1962; van der Merwe, 1964; Navdocks, 1962). Using such criteria, the saturated clays at the site of some proposed apartment structures in Gainesville, Florida, had a very high swell potential (see Table 1). Swell tests on remoulded samples of the saturated clay from the apartment site confirmed the significant swell pressures and high volume changes expected from these clays. A nearby restaurant and a service station placed in a shallow cut in this clay experienced serious heave of their floor slabs and some structural distortion due to footing heave. However, all this evidence of a serious swell problem proved misleading in one important respect.

Undisturbed samples of these clays, obtained with the standard Swedish 54 mm fixed piston sampler, showed greatly reduced swell tendency compared with the samples remoulded at their natural water content (Tables 2 and 3). This clay may have a swell sensitivity as high as 20 when at its natural void ratio.

The data in Table 3 show that the swell sensitivity in such clay will erroneously decrease to less than 1·0 when the undisturbed clay first experiences virgin consolidation, and therefore large volume strain, before unloading and determining its undisturbed swell index. When determining swell sensitivity it appears necessary to measure undisturbed swell behaviour before imposing any large strains.

Lambe and Whitman (1969) listed five cases where they compared the swelling index of undisturbed and remoulded clays. In all cases they report essentially no change due to remoulding. However, they report on only relatively recent normally consolidated or lightly overconsolidated clays. On the other hand, Holtz (1959) presented data from three clays in which he showed clearly that remoulding greatly increases their swell potential. He stated, 'It has been found that, when undisturbed and remolded specimens of the same soil are tested in consolidometers at the same moisture density conditions, the expansion of the remolded specimen will be considerably greater.' Thus some clays exhibit swell sensitivity and others do not. Bjerrum (1967) suggests the reason for this.

* Professor of Civil Engineering, University of Florida.

[1] The swelling index is the void ratio increase during 1 log cycle pressure decrease in an oedometer.

TECHNICAL NOTE

Table 1. Swell potential of Gainesville clay

Sample	Depth (ft)	Water content (%)				LI	C*	Activity (PI/C)	Swell potential		
		w_L	$w_{natural}$	w_p	PI (%)				Seed *et al.* (1962)	van der Merwe (1964)	Navdocks (1962)
C7-3U-a	12·8	106	56	42	64	0·22					$C_s = 0{\cdot}12$†
C7-4U-b	15·8	132	69	47	85	0·26					$C_s = 0{\cdot}15$
C7-4U-a	16·3	134	80	67	67	0·19					$C_s = 0{\cdot}15$
C7-5U-c	18·4	128	67	55	73	0·16					$C_s = 0{\cdot}14$
C8-1U-a	10·2	117	55	46	71	0·13	78	0·91	very high‡	very high§	$C_s = 0{\cdot}16$
C8-2U-b	13·0	100	46	46	54	0	66	0·82	very high	very high	$C_s = 0{\cdot}12$

* Per cent finer than 0·002 mm by hydrometer analysis.
† Reference does not specify whether swell index is for undisturbed or remoulded soil.
‡ Swelling potential exceeds 25% volume increase.
§ Potential expansiveness greater than 1 in./ft.

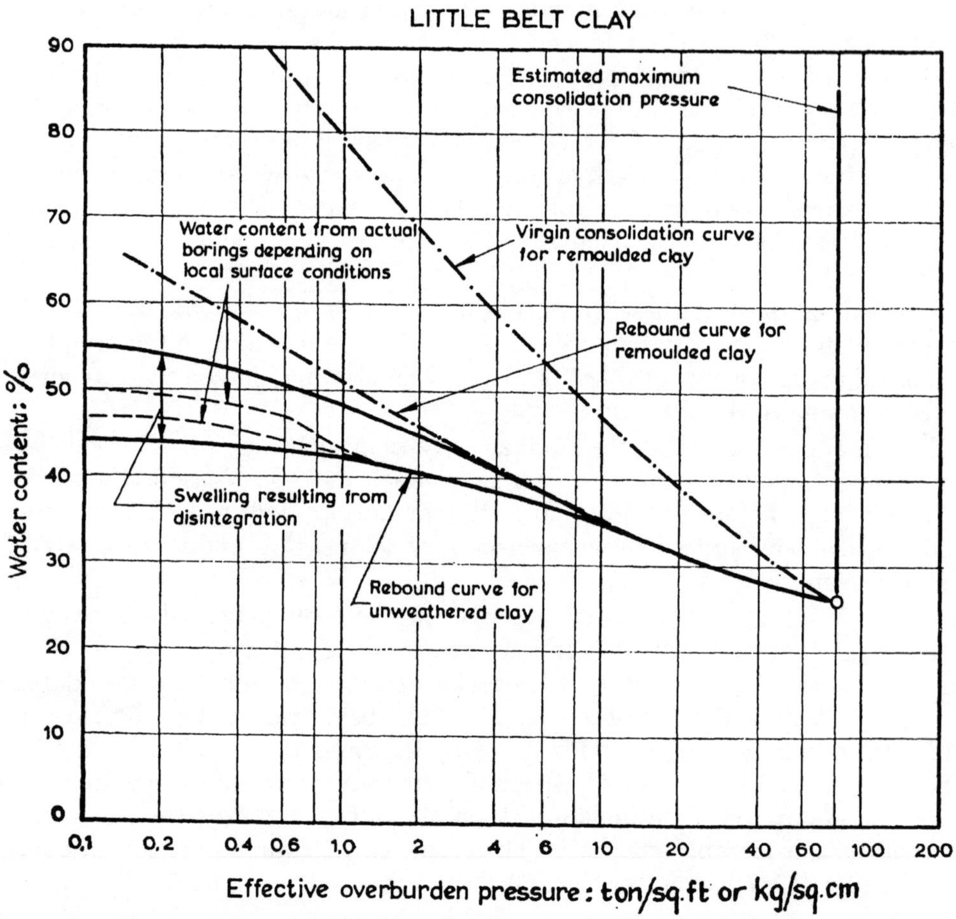

Fig. 1. Example of a clay in which swelling increases with destruction of bonds (after Bjerrum, 1967)

TECHNICAL NOTE

Table 2. Results of Gainesville clay free swell tests*

Sample	Initial w (%)	Per cent volume increase		Ratio R/U
		Undisturbed† (3 days)	Remoulded† (1 day)	
C7-SU-a	59·9	15		4·3
-b	50·8		65	
C7-4U-b	58·2	30		3·2
-b	47·6		95	
C8-1U-a	53·7	25		3·0
-a	51·9		75	
C8-1U-b	51·4	20		3·2
-c	50·7		65	
C8-2U-b	39·4	30		2·8
-b	45·7		85	

* Unrestrained specimen placed in dish of distilled water. Water contents determined before and after swell. Volume increase calculated assuming $S=100\%$.

† Undisturbed and remoulded specimens from vertically adjacent clay in the same sampler liner or adjacent liners (each 17 cm long).

Table 3. Results of oedometer swell tests on Gainesville clay

Sample	Structure*	Natural w (%) (e)	Swell pressure† (kg/sq. cm)	Swell time (days)	Swell to 0·007 kg/sq. cm (% height increase)	Average swell index C_s	Over pressure range (kg/sq. cm)	Swell sensitivity (3 days)
C8-1U-a	Remoulded	52·5 (1·497)	0·99	1 3 14	10·2 14·1 17·1	0·164	0·99–0·007	4·8
	Undisturbed	54·9 (1·490)	0·48	3	2·5	0·034	0·48–0·007	
		after oedometer consolidation to 50 kg/sq. cm and $e=0{\cdot}910$				0·19	10–0·1	0·9
C8-2U-b	Remoulded	36·1 (1·050)	1·48	1 3 14	8·5 11·4 13·3	0·100	1·48–0·007	20
	Undisturbed	45·6 (1·245)	0·22	3	0·6	0·005	0·22–0·007	
		after oedometer consolidation to 50 kg/sq. cm and $e=0{\cdot}777$				0·14	10–0·1	0·7

* Undisturbed and remoulded specimens from vertically adjacent clay in the same sample liner.

† Equilibrium vertical pressure to maintain natural void ratio when specimen allowed free access to distilled water.

TECHNICAL NOTE

Bjerrum presented a strong argument that in some overconsolidated clays there exists a considerable amount of potentially recoverable strain energy stored internally by the bending of clay particles. Bonds of various origins, formed before stress removal, then hold these particles in their bent positions after removal of the overconsolidation stresses. Later loss of these bonds as a result of weathering and/or strain changes the equilibrium and permits the clays to imbibe water and swell. Fig. 1 shows how he plotted this phenomenon using for example a highly overconsolidated, stiff, fissured clay. He believed this clay sufficiently preloaded to a high stress level for a long enough time to have developed significant bonds and locked-in strain energy. His experimental data clearly show an even greater swell after mechanical remoulding, which presumably develops a more thorough destruction of bonds than achieved by natural weathering processes. Bjerrum's observations suggest a method for identifying soils likely to have a significant swell sensitivity.

Stiff and fissured also describe the Gainesville clay. In unconsolidated, undrained compression tests this soil failed at strains of 1–2% along sharply defined failure planes, sometimes showing slickensides. The UU shear strengths ranged from 0·3 to 1·2 kg/sq. cm at both 5 and 10 lb/sq. in. confining pressures. In situ vane tests indicated strengths ranging from 1·03 to $2{\cdot}15^{+}$ (limit of equipment) with sensitivities between 2·8 and $4{\cdot}2^{+}$. The oedometer-determined preconsolidation stress, although not well defined because of insufficient pressure capability, appears to be about 25 ton/sq. ft. The present overburden is 0·5 ton/sq. ft. Heavily overconsolidated and old describe this clay. Its vane sensitivity, brittle behaviour and fissures further suggest that the clay probably has significant but breakable bonds. This Gainesville clay appears similar to the types discussed by Bjerrum.

In view of this evidence it seems reasonable to suspect that any active clay with a low liquidity index and with breakable bonds will exhibit significant swell sensitivity. The engineer should suspect any old, highly overconsolidated clay as yet unweathered by natural processes. On the other hand relatively young and normally consolidated clays will probably have little bond formation and therefore little or no swell sensitivity. Similarly active clays intermediate in bond formation should show intermediate swell sensitivity.

Before taking advantage of the possibly much more favourable swelling behaviour of an undisturbed clay with breakable bonds, the engineer should assure himself that the clay will not lose these bonds during construction or during the life of the structure as a result of such factors as natural weathering, change in groundwater level, man-induced chemical reactions or by the application of large shear strains. For example, the swell and structural problems noted for the Gainesville restaurant and service station probably resulted largely from soil structure disturbance, including deliberate compaction, by construction equipment. Prior recognition of the phenomenon of swell sensitivity will help avoid such soil engineering failures.

REFERENCES

BJERRUM, L. (1967). Progressive failure in slopes of overconsolidated plastic clay and clay shales, Third Terzaghi Lecture. *J. Soil Mech. Fdns Div. Am. Soc. civ. Engrs* **93**, Part 1, 16.

HOLTZ, W. G. (1959). Contribution to symposium on theoretical and practical treatment of expansive soils. *Colo. Sch. Mines Q.* **54**, No. 4, 95.

HOLTZ, W. G. & GIBBS, H. J. (1962). Discussion on prediction of swelling potential for compacted clays. *J. Soil Mech. Fdns Div. Am. Soc. civ. Engrs* **88**, SM4, 207.

LAMBE, T. W. & WHITMAN, R. V. (1969). *Soil mechanics*, p. 323. New York: Wiley.

NAVDOCKS. (1962). Design manual no. 7, p. 7-3-14. Department of the Navy, Bureau of Yards and Docks.

SEED, H. B., WOODWARD, R. J. & LUNDGREN, R. (1962). Prediction of swelling potential for compacted clays. *J. Soil Mech. Fdns Div. Am. Soc. civ. Engrs* **88**, SM3, 80.

VAN DER MERWE, D. H. (1964). The prediction of heave from the plasticity index and the percentage clay fraction. *Civ. Engr S. Afr.* **6**, No. 6, 103.

Journal of the
SOIL MECHANICS AND FOUNDATIONS DIVISION
Proceedings of the American Society of Civil Engineers

STATIC CONE TO COMPUTE STATIC SETTLEMENT OVER SAND

By John H. Schmertmann,[1] M. ASCE

INTRODUCTION

Settlement, rather than bearing capacity (stability) criteria, usually exert the design control when the least width of a foundation over sand exceeds 3 ft to 4 ft. Engineers use various procedures for calculating or estimating settlement over sand. Computations based on the results of laboratory work, such as oedemeter and stress-path triaxial testing, involve trained personnel, considerable time and expense, and first require undisturbed sampling. Interpreting the results from such testing often raises the serious question of the effect of sampling and handling disturbances. For example: Does the natural sand have significant cement bonding even though the lab samples appear cohesionless? When dealing with sands many engineers prefer therefore to do their testing in-situ.

Settlement studies based on field model testing, such as the plate bearing load test, often require too much time and money. This type of testing also suffers from the serious handicap of long-existing and still significant uncertainties as to how to extrapolate to prototype foundation sizes and nonhomogeneous soil conditions. A new type of test for field compressibility, involving a bore-hole expanding device or pressuremeter, is now also used in practice. The accuracy of a settlement prediction using such devices and semi-empirical correlations is not yet, to the writer's knowledge, documented in the English literature and may not yet be established. Whatever its prediction accuracy, such special testing and analysis should prove more expensive than settlement estimates based on the results of field penetrometer tests.

Presently, engineers commonly use settlement estimate procedures based on two very different types of field penetrometer tests. U.S. engineers have used the Standard Penetration Test for 20 yr. The hammer blow-count, or N-value, has been empirically correlated to plate test and prototype footing

Note.—Discussion open until October 1, 1970. To extend the closing date one month, a written request must be filed with the Executive Secretary, ASCE. This paper is part of the copyrighted Journal of the Soil Mechanics and Foundations Division, Proceedings of the American Society of Civil Engineers, Vol. 96, No. SM3, May, 1970. Manuscript was submitted for review for possible publication on January 22, 1969.

[1] Prof. of Civil Engrg., Univ. of Florida, Gainesville, Fla.

settlement performance. Because of the completely empirical nature of this method the engineer sometimes finds it not very informative or satisfying to use. Some engineers believe that it often results in excessively conservative (too high) settlement predictions. Another method, based on the Static Cone Penetration Test, has a European history of over 30 yr. In this method the quasistatic bearing capacity of a steel cone provides an indicator of soil compressibility. Settlement predictions have proven conservative by a factor averaging about 2.0.

The field penetrometer methods have the great advantage of practicality, with results obtained in-situ, quickly, and inexpensively. These advantages permit testing in volume, and thereby permit a better evaluation of any important consequences resulting from the nonhomogeneity of most sand foundations.

Perhaps the empirical nature of the present penetrometer methods represents their greatest disadvantage. The engineer does not find it easy to trace the logic and data to support these methods. Herein he will find a new approach, based on static cone penetrometer tests, which has an easily understood theoretical and experimental basis. Compared to the best procedure now in use, this new method has a more correct theoretical basis, results in simpler computations, and test case comparisons suggest it will often result in more accuracy without sacrificing conservatism.

CENTERLINE DISTRIBUTION OF VERTICAL STRAIN

Engineers have often assumed that the distribution of vertical strain under the center of a footing over uniform sand is qualitatively similar to the distribution of the increase in vertical stress. If true, the greatest strain would occur immediately under the footing, the position of greatest stress increase. Recent knowledge all but proves that this is incorrect.

Elasticity and Model Studies.—Start with the theory of linear elasticity by considering a uniform circular loading, of radius = r and intensity = p, on the surface of a homogeneous, isotropic, elastic half space. The vertical strain at any depth z = ϵ_z, under the center of the loading, follows Eq. 1 from Ahlvin & Ulery (1):

$$\epsilon_z = \frac{p}{E} (1 + \nu) [(1 - 2\nu) A + F] \quad \ldots\ldots (1)$$

in which A and F = dimensionless factors that depend only on the geometric location of the point considered; and E and ν = the elastic constants.

Because p and E remain constant, the vertical strain depends on a vertical strain influence factor, I_z. Thus

$$I_z = (1 + \nu) [(1 - 2\nu) A + F] \quad \ldots\ldots (2)$$

Fig. 1 shows the distribution of this influence factor, and therefore strain multiplied by the constant E/p, with a dimensionless representation of depth for Poisson's ratios of 0.4 and 0.5. The area between the I_z = 0 axis and these curves represents settlement. Note that maximum vertical strain does not occur immediately under the loading, where the increase in vertical stress is its maximum, $1.0p$, but rather at a depth of (z/r) = 0.6 to 0.7, where the Boussinesq increase in vertical stress is only about $0.8p$.

SM 3 SETTLEMENT OVER SAND

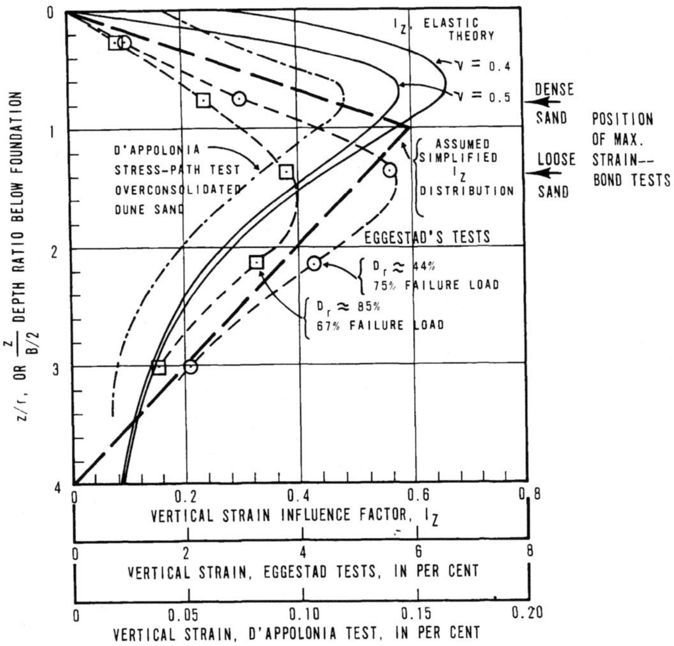

FIG. 1.—THEORETICAL AND EXPERIMENTAL DISTRIBUTIONS OF VERTICAL STRAIN BELOW CENTER OF LOADED AREA

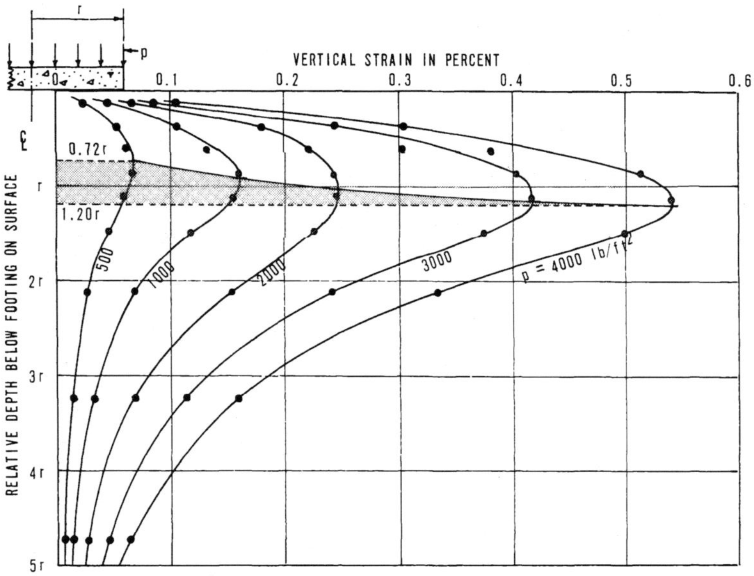

FIG. 2.—NONLINEAR, STRESS DEPENDENT FINITE ELEMENT MODEL PREDICTION OF VERTICAL STRAINS UNDER CENTER OF 10-FT DIAM, 1.25 FT THICK, CONCRETE FOOTING LOADED ON SURFACE OF NORMALLY CONSOLIDATED SAND WITH ϕ = 37°

Evidence similar to that previously given would result from considering uniformly loaded rectangular areas of least width = B. The writer obtained the following from the elastic settlement solutions tabulated by Harr (15): the maximum vertical strain under both the center and corner of a square occurs at a depth $z/B/2$ = 0.8 and 0.6 for Poisson's ratio = 0.5 and 0.4, respectively; the corresponding relative depths to maximum strain under a rectangle with L/B = 5 are 1.1 and 0.9.

Model studies using sand all show that the depth to maximum vertical strain increases compared to that indicated by elastic theory. Fig. 1 includes two representative vertical strain distributions from Eggestad's (10) tests on homogeneous sand under a rigid, circular footing of radius = r. He reports a depth to maximum vertical strain of about (z/r) = 1.5 for both loose and dense sand. Eggestad also reported the results of a similar model study by Bond (5) with depth to maximum vertical strain at (z/r) = 0.8 for dense sand and 1.4 for loose sand. Holden (16) using a uniformly loaded circular area on the surface of a medium sand with a relative density of 67%, reports maximum vertical strain at z/r = 1.1.

Vertical strain distributions have also been reported from the results of stress path tests on triaxial specimens of reassembled sand. Fig. 1 includes one from Ref. 6, from test results on a dense, overconsolidated sand.

Finite Element Computer Simulation.—A comprehensive, computer modeling technique has also been employed to study the axial-symmetric strain distribution under a circular, concrete footing resting on the surface of homogeneous sand. The finite element technique permits modeling the soil realistically, as a material with gravity stresses, nonlinear stress-strain behavior, and with stress-strain behavior dependent on effective stress. Fig. 2 presents some computer predicted, centerline strain distributions for one specific case: a 10-ft diam concrete footing, 1.25 ft thick, resting on the surface of a homogeneous, cohesionless soil with ϕ = 37°, and with unit weight = 100 lb per cu ft. (For the cases studied the vertical strain distributions were almost the same from the center line to between $0.5r$ to $0.75r$.) This model soil also has K_0 = 0.50 and Poisson's ratio = 0.48, thus approximating a normally consolidated state.

The computer-predicted settlements of this footing increase linearly to about 0.8 in. when p = 4,000 psf—a reasonable value for a real sand with ϕ = 37°. In view of the strain information in Fig. 1, the strain distributions in Fig. 2 also appear reasonable. (This is a preliminary study, done in June, 1969, by J. M. Duncan at the University of California, Berkeley, for Nilmar Janbu and the writer.) The depth to greatest vertical strain gradually increases as p increases, from about $0.72r$ at 500 psf to $1.20r$ at 4,000 psf. The same analysis, but with a 100-ft diam footing, results in a similar strain distribution, but with the depth to maximum strain remaining at about $0.72r$ while p increases from 1,000 psf to 4,000 psf. Results are also similar with a 1.0-ft diam footing, but depth to maximum strain increases from about $0.75r$ to $1.19r$, while p increases from 50 psf to 500 psf. It seems clear that the depth to maximum, centerline, vertical strain increases at the ratio of structural/gravity stresses increases. However, the increase is only over the $0.7r$ to $1.2r$ range. Both this range of depths to maximum strain, and the shape of the strain distribution curves, tend to confirm the other types of similar data presented in Fig. 1.

This computer study also showed that over the range of diameters investigated, 1 ft to 100 ft, and over the range of footing pressure investigated, 50 psf to 4,000 psf, approximately 90% of the settlement occurred within a depth = $4r$ below the footing. From a practical viewpoint, it seems reasonable to reduce exploration and computation by ignoring the static settlement of sand below $4r$.

Single, Approximate Distribution.—From the theoretical, model study, and experimental and computer-simulation results, it seems abundantly clear that the vertical strain under shallow foundations over homogeneous, free draining soils proceeds from a low value immediately under a footing to a maximum at a significant depth below the footing and thereafter gradually diminishes with depth. This is considerably different than one would expect when assuming a vertical strain distribution similar to the distribution of increase in vertical stress. Such an assumption is likely to be incorrect. The reason it is incorrect is that vertical strains in a stress dependent, dilatent material such as sand depend not only on the level of existing and added vertical normal stress, but also on the existing and added shear stresses and their respective ratio to failure shear stresses. The importance of shear in settlement has been noted repeatedly, by DeBeer (8), Brinch Hansen (13), Janbu (17), Lambe (21), and Vargas (38).

Considering the evidence in Figs. 1 and 2, for practical work it appears justified to use an approximate distribution for the vertical strain factor, I_z, under a shallow footing rather than to work indirectly through an approximate distribution of vertical stress. Why use an unnecessary and uncertain intermediate parameter? Possibly the most accurate estimate of a distribution for the strain factor for a particular problem would involve a complex consideration of the vertical distribution of changes in deviatoric and spherical stress. Each problem would then involve a special distribution. However, as shown subsequently by test cases, a single, simple distribution seems accurate enough for many practical settlement problems. The writer suggests the triangular distribution shown by the heavy, dashed line in Figs. 1 and 6 for the approximate distribution of a strain influence factor, I_z, for use in design computations for static settlement of isolated, rigid, shallow foundations. The writer uses this I_z triangle, referred to as the 2B-0.6 distribution, throughout the remainder of this paper.

The approximate distribution defines a vertical strain factor, and not vertical strain itself. Eqs. 1 and 2 show that this factor requires multiplication by p/E to convert it to strain.

This approximate distribution for the strain factor, which equals the shape of the actual strain distribution for a sand with constant modulus, applies only under the center portion of a rigid foundation. However, with knowledge of the vertical strain distribution under any point of the foundation the engineer can solve for the settlement of a concentrically loaded, rigid foundation. This is the case assumed herein. Consideration of other cases requires extension of this work.

CORRECTIONS TO ASSUMED APPROXIMATE STRAIN DISTRIBUTION

Foundation Embedment.—Embedding a foundation can greatly reduce its settlement under a given load. For example, Peck et al. (29) suggests a reduction factor of 0.50 when D/B changes from 0 to 4. D = the depth of foundation embedment and B = the least width of a rectangular foundation. Teng (34) suggests a reduction factor of 0.50 when D/B changes from 0 to 1. Meyerhof (25) suggests 0.75 for the same embedment. Yet, no major change in the 2B-0.6 I_z distribution is required to correct for embedment when using cone data.

Cone bearing values in sand soils usually start from low values at the surface and increase with depth. Thus, even with homogeneous soil, a surface foundation would have an average cone value over the 0-2B interval that can be considerably less than the average value over B-3B, which becomes the 2B interval when $D = B$. For example, if q_c increased proportional to the square root of z/B, from zero at the surface, then settlement when $D/B = 1$ computes about 0.60 the settlement when $D/B = 0$, and about 0.35 of this settlement when $D/B = 4$ (using the new method described later).

Another, usually relatively minor, correction for embedment results from the use of elastic theory. According to solutions from the linear theory of elasticity, once the depth, D, of a buried square footing exceeds about five times its least width, B, then elastic settlement reduces to one-half surface values (15). The assumed elastic, weightless material above the level of loading permits tension to relieve load and strain under that level. Sands, contrary to this, cannot sustain loads in tension. However, an arching-induced reduction in compressive stresses can replace elastic tension, with the compressive stresses due to the overburden weight of the sand.

To take some account of the strain relief due to embedment, and yet retain simplicity for design purposes, the writer proposes to retain the 2B-0.6 shape of the strain influence factor, I_z, but to adjust its maximum value to something less than 0.6. To conform to the arching-compression relief concept this adjustment should not depend solely on the D/B ratio. Instead use the ratio of the overburden pressure at the foundation level, = p_0, to the net foundation pressure increase, = $(p - p_0) = \Delta p$, or $(p_0/\Delta p)$. The following equation defines a simple, linear correction factor, C_1:

$$C_1 = 1 - 0.5 \left(\frac{p_0}{\Delta p}\right) \qquad (3)$$

However, in accord with elasticity, C_1 should equal or exceed 0.5.

Creep.—In the past it has not been common to consider the time rate of development of settlement in sand. Contrary to this, many, but not all, of the published settlement records show settlement continuing with time in a manner suggesting a creep type phenomenon.

Brinch Hansen (13) noted the importance of this creep and included a mathematical estimate of its contribution in his sand settlement analysis procedure. Nonveiler (28) also noted its importance and suggested this linear decay correction on a semilog plot:

$$\rho_t = \rho_0 \left[1 + \beta \log\left(\frac{t}{t_0}\right)\right] \qquad (4)$$

in which ρ_0 = the settlement at some reference time t_0; ρ_t = the settlement at time t and β = a constant which was about 0.2 to 0.3 in the problem investigated. The apparent creep is not completely understood and most likely arises from a variety of causes. But, the effect is similar to secondary compression in clay. Because of the simplicity of Eq. 4, the writer has adopted it as a correction factor, C_2, in this new settlement estimate procedure. Tentatively, β = 0.2 and the reference time, t_0 = 0.1 yr. The principal justification for this reference time is that it is convenient and appears to give reasonable predictions in the test cases noted subsequently. Then C_2 becomes:

$$C_2 = 1 + 0.2 \log \left(\frac{t \text{ yr}}{0.1}\right) \quad (5)$$

Shape of Loaded Area.—The various shape correction factors used when applying the theory of elasticity to the settlement of uniformly loaded surface areas suggests that the distribution of the assumed strain influence factor, I_z, also needs modification according to the shape of the loaded area. However, a correction does not appear necessary at this time.

Consider a rectangular foundation of constant, least width = B and with constant bearing pressure = p. As its length L, and L/B, increases the total load on the foundation increases and one might therefore expect a greater settlement although both B and p remain constant. However conditions also become progressively more plane strain. The full transition from axially symmetric to plane strain involves some increase in the angle of internal friction. This increased strength results in reduced compressibility, which tends to counteract the effect of a larger loaded area and a larger load. Neither behavior is well enough understood over a range of L/B ratios to permit preparing quantitative shape factor corrections. The writer assumes herein that these compensating effects cancel each other. It may be significant to note that no such correction is used with SPT empirical methods. The subsequent test cases, involving a considerable range of L/B ratios, also do not suggest an obvious need for such correction.

Adjacent Loads.—The design engineer must also deal with the practical problem of how to compute the settlement interaction between adjacent foundation loadings. This complicated problem involves a material (sand) with a nonlinear, stress dependent, stress-strain behavior. Not only do strain and settlement depend on the position and magnitude of adjacent loads, but also on their sequence of application. A later application of a smaller, adjacent load should settle less, possibly much less, than had that load been applied without the lateral prestressing effects of the first load.

In stress oriented settlement computation procedures the adjacent load problem is ordinarily handled by assuming linear superposition of elastic stresses. The analogous in a strain oriented procedure would be to superpose strains, or strain influence factors. However, any simple, linear form of superposition possibly invites serious error because of the nonlinear importance of stress magnitude and loading sequence. More research is needed to formulate design rules for this problem. Model studies, in the laboratory or by computer simulation, or both, look most promising.

The present state of knowledge requires the engineer to use conservative judgement. Obviously if two foundations are far enough apart any interaction will be negligible. The writer would consider this the case if 45° lines from the edges intersect at a depth greater than $2B_2$, when a second loading of width B_2 is placed next to an existing foundation of greater width B_1. For a 45° intersection depth also greater than B_1, assume them independent regardless of load sequence. If adjacent foundations are close enough to interact without question, say the distance between them is less than B of the smallest and they are loaded simultaneously, then the writer would treat them as a single foundation with some appropriate, equivalent width. Intermediate situations should fall within these boundaries.

CORRELATION BETWEEN STATIC CONE BEARING CAPACITY AND E_s VALUES USED IN SETTLEMENT COMPUTATIONS

Continuing the previous notations, the calculation of settlement requires an integration of strains. Thus

$$\rho = \int_0^{\infty} \epsilon_z \, dz \approx \Delta p \int_0^{2B} \left(\frac{I_z}{E_s}\right) dz \approx C_1 C_2 \Delta p \sum_0^{2B} \left(\frac{I_z}{E_s}\right) \Delta z \quad \ldots\ldots (6)$$

The last form of Eq. 6 permits approximate integration and a way of accounting for soil layering. The key soil-property variable that still remains to be determined is the equivalent Young's modulus for the vertical static compression of sand, E_s, and its variation with depth under a particular foundation.

Screw-Plate Tests.—A direct means of determining vertical E_s in sands would be to test load a plate in-situ, measure its settlement, and use Eq. 6 to backfigure its modulus. Any attempt to test at depths other than near the surface requires an excavation with its attendant load-removal stress and strain disturbances. Many sites would also require dewatering, with still further stress disturbances. To avoid such difficulties the writer used a form of plate bearing load test used in Norway (19), known as the screw-plate test. The writer's screw-plate consisted of an auger with a pitch equal to 1/5 its diameter, and a horizontally projected area of 1.00 sq ft over a single, 360° auger flight. This special auger was screwed into the ground, taking care to assure that the vertical rods remain plumb. The buried plate was loaded by using a hydraulic jack at the surface, reacting against anchored beams. Rod friction to the screw-plate seemed negligible. Elastic compression was subtracted and care was used to assure the column of rods to the plate did not buckle significantly. Sands at depths from 3 ft to 26 ft (1 m to 8 m) were tested in this way.

Fig. 3 shows photographs of the screw-plate and the load test set-up. The load was applied to the top of the column of rods, using increments in the conventional manner. The usual results consisted of a conventional appearing load-settlement curve with tangent moduli decreasing slightly with increasing pressure.

Correlation with Static Cone Bearing.—Although the screw-plate type of load test to determine sand compressibility is faster and less expensive than burying a rigid plate, it is nevertheless still too time consuming for routine investigations. For this reason data were accumulated in an attempt to see if static cone bearing capacity would correlate with screw-plate bearing compressibility. Fig. 4 presents the results of this correlation on a log-log plot. This investigation used the mechanical Dutch friction cone (32), advanced at the common rate of 2 cm per sec. Sand compressibility, in inches per ton per square foot (tsf), was taken as the secant slope over the 1 tsf-3 tsf increment of plate loading. This interval was chosen for convenience because the seating load was 0.5 tsf, almost all tests were carried to a minimum of 3 tsf, and real footing pressures commonly fall within this interval.

Note that a different symbol denotes each of 10 test sites. Four of these are in Gainesville, Florida. The remaining six are within a radius of about

SM 3 SETTLEMENT OVER SAND

FIG. 3.—UNIVERSITY OF FLORIDA SCREW-PLATE LOAD TEST: (*a*) 1.0 SQ FT SCREW-PLATE; (*b*) LOAD TEST SET-UP

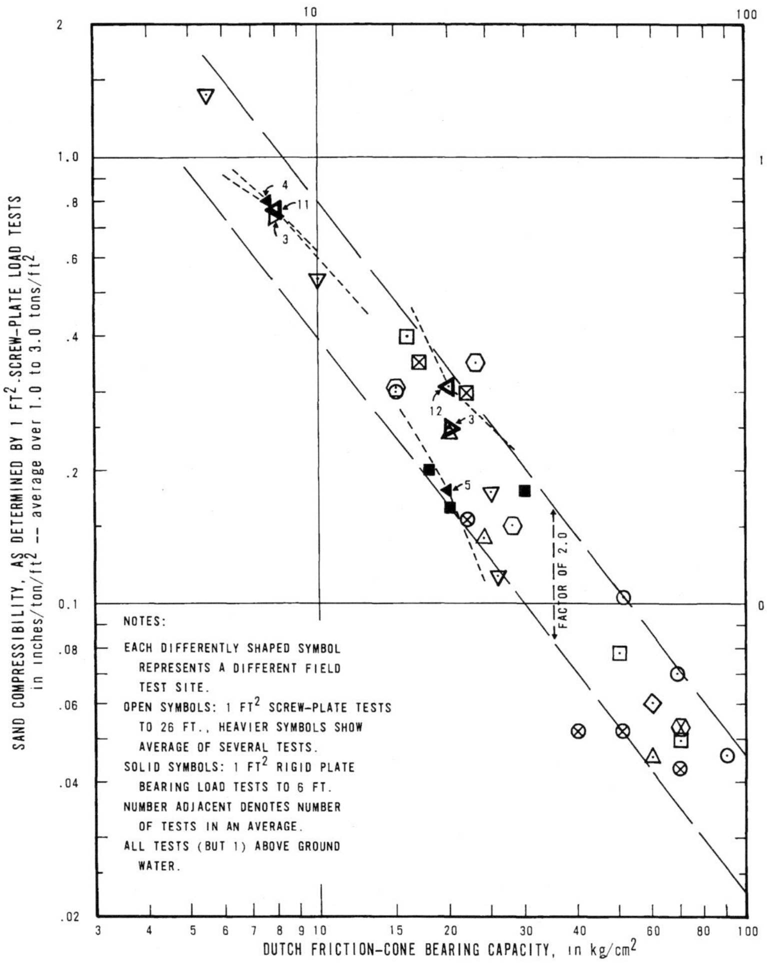

FIG. 4.—EXPERIMENTAL CORRELATION BETWEEN DUTCH CONE BEARING CAPACITY AND COMPRESSIBILITY, UNDER IN-SITU SCREW-PLATE LOAD TEST, OF SOME FINE SANDS IN FLORIDA

150 miles from Gainesville. The sands tested were above the water table, and include silty fine sand to uniform medium sand. However, most tests involved only fine sand with a uniformity coefficient of 2 to 2.5.

Fig. 4 includes 29 screw-plate tests from two research sites on the campus of the University of Florida. To condense the results from these 29, Fig. 4 shows only the average values for each group of tests at the same depth at the same site. Dashed lines indicate the spread of the data from one site. These special research tests involved only two plate depths, 2.8 ft and 6.1 ft. Nine tests were also made on 1.0 sq ft rigid, circular plates at these same plate depths at one of these sites. Again, average values and spread are indicated. The adjacent number indicates the number of individual tests in the average. The eight remaining sites account for 24 screw-plate tests at depths ranging from 3 ft to 26 ft, averaging 9.3 ft. At one of these sites data were also available from three 1-ft square rigid plate tests by Law Engineering Testing Co. Thus, the total number of individual plate tests included in Fig. 4 consists of 53 screw-plate and 12 rigid plate tests.

It appears from Fig. 4 that about 90% of these data fall within the factor-of-2 band shown. It is not surprising that a good correlation exists between compressibility and cone bearing in sands because in some ways the penetration of the cone is similar to the expansion of a spherical or cylindrical cavity, or both (2). Alternatively, if the cone is thought of as measuring bearing capacity and hence shear strength, then one can also argue, as the writer has already done, that the compressibility of sand is greatly dependent on its shear strength.

To convert screw-plate compressibility into E_s values required for Eq. 6 only required backfiguring that E_s value needed to satisfy Eq. 6 and each measured settlement. This resulted in the correlation in Fig. 5. Because the grouping of the individual points proved similar to that in Fig. 4, only the factor-of-two- band is shown (dashed lines). With this band as a guide the writer then chose a single correlation line for design in ordinary sands. Thus

$$E_s = 2\, q_c \quad (7)$$

This line was chosen because it falls within the screw-plate band, because it results in generally acceptable predictions for settlement in the subsequent test cases and also because of its simplicity. Eq. 7 permits the use of inexpensive cone bearing data to estimate static sand compressibility, as represented by E_s. Then compute settlement from Eq. 6.

Webb (40) recently reported the results of an independent correlation study in South Africa between the insitu screw-plate compressibility of fine to medium sands below the water table and cone bearing. His data include seven tests using a 6-in. diam plate (0.20 sq ft), eight tests with a 9-in. plate (0.44 sq ft) and one test with a 15-in. plate (1.23 sq ft). Cone bearing ranged between about 10 tsf and 100 tsf. He offers the following correlation equation for converting q_c to his E':

$$E' \text{ (tsf)} = 2.5\,(q_c + 30 \text{ tsf}) \quad (8)$$

Comparison of the elastic settlement formula in his paper and Eq. 6 herein shows that $E_s = C_1 C_2\; 0.6\; E'$. This assumes a constant E_s for a $2B$ depth below the screw-plate, permitting $\Sigma\, I_z\, \Delta z$ = area under $2B$-0.6 I_z distribution = $0.60B$. The average product $C_1 C_2$ used by the writer when converting his screw-plate data was about 0.88. Thus, $E_s \approx 0.53\; E'$. Webb's equation then converts to $E_s \approx 1.32\ (q_c + 30)$. Further comparison with Eq. 7 now shows the same prediction for E_s when $q_c \approx 60$ tsf, and a difference of 20% or less when q_c lies between 35 tsf and 170 tsf. Reference to Tables 1 and 2 shows that this range includes most natural sands. Such agreement supports the validity of using cone bearing data to estimate the insitu compressibility of sand under a screw-plate.

Method of Accounting for Soil Layering, Including a Rigid Boundary Layer.— The simple I_z distribution developed herein from elastic theory and model experiments assumed or used a homogeneous foundation material. But, sand deposits vary in strength and compressibility with depth. It is further assumed that the I_z distribution remains the same irrespective of the nature of any

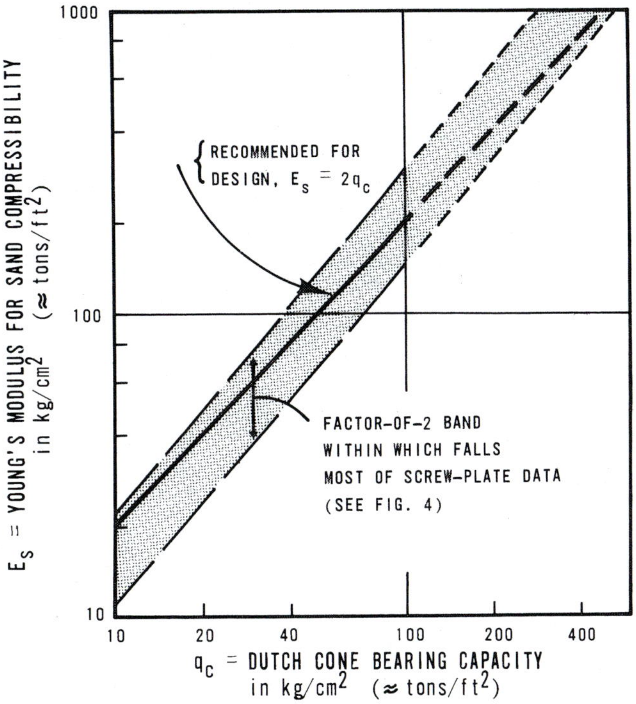

FIG. 5.—CORRELATION BETWEEN q_c AND E_s RECOMMENDED FOR USE IN ORDINARY DESIGN

such layering and that the effects of such layering are approximately, but adequately, accounted for by varying the E_s value in Eq. 6 in accord with Eq. 7.

It is possible that the above method of accounting for layering represents an oversimplification and will result in serious error under special circumstances not now appreciated. More research would be useful to define the limitations of this method and to improve it. Model studies, especially computer simulation using the nonlinear, stress dependent finite element technique, appear to have great promise for investigating such problems. This approach to layering also includes the treatment of a rigid boundary layer encountered within the interval 0 to $2B$. The 2B-0.6 I_z distribution remains the same but the soils below this boundary, to the depth $2B$, are assumed to have a very high modulus. Vertical strains below such a boundary then become negligible and can be taken equal to zero.

TABLE 1(*a*).—SETTLEMENT ESTIMATE FOR EXAMPLE IN FIG. 6 USING NEW STRAIN-DISTRIBUTION METHOD AND SOLVING EQ. 6

Layer (1)	Δz, in centimeters (2)	$\bar{q}_c$, in kilograms per square centimeter (3)	E_s, in kilograms per square centimeter (4)	z_c, in centimeters (5)	I_z (6)	$\left(\frac{I_z}{E_s}\right)\Delta z$, in centimeters per kilogram per square centimeter (7)
1	100	25	50	50	0.23	0.462
2	30	35	70	115	0.53	0.227
3	170	35	70	215	0.47	1.140
4	50	70	140	325	0.30	0.107
5	100	30	60	400	0.185	0.308
6	70	85	170	485	0.055	0.022
Total						2.266

$C_1 = 0.89$; C_2 (5 yr) $= 1.34$; $\Delta p = 1.50$; $\rho = (0.89)(1.34)(1.50)(2.266) = 4.05$ cm $= 1.60$ in.

TABLE 1(*b*).—SETTLEMENT ESTIMATE FOR THE EXAMPLE IN FIGURE 6 USING BUISMAN-DEBEER METHOD[a]

Layer (1)	Δz, in centimeters (2)	z_c, in centimeters (3)	σ'_{vic}, in kilograms per square centimeter (4)	$\bar{q}_c$, in kilograms per square centimeter (5)	$1.535\left[\left(\frac{\sigma'_{vi}}{q_c}\right)\Delta z\right]$, in centimeters (6)	z/B (7)	$\frac{\Delta\sigma_v}{\Delta p}$ (8)[b]	$\Delta\sigma_v$ ($\Delta p = 1.50$), in kilograms per square centimeter (9)	$\left(\frac{\Delta\sigma_v + \sigma'_{vi}}{\sigma'_{vi}}\right)$ (10)	log( ) (11)	$\Delta\rho$ = Column 6 × Column 11, in centimeters (12)
1	100	50	0.37	25	2.272	0.19	0.90	1.35	4.649	0.6674	1.516
2	30	115	0.435	35	0.573	0.44	0.75	1.125	3.586	0.5546	0.318
3	170	215	0.535	35	3.988	0.83	0.59	0.885	2.654	0.4239	1.691
4	50	325	0.645	70	0.708	1.25	0.47	0.705	2.093	0.3208	0.227
5	100	400	0.72	30	3.684	1.54	0.41	0.615	1.854	0.2681	0.988
6	50	475	0.795	85	0.718	1.83	0.36	0.54	1.679	0.2251	0.161
7	150	575	0.895	170	1.213	2.21	0.31	0.465	1.520	0.1818	0.221
8	100	700	1.02	60	2.610	2.69	0.26	0.39	1.382	0.1405	0.367
9	100	800	1.12	100	1.719	3.08	0.22	0.33	1.295	0.1123	0.193
10	150	925	1.245	40	7.167	3.56	0.19	0.285	1.229	0.0896	0.642
11	100	1050	1.37	66	3.187	4.04	0.16	0.24	1.175	0.0700	0.223
12	200	1200	1.52	120	3.888	4.62	0.14	0.21	1.138	0.0561	0.218
13	100	1350	1.67[c]	120	2.137	5.19	0.12	0.18[c]	1.108	0.0445	0.095
											$\rho = \Sigma$ = 6.860 cm = 2.70 in.

[a] Equation to be solved: $\rho = \Sigma\ \{1.535\ [(\sigma'_{vi}/q_c)\ \Delta z]\ \log\ [(\Delta\sigma_v + \sigma'_{vi})/\sigma'_{vi}]\}$ . . . . Eq. (9)

[b] Taken from charts based on Buisman distribution of vertical stress. For this case (rigid foundation) used stresses under DeBeer 'singular point'.

[c] Layer 13 is the last layer because stress increase at bottom of layer ≈ 10% effective overburden pressure.

Justification for the previous approach is primarily pragmatic. The computational procedure retains its simplicity despite layering. This method appears successful in the test cases noted subsequently, including the case with a rigid boundary at $0.23B$. Also, a series of model tests by the writer, using a circular, rigid, plate of 2.3 in. diam, on the surface of a dry sand with a relative density of about 25%, showed the effect of a rigid boundary on settlement to be very similar to that obtained from the $2B$-0.6 I_z distribution and the simple cut-off procedure previously suggested.

The simple conversion from cone bearing to modulus suggested herein could require modification for such effects as the magnitude of foundation pressure increase, different ground water conditions and different states of overconsolidation. This topic falls beyond the scope of the present paper. No such corrections are suggested herein. The subsequent test case comparison results suggest that the simplest approach, ignoring them, often produces acceptable prediction accuracy.

SETTLEMENT ESTIMATE CALCULATION

The following information must be gathered before a settlement estimate can be computed by the method suggested herein:

1. A static cone bearing capacity (q_c) profile over the depth interval from the proposed foundation level to a depth below this of $2B$, or to a boundary layer that can be assumed incompressible, whichever occurs first. Because the correlation with E_s is empirical and is based on q_c values obtained primarily from Dutch static cone equipment, it is desirable that the needed q_c profile be obtained with similar equipment. The Dutch cone has a 60° hardened steel point, a projected end area of 10 sq cm, and is advanced during a measurement at a rate of 2 cm per sec. The rods above the points are screened from soil friction by an outer, casing rod system. Other static cone systems may be used provided they can be correlated with the Dutch cone results or provided independent calibrations with E_s can be established for each system.
2. The least width of the foundation = B, its depth of embedment = D, and the proposed average foundation contact pressure = p. The same data is needed for adjacent foundations close enough to interact with the one for which settlement is being estimated.
3. The approximate unit weights of surcharge soils, and the position of the water table if within D. These data are needed for the estimate of p_0, which is needed for the C_1 correction factor.

With this information gathered, proceed as in the example illustrated by Fig. 6 and Table 1(a). This example is an actual pier foundation and is the first test case comparison in the next section herein.

4. Divide the q_c profile into a convenient number of layers, each with constant $\bar{q}_c$, over the depth interval 0 to $2B$ below the foundation.
5. Prepare a table with headings similar to Table 1(a) herein. Fill in columns 1, 2, and 3 with the layering assigned in step 4.
6. Multiply the values of q_c in column 3 by the factor 2.0 to obtain the suggested design in values of E_s. Place these in column 4.
7. Draw the assumed 2B-0.6 triangular distribution for the strain influ-

May, 1970 SM 3

TABLE 2.—LISTING

Number	Reference		Structure	B, in feet	L/B	D/B
(1)	(2)		(3)	(4)	(5)	(6)
1	DeBeer (9)		Belgian bridge pier	8.5	8.8	0.78
2	DeBeer (9)		Belgian bridge pier	9.8	4.2	1.0
3	DeBeer (7)		Belgian bridge pier	8.2	2.5	1.2
4	DeBeer (7)		Belgian bridge pier	19.7	2.7	0.58
5	Bjerrum (3,20)		Test fill	62	1.0	0
6	Nonveiler (28)		Grain silo	87	2.2	0.1
7	Muhs (27)		Model concrete pier load tests			
	Test: V			3.3	1.0	0
	VI			1.7	3.9	0
	XI			1.7	3.9	0
	VIII, IX			3.3	1.0	0.5
	X, XII			1.7	3.9	1.0
	XV, XVII			3.3	1.0	0.5
	XVI, XVIII			1.64	4.0	1.0
	XXXVII			3.3	1.0	0.5
	XXXVIII			3.3	1.0	0.5
	XXXIX			1.64	4.0	1.0
8	Law load test in Florida	No.				
		5	Steel plate	2.0	1.0	0.55
		6	Steel plate	2.0	1.0	1.5
		7	Concrete plate	3.0	1.0	0.3
		8	Concrete plate	3.0	1.0	1.0
		9	Concrete plate	4.0	1.0	0.17
		10	Concrete plate	4.0	1.0	0.75
9a	Tschebotarioff (37)		Liquid storage building	90	1.1	0.1
9b	Tschebotarioff (37)		Test plate	2.0	1.0	0
10	Grimes and Cantlay (12)		20 St Office Building (center of 3)	42.7	2.1	0.16
11	Webb (40)		Concrete test plate	20	1.0	0.03
12	Bogdanovic (4)		8-story apartment	79	3.6	0
13	Brinch Hansen (13)		Steel tank	184	1.0	0
14	Kumennje (19) Janbu (18)		Oil Tank	96	1.0	0
15	Meigh and Nixon (23)		Factory concrete footings	4.7	1.0	0.85
16	D'Appolonia (6)		Over 300 steel factory footings	12.5	1.6	0.64

SM 3 SETTLEMENT OVER SAND

OF TEST CASES

Soil	Approximate average 0-2B q_c, in kilograms per square centimeter	Stresses, in tons per square foot		Notes
		p_0	Δp	
(7)	(8)	(9)	(10)	(11)
Foundation at ground-water table	40	0.33	1.21 1.70	No live load Full live load
Silty to fine sand	120	0.33	1.27 1.86	No live load Full live load
Cut in sand, some clay layers	220	0.54	2.43	Probably full live load
	120	0.64	1.78	Probably full live load
Coarse silt, fine sand, ground-water table at surface	70 60	0	0.78	Nearest q_c average 2 nearest
Fine sand, 1/3 calcite (shells)	90	0.56	2.07	Rock below $D = B$
Natural fine sand, above ground-water table	135 100 100	0 0 0	2.05 2.05 3.07	 $q_{ult} \approx$ 8 tsf $\approx$ 10 tsf
Compacted moist sand embankment	180 150 70 55	0.10 0.10 0.10 0.10	5.16 5.16 3.07 2.56	$\approx$ 25-30 tsf $\approx$ 20 tsf $\approx$ 9-11 tsf $\approx$ 7-8 tsf
Compacted moist sand embankment, but water at base of pier	45 45 35	0.09 0.09 1.10	3.07 2.56 1.53	$\approx$ 8-1/2 tsf $\approx$ 7 tsf $\approx$ 4 tsf
Uniform, very fine sand above ground-water table	18 22 20 23 27 32	0.06 0.15 0.04 0.15 0.03 0.15	1.14 1.95 1.20 0.90 1.82 2.35	
	80 70	0.50 0.50	3.1 3.2	Previous structure on site
Vibrofloted sand below water table	125	0.38 to 0.59	1.42	Compressible clays below sand
Alluvial sand below ground-water table	40	0	2.0	
Variety of sands, some clay and silt	115 100	0 0	0.68 0.68	Corner III Opposite corner IV
Hydraulic fill below ground-water table	30	0	1.23	Incompressible clay below 0.23B
Fine sand, slightly organic, below ground-water table	70	0	1.33	
Gravel with flints, some fine sand	130	0.25	1.0	2 footings
Overconsolidated dune sand	120	0.44	1.70	Average size, depth and loading herein

ence factor, I_z, along a scaled depth of 0-2B below the foundation. Locate the depth of the mid-height of each of the layers assumed in step 4, and place in column 5. From this construction determine the I_z value at each layer's mid-height and place in column 6.

8. Calculate $(I_z/E_s)\,\Delta z$ and place in column 7. This represents the settlement contribution of each layer assuming that C_1, C_2 and Δp all = 1. Then determine the sum of the values in column 7.

9. Determine separately C_1 from Eq. 3 and C_2 from Eq. 5. Multiply the Σ (col. 7) by these C_1 and C_2 factors and by the appropriate Δp to obtain the

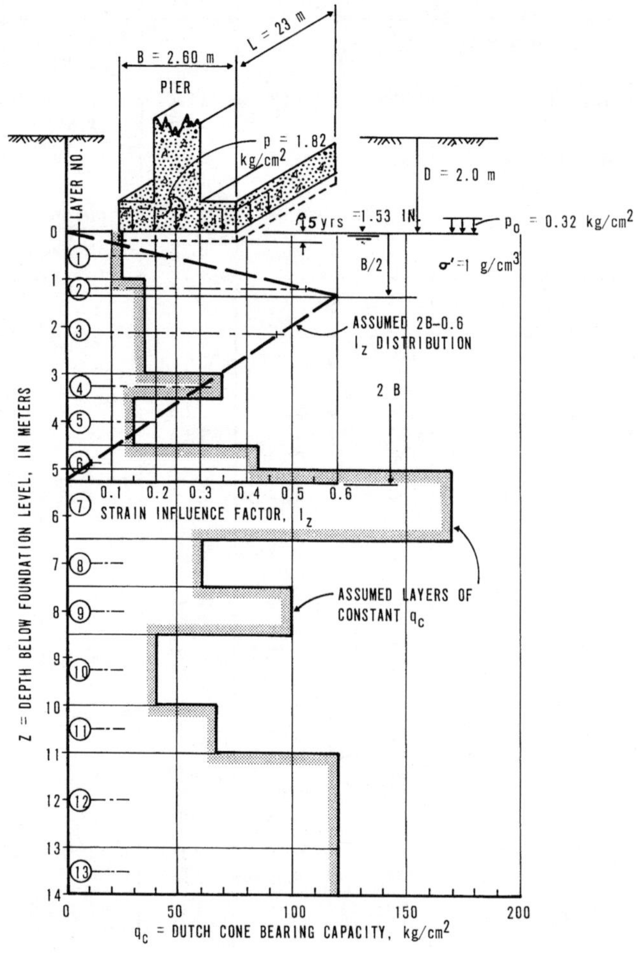

FIG. 6.—TEST CASE NO. 1 AS COMPUTATIONAL EXAMPLE

final settlement estimate for the time-after-loading assumed in the calculation of C_2.

10. Any consistent set of units may be used in this calculation procedure. Because q_c is obtained in kilograms per square centimeter, which for all practical purposes is also equal to tons per sq ft, it is convenient to use these pressure units for E_s, p_0 and Δp. If all lengths are either centimeters or inches, then the settlement will also be in centimeters or inches.

As analyzed subsequently in more detail, the Buisman-DeBeer method represents a competing method of estimating settlement from static cone data. For subsequent reference, Table 1(*b*) lists the calculations for this same example using the Buisman-DeBeer method.

TEST CASE COMPARISONS

How accurate is the proposed settlement estimate calculation procedure when compared to cases where settlements have been measured and where the requisite data (steps 1, 2 and 3) are available? The writer searched the literature for such cases and found a few with sufficient, or nearly sufficient data. Their scope should also be sufficient to demonstrate the prediction accuracy expected. Table 2 lists the pertinent data from all cases. Table 3 lists the measured and predicted (afterwards) settlements. Table 3 also includes settlements as predicted from using the Meyerhof and Buisman-DeBeer methods, which will be discussed further in the next section of this paper. The following comments supplement the information in these tables.

Belgian Bridge Piers (cases 1-4).—These make especially good test cases because of the completeness of the data supplied by DeBeer and his associates in the reference cited. Two loads are given for the first two cases, one includes dead load only and the other dead plus design live load. DeBeer kindly made these data available in a personal communication. Note that the settlements reported for all four cases are for times of 2-1/2 yr to 7 yr and thus include the settlement effects of the test loads on these bridges and the subsequent traffic live loads. The writer based the settlement calculations for cases 1 and 2 on an equivalent static loading assumed at dead plus 2/3 the design live load. For cases 3 and 4 the loadings used are as obtained from the references cited. They probably include full live load, but this is uncertain.

Norwegian Test Fill (case 5).—This fill was constructed specifically to determine, by large scale tests, what settlements should be expected at the site of a large industrial project. The top of the fill was 46 ft by 46 ft, the bottom was 79 ft by 79 ft, giving fill side slopes of about 40°. The nearest cone sounding was about 250 ft away. The second nearest was about 500 ft away in the opposite direction. Table 3 includes two computed settlements, one using only the nearest q_c profile and the other the average profile from these two nearest. L. Bjerrum kindly made several pertinent Norwegian Geotechnical Institute (NGI) internal reports available to the writer. These present more detailed site data than available in the published reference.

Settlements were measured at the base of the test fill. The value in Table 3 was the maximum under the central 46 × 46 ft area, but settlements under this area were approximately constant. The Buisman-DeBeer calculation for this case is based on stress increase under a rigid foundation rather than under the center of a uniform loading. This reduces the computed *B-D* settlement and makes their comparison with measured settlement more favorable than when using a uniform loading.

Grain Silo (case 6).—The reference details somewhat complicated foundation conditions, with abandoned, partially installed, pier foundations at one end of the silo and a tower structure adjacent to the other end. The soil was unusual in that the fine sand was reported to be about 1/3 calcite, much of it in the form of shell fragments. Rock was at a depth of $1.0B$ below the foundation level.

TABLE 3.—MEASURED AND ESTIMATED

Case Number	Measured Settlement, in inches				
	Time	Minimum	Average	Maximum	Notes
(1)	(2)	(3)	(4)	(5)	(6)
1	5 yr	1.02		1.53	
2	7 yr	0.78		0.90	Nearby fill
3	3 yr	0.24		0.32	
	5 yr	0.35		0.39	
4	several months	0.43		0.47	Cone data 85 ft from pier
	2-1/2 yr		1.10		
5	400 days			2.48	Nearest q_c (250 ft) q_c average 2 nearest
6	2 yr		10.6		
7 V	Assumed 1 day for all load tests		0.142		
VI			0.157		
XI			0.264		
VIII, IX			0.173		
X, XII			0.165		
XVI, XVIII			0.102		
XXXVII			0.236		
XXXVIII			0.185		
XXXIX			0.138		
8-No. 5	Assumed 1 day for all tests		0.27		1 load cycle
8-No. 6			0.50		1 load cycle
8-No. 7			0.30		1 load cycle
8-No. 8			0.25		1 load cycle
8-No. 9			0.51		1 load cycle
8-No. 9			0.66		6 load cycles
8-No. 10			0.50		1 load cycle
8-No. 10			0.56		several cycles
9a	Assumed 1 yr	3.0		3.7	
9b	Assumed 3 days		0.36		
10	1.7 yr		0.95 (0.38)		Not all settlement in surface sand
11	Assumed 4 days			3.25	
12-III	2 yr	1.97			Corner building
12-IV	2 yr			3.54	Opposite corner
13	0.3 yr		1.46		Measured around perimeter
	2 yr		1.73		
	7 yr		2.91		
14	5 days	4.9	6.3	7.4	Measured around perimeter
15	4 months	0.04		0.09	2 footings N = 13, N = 21
16	3-1/2 yr	0.1	0.32	0.6	Over 300 footings

SM 3 SETTLEMENT OVER SAND

SETTLEMENT FOR TEST CASES

Computed Settlement Estimate, in inches				Symbol in Figs. 7, 8
Meyerhof (7)	*B*-DeBeer (8)	Schmertmann (9)	Using Δp, in tons per square foot (10)	(11)
2.05	2.70	1.60	1.54	⊙
0.46	1.28	0.78	1.67	⦸
		0.44	2.43	⊗
0.54	0.62	0.46	2.43	⊗
0.67	1.79	0.96	1.78	⊘
		1.16	1.78	⊘
0.76	3.79	3.60	0.78	⧄
0.97	4.28	3.91	0.78	⧄
1.2	8.6	5.7	2.07	
0.62	0.130	0.159	2.05	•
1.02	0.126	0.130	2.05	
1.54	0.154	0.193	3.07	
1.02	0.236	0.237	5.16	
1.18	0.213	0.156	5.16	
0.75	0.35	0.184	2.56	
5.2	0.528	0.599	3.07	
4.4	0.437	0.499	2.56	
1.53	0.303	0.187	1.53	
1.90	0.46	0.31	1.14	▲
2.95	0.69	0.46	1.95	
2.00	0.66	0.46	1.20	
1.28	0.59	0.28	0.90	
1.89	0.83	0.65	1.82	
1.89	0.83	0.65	1.82	
2.61	1.16	0.79	2.35	
2.61	1.16	0.79	2.35	
0.9	11.3	6.2	3.1	+
1.6	1.10	0.28	3.2	
0.32	1.37	0.79	1.42	×
5.2	4.79	4.32	2.0	⬡
0.30	0.85 (corner stress)	2.21	0.68	△
0.42	1.69 (corner)	3.70	0.68	△
	6.04 (rigid)			
	7.9 (center)	1.55	1.23	⊡
0.5	6.6 (rigid)	1.79	1.23	⊡
	4.0 (perimeter)	1.94	1.23	⊡
1.1	8.4 (rigid)	5.6	1.33	⬡
	5.5 (perimeter)		1.33	⬡
0.31	0.19	0.07	1.0	▽
0.19	0.12	0.04	1.0	▽
1.05	1.22	0.97	1.70	⊙

This test case resulted in a poor, nonconservative measured-predicted settlement comparison, due perhaps to the complex nature of the foundation conditions or the unique (in these test cases) shell content in the sand, or both.

DEGEBO *Model Piers (case 7)*.—These 14 individual tests are part of an extensive program of large scale, model pier, settlement and bearing capacity tests carried out in Berlin under the direction of H. Muhs. Muhs, via personal communication, kindly made available the details of a number of these tests, including extensive static cone sounding data. The DEGEBO cone is somewhat different than the Dutch equipment. It also has the 10 sq cm, 60°, steel point, but the back-taper design is different, and electrical strain gages (30) permit a more accurate determination of point resistance. The rate of penetration used by DEGEBO may be different than the standard Dutch 2 cm per sec, but the writer treated these data as if they were obtained by the Dutch cone.

Most of these tests are in a partially saturated, or saturated, embankment compacted in layers. These are the only test cases herein which involve compacted soil. Some of these test results represent the average of two tests intended to be identical. Each series of two showed similar results. The reference cited (in German) describes more of the interesting details about this phase of DEGEBO's extensive series of pier tests.

Law Plate Load Test Research (case 8).—These 6 individual tests are part of a 1967 to 1968 research program conducted in Jacksonville, Florida, by Law Engineering Testing Company. The University of Florida participated by obtaining the static cone data.

The sand at this research site has the lowest q_c values of any of the test cases, although some of the load tests in Fig. 4 had lower. Two independent sets of relative density tests, both by the Burmister method, yielded relative densities between 50% to 60% over the 0 ft to 6 ft depth interval. It is important that had these test plates been subjected to significant dynamic loading, or to a larger number of cycles of repeated static loadings, the measured settlements would have been greater. None of the settlement prediction methods discussed herein are intended to include loadings outside the range of loads, including live loads, that are usually treated as equivalent static loadings. Ultimate bearing capacity was not clearly defined by some of these plate tests. Perhaps some of these measured settlements reported in Table 3 are at average plate pressures greater than allowable by dividing ultimate bearing capacity by an appropriate safety factor.

Heavy, Rigid Storage Building and Plate Load Test (case 9).—The computed versus measured settlement comparison a in Table 3 is for the structure itself. Here the surface sand layer extends to a relative depth of only $0.72B$ below a mat foundation. The hard clay reported below this was assumed incompressible. The reference reports ground water level at $0.23B$. Comparison b is from a plate load test at the same site, with ground water below $2B$. In neither is a time given for the measured settlements. The writer assumed times to permit calculating his C_2 correction factor.

Nigerian Office Building (case 10).—At this building, the center of a complex of three, the surface soil consisted of 32 ft of loose, medium over fine sand. The engineers had this layer compacted by vibroflotation. Then they placed the structural foundation, a 7-ft thick mat, bearing at about the depth of the water table, also 7 ft. After vibroflotation cone bearing increased to about

60 kg per sq cm for 5 ft below the mat, then increased abruptly to about 200 kg per sq cm for the next 8 ft below, and the final 13 ft remained at about 90 kg per sq cm.

The total thickness of that part of the surface sand below the mat represents a relative depth of only 0.58B. The computed settlements in Table 3 represent only the contribution of this layer. However, the measured settlement of 0.95 in. includes the contribution of cohesive layers below this sand. The per cent of the total contributed by the surface sand is not known. The authors conservatively forecast a total settlement of 3.75 in. of which they thought 1.5 in. or 40%, would be in this surface sand. Applying this percentage to 0.95 in. gives 0.38 in.

South African Load Test (case 11).—Much of the pertinent data associated with this unusually large load test can be found in the cited references. Webb kindly made available even more complete data via personal communication. The writer used the average of four cone soundings, two under and two immediately adjacent to the test plate, when calculating the settlements reported in Table 3.

Boring logs and inspection shafts showed some clayey sand layers, organic sand and even a thin rubble fill. However, the predominant soil in the upper 50 ft to 60 ft is a normally consolidated, alluvial, fine sand. The borings also showed the water table at a depth of only about 3 ft. The writer considered all sand when preparing Table 3.

The load test plate was 12 in. thick reinforced concrete cast directly on natural sand, 6 in. below its surface. The interaction of the iron ingots used to load the plate provided extra stiffening, resulting in a ratio of center/corner settlement of only 1.25. Table 3 records the center settlement.

The remaining test cases all involve a greater degree of uncertainty regarding the correct values of q_c to use in the calculations. Either the q_c profile was incomplete or it was missing and was estimated (before any settlement calculations) from other available data. Had real q_c data been obtained the real values would be somewhat different than estimated herein, and could possibly be very much different. Tables 2 and 3 nevertheless include these additional cases to show that a reasonable estimate for the q_c values usually results in a reasonable settlement estimate. These cases also provide more method comparisons for Table 3.

Belgrade Apartment House (case 12).—In this case two parallel apartment buildings, each 34 ft wide, were separated by only 11 ft. They were built and loaded simultaneously. The settlement estimate was made on the basis of a single structure with $B = 79$ ft. The q_c data extended only to a depth of about 1.0B. For the interval 1.0 to 2.0B, the writer estimated q_c at 120 kg per sq cm. Then the 1-2B layer contributes about 20% of the computed settlements listed in Table 3.

Note that two settlements are given for the same structure, they are for opposite corners. Cone soundings at the same corners showed significantly different q_c profiles. This is the way the writer recommends treating nonhomogeneity under a foundation and estimating tilt or differential settlement, or both, therefrom. Tilt due to eccentric loading is a different matter, not considered herein.

Danish Tank on Hydraulic Fill (case 13).—Careful tests in Denmark established that its relative density was about 46%. On the basis of previous correlation work in similar, but natural, sands $q_c = 30$ kg per sq cm seemed reasonable. A constant value of q_c = 30 was assumed in the settlement calculation.

An interesting aspect of this test case is that there is a relatively incompressible boundary layer at a relative depth of only 0.23 B below the tank foundation. Thus, only a small part of the 2B-0.6 I_z distribution is used in the settlement estimate for this case.

Note that the settlements were measured on the perimeter of the tank—at the edge of a uniformly loaded circular area. According to the theory of elasticity, including the effect of a rigid boundary at 0.23 B, the edge settlement of a flexible circular plate should be only about 0.5 that of a rigid plate. However, simple model tests by the writer with uniform, circular loads on dry sand, with a relative density about 25% and with a rigid boundary at various relative depths below the load level, show that approximately uniform settlement results with a rigid boundary at 0.23 B. It may actually be greater at the perimeter than at the center, by about 10%. Therefore, for this case the rigid settlement estimate can be checked approximately against measurements made at the perimeter of the tank.

The writer again was uncertain as to which point under the tank to compute the Buisman vertical stress increase for the Buisman-DeBeer settlement estimate. The results noted in Table 3 include three points. Because such a tank foundation pressure is almost perfectly uniform, and the settlements were measured along the perimeter, the subsequent comparison of prediction results is for the perimeter value only, which is also the most favorable. The same procedure was used for the case 14 tank.

Brinch Hansen (13) made a more sophisticated, and more accurate, check on the observed settlement for this tank. His method requires laboratory tests and considerable computational work.

Norwegian Tank (case 14).—This is another case where q_c data were not obtained. However, screw-plate load tests were used, perhaps for the first time, to depths of 33 ft (0.34 B). Using screw-plate determined compressibilities permits eliminating the q_c to E_s correlation (step 6). The writer then extrapolated E_s values for the remaining strain-depth interval of 0.34-2.0 B on the basis of other types of sounding data obtained at the site (see references cited). The depths and E_s values used in the computations were: 0-0.34 B:66 kg per sq cm; 0.34-1.0 B:175 kg per sq cm; 1.0-2.0 B:200 kg per sq cm.

Again the settlements reported in Table 3 are for points on the tank perimeter. The same experiments just presented show that with a uniform, loose sand foundation to relative depth 2 B, the edges settle about 80% of the settlement at the center and 90% of the settlement of a rigid foundation. However, in this case there is a significantly less compressible boundary at about 0.34 B which, as noted previously, increases the relative settlement of the perimeter. After considering these factors, it is the writer's opinion that the perimeter settlements of this tank would also approximately equal those of a rigid tank of the same size and loading.

English Factory Footings (case 15).—The foundation sands in this case, a gravel with flints and some fine sand, are much coarser than in all other cases. Static cone tests were not performed, but standard penetration tests were. The average N-value in the area of the test footing was reported as 21 before the footing excavations, reducing to 13 from the bottom of the excavation. At the Dugeness, Kent, site reported in the same reference there appears to be, in a similar gravel, a q_c/N ratio of about 10. Using this factor, the writer assumed constant q_c values of 130 kg per sq cm and 210 kg per sq cm and reports a settlement estimate for each.

Michigan Factory Footings (case 16).—The soils at this site consist of overconsolidated dune sands. Again, SPT N-value data were obtained, but there were no cone tests. Some relative density estimates were also available. On the basis of previously noted correlations the writer estimated q_c profiles assuming a high (for fine sands) q_c/N ratio of seven because of the overconsolidation. Admittedly, this could be seriously in error. The computed settlements are too high so perhaps the factor is actually greater than seven.

Because a majority of the footing load was live load, there is uncertainty regarding the Δp value to assign to the problem. The writer used the authors' figures for load. Note also that the Buisman-DeBeer calculation method is not intended to be used in overconsolidated sands (8). But, the obvious difficulty is that in many applications the degree of overconsolidation of a sand is not known and cannot be determined easily.

COMPARISON WITH ALTERNATE METHODS USING STATIC CONE TEST DATA

To help judge how the proposed new settlement estimate procedure competes with those methods already in practical use, it is also necessary to compare the test cases with the results obtained using such existing methods. A simple procedure was suggested by Meyerhof (25). A more complex procedure was first suggested by Buisman and has been somewhat modified and used extensively by DeBeer and others for about 30 years in Belgium and elsewhere (8). Recently, Thomas (36) proposed a sand settlement estimating procedure also adapting a solution from linear elastic theory. Even more recently Webb (40) suggested still another procedure which also adapts linear elastic theory.

The Meyerhof Method.—Meyerhof started with the Terzaghi and Peck (35) SPT-settlement design curves for dry and moist sands and developed approximate equations to describe them. His experience, further confirmed herein, indicated that for sands the q_c/N ratio was four, on the average. After introducing this value for the ratio he offered the following equations for the allowable net foundation bearing pressure which will produce a settlement of 1.0 in.:

$$q_a = \frac{q_c}{30}\,;\ \text{if } B \le 4\ \text{ft}, \qquad (9a)$$

$$q_a = \frac{q_c\left(1 + \dfrac{1}{B}\right)^2}{50}\,;\ \text{if } B > 4\ \text{ft}, \qquad (9b)$$

in which q_c = the average static cone bearing over a depth interval of B below the foundation.

Still following Terzaghi and Peck, he also suggested for pier and raft foundations that q_a be twice that given by Eqs. 9*a* and 9*b*. Also, another correction factor has to be applied to q_a to take account of the level of the water table. If the water table is at the foundation level or above, this factor is 0.50. If at a depth of $1.5B$ or below, the factor is 1.00. Use linear interpolation between 0 and $1.5B$.

When the foundation Δp differs from the computed q_a, then the settlement is estimated using linear interpolation or extrapolation, provided that Δp is less than one half the ultimate bearing capacity.

Buisman-DeBeer Method.—This method is explained generally in Refs. 8, 9. However, DeBeer informed the writer via personal communication of two important aspects of this method not noted in these references. These additional aspects were used to arrive at the Buisman-DeBeer settlement estimates reported in Table 3. Table 1(*b*) presents a listing of the computations using this method, with test case 1 as the example.

First, when considering rigid foundations such as the piers in test cases 1 to 4, the Buisman formula (8) for vertical stress increase is applied to the singular point of the foundation. DeBeer defines this point as that where the stress distribution is nearly independent of the distribution of contact pressure under the footing. Thus, the settlement of this point will be almost the same under an assumed uniform distribution as under the true distribution of a rigid foundation. In this way, at this point, DeBeer estimates the settlement of a rigid foundation using an assumed uniform contact pressure. DeBeer reports the singular point for an infinitely long footing at about $0.29B$ from its centerline. The writer assumed its location at $0.25B$ for a square and circular footing.

The second modification is that all vertical strain, and therefore contribution to settlement, is assumed to be zero below the point at which the Buisman vertical stress increase becomes less than 10% of the existing overburden vertical effective stress. This depth limit was included, where applicable, in the Buisman-DeBeer calculations. However, in some cases the cone data were not available to the 10% limit depth. In these cases (nos. 2, 3, 4, 7, 16) the Buisman-DeBeer settlements reported in Table 3 are too low by unknown, but probably minor amounts.

Recently, others have proposed at least three modifications in the Buisman-DeBeer procedure for evaluating E_s, their compression modulus, from static cone data. Vesić (39) suggests a simple modification which includes a correction for relative density. However, reliable relative density data are rarely available in practical work. Furthermore, the always-possible cementing in granular soils makes relative density of questionable value as an indicator of compressibility in some natural deposits. Schultze (33) suggests an empirical formula to evaluate E_s which would add considerably to the complexity of prediction calculations. Both these suggestions evolved from research work in large sand bins. While they may prove valuable, there is at present no test-case evidence that the writer is aware of that demonstrates that either suggestion will systematically improve settlement prediction accuracy without sacrificing necessary conservatism. Because of this, and to simplify this presentation, neither modification was used in the Buisman-DeBeer settlement estimates noted herein.

A third modification has been suggested by Meyerhof (25). On the basis of settlement measured-predicted comparisons, mostly from Belgian bridges, he noted that predictions were generally conservative (too high) by a factor of two. He recommended increasing allowable contact pressures by 50% for the same computed settlement. A few trial computations show this is roughly equivalent to increasing the Buisman-DeBeer modulus, E_s, by 28%. Without this correction E_s = 1.5 q_c in this method. With this correction it would equal about 1.9 q_c. The writer, using an independent approach and data, arrived at nearly the same E_s = 2.0 q_c. Both E_s definitions are the same although used in different formulas. Because Meyerhof's suggestion is not yet in common use it has not been used in the computations herein.

Although some of the published test cases include settlement predictions using the Buisman-DeBeer method, the writer has recalculated them and all results presented in Table 3 are from his calculations. Table 1(*b*) is an example. This was necessary so that all methods would be compared using the same assumed q_c data, layering and Δp loadings.

Long experience has proven that the *B-D* method gives a conservative answer. Its use permits the rapid, economical determination of an upper bound settlement which an engineer can use with considerable confidence. Any competing method must be weighed against this very useful feature.

Thomas Method.—This method involves the use of an independent, laboratory correlation from q_c to E_s, combined with the settlement formula from elastic theory and the geometrical influence factors from this theory. A discussion by Schmertmann (31), using many of the test cases also used herein, points out that this method tends to seriously underestimate settlement. The difficulty may be that the laboratory q_c to E_s correlation experiments did not adequately model the stress-strain environment found under footing and raft foundations.

Because this method is too new to assess field experience performance, and from the above many need further research and revision before it can be considered conservatively reliable, it is not considered further herein.

Webb Method.—Webb also used the insitu screw-plate test to obtain a correlation between cone bearing and sand compressibility. As already noted, these independent correlations check well.

Although similar in concept, Webb's method and the new one proposed herein differ in an important way. The new method uses the 2B-0.6 I_z distribution to estimate vertical strain and settlement. Webb's method still requires the extra computation of vertical stress increase (he recommends Boussinesq).

Webb's method is also too new to assess any field experience with its use. His very recent paper was received too late to include test case comparisons herein without a major revision of this paper. If desired, the reader can use the data in Tables 2 and 3 to make his own comparisons.

Settlement Comparisons.—On the basis of the test cases presented in Table 3 it seems obvious that the Meyerhof procedure produces the least accurate comparisons of the three considered. The settlement of small foundations appears greatly overestimated and that of large foundations underestimated. This method should be discarded in its present form. Remember that this method is based on the Terzaghi-Peck SPT method with a q_c/N ratio taken = 4. Data presented subsequently shows that four for this ratio should not usually be grossly in error. This suggests the Terzaghi and Peck design curves may be in error, especially for very small and very large foundations.

Figs. 7 and 8 present graphs showing how the predicted settlements using the Buisman-DeBeer and new methods compare with those measured. The abscissa is the predicted settlement to a log scale. The ordinate is the correction factor needed to change the predicted settlement to the settlement actually measured. The symbols in Figs. 7 and 8 can be matched to the test

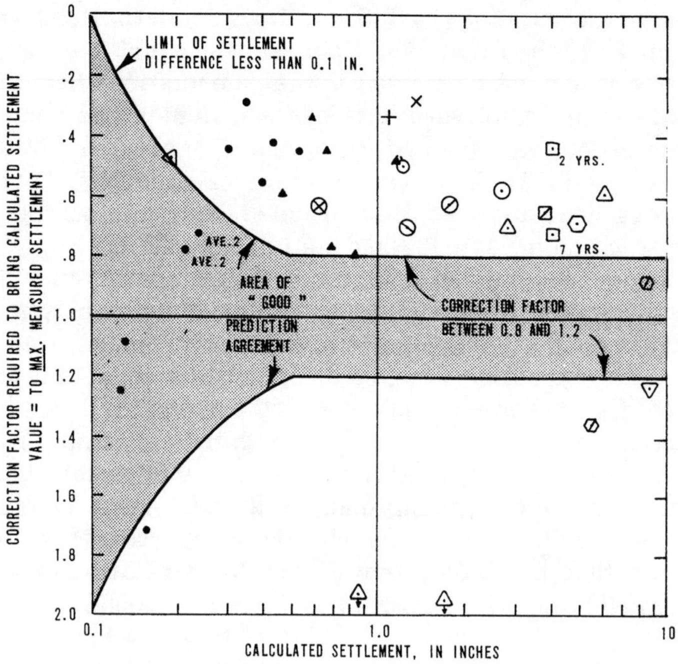

FIG. 7.—SETTLEMENT PREDICTION PERFORMANCE FROM TEST CASES, USING BUISMAN-DeBEER METHOD

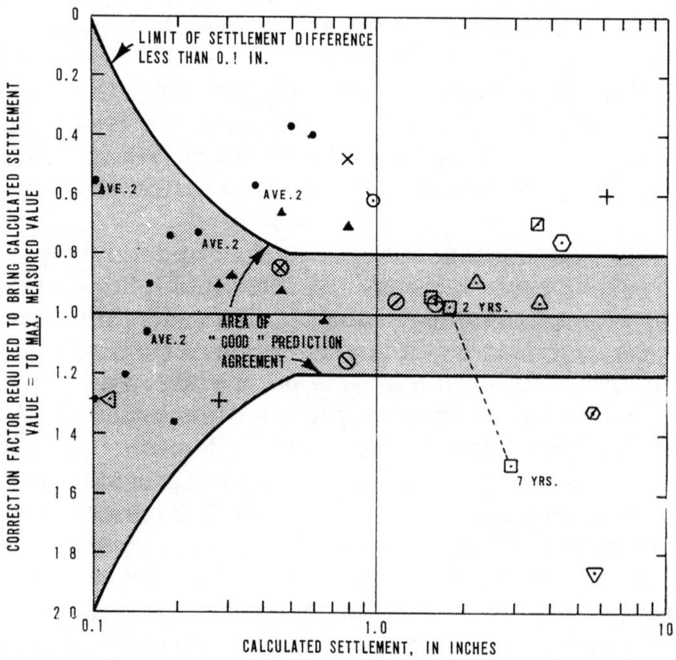

FIG. 8.—SETTLEMENT PREDICTION PERFORMANCE FROM TEST CASES, USING NEW STRAIN FACTOR METHOD

cases by the last column in Table 3. To maintain a conservative outlook the predicted settlements are compared with the maximum measured values.

If good prediction-measured agreement is defined as within 0.1 in. (0.25 cm), or requiring a correction factor within the 0.8 to 1.2 interval, then it is apparent that there are more instances of good agreement using the new method. In Fig. 7 the agreement would be considered good for seven of the 37 points plotted, while in Fig. 8 it would be 21 out of 36.

Considering relative conservatism, and defining conservative as prediction exceeding measured, Fig. 7 shows five points on the unconservative side of the good agreement range. These involve four of the test cases, including one of the DEGEBO load tests. Fig. 8 has three points on the unconservative side of good agreement, involving three test cases.

Fig. 7 also shows that most of the Buisman-DeBeer comparisons fall within a correction factor band of 0.4 to 0.8. This checks, approximately, DeBeer's statement (8) that this method has proven, on the basis of measurements from over 50 Belgian bridges, to yield a mean prediction-measured settlement ratio of two, which inverts to a correction factor of 0.5. The present test cases include only four of these bridges. These data also check Meyerhof's suggestion (25) which, as noted previously, in effect would increase E_s from 1.5 q_c to 1.9 q_c without sacrificing essential conservatism. Were this done and a new Fig. 7 prepared using the new, reduced settlement predictions, there would still be only five points on the unconservative side of good prediction agreement. These points would, of course, then be more unconservative. In comparison to the 0.4 to 0.8 band in Fig. 7, Fig. 8 shows that most of the new method comparisons fall within the 0.6 to 1.2 band, also a factor of 2.0.

Summarizing, it is the writer's opinion, based on the test cases presented, that the strain-distribution method presented herein results in more accurate settlement predictions than the unmodified Buisman-DeBeer method. While the new method is less conservative, the results are no more often on the unconservative side of good prediction-measured agreement than with the Buisman-DeBeer method. The new method thus retains the "upper bound" feature of Buisman-DeBeer. However, a simple modification of the Buisman-DeBeer method, as suggested by Meyerhof, results in the B-D method producing results similar to those achieved using the new method proposed herein.

The new method has the advantage of requiring simpler computations [compare Tables 1(*a*) and 1(*b*)] and probably results in a more accurate distribution of vertical strain below the center of an isolated foundation. The Buisman-DeBeer method has the present advantage of more conveniently, though perhaps inaccurately, accounting for the interaction of adjacent loads by assuming stress superposition, plus an experience base of 30 yr.

Besides the difference in distribution of vertical strain, the Buisman-DeBeer and new methods also respond differently to the magnitude of the pressure increase Δp. For example, using the new method a 50% increase in Δp results in a somewhat greater than 50% increase in predicted settlement. Such overlinear behavior results from C_1 increasing when Δp increases (see Eq. 3). In the Buisman-DeBeer method the effect of changing Δp is more complicated [see Eq. 9 in Table 1(*b*)]. The effect is linear on a log-Δp scale, and therefore underlinear. For example, the problem in Table 1(*b*) yields a settlement prediction of 1.96 in. if Δp = 1.00 instead of 1.50 kg per sq cm, using a 10% limiting depth of 1200 cm. In this case a 50% increase in Δp results in only a 38% increase in the predicted settlement.

It is unusual for static load tests in sands to exhibit underlinear load settlement behavior, usually it is approximately linear a low pressure and becomes progressively more overlinear as bearing capacity failure is approached. This may be a further indication of some significant theoretical inaccuracy in the Buisman-DeBeer method.

At this point it is well to note again that both methods ignore at least one effect of layering in E_s values. The Buisman-DeBeer method does not include a correction for changes in the profile of vertical stress increase resulting from layering. The new strain-distribution method does not include a correction for changes in I_z resulting from layering.

TEMPORARY USE OF STANDARD PENETRATION TEST DATA

Although used world wide, presently the static cone penetration test is not used extensively in the United States. An engineer may not be able to specify this type of test on his project because the necessary equipment is not available. On the other hand, use of the SPT is common and the equipment is readily available. It is therefore of interest to note any empirical correlation that may exist between q_c and N.

Many investigators have explored this correlation. Meyerhof (24) suggested that $q_c/N = 4$. Others are noted by Sanglerat (30) and Schultze (33). The writer's experience with this correlation in granular soils, limited mostly to uniform fine sands but including some silty and medium sands, is summarized by the data in Fig. 9. The mean values of q_c/N fall in the range of 4.0 to 4.5, which for fine sands checks Meyerhof's suggestion. But there is a great spread around the means. This should be expected. Both types of tests, but particularly the SPT (11,26), are subject to error. The many sites, testing laboratories, drillers and types of equipment involved in the writer's data accentuate the variability in SPT results. However, in all cases N was to be determined in substantial accord with ASTM D1586. It should be noted that at some individual sites, with only one laboratory, driller and piece of equipment involved, the q_c/N correlation spread was similar to that presented for all sites. At other sites the spread was much less.

It is also quite clear from the writer's experience, and that of others, that the q_c/N ratio varies with grain size and perhaps with gradation. The finer grained the soil, the smaller the q_c/N ratio, reaching as low as about 1.0 for some clays and as high as 18 (22,23), for some gravels.

If an engineer wishes to use the settlement estimate procedure of Buisman-DeBeer, or the new one suggested herein, but he has only SPT N-values, then he must convert these as best as he can to q_c values. This conversion should ordinarily be conservative, with the q_c values on the low side of reality. Obviously, in view of the potential scatter demonstrated by the data in Fig. 9, it is much more desirable, and should lead to less expensive design, to have direct determination of q_c. As a temporary expedient the writer recommends the following q_c/N ratios which are usually conservative:

Soil Type	q_c/N
Silts, sandy silts, slightly cohesive silt-sand mixtures	2.0

Clean, fine to med. sands & slightly silty sands	3.5
Coarse sands & sands with little gravel	5
Sandy gravels and gravel	6

Assume these ratios are independent of depth, relative density, and water conditions. The writer also suggests that as many N-values as possible be

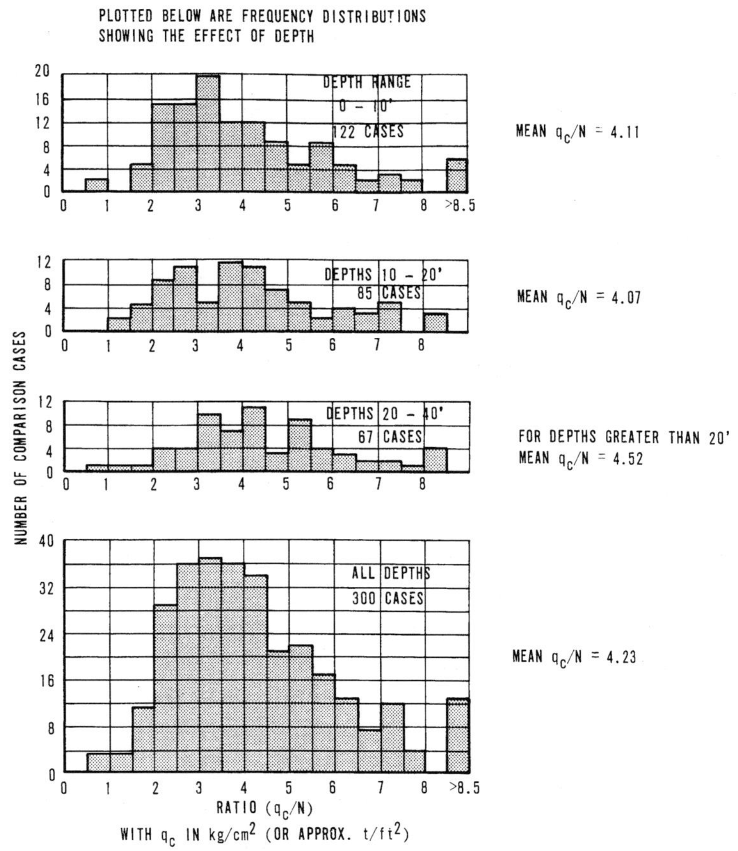

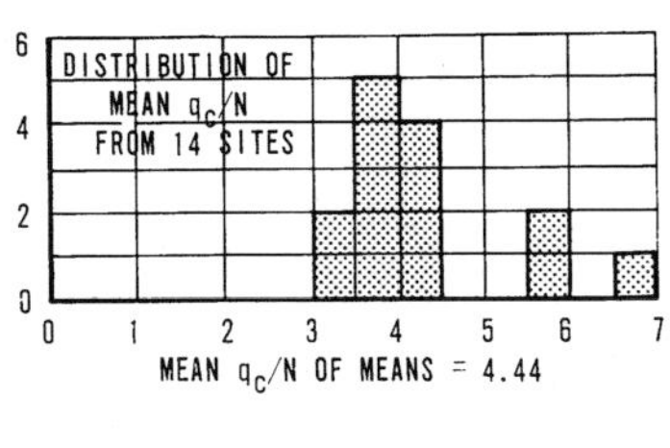

EFFECT OF MAGNITUDE OF SPT N-VALUE

RANGE	LEAST SQUARES LINE	CORREL. COEFF.
0<N<10	$q_c = 4.86\ N$	0.43
0<N<30	$q_c = 4.13\ N$	0.72
ALL N	$q_c = 18.3 + 2.9\ N$	0.80

FIG. 9.—DATA FOR CORRELATING N AND q_c IN SILTY TO MEDIUM SANDS (Comparison holes 3-10 ft apart; All q_c by University of Florida; N by 7 firms at 14 sites, 13 of which in Florida; All N are uncorrected.)

obtained to minimize, by averaging, the large correlation error possible with only few data.

CONCLUSIONS

1. A new method is presented herein for the systematic computation of the static settlement of isolated, rigid, concentrically loaded shallow foundations over sand. The computations involved are simple and can be done in the field with a slide rule. The method employs elastic half-space theory in a simplified form and uses the static cone bearing capacity as a practical means for determining in-situ compressibility, E_s.

2. The proposed method includes a simplified distribution of vertical strain under a foundation, expressed in the form of a strain influence factor, I_z. This distribution of I_z results in centerline strains showing better agreement with available data than when computed on the usual basis of increases in vertical stress.

3. The test case comparisons presented herein, from 16 sites in 10 countries and including considerable scope in geometry, loading and soil parameters, demonstrate the accuracy of the strain-distribution method. It appears from these cases to be the most accurate of the three methods compared herein which use static cone data. Yet, it yields a conservative solution as often as the Buisman-DeBeer method.

4. A simple modification to the existing Buisman-DeBeer procedure, suggested by Meyerhof, would result in accuracy and conservatism comparable to that from the new procedure developed in this paper. This would change the important estimate of E_s from $= 1.5\ q_c$ to $= 1.9\ q_c$, which is in agreement with the writer's independent development, using screw-plate load tests, of his $E_s = 2\ q_c$. Although the two E_s values have the same definition, they are used in very different formulas. Thus, this research confirms the conservative validity of the long-used 1.5 factor. Webb's recent work adds to this confirmation.

5. The new method is simpler than the Buisman-DeBeer method of computation. It does not require computation of the below-foundation distribution of effective overburden stress and vertical stress increase.

6. On the very limited basis of single test cases, the test case comparisons point out the possibility that modifications to the new procedure may be needed for some soil conditions. Very shelly sands (case 6) may have greater compressibility, and overconsolidated sands (case 16) less compressibility than when computed from Eq. 7.

7. It is possible, but with reduced accuracy, to use the proposed settlement calculation procedure in conjunction with standard penetration test data. Correlation data are presented to permit approximate, usually conservative, conversion from N to q_c values. Such conversion is recommended only as a temporary expedient until cone data can be used directly.

ACKNOWLEDGMENTS

The National Science Foundation provided much of the financial assistance needed to accomplish this work through their Grant No. GK-92. The University of Florida Engineering and Industrial Experiment Station also provided significant assistance. Many engineers helped by providing valuable data relating to the test cases developed herein. Their help is noted in each case.

SM 3 SETTLEMENT OVER SAND

The following University of Florida personnel also assisted the writer with the extensive field work required to accumulate the cone-screw-plate and cone-STP data correlations: R. E. Smith, and K. DiDonato, Jon Gould, K. H. Ho and Billy Prochaska. Anne Topshoj, performed the special sand model tests referred to herein. W. Whitehead provided general assistance.

APPENDIX I.—REFERENCES

1. Ahlvin, R. G., and Ulery, H. H., "Tabulated Values for Determining the Complete Pattern of Stresses, Strains, and Deflections beneath a Uniform Circular Load on a Homogeneous Half Space," *Highway Research Board Bulletin*, No. 342, 1962.
2. Bishop, R. F., Hill, R., and Mott, N. F., "The Theory of Indentation and Hardness Test," *The Proceedings of the Physical Society*, 1, May 1945, No. 321, p. 147.
3. Bjerrum, L., "Development of an Industry on a Silty Sand Deposit," author's notes for a series of two lectures presented at MIT in March, 1962. Also see Norwegian Geotechnical Institute internal reports 0.728-1,3,5.
4. Bogdanovic, L., Milovic, D., and Certic, Z., "Comparison of the Calculated and Measured Settlements of Buildings in New-Belgrade," *Proceedings European Conf. on Soil Mechanics and Foundation Engineering*, Vol. I, Wiesbaden, 1963, pp. 205–213 (building 7).
5. Bond, D., "The use of Model Tests for the Prediction of Settlement under Foundations in dry Sand," thesis, presented to the University of London, at London, England, in 1956, in partial fulfillment of the requirements for the degree of Doctor of Philosophy.
6. D'Appolonia, D. J., D'Appolonia, E. E., and Brissette, R. F., "Settlement of Spread Footings on Sand," *Journal of the Soil Mechanics and Foundations Division*, ASCE, Vol. 94, No. SM3, Proc. Paper 5959, May, 1968, pp. 735–760 (see Fig. 21).
7. DeBeer, E., "Settlement Records of Bridges Founded on Sand," *Proceedings Second International Conference on Soil Mechanics and Foundation Engineering*, 1948, Vol. II, The Netherlands, p. 111.
8. DeBeer, E. E., "Bearing Capacity and Settlement of Shallow Foundations on Sand," *Proceedings of a Symposium on Bearing Capacity and Settlement of Foundations*, Duke University, 1967, Lecture 3, pp. 15–33.
9. DeBeer, E., and Martens, A., discussion of "Penetration Tests and Bearing Capacity of Cohesionless Soils," by G. G. Meyerhof, *Journal of the Soil Mechanics and Foundations Division*, ASCE, Vol. 82, No. SM4, Proc. Paper 1079, Oct., 1956, pp. 1095–7.
10. Eggestad, Aa., "Deformation Measurements below a Model Footing on the Surface of dry Sand," *Proceedings of the European Conference on Soil Mechanics and Foundation Engineering*, Vol. 1, Wiesbaden, Germany, 1963, p. 233.
11. Fletcher, G., "Standard Penetration Test: Its Uses and Abuses," *Journal of the Soil Mechanics and Foundations Division*, ASCE, Vol. 91, No. SM4, Proc. Paper 4395, July, 1965, p. 75.
12. Grimes, A. A., and Cantlay, W. G., "A twenty-story block in Nigeria founded on loose sand," *The Structural Engineer*, Vol. 43, No. 2, February, 1965, pp. 45–57.
13. Hansen, J. Brinch, "Improved Settlement Calculation for Sand," *Danish Geotechnical Institute Bulletin* No. 20, 1966, pp. 15–20.
14. Hansen, J. Brinch, "Stress-Strain Relationships for Sand," *Danish Geotechnical Institute Bulletin* No. 20, 1966, p. 8.
15. Harr, M. E., *Foundations of Theoretical Soil Mechanics*, McGraw-Hill Book Co., Inc., New York, 1966, p. 81.
16. Holden, J. C., "Stresses and Strains in a Sand Mass Subjected to a Uniform Circular Load," *Departmental Report No. 13*, Department of Civil Engineering, University of Melbourne, Melbourne, Australia, 1967, 362 pp., p. 164.

17. Janbu, N., "Principal Stress Ratios and Their Influence on the Compressibility of Soils," *Proceedings of the 6th International Conference on Soil Mechanics and Foundation Engineering,* 1965, Vol. I, p. 249.
18. Janbu, N., "Settlement Calculations based on the Tangent Modulus Concept," *Bulletin 2,* The Technical University of Norway, Trondheim, Norway, 1967, p. 32.
19. Kummeneje, O., "Foundation of an Oil Tank," *Norwegian Geotechnical Institute Publication* No. 12, Oslo, 1956 (in Norwegian). A later reference to screw-plate testing is: Kummeneje, O., and Eide, O., "Investigation of Loose Sand Deposits by Blasting," *Proceedings, 5th International Conf. on Soil Mechanics and Foundation Engineering,* Vol. I, Paris, 1961, p. 491.
20. Kummeneje, O., and Eide, O., "Investigation of Loose Sand Deposits by Blasting," *Proceedings 5th International Conference on Soil Mechanics and Foundation Engineering,* Vol. I, Paris, France, 1961, p. 493.
21. Lambe, T. W., "Stress Path Method," *Journal of the Soil Mechanics and Foundations Division,* ASCE, Vol. 93, No. SM6, Proc. Paper 5613, Nov., 1967, pp. 309–331.
22. Meigh, A. C., Discussion in Vol. III of the *Proceedings, European Conference on Soil Mechanics and Foundation Engineering,* Wiesbaden, Germany, 1963.
23. Meigh, A. C., and Nixon, I. K., "Comparison of In-Situ Tests for Granular Soils," *Proceedings Fifth International Conference on Soil Mechanics and Foundation Engineering,* Vol. I, Paris, France, 1961, p. 499 (Purley Way site).
24. Meyerhof, G. G., "Penetration Tests and Bearing Capacity of Cohesionless Soils," *Journal of the Soil Mechanics and Foundations Division,* ASCE, Vol. 82, No. SM1, Proc. Paper 866, Jan., 1956, p. 5.
25. Meyerhof, G. G., "Shallow Foundations," *Journal of the Soil Mechanics and Foundations Division,* ASCE, Vol. 91, No. SM2, Proc. Paper 4271, March, 1965, p. 25.
26. Mohr, H. A., discussion of "Standard Penetration Test: Its Uses and Abuses," by Gordon F. A. Fletcher, *Journal of the Soil Mechanics and Foundations Division,* ASCE, Vol. 92, No. SM1, Proc. Paper 4594, pp. 196–196.
27. Muhs, H., "Die zulässige Belastung von Sand auf Grund neuerer Versuche und Erkenntnisse," *Bau-Technik,* Heft 10-11, Oct.-Nov., 1963, pp. 130–147.
28. Nonveiler, E., "Settlement of a Grain Silo on Fine Sand," *Proceedings, European Conference on Soil Mechanics and Foundation Engineering,* Vol. I, Wiesbaden, Germany, 1963, pp. 285–294.
29. Peck, R. B., Hanson, W. E., and Thornburn, T. H., *Foundation Engineering,* John Wiley & Sons, Inc., New York, 1953, p. 243.
30. Sanglerat, G., *le pénétromètre et la reconnaisissance des sols,* Dunod, Paris, France, 1965.
31. Schmertmann, J. H., discussion of Thomas, (1968), *Geotechnique,* June, 1969.
32. Schmertmann, J., "Static Cone Penetrometers for Soil Exploration," *Civil Engineering,* Vol. 37, No. 6, June, 1967, pp. 71–73.
33. Schultze, E., and Melzer, K., "The Determination of the Density and the Modulus of Compressibility of Non-cohesive Soils by Soundings," *Proceedings 6th International Conf. on Soil Mechanics and Foundation Engineering,* Montreal, Canada, 1965, Vol. I, p. 357.
34. Teng, Wayne C., *Foundation Design,* Prentice-Hall International, Inc., Englewood Cliffs, N.J., 1962, p. 119.
35. Terzaghi, K., and Peck, R. B., *Soil Mechanics in Engineering Practice,* John Wiley & Sons, Inc., 2nd edition, New York, 1967, p. 491.
36. Thomas, D., "Deepsounding Test Results and the Settlement of Spread Footings on Normally Consolidated Sands," *Geotechnique,* London, Vol. 18, Dec. 1968, pp. 472–488.
37. Tschebotarioff, G. P., *Soil Mechanics, Foundations, and Earth Structures,* McGraw-Hill Book Company, Inc., New York, 1951, pp. 357, 379, 232.
38. Vargas, M., "Foundations of Tall Buildings on Sand in Sao Paulo (Brazil)," *Proceedings 5th International Conf. on Soil Mechanics and Foundation Engineering,* Vol. I, Paris, France, 1961, p. 842.
39. Vesić, A. S., "A Study of Bearing Capacity of Deep Foundations," final report, project B-189, Georgia Institute of Technology, March, 1967, p. 170.
40. Webb, D. L., "Settlement of Structures on deep alluvial sand sediments in Durban, South Africa," *British Geotechnical Society Conference on In-Situ Investigations in Soils and Rocks,* Session III, paper 16, London, England, 13–15 May, 1969, pp. 133–140.

SM 3 SETTLEMENT OVER SAND

APPENDIX II.—NOTATION

The following symbols are used in this paper:

A = constant in elastic strain equations, depending only on geometry of point considered;
B = least width of a rectangular foundation, diameter of circular foundation;
C_1 = correction factor to approximately account for depth of embedment effects;
C_2 = correction factor to approximately account for creep type settlement;
D = depth of embedment of a foundation = vertical distance from shallowest adjacent ground level of base of foundation;
D_r = relative density, void ratio basis;
E = Young's modulus in a linearly elastic media;
E_s = equivalent Young's modulus for granular soil in compression;
F = similar to A above;
GWT = abbreviation for ground water table (level);
H = depth below foundation to an assumed incompressible boundary layer;
I_z = influence factor for vertical strain;
L = length of a rectangular foundation;
N = blow-count in the standard penetration test (uncorrected);
p = average pressure of foundation against soil;
p_0 = overburden pressure at foundation level;
Δp = average net increase in soil pressure at foundation level, = $(p - p_0)$;
q_a = allowable, net average foundation pressure to produce an estimated settlement of 1.00 in. (Meyerhof method);
q_c = static, Dutch cone bearing capacity, in kilograms per square centimeter;
r = radius of a circular foundation;
SPT = abbreviation for standard penetration test;
t = time;
t_0 = a reference time (0.1 yr used herein);
z = depth below foundation level;
β = constant designating semi-log linear creep rate;
γ' = effective unit weight of soil;
ϵ_z = vertical strain;
ν = Poisson's ratio;
ρ_t = settlement at time = t;
ρ_0 = settlement at reference time;
$\Delta\sigma_v$ = increase in vertical stress below D, due to Δp; and
σ'_{vi} = initial vertical stress, at depth D, due to surrounding surcharge at time of loading foundation.

FOURTH PANAMERICAN CONFERENCE

J. H. SCHMERTMANN [11]--deMello in his state-of-the-art paper has achieved for the profession a new level of thorough review and insight into the significance of SPT data. He notes, in scholarly detail, many unresolved problems towards achieving a better understanding of SPT data. On page 14 deMello discusses the state of knowledge of the importance of the sampler side wall friction contribution to the blowcount, or N-value. Apparently much confusion exists on this subject, with a consequent diversity of opinions. On page 44 he notes a general agreement that lateral stresses affect N-values, but also notes a lack of quantitative data. Herein this discusser examines further the effects of side wall friction and lateral stress, using new insight made possible by new experience with the static friction-cone penetrometer.

Many engineers have noted that the blowcounts for each successive 6 inches of penetration, from 0 to 18 inches, usually successively increase in soils without gravel. They usually find a notably low count for the first 6 inches, and commonly attribute this to "seating" in the loosened soil at the bottom of the borehole. Therefore, the ASTM SPT method counts only the blows from 6 to 18 inches. Figure S-1 shows an example of the successive increase in blowcount in a silty marl in southern Florida.

At this site the engineers had four series of parallel borings using a 5 foot long, 2 inch I.D., SPT-type sampler driven in five 1 foot increments, borings using the standard SPT sampler with blows recorded in 6 inch increments from 0 to 18 inches, and soundings using the Begemann static, mechanical friction-cone penetrometer. Consider only the relatively uniform, 30 foot layer of silty marl found at a depth of 40 feet. Note the stair-step progression of blowcounts with each additional increment of penetration--very dramatic for the 5 foot sampler, but equally noticeable with the standard sampler. The comparative uniformity of the cone bearing profile shows that the consistent stair-step pattern probably results from the way in which the sampler mobilizes penetration resistance rather than a strength pattern within the soil.

A former University of Florida graduate student, Zaid Awkati, made a thought provoking observation when he compared SPT and friction-cone penetrometer data in mostly sandy clays at a research-consulting site in Gainesville. He noted that N-values appeared to correlate better with friction along the friction-cone penetrometer sleeve than with cone end bearing.

---

[11]Department of Civil Engineering, University of Florida, Gainesville, Florida, USA

DISCUSSION–SESSION I

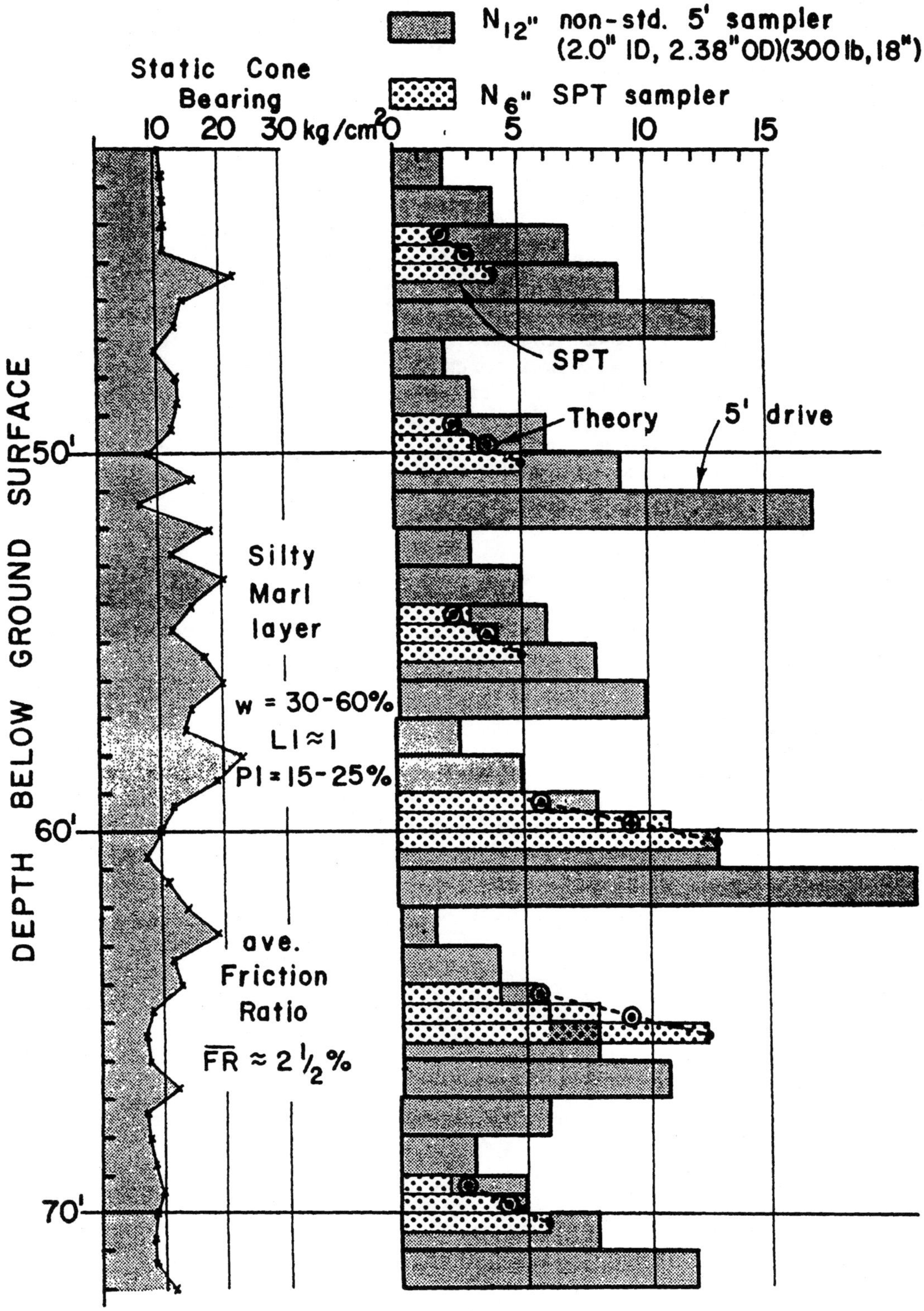

FIG. S-1 EXAMPLE OF STAIR-STEP BLOWCOUNT PROGRESSION

FOURTH PANAMERICAN CONFERENCE

All of the above, plus deMello's observations on page 24 regarding the importance of increasing clay sensitivity reducing N-values, suggests that side wall friction may have great importance in generating SPT blowcounts.

The Friction-Cone SPT Formula

The reader can find descriptions in the literature of the two most common types of static friction-cone penetrometer tips--the Dutch mechanical tip (Begemann, 1965) and the Fugro tip (deRuiter, 1970). Both of these tips permit the measurement of friction along a special friction sleeve in the immediate vicinity above the cone, called the "local friction". Begemann showed that the slope of an arithmetic graph of local friction vs. cone bearing remains approximately constant for a given soil, and that this slope varies with soil type. The soil mechanics group at the University of Florida has both types of friction-cone tips and now has experience with them in laboratory and field research and in practice. We confirm Begemann's findings, which we express by the concept of a friction ratio. Use of this concept seems to help explain some aspects of SPT N-value behavior.

Figure S-2 shows the SPT sampler advancing, at constant velocity, into the soil at the bottom of a borehole. The total resistance to this advance equals the sum of end bearing and side wall friction resistances:

$$F = F_{end} + F_{\substack{inside\\+outside}} = F_e + F_s \qquad (1)$$

Note that the horizontally projected end area of the SPT sampler equals 10.7 cm^2--nearly the same as the 10.0 cm^2 of the above penetrometer tips. Now assume that the end bearing resistance on the sampler equals that measured by the static cone penetrometer at the same depth in the same soil, q_c:

$$F_e = A_e \, q_c \qquad (2)$$

Assume further that the inside and outside sampler wall friction both equal the local friction measured along the friction sleeve of a static friction-cone penetrometer, f_s:

$$F_s = \pi \, (D_i + D_o) \, L \, f_s \qquad (3)$$

Making use of the friction ratio concept:

$$FR = \frac{f_s}{q_c} = f \text{ (soil type)} \qquad (4)$$

permits expressing sidewall friction in terms of end bearing:

$$F_s = \pi \, (D_i + D_o) \, L \, (FR) \, q_c \qquad (5)$$

Putting (2) and (5) into (1) gives:

$$F = [A_e + \pi \, (D_i + D_o) \, L \, (FR)] \, q_c \qquad (6)$$

DISCUSSION–SESSION I

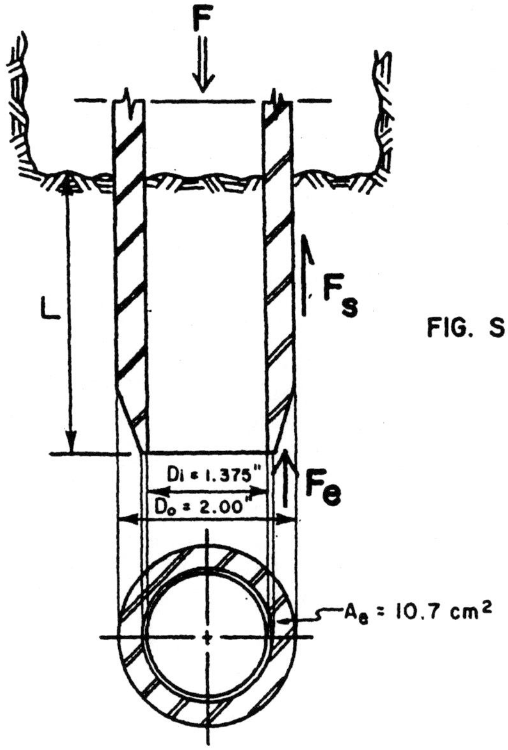

FIG. S-2 STATIC SIMULATION OF SPT SAMPLER PENETRATING SOIL AT BOTTOM OF A BOREHOLE

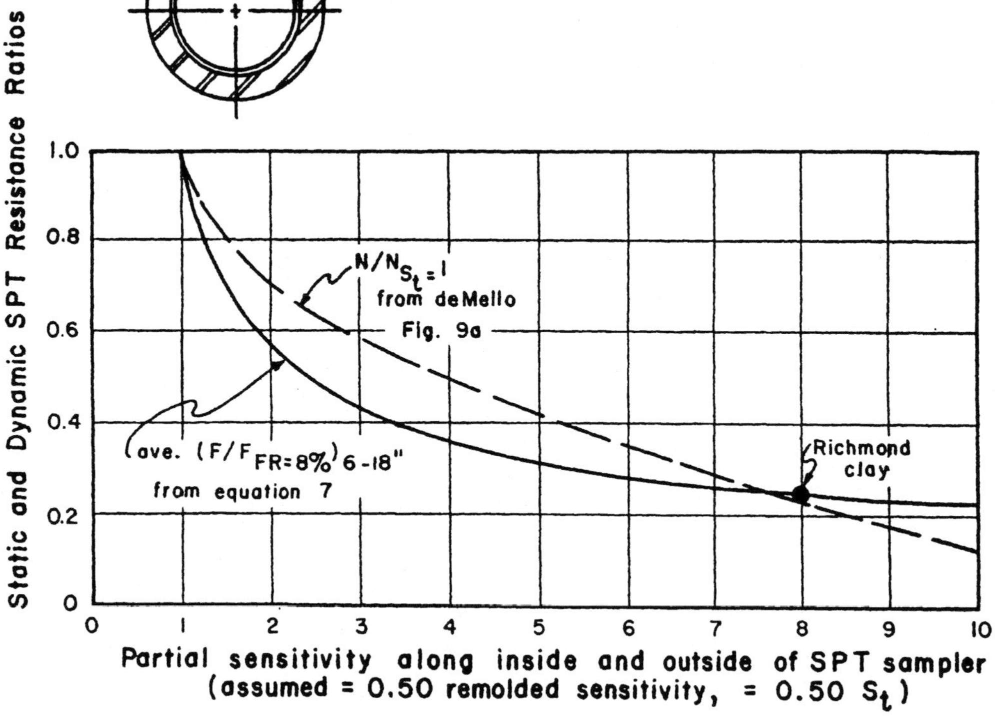

FIG. S-3 COMPARISON BETWEEN deMELLO'S AND EQ. 7 PREDICTIONS OF EFFECT OF CLAY SENSITIVITY ON SPT RESISTANCE TO PENETRATION

FOURTH PANAMERICAN CONFERENCE

Converting equation (6) to the dimensions of the SPT sampler, expressing FR in percent, q_c in kg/cm^2, the average resistance $\bar{F}$ over the penetration interval ΔL in kg., the average penetration depth of $\bar{L}$ during ΔL in inches, and assuming the blowcount dynamic resistance proportional to the static work input over ΔL, gives:

$$\Delta N_{SPT} \sim \Delta L\ [\bar{F}] = \Delta L\ [10.7 + 0.684\ (FR)\ \bar{L}]\ q_c \quad . . (7)$$

referred to herein as the "friction-cone SPT formula".

Formula Predictions

Now use equation 7 to compute the <u>relative</u> static, and therefore also the proportional dynamic, sampler resistances for the 0-6 inch, 6-12 inch, and 12-18 inch sampling increments. Normalize by obtaining the ratio of the average 0-6 inch, and 6-12 inch resistances to a reference average resistance over 12-18 inches. As shown by equation 7, these ratios depend on the friction ratio of the soil penetrated. Table S-I presents the theoretical ratios obtained by use of equation 7.

Now to the question: How do actual ratios compare with these predicted? This discusser searched his files for SPT data that met these four requirements:

1. Blowcounts recorded for each of the three 6 inch intervals.
2. SPT borings continuously stabilized with the use of drilling mud, circulated such that no coarse particles (gravel) likely left in the bottom of the borehole.
3. SPT work performed under the immediate supervision of engineers.
4. Parallel friction-cone penetrometer test data to provide assurance that the soil layer has no extreme end bearing fluctuations and to provide the needed friction ratio data.

This search found 33 SPT tests in sands with an average friction ratio of about 1%, 25 tests in the same silty marl as in figure S-1, with an average friction ratio of about 2.5%, and 33 tests in a very stiff, overconsolidated Gainesville clay, with an average friction ratio of about 4.0%. The next step noted each 0-6 inch, and 6-12 inch sampling interval blowcount and compared it to each corresponding blowcount for the 12-18 inch interval. Table S-I includes the average ratios from these field results. Note the excellent comparison between the theoretical (using equation 7) and the measured ratios at all three levels of friction ratio.

DISCUSSION–SESSION I

TABLE S-I

FRICTION RATIO = FR		$(\Delta N_{6''}/\Delta N_{12\text{-}18''})$		
		0 to 6"	6 to 12"	12 to 18"
1.0%	Eq. (7)	0.61	0.80	1.00
		0.63	0.81	--
		33 tests, loose-med sands, + and - GWT		
2.5%	Eq. (7)	0.43	0.72	1.00
		0.46	0.72	--
		25 tests, silty marl, $\bar{w}$ = 45%, LI = 1		
4.0%	Eq. (7)	0.37	0.68	1.00
		0.44	0.66	--
		33 tests, v. stiff clay, S_t = 4, LI = 0		
8.0%	Eq. (7)	0.29	0.64	1.00

Using equation 7 also permits predicting the 6 inch interval blowcount variation for the individual SPT tests diagrammed in figure S-1. Still considering the 12-18 inch blowcount value as the reference, the open circles show the result of predictions for the 6-12 and 0-6 inch interval blowcounts. Again, note the good observed-prediction agreement.

The friction-cone SPT formula, equation 7, also suggests a way to evaluate the effects of remolding sensitivity. deMello presents his estimate of the effects of such sensitivity on N-values in his figure 9a, on page 24. Figure S-3 includes a copy of his curve with a slight modification to force the curve through the 1.0 - 1.0 point. Now use equation 7 to make a similar prediction assuming end bearing remains constant and that partial sensitivity only reduces the friction ratio. For example, a partial sensitivity of 4 would reduce the friction ratio from 8% to 2%. Insensitive clay has a friction ratio of about 8%. For comparison, figure S-3 also includes a curve of the ratio of static resistances, according to equation 7, for different partial sensitivity reductions from a reference friction ratio of 8%. Note the similarity between the above and deMello's curve.

deMello informed this discusser in a private communication that he based his curve in figure 9a on about three field comparisons in Brazilian clays, with the shape of the curve guided by theoretical considerations. In deMello's reference 45, L. Casagrande notes a case in which the actual undrained strengths in a Richmond clay, with N from 4 to 8, measured four times those expected from the Terzaghi and Peck correlation in reference 300. As deMello points out, Terzaghi and Peck make no correction for partial sensitivity. Assuming their correlation

FOURTH PANAMERICAN CONFERENCE

valid for insensitive clay, this Richmond clay has a blowcount sensitivity ratio, $(N_{St}/N_{St=1})$, = 1/4 or 0.25. L. Casagrande also reported this Richmond clay had a sensitivity varying from 10 to 22. Assuming an average sensitivity of (10 + 22)/2 = 16, and a partial sensitivity along the SPT sampler of 1/2 complete remolding sensitivity (a factor deMello suggests in paragraph 2.9.5) the partial sensitivity then equals about 8. Note that the 0.25, 8 point for this Richmond clay plots close to both curves in figure S-3.

In his paragraph 2.9.5 deMello also suggests the likelihood of a sensitivity correction to N-values even greater than in figure 9a. The analysis made herein likely also errs on the too-low side. It assumes that cone, and therefore SPT sampler, end bearing remains the same regardless of sensitivity, while it most likely decreases significantly with increasing sensitivity.

Although based on the limited field support data noted above, the seeming success of the friction-cone SPT formula approach to predicting the effect of sensitivity on N-values further supports the possible usefulness of equation 7, and therefore supports the assumption of a proportional relationship between SPT sampler dynamic and static resistances.

Some Predictions

1. From the evidence presented it appears that equation 7 yields a helpful indication of the relative contribution of end bearing and side wall friction to the SPT N-value results. Using this equation also permits a prediction of the relative contributions of end bearing and side wall friction to the N-value for different values of friction ratio, and therefore for different soil types. Table S-II presents the equation-predicted percentage contributions. This work demonstrates the great importance of side wall friction, and that this importance increases dramatically with increased friction ratio.

TABLE S-II

FR	TYPICAL SOIL TYPES	PART $N_{6-18''}$ DUE TO	
		end bearing	side friction
1%	sand, above and below ground water	56%	44%
2 1/2%	silty marl, silt, clayey sand	34%	66%
4%	sandy clay, clay with S_t = 4	25%	75%
8%	insensitive clay, peaty soils	14%	86%

DISCUSSION–SESSION I

2. Theory and field comparisons presented herein suggest that N remains, at least approximately, proportional to F, and therefore to q_c, as assumed in equation 7. Lateral stress effects on q_c values would then have similar effects on N-values. The University of Florida has a research study, sponsored by the National Science Foundation, which includes investigating the effects of varying lateral stress on cone bearing. For this purpose the students use a unique test chamber, in which they subject a volume of soil 4 feet in diameter and 4 feet high to one-dimensional compression, normally and overconsolidated, with carefully controlled and measured lateral and vertical stresses. To date the tests include only one type of dry sand. A special traveling hopper and sand sprinkling system permits control and duplication of sand densities.

   Figure S-4 presents the results of 12 chamber tests performed to date on 1-D, one cycle, load-unload overconsolidated volumes of this sand. Note the different overconsolidation ratios used, two levels of relative density, and the use of both electrical and mechanical friction-cone penetrometer tips. These data clearly show a dramatic increase in cone bearing with increased lateral stress, while holding vertical stress and density constant. Roughly, doubling lateral stress increases cone bearing by 60%. These data lead to the prediction that SPT N-values would likely increase with increasing lateral stress at a given test depth. However, the stress relief effect of the borehole may reduce this effect on N-values compared to q_c values.

<u>Possible Significance of This Discussion</u>

1. Static penetration resistance appears useful when defining the components of dynamic penetration resistance of the SPT sampler.

2. The suggested friction-cone SPT formula should help define the interrelationships between different SPT-type samplers. Such will differ with different friction ratios, and therefore with the soil type sampled.

3. The low blowcount value often observed for the 0-6 inch sampler penetration into soils without gravel perhaps does not reflect a "seating" in loosened soil in the bottom of the borehole but instead reflects the minor importance of the end bearing resistance.

4. The SPT N-value reflects a varying blend of end bearing and side wall friction effects, depending on the soil type sampled. In contrast, the static friction-cone penetration test (CPT) more clearly separates the end bearing and local friction effects. The CPT's ability to accomplish this separation, plus the concept of a

FOURTH PANAMERICAN CONFERENCE

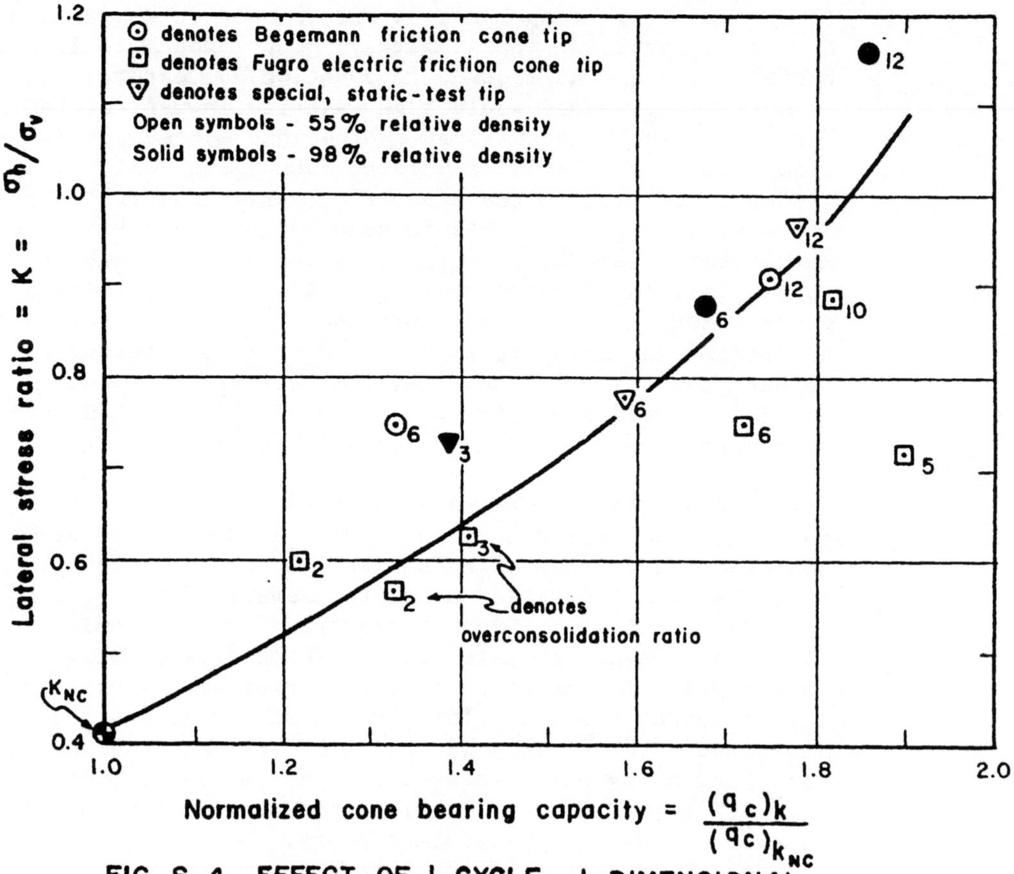

FIG. S-4 EFFECT OF 1-CYCLE, 1-DIMENSIONAL OVERCONSOLIDATION OF A SUBANGULAR, MEDIUM, DRY QUARTZ SAND ON THE STATIC CONE BEARING CAPACITY IN A 4-FT DIAMETER, 4-FT HIGH TEST CHAMBER

friction ratio, makes possible this improved interpretation of SPT data.

5. For quantitative data on a soil's strength properties, why not use the CPT method in the first place? deMello on page 65 also recommends direct use of the CPT for settlement estimates.

References

Begemann, H. K. S., 1965, "The Friction Jacket Cone as an Aid in Determining the Soil Profile", Proceedings, 6th International Conference on Soil Mechanics & Foundation Engineering, Vol. I, pp. 17 - 20.

deRuiter, J., 1971, "Electric Penetrometer for Site Investigations", Journal of the Soil Mechanics & Foundations Division, ASCE, February, pp. 457 - 472.

DISCUSSION–SESSION I

G. C. SANGLERAT[12]. --

I fully agree with the conclusions of panelist Schmertmann. More specifically the equation giving the N value as a function of static cone resistance q_c, and of the friction ratio, FR (friction on sleeve resistance to cone), is applicable in any type of soils. This equation can be formulated as follows :

$$N(SPT) = (A + B \times FR\ \%)\ q_c$$

A and B being two constants.

On the other hand, many attempts have been made to correlate static cone resistance with N values by the relation

$$q_c = n\ N$$

n varying from 2 to 8 or more depending on the type of soil (clayey to sandy). Using Schmertmann's formula for N, we can write

$$n = \frac{1}{A + B \times FR\ \%}$$

But since FR is a characteristic of soil type :

where FR is low for cohesionless soil
and FR is high for cohesive soil

one can conclude that :

n should be high for sand
n should be low for clay or peat

This is well confirmed by practical results of various authors. The fact that the friction ratio FR is a characteristic of the soil type has been verified in France with a penetrometer entirely different from those employed by Schmertmann (Fugro and Gouda devices).

In France I used the Andina static-dynamic device with friction sleeve.

Tests performed in France since 1970 show that the friction ratio FR is independent of the rate of penetration

quasi static of 2 cm/sec (FR_2)
quasi static of 1 cm/sec (FR_1)
and truly static of 0 cm/sec (FR_0)

For all practical purposes then FR_0 is equivalent to FR_2 or slightly smaller.

12Lyon, France

EFFECTS OF IN SITU LATERAL STRESS ON FRICTION – CONE PENETROMETER DATA IN SANDS.

by John H. Schmertmann, Ph.D., P.E., professor of civil engineering Univeristy of Florida.

The static cone penetration test has received wide acceptance in Holland and elsewhere as a means of evaluating the in situ engineering properties of sand deposits. We all know that the magnitude of vertical stress, or the overburden pressure, has an important effect on the static cone bearing capacity, g_c, and that the engineer must include this effect when interpreting q_c for engineering properties. However, the magnitude of the in situ lateral stress remains not considered in such evaluations. Apparently, the difficulty of measuring in situ lateral stresses plus the lack of research as to their significance to cone penetrometer data makes it difficult and perhaps impossible for engineers to consider their effect at present.

In an effort to provide qualitative insight and perhaps quantitative data on this subject, I initiated both laboratory and field research at the University of Florida with the specific aim of evaluating the effects of in situ lateral stress on friction cone penetrometer data in sands. The results are both interesting and of practical importance, as I now hope to show you.

U.F. LABORATORY STATIC CONE TESTING CHAMBER.

In this specially constructed triaxial chamber we prepare an "element" of sand which has a diameter of 4.0 ft (34 cone diameters) and a height of 4.0 ft.
This "element", or "sample", we surround at its sides and bottom by a rubber membrane with the sides and bottom loaded independently by water pressure. We rain in (pluvial compaction) the sand and can obtain uniform and repordiceable densities of dry sand. The chamber is especially designed to permit K_0 consolidation loads via a bottom piston and insert the penetrometer tip through an axial access hole in the top chamber lid.

Thus far we have performed about 100 static cone tests with this chamber, using both normal consolidation and overconsolidation conditions, and have reached some interesting conclusions with respect to the importance of lateral stress.
Fig. 1 herein presents an earlier summary of these data.
From this figure it is quite apparent that increasing the lateral stress while holding density and vertical stress constant, causes an important increase in q_c. Roughly, doubling the lateral stress increases q_c by 60%.

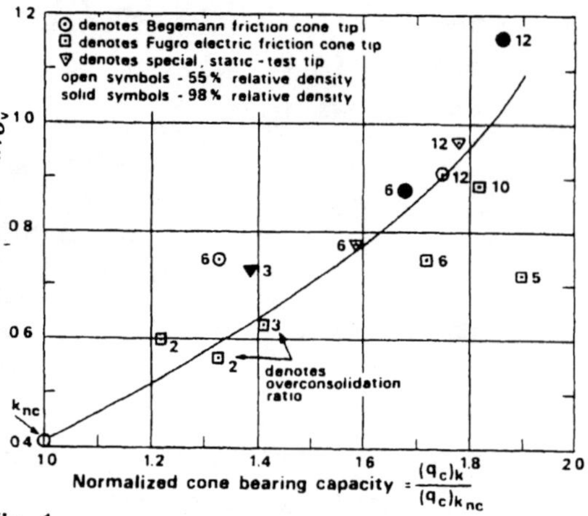

fig. 1

U.F. FIELD CUBE TESTS.

After establishing the uniformity of soil condition at a nearsurface deposit within the grounds of the University of Florida, we divided this area into a large number of squares, 5 foot on a side. Then, one at a time, we excavated trenches, 5 to 6 feet deep, around the borders of a square and inserted inflatible air mattresses into these trenches, two mattresses for each side. With a common manifold to all 8 air bags we then carefully controlled the lateral stress on the sides of such a field test "cube". We then performed static cone bearing, pressure meter, and buried rigid plate bearing tests within these cubes, using various levels of air-bag lateral stress to evaluate lateral stress effects.
In its original condition the sand at the site was at a relative density of about 21%. We performed a series of tests at this density. We then subjected the site to compaction using a vibratory roller and thereby increased its relative density to an average of about 85%. We then performed a similar series of cube tests at the higher density. Thus these field cube tests permitted some evaluation of the separate effects of both density and lateral stress.
Fig. 2 shows a cross-section of such a cube, illustrated with the provision of a plate load test within the cube.
Fig. 3 shows the load-settlement curves obtained

from such rigid plate load tests and illustrates the effect of varying lateral stress at the two density levels.

Fig. 4 shows my interpretation method for evaluating these load-settlement curves to obtain elastic modulus and yield point parameters.

Fig. 5 then shows the relative magnitude of the elastic yield pressure and modulus as determined at two relative density and two lateral stress (K') conditions.

Finally, fig. 6 introduces the cone penetrometer data obtained from other similar test cubes. These are averages from a large number of tests using the Delft mechanical friction cone, but there was not much scatter.

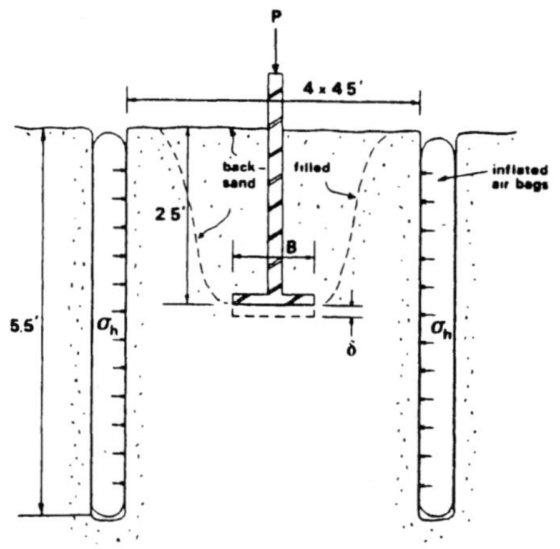

fig. 2

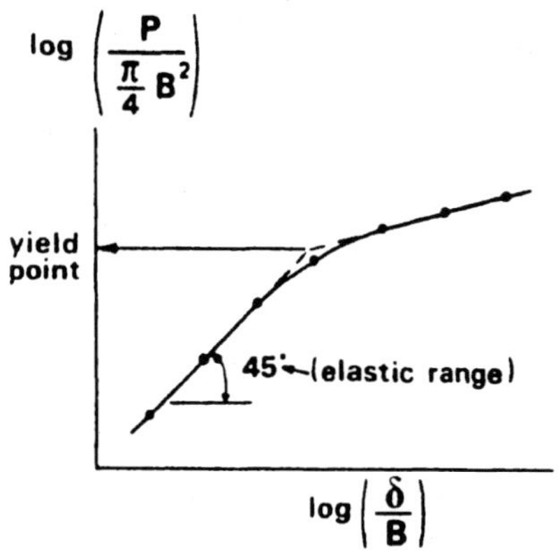

Test interpretation for yield pressure

fig. 4

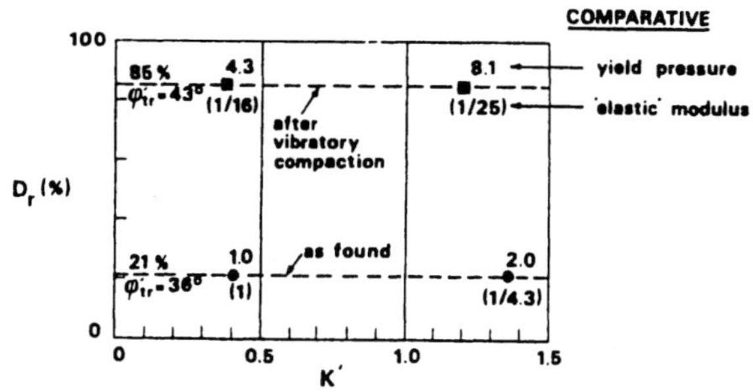

fig. 5

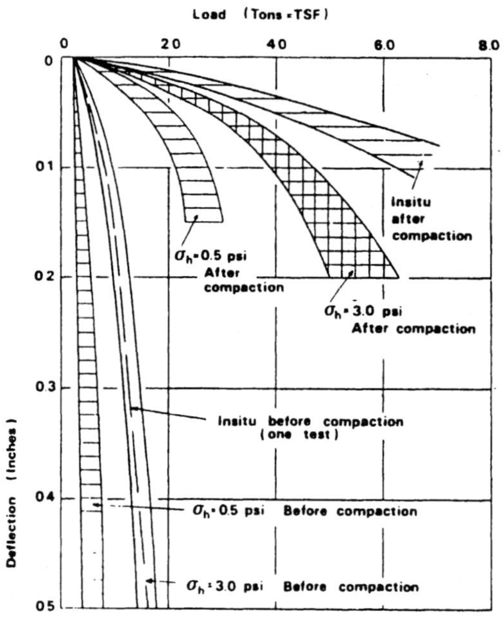

fig. 3

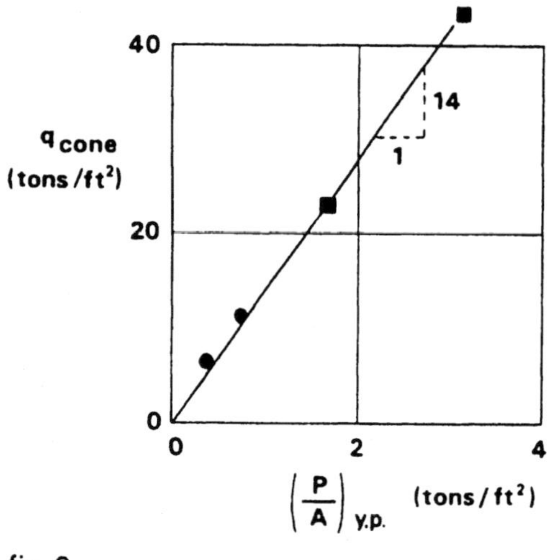

fig. 6

From these field tests it seems that if an engineer wishes to interpret for the yield point, then q_c is a good measure of this yield point (factor of 14) at both density conditions and both lateral stress conditions at each density.

However, also considering fig. 5, and noting that the q_c values at the lower lateral stress are about one half of those of the higher lateral stress, it seems that q_c reflects elastic modulus changes due to lateral stress when the sand was at high density, but would underestimate the elastic modulus by a factor of about 2 when the sand is at low relative denisty.

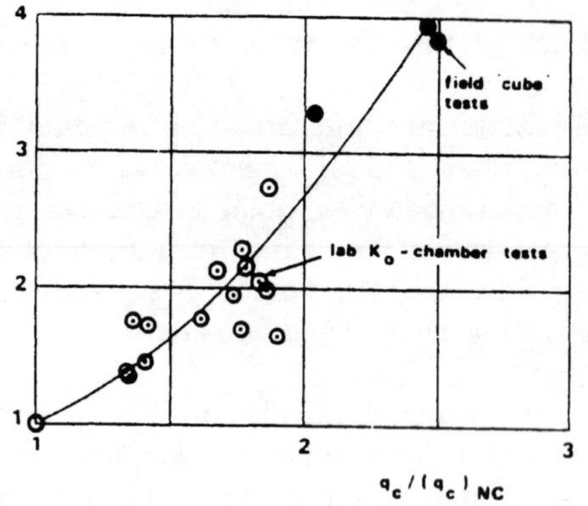

tig. 7

SUMMARY.

Fig. 7 presents, in normalised coordinates, a summary of both the laboratory chamber and field cube tests and shows the consistent importance of evaluating lateral stress effects when interpreting cone bearing data. I have no doubt that many instances of high q_c values in sand have meant not high density, but high in situ lateral stresses due perhaps to significant overconsolidation. On the other hand, unusually low values may be due not to low density but perhaps to low lateral stresses, say by adjacent excavation.

JOURNAL OF THE CONSTRUCTION DIVISION

Dewatering Case History in Florida

By John H. Schmertmann,[1] F. ASCE

Introduction

Dewatering occasionally becomes a surprisingly difficult phase of construction. This paper describes the dewatering of a small foundation in Southeast Florida in 1959-1960 where the difficulty proved an uncomfortable and expensive surprise to both engineers and contractor. Although the contractor bid the dewatering at $20,000, it actually cost about $130,000, not including lost-time costs. He expected to do it in 2 weeks but it actually took 7 months! At one point, almost 3 months after starting, wellpoint pumping costs ran $700 per day with no progress and threatened to put the contractor out of business. Note that all costs mentioned herein are in 1960 dollars.

The writer has, over the 14 yr since working on this dewatering problem, noted that the construction and the soil engineering literature includes few case histories of such problems. This job seems as instructive today as at the time.

The project involved dewatering the excavation for the foundation of Structure 37A on Canal 14 of the U.S. Army Corps of Engineers' Central and Southern Florida project—a minor structure in this vast project for drainage and flood control. It is located between Ft. Lauderdale and Pompano Beach, about 1 mile (1.6 km) from the Intercoastal Waterway. E and I, Inc., then of Ft. Lauderdale, Fla., had the contract to build Structure 37A.

The Structure 37A design called for a concrete weir to regulate flow in a relocated canal, under construction simultaneously. The Corps of Engineers planned to found the weir on a rectangular concrete slab, 42 ft × 58 ft (13 m × 18 m) with sheet pile cutoff walls going out at 45° from each corner. Fig. 1 shows a simplified plan. This plan also shows the 100-ft × 110-ft (31-m × 34-m) outline of the header for the stage 1 wellpoint system, which approximates the edge of the excavation for 37A at the time of the maximum wellpoint installation. Fig. 2 presents a photograph of the excavation at this time. The ground water was initially at an average elevation of about +1.0 ft. Only about 50 ft (15 m) away, and roughly parallel to the south side of the excavation,

Note.—Discussion open until February 1, 1975. To extend the closing date one month, a written request must be filed with the Editor of Technical Publications, ASCE. This paper is part of the copyrighted Journal of the Construction Division, Proceedings of the American Society of Civil Engineers, Vol. 100, No. CO3, September, 1974. Manuscript was submitted for review for possible publication on October 3, 1973.

[1] Prof. of Civ. Engrg., Univ. of Florida, Gainesville, Fla.

was an existing canal with tidal water at about El. +1.0 ft ± 0.5 ft. The original design called for the base of the weir slab at −14.5 ft. Thus, the initial dewatering problem involved lowering the water level a minimum of 15.5 ft (4.7 m) over an area of approx 100 ft × 110 ft (31 m × 34 m).

SOILS AND GEOLOGY

The contract documents included the results from six, non-SPT, split spoon sample borings made by the U.S. Army Corps of Engineers. The writer based

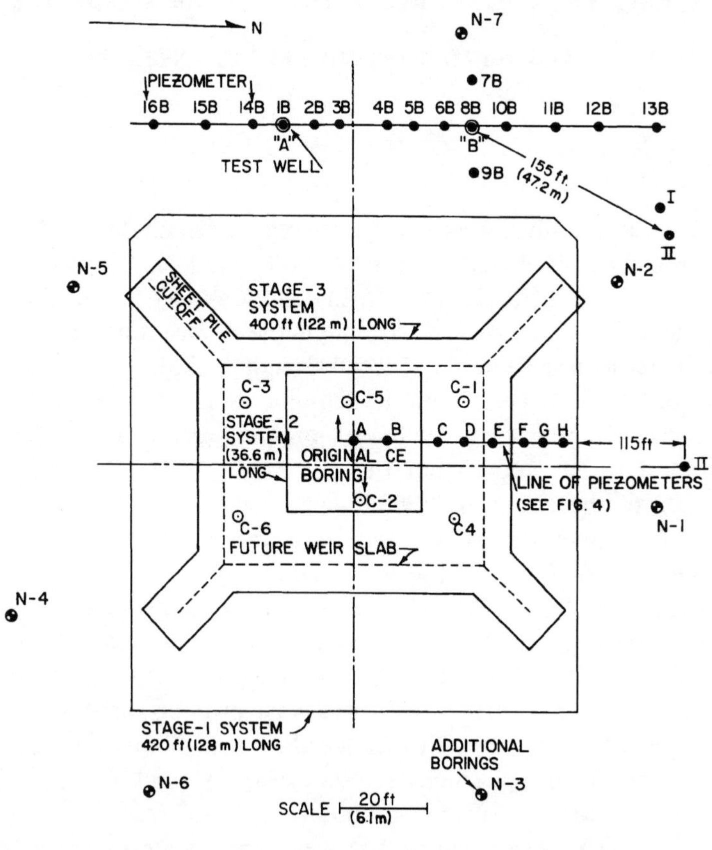

FIG. 1.—Site Plan Showing Borings, Weir Slab, Wellpoint Stages, Pump Test Wells, and Piezometer Locations

the soil profile shown in Fig. 3 on these initial borings. Points C1-C6 in Fig. 1 show these borings all located within the boundaries of the weir foundation slab.

The contract documents did not explain the geology of the site, which in retrospect might have allowed the engineers involved to anticipate this dewatering problem. Except for the surface organics, all layers explored in this project were marine coastal sediments deposited since mid-Pleistocene times. The site lies within the irregular surface-contact boundary area between the overlying Miami Oolite formation and the underlying mid to late-Pleistocene, very variably cemented, quartz sand, and shell deposits known as the Ft. Thompson formation. Interfingering and unconformities are common in this boundary area. The

CO3 DEWATERING CASE HISTORY

FIG. 2.—Looking NE Across Excavation During Well Pumping Tests

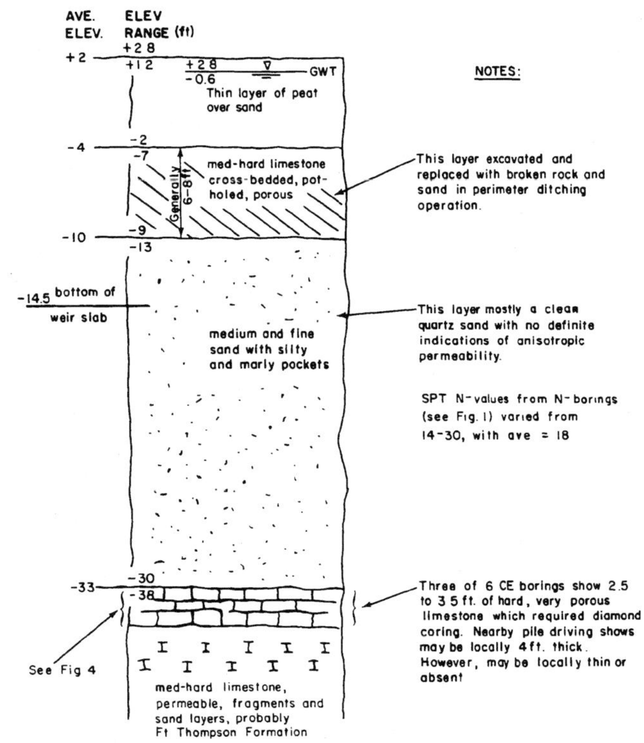

FIG. 3.—Average Geologic Section from Crops of Engineers' Borings

limestone from average −4 ft to −10 ft (see Fig. 3) was oolitic and probably Miami Oolite. The limestone below −30 ft had the well-cemented and vuggy character believed typical of the top of the Ft. Thompson. This character probably developed from solution, erosion, and secondary calcite and silica deposition during periods of much lower Pleistocene sea levels. Its many small, interconnected cavities account for its very high permeability, which might have been anticipated from this geology and also from the previous dewatering difficulties encountered by Prugh (3) in the Ft. Thompson formation in Miami.

Attempt with Wellpoints

The medium to fine sand layer, extending about 16 ft (4.9 m) below the bottom of the intended maximum excavation depth, and the comparatively small size of the excavation made it seem that using wellpoints would provide an economical method for dewatering. Three preliminary single wellpoint tests on August 13, 1959, using 1-1/2-in. (38-mm) points and tip elevations between −23 ft and −32 ft, each yielded 30 gpm-35 gpm (0.11 m^3/min-0.13 m^3/min), which the contractor thought excellent production. With this encouragement, the contractor rented a wellpoint installation. This installation finally mushroomed to 452 points pumping about 8,000 gpm (30 m^3/min) out of the three stages shown in Fig. 1. But he could not lower the water level at the center of the excavation below El. −10.8 ft despite an average of one point per 5.4 sq ft (0.50 m^2) of slab area!

A temporary system of 107 points appeared successful on August 31. The contractor then followed this with a 200-point initial line for stage 1, which included the up and downstream riprap areas protecting the canal bottom near the weir. This did not succeed in lowering sufficiently the water level under the slab. The contractor then thought that the upper limestone layer (see Fig. 3) was more porous than estimated. To prevent flow over the wellpoints he ditched through this limestone on September 13 and backfilled with relatively impervious soil as a cutoff. Only minor improvement resulted. He then added 36 points and also considerably shortened the stage 1, line, but he thereby only lowered the water level an additional 2 ft at the excavation center. Still suspecting lateral flow above the points, on October 9 he installed another 100 points in an open intercepting ditch just inside the shortened stage 1 with the tips 9 ft (2.7 m) higher at El. −15 ft. The pumpage from these additional 100 proved so small that he soon removed them. The first (of three) quicksand boils developed during this removal. Everyone now realized they had a serious problem. The wellpoint subcontractor then thought that more points were needed inside the stage 1 system to bring down the center crown in the water level. Accordingly, he installed the stage 2 and stage 3 systems, bringing the total to 452 points and an improvement of almost 4 ft (1.2 m). The water level still remained, however, almost 5 ft (1.5 m) above the El. −15.5 ft originally planned and the contractor now also had the foundation area cluttered with the stages 2 and 3 systems (see Fig. 2).

Then something happened that indicated clearly the situation was almost out of control with a danger of destroying the whole foundation soil layer. The contractor decided to put in an open-pipe piezometer to check water pressure in the rock under the center of the excavation. When the jetted pipe hit rock at El. −30 ft the second boil, approx 3 ft (0.9 m) in diameter and flowing an estimated 2,500 gpm (9.5 m^3/min), developed in about 1 min and was stopped by surcharging only after considerable difficulty. That this could happen in the middle of all those wellpoints seemed especially frightening. The contractor and his wellpoint subcontractor now sought the advice of the Corps of Engineers and subsequently brought in the writer as a consultant.

INVESTIGATING PROBLEM LAYER: UNDERLYING ROCK

Evidence abounded to indicate the rock below −30 ft controlled this dewatering problem. The six Corps borings showed this rock to be porous limestone, with the possibility of a particularly porous top cap. The two large quicksand boils clearly showed that this rock could continuously supply large quantities of water to a limited area. It seemed impossible for the sand alone to supply such quantities. The writer witnessed the third boil, an equally large one, during his first day at the site. The ineffectiveness of the September 13 ditching, the poor performance of the intercepting temporary wellpoint line of October 9, and the results from adding the stage 2 and 3 systems all suggested that the problem flow was not lateral, but vertical.

However, the nature of the seepage flow pattern required proof beyond doubt. This was particularly so because the contractor now planned to use the "relief for unusual conditions" clause in his contract. He hoped to demonstrate conclusively that he had encountered an unusual condition in that part of Florida. He eventually won this point and financial relief.

First, the contractor installed the line of open ended pipe piezometers shown in Fig. 1. Readings from these, plus the computed heads at the wellpoints in each stage, guided the construction of the flownet shown in Fig. 4. The flownet construction includes the idealizations of a uniform and isotropic sand layer and a constant top of underlying rock at El. −30 ft. This flownet, showing vertical flow from the underlying rock as the primary source of the water pumped from the wellpoints, offers reasonable agreement with the piezometer readings. Trial flownets based on lateral flow did not yield such agreement. Furthermore, the ratios of the flow into each stage as computed from the net and as measured in the discharge lines were in reasonable agreement. In order of stages 1, 2, and 3, the measured ratios were approx 6:1:3.5 This compares with 4:1:3 computed from Fig. 4. In view of this additional evidence from the piezometers and flownet, there now was no longer any doubt that the rock was the source of the water, the pressure, and the trouble.

When looking at the flownet in Fig. 4, an engineer might think that as long as so much effort had already gone into a wellpoint installation why not complete the job by perhaps doubling the number of points, with additional headers between the existing stages to lower the water table "crowns" to the required −15 ft elevation? But, when the sand excavation also reached the required −14.5 ft the average vertical hydraulic gradient would approach 1.0, which means the foundation would likely become a fully quick 10,000-sq ft (930-m^2) quicksand "tank." Any wellpoint installation in the sand could not succeed without first somehow reducing the water pressure in the undelying rock. It now remained to determine the pumping rate from the underlying rock needed to achieve the required pressure reduction.

TABLE 1.—Estimated Average Hydraulic Gradient Initiating Boils

	Head Loss, ΔH			Vertical Flow		
Date (1)	Estimated H in rock (2)	Surface of excavation (3)	ΔH, in feet (4)	Rock elevation estimated (5)	ΔL, in feet (6)	$\bar{i}_v$, $= \Delta H/\Delta L$ (7)
October 20	−1	−12	11	−30	18	0.61
October 29	−1	−7.5	6.5	−30	22.5	0.30
November 5	−1	−10	9	−30	20	0.45

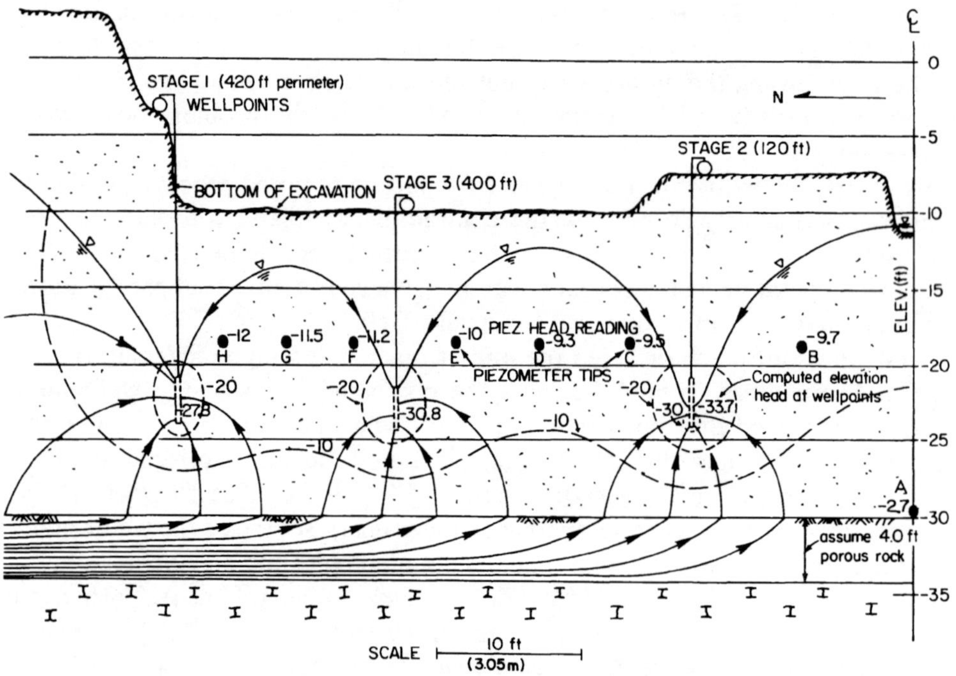

FIG. 4.—Flownet for Conditions During Pumping from Maximum Wellpoint System

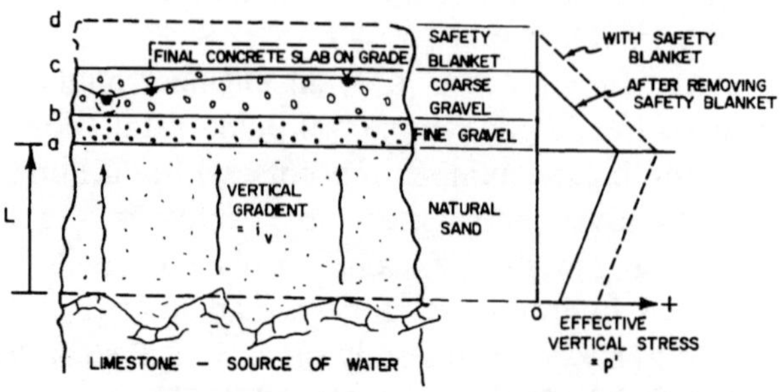

FIG. 5.—Schematic Diagram of Gravel Blanket Scheme

Determining the required pressure reduction was aided by the evidence available from the three quicksand boils. Table 1 summarizes this evidence.

From this evidence, and considering the washing-provocation of the last two boils, the writer decided that the average vertical hydraulic gradient, $\bar{i}_v$, should not exceed 0.4. This compares with about 0.9 for a theoretical quick condition if the quartz sand layer was loose and homogenous. With the dewatered excavation water level at El. −15 ft, and the highest rock at −30 ft, the head in the rock then could not exceed −9 ft. The required head reduction in the rock then equals $[+1 - (-9)] = 10$ ft (3.0 m).

The contractor now needed an estimate of the rock pressure relief pumping required, based on the Corps borings and the available piezometer information. The borings indicated that the porous zone in the rock might be limited to a harder cap rock only approx 4 ft (1.2 m) thick. The flownet of Fig. 4 shows all the water flow through this cap. Assuming this the case, then the readings from the only two piezometers in the rock, nos. A and II and both in the section of the flownet, indicated a permeability for the cap rock of about 6,000 $\times 10^{-4}$ cm/s. Such extremely high permeability may surprise some readers. But Prugh (3) reported 10,000 $\times 10^{-4}$ cm/s-32,000 $\times 10^{-4}$ cm/s for the Ft. Thompson formation at one site in Miami, and Shea (4) had reported limestone with up to 400,000 $\times 10^{-4}$ cm/s at other sites of the Central and Southern Florida project. So the 6,000 $\times 10^{-4}$-cm/s permeability value appeared reasonable and possibly even low. Using a higher permeability of 40,000 $\times 10^{-4}$ cm/s, plus the assumed 4-ft (1.2-m) cap source, and assuming the relief wells installed immediately outside the stage 1 system and into the rock, resulted in a computed total pumping requirement of about 20,000 gpm (76 m^3/min).

A rate of 20,000 gpm (76 m^3/min) was feasible, though expensive. This computation justified the further investigation of the character of the rock. Accordingly, the contractor had seven additional borings made outside the stage 1 header, to avoid starting boils (see Fig. 1), with the objective of obtaining more information about rock permeability and thickness. He also constructed a large well pumping test for further quantitative information.

Elevation of the lower rock surface in the additional borings varied from −28.0 ft to −33.1 ft, which agreed with the Corps borings. Each boring ended in porous rock, even though three of them penetrated 22 ft (6.7 m) into the rock. Fourteen recharge tests, which provide a semiquantitative measure of permeability, demonstrated that the high-permeability rock layer was not limited to a harder cap rock but had a thickness of at least 22 ft (6.7 m).

The subsequent well-pumping tests constituted a most important part of the investigation of the character of the underlying rock. Fig. 1 includes the layout of the two wells, A and B, and the associated piezometers. These wells consisted of 12-in. (300-mm) diam steel casing driven 6 ft (1.8 m) and 8 ft (2.4 m), respectively, into the rock to prevent the movement of sand from the layer above the rock, and an open hole 2 ft (0.6 m) and 6 ft (1.8 m), respectively, into the rock below the casing. The contractor made these holes by punching into the rock below the casing with a steel rail with a chisel point and blowing out the rock fragments with compressed air. The line of piezometers noted in Fig. 1 consisted of 1-1/2-in. (38-mm) wellpoints jetted to good contact with the rock and, in some cases, driven into the rock. All wellpoints flowed freely during the cleaning of the 12-in. (300-mm) wells with compressed air, and thereby demonstrated flow continuity with the test wells. Fig. 2 also shows test well A in the foreground.

In the first two tests the contractor only pumped well B with a pumping rate of 2,250 gpm (8.3 m^3/min). The piezometers showed a drawdown of only 2.5 ft (0.76 m) and a radius of influence of only about 75 ft (23 m). In the final two tests both wells were pumped. Well B was pumped at 2,250 gpm (8.5 m^3/min), with a drawdown of 3.4 ft (1.0 m), and in addition the pumpage from A was 3,100 gpm (11.7 m^3/min), with a drawdown of about 11 ft (3.4 m), for a total of 5,350 gpm (20.2 m^3/min). Each series of two B and AB tests gave very similar results.

While these piezometer data show that the flow into the wells was not radially uniform but followed preferred channels which did not intersect all the piezometers, the writer nevertheless attempted to estimate average rock permeability from these data. In porous, channeled limestone such an estimate does not represent a physical reality valid for small volumes within the rock, but only a large-volume average behavior needed for subsequent use in well formulas. The average rock permeabilities computed from estimated average drawdown curves were $k_B = 36{,}000 \times 10^{-4}$ cm/s and $k_{AB} = 21{,}000 \times 10^{-4}$ cm/s.

Using a value of $k = 30{,}000 \times 10^{-4}$ cm/s and a porous rock thickness of 25 ft (7.6 m), the pumpage for a circle of 12-in. (300-mm) wells placed just outside the stage 1 header computed at about 100,000 gpm (380 m^3/min) just to reduce the head in the rock at the center of the excavation the specified 10 ft (3.0 m). It seemed probable that the permeable rock thickness exceeded 25 ft (7.6 m), and that even greater pumpage would actually be required. Thus, even to achieve the 10-ft (3.0-m) minimum pressure relief in this rock required an enormous and far too expensive dewatering installation. The problem needed an entirely different solution.

Gravel Blanket Scheme

Engineers from the Corps of Engineers (CE), the contractor, and the writer now had a conference at CE Division Headquarters in Jacksonville, Fla., to consider alternate dewatering schemes, which included the following:

1. Using grout to form an underlying blanket to greatly reduce the permeability of the upper 5 ft (1.5 m) of the underlying rock.
2. Sheet pile cofferdam to rock and a grout curtain to impermeable rock and thereby cut off the water supply to the porous rock.
3. Sheet pile cofferdam to rock with tremie seal concrete foundation mat.
4. Excavating the porous limestone and replacing with sand.
5. Forming an impervious rock blanket in the upper 5 ft of rock by using explosives to reduce permeability.
6. Overlying gravel blanket as explained subsequently.

On the basis of preliminary engineering and economic considerations only schemes 1, 3, and 6 received detailed study.

All the engineers involved considered scheme 3 the conventional and safe approach. But, the cost would exceed $100,000. They thought scheme 1 more risky, but an independent grouting-dewatering contractor bid $90,000. Scheme 6 was untried in the experience of all the engineers concerned, but estimated to cost only \$50,000. Because of the engineering interest in this scheme, plus the possibility of saving \$40,000–50,000 or more, the Corps allowed the contractor to try the gravel blanket scheme.

Theory of Gravel Blanket.—Fig. 5 presents a schematic diagram to aid explaining the gravel blanket method. Please refer to Fig. 5 and the following steps:

1. Remove existing wellpoints and other equipment and allow the excavation to fill with water.
2. Excavate the sand to elevation *a*.
3. Place, underwater, a high-permeability layer (gravel) of thickness *a-d* over the bottom of the excavation. If needed, first place a filter layer of coarse sand or finer gravel *a-b*.
4. At the appropriate elevation during the gravel placement in step 3, place a perforated pipe water collection system around the bottom perimeter of the excavation, outside the planned foundation slab.
5. Pump from the aforementioned collector pipe system until the water level in the excavation reaches the position shown within the gravel blanket. Continue the pumping to maintain this position.
6. Install piezometers and check the distribution of vertical effective stress over distance L.
7. Remove the thickness, *c-d,* of safety gravel and construct the foundation on the dry gravel at elevation *c*.

Quicksand action, boils, or a general hydraulic uplifting of a plug of overburden soil can only occur when the water pressure at some depth equals or exceeds the total vertical overburden pressure of the soil at that depth (zero or negative vertical effective stress). The success of the foregoing method hinges on the ability to everywhere, at all times and all depths, maintain a positive vertical effective stress. As also shown in Fig. 5, effective stresses must be great enough when including the weight of the safety gravel blanket to remain positive after its removal. Step 6 permits the engineer to check that a sufficient margin of effective stress exists before removal to permit this removal.

The gravel blanket scheme had a well-established base in the technical literature, although it happened to be new to all the parties in this dewatering project. It can be considered a special adaptation of a horizontal, weighted, graded, and inverted filter acting as both a filter and a drain. For examination of related seepage and filter theory the reader may refer to Cedergren (1), Taylor (5), Terzaghi (6), and Terzaghi and Peck (7).

The problem now reduces to one of proper pressure balancing. The factors that influence this pressure balance are the unit weights of the soil layers involved, the water pressure head in the underlying rock, and the excess hydrostatic pressure-flow distance, L.

At this point the Corps of Engineers helped the dewatering effort with a minor redesign of their weir to permit raising the bottom of the foundation slab 1 ft to -13.5 ft, and thereby increasing L by 1 ft. Later events proved this help of critical importance.

Based on experience and simple lab density tests on jar samples from the Corps and N-borings, the writer estimated the saturated unit weight of the natural quartz sand at between 120 pcf and 125 pcf (1,900 kg/m^3 and 2,000 kg/m^3), and assumed that the blanket gravel would also have about this unit weight. According to piezometer II (see Fig. 1) after a 5-1/2-in. (140-mm) rainstorm the head in the underlying rock could reach +2 ft, the elevation of the water in the adjacent channel. The head could also be as low as −2.7 ft, read by piezometer A in the center of the excavation. The highest elevation for the top of the underlayer rock determined by the Corps borings under the slab

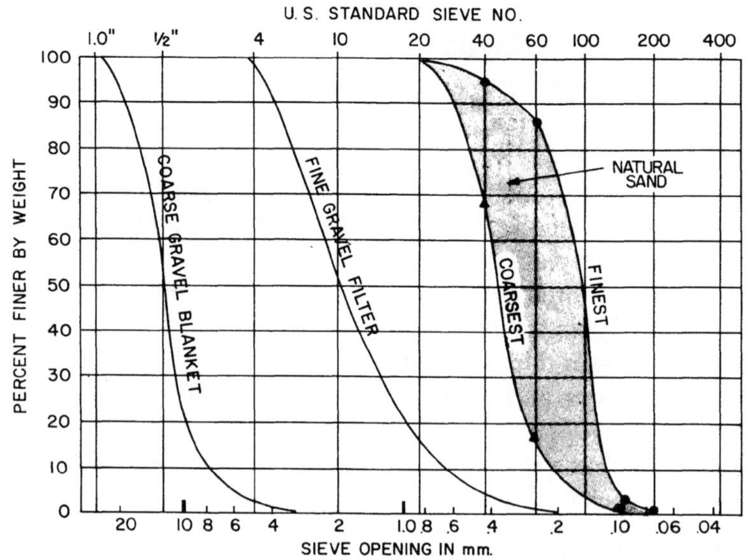

FIG. 6.—Results of Sieve Analyses of Natural Sands and of Gravels Used in Blanket

FIG. 7.—Excavation Immediately after Dewatering by Sump Pumping from Gravel Blanket—Safety Blanket Still in Place

was −30 ft, with the average at −33 ft. The corresponding information from the outlying N-borings was −28 ft and −30.7 ft.

Vertical seepage head loss concentrated in the lower half of the sand layer would aid the weight balancing by increasing minimum values of effective stress, and concentration in the upper half would decrease minimum effective stress. The comparative permeabilities of the sand at different depths below the blanket became of great importance and sieve analyses were made to estimate this. Fig. 6 shows these and other sieve results. The coarsest and finest sands from the sand layer had grain diameter ratios of about 2.4 at D_{60} and about 1.7 at D_{10}, indicating, if permeability varies as D^2, that the upper and finer grained part of the sand layer may have a permeability as low as one-fourth that of the lower part.

The writer now investigated different combinations of elevation of the top of the rock underlayer, pressure head at the top of the rock, unit weight of the sand and gravel, and relative permeabilities between the upper and lower half of the sand layer, considering the then-known extreme and estimated most likely values for each parameter. It became clear that if all parameters combined most unfavorably the final excavation could not reach El. −13.5 ft, with the water level at about −14 ft, without a portion of the sand layer first reaching a condition of zero effective stress. However, if only the most likely values combined unfavorably the dewatering could be accomplished with a 100-psf (4.8-kN/m^2) margin of effective stress safety, considered the minimum acceptable. Only the event would demonstrate how these parameters combined.

One of the attractive features of the blanket scheme was the ease of providing extra safety by means of an extra, temporary, weight of gravel blanket to provide a margin of safety against the most unfavorable combination, as well as any other unforeseen unfavorable effects. This protection permitted the subsequent use of the "observational approach" as described, for example, by Peck (2). Then, if the step 6 piezometer check proved the safety blanket could be removed and still maintain a minimum of 100 psf (4.8 kN/m^2) effective stress, the contractor could proceed with step 7. If not, then the contractor would first install a supplementary pressure relief system of 12-in. (300-mm) wells into the rock underlayer, as required. Someone had to choose the required thickness of the safety blanket—the thicker the safer but also the more expensive. The writer chose 3.0 ft (0.9 m)—a choice that luckily turned out adequate, but barely so.

Construction of Gravel Blanket.—The contractor began by implementing step 1. For step 2 he needed to know the steepest permissible excavation slopes and how to protect these slopes against seepage breakout after dewatering. For the step 4 design of the pipe collector system and pumping machinery he needed to know the maximum pumping rate expected. Flownet analysis helped answer this question and also provided a method to estimate seepage flow, pressure conditions, and stability for various assumed slopes. The maximum recommended slope was 2-1/2 horizontal to 1 vertical with a 2-ft (0.6-m) cover of fine gravel.

As planned, the contractor excavated the natural sand underwater to El. −17.5 ft and to 3 ft (0.9 m) outside the perimeter of the slab. He also sloped the sand side slopes underwater. It proved too difficult to get the maximum 2.5 on 1 side slopes, so the contractor settled for about 2.7 to 3 on 1 slopes.

Finding suitable gravel for the blanket provided an unexpected problem. Blanket gravel with a saturated unit weight of greater than 100 pcf (1,600 kg/m^3) simply was not available in the entire limestone Florida peninsula. Yet, to sacrifice 20 pcf (320 kg/m^3) for 7 ft (2.1 m) meant losing 140 psf (6.7 kN/m^2), which was not acceptable in view of the careful pressure balance and small effective stresses expected. Although expensive, the writer considered it necessary to import gravel by truck from Chattahoochee, Fla.—almost 500 miles (900 km) away. Tests showed the saturated unit weight of this gravel averaged about 123 pcf (1,970 kg/m^3). Fig. 6 also shows its average grain size distribution.

Because the blanket gravel had to be coarse to provide a high permeability and thereby permit seepage collection through a relatively thin blanket layer by a horizontal drain system 3 ft (0.9 m) outside the perimeter of the slab, the contractor also had to provide a fine gravel filter between the coarse gravel and the natural sand. This filter then also became part of the blanket. Again, the contractor could not find suitable fine gravel in peninsula Florida. Either it was carbonate rock and too light, or it was too fine for proper filter action. The minus no. 4 sieve fraction of the Chattahoochee gravel had suitable weight and a suitable grain size distribution, as shown in Fig. 6. Unfortunately, the contractor could only obtain this special gravel in limited daily quantities. Consequently, he reduced the originally planned 2-ft (0.6-m) thick layer to 1 ft (0.3 m), and special care was used, via skin diving enthusiasts in the contractor's crew, to insure a minimum thickness of 6 in. (150 mm).

The contractor planned to use 14-in. (350-mm) pipe for the perimeter drain system, but he had to use 12 in. (300 mm) because a then current steel strike made 14 in. unavailable. To make the pipe pervious the contractor cut numerous 1/4-in. (6-mm) wide horizontal slits in this pipe with torches and then wrapped it with 1/4-in. (360-mm) screening. When the gravel bed reached −15.5 ft he installed this pipe collection system underwater, again with the aid of skin divers. The perimeter pipes led to an open cage sump. He provided pumping capacity of 5,000 gpm (19 m^3/min), double the expected seepage into the excavation, by using two pumps, one a standby, mounted on a pile supported wood platform directly over the sump. Fig. 7 includes this pumping installation. Seepage never exceeded 2,500 gpm (9.5 m^3/min) and the standby remained idle.

Final Dewatering

After raising the coarse gravel blanket to El. −10.5 ft, the contractor eagerly started the dewatering by pumping out the lake now filling his excavation. He did this on February 27, 1960, lowering the water level at the rate of about 1 ft (0.3 m) per hr.

Unexpected Slope Boils.—Between a water level of −6 ft to the final −14 ft, many small and unexpected boils occurred along the slopes at the level of the water in the excavation or at the toe of the slope when the water was below the toe. The contractor stopped each boil by surcharging with gravel borrowed from the top layer (the safety layer) of the gravel blanket. Most of these boils could be stopped by the shoveling action of one or two men. A few required the aid of mechanical equipment. The frequency of boil development gradually diminished as the water pressure under the slopes gradually diminished and no further slope boils developed after approx 24 hr from the start of dewatering. No boils developed inside the perimeter drain system, so the blanket remained intact under the foundation slab area. Fig. 7 also shows the dewatered excavation on February 28. The surface elevation of the gravel blanket had now been reduced to −11 ft to provide the surcharge gravel to stop the slope boils, reducing the thickness of the safety blanket 0.5 ft (0.15 m) and the margin of safety weight it provided.

The slope boils surprised everyone. They resulted from the most important mistake made implementing the gravel blanket scheme. It was intended that adequate safety to protect against the unexpected be provided by the extra, temporary safety weight gravel. This proved the case.

The boils suggested that an unexpected low permeability layer existed below the 1 ft (0.3 m) [reduced from 2 ft (0.6 m) because of short supply] of fine gravel filter covering the slopes, impeding the expected drawdown and pressure relief in the sand behind the slopes. When a boil started, it was generally first detected by the flow of a grey, silty suspension coming up through the brown-orange gravel. This observation suggested that a very silty and therefore low permeability layer existed near the undersurface of the slope-protecting gravel filter. A careful review of the boring logs and the samples showed no indications that such a silty layer existed anywhere in the sand before the excavation. Putting the evidence together suggested that the silty layer formed due to settlement of silty fines put in suspension by the agitation of underwater shaping of the natural fine sand to form the bottom and sides of the flooded excavation.

The contractor subsequently reported his skin divers had trouble seeing underwater during construction because of the fines in suspension and that he failed to clean out the excavation by periodic pumping-out of the silty suspension therein. Hindsight shows that such cleaning would probably have prevented most, if not all, of future difficulties. The writer also noticed a silty surface scum, perhaps 1/8 in. (3 mm) thick, on top of the gravel blanket after the dewatering, indicating there was considerable silt in the water at this time. If a silty layer existed for the aforementioned reason, then this explained the slope boils. Then similar low-permeability silty layer should exist under the blanket in the slab area. Unfortunately, this proved true.

Blanket Piezometer Tests (Step 6).—The excavation was not dry to the gravel at −11 ft, with the water at −14 ft (see Fig. 7). The slopes and bottom appeared stable and even supported a small bulldozer. The delighted contractor wanted to remove immediately the final 2.5 ft (0.76 m) of dewatered safety gravel. He had now been working on this dewatering job for 177 days after first expecting it to take only 15 days, and he had spent many times his estimated cost. With the prize seemingly so obviously within grasp he wanted to end this nightmare quickly. But it was only with great difficulty that he was persuaded to await the planned piezometer tests before removing the remaining safety blanket gravel.

The piezometer tests started that day, using wellpoints for piezometers, with all but 6 in. (150 mm) of screen length (No. 60 screenings) covered. The contractor took five piezometer readings in a set, one in the center of each quadrant of the slab area and one in the center of the entire slab. He then took sets at different elevations. Fig. 8 presents these piezometer test data (along with others that will be explained subsequently). In Fig. 8 the reader can compare these piezometer data with the maximum piezometric water level heights permitted at different depths for an effective stress of not less than 100 psf (4.8 kN/m^2) after removing the safety gravel blanket. From the very high hydraulic gradient under the gravel blanket, it was immediately apparent that a low permeability zone existed immediately below the gravel blanket and resulted in water pressures too great to permit immediate removal of the safety blanket. In fact, the contractor had a barely stable excavation bottom with the gravel at −11 ft! The minimum effective stress dipped to only 50 psf (2.4 kN/m^2).

The piezometer data indicated (see Fig. 8) the low-permeability layer only about 1.5 ft (0.3 m) thick, but the piezometer readings also indicated it had a permeability only 1/17 that of the underlying sand. It seems likely that this layer also resulted from the settlement and mixing of silt fines during the time the excavation was flooded and shaped and awaiting the start of the placement of the gravel blanket. As noted previously, this would also explain the silty layer and boil trouble with the slopes.

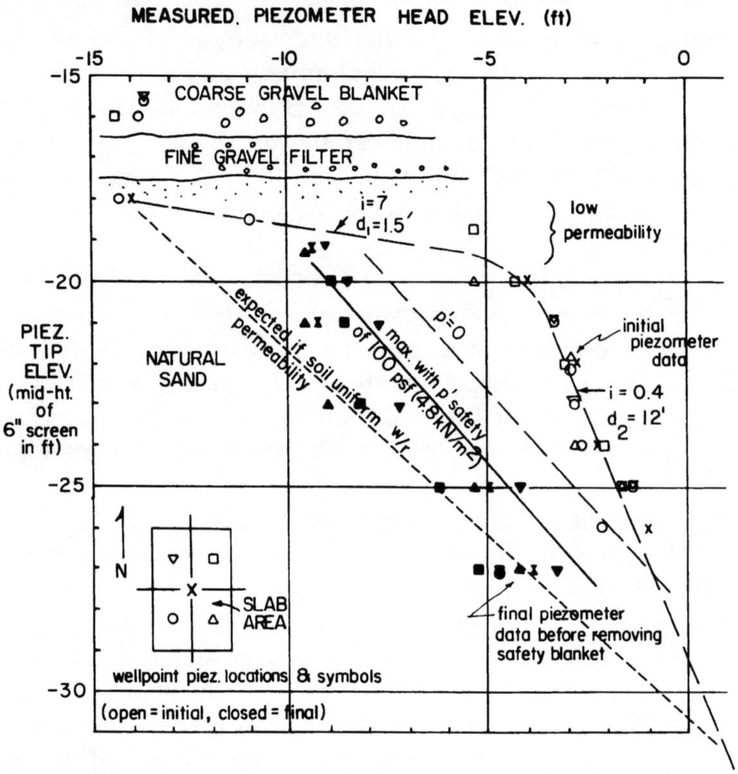

FIG. 8.—Actual and Required Piezometer Readings for Effective Stress Control in Natural Sand Layer

Introducing Unusual Wellpoint System.—To carry the gravel blanket scheme to completion a way had to be found to increase the permeability of the silty layer immediately below the gravel. The writer suggested a grid of self-pumping wellpoints with tip screens just below the silty layer and cutoff at El. −16 ft (within the gravel). The contractor tested this idea by installing four 2-in. (50-mm) diam wellpoints at the corners of a 10-ft (3-m) square, with a piezometer in the center. The results showed significant, but not enough, pressure relief under the gravel. A flownet study indicated that 50 points on 7-ft (2.1-m) centers over the slab area might do the job. The contractor installed these 50, but piezometer checks showed the pressure reduction was not quite adequate. He then placed an additional 70 wellpoints between the first 50. Piezometers then placed at the most unfavorable locations between wellpoints gave the data also shown in Fig. 8. These readings indicated that now a minimum 100-psf (4.8-

CO3 DEWATERING CASE HISTORY

kN/m²) effective stress would remain after safety gravel blanket removal.

Fig. 9 shows how this unusual wellpoint system looked and functioned. After installing these wellpoints the pumping rate from the horizontal 12-in. (300-mm) pipe drain system in the gravel increased approx 1,000 gpm (3.8 m^3/min) or 8 gpm (0.03 m^3/min) per wellpoint. This entire supplementary wellpoint system was below the final gravel surface and remained in place under the completed weir structure.

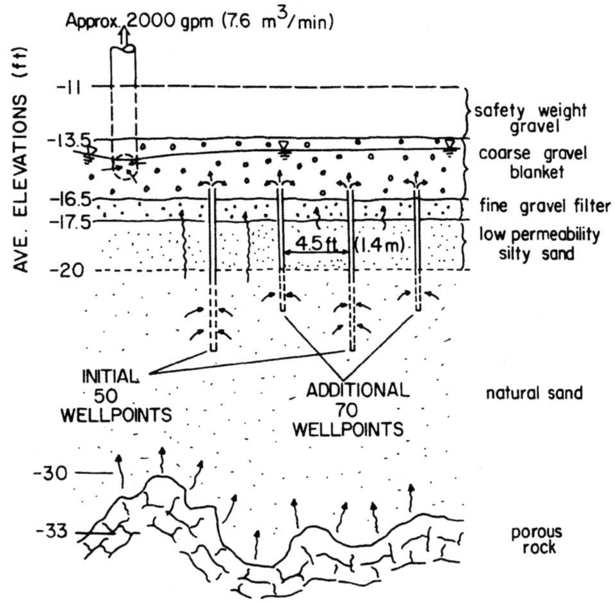

FIG. 9.—Self-Pumping Wellpoint Field Used for Pressure Relief under Gravel Blanket

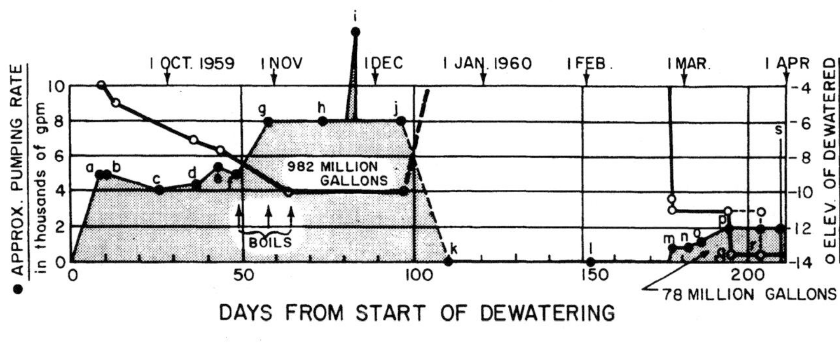

FIG. 10.—Summary of Job Pumping and Dewatering History

The contractor now drove the lines of cutoff sheeting and removed the safety blanket over the south half of the slab area and completed without incident his forming, steel layout, and pouring of the south half of the weir foundation slab. This took about 1 week. He then followed a similar procedure for the north half. However, while he was completing his forming on the third day after removing the north half safety blanket, a boil developed in the NW quadrant of the slab! The contractor finally overcame this last problem by a combination of a temporary surcharge of gravel, several additional relief wellpoints in the area, and allowing a few inches of water within the forms and pouring the final concrete into and displacing this water. To this point the excavation had remained stable for about 2 weeks after first removal of part of the safety gravel.

Looking at the data in Fig. 8 with the advantage of hindsight one can see that conditions under the NW slab quadrant were least favorable and marginal and a few more wellpoints should have been placed there before removing the safety blanket.

Once the contractor succeeded in placing the weight of the concrete slab he passed the most critical point in the pressure balancing. He had no further trouble with the dewatering and quickly completed the structure. The final cost of implementing the gravel blanket scheme reached about $60,000 or only $10,000 over the initial estimate and probably still only one-half that of a tremmied mat. To the writer's knowledge this foundation has performed successfully since 1960.

Summary and Conclusions

1. Fig. 10 summarizes by graph the history of pumping methods, pumping rates, and center-excavation water levels for this dewatering job. The history of dewatering at structure 37A is as follows: (a) Stage 1 (longer) completed, 220 wellpoints, tips at −30 ft; (b) ditched and backfilled upper limestone; (c) shortened stage 1, added 36 wellpoints, tips at −24 ft; (d) put in temporary stage just inside stage 1, 100 wellpoints, tips at −15 ft; (e) stage 2 completed, 50 wellpoints, tips at −24.5 ft; (f) removed temporary 100-wellpoint line; (g) stage 3 completed, 146 wellpoints, tips at −24 ft, consultant contacted; (h) seven new N-borings completed; (i) 12 in. well pumping tests; (j) decision to attempt gravel blanket scheme; (k) removed all wellpoints, flooded and started underwater excavation; (l) started filter and drain construction; (m) pumped flooded excavation to below top of gravel blanket (at −10.5 ft); (n) first tests for relief wellpoints; (o) initial 50, 2 in., No. 60 screen, relief wellpoints installed; (p) additional 70 relief wellpoints installed; (q) removed south half of safety gravel; (r) removed north half of safety gravel; and (s) foundation slab completed, critical point now past.

2. The attempt to dewater using wellpoints reached a maximum of 452 points pumping 8,000 gpm (30 m^3/min) without success. The contractor pumped about 982,000,000 gal (3,700,000 m^3) of water before abandoning the attempt with wellpoints. Dewatering the excavation would not have succeeded even with an infinite number of wellpoints pumping from the sand layer under the bottom of the excavation.

3. Additional borings, recharge tests, and a single and dual large diameter well pumping test, carefully monitored with piezometers, confirmed the pressure of a very high permeability limestone layer under the sand. They also showed that to relieve adequately the water pressure in the rock layer would require a prohibitively large and expensive well installation.

4. Careful application of the principles of soil mechanics and ground-water seepage, using the observational approach, permitted the use of a dewatering scheme, herein referred to as the "gravel blanket scheme," which did not require reducing the water pressure in the underlying rock.

5. Despite some unexpected problems, which were solved as they occurred, the gravel blanket scheme proved successful. Total dewatering pumpage was only about 1/13 that of the unsuccessful attempt with conventional wellpoints at a cost only about one-half that of the conventional alternative of sheet piling and a tremmied slab.

6. As shown by this case history of a small job, dewatering problems can involve considerable technical challenge and the application of soil mechanics principles can improve results and save money.

APPENDIX I.—REFERENCES

1. Cedergren, H. R., *Seepage, Drainage and Flow Nets*, John Wiley and Sons, Inc., New York, N.Y., 1967, pp. 174, 197, 455.
2. Peck, R. B., "The Observational Method in Applied Soil Mechanics," Nineth Rankine Lecture, *Geotechnique*, London, England, Vol. 19, No. 2, June, 1969, pp. 171-187.
3. Prugh, B. J., "Dewatering Miami's Biscayne Aquifer," *Journal of the Soil Mechanics and Foundations Division*, ASCE, Vol. 83, No. SM3, Proc. Paper 1299, July, 1957, pp. 1299-1-1299-15.
4. Shea, P. H., "Unusual Foundation Conditions in the Everglades," *Transactions*, ASCE, Vol. 120, Paper No. 2718, 1955, pp. 92-102.
5. Taylor, D. W., *Fundamentals of Soil Mechanics*, John Wiley and Sons, Inc., New York, N.Y., 1948, pp. 156, 130.
6. Terzaghi, K., *Theoretical Soil Mechanics*, John Wiley and Sons, Inc., New York, N.Y., 1943, pp. 257-262.
7. Terzaghi, K., and Peck, R. B., *Soil Mechanics in Engineering Practice*, 2nd ed., John Wiley and Sons, Inc., New York, N.Y., 1967, pp. 172, 60-62.

APPENDIX II.—NOTATION

The following symbols are used in this paper:

D = sand grain diameter as determined by sieving;
D_{10} = that value of D for which sand contains 10% smaller than D, by weight;
d_1 = thickness of silty sand layer at surface of natural sand in flooded excavation;
d_2 = $L - d_1$;
H = head of water, expressed as vertical distance between water level in piezometer and arbitrary reference elevation datum;
$i = \Delta H/\Delta L$ = hydraulic gradient (dimensionless);
i_v = component of hydraulic gradient acting in vertical-up direction;
$\bar{i}_v$ = average i_v over ΔL;
k = soil mechanics coefficient of permeability;
L = length of shortest vertical flow path from limestone source, through natural sand to gravel blanket (see Fig. 5);
p' = effective vertical normal stress;
ΔH = loss in head between two points on flow path; and
ΔL = flow distance between two points on flow path.

JOURNAL OF THE GEOTECHNICAL ENGINEERING DIVISION

Slope Effect on House Shrink-Swell Movements

By John H. Schmertmann,[1] F. ASCE and David K. Crapps,[2] M. ASCE

Introduction

Many houses in Gainesville, Fla. have experienced damaging differential vertical movement of their footing foundations placed over a calcium montmorillonite clay. This clay can, and does, both shrink and swell depending on many site, construction, and weather variables. During their engineering investigations of some of these damaged houses the writers have noticed a preference for the upslope part of a house to move upward relative to the downslope part of a house, and a trend showing the relative magnitude of differential movement increases with the steepness of the slope. This paper offers some evidence to document these findings and suggests a reason for this behavior based on the interrupted natural drainage of the slope.

Gainesville Clay

The subject clay belongs to the Hawthorn Formation, a miocene, probably alluvial deposit found in much of north central and central Florida. Sowers and Kennedy (5) include this clay in their general description of active clays in the southeastern part of the United States. It occurs close enough to the ground surface to affect house foundation behavior in the western half of Alachua County, and extends into Columbia County (Lake City, 40 miles north of Gainesville), and into Marion County (Ocala, 35 miles south of Gainesville). Depths to the top surface of this clay of up to 6 m (20 ft) have caused detrimental movements. The Ocala limestone bedrock occurs at depths of about 5 m–15

Note.—Discussion open until May 1, 1981. To extend the closing date one month, a written request must be filed with the Manager of Technical and Professional Publications, ASCE. This paper is part of the copyrighted Journal of the Geotechnical Engineering Division, Proceedings of the American Society of Civil Engineers, Vol. 106, No. ST12, December, 1980. Manuscript was submitted for review for possible publication on October 2, 1979.

[1]Principal, Schmertmann & Crapps, Inc., Gainesville, Fla.
[2]Principal, Schmertmann & Crapps, Inc., Gainesville, Fla.

DECEMBER 1980 GT12

m below ground surface. The piezometric water level lies within this rock, but a perched water condition can occur locally and temporarily over the clay.

The first writer has described some of the ordinary soil classification parameters of this clay, as well as some of its swelling properties (4). Fig. 1 shows the Atterberg limits from this clay from 24 sites, 19 in the Gainesville area, three in Lake City, and two in Ocala, plotted in relationship to the Casagrande A-line. When relatively free of sand the clay often has a plasticity index of over 100%. However, one often finds it mixed with sand pockets, lenses, and sand-filled cracks, which reduce its overall plasticity, and increases its mass permeability. X-ray and other studies show calcium and magnesium montomorillonite for over 90% of the $-2\ \mu$ fraction. The clay has a combined Ca and Mg base exchange capacity of about 50 meq/100 g. The -2μ fraction typically constitutes 60%–80% of the grains, by weight, and this fraction has an activity between 0.8 and 2.1. The natural water content of the clay usually falls between 35%–70% and liquidity indices between −0.1 and +0.3. The shrinkage limit of undisturbed samples ranges from 15%–25%.

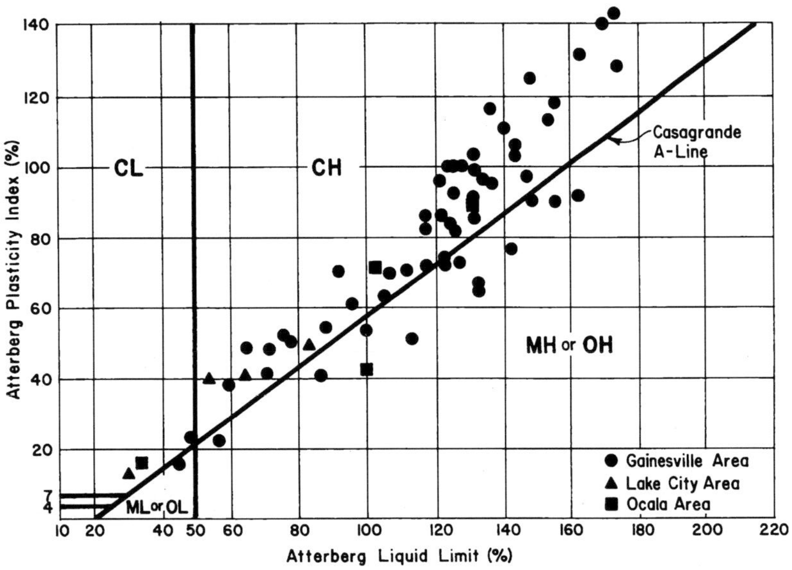

FIG. 1.—Plasticity Test Results from 24 House Sites over Gainesville Clay

The clay typically occurs in relatively thin layers and pockets, often irregularly interspersed with sand and clayey sand. It usually, but not always, has a sand or clayey sand cover, or both. Water accounts for about 60% of the volume of the clay and it has a large shrink and swell potential. Fig. 2 shows an idealized typical situation of a sand cover over a 5-ft (1.5-m) layer of the clay and shows how much the top surface of the clay could move up due to swell or down due to shrinkage and thus also shows its potential for causing vertical wall movements.

The writers have measured average vertical heave movement rates of one story concrete block walls as high as +2 mm/day, +0.8 mm/day, and +0.4 mm/day over 1-day, 10-day and 50-day intervals, respectively, and settlement rates as high as −0.6 mm/day and −0.5 mm/day over 10-day and 50-day intervals respectively—all probably due only to the swelling and shrinking of the underlying Gainesville clay. The +2 mm/day, +0.8 mm/day, and −0.5 mm/day occurred at the same house during, respectively, very wet weather with 80 mm of rain in 1 day and 160 mm in 10 days, and drought with only 6 mm of rain in 50 days. Thus, Gainesville engineers, contractors, and house owners must deal with an active clay that can either cause large swelling heaves during periods of very wet weather or if one disturbs its equilibrium in the direction of increased supply of water, and large drying settlements during droughts or if one disturbs its equilibrium in the direction of drying. For reference, Gainesville has a 70-yr

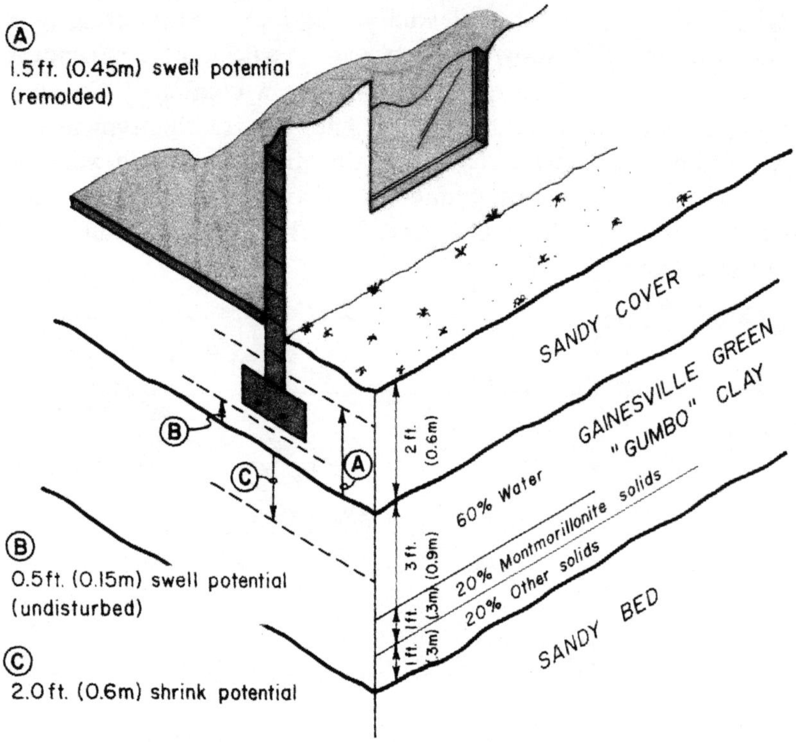

FIG. 2.—Idealized Example of Vertical Movement Potential of Top of 5-ft Layer of Gainesville Clay

average annual rainfall of 55 in. (1,400 mm) with monthly extreme averages of 2 in. (50 mm) in November, and 8 in. (200 mm) in August.

Measuring Apparent Accumulated Vertical Movement of Houses Founded Over Gainesville Clay

The typical, nonengineered, house construction observed in the Gainesville area involves unreinforced, or only lightly reinforced, concrete block (CCB) and sometimes brick walls carried by continuous wall footings of poured concrete typically 18 in. (450 mm) wide and 8 in. (200 mm) deep with two longitudinal #5 steel reinforcing bars. The masonry wall begins directly over the center of the 8-in. × 18-in. rectangular footing. The depth to the bottom of this footing usually ranges between 1 ft–2 ft (0.3 m–0.6 m). Roof trusses usually span to the outside walls, with interior bearing walls usually only where a multilevel house floor system changes floor level. When the ground slopes the wall footings going down the slope usually have vertical steps. The contractor typically places a 4 in. (100 mm) concrete slab between the outside walls, unreinforced except

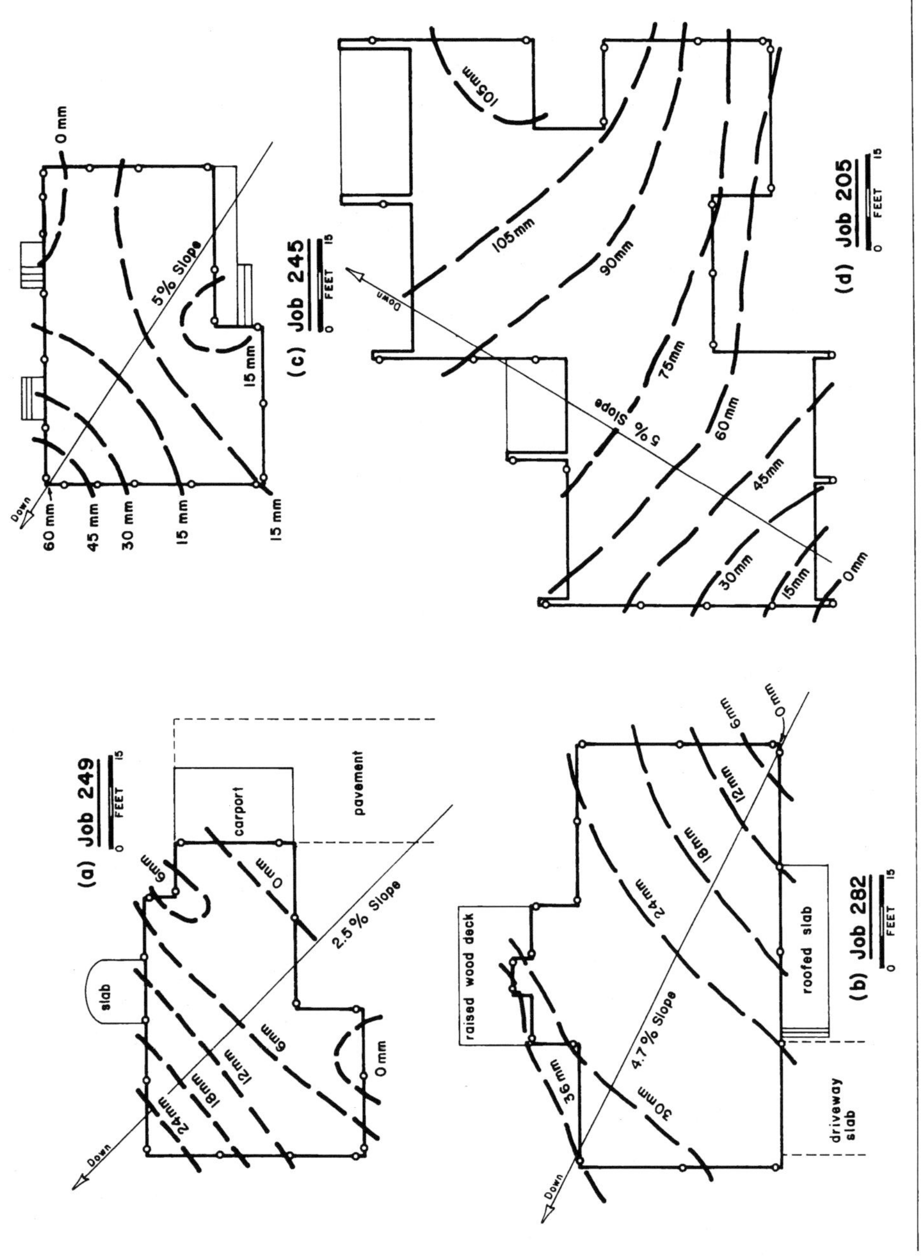

FIG. 3.—Plan of Houses in Gainesville Showing Contours of Vertical, Post-Construction, Wall Footing Movements Arbitrarily Shown as Settlement from Highest Point

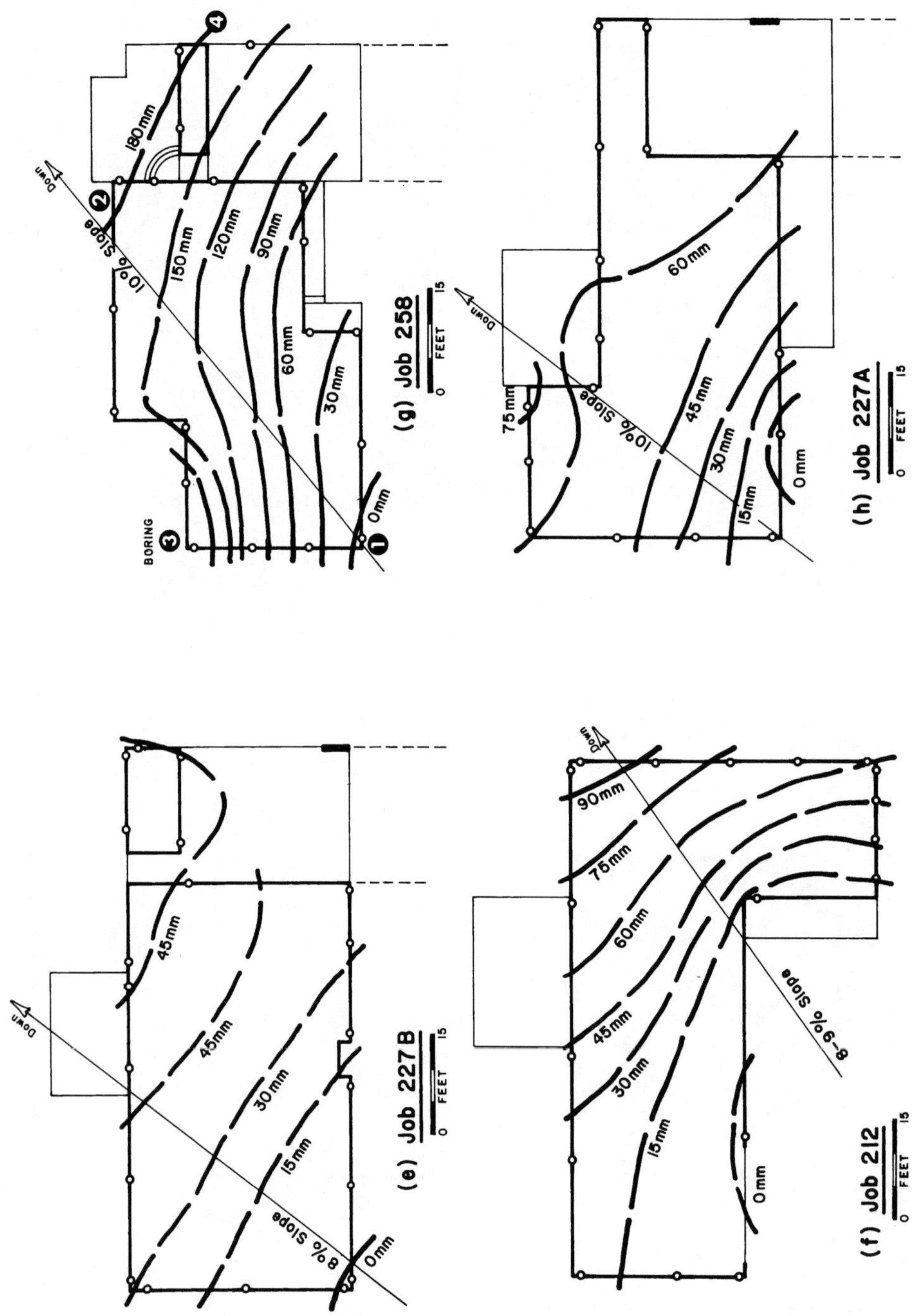

FIG. 3.—*Continued*

for a light wire mesh, and poured over compacted sand and a plastic moisture barrier and directly against the CCB walls with no special precautions either to insure slip or to insure bonding with the concrete block.

Contractors use a tightly stretched, horizontal cord to control the laying of

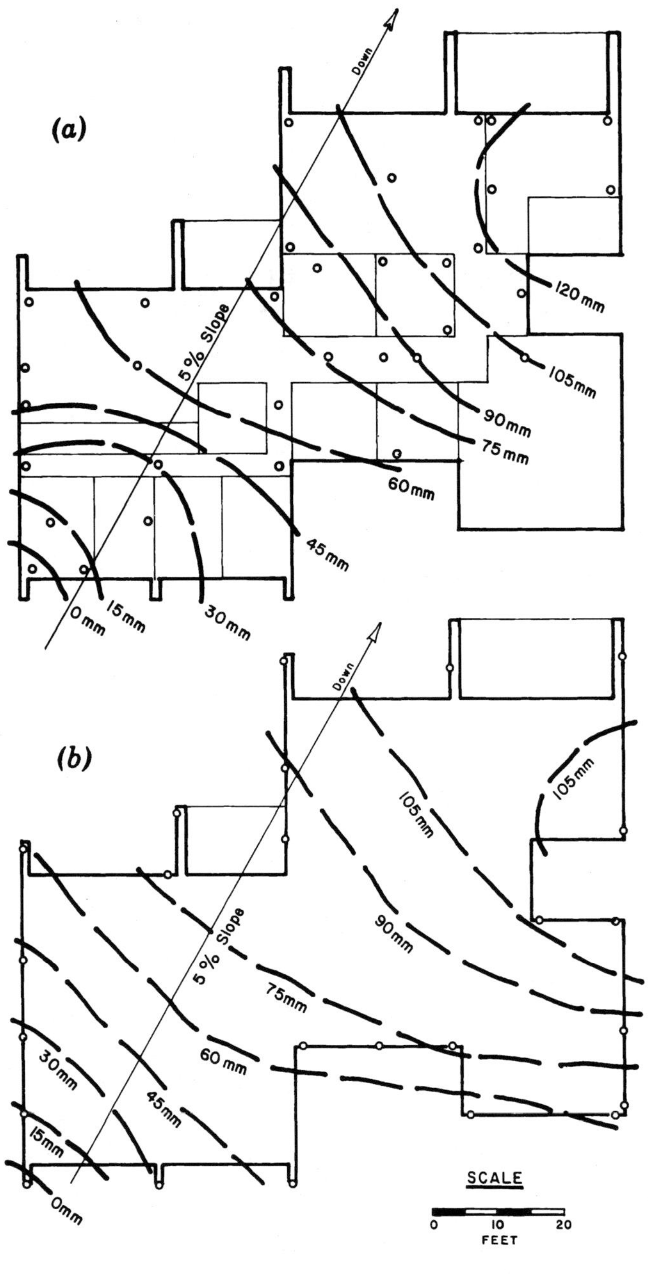

FIG. 4.—Example Comparison Between Vertical Movement Contours Based on: (*a*) Interior Slab-on-Grade Measurements; (*b*) Exterior Wall Joint Measurements

the longitudinal mortar joints in CCB or brick walls. M. A. Johnson (1,2), and the writers made check measurements in 1978 of the horizontal accuracy to which local contractors load the block or brick joints on 11 houses, one apartment building, and one office building under construction in Alachua County. The number of points checked around the outside walls of a structure varied from 16–26, with an average of 20. The difference between the highest and lowest points on opposite walls in each of these surveys varied from 0.29 in. (7.4 mm)–1.46 in. (37.1 mm), with an average of 0.73 in. (18.5 mm). The maximum mortar joint elevation gradient between opposite walls varied from 0.77–3.85 in./100 ft (19.3 mm/30 m–96.1 mm/30 m), with an average of 1.69 in./100 ft (42.3 mm/30 m). When comparing with the lowest point from a survey, the average mortar joint elevation difference equaled approx 0.30 in. (8 mm), with a standard deviation of approx 0.25 in. (6 mm). Approx 95% of the points fell within an elevation difference of ±0.4 in. (10 mm) from the average. These data did not suggest any obvious correlation between, or pattern of, construction deviations from horizontal and the magnitude or direction of the slope of the natural ground around and under the house.

The preceding background knowledge helps permit an estimate of the amount of post-construction differential movement that has taken place in a house. The writers choose the lowest convenient horizontal mortar joint in the masonry walls and obtain comparative elevations on this joint around the house perimeter. Comparing these elevations permits the drawing of contours of equal elevation, after correcting for any changes in the joint surveyed because of changes in joint necessitated by the slope of the ground surface surrounding the house. Then, by comparing these elevation contours, and keeping in mind construction tolerances, the writers estimate the amount and pattern of relative heave or settlement, or both, that had taken place since construction. Fig. 3(*a–h*) present eight examples of the plan of a house with indicated contours of relative settlement, in millimeters, with the highest contour as the arbitrary reference. Ground slope steepness increases from Fig. 3(*a–h*).

Each example in Fig. 3 shows the plan location of the level measurement points. The reader may wonder why the contours shown cut across the center of the house when based on comparative elevations only around the outside walls. The writers justify this because their experience with six comparisons has shown that relative elevations on the interior floor slab over the house area usually produce elevation differences that nearly match those determined from only the wall perimeter data. Two of these comparisons involved houses shown in Fig. 3(*d*) and Fig. 3(*g*). Fig. 4(*a*) shows the house from Fig. 3(*d*) with contours of relative movement based only on comparative elevations on the slabs at the points shown. Fig. 4(*b*) for convenience repeats Fig. 3(*d*). Comparing the two sets of contours in Fig. 4 will show their similarity

The Fig. 3 examples clearly show a definite pattern of apparent increasing differential movement as one moves across a house and down the dip of the original slope, the magnitude of which usually greatly exceeds the aforementioned observed construction tolerances.

Relationship Between Ground Slope and Apparent Vertical Movements

Each of the plans in Fig. 3 also shows the approximate dip direction and magnitude of the slope of the natural ground surface on which the contractor built the house, as a percentage. Note that all the contours of equal relative movement occur approximately perpendicular to the line of slope dip, and that the downslope part of the house has apparently moved downward relative to the upslope part. Note also that the magnitude of the post-construction relative differential movement increases with increasing magnitude of the slope.

The writers prepared Fig. 5, which shows most of the available points (as explained subsequently) relating to the slope of the natural ground surface,

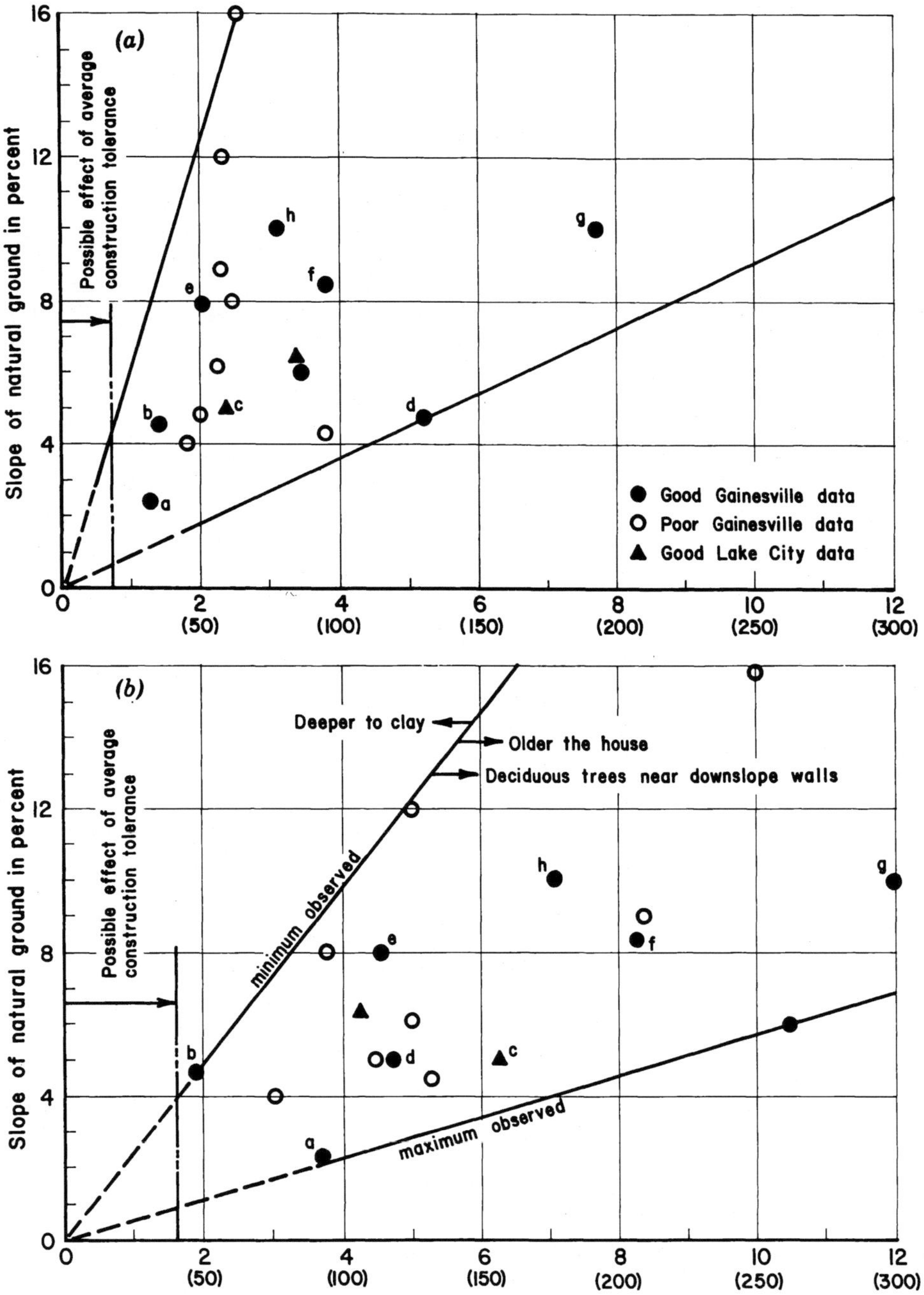

FIG. 5.—Effect of Natural Ground Slope on Apparent Maximum Differential Vertical Movement of House Walls over Gainesville Clay: (*a*) Movement Between Upslope and Downslope Masonry Walls, in inches (millimeters); (*b*) Gradient Between Upslope and Downslope Masonry Walls, in inches/100 ft (millimeters/30 m)

as a percentage, and the maximum apparent differential vertical movement between upslope and downslope walls since construction. In all cases the downslope part of the house experienced a relative settlement or the upslope, a relative heave. These data alone do not permit determining which. Fig. 5(*a*) shows the maximum apparent differential movement between the upslope and downslope walls of a house, measured in the approximate direction of the dip of the slope. This figure shows considerable scatter—e.g., a 2-in. (50-mm) differential movement could occur with slopes ranging between 2% and 13%. In an effort to reduce the scatter by normalizing, the writers plotted slope versus the maximum gradient of apparent movement in Fig. 5(*b*). They measured the maximum apparent differential movement between walls again approximately along the dip of the slope and then divided this movement by the distance between the wall points, in feet and multiplied the result by 100 ft to give the gradient. This method of plotting should more clearly show the effect of slope. It provides a method for compensating differences between large and small houses and between the same size houses positioned downslope in their long direction compared to others positioned downslope in their short direction.

Fig. 5 also includes the average construction tolerances noted previously. The figure shows these as positive for possible direct subtraction from the measured apparent movements to give a net, after construction movement. However, because the construction tolerances did not seem associated with slope, they could also show as negative and add to the measured movements. For all the cases presented in Fig. 5 the measured apparent movements exceed the average construction tolerance and thus strongly suggest that post-construction movements of the type described actually did occur.

Fig. 5(*b*) seems to reduce the scatter and shows more clearly the trend of increased differential movement gradient with an increase in natural slope gradient. The letter identification adjacent to some of the points in Fig. 5 matches the cases in Fig. 3 and allows the reader to check how the Fig. 3 data translated to the Fig. 5 points. In addition, the solid points indicate 10 houses where the writers obtained relative house movement data that showed the entire house involved with the apparent vertical movement gradient in approximately the same direction as the dip of the slope. Fig. 3 gives eight examples of such data. The open points in Fig. 5 present data from eight additional houses where the data required more interpretation. For example, perhaps only a part of the house settled in the pattern expected, or other factors entered into the clay behavior and distorted the apparent movement pattern. Nevertheless, the eight houses that produced such more confused data still fit well within the overall pattern suggested by the 10 houses that produced less ambiguous data.

The writers also had available six houses, in addition to the aforementioned 18, in their pool of observed house-on-slope-with-clay movement data wherein the relative movement did not seem to correlate well with the dip of the slope. Two houses even appeared to move counter to the trend shown in Figs. 3–5. Of course, many other factors, such as those noted subsequently, can influence the position of a point on Fig. 5. Furthermore, these and other factors can produce a differential vertical movement across a house set on level ground, or overwhelm the effect of only slope and tend to conceal from or confuse the engineer as to the effect on only slope.

Although not documented herein to avoid confusing the main theme of this paper, the writers' experience indicates the following as some of the more common factors that can influence the position of a house point in Fig. 5. The deeper the clay layer, or the thinner the layer, the less the differential movement for a given ground slope. The older the house the more opportunities it has had in its history to have the underlying clay shrink or swell because of weather fluctuations, and therefore the higher the probable differential vertical movement for a given slope.

Trees can have an important and sometimes dominant effect on house vertical movements over clay, regardless of the slope. During times of drought the root systems of leafy trees and other large plants will take water out of the Gainesville clay and thus, produce settlement. Such trees; notably sweetgum, various oaks, hickory, magnolia, and ligustrum, when near the downslope walls of a Gainesville house can increase the movement or gradient when plotted in Fig. 5. Ten of the 18 houses in Fig. 5, including five of the solid points, had such downslope trees. Such trees when near the upslope walls can decrease the movement or gradient in Fig. 5. Also, the removal of trees before house construction can cause the house to heave as the clay swells in rebound response to the removal of the tree root suction (the writers have measured several cases of a response heave of about 2 in. (50 mm) in less than 1 yr after removing nearby trees). Thus, the presence or removal of trees and large plants can distort the house vertical movement pattern that might result from only the effect of ground slope.

Considering the aforementioned, and many other possible conditions that effect local shrink and swell movements, the trend shown in Fig. 5 seems quite strong and in the opinion of the writers reflects a major influence of the slope itself on such movements. The next section hypothesizes about why this might, or even should, occur.

Hypothesis for Differential Movements Related to Ground Slope

As part of their engineering studies of houses on the Gainesville clay, the writers have made some long-term (more than 6 months) precision level measurements, always using a special benchmark into rock for each house, to discover the pattern of the continuing movements of houses prior to selecting remedial measures. These measurements involve a precision of 0.1 mm (1/250 in.) and an accuracy of about ± 0.3 mm (1/100 in.), and thus, permit determining settlement or heave movements over periods of time as short as a few days or a week. The reader should not confuse these level measurements with the less precise data (about $\pm 1/10$ in. or 2 mm) used for the Fig. 3 comparative mortor joint elevations made without a special benchmark. Inspection of the precision level data helped develop the writers' understanding of the general effect of slopes on movements. We noticed that upslope walls usually moved in the heave mode during normal weather, heave at a reduced rate during moderate droughts, or even in the settlement mode during severe droughts—but at a rate of less than the downslope walls, thus maintaining the relative heave movement. Downslope walls usually moved in the settlement mode, although they sometimes heaved slightly during periods of very wet weather—but at a rate less than the heave of the upslope walls. Such observations, together with the apparent quantitative effect of slope shown in Fig. 5, provide the basis for the hypothesis presented in this section.

An active clay will shrink when and where the house construction produces increases in effective stress compared with the previous natural effective stress with which the clay maintained an equilibrium. Heave results from construction that decreases effective stresses. Changing pore water suction pressures represents one way to change effective stresses. For example, increasing water availability will decrease suction and allow clay swell. Decreasing water availability will increase suction and induce clay shrinkage. The writers believe that the typical Gainesville area house construction on a slope increases the upslope water availability, producing heave of the upslope walls, and reduces the downslope availability and thus produces settlement of the downslope walls. The upslope heave combined with the downslope settlement produce the total vertical differential movement apparent in data such as shown in Fig. 3.

To produce an approximately level area for the house the Gainesville area

FIG. 6.—Examples of Upslope Cut and Downslope Fill Method of Construction often Used by Contractors to Accommodate House to Slope: (*a*) Upslope Cut into 10% Slope; (*b*) Upslope Cut into 5% Slope; (*c*) Fill under Upslope House and Cut under Downslope House, Houses 24 ft (7 m) Apart of 8% Slope; (*d*) Example of Downslope fill with House on 5% Slope

contractor without engineering advice ordinarily excavates for the upslope part of the house and fills in the downslope part. Fig. 6 shows photo examples of such typical upslope cuts and downslope fills. As, e.g., Jones (Ref. 3, p. 224) clearly points out, "Structures should be located to have consistent depth of either cut or fill beneath them, and never upon mixed cut and fill." But, a sloping lot makes this difficult to achieve economically, even if the contractor knew of the underlying clay problem. The writers know of no local houses, built on slopes, that have a consistent depth of either cut or fill.

The aforementioned cut and fill can produce a major clay reaction. Consider Fig. 7 which shows the cross section of a house constructed on a typical slope, in this case underlain by an active clay. The excavation prior to constructing the upslope walls, together with the additional excavation required to build the wall footings, produces a complete or partial dam tending to prevent or reduce the natural downslope movement of the infiltrated rain water over the surface of the clay. Rain water from the roof also tends to pond behind the upslope walls. This would produce an increase of water available to the clay in the vicinity of the upslope wall footings and tend to cause the underlying clays to swell and the footing to heave. Because of greater water availability and greater stress relief, the clay and its overlying footing would heave more outside than inside the walls and thus also tilt the walls downslope.

The downslope walls have footings placed much higher above the clay, possibly even in the fill, and water can more freely flow under this footing. However, the amount of water available for flowing across the underlying clay has now

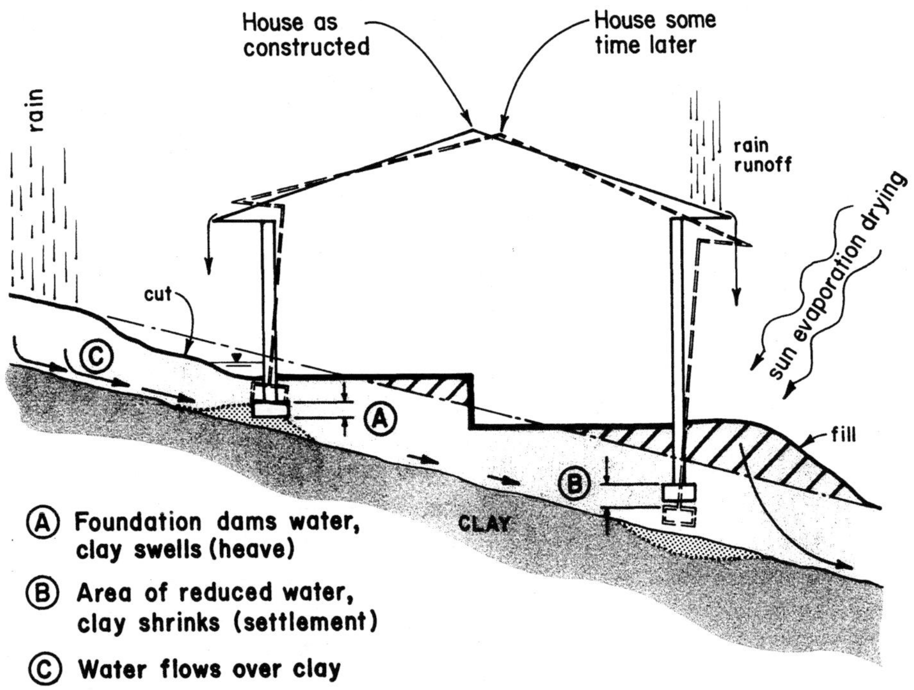

FIG. 7.—Hypothesis for Heave, Settlement, and Wall Tilt Tendencies of House Built on Slope with Shrink-Swell Clay

decreased because of the upslope damming action, which tends to force the surface water around the sides of the building. Also, the roof catches water that previously reached the covered area and could have flowed across the clay, but which now gets dumped downslope and does not easily flow back under the footing. The clay under the downslope footing thus may experience a reduction in available water and as a result shrink and cause the downslope walls to settle. Greater shrinkage can occur outside the footings compared to inside and thus these walls also tend to tilt downslope.

The aforementioned changes in the slope drainage produce a net effect of differential vertical movement of relative settlement down the dip of the slope and across the house, as measured and described by Figs. 3–5. The writers also commonly measure the described downslope tilt of the house. Typical wall tilt values range from 1–4 horizontal to 200 vertical.

Water Content Check of Proposed Hypothesis.—In early 1980 the writers had the opportunity to make a water content check of the proposed hypothesis at one site in Gainesville. This involved the house shown in Fig. 3(*g*) which also shows the locations of our splitspoon borings around the house. The driller obtained continuous samples over the 0 ft–12 ft (3.7 m) depth interval and saved the samples in 6-in. (150-mm) increments. The writers then inspected the samples and obtained a representative water content from each 6 in., as well as less frequent but also representative Atterberg limit tests. From this information it seemed apparent that a typical Gainesville clay layer of from 5.5 ft–7 ft (1.7 m–2.3 m) thickness existed within the surface 10-ft (3-m) depth and that the past shrink-swell behavior of this layer most likely caused the damaging differential movement of this house. Movement surveys of settlement plates placed at 5-ft (1.5-m), 10-ft (3.0-m) and 15-ft (4.6-m) depths between borings 2 and 3 showed that over 90% of the movement took place above the 10-ft (3.0-m) depth.

At the time of testing the house had existed at the site for 13 yr, and had experienced approx 8 in. (200 mm) of differential movement. If the writers have the correct hypothesis, and if the clay had the same water content distribution at all points under the house before building the house, then the present upslope water contents should exceed the present downslope water contents. We therefore compared the average water content in the clay layer at each of the four locations. We also compared the Atterberg limit data to determine if we probably had the same clay at all four locations. Table 1 summarizes these data and gives the averages, the number of tests involved in each average, and some standard deviations.

Note that Table 1 lists the borings in order of progressively greater downslope position around the house. One can see immediately that the average water content of the clay layer decreases steadily when moving downslope, with a surprising approx 4% decrease for each 1-ft (0.3-m) decrease in elevation. This checks the hypothesis prediction. On the other hand the average liquid limits vary from only 118.6%–122.1%, with no downslope order. Visual comparisons and these limit data indicate we have the same clay at all four locations. The plasticity index and therefore also the plastic limit data show more variation in the averages but the writers believe that this results primarily from differences in plastic limit technique between two inexperienced student technicians.

The reader can also check to see that the average water content differences listed in Table 1 can, when distributed over a 5.5-ft–7-ft layer of assumed saturated clay, more than account for the measured apparent differential vertical movements of house Fig. 3(*g*). We need no other explanation than that of induced water content differences to reach and even exceed the magnitude of the apparent movements.

We believe that this field check supports the proposed hypothesis, but we also must admit that trees near the downslope walls of this house probably caused some of the water content decrease.

Possible Factors Other than Drainage and Water Availability.—Some construction factors other than drainage and water availability could also, but probably did not, contribute significantly to the type of differential movement noted in Figs. 3–5.

DECEMBER 1980 GT12

TABLE 1.—Comparative Test Data from Four Borings Around House in Fig. 3(g)

Boring	Comparative surface elev.	clay depth (thickness) ft	average *% (± one standard deviation) No. tests: natural water content w	liquid limit LL	Plasticity Index PI
1	reference	0.5 - 7.5 (7)	62.5 (53.3 - 71.7) 13	121.8 5	Student A 66.8 5
3	-1.5 ft	4.5 - 10 (5.5)	54.6 (43.2-66.0) 11	118.6 5	A 64.4 5
2,2R	-4.0 ft	2-7.5 (5.5)	46.3 (33.8-58.8) 15	122.1 8	A 62.8 4 Student B 85.2** 4
4	-6.0 ft	2.5 - 8.5 (6)	39.5 (35.1-43.9) 12	120.6 4	B 78.0** 4

*All test samples dried at 110° F due to student error. All values too low compared to drying at std. 105° C, but acceptable for comparison purposes.

**Probably reflects inexperienced, A and B grad student's different technique for PL rather than difference in clay.

GT12 SLOPE EFFECT

The upslope cut and downslope fill method of accommodating a house to a slope will usually result in a reduced thickness of sandy soil or fill between the footings and the active clay under upslope compared to downslope footings. The reduction in overburden pressure upslope would encourage swell and the increase in downslope overburden pressure would restrain swell. This behavior would then contribute to the reported upslope-downslope differential movement, as further detailed in the following.

The writers have measured the heave movements of 5-ft, 10-ft, and 15-ft (1.7-m, 3.0-m, and 4.6-m) deep settlement plates placed into undisturbed clay and 1 ft–5 ft (0.3 m–1.7 m) from the CCB walls on three Gainesville houses on which we also obtained vertical wall movements. Our analysis of these field data indicate a swelling pressure of about 1 ton/sq ft (95 kPa) for the clay at a 15-ft–20-ft (4.5-m–6.1-m) depth. They further indicate that for the case with 3-ft (0.9-m) more downslope sandy soil cover over the clay than upslope cover, or about the maximum we have observed, that the upslope clay will swell about 50% more than the same thickness of the clay downslope. This suggests that this behavior could contribute in an important way to the upslope-downslope differential movement described previously. However, it requires water availability to produce heave and the relative water starvation often causes downslope drying and settlement rather than swelling and heave. Also, the artificial introduction of water near the downslope walls almost always causes them to heave and thus confirms their water-starved condition.

If the upslope excavation entered the clay the associated disturbance of the swell-sensitive clay structure could greatly increase the swell of the disturbed clay (4) compared to the undisturbed clay under the downslope part of the house. However, the Gainesville and Alachua County building inspectors usually discourage footing construction directly on the clay and the writers have found few such cases. The remaining downslope cover of possibly loose natural sand over the clay could cause compressibility settlement of the downslope footings. Also, uncompacted or poorly compacted fill under these footings might also produce such settlement. However, the building inspectors rarely allow footings on unengineered fill, and usually require compaction of the natural sands under the footings. The writers have found few such possible instances of settlement of downslope fill or sand (using hand cone penetration test and auger sampling methods to estimate compressibility). The clay itself has a high preconsolidation stress (4) and will not consolidate significantly under house footing stresses.

The writers believe that the aforementioned factors usually did not contribute significantly to the observed house differential movements and that the effects of drainage and water availability dominate in the explanation of how Gainesville area slopes contribute to these movements.

PRACTICAL APPLICATIONS

This paper demonstrates, to the writers' knowledge for the first time in the geotechnical literature, how house construction on a slope that includes shrink-swell clay can induce a damaging upslope heave or a downslope settlement, or both, and that the relative movement increases with the magnitude of the slope. The writers suggest that this house distortion pattern results primarily from the changes in near-surface drainage and water availability that, at least in the Gainesville, Fla. area, typically results from the cut and fill site preparation that contractors use to accommodate the house to the slope. The writers have described Gainesville area active clay, weather, and construction conditions. The reader must judge if the phenomenon described applies to his or her environment and house construction methods.

The paper also includes a description of a method the writers developed and found very practical and useful to estimate the magnitude and pattern, but not the direction, of relative movement that a house has experienced. The method uses comparative elevations along horizontal mortar joints in masonry walls, on floor slabs, or both, after calibration for area construction tolerances.

Once engineers recognize the slope behavior described herein they will doubtless devise methods to eliminate or reduce damage from this cause. Although it falls beyond the intended scope of this paper, the interested reader may appreciate a few comments concerning the writers' experience dealing with this problem. This experience involves almost exclusively houses from 5 yr–30 yr old where the owners wished to repair present and prevent future crack damage—at an affordable price.

First, we can offer some comments about what did not solve the problem. About 50% of the houses investigated by the writers had previous footing underpinning work done by contractors—almost always without engineering recommendation or guidance. Such underpinning also almost always made the problem worse because of local clay and drainage disturbance around the added, deeper footings. Such underpinning failed because it did not address the basic problem—the clay.

The writers have had encouraging success with improving upslope drainage and wetting downslope areas, and with the selective removal of trees and other large plants. We have not yet tried to isolate the clay under a house by perimeter horizontal or vertical mechanical moisture barriers, nor have we had the opportunity to try chemical treatments such as lime slurry injection. The Gainesville clay reacts favorably with lime and we hope to try lime columns and injection in the near future.

CONCLUSIONS

The writers offer the following conclusions from their studies on Gainesville, Lake City, and Ocala, Fla. houses built over a near-surface, active clay layer and on natural ground slopes to 16%:

1. Comparative elevations of nominally horizontal mortar joints in brick or concrete block walls, or the surface of slabs, or both, can provide useful information about the magnitude and pattern of post-construction vertical movements. Determining the direction of such movements requires additional data.
2. One must consider typical construction tolerances, about ±0.4 in. (100 mm) for recent Gainesville construction, when interpreting the aforementioned comparative elevations.
3. The writers hypothesize that houses build on slopes over clay that can both shrink and swell, and with cut and fill to accommodate house construction to the slope, tend to swell on their upslope walls and settle along their downslope

walls. This probably occurs primarily because of associated changes in near-surface drainage and consequent water availability, and secondarily because the upslope cut encourages more clay swell than the downslope fill. Other factors, such as nearby trees, can distort and confuse the movement pattern of a house over clay and on a slope.

4. Comparative clay layer water contents at one house location on a lot with a 10% slope showed a 4% decrease in average clay layer water content for each 1 ft decrease in elevation around the house. This large difference tends to support the proposed hypothesis.

5. Houses subjected to the slope related vertical movements also have a tendency to tilt down hill.

6. The present evidence suggests that an increased magnitude of natural ground slope tends to increase the magnitude of the aforementioned differential movements.

7. Underpinning an existing house with footings does not solve the differential movement problem resulting from the slope and the clay. The writers have had some success with modifying drainage and with the selective removal of trees.

Acknowledgments

While all of the case histories and data therefrom presented herein came from the writers' consulting work over a period of about the last 5 yr, the writers also wish to acknowledge important help from the University of Florida's Department of Civil Engineering. John Davidson acted as the Principal Investigator on two Engineering and Industrial Experiment Station sponsored projects, W-18 and W-32, that included measuring house movements due to shrink-swell clay behavior. He supervised graduate students, M. Johnston, and Y. Taesiri, who did their masters research on these projects. W. J. Whitehead, Assistant in Civil Engineering, provided important assistance in the field work, such as soil sampling and the placement of the special benchmarks for precision leveling. D. Earl Jones, Jr., Chief of the Planning & Engineering Branch, Housing and Urban Development (HUD), Washington, D.C., kindly reviewed this paper and made many useful suggestions intended to add to its possible practical value. Richard Furman prepared the figures for publication. The Gainesville office of Universal Engineering Testing Co. contributed the borings shown in Fig. 3(*g*) and Table 1.

Appendix.—References

1. Davidson, J. L., "Shrink-Swell Clay Damage to Residential Structures, Preliminary Study," *Research Report 245*W32*, University of Florida Engineering & Industrial Experiment Station, June, 1979.
2. Johnson, M. A., "The Identification of Sinkhole Related Damage to Residential Structures," thesis presented to the University of Florida in 1978, in partial fulfillment of the degree of Master of Engineering.
3. Jones, D. E., Jr., "The Expansive Soil Problem," *Underground Space*, Great Britain, Vol. 3, No. 5, 1979, pp. 221–226.

DECEMBER 1980 GT12

4. Schmertmann, J. H., "Swell Sensitivity," Technical Note, *Geotechnique*, Dec., 1969, pp. 530–33.
5. Sowers, G. F., and Kennedy, C. M., "High Volume Change Clays of Southeastern Coastal Plain," *Proceedings of the 3rd Pan American Conference on Soil Mechanics and Foundation Engineering*, Vol. II, 1967, pp. 99–120.

DILATOMETERS SETTLE IN

As U.S. use increases, this in situ testing tool imported from Italy is praised as being fast, accurate, and economical.

JOHN H. SCHMERTMANN

Skepticism was my reaction when I heard about the dilatometer test DMT some ten years ago. This geotechnical in situ testing tool was developed and patented by an Italian professor and engineer named Silvano Marchetti. After eight years of research and practical experience with the dilatometer in the U.S., my skepticism has greatly diminished. The DMT delivers a surprising combination of accuracy, reproducibility, ruggedness, economy and versatility. Professor Marchetti has succeeded in combining the simplicity and economy of a penetration test with the sophistication of a test that also measures an insitu stress and modulus. We have found the DMT most useful for the rapid calculation of expected settlements in many types of soil, but it has many other uses.

For those unfamiliar with the DMT (See photo), the technician operator uses a drill rig or other suitable equipment to push or drive the blade into the soil to be tested, stopping at suitably spaced test depth intervals as close as 150 mm. At each depth the operator uses gas pressure to expand and deflate a stainless steel membrane horizontally in the soil and obtains four data readings, typically taking 1 to 2 minutes total. The data are then interpreted to obtain the following information:

- Soil type, in situ lateral stress, one-dimensional compression modulus M, and the effective overconsolidation ratio in all soils
- The excess hydrostatic pore pressure, undrained shear strength, and coefficients of consolidation and permeability in clays
- The friction angle and insitu water pressure in sands
- All for use in any rational design procedures using these soil properties. The special equipment costs vary from $9-15,000 and the commercial cost per test typically ranges from $15-35, including data reduction and drill rig time.

QUICK AND ACCURATE

Same-day calculation of ultimate settlements in a wide variety of soil types are often possible with the DMT because of the rapid accumulation of relevant modulus and stress history data. These rapid calculations are sufficiently accurate for most practical work and I find this capability perhaps the most generally useful advantage of the DMT. I have knowledge of 30 case histories of DMT settlement analysis predictions vs measurements. My experience shows we may expect a blind (no adjustments for local experience) calculated/measured ultimate settlement prediction accuracy within a ratio of 0.67 to 1.50 in all penetrable soils, with a mean of 0.95 and a coefficient of variation of 25%. This compares well with the much slower and more expensive conventional soil sampling and laboratory methods.

I've written in detail about the results in 16 of those case histories; the following three examples illustrate my point with predictions prior to construction.

The Sunshine Skyway Bridge Main Piers. At this recently constructed bridge site, the DMT

DMT tests at the Sunshine Skyway bridge piers were more accurate than lab tests; the results from the latter indicated the need for a more costly foundation design.

DMT equipment showing control and readout box, blade and membrane, the pneumatic-electric cable and a tank source of gas pressure.

proved more accurate than lab consolidation tests. If we had relied on the lab tests alone, we might have asked for a more costly foundation design for the piers.

The bridge, across Tampa Bay in Florida, may have the current world record span for cable stayed, post tensioned, concrete box girder construction. Each main pier (see photo) is carried by forty-four 1.5 m diameter drilled shafts founded within and above a hard, calcium montmorillonite, Miocene clay with many limey, cemented zones.

The geotechnical testing constraints included a fast-tracked schedule and a difficult soil to sample at a difficult location. The settlement of the main piers had to fall within acceptable vertical and tilt tolerances so as to mesh successfully with the very detailed segmental construction procedure. We elected to use in situ testing methods to determine elastic and compression moduli, and included Marchetti DMTs for this purpose, supplemented with a few laboratory consolidation tests performed on samples obtained using 4 in. double tube core barrel methods.

There were approximately 1000 DMTs performed for the entire project. We did 35 of these in the calcareous hard clays supporting the two main piers. All 35 required predrilling and using the SPT hammer to drive the blade because the strength of the soil prohibited quasi-static pushing. The DMT indicated average 1-D compression moduli of M = 200 MPa below the drilled shaft foundations, while the lab tests gave approximately M = 50 MPa. We suspected the lab results fell on the low side of reality because of sample disturbance and laboratory equipment compressibility effects. We also thought the DMTs were low because we had to hammer-drive the blade.

Since we did not know the ground truth, and felt this was a very important settlement prediction, we decided to compromise on the conservative side. We used the value of 100 MPa for our settlement calculations. The bridge has just been completed and the engineers have measured a maximum combined immediate (elastic) plus consolidation (volume change) pier settlement of 18 mm vs. our predicted 43 mm for the loads applied. It appears that the ground truth M = approximately 43/18 x 100 = 240 MPa. In this case the DMT proved much more accurate than the conventional lab methods. Had we relied on the lab results, we would have predicted 86 mm of settlement for the load applied and 150 mm for the design maximum loads.

Jacksonville Power Plant. This case history involved the ground improvement quality control for the soils under a major power plant near Jacksonville, Fla. The designers elected to save the cost of a pile foundation by using a combination of deep compaction grouting and shallower dynamic compaction. The goal was to produce a 50 ft layer of sand and silt strong enough to prevent differential settlements of more than 6 mm between major components of the power block. Hayward Baker Co. (now GKN-Hayward Baker Co.) guaranteed maximum differential settlement results and were most anxious to provide the best quality control possible. They utilized more than 900 Dutch cone (CPT) soundings in conjunction with more than 30 DMT soundings. The results proved acceptable, but not without some difficulty.

The designers set the quality standards in terms of minimum CPT profiles, which proved difficult to achieve in a trial section of the compaction area. In contrast, the DMT soundings showed M values adequate to meet the 6 mm maximum differential settlement requirement. The engineers, after considerable study, agreed to use the DMTs as a calibration for the CPT soundings. They considered a minimum equivalent M-profile as an alternate acceptance criteria when the CPT soundings did not meet the initially set requirements. The contractor invoked this alternate requirement many times during the ground improvement work and it proved successful.

In my opinion, DMT modulus values are probably more accurate than any values inferred from CPT results. This example is a case in point, but there are at least two reasons for my conviction: The DMT blade causes less soil structure disturbance than the CPT cone, and the DMT actually measures a modulus while the CPT measures only a bearing capacity.

Ontario Orange Juice Tanks. Here, John Hayes, President of Site Investigation Services, Ltd., Peterborough, Ontario, used the DMT M values to predict the settlements under a closely spaced group of four tanks on a common mat foundation (see photo). The initial DMT soundings predicted excessive settlement for the location of first choice. After further DMT sounding exploration, Hayes suggested moving to a location with settlement predictions of acceptable magnitude. He reports that the tanks have now been built, 100% load tested, and are now in service. The measured settlements of the compressible surface sand layer, with some contribution from underlying clay layers, have almost reached the predicted magnitude of 30 mm.

The Ontario experience shows

FIGURE 1
DMT SOUNDINGS

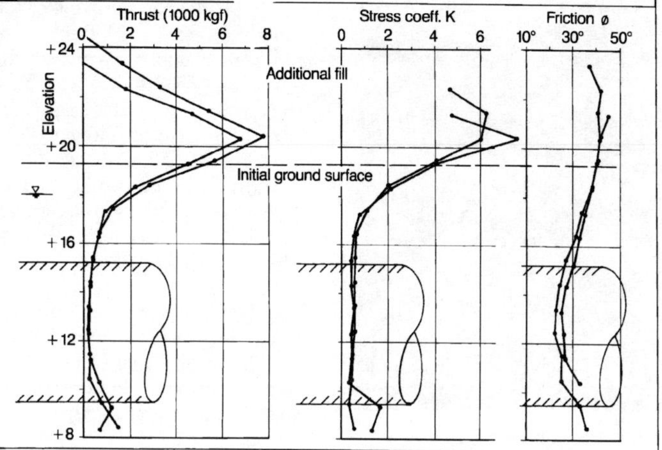

Results of two soundings through compacted trench backfill to an existing rigid pipe.

how effective use of DMT soundings can provide an economical solution to a foundation problem. Hayes reports that as of Sept. 1987, he used the DMT for settlement predictions in approximately 40 projects and has had settlement measurements to check his results in eight cases. All were acceptably accurate.

SPECIAL APPLICATIONS

The Marchetti dilatometer has potential use for other types of problems. The following examples touch on a few possibilities.

Pipe Trench Backfill Pressure—Here the problem involved attempting to compute the soil pressure on a 1.8 m diameter reinforced concrete pressure pipe that had an unexpected compacted fill placed over the pipe. The original design was via the Marston calculation method, which uses the lateral stresses and internal friction of the backfill as the key soil parameters involved in making such a calculation (see Fig 1).

The DMT provides a very practical and suitable tool to investigate such a problem. The blade could be inserted vertically through the trench backfill immediately adjacent to the pipe, going past the pipe and into the underlying natural bedding soil. The blade and its membrane were kept parallel to the axis of the pipe so as to measure the soil pressures in the proper direction perpendicular to the axis of the pipe and trench. Fig. 1 also shows the total thrust to push the blade into the soil, and the lateral stress coefficient K and plane strain friction angle results obtained from the routine DMT data reduction.

In this case use of the Marston equation with these data showed that the computed vertical soil weight carried by the pipe did not exceed the original design calculations and that the surcharge was acceptable without reducing the pressure capacity of the pipe. The DMT data incidentally also showed the absence of any of the specified compaction of the backfill around the pipe.

Pressures Against Existing Walls-The question occasionally arises as to what pressure an existing wall actually carries vs. its design. Perhaps the wall shows signs of distress, or may need to carry an unanticipated surcharge. The DMT can provide a means of measuring existing pressures, including water pressures.

For example, a vertical wall in a buried utility vault showed signs of distress after its height was increased from 18 to 30 ft due to adjacent roadway construction. The question arose as to whether this distress resulted from the unanticipated magnitudes of earth pressure. The engineers involved decided to check the actual earth pressures on the wall with two DMT soundings 5 ft from the wall and 30 ft apart along a portion of the wall showing distress.

A simplified cross section of the wall shows the DMT-determined lateral stresses against the wall, in both the surface backfill layer and the underlying natural residual silty soils. These pressures exceeded the backfigured capacity of the wall and explained its distress.

The speed of performing the DMT soundings was an important advantage at this site. Heavy road traffic had to be blocked to allow the soundings through the pavement and parallel to the wall of the vault. One rig performed both soundings, and blocked only one lane of traffic for about six hours.

Lateral Movement of Piles—The DMT approximately models the problem of the lateral movement of a pile. This is because the blade insertion wedges the soil apart in the horizontal direction and the measured lateral stress and modulus are also in the horizontal direction. Furthermore, the DMT provides detailed data right from the ground surface, where it is the most important for this problem.

New and practical methods for calculating lateral pile movements using the DMT and the p-y curve method are described in several recent paper. Alternatively, one can use the DMT to obtain Young's modulus data and utilize elastic methods that have the advantage of including group effects.

YOUR MOVE

After eight years of experience with the DMT, I believe it represents a big improvement in the practical soil testing tools available to the practicing geotechnical engineer. Almost all the practicing geotechnical engineers I know who have tried the DMT have found it useful. The technical literature now has well over 100 papers that contain many examples in addition to those mentioned here.

A caveat: one should only use the DMT or any other penetrometer test in appropriate soils — for example not in rocky fills and not in very compact and stony glacial tills. Generally, problem soils are finer and weaker where one can easily use the DMT.

John H. Schmertmann is a consulting engineer with his own firm, Schmertmann & Crapps, Inc., and part owner of GPE, Inc., which markets the dilatometer in North America. Both firms are in Gainesville, Fla.

Reprinted from Civil Engineering, March 1988

Density Tests above Zero Air Voids Line

By John H. Schmertmann,[1] Fellow, ASCE

Abstract: To compute the degree of saturation, S, one must have the results from tests for water content, wet unit weight, and specific gravity. Each has inherent variability, which results in a variability in S. This variability results in a percentage of S plotting above the zero air voids line (ZAVL). After first showing that Gaussian normal distributions apply to the test results, the writer proposes typical low, average, and high values for test and soil plus test variability and then uses both numerical and analytical methods to predict the percent of S above the ZAVL. He then presents the density test results from three large cohesive fills to support the analytical study. Knowing the percent of tests above the ZAVL may provide the engineer with an additional quality check of a cohesive fill. Arbitrarily rejecting all tests above the ZAVL will distort the perceived character of the fill, which could result in action detrimental to the engineering behavior of the fill and might provide the basis for a claim for extra compensation by the contractor.

Introduction

True or false?—All valid field density tests have a degree of saturation, S, equal to or less than 100% and must therefore plot on or below the zero air voids line (ZAVL). Many engineers would answer true and some reject all tests that plot above this line. The writer will argue for an answer of false, and that tests above the ZAVL have validity and may have a practical value in evaluating the quality of compaction control.

Of course, from a physical viewpoint one cannot have a compacted soil with a degree of saturation, S, greater than 100%. But the writer recently reviewed compaction records wherein 26% of a large number of tests did indeed plot above the ZAVL. Does this prove incompetent testing? The writer will show it does not. If not, how could it happen? The answer involves soil and testing variability. Furthermore, rejecting tests that plot above the ZAVL distorts any statistical findings from the field density testing and may have important engineering and economic consequences.

Although this paper focuses on the field density testing of compacted cohesive fills because of their economic importance, the concepts discussed also apply to all tests that involve a calculation for the degree of saturation. For example, calculating the value of S for the soil specimen in an oedometer test often produces $S > 100\%$, say 105%. What do we do then? Some practicing engineers will report 105%, and others will report only 100% even though they calculate 105% because 105% looks foolish. A few research engineers might even reject the results from such an oedometer test. The concepts in this paper will show that one should report the 105% and use the test because it probably provides information of equal value to a test with a calculated $S < 100\%$.

[1]Prin., Schmertmann & Crapps, Inc., 4509 N.W. 23rd Ave., Ste. 19, Gainesville, FL 32606.

Note. Discussion open until December 1, 1989. To extend the closing date one month, a written request must be filed with the ASCE Manager of Journals. The manuscript for this paper was submitted for review and possible publication on July 28, 1988. This paper is part of the ***Journal of Geotechnical Engineering***, Vol. 115, No. 7, July, 1989. ©ASCE, ISSN 0733-9410/89/0007-1003/$1.00 + $.15 per page. Paper No. 23685.

VARIABILITY IN TESTS THAT DETERMINE SATURATION

Eq. 1 gives the formula for the degree of saturation, S, as determined from the results of the three separate tests usually used to measure saturation

$$S = \frac{\gamma w G}{(1 + w)G\gamma_w - \gamma} \qquad (1)$$

where γ_w = unit weight of water; γ = wet unit weight; w = water content; and G = specific gravity of soil solids. Each test may involve subtests, such as separate subtests for weight and volume to determine γ, but for the purposes of this paper the writer includes all such subtests within the word *test* for γ, w, and G. No one can perform each of these tests perfectly every time. Each test result includes some deviation, denoted $\Delta\gamma$, Δw and ΔG, from the "true" value and these deviations will vary from test to test. The variability of the results obtained from such testing then leads to values of S that also deviate from the true value, as expressed by Eq. 2.

$$S \pm \Delta S = \frac{(\gamma \pm \Delta\gamma)(w \pm \Delta w)(G \pm \Delta G)}{(1 + w \pm \Delta w)(G \pm \Delta G)\gamma_w - (\gamma \pm \Delta\gamma)} \qquad (2)$$

Furthermore, the true values themselves have an additional variability due to the variability of the soil itself in any mass of soil such as a compacted fill.

The writer herein will use only three common statistical terms. The mean or average of a group of replicate tests usually gives the best estimate of its

TABLE 1. Writer's Values of Coefficients of Variation for the Tests in Common Use to Compute the Saturation of a Compacted Fill

Test (1)	V_{low} (%) (2)	V_{avg} (%) (3)	V_{high} (%) (4)
γ	1	2	4
w	5	10	20
G	0.5	1.0	2.0

TABLE 2. Writer's Values of Coefficients of Combined Variation for Tests and Soil in a Compacted Fill

Test (1)	V_{low} (%) (2)	V_{avg} (%) (3)	V_{high} (%) (4)	V From Harr (1987) (%) (5)	V^c From Lee et al. (1983) (%) (6)
γ	1.5	3	6	3	3
w	7.5	15	30	13, 20	15
G	1.0	2	4	2	
Computed S	7%[a]	14%[a]	28%[a]	10%[b,c]	20%

[a]From Table 3.
[b]In lab.
[c]Tests only, not soil.

correct value. An overhead bar denotes the mean, such as $\bar{S}$ for the mean degree of saturation. The standard deviation, denoted σ, provides a convenient measure of deviation around the mean. The coefficieint of variation, denoted V, equals the standard deviation divided by the mean. Engineers usually express V as a percentage. It provides a convenient, normalized way of expressing variability. Then $\bar{\gamma}$, $\bar{w}$, $\bar{G}$, and $\bar{S}$ denote the mean values from each test and the computed saturation, and σ_{γ}, σ_w, σ_G, and σ_S denote their standard deviations. Finally, $V_{\gamma} = \sigma_{\gamma}/\bar{\gamma}$, $V_w = \sigma_w/\bar{w}$, $V_G = \sigma_G/\bar{G}$, and $V_S = \sigma_S/\bar{S}$ denote their coefficients of variation.

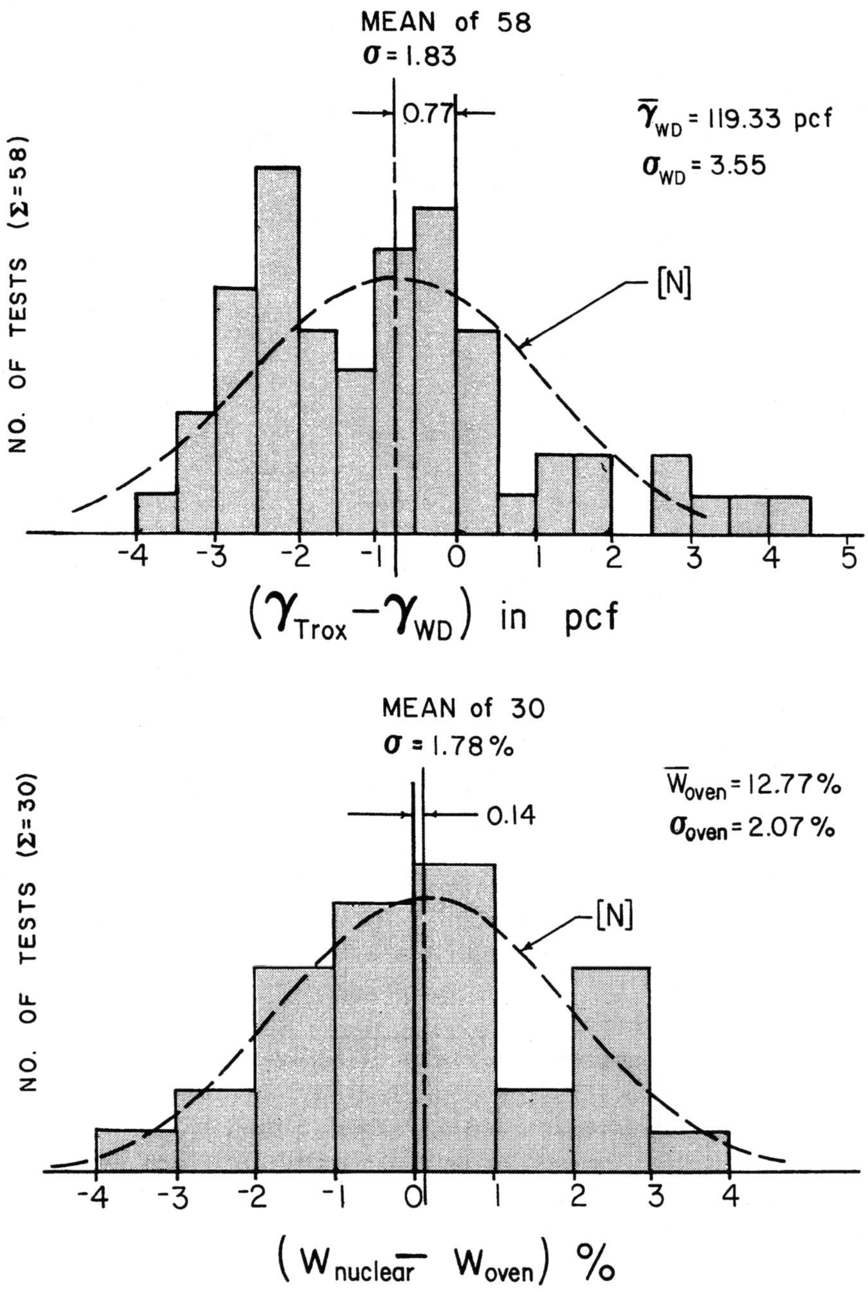

FIG. 1. Examples of Variability in Tests Used to Compute Degree of Saturation, with Matching Normal Distributions

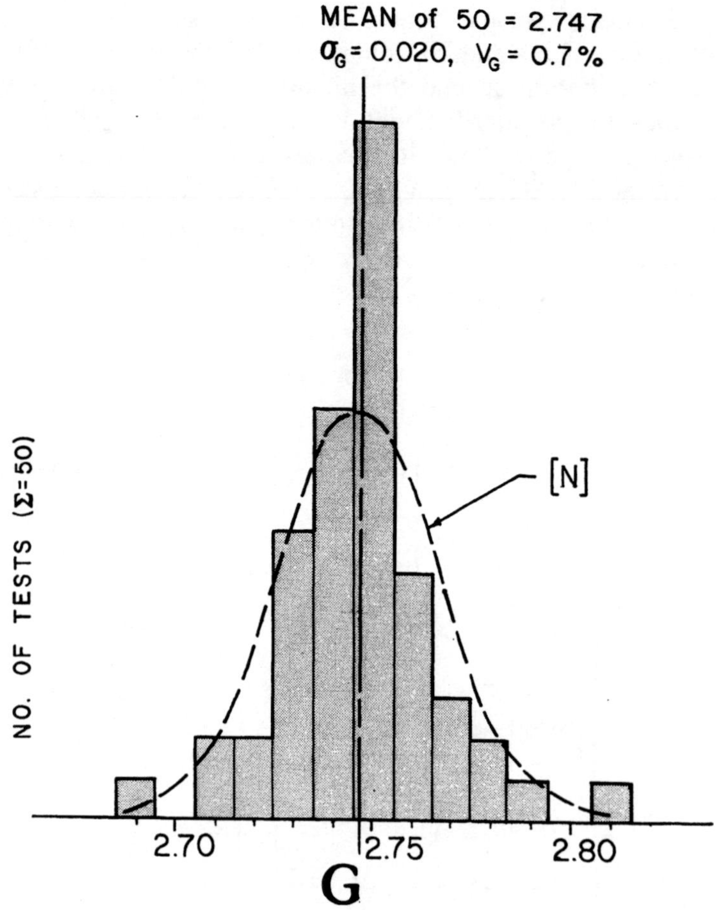

FIG. 1. ***(Continued)***

After considerable study of the literature and his own files, the writer made estimates of what he would consider low, average and high values of V_γ, V_w, and V_G. Tables 1 and 2 list these estimates, with Table 1 for only the tests and Table 2 for the combined variability of tests and soil. The low estimates assume a fill with less than average variability and/or testing with better than average variability. The high estimates assume a fill with more than average variability and/or worse than average testing variability. The values in Tables 1 and 2 are important to the quantitative use of this paper. To help support the writer's average values, Table 2 includes "representative" values from Harr (1987) and "recommended standard" values from Lee et al. (1983).

The writer next investigated the distribution of variability in the tests for γ, w, and G. Figs. 1(*a*)–(*c*) show single-study examples of the distribution of the deviations or errors associated with each of the three tests considered in this paper. Fig. 1(*a*) shows a group of 58 pairs of Troxler nuclear probe and Washington Densometer field tests, comparing wet unit weight (adapted from LaCroix and Horn 1973); Fig. 1(*b*) a group of 30 pairs of field (nuclear) and lab (oven) tests, comparing water content w; and Fig. 1(*c*) shows one lab's series of 50 tests for specific gravity, G, on samples from an existing cohesive fill. Had each of these studies included an infinite number of comparisons, then each of these distributions would have given a smooth curve of some shape. But what shape? The ordinary Gaussian normal distribution gives the most convenient shape in terms of the mathematics for this paper, and for subsequent use by the reader. Figs. 1(*a*)–(*c*) include, superposed, a dashed line showing the normal distribution curve that matches with the mean and standard deviation obtained from the actual test comparison data.

The comparisons in Figs. 1(*a*)–(*c*) vary from perhaps good in Fig. 1(*c*) to perhaps only fair in Fig. 1(*a*). It would be unreasonable to expect every comparison of the type shown to give good comparisons. The writer believes on the basis of general impressions from his experience and from the literature, as well as from specific comparisons such as in Fig. 1, that using normal distributions to describe the variability in γ, w, and G in test results give acceptable approximations well-supported by experience. Furthermore, this paper will show subsequently that the use of the normal distribution assumption appears to predict results that match well with at least some field experience.

The assumed coefficients of variation listed in Table 2 and the assumed normal distribution for each test's variability and the soil variability now provide a basis for calculating the variability in S when computed by Eq. 1.

RESULTING VARIABILITY IN COMPUTED S

Numerical

One conceptually simple, but computationally long, way to calculate the variability in computed S requires using Eq. 1 for a large number of trials and for each trial varying the γ, w, and G test values around their assumed mean values in a normally distributed random manner using the coefficients of variation in Table 2. Next, review all these calculations for S and study their variation. After the writer performed exploratory 50 to 100 trial cases and obtained encouraging results, his statistical consultant made a similar but more comprehensive study, initially with 1,000 and then with 5,000 trials each to make sure we had enough trials for an accurate result. He did this for the many combinations of γ, w, and G shown in columns 1, 2, and 3 in Table 3. The 1,000 and 5,000 trials produced nearly the same results. Columns 5, 6, and 7 in Table 3 list the results from the presumed most accurate, 5,000 trial cases.

Analytical

The field and lab tests for water content, unit weight and specific gravity usually have very little, if any, interdependence. Fig. 1 and the associated discussion showed that the results from repeated tests varied approximately symmetrically around their means in a normally distributed, random manner. After assuming independence and randomness one can use the first two terms of the Taylor series to obtain a first-order approximation of the variability in the degree of saturation, S, when using Eq. 1. Eq. 3 gives the differential equation for the first-order approximation and Eq. 4 the result when applying Eq. 3 to Eq. 1.

$$\sigma_s \cong \left[\left(\frac{\partial S}{\partial w}\sigma_w\right)^2 + \left(\frac{\partial S}{\partial G}\sigma_G\right)^2 + \left(\frac{\partial S}{\partial \gamma}\sigma_\gamma\right)^2\right]^{1/2} \quad (3)$$

TABLE 3. Results of Studies to Determine Expected Variability in S and $\%S$ above ZAVL (G = 2.75)

Assumed				From 5,000 Trials			From [N] Analytical		
γ_d[a] (pcf)[b] (1)	$\bar{\gamma}$ (pcf) (2)	$\bar{w}$ (%) (3)	V (Table 2) (4)	$\bar{S}\%$ (5)	σ_s (6)	Percentage trials $S > 100\%$ (7)	$\bar{S}\%$ (8)	σ_s (9)	Percentage $S > 100\%$ (10)
110	125.67	14.25	Average	69.77	9.45	0.1	70	9.35	0.1
	126.80	15.27		75.30	9.82	0.7	75	9.84	0.6
	127.92	16.29		80.05	10.45	3.1	80	10.32	2.6
	129.04	17.31		84.98	10.89	9.0	85	10.77	8.2
	130.16	18.33		89.93	11.19	18.3	90	11.22	18.7
	131.29	19.35		95.22	11.75	33.9	95	11.65	33.4
	132.40	20.36		100.27	12.17	50.6	100	12.06	50.0
120	133.14	10.95	Average	70.38	10.77	0.6	70	10.45	0.2
	134.08	11.73		75.25	11.33	2.1	75	11.04	1.2
	135.01	12.51		80.53	11.93	6.2	80	11.61	4.3
	135.95	13.29		85.34	12.62	12.3	85	12.17	10.9
	136.88	14.07		90.16	13.08	21.7	90	12.71	21.5
	137.83	14.86		95.52	13.08	35.8	95	13.24	35.4
	138.76	15.64		100.64	13.89	50.1	100	13.76	50.0
130	140.59	8.15	Average	70.91	12.82	2.3	70	12.11	0.7
	141.35	8.73		75.72	13.34	4.3	75	12.84	2.6
	142.10	9.31		81.02	13.93	9.2	80	13.55	7.0
	142.86	9.89		86.10	14.67	16.3	85	14.25	14.6
	143.61	10.47		90.81	15.73	26.1	90	14.94	25.2
	144.37	11.06		96.24	16.33	37.5	95	15.62	37.4
	145.13	11.64		101.59	16.90	50.8	100	16.29	50.0
120	133.14	10.95	High	71.62	23.28	10.2	70	20.90	7.6
	134.08	11.73		76.89	24.10	15.4	75	22.08	12.9
	135.01	12.51		81.96	26.07	22.0	80	23.22	19.5
	135.95	13.29		86.57	26.66	27.1	85	24.33	26.9
	136.88	14.07		92.14	28.48	35.1	90	25.41	34.7
	137.83	14.86		97.41	30.47	42.2	95	26.45	42.5
	138.76	15.64		101.77	31.49	47.8	100	27.52	50.0
120	133.14	10.95	Low	70.24	5.20	0.0	70	5.22	0.0
	134.08	11.73		75.07	5.56	0.0	75	5.52	0.0
	135.01	12.51		80.19	5.75	0.1	80	5.80	0.0
	135.95	13.29		85.17	6.08	1.0	85	6.08	0.7
	136.88	14.07		90.11	6.32	5.9	90	6.35	5.7
	137.83	14.86		95.15	6.62	23.2	95	6.62	22.5
	138.76	15.64		100.18	7.06	50.1	100	6.88	50.0

[a] γ_d = dry unit weight, $\gamma/(1 + w)$.
[b] 1.00 pcf = 0.157 kN/m^3.

$$\sigma_s \cong \frac{\{[G\gamma(G\gamma_w - \gamma)\sigma_w]^2 + [w\gamma^2\sigma_G]^2 + [wG^2\gamma_w(1 + w)\sigma_\gamma]^2\}^{1/2}}{[(1 + w)G\gamma_w - \gamma]^2} \quad \ldots\ldots\ldots\ldots (4)$$

Applying Eq. 3 to the cases in Table 3 gives the results for σ_s listed in column 9 of this table. The writer now made the additional assumption that the deviations in S also have the Gaussian normal distribution. Fig. 2 presents an actual distribution of S from field data and shows that the equivalent normal distribution gives a reasonable approximation in this case. If one accepts the normal distribution approximation, it then becomes a simple mat-

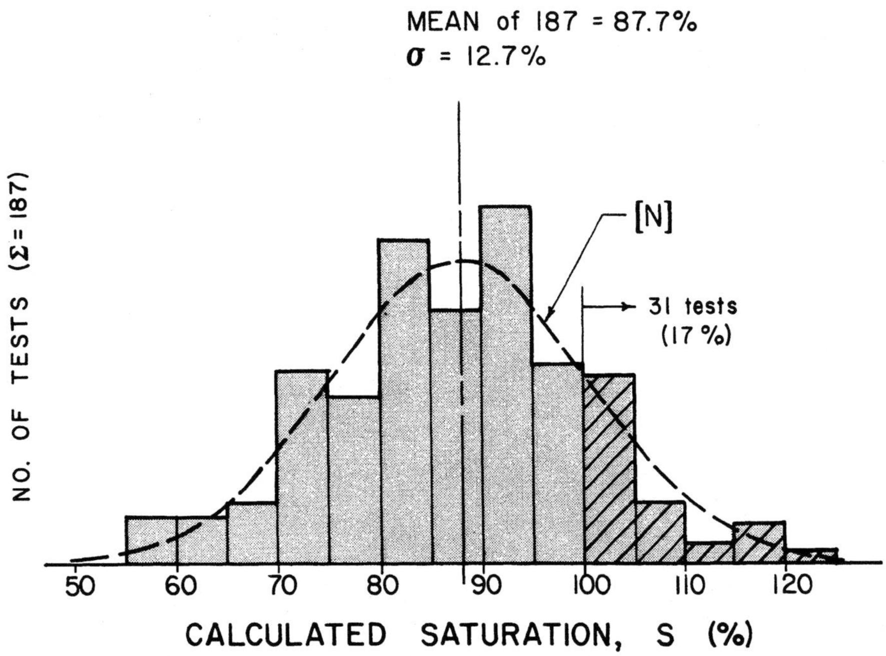

FIG. 2. Field Data Showing Approximate Normal Distribution for Calculated Values of Degree of Saturation

ter to determine the number of standard deviations between the mean S and $S = 100\%$ for each case, and then to enter the appropriate table of normal distribution areas and obtain the expected percentage of tests that will compute with $S > 100\%$. Column 10 of Table 3 lists the results.

Comparing columns 7 and 10 in Table 3 shows that they have nearly the same values for the percentage $S > 100\%$. This confirms the accuracy of using the assumption of a normal distribution for S in the analytical solution. The writer's consultant also performed an analytical study using the second order approximation from the Taylor series so as to check the adequacy of the first order approximation used to obtain Eq. 4. This produced equations much more complex than Eqs. 3 and 4 herein, but with only negligible changes in the results. All this demonstrates that one can use the convenience of the normal distribution of S and the relatively simple Eq. 4 when determining the percent of field data points expected to fall above the ZAVL.

CALCULATIONS WITH $S > 100\%$

The studies described earlier produced a perhaps surprising result: many of the calculations gave $S > 100\%$ and therefore would plot above the ZAVL. Fig. 3 attempts to show why one should expect some calculations with $S > 100\%$, and that this percentage should increase with increasing $\bar{S}$ and variability.

Fig. 3 shows the typical γ_d-w-ZAVL graph used to plot the results of field density tests. Consider some test point X shown in its most likely correct position (using $\gamma_d = 120$ pcf, $w = 14\%$, and $S \approx 90\%$ in this illustration). The bell-shaped curves shown represent the normally distributed spread or variability only in w, one curve for a $V_w = 5\%$ and another for $V_w = 20\%$. If the total (shaded) area under each curve represents all the tests, then the

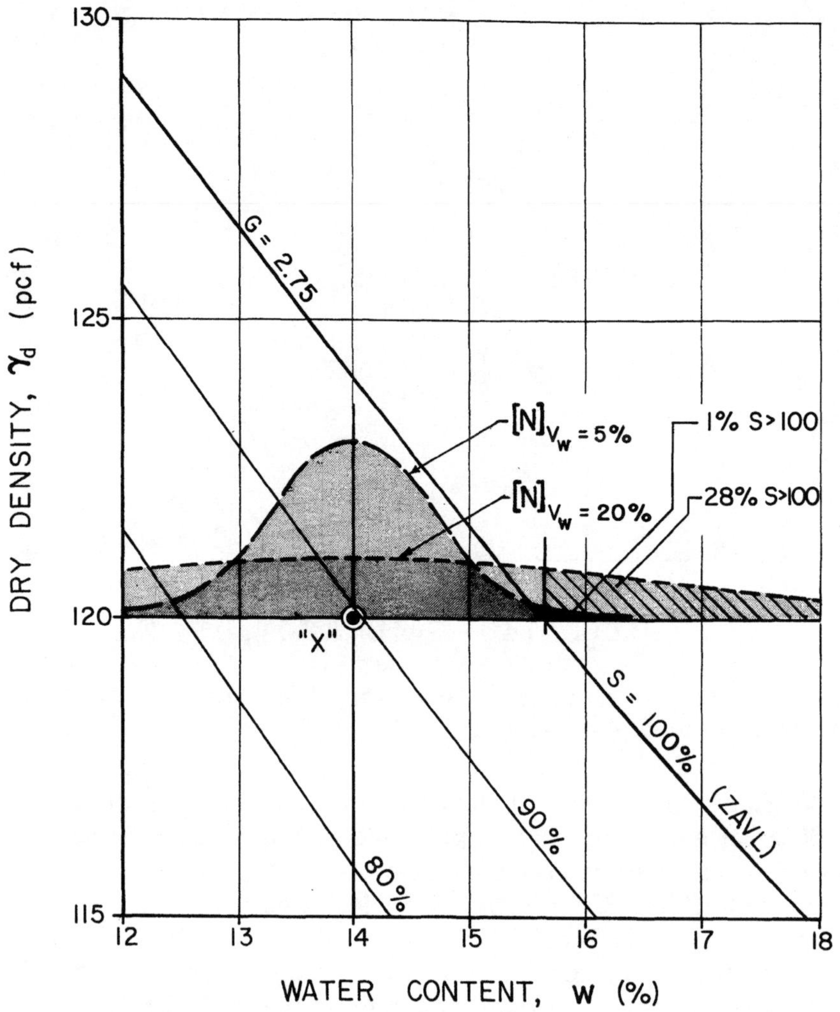

FIG. 3. Example to Illustrate How Variability in Results from Test for Water Content Will Produce Computed Points above ZAVL, and How Increasing Variability Increases Percentage of Such Points

cross-hatched area above the ZAVL represents all of these tests expected to compute above the ZAVL. The ratio of areas gives the percentage above the ZAVL due to V_w only. As illustrated, the higher V_w the greater this percentage, with 1% $S > 100$ for $V_w = 5\%$ and 28% for $V_w = 20\%$.

Fig. 4 presents the results from column 7 in Table 3 for the percentage of the 5,000 trials which, by count, would plot above the ZAVL (• points), and also the nearly matching analytical results from column 10 (○ points). The solid lines in this figure present the above-ZAVL percentage for the high/avg/low variability cases from Table 2, considering only a dry unit weight of 120 pcf. To illustrate sensitivity to γ_d, the dashed curves show the average variability case using $\gamma_d = 110$ and 130 pcf. The writer also made a sensitivity check for the effect of varying G and determined that varying G over the normal range of 2.6 to 2.9 had a negligible effect on the positions of the Fig. 4 curves calculated using $G = 2.75$.

Note that the curves in Fig. 4 have the expected shape: varying from near zero at low $\bar{S}$ because only the very tail end of the variability distribution

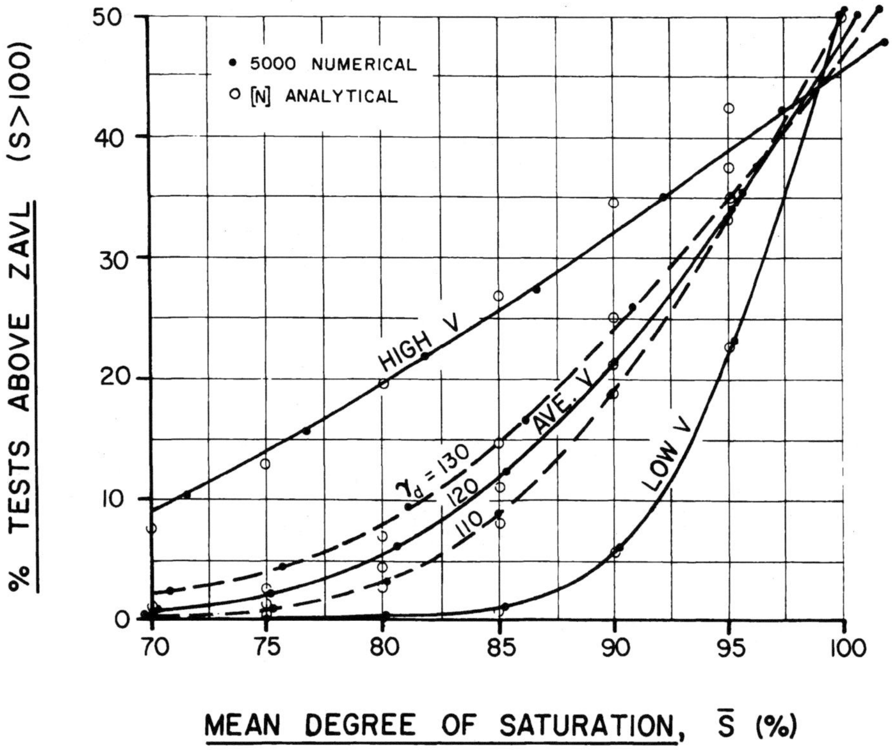

FIG. 4. Numerical and Analytical Values of Percentage of Field Tests Expected to Plot above ZAVL when Using Assumed Test + Soil Variabilities in Table 2 and G = 2.75

would extend past 100%, to near 50% when $\bar{S}$ = 100% because any approximately symmetrical distribution of variability around the mean would produce approximately one half the computed S values higher than the mean of 100%.

S > 100% FROM FIELD MEASUREMENTS

The reader might now like to see how some actual field density test data compare with the numerical predictions presented in Fig. 4. For this purpose the writer has prepared Table 4 listing the results from the density check testing in three large cohesive fills herein identified as C, D, and E. The fill C had various subgroups of data with different mean values for the degree of saturation. Fig. 5 contains only the curves from Fig. 4 to reduce confusion. The solid points in Fig. 5 plot the data in columns 8 and 9 in Table 4.

One can see in Fig. 5 that the field examples of S > 100% fit well within the numerically predicted curves. If the engineer had the curves in Fig. 5 available as a reference, and plotted any of these field test results on the figure, and made some adjustment in the position of the curves for the different γ_d, he or she might then conclude that sites C, D, and E had approximately average testing + soil variability. In this way a plot such as Fig. 5 could have provided an additional tool for quality control or quality assurance for the density check testing at these three fills.

TABLE 4. Average S Data Computed from Field Density Tests in Three Cohesive Fills

ID (1)	Geology (2)	Classification (USCS, AASHTO) (3)	Number of tests (4)	$\bar{\gamma}_d$ (pcf) (5)	$\bar{w}$ (%) (6)	$\bar{G}$ (7)	$\bar{S}$ (%) (8)	Percentage of tests with $S > 100\%$ (9)
C	Glacial	SM SC	1,351 (Includes C1–C6)	122.6	12.63	2.71	91.0	25.7
C1			220	126.7	10.61	2.73	84.5	8
C2			143	123.4	12.14	2.70	90.4	21
C3			40	127.3	11.37	2.73	92.7	29
C4			205	125.1	11.80	2.69	93.7	30
C5			144	123.0	13.17	2.69	97.8	47
C6			61	119.7	14.72	2.65	102.9	60
D	Residual	A4–A6	658	108.0	18.4	2.82	82.4	4.7
E	Residual		762	109.8	17.6	2.70	89.0	15.6

EFFECTS OF BIAS ERROR

All the previous discussion in this paper has assumed that the variability of results from a particular type of soil parameter test distribute randomly around a correct mean value. It always remains possible that the variability will distribute randomly around an incorrect mean value. The difference be-

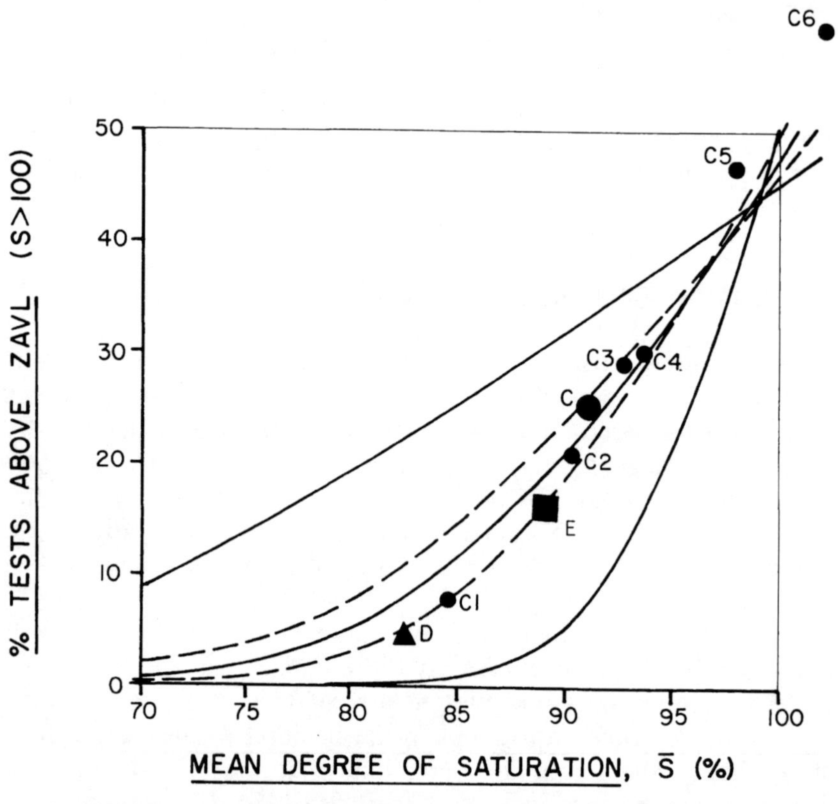

FIG. 5. Table 4 Field Density Case Examples from Cohesive Fills Plotted on Fig. 4 Theoretical Curves for Expected Percentage above ZAVL

TABLE 5. Approximate Change in S Due to Test Bias

Test (1)	Bias in mean (2)	Change in $\bar{S}$ (3)
γ_d	±2 pcf	±5%
w	±1%	±6%
G	∓0.05	±4%

tween an incorrect mean and the correct mean represents an error in the mean itself, or a bias error. For example, the C6 group of tests in Table 4 computed to have an $\bar{S}$ = 102.9%, which physically cannot occur and therefore indicates a bias error somewhere (probably a too-low G in this case).

A bias error in either the unit weight, water content or specific gravity test will change $\bar{S}$ and thus also the percent of points above the ZAVL. Table 5 shows the approximate expected change in the computed mean degree of saturation resulting from various assumed bias errors.

Another type of bias, the sampling error, can result from a biased selection of locations for the field density tests. For example, the preferential testing of locations deemed most likely to have a low γ_d. A sampling bias affects the representativeness of the test results versus the entire fill. But, as discussed subsequently, it does not directly affect using the proposed method to evaluate testing variability.

The curves in Figs. 4 and 5 result from using constant low, average, or high values of fill plus test variability. Bias can change the value of $\bar{S}$, but the $\%S > 100$ will change approximately along a curve of constant variability. Bias will therefore not affect the relative position of a point between the curves of low/average/high variability and thus will not affect one's evaluation of the relative variability of the field density testing. The engineer should obviously try to minimize bias in all three tests, but for reasons other than judging test variability.

SOME CONSEQUENCES OF REJECTING TESTS ABOVE THE ZAVL

As noted in the introduction of this paper, some engineers reject all field density check tests that produce a calculated value of $S > 100\%$ because in their view all such results have obvious error. They typically do this rejection despite the lack of any other evidence to indicate a non-representative test. Sometimes the rejection occurs in a more subtle form by the technicians performing the testing. Just knowing that the engineers reviewing their work do not like to see test results with $S > 100\%$ encourages them to find one way or another to suppress such results. However, this paper has demonstrated that engineers and technicians should expect a certain percentage of all random tests to compute with $S > 100\%$, that this percentage increases with the mean degree of saturation $\bar{S}$, and that any such tests have as much validity as the tests that plot below the ZAVL. Arbitrarily rejecting such tests may sometimes have minor but other times may have important and perhaps unsuspected consequences for both the engineer and contractor.

The consequences depend on whether one uses the field density tests for quality control (QC) or quality assurance (QA). If for QC, then the engineer reviews each test one at a time. Each rejected test presumably requires a

TABLE 6. Required Changes in Water Content to Hold %S > 100% to 1%

γ_d (pcf) (1)	V (%) (2)	$\bar{w}$ (%) (3)	$\bar{S}$ (%) (4)	$\Delta\bar{w}$ (%) (5)
110	Average	15.3	75	0
		12.3	85	−2.1
		19.4	95	−4.6
120	Average	11.7	75	−0.3
		13.3	85	−2.3
		14.9	95	−4.0
130	Average	8.7	75	−0.8
		9.9	85	−2.3
		11.1	95	−3.9
120	High	11.7	75	−5.0
		13.3	85	−7.5
		14.9	95	−10.4
120	Low	11.7	75	0
		13.3	85	0
		14.9	95	−1.7

replacement test that passes with $S < 100\%$. To reduce the percentage of tests with $S > 100\%$ to an acceptably small amount, say 1%, and taking G as a constant, the Contractor must reduce either soil variability or w or γ_d, or combinations, with consequent effects on the engineering properties of

TABLE 7. Perceived Changes in Fill Due Only to Rejecting Tests with S > 100%

		$\bar{S}$ (%)		(Perceived − Actual)		
γ_d (1)	V condition (2)	Actual[a] (3)	Perceived (4)	$\Delta\bar{S}$ (%) (5)	$\Delta\gamma_d$ (pcf) (6)	$\Delta\bar{w}$ (%) (7)
110	Average	75.30	75.07	−0.23	0	−0.03
		84.98	82.94	−2.04	−0.22	−0.26
		95.22	88.74	−6.48	−0.33	−1.03
		100.27	90.44	−9.83	−0.33	−1.61
120	Average	75.25	74.65	−0.60	−0.12	−0.05
		85.34	82.27	−3.07	−0.36	−0.27
		95.52	87.88	−7.64	−0.72	−0.71
		100.64	89.67	−10.97	−0.96	−1.10
130	Average	75.72	74.28	−1.44	−0.26	−0.06
		86.10	81.45	−4.65	−0.65	−0.20
		96.24	86.33	−9.91	−1.17	−0.53
		101.59	8.18	−13.41	−1.56	−0.73
120	High	76.89	69.59	−7.30	−0.96	−0.50
		86.57	74.19	−12.38	−1.32	−0.98
		97.41	77.54	−19.87	−1.68	−1.78
		101.77	78.77	−23.00	−1.92	−2.14
120	Low	75.07	75.07	0	0	0
		85.17	85.00	−0.17	0	−0.13
		95.15	92.49	−2.66	−0.24	−0.25
		100.18	94.57	−5.61	−0.36	−0.59

[a]As selected from the 5,000 trial cases in Table 3.

the fill. The engineer could also contribute by reducing the variability in the testing for γ and/or for w and/or for G. Most likely the engineer will hold the contractor to the dry density requirements and thus force him to produce a drier fill. In wet areas this could cause much unnecessary extra drying work.

The writer prepared Table 6 to documant how much extra drying one would need to reduce the $S > 100\%$ tests to 1% of all random tests. The cases come from Table 3 and the writer has assumed a normal distribution for S and that γ_d and G remain constant. Note the similarity to the concept illustrated in Fig. 1 except that the writer now reduces the mean water content with constant V_w to achieve the $S < 100\% = 1\%$ of tests versus by the reduction in V_w shown in Fig. 1. Table 6 predicts that very substantial reductions in $\bar{w}$ must occur if the contractor can only have 1% of tests with $S < 100\%$ and with all other factors held constant.

If for QA, then the engineer reviews the tests in large batches and the prior rejection of all tests with $S < 100\%$ has statistically predictable consequences on how he or she perceives the fill after this censoring of results versus its actual character. Consider the simplified case where the engineer has specified, and the contractor has actually achieved, mean values of compaction at exactly the optimum compaction values of $\bar{\gamma}_d = 120$ pcf and $\bar{w} = 13.3\%$, at $\bar{S} = 85\%$. Column 7 in Table 3 shows that one should expect 12.3% of all the random density tests to have a computed $S > 100\%$. The engineer now rejects these 12.3%. Reviewing the same 5,000 trial numerical study that produced the 12.3% and rejecting this 12.3% leaves 4,385 trials or tests the engineer now has remaining to evaluate the fill. These 4,385

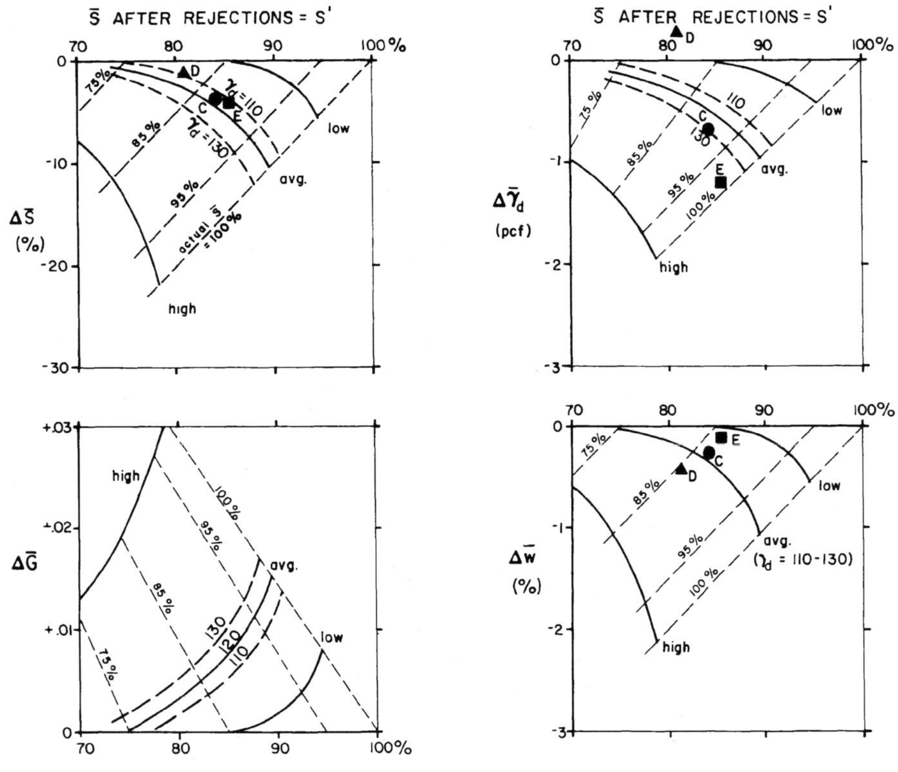

FIG. 6. Changes in Perceived Properties of Fill Due to Rejecting All Tests with $S > 100\%$, Analytical Results with Superposed Field Cases from Table 4

gave computed mean values of $\gamma_d' = 119.6$ pcf, $w' = 13.0\%$ and $S' = 82.3\%$. The engineer incorrectly thinks he or she has a fill undercompacted by 0.4 pcf and may require the contractor to increase compactive effort. He or she also incorrectly thinks the fill too dry by 0.3% water content and may ask the contractor to reduce drying or require the contractor to add water. Such actions could be detrimental to the fill's engineering behavior. The contractor has to do additional work and may make less profit and have a basis for a claimed extra or even a lawsuit. Here again the engineer has, by the rejection of tests with $S > 100\%$, produced a distorted picture of the fill that may produce unfavorable reactions.

Table 7 presents these and some other comparisons from the 5,000 trial numerical studies in Table 3 to illustrate further the consequences of rejecting or censoring-out the trials (tests) where $S > 100\%$. In all cases the rejections reduce the actual values of $\bar{S}$, $\bar{\gamma}_d$, and $\bar{w}$ (but increase $\bar{G}$) from the actual to the censored values. One can also think of these reductions (or increases) as corrections to the censored values to obtain the correct values.

Fig. 6 illustrates the analytical results in Table 7, and shows the effects on the measured average properties of a fill due to rejecting data with $S > 100\%$. The effects increase with increasing actual $\bar{S}$ and with increasing V. The rejections depress the perceived $\bar{S}$ to the point that the censored S' tends to reach a limit and it becomes difficult or impossible to recognize the higher actual $\bar{S}$. In some cases a fill actually above optimum $\bar{w}$ will have a perceived $\bar{w}$ below optimum. Fig. 6 also includes some field data points from the Table 4 cases and once again they provide some support for the findings developed numerically in this paper.

CONCLUSIONS AND SUGGESTIONS FOR USING TEST DATA WITH $S > 100\%$

1. Do not reject tests simply because they compute above the ZAVL. To do so not only destroys a possible quality assurance tool, but you might (ironically) force the contractor to produce a fill having a higher mean degree of saturation than intended by the specifications. The cost of extra work could trigger a claim by the contractor. A higher $\bar{S}$ could also produce a fill that was weaker, more compressible and more susceptible to developing high pore pressures during construction.

2. An engineer should expect a significant percentage of uncensored field density data to plot above the ZAVL, as indicated in Fig. 4. Consider the average curve percentages as routine and not as evidence of poor fill and/or testing methods.

3. The QC engineer, working with field density check tests one at a time, and who rejects those with $S > 100\%$ will likely produce a drier and less dense fill than intended. The QA engineer, evaluating batches of tests, who rejects those tests with $S > 100\%$ will likely perceive a fill drier and less dense than the actual fill.

4. As the fill density testing progresses keep a statistical count of the percent above the ZAVL using all the tests performed, as well as a computed current $\bar{S}$ and $\bar{\gamma}_d$, and plot the results into Fig. 5.

5. Note the relative location of your plotted point(s) with respect to the high/average/low variability curves presented by the writer and decide on the probable overall combined variability of the fill and of the performance of the density

testing. You can prepare your own Fig. 5 curves using Eq. 4 if you disagree with the writer's high/average/low variability values. The more tests you have available the better the validity of this quality check.

6. If this proposed $\%S > 100$ check indicates a too-high variability and you wish to try to locate and eliminate its source, you need to consider both the soil and the testing. Checking the testing requires one or more series of calibration or comparative tests. These tests should aim at determining the job-operative values of V_γ, V_w, and V_G to detect which one(s) are too high, perhaps by comparison with Table 1. The offending test(s), then require improved methods or a change to a less variable type. Note that the writer has found in various trials that V_γ and V_w contribute approximately equally to V_s, with the V_G contribution usually negligible. One should also make calibration or comparative tests to check for and reduce bias.

7. The supplemental quality check tool described herein should prove most sensitive and therefore most useful when there is a significant spread in the expected percentage above the ZAVL between the low and high variability curves. This may limit the usefulness of Fig. 5 to soils wherein the value of $\bar{S}$ falls between approximately 80 to 95%, for example in cohesive fills at or above the optimum water content for compaction. In some areas this will include many fills.

8. Reduce bias errors to as near zero as practical. Bias changes $\bar{S}$ and therefore the expected $\%S > 100$. However, bias does not directly affect the check for variability described earlier.

9. The 5. suggested variability check of course does not by itself adequately check a fill but merely adds a possible supplement to other checks. The engineer also needs to check such things as matching test results with the proper compaction curves, adequate numbers and distribution of tests, staying within the dry density and water content tolerances, and staying within any gradation or plasticity requirements. The engineer also should always consider the broader question of how the field density acceptance criteria for a fill relate to the desired engineering properties.

ACKNOWLEDGMENTS

Carl Schmertmann provided the numerical computer studies, performed the second order Taylor series analysis, and generally acted as a statistical consultant. Paul Bullock and John Shoucair performed the tabulations and statistical evaluations for the case C and D measurements summarized in Table 4. G. Beacher supplied the data for case E.

APPENDIX I. REFERENCES

Harr, M. E. (1987). *Reliability-Based Design in Civil Engineering*. McGraw-Hill Book Co., New York, N.Y., 30.

LaCroix, Y., and Horn, H. M. (1973). "Direct determination and indirect evaluation of relative density and its use on earthwork construction projects." *STP 523*, Am. Soc. for Testing and Materials, Philadelphia, Pa., p. 274.

Lee, I. K., White, W., and Ingles, O. G. (1983). *Geotechnical engineering*. Pitman, Boston, Mass., p. 62.

APPENDIX II. NOTATION

The following symbols are used in this paper:

G = specific gravity of soil solids;
S = degree of water saturation, as a percentage;
V = σ/mean = coefficient of variation, as a percentage;
w = water content, dry weight basis, as a percentage;
Δ = change in a test or computed result;
γ = total or wet unit weight, or density, F/L^3;
γ_d = oven dry unit weight;
γ_w = 62.4 pcf, 9.81 kN/m^3, unit weight of water; and
σ = standard deviation.

Reprinted from Grouting, Soil Improvement and Geosynthetics
Proceedings, GT Div/ASCE
New Orleans, LA/Feb. 25-28, 1992

A Design Theory For Compaction Grouting

John H. Schmertmann[1], F. ASCE
James F. Henry[2], F. ASCE

Abstract

The authors present a new theory and design method for compaction grouting to protect overhead construction against local zones of sinkhole settlement. The theory uses shear enhancement along the grout columns, due to lateral stress and density increases, to support the soils above and between the columns. Unique laboratory model tests help support the design theory. The paper includes a brief parametric study and case history example.

Introduction

Ground improvement by compaction grouting provides a method for reducing and perhaps eliminating the risk of structural damage due to the sinkhole-type slow or sudden collapse of subsurface voids. Henry (1986), Schmertmann et. al. (1986), Welsh (1988) and Henry (1989) present a variety of case histories that describe the apparent successful application of compaction grouting for this purpose. Some progressive engineer-contractors have started to develop rational design methods, for example Gambin (1991) for settlement reduction. This paper provides another concept and design method, as well as some findings from a lab study and a parametric study, and a case history.

The first author had the problem of protecting the bottom-liner system of a landfill against underlying local loss of support due to sinkhole action. Figure 1 illustrates the problem. The remedial actions taken included a compaction grout "mat" to greatly reduce the possibility of such local loss of support ever occurring. The Florida Dept. of Environmental Regulation (DER) review professionals wanted a rational design procedure for the "mat". The first author

[1]Principal, Schmertmann & Crapps, Inc., 4509 NW 23rd Ave., Suite 19, Gainesville, FL 32606

[2]V.P., Hayward Baker Inc., 6850 Benjamin Road, Tampa, FL 33614

thereupon developed the design method described herein. The second author's company later installed this compaction grout mat.

The scope of the paper includes a description of the developed design method, a parametric study using the method, as well as its use in an example calculation for the aforementioned compaction grouting project. It also includes a description of a unique laboratory test program designed to provide a preliminary confirmation of the new design analysis concepts and the values for the parametric study. All notations are defined when introduced.

New Design Concept and Analysis Method

The Concept

Using Figures 1 and 2, the concept can be summarized:

1. High pressure injections of low-slump cement mortar form a geometric pattern of grout columns on a layer with good bearing.

2. Multiple injections at suitable depth intervals form each grout column to a height h. Each helps to displace and compact the soil in the area s^2 between columns and increases the lateral pressure, denoted σ_h, and soil friction, denoted τ, at the column/soil boundaries. The lateral pressure increase eventually partly dissipates to a fraction, denoted α, of its initial value due to creep and other effects. The increase in density and stress increases the potential column/soil shear force to a value denoted T, and the top bearing to a value denoted R.

3. The column spacing s, height h, and avg. diameter d are chosen in suitable combinations until the overburden soils can be entirely supported by T and R. See Fig. 2. Any surcharge, denoted q, generates additional ΔT and ΔR support at the grout columns.

4. Removing all or part of the overburden load γz and surcharge q over voids by arching to the grout columns also removes all or part of the risk of surface settlement.

The Design Analysis

The analysis aims at achieving a certain factor of safety F, or alternatively when F < 1, a stress reduction ratio S_R. Use only effective forces and stresses. F comes from a comparison of vertical forces and S_R from a comparison of vertical stresses, as follows per square in the grid of grout columns:

Weight requiring support: Equation (1) expresses this $F_v\downarrow$ vertical downward force.

$$F_v\downarrow = (\gamma z+q)s^2 \qquad (1)$$

Forces Available for Support, $F_v\uparrow$: These vertical upward

COMPACTION GROUTING DESIGN

forces consist of T and R plus ΔT and ΔR from any surcharge. The following equations (2)-(8) and Fig. 2 describe these forces.

$$T = \tau \pi d h \qquad (2)$$

$$\tau = \sigma_h \tan \phi_g \qquad (3)$$

(wherein ϕ_g = soil/column friction angle)

$$\sigma_h = [K_o + \alpha(K_g - K_o)](\sigma_v) \qquad (4)$$

(see below)

$$\Delta T = K_o q \tan\phi_g \pi d h \qquad (5)$$

$$R = \sigma_{vc}\ \pi d^2/4 \qquad (6)$$

(wherein σ_{vc} = grouting pressure at top of columns)

$$\sigma_{vc} \simeq \sigma_{hc} \qquad (7)$$

$$\Delta R = \theta q \pi d^2/4 \qquad (8)$$

(wherein θ = stress concentration factor on top of columns)

Equation (9) defines the factor of safety.

$$F = \frac{F_v\uparrow}{F_v\downarrow} = \frac{T+\Delta T+R+\Delta R}{(\gamma z+q)s^2} \qquad (9)$$

When F < 1 we recommend using the stress reduction factor S_R in (10) with σ denoting the average vertical normal stress acting between and at the base of the columns:

$$S_R = \frac{\sigma_{after}}{\sigma_{before}} = \frac{[s^2(\gamma z+q) - (T+\Delta T+R+\Delta R)]/(s^2-\pi d^2/4)}{(\gamma z+q)} \qquad (10)$$

<u>Lateral stress, σ_h</u>: The use of the lateral stress coefficient $K = \sigma_h/\sigma_v$ provides a convenient and conventional normalized way of expressing lateral stress. Using K_o and K_g to denote this coefficient before and after grouting, and α to denote the fraction of the boundary stress increase during grouting that engineers can rely upon, after grout curing and soil creep, for the long term lateral stress in design, the stress $\sigma_h = K\sigma_v$ wherein

$$K = K_o + \alpha(K_g - K_o) \qquad (11)$$

Let us now review expected values for K_g in cohesionless soils. The writers believe that the insitu pressuremeter test limit pressure, p_L, may provide a good model for the grout/soil boundary lateral pressure required to form a compaction grout column. A pressuremeter test lifting off at p_o and carried to p_L would provide a direct measurement of the lateral stress increase. The Marchetti (1980) dilatometer p_o value also provides a direct measurement of the lateral pressure needed for an expansion somewhere between plane strain and axisymmetric conditions. The Vesic (1972) cavity expansion theory can also be used to estimate the lateral expansion pressure. Pressuremeter tests give equivalent K_g values by their p_L/p_o ratios, which typically fall in the range of 3 to 10. The K_D index in the DMT also expresses this ratio, which typically equals 1.5 in very loose, underconsolidated sands to 4 in normally

GROUTING/SOIL IMPROVEMENT AND GEOSYNTHETICS

consolidated, medium dense sands. The factor F_q' expresses approximately the K_g ratio in the Vesic cylindrical cavity expansion theory. This factor varies from approximately 2 to 8 over the same range of cohesionless soils. Based on these tests and theories, the writers expect K_g to vary from 1.5 to 10. K_g should increase with density and with the lateral grout displacement ratio β, defined in eqn. (12).

$$\beta = \left(\frac{\pi d^2}{4}\right)/s^2 \ (100\%) \qquad (12)$$

For future use, the following defines an overall grout volume strain:

$$\epsilon_v = \frac{\pi d^2 h}{4 s^2 z} \ (100\%) = \beta\left(\frac{h}{z}\right) (100\%) \qquad (13)$$

If the total grout volume determines most of the cost of compaction grouting, ϵ_v can be used as an approximate measure of total cost.

Note that α in eqn. (11) should fall between 0 and 1. Using $\alpha = 0.5$, $K_o = 0.5$, and the aforementioned expected range in K_g, Table 1 gives the writers' current suggested design values of K and tan $\emptyset_g$, based in part on the subsequently described experimental results. Soils with significantly different behavior, such as carbonate sands or lower permeability soils, may have significantly different (usually lower) K and tan $\emptyset_g$ values.

Table 1 - Suggested Design Values For K and Tan $\emptyset_g$ In Hard Mineral, Cohesionless Sands & Silts*

	K over tan $\emptyset_g$				
soil (1)	β = 5% (2)	10% (3)	15% (4)	20% (5)	25% (6)
very loose (contractant)	1.0 / 0.60	1.3 / 0.65	1.4 / 0.68	1.5 / 0.70	1.5 / 0.70
loose (neutral)	1.5 / 0.70	2.0 / 0.73	2.3 / 0.75	2.5 / 0.77	2.6 / 0.78
med-dense (dilatant)	2.5 / 0.80	3.3 / 0.82	3.7 / 0.83	4 / 0.84	4.2 / 0.85

* assumed permeable enough that significant pore pressures do not develop to dissipate after grouting

Laboratory Experiments

While the design theory described in Section 2. seems simple and conservative, the writers wanted to provide some experimental verification. Such experiments began at the University of Florida in October 1990 and were completed in June 1991, as described by Chen (1991). This lab research on compaction grouting is pioneering and the results are informative but preliminary.

Figure 2 illustrates the experimental setup, with the plan and section through the two similar sandboxes used in the experiments. The experiments involved a small scale model of the mat compaction grouting idea. Chen made the 1/2" plywood, reinforced, boxes large enough to hopefully produce negligible boundary effects. To further make such effects negligible he covered the sides and bottom of each box with a double layer of plastic sheeting with vaseline grease between the sheets. He injected 4 grout columns in a square pattern in the center of each box and instrumented one box with a load cell under each column and a stress cell at the "rock" and centered between the 4 columns. The second box, used only for the replicate "B" tests, had only the centered stress cell.

Typical Experimental Sequence

Still referring to Figure 2, plus the photos in Fig. 3, an experimental testing sequence and program consisted of the following:

1. Set up the box-bottom load and stress cells in the appropriate pattern and check that they function properly. See Fig. 3(a).

2. Fill the box with 15" of dry sand, using gentle pouring from buckets, to impose a bottom overburden pressure of 0.75 psi.

3. Place a surcharge q consisting of 19" of water (= 0.69 psi). This calibrates the stress cell. See Fig. 3(b).

4. Remove the water surcharge and inject 4 grout columns directly over the load cells and to the average height and diameter desired. The injection process involved a model grout pipe (1/2" I.D.) and air pressure injection. See Fig. 3(c). The grout equipment and injection technique took a long time and many trials to develop. Let the grout cure a minimum of one day.

5. Repeat the water surcharge and measure the stress cell Δp and load cell ΔP responses.

6. Partly dig out and expose the grout columns. Calibrate the load cells under each with axial dead loads directly on the exposed columns. See Fig. 3(d). Measure the size, shape and weight of each of the 4 grout columns as actually injected.

7. Compare the load and stress cell incremental readings due to the surcharge loadings, with the similar readings before the

grouting. The R_a defines the effectiveness of the column-induced arching at the center point between columns. R_a should vary between 0 and 1. The smaller R_a, the more effective the grout-induced arching.

$$R_a = \frac{\Delta p/q(\text{after grouting})}{\Delta p/q(\text{before grouting})} \quad \ldots\ldots\ldots\ldots\ldots \quad (14)$$

Experimental Results

Table 2 provides a listing of the experimental conditions and results. The photographs in Figure 3 show the testing equipment and some of the test steps. See Chen (1991) for the details.

Table 2 - Summary of Experimental Results (z = 15", D_r* ≃ 20%, q = 0.69 psi)

Test No.	s (in)	avg. h(in)	β(%)	measured		interpreted**	
				R_a	ρ	ρ (Fig. 5)	m_3
(1)	(2)	(3)	(4)	(5)	(6)	(7)	(8)
1A	10	9.5	7.6	0.57		0.80	0.21
B	10	9.1	3.7	.34		0.93	.79
2A	10	6.0	7.2	0.67		0.90	0.26
B	10	5.75	7.8	.45		.90	.45
3A	10	4.5	9.6	0.70		0.90	0.22
B	10	4.75	6.8	.46		.92	.66
4A	10	8.25	8.5	0.43	0.80	0.80	0.35
B	10	5.4	11.4	.25		.82	.61
5A	10	10.0	9.5	0.75		0.70	0
B	10	12.75	8.7	.23		.70	0.23
6A	20	7.8	1.6	0.80	0.96	0.99	0.90
B	20	6.0	2.7	.78		.98	.75
7A	20	8.3	3.6	1.06	1.26	-	-
B	20	9.1 (z=17")	2.8	.69		0.97	0.67
8A	10	8.1	13.1	0.28	0.41	0.60	0.30
B	10	9.4	14.0	.33		0.50	.15

*D_r denotes relative density, est. range 10-30%

** Using
$K = 1.5$
$\theta = 1.5$
$K_o = 0.5$
$m_2 = 0$
$\tan\emptyset_g = 0.75$
and calc. R_a = meas. R_a

Let us now review some of the experimental results. From the design concept presented herein one would expect that, if all else the same, the proportion of surcharge stress reaching the stress cell between the grout columns would decrease with increasing column displacement β. Figure 4 presents a graph of β vs. R_a. It clearly shows the expected trend, although with considerable experimental scatter due to experimental errors, different h/z ratios, etc. The results in Figure 4 seem to support the overall design concept.

After grouting, the surcharge induced stresses preferentially arch to the columns and their now-stiffened surrounding sand. This results in a below-average stress increase at the center stress cell. As shown in Table 2, four of the tests had the necessary grout column load cell force measurements to permit calculating the center-stress reduction factor ρ defined by equation (15).

$$\rho = \frac{\Delta p(\text{center, at stress cell})}{\sigma(\text{avg. betw. cols, see eqn. (10)})} \quad \ldots\ldots\ldots\ldots (15)$$

Figure 5 shows the results. After recognizing that $\rho = 1$ along the axes, the writers estimated contours of $= \rho$ values based on the four available tests and later modified them into Figure 5 to better match the results from all the tests.

The grouting produces primarily a net radial expansion strain. It may or may not mobilize vertical shear strain to start to mobilize T. Also, the bottom of the model box did not move downward to simulate sinkhole-type or other settlement action. The surcharging alone, without a settlement, does not develop enough column/sand vertical shear strain to fully mobilize ΔT. The analytical study of the experiments therefore included mobilization factors m_2 and m_3 for T and ΔT, respectively.

After introducing the above m_2 and m_3 factors, the writers used trials with reasonable values for the variables involved until each calculated R_a matched the measured R_a in Table 2. Values of ρ were selected from the contours in Figure 5. Table 2 presents the results using constant values for m_2, K, θ, K_o, and tan $\emptyset$, the Figure 5 values for ρ and values for m_3 between 0 and 1. These results now helped with our parametric study of the field design problem, for which any sinkhole subsidence of concern was assumed to produce enough strain to fully mobilize T and ΔT (m_2=m_3=1).

Parametric Study

The writers used a calculator program incorporating equations (1) through (13), set up dimensionless variables, and then produced an example design-guide chart. This parametric study uses the loose category in Table 1 and the K and tan $\emptyset_g$ values listed therein, plus the constant K_o and θ values listed under Table 2.

Figure 6 presents the results in the form of a dimensionless chart that should prove useful for insight and preliminary design. The writers offer the following comments based on these results:

GROUTING/SOIL IMPROVEMENT AND GEOSYNTHETICS

1. The total grout volume (ϵ_v) required at constant column spacing (s/z) to achieve a given F (or S_R) decreases with increasing β. See example F = 1 and ϵ_v = 3% comparison a in Figure 6. This occurs because of the Table 1 increases in K and tan ϕ_g with β.

2. The total grout volume required decreases approximately linearly with decreased column spacing -- at constant F. See b in Figure 6. However, this comes at the expense of more grout columns.

3. At the same β, increasing the grout column height to h/z > approximately 0.5 produces very little improvement in F or S_R.

4. For a given column spacing and height, the total grout volume ϵ_v increases in proportion to F.

5. It requires less grout volume to support an equal surcharge vs. the overburden at the time of grouting. See c in Figure 6.

Case History Example

As noted in the Introduction, Figure 1 provides a schematic for a case history. An extensive SPT boring and cone penetration testing program defined 2.65 acres of a 9.79 acre site that appeared to have a relatively high risk for sinkhole action. Hayward Baker Inc. compaction-grouted 2.95 acres in 1990. Parker (1991) describes this project in more detail.

The design engineers chose a square pattern for the primary compaction grout columns. They also chose a primary spacing s = 20 ft and a lateral volume displacement ratio β = 15%, both based on experience, and then used the theory herein to calculate the required height of grout columns for F = 1.0 from eqn. (9) when the landfill reached its full surcharge height at elev. +168. To summarize: for the design calculations using eqns. (1) to (9), made prior to the lab research described herein, the engineers considered the soils as equivalent to loose sand and used K_g = 4, K_o = 0.4, α = 0.58, q = 6000 psf from the landfill, avg. effective γ = 64 pcf, z = 55 ft, tan ϕ_g = 0.76, and θ = 2.7. For F = 1.0 they obtained h/z = 0.20, or h = 11 ft. The required delivered minimum grout pressure = $K_g\sigma_{vz}$ = 98 psi. Calculating the grout column base compressive stress gives p_c = 260 psi without the landfill surcharge, and an eventual p_c = 440 psi with full surcharge. The engineers required a minimum 2-day grout unconfined compressive strength = 200 psi to obtain the minimum 440 psi needed after full surcharge. Table 3 summarizes the design, planned vs. executed.

The Florida DER required performing some type of field verification testing to demonstrate achievement of the grouting objectives. The engineers used the SPT blowcount N-values for this purpose and established minimum after-grouting N-values for sand and silt at various depths. In all cases from 38 check borings the N-values equalled or exceeded the minimum values established prior to the compaction grouting -- see Parker (1991) for details.

Table 3 - Example Landfill Liner Grout Protection Design vs. Built

	Design	Built
No. primary grout holes (s = 20′)	289	311
Avg. depth z (ft)	55	59.3
Area grouted (acres)	2.65	2.95
Total grout volume (yd^3)	6,600	9,390
% grout, ϵ_v	3.0	3.33
Average h/z	0.20	0.25
β (%)	15	13.4
F with full surcharge, using		
pre-research parameters	1.00	1.01
post-research (this paper)	0.86	0.98

Table for Conversion to SI units:

1 inch	= 25.4 mm
1 foot	= 0.305 m
1 yd	= 0.765 m^3
1 acre	= 4,047 m^2
1 psi	= 6.89 KPa

Conclusions

1. The design concepts and analyses in this paper provide a rational basis for the design of a grout-soil "mat" that partly or completely supports the overlying soils between the grout columns.

2. The new design concept uses the lateral stress increase and enhanced soil shear arching ability that results from the lateral volume displacement from the compaction grouting.

3. The laboratory experiments and a case history described herein provide some confirmation of the new design concept.

4. A parametric study suggests that the following can improve the economy of compaction grouting:
   - o Reduce spacing (but increase No. grout columns).
   - o Increase column diameters.
   - o Use shorter columns.

For example: If q = 0, changing s/z = 0.5, β = 5% and h/z = 0.75 to s/z = 0.25, β = 15% and h/z = 0.25 increases F from 0.8 to 2.3 at the same ϵ_v = 3.75%, but uses nine times as many grout columns.

5. A case history using the design theory proved practical to construct and matched well with experience on similar projects.

GROUTING/SOIL IMPROVEMENT AND GEOSYNTHETICS

REFERENCES

Chen, Hung-Yun, (1991). "Laboratory Verification of a Compaction Grouting Technique to Minimize Settlement Induced by Local Zones of Sinkholes", Masters Thesis presented to the Graduate School of the University of Florida, 145 pp.

Gambin, M.P., (1991). "Lateral Static Densification at Monaco: Design, Construction, and Testing," Deep Foundation Improvements: Design, Construction, and Testing, ASTM STP 1089, Melvin I. Esrig and Robert C. Bachus, Eds., ASTM, Philadelphia, pp. 248-265.

Henry, J.F., (1986), "Low Slump Compaction Grouting for Correction of Central Florida Sinkholes", presented at a conference titled "Environmental Problems in Karst Terrain and Their Solutions", Bowling Green, KY, Oct. 28-30, 6 pp.

Henry, J.F., (1989), "Ground Modification Techniques Applied to Sinkhole Remediation", Proceedings Third Multi-disciplinary Conf. on Sinkholes, Florida Sinkhole Institute, pp. 327-332.

Marchetti, S., (1980), In Situ Tests by Flat Dilatometer", JGED, ASCE, Vol. 106, GT3, March, p. 299.

Parker, L.W., (1991), "Compaction Grouting Stabilizes Landfill Foundation", Prepared by Jones Edmunds & Assoc., Gainesville, FL, presented and distributed at ASCE Convention, Orlando, 26 pp.

Schmertmann, J.H., Baker, W., Gupta, R., and Kessler, K., (1986), "CPT/DMT QC OF GROUND MODIFICATION AT A POWER PLANT", Va. Tech., Blacksburg, VA, ASCE Geotechnical Spec. Publ. No. 6, pp. 303-321.

Welsh, J.P., (1988), "Sinkhole Rectification by Compaction Grouting", Proceedings, Geotechnical Aspects of Karst Terrains, Nashville, Tennessee, May.

Vesic, A.S., (1972), "Expansion of Cavities in Infinite Soil Mass", Journal SMFED, ASCE, Vol. 98, No. SM3, pp. 265-290.

ACKNOWLEDGEMENTS

HAYWARD BAKER Inc. provided the financial support for the research at the Univ. of Florida (UF), Gainesville, FL. H-Y Chen, a master's degree student in Civil Engineering working under Professor Frank Townsend, performed the laboratory research. Professor David Bloomquist provided valuable help with the experiments. Several student assistants also provided valuable help, notably Chris Boston. For the case history: The first author acted as a consultant to the designer Jones Edmunds and Assoc., Gainesville, FL. Their geotechnical engineer, Lee Parker, provided continuous and very competent field supervision during the grouting.

COMPACTION GROUTING DESIGN

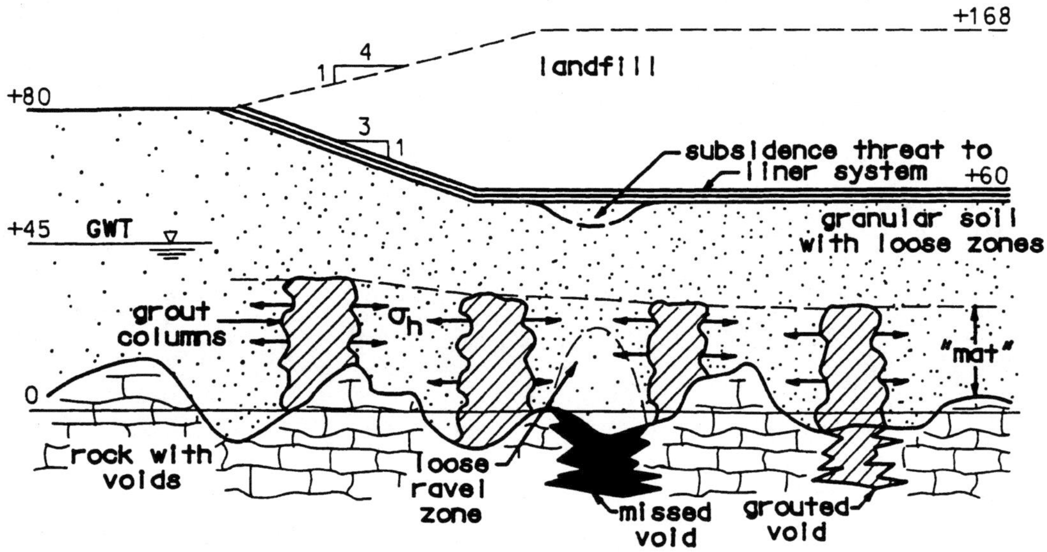

FIGURE 1 - COMPACTION GROUT SOIL BLANKET TO PREVENT VOID AND RAVEL INDUCED LINER SUBSIDENCE (not to scale)

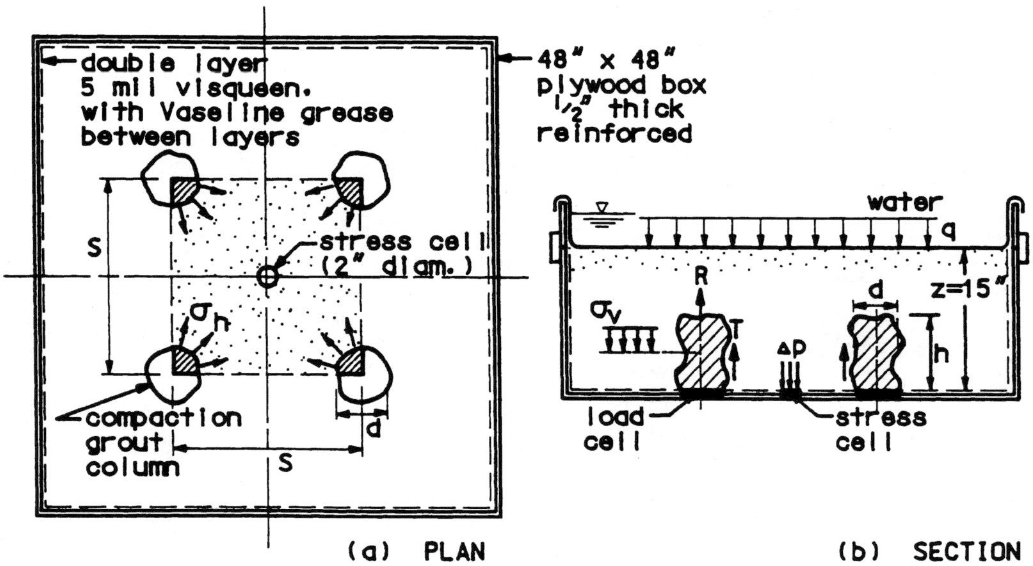

FIGURE 2 - EXPERIMENTAL MODEL TEST BOX AND SYMBOLS USED

GROUTING/SOIL IMPROVEMENT AND GEOSYNTHETICS

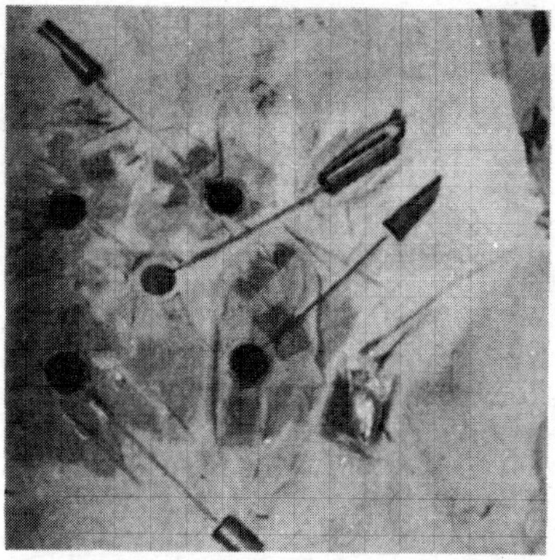

(a) Laying out the pattern of 4 column load cells & center stress cell on bottom of box.

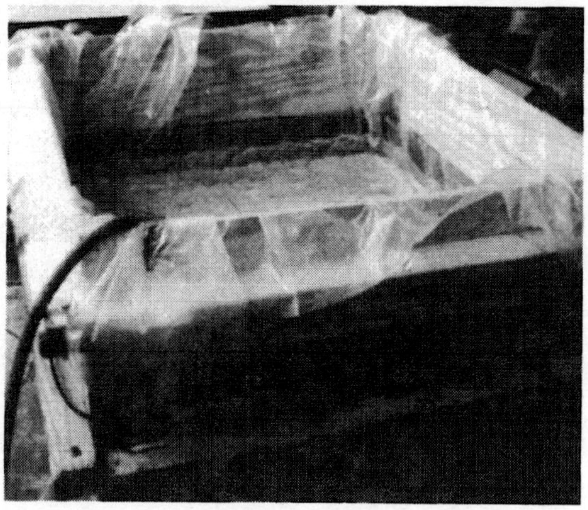

(b) Box extension and plastic liner placed for water surcharge test.

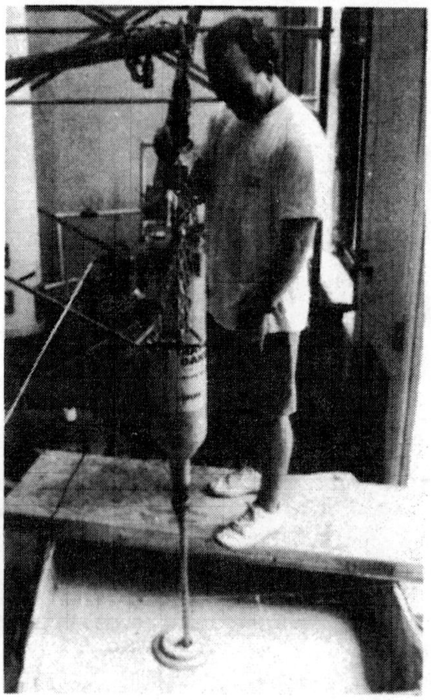

(c) Injecting grout columns by air pressure from suspended tank. Using load cell and visual measurement for flow control.

(d) Grout columns exposed after test and load cells calibrated with dead weights.

FIGURE 3

PHOTOS ILLUSTRATING THE LABORATORY RESEARCH

COMPACTION GROUTING DESIGN

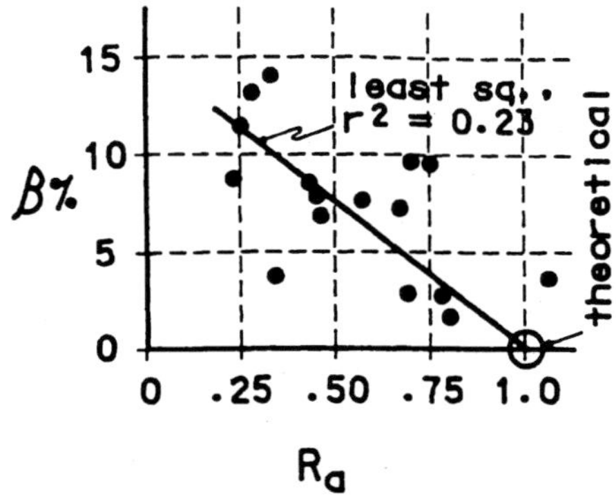

FIGURE 4 – EXPERIMENTAL DATA SHOWING DECREASING R_a WITH β INCREASING (all tests)

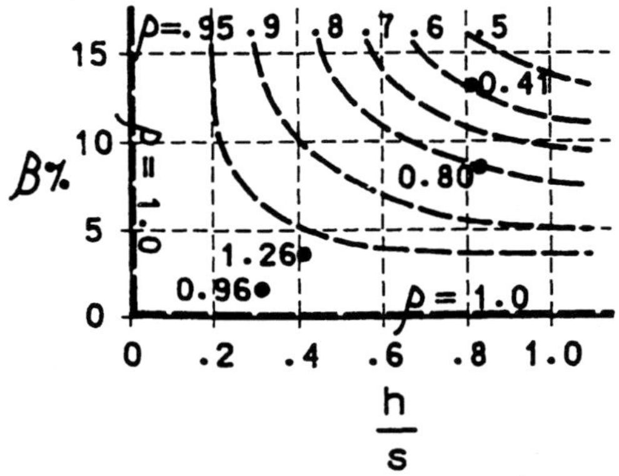

FIGURE 5 – EXPERIMENTAL DATA AND ESTIMATED ρ CONTOURS

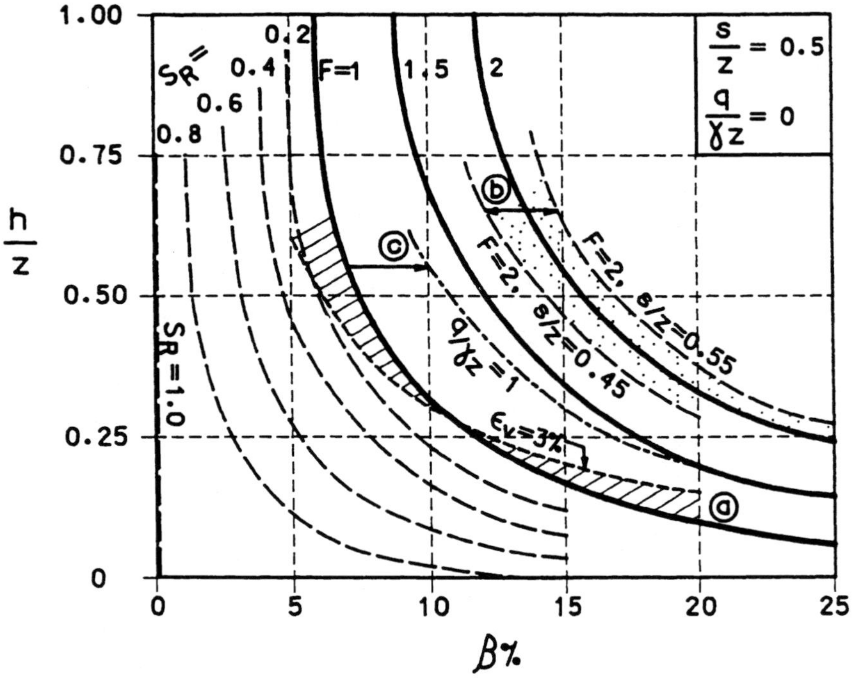

FIGURE 6 – DESIGN GUIDE RESULTS FROM PARAMETRIC STUDY

GROUTING/SOIL IMPROVEMENT AND GEOSYNTHETICS

List of Symbols

d = avg. diameter of grout columns. Fig. 2.
h = height of grout columns. Fig. 2.
m_2, m_3 = partial mobilization factors due to non-yielding box base.
p_c = compressive stress at base of grout columns.
p_L = limit pressure from pressuremeter test.
p_o = first dilatometer reading after blade insertion, total stress. Also liftoff pressure in pressuremeter test.
q = surcharge pressure.
s = side spacing of square pattern of grout columns.
u_o = pore water pressure.
z = total depth of soil above base of grout columns. Fig. 2.
D_r = relative density.
F = factor of safety of grout column support of overburden and any surcharge. Eqn. (9).
F_q' = lateral stress factor for cohesionless soil in Vesic (1972).
$F_v\downarrow$ = forces acting downward within s^2. Eqn. (1).
$F_v\uparrow$ = forces acting upward within s^2. Eqn. (1).
K = long-time effective lateral stress coefficient along grout columns after grouting. Eqn. (11).
K_D = Marchetti DMT lateral stress index = $(p_o-u_o)/(\sigma_v-u_o)$.
K_g = lateral stress coefficient at the grout/soil interface at the completion of the injection of the grout columns.
K_o = initial lateral stress coefficient and incremental value during surcharge loading.
P = end bearing force on base of grout column.
R = end bearing support force over top of columns. Eqn. (6).
ΔR = additional R due to surcharge. Eqn. (8).
R_a = arching effectiveness ratio, surcharge loading. Eqn. (14).
S_R = average normal stress reduction ratio on plane of base of grout columns. Eqn. (10).
T = potential total shear force due to τ per s^2 area. Eqn. (2).
ΔP = change in P after grouting due to surcharge q.
ΔT = additional potential shear force due to surcharge. Eqn. (5).
α = time reduction factor in grouting induced ΔK, = $\alpha[(K_g-K_o)]$.
β = grout lateral volume strain between columns. Eqn. (12).
ϵ_v = grout volume strain over entire depth z. Eqn. (13).
γ = avg. eff. unit weight of soil over z, (z-h/2) or (z-h).
σ_h = horizontal stress at grout columns. Eqn. (4), Fig. 2.
σ_v = vertical stress. Fig. 2.
σ_{vc} = grouting pressure at top of columns. Eqn. (7).
σ_{vz} = change in P after grouting at bottom of columns.
Δ_p = measured vertical stress increase on pressure cell, centered between and at base of grout columns, due to q.
ϕ_g = soil/column friction angle after grouting.
τ = avg. column/soil shear stress potential along vertical sides of grout columns. Eqn. (3).
ρ = stress reduction factor during q. Eqn. (15).
θ = stress concentration factor, surcharge on top of columns.

CONICAL TEST LOAD MEASURING COMPRESSIBILITY

By John H. Schmertmann,[1] Fellow, ASCE

INTRODUCTION

During the past 40 years the writer has occasionally found it practical and useful to perform a type of load test, herein referred to as the "conical test load" or CTL, to measure soil compressibility on a scale between in situ plate and embankment tests. The literature may have only one case history describing the use of a conical test load. The writer recommended the CTL described by Welsh (1983) and used for testing the compressibility of a sanitary landfill before and after dynamic compaction. R. Lukas (1986) briefly mentions the same CTL. This paper describes the concept of the CTL, its special advantages, and gives another case history example.

CONCEPT

The CTL involves the construction of a cone-shaped mound of soil on the ground over a center plate and measuring the settlement of the plate due to the weight of the mound. Based on the layering of the ground under the plate, plus the expected CTL-induced stress increases in each layer, the engineer can estimate the compressibilities of the underlying soil layers affected. The vertical stress is a principal stress along the axis of the cone, and therefore under the plate. The engineer can then use the compressibility information to calculate the settlements of interest.

As given by (1), Harr and Lovell (1963) used calculus to solve for the vertical stress increase σ_z under the axis of a cone-shaped, vertical loading on the surface of a homogeneous, isotropic, elastic half-space

$$\sigma_z = p\left\{1 - \frac{1}{\left[\left(\frac{a}{z}\right)^2 + 1\right]^{1/2}}\right\} \qquad (1)$$

in which p = the maximum vertical contact pressure under the cone axis, a = its base radius, and z = the depth below the base. The reader might also find use for (2), from Schiffman (1963) and also based on isotropic elasticity

$$E = \frac{p}{\Delta}a(1 - \nu^2) \qquad (2)$$

Eq. (2) permits estimating a Young's modulus E if one knows the cone axis settlement at the ground surface, Δ, and Poisson's ratio ν.

---

[1]Prin., Schmertmann and Crapps, Inc., Gainesville, FL 32606.

Note. Discussion open until October 1, 1993. To extend the closing date one month, a written request must be filed with the ASCE Manager of Journals. The manuscript for this paper was submitted for review and possible publication on December 19, 1990. This paper is part of the ***Journal of Geotechnical Engineering***, Vol. 119, No. 5, May, 1993. ©ASCE, ISSN 0733-9410/93/0005-0965/$1.00 + $.15 per page. Paper No. 1035.

ADVANTAGES

The writer recommends the conical shape in part because of its simplicity. Contractors can easily construct an approximate cone-shaped mound—most soils repose in this shape when dumped symmetrically around a vertical axis. The dumped soil, usually sand, also falls in an loose to very loose state. This minimizes any internal arching and helps achieve a vertical stress loading similar to that assumed when deriving (1) and (2).

Besides the use of a simple dead loading, the CTL has the advantage of bigness. This test applies total loads, W, typically in the range of 400–1,000 kips, that significantly stress a large volume of ground to depths not easily attained by other methods. Typical cone heights range from 8 to 16 ft and diameters from 30 to 50 ft. Despite bigness, the stable configuration of the dead loading enhances safety and permits long-time loading with minimal attention.

Again despite bigness, the advantages of the CTL include the speed and low cost of constructing a test. Assuming the often common availability of fill soil that one can temporarily divert, and a front-end-loader type of machine with enough reach to build a cone to the required height, the CTL requires very little extra equipment. One needs a plate, some short lengths of casing to build up from the plate to provide access and protection for a survey level rod, a survey level rod with sufficient extensibility to project above the maximum cone height anticipated, and a surveying level to obtain comparative elevations of the plate with respect to some distant benchmark. Using this measurement system requires that the plate area inside the casing be kept clean. This requires attaching the casing to the plate, usually by screwing on to a coupling welded to the plate. The loose soil-to-casing friction appears to have a negligible effect on the results. Two men, an engineer or technician and a machine operator, can usually perform the load buildup part of a CTL in half a day.

The conical shape of the load promotes rapid consolidation settlement. For small ratios of compressible layer thickness to CTL diameter, H/d, we have approximate one dimensional compression and drainage under the CTL axis. The larger the H/d ratio the further the deviation from 1-D conditions and the more rapid the primary consolidation because of a greater contribution by radial drainage.

Table 1 lists the time factors for 50% and 90% consolidation settlement under the plate for several H/d ratios, determined as follows: Davis and Poulos (1972) provided the time factors for a uniform cylindrical loading and its superposed vertical and radial pore pressure dissipations. To approximate the conical loading, the writer and S. Subramanian, a graduate student in civil engineering at the University of Florida, superposed 3–6 cylindrical load steps, the minimum needed to achieve three-place accuracy in the results. Each had the same step increment of pressure increase but a different diameter. The sum of all the steps matched the CTL maximum pressure p and total load W. Each step had its effect weighted according to its generation of pore pressure, determined by using the methods described in Davis and Poulos (1972). We then superposed the three to six steps to arrive at their combined effect as listed in Table 1.

Note that Table 1 applies to a consolidating layer with its top surface at the base of the CTL. The presence of an intermediate layer under the CTL makes the drainage conditions more 1-D and increases the time factors. See Poulos and Davis (1972) for the effects of an intermediate layer under a uniform, circular, surface loading.

TABLE 1. Time Factors for Combined Vertical and Radial Primary Consolidation under Center of Permeable CTL [Based on Davis and Poulos (1972) Method, Cylindrical Modified to Conical Loading]

	Drainage Top and Bottom		Drainage Top Only	
H/d (1)	T_{50} (2)	T_{90} (3)	T_{50} (4)	T_{90} (5)
(*a*) 1-D, Uniform Surcharge				
0	0.05	0.20	0.21	1.00
(*b*) Anisotropic c_h/c_v = 10 Decreases T by Additional Factor of Approximately 3 to 5				
0.25	0.03	0.17	0.10	0.69
0.50	0.02	0.145	0.05	0.395
1.0	0.009	0.12	0.023	0.21
2.5	0.0015	0.055	0.003	0.095

Note: T = conventional time factor = ct/H^2, for primary consolidation settlement of 50 and 90% of ultimate primary consolidation settlement; H = thickness of consolidating layer, from surface and directly under axis of CTL; d = base diameter of CTL; and c = isotropic coefficient of consolidation.

The reader can see from Table 1 that the pore pressure dissipation takes place more rapidly under the center of a conical loading versus under a uniform surcharge loading. This more rapid rate should sometimes allow the engineer to use the CTL to measure not only immediate compressibility but also time-delayed compressibility in soils otherwise considered too impermeable to test in situ by surcharge methods in the time available.

EXAMPLE

The example described subsequently illustrates the CTL method, the results, and its possible special usefulness. The project involved the placement of a large industrial machine foundation on a 20-ft thickness of construction debris fill. The engineering team, which included the writer, had the surface 5 ft excavated and the remaining 15 ft treated with dynamic compaction using an 18-kip, 78-in. diameter steel weight dropped 30–35 ft. The overall intensity of treatment achieved = 6,900 ft-lb/cu ft of debris fill. This treatment produced an average settlement of 1.6 ft over a 90,000-ft^2 treatment area. The contractor then placed an average 6.5-ft thickness of select granular backfill, compacted to a minimum of 95% and a minimum average of 98% modified Proctor (*ASTM D1557*).

Subsequently, 16 mechanical, mantle cone (CPT) soundings (*ASTM D3441*) probed this granular backfill. As expected, this backfill proved very strong, with many q_c values exceeding 500 tsf. Ten of the 16 succeeded in penetrating this backfill, and two almost succeeded in fully penetrating the underlying debris fill. Fig. 1 shows the CPT sounding log for one of these two soundings. This sounding had the weakest q_c profile in the debris fill. We also located the subsequent CTL over this sounding.

A problem developed at this stage of construction. The combined fill compressibility estimated from the q_c profiles in the two fills matched approximately that given to the structural engineers for preliminary foundation design. However, we did not have much confidence in the accuracy of our

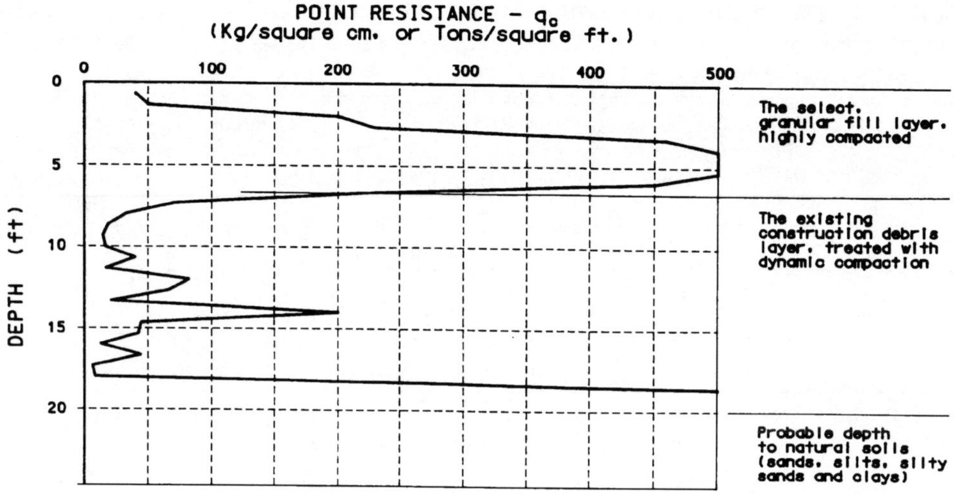

FIG. 1. Mechanical CPT Log through Fills at Location of Example CTL

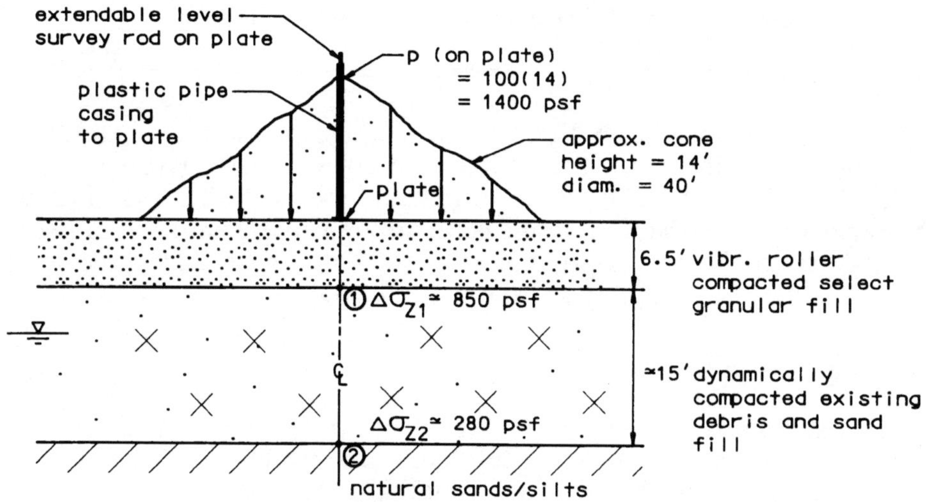

FIG. 2. Example CTL with $W \simeq$ 585 kips

q_c-compressibility correlation for a debris fill. The writer also estimated compressibility of the debris fill using elevations for the top of the debris fill before and after placement of the granular backfill. The apparent settlements divided by the pressure of the 6.5 ft of overlying fill gave a compressibility greater than the design value by a factor of 10. However, the before and after elevations also included much uncertainty. The engineers decided to resolve the problem with a CTL test.

In this case the CTL provided enough load over a large enough area for the effects to penetrate the backfill and significantly test the compressibility of the underlying debris fill. Fig. 2 shows a section through the CTL used at this site. It also gives the stress increases at the top and bottom of the debris layer as estimated by using (1) and corrections for the layered system. Because an available elastic solution for the stress distribution in a layered system under a conical loading does not appear to exist, the writer used an equivalent, uniform, circular loading and the layered solutions in NAVFAC (1982) to estimate the layering corrections. We chose the size of the CTL

(a) (b) (c) (d)

FIG. 3 Construction of Example CTL and Method of Plate Settlement Measurement: (*a*) Initial Layout: Plate with Starter Casing, Guide Circles; (*b*) Building Cone to Initial h = 6 ft; (*c*) Building cone to h = 12 ft, Helper Keeping Casing Plumb; (*d*) Settlement Measured by Level on Rod Placed on Plate and Protected by Casing: H = Final 14 ft and Measurements Taken versus Time

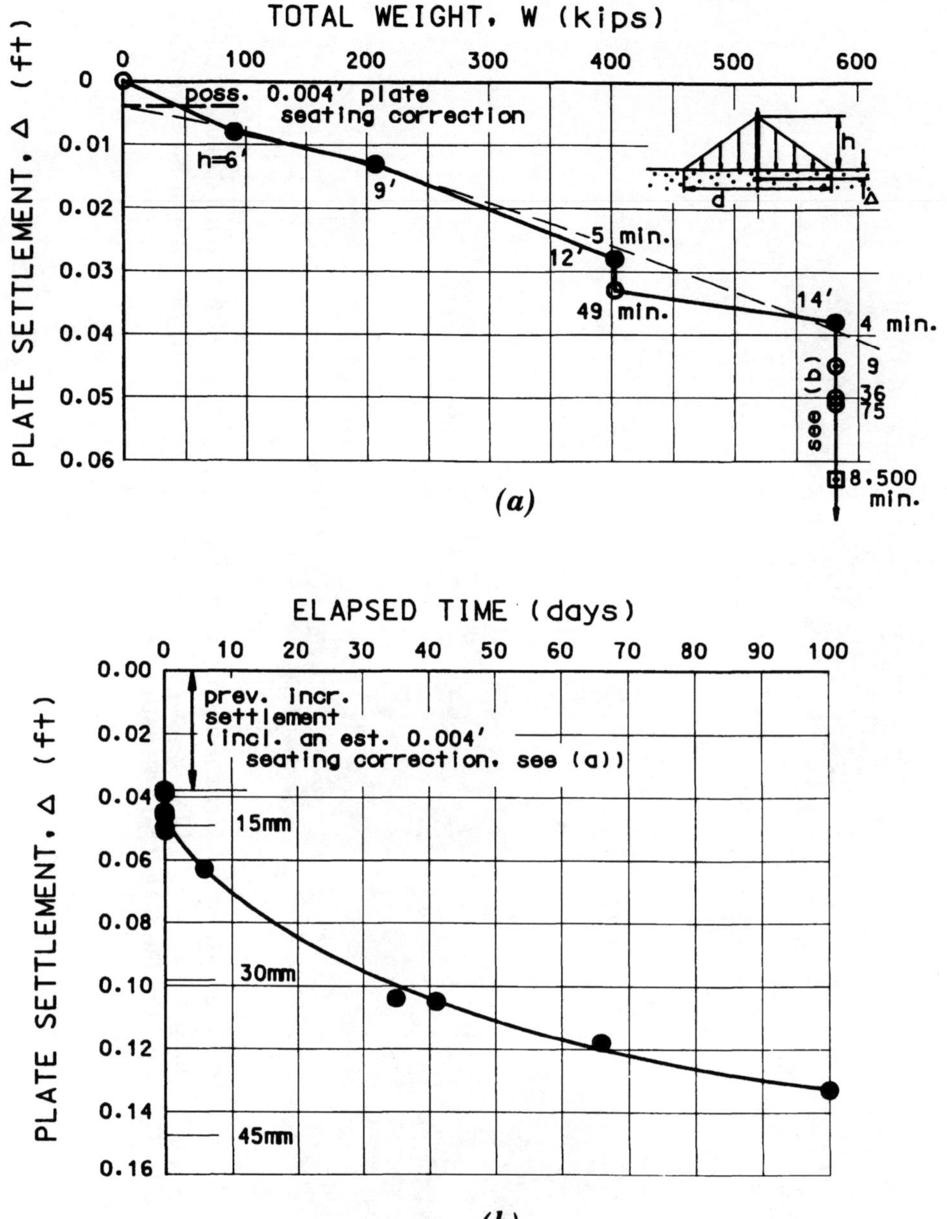

FIG. 4. Settlement-Load-Time Results from Example CTL

such that the stress increases under its vertical axis matched approximately with those expected in the debris fill under the future machine foundation.

Fig. 3 shows four photographs taken during this example CTL. It took 1 hour to locate the test point, level the fill surface, place the 17 × 23 × 1 in. steel plate, and draw a series of concentric circles in spray paint around the plate to help the front-end-loader operator get an axisymmetric placement of soil around the vertical axis above the plate. This test imposed the load in four increments, raising the height of fill along the axis to 6, 9, 12, and 14 ft, while increasing the base diameter to 12, 24, 36, and 40 ft, respectively. Push-cylinder samples of the fill gave its unit weight as approximately 100 pcf. The operator only had to travel 100–300 ft to the fill stockpile. Applying all four increments took about 5 hours. Unrelated environmental permitting problems then delayed construction. This allowed leaving the CTL in place to obtain additional data versus time at nearly constant load.

Fig. 4 shows the results from this CTL. Fig. 4(*a*) shows a reasonable and consistent, approximately linear response to the total weight of the soil cone. Fig. 4(*b*) shows the settlement-time behavior of the plate under the maximum h = 14 ft. Fig. 4(*b*) shows a gradual reduction in the rate of settlement for the first 100 days. A least-squares, hyperbolic-curve fit and extrapolation predicted a CTL settlement of 0.16 ft in five years. The five-year compressibility of the combined fill layers fell between the values obtained from the aforementioned q_c and the before-and-after elevation methods.

PRACTICAL APPLICATIONS AND CONCLUSIONS

1. The CTL provides a simple and inexpensive method to apply a large test load over a large area to measure in situ ground compressibility. One can usually impose on the soil layer(s) of special interest, the stress increase magnitudes expected from the proposed structure.
2. Because of the special geometry of the test loading, consolidation settlement in saturated soils will occur more rapidly than under a blanket loading.
3. If convenient, a CTL can remain in place without attention for a long time, to permit measuring time effects. Using the CTL may permit practical testing for time delayed compressibility.
4. The CTL can provide a method for testing the in situ compressibility of natural earth materials or fills otherwise difficult to sample and test at a smaller scale. The CTL can also provide the settlement data for an acceptance criteria for ground-improvement methods used in such materials or fills.

APPENDIX. REFERENCES

Davis, E. H., and Poulos, H. G. (1972). "Rate of settlement under two- and three-dimensional conditions." *Geotechnique*, London, England, 22(1), Mar., 95–114.

"Department of the Navy Design Manual 7.1." (1982). *Soil Mech.*, 179.

Harr, H. E., and Lovell, C. W. (1963). "Vertical stresses under certain axisymmetrical loadings." *Highway Res. Record*, 39, 68–77.

Lukas, R. G. (1986). "Dynamic compaction for highway construction, Vol. I: Design and construction guidelines." *FHWA/RD Final Report 86/133*, Jul., 30.

Schiffman, R. L. (1963). "Discussion to Harr and Lovell." *Highway Res. Record*, 39, 81.

Welsh, J. P. (1983). "Dynamic deep compaction of sanitary landfill." *Proc., 8th European Conf. on Soil Mech. and Foundation Engrg.*, 1, 319–321.

CPT '63(+) - A History in the USA from a University of Florida Viewpoint

John H. Schmertmann
Professor Emeritus, University of Florida
Principal, Schmertmann & Crapps
Gainesville, Florida, USA

1. INTRODUCTION

The following gives my recollection of my pre-banquet talk to the attendees of CPT '95 and their guests. It presents some history of the development of the CPT in the USA as I experienced it through our efforts at the University of Florida (UF), in Gainesville, Florida. This is necessarily a personal story with many references to my own work, but I believe the organizers of CPT '95 wanted such a story when they invited me to give this talk. As explained later, my CPT '63+ story began with a visit to the Swedish Geotechnical Institute (SGI) in 1963.

Over the years since my visit in 1963 I have become a proponent of insitu testing. Like many engineers in my age group, I started my career with a strong laboratory testing orientation. My bachelor thesis at MIT, and my Master and Doctorate theses at Northwestern involved primarily laboratory work. On my first job I had charge of a consulting engineering laboratory for three years. However, I now believe that most geotechnical investigation can and should be done primarily in the field.

As an example: *Figure 1* shows the Skyway bridge across Tampa Bay, costing c. $300M 10 yrs. ago, for which my partner Dr. Crapps and I had primary responsibility for geotechnical investigation, geotechnical design and construction inspection of the bridge foundations. We did this using almost entirely insitu testing (Handy et. al., 1985, Sonnefeld et. al., 1985). I can report that the bridge performed even better than predicted, and much better than predicted from laboratory tests (Schmertmann, 1988).

Figure 1

Of course, "BCPTs", meaning static cone penetration tests before the use of the abbreviation "CPT", occurred in the USA before 1963. Terzaghi used an early form of BCPT in his consulting work on the NY City subway in 1928 (Terzaghi and Peck, 1948), but it did not appear to stir any interest. Tschebotarioff (1951) tried to introduce the Dutch CPT into USA practice in 1948-49. He failed, in part because the Dutch manufacturers of the equipment he imported had too many restrictions on its use. Hvorslev designed special CPT equipment for the U.S. Army Waterways Experiment Station (WES) and used it extensively for research in the deep loose sand deposits along the Mississippi River in the 1950's. This work got publicized in 1953 (Hvorslev) and 1961 (Shockley et.al.). All this BCPT work, and possibly other unknown to me, appeared to have had little or no success in introducing BCPT's into USA practice.

2. SCANDINAVIAN BACKGROUND AND 1963 SGI VISIT

As some of you might know, I consider myself at least part Scandinavian. My wife, Pauline, who is in the audience, had a Swedish Grandmother born only a 14 kroner bus ride from this building. My Stepfather was born in Christiania (now Oslo) and started his working life as a seaman on a Norwegian whaling ship working off the coast of Africa. But my real Scandinavian conversion came after a famous Dane-turned-Norwegian invited me to come for one year to the Norwegian Geotechnical Institute. Of course I mean Dr. Laurits Bjerrum. Our family spent 14 months in Norway and Pauline gave birth to our youngest son in an Oslo suburb hospital. After returning to the USA we also had a lovely Norwegian girl, with the marvelous name of Siri Klockerstuen, live with us for one year as an exchange student.

In 1963 Dr. Bjerrum suggested I visit the SGI as the guest of its then director, Dr. M.J. Osterman and its secretary, N. Flodin. Among the many things SGI engineer T. Kallstenius showed me was the SGI's BCPT equipment. This included the one-tonn Nilcon combination CPT and field vane shear test, which I immediately fell in love with.

3. 1963(+) AT U.F.

I got back to the USA in 1963, ordered the Nilcon unit for the University of Florida and used it in a San Francisco Bay mud project during the summer of 1964. It produced excellent results -- to my knowledge for the first time showing the linear increase in s_{uv} with depth in this clay and the projected Jakobsen origin cohesion (see Schmertmann, 1976).

However, when I returned to the sands of Florida it became obvious that the equipment did not have enough penetration thrust. We did not really have a practical CPT until 1965 when we got a 10-tonn unit from the Goudsche Machine Fabriek. I discovered, as had Tschebotarioff, that the Dutch are sometimes difficult businessmen. Despite considerable effort, there was no way that I could beg a price discount for the university. We paid full price. On the other hand we found it very well made and it seems to last forever - a consultant in the Washington, DC area recently purchased it and now operates this same equipment 30 years later!

In 1965 I already knew that the CPT would provide a valuable tool in the practice of geotechnical engineering. I decided to promote the CPT for use by USA practitioners. Considering the previously mentioned lack of success, it seemed apparent that I needed to make a large effort and use a systems approach as shown in *Table 1.* The delivery of site investigation and geotechnical design services involves engineer and client education, field equipment availability and suitability, citeable literature (few civil engineers want to be first), reference standards and/or specifications, the blessing of governmental agencies, design manuals, and realistically the potential for increasing rather than diminishing profits. My whole CPT '63(+) experience perhaps constitutes a case history on how to accomplish a major innovation or change in the practice of civil engineering - perhaps the most conservative and slow-to-change of the engineering professions.

Table I

CPT → USA

A SYSTEMS APPROACH

- EDUCATION
- LECTURES
- RESEARCH— PAPERS
- CONSULTING DEMONSTRATIONS, PROFITS
- STANDARD FOR SPECIFICATIONS
- MANUAL FOR DESIGN

On the educational front: I started with UF

Table II

FUGRO FELLOWS

1974	Z. AL-AWKATI
1975	M. TUMAY
1976	R. WOODS
1977	R. CAMPANELLA
	M. ARNOLD
1978	K. SENNESET

CONTINUED AT LSU

students and in 1967 introduced perhaps the first insitu testing course in North American, where these future engineers got hands-on experience with different insitu testing methods -- of course including the CPT. The entire course then consisted (and still consists) of one afternoon per week of performing a different insitu test and then preparing a report. We also stimulated interest and spread the word via a series of post-doctoral students, listed in *Table 2*, that came to UF to learn about and from what we were doing. I am proud say we have one in the audience today, Dr. Tumay. Fugro, Holland sponsored this post-doctoral program largely through the efforts of Director Jaap deRuiter. I'll have more to say about Jaap later on.

I also talked to engineering groups about the CPT whenever I had the opportunity. I got the attention of local consultants by taking the occasional job from them (especially from local architects who immediately saw the advantages of the CPT method), but later I helped them by providing specialty CPT subconsultant services when they requested such services. These roles have now reversed, which I consider a sign of success.

In approximately 1973 I started with an American Society for Testing and Materials (ASTM) subcommittee to develop a CPT standard, which got published in 1975. Incidently, I found it relatively very easy to get an ASTM standard accepted for something not yet in use and that few people in the USA had any experience with. I had everyone in the USA who knew much about it on my committee. Everyone else didn't know enough to raise the usual host of objections to arriving at the consensus needed to achieve an ASTM standard. This standard, D3441, has now served well for 20 years and has helped others to write similar standards or reference procedures for the CPT. D3441 is currently under major revision to better accommodate the needs of electric, piezoocone, and environmental testing.

In addition, with the help of Jaap de Ruiter and Fugro, I wrote a design manual and made a videotape about the CPT for the Federal Highway Administration (Schmertmann, 1978). This added to CPT credibility.

Of course, we also did research at the University of Florida, both in the field and in the laboratory (large chamber testing). The published results stimulated the use of the CPT for problems of interest to both practitioners and academics, especially for determining the settlement and relative density of sands.

We also had to show the local engineers how they could use their SPT drill rig equipment to perform the CPT and, given the usual much greater daily footage achievable with the CPT, how the CPT could earn money for them. We then showed how a special cone truck (we got our first in 1971) could earn even more money! All of the above were done more or less simultaneously and with a rough coordination based on expediency. I believe that all of the above factors in the system played a part in the success we have achieved in making a perhaps substantial wedge in the former almost complete dominance of the SPT for geotechnical investigation and design in North America. Of course, the arrival of the electric CPT, the piezocone, and the computer, plus all the CPT-related offshore and environmental business, has helped greatly to expand the demand for CPT work. I have prepared the chronological *Table 3* to give the reader at least an approximate but still incomplete accounting of the progression of CPT events as we experienced them at the

Table IIIA

SOME KEY UF-USA CPT DATES

1964/65	SWEDISH/DUTCH EQUIPMENT TO U.F.
1965	STARTED FIELD RESEARCH
1967	STARTED INSITU TESTING COURSE
1970	STARTED CHAMBER TESTING
1970	SAND SETTLEMENT PAPER
1973	STARTED "CPT"
1974	STARTED FUGRO FELLOWSHIP

Table IIIB

MORE DATES

1975	ASCE RALEIGH CONFERENCE
1975	ASTM D3441 PUBLISHED
1978	FHWA MECH. & ELEC. CPT 'GUIDELINES' PUBLISHED
1981	HOGENTOGLER CPT TRUCKS FOR SALE (IN USA = 50+)
1984	HOGENT. 'GUIDELINES' FOR ELECTRONIC CPT
1986	ASCE VA. TECH CONF.
1988	ISOPT-1 AT DISNEY WORLD

University of Florida.

4. "CPT"

Perhaps a small item in the success of the CPT, but of special interest to this CPT '95 Symposium, concerns the history of the abbreviation "CPT". As best I can tell the "CPT" is now 22 years old. As I recall, Jaap deRuiter and I had a discussion in 1972 that included the subject of how to best refer to the Dutch static cone penetration test in some abbreviated form, both for convenience and to compete with the "SPT" in common usage as an abbreviation for the standard penetration test. We considered "DSCPT" for Dutch static cone penetration test, "DCPT" for Dutch cone penetration test, and others. I suggested, "Why not just CPT?" and he agreed that sounded pretty good. So we went "Dutchless" with the bare "CPT".

According to my records I first used "CPT" in print in 1973 (Schmertmann, 1973) and then several times in the more widely distributed ESOPT Proceedings (Schmertmann, 1974). Nobody seemed to object so I used CPT freely in my 1975 Raleigh conference paper (Schmertmann, 1975) and, as shown in *Table 4*, even expanded the abbreviations to include VST for vane shear test, PMT for pressuremeter test, BST for borehole shear test. The list has grown since and continues to grow. We will soon need a dictionary!

Table IV

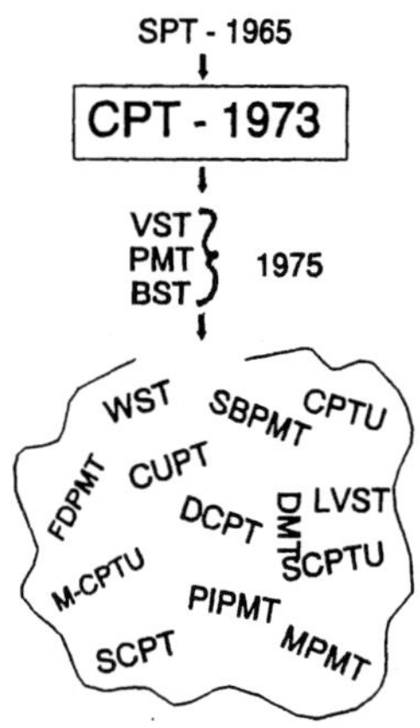

5. TECHNICAL SUMMARY

Perhaps I can begin to end this before-banquet, and perhaps more importantly before-drinks, talk with a confession. First, it's my first before-banquet talk. In fact, as I told Prof. Massarsch, it's the first I ever heard of. So, I am somewhat at a loss of how much technical content to include. However, on the assumption that some of you will be disappointed if I don't make at least a few technical comments, I thought you might find it of interest for me to list and comment upon what I think we learned from the CPT at the University of Florida after all those years of

Table V

WHAT DID WE LEARN AT U.F. ABOUT THE CPT?

- HORIZONTAL STRESSES DOMINATE
- USEFUL TO PREDICT SAND SETTLEMENT
- EXCELLENT FOR SOIL PROFILING
- NEED SYSTEMS APPROACH TO 'SELL'
- PORE PRESSURE EFFECTS MINOR IN SANDS
- CHAMBER CALIBRATION TESTS USEFUL
- HELPS TO UNDERSTAND SPT

effort. Please refer to *Table 5*.

• We learned almost immediately that horizontal stresses dominate the cone bearing capacity and I presented this at FUGRO (Schmertmann, 1972), in a paper which stands out because of it's brevity (2 1/2 pages) and as an island of English in a sea of Dutch.

• We then discovered a useful correlation between the CPT and predicting the settlements of shallow foundations in North Florida sands, as detailed in my widely used sand settlement paper (Schmertmann, 1970).

• We discovered for ourselves, as has everyone else using the CPT in the field, that even the mechanical CPT provides an excellent tool for soil profiling and site investigation. This has been improved even further by the electric CPT and piezocone testing.

• As I mentioned before, we found it necessary to use a systems approach to sell the CPT to practicing engineers. One might also consider it as a "carrot and stick" approach -- the carrot of better engineering and possibly better profits, the stick of the potential loss of consulting work if you don't have the CPT available.

• We learned from our extensive calibration test efforts that pore pressure effects had only a minor influence in sands and that we did not have to saturate them. We could test them dry. We also produced useful calibration charts for relative density -- they provided a reference set of values that engineers found useful in ground improvement specifications. They have now been largely replaced by the more extensive Italian chamber test results, for which attendees ISSMFE President Mike Jamiolkowski and theme lecturer Dr. G. Baldi deserve a lot of the credit.

• Finally, at least in the present listing, we learned that the end bearing and side shear on the CPT friction cone provided a very useful model toward a better understanding of how to interpret quantitatively the SPT blow count. I detailed this in Schmertmann (1979). Anything useful about the SPT is a 'big deal' in the USA.

6. STOCKHOLM CITY HALL

When I visited Stockholm in 1963 I discovered something about the Stockholm City Hall building that intrigued me and which I often mentioned to my students. I assumed that all Swedish geotechnical engineers knew this, but when I checked with Professor Massarsch I found that many probably did not. This had me worried because I thought I might have been mistaken. However, when I checked with Professor Broms he confirmed what I had heard in 1963. Because I thought this apparently little known fact might interest and intrigue you also, and because it shows the marvelous ingenuity of Swedish geotechnical engineers, I close with this item:

Figure 2

My daughter Joy, a graduate biologist and

an ardent environmentalist, sometimes considers civil engineers "the enemy". I often tell her that we don't fight Mother Nature, but rather try to help. The Stockholm City Hall, shown in *Figure 2*, provides a good example. By walking around the outside of this large and beautiful castle-like structure you will see surrounding bedrock. The City Hall sits on bedrock as befits such a structure. But, where the architects wanted bedrock Mother Nature supplied only many meters of soft river deposits. The engineers got around this problem by supporting the rock on piles. We have all heard of piles supported by rock, but this is the only example I know of with rock supported by piles! Perhaps the site investigation for the pile design included BCPT's?

7. LIST OF REFERENCES

Handy, R.L., Schmertmann, J.H., and Lutenegger, A.J.(1985). BOREHOLE SHEAR TESTS IN A SHALLOW MARINE ENVIRONMENT, *Strength Testing of Marine Sediments: Laboratory and In-Situ Measurements, ASTM STP 833*, R. C. Chaney and K. R. Demars, Eds., American Society for Testing and Materials, Philadelphia, pp. 140-153.

Hvorslev, M.J. (1953). CONE PENETROMETER OPERATED BY ROTARY DRILLING RIG, *Proc. 3rd ICSMFE*, Vol. 1, p. 236

Schmertmann, J. H. (1970), STATIC CONE TO COMPUTE STATIC SETTLEMENT OVER SAND, ASCE, *Journal of the Soil Mechanics and Foundations Division*, May, pp. 101-1043.

Schmertmann, J. H. (1975). MEASUREMENT OF INSITU SHEAR STRENGTH, ASCE *Proceedings of a Conference on insitu measurement of soil properties*, Vol. II, State-of-the-Art Report for Session III, June, pp. 57-138. Discussions and closure to p. 179.

Schmertmann, J. H. (1976). THE SHEAR BEHAVIOR OF SOIL WITH CONSTANT STRUCTURE, *Laurits Bjerrum Memorial Volume*, Norwegian Geotechnical Institute, pp. 65-98.

Schmertmann, J. H. (1979). "STATICS OF SPT", ASCE, *Journal of the GED*, GT5, May, pp. 655-670.

Schmertmann, J. H. (1988). DILATOMETERS SETTLE IN, *Civil Engineering*, March, pp. 68-70.

Schmertmann, J. H. (1974). CPT TO HELP UNDERSTAND SPT, *Discussion to ESOPT I*, June, Stockholm, Vol. 2.1, p. 146.

Schmertmann, J. H. (1972). EFFECTS OF IN-SITU LATERAL STRESS ON FRICTION-CONE PENETROMETER DATA IN SANDS, *Fugro Sondeer Symposium* (Holland), October, pp. 37-39.

Schmertmann, J. H. (1973). SETTLEMENTS OF STRUCTURES ON GRANULAR SOILS -- ESTIMATING PROCEDURES AND CASE HISTORIES", ASCE NYC Met. Section Seminar on Settlement of Structures, *Proceedings*, May, Part II, 26 pp.

Schmertmann, J. H. (1974). PENETRATION TESTING IN THE USA, State-of-the-Art Report, *Proceedings European Symposium on Penetration Testing (ESOPT)*, Stockholm, June, Vol. 1, pp. 217-218.

Schmertmann, J. H. (1978). GUIDELINES FOR THE CONE PENETRATION TEST PERFORMANCE AND DESIGN, U.S. Department of Transportation, *FHWA-TS-78-209*, July, 145 pp.

Shockley, W.G., Cunny, R.W. and Strohm, Jr., W.E. (1961). INVESTIGATIONS WITH ROTARY CONE PENETROMETER, *Proc. 5th ICSMFE*, Vol. 1, p. 523.

Sonnenfeld, S.L., Schmertmann, J.H. and Williams, R. C. (1985). A BRIDGE SITE INVESTIGATION USING SPTS, MPMTS, AND DMTS FROM BARGES, Strength Testing of Marine Sediments: Laboratory and In-Situ Measurements, *ASTM STP 883*, R. C. Chaney and K. R. Demars, Eds., American Society for Testing and Materials, Philadelphia, pp. 515-535.

Terzaghi, K. and Peck, R.B. (1948). *SOIL MECHANICS IN ENGINEERING PRACTICE*, p. 278. In second edition, 1967, see p. 321.

Tschebotarioff, G. P. (1951). *SOIL MECHANICS, FOUNDATIONS, AND EARTH STRUCTURES*, McGraw-Hill, p. 357. In second edition, 1973, see p. 44.

OBSERVATIONS FROM OSTERBERG CELL TESTS OF BORED PILES

by John H. Schmertmann, Professor Emeritus, University of Florida,
Principal, Schmertmann and Crapps, Inc., Gainsville, Florida,
and
John A. Hayes, president, LOADTEST Inc, Gainsville, Florida

Introduction

The availability of the Osterberg cell (O-cell) testing method, permitting the application of the very large loads with an automatic separation of side shear (SS) and end bearing (EB), will likely have a major impact on the design and construction of bored piles (drilled shafts). This paper presents data in the form of summarized case histories from the files of LOADTEST Inc which show that engineers often grossly underestimate the capacity of well-constructed bored piles and that contractors sometimes use construction methods that grossly reduce capacity.

We will start with a brief review of the O-cell test method. It involves placing a calibrated hydraulic jacking device, the Osterberg cell, at or near the bottom of a pile and then expanding it by applying hydraulic pressure. The shaft above the cell moves upward, resisted by side shear. The bottom moves downward, resisted primarily by end bearing. An array of telltales, LVDTs, and electronic gages provide the pile movement and compression data. LOADTEST Inc. uses the ASTM D1143 Quick Load Test Procedures, and presents the results in the form of upward SS and downward EB load-movement curves with creep or yield limit loads indicated. Combining these data allows construction of a load-movement curve that simulates an equivalent, conventional 'top-loaded' curve with an indicated creep or yield limit load.

Ordinarily, only one of the resistance components, shaft shear (SS) or end-bearing (EB) depending on the design of the test, reaches an ultimate or 'failure' loading, which we herein denote 'Q'. A top-loaded test would then have a combined SS + EB ultimate resistance that exceeds 2Q by some unknown amount. In the following discussion, we have conservatively taken 2Q as equal to the ultimate resistance.

Measured vs. Estimated Capacities

A review of the test results from LOADTEST, Inc. files indicates that, in most cases, the measured capacity of the test shaft exceeds the designer's estimated capacity. Ironically, the ratio of measured to estimated capacity (M/E) tends to increase as the strength of the supporting strata improves. Fig. 1 provides a summary of the M/E ratio data from 25 projects where we have information regarding the engineer's estimated capacities. Tables 1.1, 1.2, and 1.3 provide more detail about some of these projects in soils, intermediate materials, and bedrock, respectively.

We based the horizontal positioning on the graph within the material classifications shown on a subjective assessment of material characteristics. The data points with an upward arrow indicate tests in which neither component reached an ultimate resistance and, therefore, the test would plot at some unknown higher M/E ratio had the O-cell had enough capacity to reach an ultimate.

The O-cell testing shows that in "normal" soil conditions (sand, silt, and clay strata) the engineers usually provide reasonably good estimates, with M/E ranging from about 0.7 to 2.

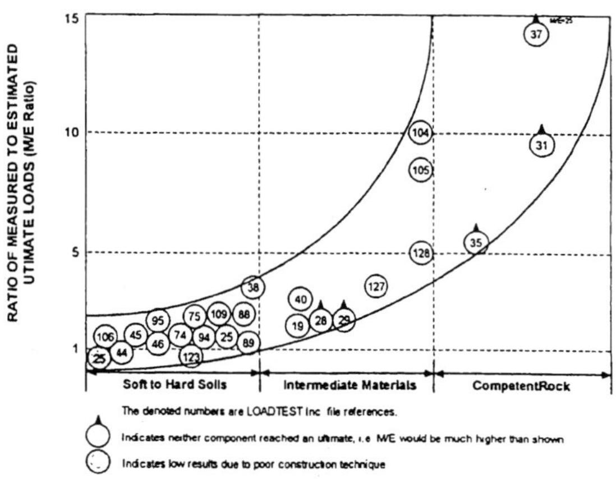

Fig. 1 Ratio of Measured to Estimated Ultimate Load

However, in intermediate geomechanical materials (soft rocks, weathered rocks, coarse dense soils, and glacial tills), the measured/estimated (M/E) ratio generally increases into the 2 to 5 range. This no doubt reflects the difficulty in measuring and estimating the strength parameters of these materials, coupled with natural engineering conservatism. In unweathered competent rock formations, the M/E ratios in Fig. 1 become even higher, from 7+ to 25+. Paradoxically, it would appear that geotechnical engineers have tended to increase the "factors of safety" applied to bored pile foundations as the competency of the founding formation improves. However, e suspect that, to some extent, the high M/E ratios in competent rock have resulted from the inability (prior to the O-cell) to test the high capacity of these foundations and to generate the experience essential to a confident design with smaller margin of safety. Or, in other words, the high M/E ratios reflect "factors of uncertainty" which engineers can reduce only after improving the understanding by using better testing methods that provide more definitive information from the field.

The Influence of Construction

Although geotechnical engineers have made, and continue to make, great strides in understanding soil and rock mass behavior under load, our understanding of the interaction between foundations and the surrounding soil or rock mass continues to be uncertain. This is especially true for bored pile foundations, which are built within the soil or rock mass. The in-situ construction process introduces many disturbing forces, and, sometimes, materials, which can have a major effect on the subsequent behavior of the bored pile under load. In some cases, the impact of the construction process can reduce actual bored pile capacities to 5% to 50% of the capacity achieved with better construction methods. Tables 2.1 and 2.2 provides examples separated on side shear and end bearing, respectively.

The O-cell test method readily provides evidence of any abnormality related either to side shear or to end bearing capacity. This is an important distinction as compared to top load testing methods. A top load test may indicate apparently reasonable test capacity when, in fact, either the side shear component or the end bearing component could be seriously compromised by construction technique. The O-cell test by contrast relies on, and measures, both end bearing and side shear capacities and a deficiency in either soon becomes obvious. The examples in Tables 2.1 and 2.2 illustrate how the O-cell test method has provided the comparative data we need to help our understanding of the impact of construction technique on bored pile capacity.

Side Shear. In Example SS 23, the contractor constructed the bored pile by "thinning" the soil with water and bentonite when excavating the shaft. He employed a drilling auger to churn the sand-water-bentonite mixture and placed the concrete by lowering it through the "thinned" soil slurry using a cleanout bucket. The process resulted in highly disturbed soils both along the shaft side walls and at the shaft base. The contractor later changed the construction procedure to lower the sand content in the slurry to about 8% and placed the concrete by pumped tremie methods. The testing showed that the side shear increased from 15 kPa to 70 kPa as a result of these improved construction procedures.

In Example SS 27, the contractor attempted to drill the shaft "dry," with no slurry to counteract hydrostatic groundwater pressure. The residual soils were impermeable enough that seepage became evident only when drilling near the shaft base. Disturbance to the surrounding soils was massive, however, resulting in an ultimate load in side shear of no more than 440 kN. A second test shaft constructed using water as a stabilizing fluid resulted in an ultimate load of 6,100 kN in side shear.

In Example SS 108, a rock socket test shaft was drilled "dry" with a casing extending to the bottom of the shaft but with no water inside the casing. The hydrostatic head outside the casing was 12.2 m. The 3.7 m deep rock socket was then filled with concrete followed by pulling the casing back about 4.0 m. Virtually no side shear bond developed between the shale and concrete. The test shaft was reconstructed with casing extending only to the top of the rock socket and with water inside the casing to counteract hydrostatic pressures. Concrete was placed by tremie methods. The side shear capacity of the reconstructed shaft exceeded the 10,000 kN capacity of the O-cell.

In Example SS 93, the test shaft was drilled through 30.5 m of saprolite clay using a rotating core barrel type casing technique. The material inside the casing was removed with a clam type digging tool. The casing was pulled in sections as the concrete was placed. The O-cell was set at a depth of 15 m. A special "rifling" tool was used to roughen the side walls of the section of shaft above the O-cell. The test results showed that the lower smooth-walled section of the shaft and the base failed for an ultimate SS load of 7,100 kN at a movement of 84 mm. In contrast, the upper rifled

Fulcrum The Newsletter of the Deep Foundations Institute ***Winter '96-'97***

section sustained a shear load of 10,000 kN at a movement smaller than 2 mm.

End Bearing. Examples EB 128, EB 141 and EB 142 illustrate the impact on end bearing capacity of loose material at the base. In all three cases, the test shafts were drilled "dry" (i. e., no stabilizing fluid) making it possible to visually inspect the base of the shaft before placing the concrete. Rough estimates of the depth of loose material at the base were made by "sounding" with a weighted tape and by visual observations.

Fig. 2 shows an end bearing load-movement curve from EB 128. This extreme example dramatizes the point. The testing shows that for all practical purposes the base capacity was lost (a movement of 25 mm occurred for a load of only 250 kN). It should be noted that, in all of these examples, a typical top-loading test would not have revealed a significant problem with the test shaft capacity because the side shear capacity would have masked the end bearing deficiency.

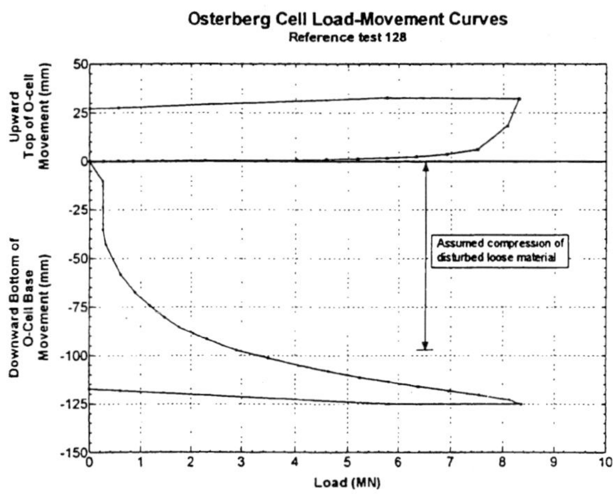

Fig. 2 Measured O-Cell movement versus Load

Such experiences have led us to question the potential for bottom-condition problems when the use of water or drilling mud prevents a simple visual inspection. As we have seen when we inspect such shaft bottoms with a special video camera system designed for this purpose, it often requires considerable effort to provide a "clean" bottom under wet conditions. These experiences also reinforce the importance of performing a loading test to determine the effect of construction technique as well as to assess side shear and end bearing parameters.

Conclusions

Our experience shows that, generally, the stronger the supporting material, the larger the ratio between the measured shaft capacity and the prior engineering estimate of capacity. This ratio often exceeds a value of 10 in rock sockets!

The techniques of shaft construction play an important part in subsequent shaft capacity. Improper hydrostatic balance, not desanding a slurry, poor bottom cleaning technique, failure to roughen side walls, dropping concrete through water, and premature casing withdrawal can all seriously reduce capacity. A conventional top-loading test may mask such defects, but they become obvious in an O-cell test.

TABLE 1.1 CASE HISTORIES IN SOILS

FILE REFERENCE	38	46	106
LOCATION	Truth or Consequences, NM	Los Angeles Coliseum, Los Angeles, CA	Owensboro, KY
PURPOSE	Test of 2.2 m bored pile for remedial foundation system	Tested 28 (1.1 m diameter) bored piles to allow a Factor of Safety of 1.5	Isolate shear and end bearing characteristics
METHOD	Cased to 3.0 m, then drilled under polymer slurry to 22.3 m. O-cell placed 4.0 m above bottom. Concrete to 5.8 m from surface	Drilled uncased and dry to 19.8 m. O-cell placed at bottom	1.5 m diameter bored pile under water to 23.5 m. One O-cell placed at 17.7 m and another one at the bottom (multi-level)
CONDITIONS	Interbedded cobbly sand/sandy gravel N = 20 - 86 GWT at 13.4 m intermediate material	Interbedded sand and gravel, compact to dense	Clay to 3.0 m, then fine to coarse sand with silt layers N = 7 to 30 along shaft, N = 30 at base
RESULTS	Tested to 3,770 kN Estimated Ultimate Load = 10,300 kN M/E = 3.7	Tested to 10,700 kN M/E ≈ 1.2	Tested to 11,300 kN EB = 3,100 kN+ SS = 8,200 kN M/E = 1.3

GWT denotes groundwater table
N denotes standard penetration test blow count

Fulcrum The Newsletter of the Deep Foundations Institute ***Winter '96-'97***

TABLE 1.2 CASE HISTORIES IN INTERMEDIATE MATERIALS

FILE REFERENCE	19	28	29	127
LOCATION	Forest Glen, MD	Owensboro, KY	Owensboro, KY	East Milton, MA
PURPOSE	Determine end bearing and side shear compon-ents for design of a 0.9 m diameter bored pile	Determine end bearing and side shear to confirm design parameters of a 1.8 m diameter rock socket	Determine end bearing and side shear to confirm design parameters of a 1.8 m diameter rock socket	Determine end bearing and side shear to allow shorter shafts if possible
METHOD	Cased to 5.5 m then drilled "dry" to 15.5 m. O-cell at bottom, concrete placed in 2 stages to get ultimate shear and end bearing. First stage with 1.5 m socket. Second stage with 9.4 m socket.	Cased to top of rock at 27.4 m, then drilled under polymer slurry to 45.7 m. O-cell at bottom of shaft.	Cased to top of rock at 27.4 m then drilled under polymer slurry to 33.5 m, O-cell at bottom of shaft	Cased to a depth of 2.8 m then drilled "dry" to 9.5 m. O-cell at bottom of rock socket
CONDITIONS	Clay to 4.6 m, then weathered schist to 19.8 m. Unconfined compression strength 500 psi to 2500 psi.	Water to 18.9 m, then sand and gravel to 27.4 m on shale and limestone.	Water to 18.9 m, then sand and gravel to 27.4 m. Then shale and limestone with coal seams.	Earth Fill to 1.5 m then shale to 15 m+. Unconfined Compressive Strength = 17.2 MPa to 20.7 Mpa. RQD = 17
RESULTS	SS Stage 1, averaged 720 kPa at EB = 8,600 kN Total SS exceeded 17,800 kN. M/E > 2	SS = 24,500 kN at 7 mm. EB = 26,700 kN at 81 mm. M/E > 2. Ultimate load not reached	SS = 25,800 kN at 7 mm. B = 27,600 kN at 34 mm. M/E > 2 Ultimate load not reached	SS = 27,600 kN at 13 mm. EB = 28,000 kN at 20 mm. M/E = 3.3

Movement is downward for EB and upward for SS

TABLE 1.3 CASE HISTORIES IN BEDROCK

FILE REFERENCE	35	31	129
LOCATION	Decatur, AL	Burgin, KY	Stillwater, MN
PURPOSE	Verify design parameters, leading to adjustments where feasible for a 0.9 m diameter shaft	Proof test on a 0.9 m production shaft with 1,800 kN design load)	Pre-design test to determine shear parameters for axial test on 1.2 m diameter shaft. (Deep-seated lateral load test also carried out).
METHOD	Cased to top of rock socket at 17.7 m then cored to 22.6 m under water. 26,700 kN O-cell placed at 22.1 m. Concrete-filled rock socket to 17.7 m level (i.e., 4.4 m socket)	Cased to top of rock at 3.0 m, then cored "dry" to 6.1 m. O-cell placed on bottom. Concreted to form 2.0 m socket.	Cased to 15.0 m, then drilled to 55.0 m. O-cell placed 4.5 m above bottom (i.e., 3.5 m socket above O-cell)
CONDITIONS	Fill sand and clay to 8.2 m, then limestone with some voids and clay seams. Uncon-fined Compr. Strength of limestone = 100 - 150 MPa GWT at 1.8 m	Earth fill and sand to 3.0 m, then limestone. Unconfined Compr. Strength = 100 - 140 MPa	Overburden to 15.0 m then dolostone over sandstone. Unconfined Compr. Strength = 50 - 58 MPa
RESULTS	SS = 22,200 kN at 18 mm EB = 22,700 kN at 36 mm M/E > 5.5	SS = 8,000 kN at 2 mm EB = 8,000 kN at 4 mm M/E > 9.3	SS = 21800 kN at 33 mm SS + EB = 21,800 kN, in low-er socket at 15 mm

GWT denotes groundwater table
Movement is downward for EB and upward for SS

TABLE 2.1 IMPACT OF CONSTRUCTION TECHNIQUE — SIDE SHEAR

File Reference	SS 23	SS 27	SS 108	SS 93	SS 93
Diameter	990 mm	910 mm	910 mm	1520 mm	1520 mm
Length of Test Shaft	5.8 m	18.3 m	3.7 m	15.2 m	15.2 m
Subsurface Conditions	Sand	Residual soils (silt/sand)	12.2 m overbur-den over shale (rock socket)	Sapro-litic Clay	Sapro-litic Clay
G/W Head	2.1 m	10.7 m	15.2 m	15.8 m	0.6 m
Construc-tion Method	Heavy sand slurry plus improper concrete place-ment	Drilled "dry" with no fluid to countera ct hydro-static pressure	Drilled "dry" with casing extended to bottom of rock socket at 15.8 m	Drilled with corebarr el-type casing (no rifling)	Drilled with core-barrel Side walls rifled
First Test Result	SS = 300 kN	SS = 400 kN	SS = 400 kN	SS = 710 0 kN at 84mm	—
Remedial Test Result	SS = 1,600 k N	SS = 6,100 kN at 6 mm	SS = 10,700 kN	—	SS = 10,700 kN at 2 mm

TABLE 2.2 IMPACT OF CONSTRUCTION TECHNIQUE — END BEARING

File Reference	EB 128	EB 141	EB 142
Diameter	610 mm	860 mm	1220 mm
Length of Shaft	6.7 m	13.1 m	13.7 m
Subsurface Conditions	Decomposed granite	Dense to very dense silty sand (N = 30 to 85)	Weathered rock, silty sand (N = 30 to 100+)
G/W Head	—	0.6 m	6.1 m
Construction Method	Drilled dry. Cleaned out only with auger (Est-imated 150 mm to 200 mm loose material at base)	Drilled dry. Cleaned out only with auger (Esti-mated 50 mm to 80 mm loose material at base)	Drilled dry. Cleaned with clean-out bucket. Seep-age at base
Test Result	SS = 8,000 kN at 33 mm100 mm compression of loose base material	SS = 4,000 kN at 2.5 mm) EB = 2,200 KN at 38 mm. 32 mm compression of loose base material	EB = 2200 kN at 127 mm Estimated 115 mm compres-sion of loose base material

TABLES 2.1 and 2.2: SS = Side Shear Component; **EB** = End Bearing Component; **G/W** = Groundwater Table

Discussion of "Time for Development of Internal Erosion and Piping in Embankment Dams" by Robin Fell, Chi Fai Wan, John Cyganiewicz, and Mark Foster

April 2003, Vol. 129, No. 4, pp. 307–314.
DOI: 10.1061/(ASCE)1090-0241(2003)pp. 129:4(307)

John H. Schmertmann, F.ASCE[1]

[1]Professor Emeritus, Dept. of Civil and Coastal Engineering, Univ. of Florida, Gainesville, FL 32611. E-mail: schmert@ufl.edu

The writer wrote a discussion to Foster et al. (2000a,b) that included a brief presentation of two time-to-failure piping examples involving small embankment dams from first filling and from a trigger event. The times involved 2 years and 3 d in one example and 3 months and 7 h in the other. In both cases the failure review boards concluded that a partial pipe had formed previously, which the triggering event completed to the upstream face. These support the authors' finding that many dams stood successfully for a relatively long time and then failed quickly by piping following a triggering event.

It seems that the authors may have missed an important practical point concerning the role of the coefficient of uniformity, C_U, of an internally stable (no selective erosion of fines out of a coarse matrix) cohesionless soil versus its resistance to piping. The writer reviewed (Schmertmann 2000) 115 flume piping tests with sands, almost all performed at the Delft Hydraulics and Geotechnics Laboratories and at the University of Florida, using flumes with horizontal direct piping path lengths varying from 0.8 to 12.0 m. Fig. 1, a copy of Fig. 6 from Schmertmann (2000), shows the dramatic importance of C_U on the gradient needed to develop and maintain piping in cohesionless soils in the laboratory and presumably also within or under a dam. Put another way, the factor of safety versus the propagation of a pipe, through any point in a given cohesionless soil, depends on C_U as a major variable. The reference does not discuss the rate-of-propagation of a pipe. However, it seems obvious that with all else equal higher gradients with constant C_U will increase the rate of propagation. Also, for a constant gradient a lesser C_U will increase the rate.

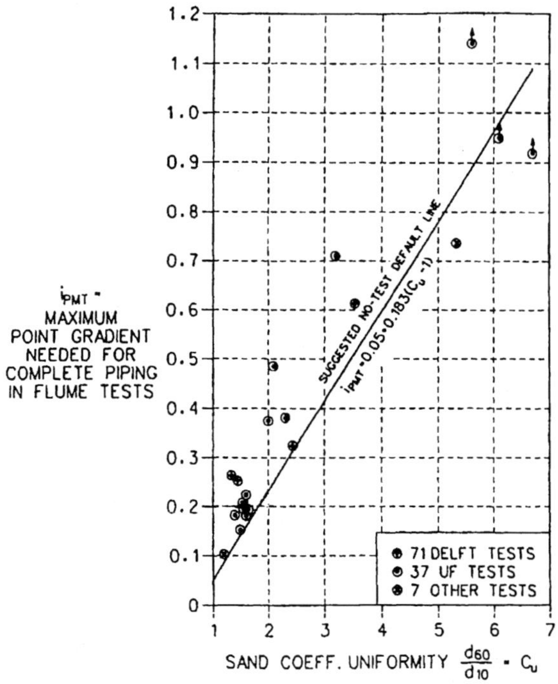

Fig. 1. Horizontal piping gradients with no-test default line

The authors appear to agree with the importance of gradient by noting that for the vast majority of their cases of piping failure they found the reservoir at or near its highest levels. But, the importance of C_U versus piping by backward erosion seems incorrect—at least in the top-left item in their Table 5, where they include uniform and well-graded cohesionless soils in the same risk category.

References

Foster, M. A., Fell, R., and Spannagle, M. (2000a). "The Statistics of embankment dam failures and accidents." *Can. Geotech. J.*, 37(5), 1000–1024.

Foster, M. A., Fell, R., and Spannagle, M. (2000b). "A method for estimating the relative likelihood of failure of embankment dams by internal erosion and piping." *Can. Geotech. J.*, 37(5), 1025–1061.

Schmertmann, J. H. (2000). "The no-filter factor of safety against piping through sands." *Geotechnical Special Publication No. 111*, ASCE, 65–132.

Closure to "Time for Development of Internal Erosion and Piping in Embankment Dams" by Robin Fell, Chi Fai Wan, John Cyganiewicz, and Mark Foster

April 2003, Vol. 129, No. 4, pp. 307–314.
DOI: 10.1061/(ASCE)1090-0241(2003)129:4(307)

Robin Fell[1]; Chi Fai Wan[2]; John Cyganiewicz[3]; and Mark Foster[4]

[1]Professor, School of Civil and Environmental Engineering, Univ. of New South Wales, Sydney NSW 2052, Australia. E-mail: r.fell@unsw.edu.au

[2]Research Engineer, School of Civil and Environmental Engineering, Univ. of New South Wales, Sydney NSW 2052, Australia. E-mail: wan@civeng.unsw.edu.au

[3]Senior Geotechnical Engineering Specialist, U.S. Bureau of Reclamation, P.O. Box 25007, Denver, CO 80215. E-mail: jcyganiewicz@do.usbr.gov

[4]Associate, URS Australia, 116 Miller Street, North Sydney NSW 2060, Australia. E-mail: mark_foster@urscorp.com

The writers welcome this discussion and acknowledge that the coefficient of uniformity has an influence on the gradient needed to develop and maintain piping in cohesionless soils. The scenario our paper considers is that erosion has initiated, and filters have not stopped it continuing. We are then considering the time from when the resulting concentrated leak forms to when breach occurs. During this time we would expect the reservoir level to remain much the same, and average gradients not change. We accept that cohesionless soils with a low C_u will erode more rapidly than one with a high C_u—maybe several times more rapidly. However, recent work done at the University of New South Wales (Wan and Fell 2004) shows that the range of rates of erosion of soils varies over at least 5 orders of magnitude, with most clays eroding a hundred or a thousand times more slowly than cohesionless silts and silty sands.

Given this, the relatively small difference in the rate of erosion with coefficient of uniformity for cohesionless soils is not significant in the context of Table 5 of the paper, or the accuracy of our method, and we do not believe it is necessary to differentiate between sands with different C_u.

Reference

Wan, C. F., and Fell, R. (2004). "Investigation of rate of erosion of soils in embankment dams." *J. Geotech. Geoenviron. Eng.*, 130(4), 373–380.

Estimating Slope Stability Reduction due to Rain Infiltration Mounding

John H. Schmertmann, F.ASCE[1]

Abstract: Rainwater can infiltrate into slopes and earth dams. Given sufficient infiltrating rainfall intensity, frequency, and time, the phreatic line within the slope or dam will "mound" above its otherwise steady state position. Such a mound will increase downslope seepage and reduce downslope stability. This paper develops an approximate method for solving this transient seepage "mounding problem" and the resulting reduction in stability. It shows the following variables will reduce stability: increased rainfall intensity and duration, decreasing permeability—especially over the 10^{-2}–10^{-4} mm/s range—and its anisotropy, increasing the slope cotangent, decreasing the depth to an impermeable underlayer, or increasing the depth to a permeable underlayer. Increasing size and effective porosity has only a time effect. In special circumstances the mounding effect can reduce slope stability by 50% and the paper includes a slope case history example.

DOI: 10.1061/(ASCE)1090-0241(2006)132:9(1219)

CE Database subject headings: Slope stability; Rainfall; Permeability; Dams, earth; Infiltration.

Introduction

Engineers know that a period of very heavy rainfall can add to the saturation of earth slopes and trigger "mudslides." The same can happen to earth dams. The author calls this the "mounding problem" and herein calculates mound formation and its possible importance.

High intensity rainfall over a sufficient length of time can reduce the stability of slopes, earth dikes, levees, and dams (herein collectively referred to as "slopes") in several ways, such as raising upstream pond levels, overtopping, erosion, piping, and slope seepage and pore pressure effects. This paper attempts to determine if and when rainfall infiltration into a slope might produce significant "mounding." The writer develops an approximate solution for this transient mounding and slope stability problem and identifies the effects of the variables investigated.

Fig. 1(a) shows schematically a picture of an increase in the elevation of the saturation or phreatic line within a dam. It shows a "mound" as that part above the assumed linear steady state phreatic line from the upstream to downstream water levels within a homogeneous dam. Mounding can aggravate all the effects noted above and thus contribute to a failure. It may occasionally make an important contribution that goes unrecognized because engineers do not suspect it. This paper attempts to provide the engineer with a tool to estimate the possible importance of mounding in reducing stability.

The "mounding problem" involves the complex interaction of variables, including soil properties that an engineer can determine or control only approximately. A rigorous solution seems unnecessary and perhaps impossible. The simplified and approximate solution explained and used in this paper may sometimes prove adequate. It considers the following problem variables: the size and geometry of the slope; the coefficient of permeability and its anisotropy; the rainfall intensity, duration, and sequencing; the location of the nearest horizontal impermeable or permeable layer; and the seepage effect of the mound and its negative effect on downslope stability.

This paper presents no new experimental or field data. Rather, it rearranges in a possibly new way the observational, experimental, and computational work of others regarding the well-known effects of sudden drawdown in dams and applies this to the mounding problem. It includes a case history of a slope failure by mounding.

Mounding doubtless occurs—as in the case history, or a perched water table mound above a relatively impermeable layer of limited extent, or the mound of rising ground water level in a new fill. The questions here: when, how fast, and with what stability consequences does it form within a slope? The writer's answers may also have more general practical applications.

Phreatic Line Position Analysis

Consider the mounding problem as the reverse of the transient seepage sudden drawdown problem used for dam upslope stability analyses. Instead of the upslope pond level dropping faster than the dam can drain, which produces a transient elevated phreatic line and reduced upslope stability, rain (or any other water) infiltration enters a slope faster than it can drain and produces a transient elevated phreatic line (the mound) and reduced downslope stability.

The writer made extensive use of the approximate full and partial drawdown analysis presented by Newlin and Rossier (1967), based in part on original work by A. Casagrande. But, instead of the slope or dam draining water it accumulates water.

[1]Professor Emeritus, Univ. of Florida, Gainesville, FL 32611. E-mail: schmert@ufl.edu

Note. Discussion open until February 1, 2007. Separate discussions must be submitted for individual papers. To extend the closing date by one month, a written request must be filed with the ASCE Managing Editor. The manuscript for this paper was submitted for review and possible publication on June 3, 2005; approved on February 6, 2006. This paper is part of the ***Journal of Geotechnical and Geoenvironmental Engineering***, Vol. 132, No. 9, September 1, 2006. ©ASCE, ISSN 1090-0241/2006/9-1219–1228/$25.00.

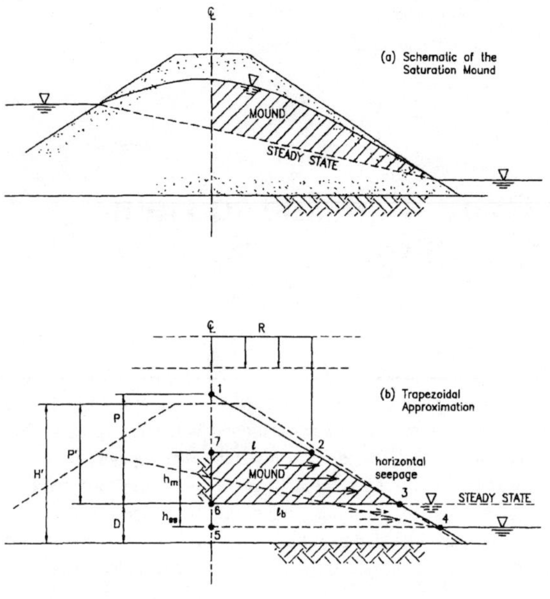

Fig. 1. Schematic of saturation mound and trapezoidal approximation

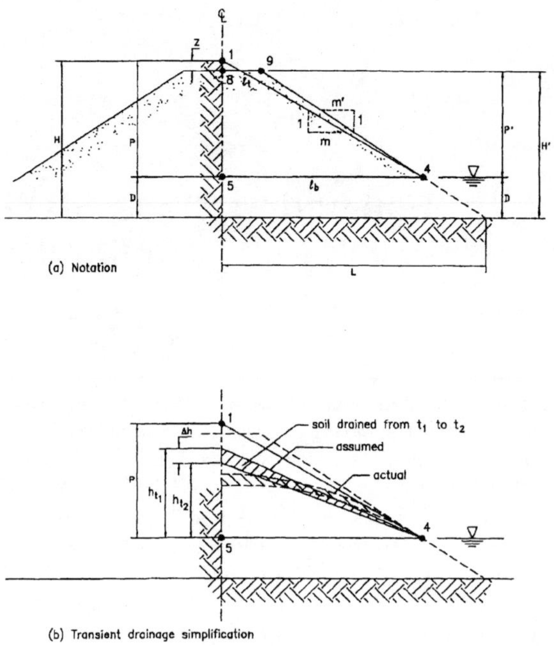

Fig. 2. Newlin and Rossier (1967) transient drainage analysis method

For convenience in utilizing this important reference think of accumulation as negative drainage. This maintains much of their notation, including their use of U to denote the relative drainage in the drawdown problem. In this paper $M=(1-U)$ with "M" denoting the relative mounding.

Fig. 1(b) shows the simplified downslope of a dam as an equivalent triangle, in drainage rate as discussed subsequently, and the mound approximated as a horizontal trapezoid above an underlying trapezoid for the steady-state condition. The actual phreatic surfaces of rainwater mounds, or after sudden drawdown, will look similar to the hyperbolic curves shown with dashed lines in Fig. 2(b). See Newlin and Rossier for a more detailed description of these curves. In Fig. 1(b) U=area 1-2-7 divided by area 1-3-6. Note that their work assumed isotropic permeability, or $k_h=k_v$, while this paper includes the effects of anisotropy. This requires the use of a transformation factor f, and an equivalent permeability k, and Eqs. (1) and (2) to convert anisotropic flow conditions to isotropic

$$k=\sqrt{k_h k_v} \tag{1}$$

$$f=\sqrt{\frac{k_h}{k_v}} \tag{2}$$

Fig. 2 illustrates the simplified, equivalent triangle, transient drainage analysis method used by Newlin and Rossier. Fig. 2(a) gives the notation. Eq. (3) gives $U=(P-h/P)$ from Fig. 2(b), which also$=[(P-h_{\mathrm{md}})/P]^2$ from Fig. 1(b). Eq. (4) presents the equation they developed, and verified using flow nets and parallel plate seepage experiments, for their time factor T needed to reach various degrees of drainage U. Eq. (5) gives their definition for the time factor T, which depends on the equivalent isotropic permeability k, the effective prorosity$=n_e$, the base length of the slope L, a calibration factor C for which they chose 2.0, and the total time for drainage t.

$$U=(P-h)/P=[(P-h_{\mathrm{md}})/P]^2 \tag{3}$$

$$T=\frac{P}{(H-P)}\ln\left[\frac{1-\left(\dfrac{PU}{2H-P}\right)}{1-U}\right] \tag{4}$$

$$T=\frac{2kHt}{n_e L^2} \tag{5}$$

In their discussion of the accuracy of Eqs. (4) and (5) they said: "The solution is simple and yields results with a degree of precision well within the range that soil characteristics can be determined. The accuracy of this approach can be favorably compared to that obtained from trial and error sketching of steady state flow nets." They cautioned against using these equations with $h/P<0.15$ or >0.85. According to their Fig. 7, one may therefore use these equations between $U=0.03$ and 0.76, or $M=0.97-0.24$. They also noted that they only established the validity of the equations over m (cotangent of slope angle) values of 2–3, but they thought "it not unreasonable" to expect validity over a greater slope range. For the less rigorous purposes of this paper the author assumed Eqs. (4) and (5) valid for all U and m.

In the case of a trapezoidal dam section, to use the Newlin and Rossier analysis the engineer must first convert to the equivalent triangle. Refer to Figs. 1(b) and 2(a). A separate analysis, similar to that used by Newlin and Rossier, yields Eqs. (6) and (7) for the subject conversion. When used for input into Eqs. (3)–(5) they produce approximately the same rate of drainage for the simplified Newlin and Rossier triangle as for the trapezoidal dam

Table 1. M_e versus Log of R_e for Reference Slope

	$\log(R_e)$ (mm/day)				
M_e	D/H=0	0.2	0.4	0.6	0.8
0.998	2.983	3.063	3.130	3.188	3.240
0.99	2.627	2.708	2.776	2.835	2.886
0.98	2.468	2.550	2.619	2.678	2.731
0.97	2.370	2.454	2.525	2.585	2.638
0.94	2.193	2.281	2.355	2.417	2.472
0.90	2.039	2.139	2.219	2.282	2.340
0.85	1.907	2.011	2.094	2.165	2.225
0.80	1.795	1.910	1.995	2.068	2.134
0.70	1.588	1.725	1.829	1.913	1.983
0.60	1.406	1.557	1.701	1.771	1.848
0.50	1.187	1.388	1.527	1.632	1.716
0.40	0.951	1.206	1.366	1.482	1.574
0.30	0.667	0.996	1.181	1.311	1.410
0.20	0.290	0.733	0.948	1.091	1.200
0.10	−0.364	0.337	0.591	0.752	0.867
0.05	−0.958	−0.012	0.264	0.431	0.551

Note: Use linear interpolation.

$$z = P'\left[\frac{1}{2\left(\frac{\ell_b}{\ell_t}\right) - 1}\right] \quad (6a)$$

$$P = P' + z \quad (6b)$$

$$H = H' + z \quad (6c)$$

$$m = \frac{m'P' + \ell_t}{P' + z} \quad (7)$$

Eq. (7) also results from the equivalent triangle simply bisecting ℓ_t, and in almost all cases using $m=m'$ will produce acceptable results.

The writer solved Eqs. (4) and (5) for the rate of total drainage along the top of the equivalent length ℓ, shown in Fig. 1(b), of the idealized trapezoidal mound with the same U. The phreatic line would then maintain a constant h_m and U if the rainfall infiltration over the length l just equaled the aforementioned rate of drainage. He then solved for this equilibrium rainfall infiltration rate, R_e, for a reference slope with an equivalent isotropic permeability $k_r=10^{-3}$ mm/s, a transformation factor $f=1$, a slope $m=1$, and the depths to the nearest horizontal impermeable layer, $(D/H)=0-0.8$. Table 1 presents the results with the equilibrium $M_e=(1-U_e)$ as a function of the log of R_e in units of millimeters/day, for $(D/H)=0$, 0.2, 0.4, 0.6, and 0.8. The engineer can interpolate linearly in Table 1 to obtain the subsequent $(\Delta F/F_0)$ stability ratios with a computational error of less than ±0.01. Note that (D/H) in Fig. 2(a) has a maximum possible value of 1.0, and as shown subsequently, that R_e or M_e do not depend on size P or effective porosity n_e. As also shown subsequently the larger the value of M_e the more important for the engineer to evaluate it accurately for its stability effect. M_e of less than approximately 0.4 has a negligible stability effect.

Any infiltration rate QR, where Q=the fraction of R that actually infiltrates into the slope, has the potential to increase (or decrease) the mound height h_m and increase (or decrease) M. Various trials with the aforementioned solution for R_e as a function of M showed that the engineer can obtain the value of R_e at a given M from Eq. (8) for any other combinations of m, f, and k, herein denoted R_e'

$$R_e' = R_e\left(\frac{f}{m}\right)^2\left(\frac{k}{k_r}\right) \quad (8)$$

Eq. (9) gives the normalized change in position of the idealized mound height, $\Delta h_m/P$, as a function of the time increment involved, the excess infiltration over drainage rate $(QR-R_e')$, the transient drainable height P of the dam above tailwater and its effective porosity n_e. Eq. (10) expresses this same change in mound height as a function of the average $\bar{M}$ over the change in M. Equating Eqs. (9) and (10) gives Eq. (11). As shown subsequently, numerical, step integration of Eq. (11) will give the relationship between time and the net of infiltration and drainage within the slope to raise or lower the phreatic line and thus the mound

$$(\Delta h_m/P) = \frac{(QR-R_e')}{n_e P}\Delta t \quad (9)$$

$$(\Delta h_m/P) = \frac{\Delta M}{2\sqrt{1-\bar{M}}} \quad (10)$$

$$\Delta t = \frac{\Delta M}{2\sqrt{1-\bar{M}}}\left(\frac{n_e P}{QR-R_e'}\right) \quad (11)$$

Rain Infiltration and Delay

Considering only a rain source, the maximum infiltration rate occurs when all the rain enters the slope and the infiltration ratio $Q=1$. However, many factors can reduce Q, such as runoff, evaporation, and transpiration. The minimum $Q=0$ with rain falling, assuming no swelling or shrinking of the slope (positive or negative storage). In the absence of superior information, the writer suggests using $Q=1/2$ and assuming that all QR rainwater will infiltrate until reaching the maximum rate of infiltration given by ik_v, where $k_v=k/f$ and i_v=vertical seepage gradient for gravitational flow=1. Rainfall rates exceeding this maximum and that cannot pond will run off the slope and not contribute to mound formation.

The writer further assumed that water drainage below the phreatic line occurs under saturated conditions, while rainwater moving downward to the phreatic line flows under constant n_e, partially saturated conditions until it merges with the phreatic line. Thus one has partially saturated permeability above the phreatic line and fully saturated permeability below. For the example delay times presented subsequently, the writer assumed a constant ratio of $[k(\text{above})/k(\text{below})]=1/3$. See "Practical Comments and Suggestions" for more discussion of partial saturation.

When rain falls on the surface of a slope it takes time for this water to flow down and reach the phreatic line and contribute to the saturated mound. The idealized vertical distance in Fig. 1(b) varies from $(P-h_m)$ to 0, depending on where the rain strikes the slope surface. The average=$(P-h_m)/2$. Eq. (12) gives the average value of this delay time δt. It occurs and changes, depending on the current h_m value, with every significant change in rainfall intensity. It now remains to include the seepage effect of h_m on downslope stability

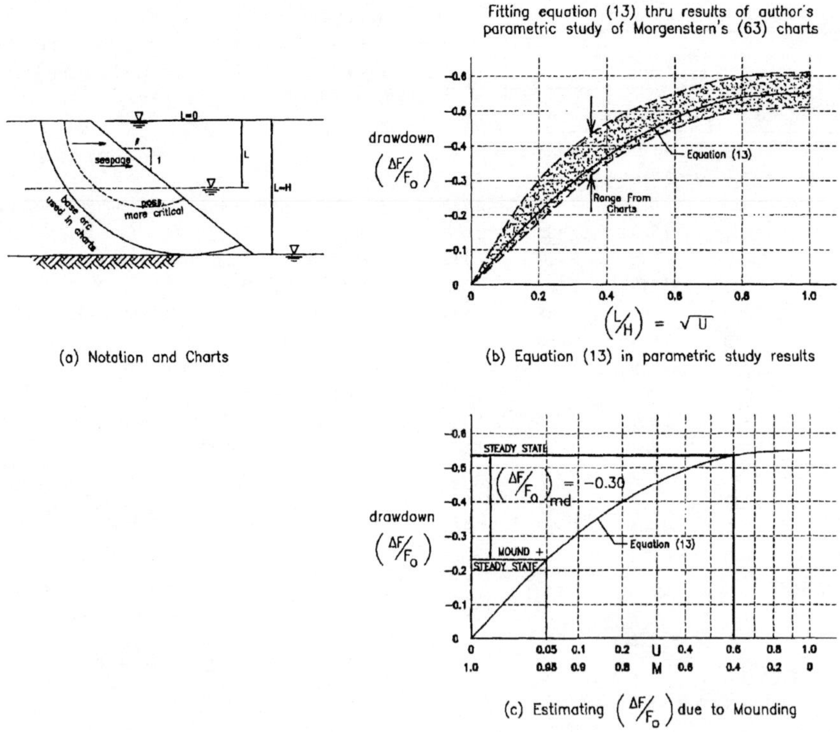

(a) Notation and Charts

(b) Equation (13) in parametric study results

(c) Estimating $\left(\frac{\Delta F}{F_0}\right)$ due to Mounding

Fig. 3. Using Morgenstern (1963) to estimate $(\Delta F/F_0)$ due to sudden drawdown and mounding

$$\delta t = \frac{1.5fPn_e\left(1-\frac{h_m}{P}\right)}{k} \tag{12}$$

Mound Seepage and Reduced Downslope Stability

Morgenstern (1963) made a useful investigation of the effect on slope stability of various levels of sudden drawdown "L" from maximum pond level "H." Fig. 3(a) shows the geometry and his notation. He showed the reasonableness of assuming perfectly horizontal seepage flow over the drawdown L and no seepage effects below L. He presented his results in a series of charts of the shear stability factor of safety "F" versus the ratio (L/H) for slopes with β (or m) from 2 to 5, $\phi'=20-40°$, and $(c'/\gamma H)=0.0125-0.050$. The writer made a parametric study of these charts and applied the results to the mounding problem as described below.

Using different combinations of (L/H) from 0 to 1, $(c'/\gamma H\tan\phi')$ from 0.02 to 0.12, and m from 2 to 5, gave $(-\Delta F/F_0)$ versus (L/H) results within the stippled band shown in Fig. 3(b). Note that $F_0=F$ when $(L/H)=0$, and that the band includes some values from more critical, nonbase-tangent circles using the example interpolation procedure given by Morgenstern. Considering the approximate nature of the aforementioned analysis methods, it seemed acceptable to represent the band by the single, parabolic curve given by Eq. (13). Fig. 3(b) shows how this curve fits within the band and Fig. 3(c) shows its use (explained subsequently) to determine $(-\Delta F/F_0)$ due to mounding.

$$-\left(\frac{\Delta F}{F_0}\right)=0.625\left(\frac{L}{H}\right)^2-1.175\left(\frac{L}{H}\right)+0.550 \tag{13}$$

The trapezoidal approximation of the mound in Fig. 1(b) leads to Eq. (14), which permits the substitution of $\sqrt{U}$ or $\sqrt{1-M}$ for (L/H) on the x axis of Fig. 3(c). Applying Eq. (13) to the steady state and steady state plus mound conditions gives the $(L/H)=\sqrt{U_{ss}}$ and $\sqrt{U_{ss+md}}$ positions along the x axis (see below for more discussion of U_{ss+md}). Projecting vertically to the curve and then horizontally to the y axis, and taking the difference in y, gives the $(-\Delta F/F_0)$ relative stability reduction due to the mound. But, because U_{ss} can vary greatly in different situations, the writer took $(L/H)=1$, $U_{ss}=1$, and $(\Delta F/F_0)=0$ from Eq. (13) as the reference steady-state condition. With infiltration over the full range of $M=0$ to $M=1$, $(\Delta F/F_0)$ due to mounding equals the "maximum" value of -0.550

$$\left(\frac{L}{H}\right)_{\text{Morg.}}=\sqrt{U}=\sqrt{1-M} \tag{14}$$

A downslope stability analysis will most likely already include any steady-state seepage effect. If so, it still remains to separate the mounding and steady-state effects. The writer does this in Fig. 3(c) using U_{ss}, a combined U_{ss+md}, and the Morgenstern (L/H) in Fig. 3(a), as explained below.

Determine $U_{ss}=[(\text{area 1-3-6})/(\text{area 1-4-5})]$ directly from the simplified geometry in Fig. 1(b). However, the stability-related U_{ss+md} used in Fig. 3(c) combines U_{ss} with a transient drainage-related definition of $U_{md}=[(\text{area 1-2-7})/(\text{area 1-3-6})]$. A study of the geometry involved will show that Eq. (15) gives U_{ss+md}. Then

Table 2. Approximate Percent of Total R_e due to Mound Lateral Drainage with Permeable Horizontal Boundary Layer (at all M)

D/H	m/f	$\left(\frac{\text{lateral}}{\text{lateral+vertical}}\right)$drainage (%)
0	1	25
	2	5
	4	0
0.2	1	55
	2	20
	4	5
0.4	1	80
	2	50
	4	20
0.8	1	100
	2	95
	4	90

$M=(1-U)$ for use in Fig. 3(c). As an example in Fig. 3(c), let $U_{ss}=0.60$ and $U_{md}=0.083$. Using Eq. (15) gives $U_{ss+md}=0.05$ and the matching $M=0.95$. Then Fig. 3(c) shows how to determine the normalized stability reduction ratio of $(\Delta F/F_0)=-0.30$ due to only the transient mound. To compute the value of ΔF due only to the mound, multiply the $(\Delta F/F_0)$ reduction ratio by the F_0 obtained from the Morgenstern (1963) charts when using $(L/H)=0$ in the charts

$$U_{ss+md} = U_{ss}U_{md} \tag{15}$$

Note in Figs. 3(b and c) that approximately 80% of the instability effects of full sudden drawdown, and therefore also of full mounding, occur as a result of the $(L/H)=0-0.5$ first half of the drawdown. For mounding this means from $M=0.75$ to 1.

Underlying Permeable Layer

The preceding considered only a horizontal impermeable lower seepage boundary. While beyond the scope of this paper to present details, the writer has made an estimate of the mounding and stability effects of an underlying horizontal boundary layer with infinite instead of zero permeability. Perfectly vertical drainage has no stability effect in a purely frictional soil and assumed for all soils. After separating the total seepage into lateral (horizontal) and total (lateral+vertical), Table 2 gives the percent of total assigned to lateral.

As seems reasonable, Table 2 shows that a permeable underlayer immediately under the slope $(D/H=0)$ will capture most of the mound drainage and make it unlikely a significant mound can form, and also that such a layer, if deep enough $(D/H=0.8)$ will have little or no effect. As a rough guide, consider the $(\Delta F/F_0)$ effect proportional to the percent of R_e due to lateral flow. For example if $(\Delta F/F_0)=-0.30$ but lateral drainage equals only 50% of the total, then $(\Delta F/F_0)$ "drops" to −0.15.

Parametric Results

The writer combined Eqs. (8)–(14) to make some parametric studies using the following steps:

1. For a given slope select the approximate (simplified) values of P, n_e, H, D, k, f, m, and QR versus time t.
2. Calculate R_e' from Eq. (8) after assuming an incremental ΔM, starting from $M_0=0$. Then the average $\bar{M}$ over the first ΔM increment$=(M_0+\Delta M/2)$. Obtain R_e by linear interpolation in Table 1.
3. The $\Sigma(\Delta t)$ values need to include the extra delay time δt whenever QR changes significantly. Use Eq. (12) for an estimate of each delay time.
4. Use Eq. (11) and the above R_e' to calculate a Δt time increment matching the selected ΔM and the resulting $\bar{M}$.
5. Select another ΔM and repeat Steps 2 and 3 to get the next increment of Δt to add to the current $\Sigma(\Delta t)$ that then matches the current $M_0+\Sigma(\Delta M)$. Continue until QR changes, the $\Sigma(\Delta t)$ reaches a desired value, or M approaches an asymptotic value M_e. The $\Sigma(\Delta t)$ approaches infinity when R_e' approaches QR.
6. Use Eqs. (13) and (14) or Fig. 3(c) with the applicable M values to make an estimate of the relative slope stability reduction due to mound formation.

However, the engineer does not have to use the above numerical procedure to calculate M_e and $\Delta F/F_0$ resulting from a continuous QR infiltration rate over the time needed to reach equilibrium. Tables 3–5 provide some parametric results. Table 6 provides an example of the use of the aforementioned step procedure. The table indicates the steps by [number]. Fig. 4 includes the plotted results from this example. The note at the bottom of Table 4 gives the direct calculation steps to avoid numerical integration for the equilibrium case.

The parametric study attempted to get some further insight into mounding behavior, mainly the effect of the variables investigated, and to try to determine which combination of variables would produce significant downslope stability reduction. The writer investigated the following range of variables: k from $1/625$ to 125×10^{-3} mm/s, P from 5 to 30 m, n_e from 0.1 to 0.2, continuous QR from 0.12 to 10.0 mm/day over infinite and 30 day time intervals, m from 1 to 8, f from 1/4 to 7, and (D/H) from 0 to 0.8. As shown in Table 3, the study involved two case examples: No. I—a "typical" case with $P=10$ m, $k=1/625$ to 5×10^{-3} mm/s, $n_e=0.10$, $f=\sqrt{2}$, $m=2.5$, and $D/H=0.1$, and No. II, the "maximum" case, with $P=5$, $k=1/125$ to 125×10^{-3},

Table 3. Example Cases for Parametric Study

Example	P (m)	n_e	$k/10^{-3}$ mm/s	f	m	D/H	QR_1 (mm/day)	Δt_1 (days)	QR_2 (mm/day)	Δt_2 (days)
I (typical)	10	0.1	1/625–5	$\sqrt{2}$	2.5	0.1	2[a]	∞	10[a]	30
II (maximum)	5	0.1	1/125–125	1	4	0	2[a]	∞	0 or 10	30
	5–20	0.1–0.2	1	1/4–4	1–8	0–0.8	2 or 8	∞	0 or 10	30
	(see Table 5)		(see Table 4)		(see Table 5)					

[a]Or less, as controlled by k_v.

Table 4. Parametric Study for Effect of Permeability

Example	$k/10^{-3}$	QR_1 (mm/day)	M_e[c]	Initial delay δt_i (days)	Total[b] $t_{0.9}$ (days)	QR_2 (mm/day)	M_{30}	$(-\Delta F/F_0)$ at M_e[c]	$(-\Delta F/F_0)$ at M_{30d}
I (typical)	1/625	0.033[a]	0.778	42y	89y	—	—	0.135	—
	1/125	0.163[a]	0.777	8.4y	18.2y	—	—	0.134	—
	1/25	0.815[a]	0.775	610	1,310	—	—	0.133	—
	1/5	2	0.608	125	380	4.07[a]	0.73	0.059	0.108
	1	2	0.286	25	200	10	0.56	0.003	0.032
	5	2	0.121	5	—	10	0.265	0	0.002
II (maximum)	1/125	0.23[a]	0.991	3.0y	7.3y	—	—	0.443	—
	1/25	1.15[a]	0.991	220	530	—	—	0.443	—
	1/5	2	0.942	43	230	5.76[a]	0.991	0.303	0.443
	1	2	0.654	9	140	10	0.925	0.075	0.247
	5	2	0.349	2	90	10	0.61	0.009	0.060
	25	2	0.107	1/2	50	10	0.365	0	0.011
	125	2	0	1/10	—	10	0.065	0	0

[a]Maximum values as controlled by $k_v(i_v=1)$=maximum vertical infiltration rate. Note $k_v=k/f$.

[b]t for $M=0.9M_e$.

[c]For continuous QR_1 over infinite time the author calculated M_e directly, without times, by putting $R_e'=QR_1$ into Eq. (8), the resulting R_e into Table 1 to interpolate for M_e, followed by using Eqs. (13) and (14) to solve for $(\Delta F/F_0)$. Shown to 0.000 for comparisons only.

Table 5. Other Parametric Results with Constant $k=10^{-3}$ mm/s

	"Maximum" Example II variable: P (m)	n_e	M_e	Initial delay δ_i (days)	Total $t_{0.9}$ (days)	$(-\Delta F/F_0)$
	Using QR_1=8 mm/day					
(a)	5	0.2	0.918	17	115	0.264
	5	0.1	0.918	9	55	0.264
	10	0.1	0.918	17	115	0.264
	20	0.1	0.918	35	230	0.264
(b)	f=0.5		0.992	4	50	0.449
	1		0.918	9	55	0.264
	2		0.654	17	55	0.075
	4		0.383	35	55	0.013
(c)	m=1		0.383	9	20	0.013
	2		0.654	9	40	0.075
	4		0.918	9	55	0.264
	8		0.992	9	55	0.449
(d)	D/H=0		0.918	9	55	0.264
	0.1		0.905	9	55	0.247
	0.2		0.888	9	55	0.226
	0.4		0.855	9	55	0.193
	0.8		0.782	9	55	0.138
(e)	QR_1=0.5 mm/day		0.383	9	390	0.013
	1.0		0.508	9	210	0.033
	2.0		0.654	9	140	0.075
	4.0		0.805	9	90	0.153
	8.0		0.918	9	55	0.264
	16.0		0.974	9	32	0.376

n_e=0.10, f=1, m=4, and D/H=0. Tables 4 and 5 present the results of this study.

Table 4 shows that decreasing permeability can dramatically increase mound height and decrease downslope stability. But, it also increases the time needed for the decrease. The long times needed may diminish the appreciation for the eventual mound formation. Table 5(a) shows that P (size) and n_e (effective porosity) have no effect on stability at infinite time. Increasing P or n_e each proportionally increases the times to achieve mound equilibrium and therefore reduces the instability effects at a given time. Part (b) shows that increasing f (anisotropy) greatly increases comparative stability. Part (c) shows that increasing m (the flatness of the slope) greatly decreases comparative stability. Of course, by itself flatter also means more stable slopes before the decrease. Part (d) shows that increasing (D/H) has a more modest effect on reducing the mounding effect on stability. Part (e) shows, as expected, that increasing the rainfall infiltration intensity (QR_1) decreases stability and the time to reach equilibrium.

Fig. 4 gives a graphical presentation of the results in Table 4 for both the "typical" and "maximum" examples. It shows that mounding occurs in soils with $k<10^{-2}$ mm/s, but it takes years for the highest mound to form in soils with $k<10^{-4}$ mm/s. This suggests a practical range of concern of $10^{-2}-10^{-4}$ mm/s, typically found in very fine sand and silt soils. More permeable soils drain too fast, and less permeable take too long to infiltrate. Points (a)–(b) in Fig. 4, for both examples I and II at $k=10^{-5}$ mm/s, show the length of time (15 and 7.5 years) needed for the mound-induced stability reduction to reach these points with a continuous infiltration of 2 mm/day. An additional heavy rainfall event of limited duration, such as illustrated for 10 mm/day for 30 days, has a significant effect over approximately the same k range of 10^{-2} to 10^{-4} mm/s.

The above findings from continuous and therefore also average infiltration rates do not tell the whole story. Consider example II with $k=5\times10^{-3}$ mm/s. As shown in pt. 1 in Fig. 4, continuous infiltration at 2 mm/day produced an equilibrium stability reduction ratio of −0.009. Then 30 days with a continuous 10 mm/day, =300 mm total, "increased" the ratio to

Table 6. Numerical Example (in Fig. 4, Maximum, $k/k_r=5$)

Input										
QR (mm/day)	For (days)	M_i	ΔM	M_{i+1}	$\bar{M}$	$\log R_e$ (Table 1)	R_e' (mm/day)	$\Sigma\Delta t$ (days)	δt (days)	$-\Delta F/F_0$
2	∞	0	—	—	—	—	—	0	—	0
		0	0.1	0.1	0.05	−0.364	0.14	15.4	1.7	—
		0.1	0.1	0.2	0.15	0.290	0.61	34.9	—	—
		0.2	0.1	0.3	0.25	0.667	1.45	87.5	—	—
		0.3	0.03	0.33	0.315	0.752	1.77	126.2	—	—
		0.33	0.01	0.34	0.335	0.781	1.89	153.4	—	—
		0.34	0.005	0.345[a]	0.343	0.795	1.95	183.71	—	0.008
(150)	2	0.345	0.005	0.35	0.348	0.809	2	185.1	1.40	—
144[b]		0.345	0.15	0.5	0.425	1.187	5	185.5	—	—
		0.5	0.2	0.7	0.6	1.588	12	186.1	—	—
		0.7	0.1	0.8	0.75	1.795	20	185.5	—	—
		0.8	0.1	0.9	0.85	2.039	34	187.07	—	—
		0.9	0.01	0.91	0.905	2.078	37	187.14[c]	—	0.254

Note: Input $P=5$ m, $n_e=0.1$, $f=1$, $m=4$, $D/H=0$.

[a]By direct calculation=0.349 (see footnote c in Table 4).

[b]Limited by k_v (see footnote c in Table 4).

[c]Reached desired value of (183.71+1.40+2=187.11).

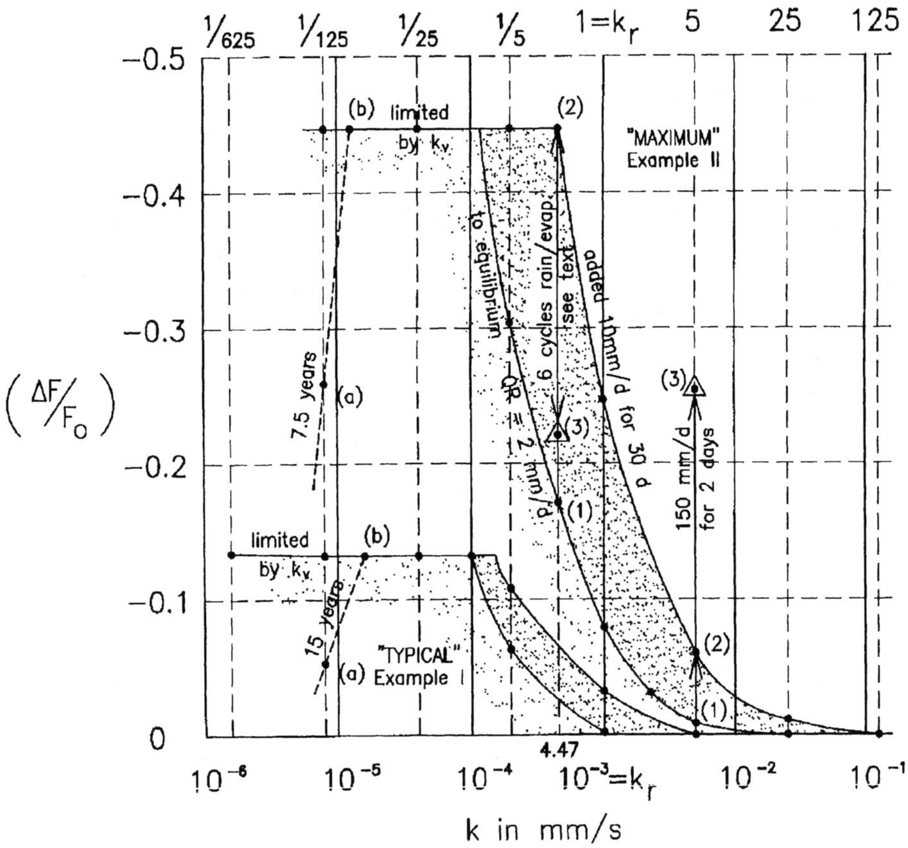

Fig. 4. Examples of effect of permeability (see Tables 4–6)

pt. 2=−0.060. But, if that 300 mm had occurred over only 2 days of continuous infiltration the ratio would have "increased" to pt. 3=−0.254. Now consider the same 1-2-3 pts. when $k=4.47\times10^{-4}$ mm/s. For pt. 3 the writer again applied the total 300 mm in 30 days, but in six cycles. Each consisted of 2 days of 25 mm/day continuous infiltration, followed by 4 days of −2 mm/day continuous evaporation. Note that for this k the maximum infiltration rate before runoff=12.9 mm/day and the stability reduction ratio "decreases." Instead of changing from −0.171 (pt. 1) to −0.443 (pt. 2), it changes from −0.171 to only −0.231 (pt. 3). To obtain these results the writer used Eq. (11) numerically with suitably small trial increments of both + and $-\Delta M$ so that Δt would match the 2 and 4 day intervals in each cycle of QR. The engineer can use this procedure for any infiltration and interval pattern.

From the above examples it seems clear that to evaluate the potential stability reduction of mounding the engineer should use, when available and appropriate, more detailed infiltration data for calculations rather than only averages over long time periods.

It seems interesting that mounding behavior appears to join other engineering behaviors that also occur most noticeably in soils with a similar 10^{-2}–10^{-4} mm/s range in k. For example, liquefaction occurs when pore pressures can generate due to earthquake cyclic–loading-induced soil structure collapse in saturated cohesionless soils. But the pore pressures must also not dissipate too quickly—as in silts and fine sands. Ice lense formation offers a static example in soils with a similar k range. The soil must have a permeability high enough for water to flow to the lense but low enough to generate the capillary and osmotic gradients to drive the flow. Those soils with the most potential for erosion and piping also have a similar range in permeability.

Case History

In a recent publication, Blatz et al. (2004), details a case history of essentially complete mounding [$M=1$, see their Fig. 18(a)] and slope failure following an extreme (for the location) rainfall event 30 years after construction. Their pre- and postfailure factors of safety F for the slope=2.0–2.2, with a minimum of 1.04 around the time of failure. According to Eqs. (14) and (15), or Fig. 3(a), changing from a steady state of $M\approx0.4$ or less (which applies to this case) to $M=1$ should produce an approximate 55% reduction in the factor of safety. Applying this to the above F=2.0–2.2 gives F=0.90–0.99, which checks the failure.

One can even use the Morgenstern charts to check the values of F directly. The average slope cotangent m=3.5. Using the charts with their $\phi'=22°$, and averaging between m=3 and 4 and $c/\gamma H$=0.0125 and 0.025, gives F=2.1 for no mounding (M=0) and 1.0 for complete mounding (M=1), which also checks. This use of $(c/\gamma H)=0.019$ implies a small value of c' along the failure although Blatz et al. assumed $c'=0$. However, their Fig. 9 suggests a possible small c'. It also seems likely that a stiff, weathered silty clay would have some effective cohesion.

Thanks to additional rainfall, temperature, and failure timing information kindly supplied by Blatz (personal communication, March 2005), the writer reconstructed the following approximate timeline. The Virden weather station recorded above-freezing temperature conditions that began April 16, 1999. The unusually heavy rains for the area began on May 3 and ended on June 8, with an average of 5.65 mm/day for this 36 day interval. With a zero date of April 1, the Blatz et al. Fig. 16 suggests the failure occurred on approximately May 20, or 34 days after thaw began and 17 days of average 7.5 mm/day after the rains began. However, the exact dates of the failure remain unknown. The reports indicate it occurred sometime within 1 week after the rains ended. This places failure about 60 days after the thaw and 40 days after the rains began. For the subsequent analysis the writer assumed 15 days of 2.5 mm/day melt followed by 20 days of 10 mm/day of continued melt and rain infiltration, and complete mound formation and failure in a total of 60 days.

Using Eqs. (11) and (12) to back-figure a set of parameters that would match approximately with the above timeline, and starting with an initial (after 30 years) Fig. 1(b) steady-state condition of h_{ss}=2.5 m and M_{ss}=0.35, gave the following reasonable (but not unique) results: P=11 m, n_e=0.025, $k=10^{-4}$ mm/s, f=1/3, m=3.5, and D/H=0.2 to an impermeable boundary. These results indicate the possibility of a much higher k (by a factor of 500) than Blatz et al. obtained from laboratory tests and perhaps also a much lower effective porosity.

The much higher k suggests that rain infiltration only filled the cracks and fissures in the weathered, surface, silty clay layer. The low n_e also suggests little or no filling of air voids in the soil matrix. Based on the lab k values, this would take years! The f=1/3 suggests that vertical cracks dominate over horizontal—not unreasonable because of common shrinkage cracking near the surface. In support, Zhang et al. (2000) describe another case history, including field infiltration rate measurements, where surface weathering and cracking appeared responsible for a ratio of (field k/lab k) greatly exceeding the above 500 in the present case history.

Use of the methods presented in this paper can produce results that match some of the reality in this case history. It does not include soil suction effects and any other complications from the applications of partially saturated soil mechanics. The results imply negligible practical effects from these complications in this case. Fig. 3(c) may show why—actual rain-induced failures often involve the near-saturation of the slope and most of the stability loss due to positive pore pressure effects occurs during the final stages of saturation (high M values). Perhaps at this important stage we often do have negligible partial saturation effects. See "Practical Comments and Suggestions" for more discussion.

Methods and Conclusions

The writer developed an approximate solution for the transient accumulation of rainwater in a slope and its stability reduction effect (the "mounding problem"), and used a sequence of approximations and assumptions. Their cumulative effect remains unknown at this time (March 2006). However, the results presented seem reasonable and fit in with the writer's experience and the case history. Tables 3–5 and Fig. 4 show examples of the parametric results obtainable from the use of the equations developed. The writer used the following methods to obtain the subsequent conclusions:

Methods

1. Considering mounding as negative drawdown;
2. Using an existing analysis for transient drainage after full and partial sudden drawdown, combined with an existing paper giving the results of full and partial sudden drawdown on dam stability;
3. Giving the mound an equivalent triangular shape for drain-

age analysis and an equivalent trapezoidal shape for rainfall infiltration and stability analyses;
4. Establishing an equilibrium rainfall infiltration rate for a reference slope and using it to determine changes in the position of the mound for other slopes;
5. Normalizing the drawdown stability results;
6. Separating the mounding effect from the combined steady state plus mounding effects; and
7. Determining in what cases significant mounding may occur and providing a theory and computational method to evaluate its effect on downslope stability.

Conclusions

1. Mounding can reduce stability 10% in some cases and 50% in special cases;
2. A permeable underlayer can greatly reduce the likelihood of significant mounding;
3. Soils within the effective permeability range of $k=10^{-2}-10^{-4}$ mm/s can mound significantly. Above 10^{-2} they drain too quickly. Below 10^{-4} the mound may take too long to form;
4. Low mounds ($M<0.4$) have a negligible instability effect. Higher mounds have a progressively greater effect with approximately 80% of the total effect occurring between $M=0.75$ and 1.0;
5. Slopes with soils in the lower end of the above permeability range have the potential to form mounds with the greatest instability effect. They also take years to form and this may diminish the appreciation of their eventual formation;
6. Increasing the size (P) of a slope or its effective porosity (n_e) does not change the equilibrium mound position for a given continuous infiltration rate. Each does proportionally change the time involved to achieve equilibrium;
7. The details of rainfall rates, duration, and sequencing can have an important effect;
8. There exists a significant delay time between rainfall and resulting changes in the mound. It varies from hours to years;
9. The study of the case history supports the possibility that soils with matrix permeabilities less than 10^{-5} mm/s may have much higher near-surface permeability, and much lower effective porosity in the field, due to the selective rain infiltration of cracks and fissures; and
10. Many variables affect mounding, making it difficult to evaluate by "engineering judgment." Such judgment might improve with the application of the methods and findings presented in this paper.

Practical Comments and Suggestions

1. Under certain conditions, mainly relatively unprotected, low slopes or dams made of silty soils and in areas of heavy rainfall, mounding represents a significant factor in reducing downslope stability and can lead to failure.
2. Partial saturation effects: We have, by definition, a partially saturated slope above the phreatic line. Its permeability and shear strength varies with its degree of saturation due to capillary and suction effects. Blatz et al. (2004) provide an example stability analysis that includes these effects. The writer's analysis of this case suggests that mounding caused the failure and one could ignore the initial suction and strength losses. But, in other cases these losses themselves might induce a failure without the infiltration reaching the mound or where a significant mound does not exist, possibly because of insufficient time to form. For those cases where a substantial mound can form it seems likely that the stability reduction due to mounding will add to that due to loss of suction, making the mounding case the most critical. Assuming no strength due to suction, plus mounding effects, produces a conservative result.
3. Alternating rain and no-rain intervals: As shown by the example in Table 6, the engineer can use Eq. (11) with any Δt time intervals and values of QR. For example, use daily rainfall amounts and 1 day time intervals. The mound height h_m would increase and decrease according to whether the net $(QR-R_e')$ turns positive or negative. Note that during time intervals without rainfall QR can become negative due to evaporation effects.
4. Reducing mounding effects: As discussed in "Underlying Permeable Layer," horizontal permeable layers under a slope can greatly reduce mound formation and therefore its effects. A horizontal toe drainage layer would have a similar but partial effect. Of course, the engineer can also reduce the value of the infiltration factor "Q." Effectively paving the top of a slope or dam would reduce Q to 0 and prevent mound formation. Sloping the crest would also help by directing some of the rainfall away from the downslope. Planting vegetation might also help by encouraging transpiration, although at the same time it slows any runoff velocities and allows more time for infiltration.
5. Homogeneous and isotropic slopes: Any real slopes or dams will likely have neither constant nor easy to estimate homogeneous conditions and anisotropic permeability. The engineer should use appropriate judgment when using the simplified analysis methods developed in this paper.
6. Implications for piping and erosion: The writer has limited the scope of the paper to the effect of mounding on slope stability. However, mounding by its nature induces greater vertical and horizontal gradients which in turn might produce internal piping, or increase the size and progression rate of existing pipes (see Schmertmann 2004). Mounding also makes it more likely that a seepage breakout will occur on the downslope portion of the dam (or slope) below the phreatic line. This in turn will make downslope surface erosion more likely and perhaps will indicate the need for slope protection.
7. Detecting mounding and predicting possible failure: Detecting mounding requires piezometers or other suitable field instruments. Its transient nature also requires that the engineer read them often after significant infiltration events to try to detect their maximum size following such events before the mound can reduce significantly. Such data may then produce a basis for estimating the magnitude of the event that could produce failure and its likelihood.
8. Other applications: The writer hopes that the mound analysis methodology developed herein will lead to checking and refinements and perhaps broader and innovative applications. A few thoughts: Leaky pipes, irrigation, snowmelt, thawing ice lenses, runoff from uphill sources, the clogging of a drain, or shutting off a near-surface well or drainage layer might produce increased mounding. Applying the methodology to axisymmetric and other three-dimensional geometries might lead to estimates of the mounding to produce perched water tables and the rise of groundwater levels following various types of fill construction.

Acknowledgments

T. William Lambe and Francisco Silva-Tulla reviewed some of the writer's initial work on the "mounding problem" in conjunction with our common investigation of a tailings dam failure in Florida. They encouraged the effort to produce a Technical Note, which then grew into this paper. The writer also thanks Loadtest, Inc., Gainesville, Fla., for its support in preparing this paper, and Suzanne Terrell and John Graman particularly for word processing and graphics, respectively.

Notation

The following symbols are used in this paper:

C = calibration factor in Eq. (5) chosen=2 by Newlin and Rossier (1967);
c' = effective stress cohesion;
D = depth between tailwater and assumed impermeable or permeable horizontal boundary layer elevations (includes h_{ss} when $\neq 0$;
F = plane strain, downslope stability factor of safety;
F_0 = downslope stability factor of safety with no seepage;
f = anisotropic permeability transformation factor [Eq. (2)];
H = elevation difference between dam crest and bottom of most critical failure circle (Morgenstern 1963). Shown at maximum drawdown in Fig. 3(a);
H = P+D [Fig. 2(a)];
H' = H-z [Fig. 2(a)];
h = height of mound [triangular simplification in Fig. 2(b)];
h_m = part of h due to mound [trapezoid simplification in Fig. 1(b)];
h_{ss} = part of h due to steady-state seepage [trapezoid simplification in Fig. 1(b)];
i_v = vertical seepage gradient;
k = equivalent isotropic coefficient of permeability (millimeters/second);
k_h = horizontal coefficient of permeability;
k_r = reference permeability of 10^{-3} mm/s;
k_v = vertical coefficient of permeability;
L = length of equivalent drainage triangle at its impermeable base [Fig. 2(a)];
L = sudden drawdown height (Morgenstern 1963) [Fig. 3(a)];
ℓ = horizontal length over which rainfall infiltrates into mound [Fig. 1(b)];
ℓ_b = horizontal bottom of dam [Fig. 2(a)];
ℓ_t = horizontal top of dam [Fig. 2(a)];
M = degree of transient mounding=$1-U$;
$\bar{M}$ = average M over increment of ΔM;
M_e = M at equilibrium with constant $QR=R_e'$;
M_{ss} = preexisting M due to steady-state seepage;
M_0 = M when $t=0$;
M_{30} = maximum M after 30 days of increased QR;
m = cotangent of slope of equivalent downslope triangle [Fig. 2(a)];
m' = cotangent of downslope of dam [Fig. 2(a)]; equals β in Morgenstern (1963);
n_e = effective porosity above phreatic line;
P = maximum mound height above h_{ss} [Fig. 1(b)] or above tailwater [Fig. 2(a)];
P' = P-z;
Q = fraction of rainfall that infiltrates into slope;
R = rainfall rate (millimeters/day);
R_e = equilibrium QR when water infiltration rate into mound equals drainage rate, for reference downslope;
R_e' = R_e for other downslopes;
T = Time factor [Eq. (5)];
t = time;
U = ratio of present to maximum ($t=\infty$) transient drainage during sudden drawdown;
z = P-P';
γ = total unit weight of dam soil;
γ_w = unit weight of water;
Δt = time increment;
δt = time delay due to rainfall infiltration flow to phreatic line; and
ϕ' = effective stress friction angle.

References

Blatz, J. A., Ferreira, N. J., and Graham, J. (2004). "Effects of near-surface environmental conditions on instability of an unsaturated soil slope." *Can. Geotech. J.*, 41, 1111–1126.

Morgenstern, N. (1963). "Stability charts for earth slopes during rapid drawdown." *Geotechnique*, 13(2), 121–131.

Newlin, C. W., and Rossier, S. C. (1967). "Embankment drainage after instantaneous drawdown." *J. Soil Mech. Found. Div.*, 93(6), 79–95.

Schmertmann, J. H. (2004). "Discussion of 'Time for development of internal erosion and piping in embankment dams' by Fell, R., et al." *J. Geotech. Geoenviron. Eng.*, 130, 980–980.

Zhang, J., Jiao, J. J., and Yang, J. (2000). "In situ rainfall infiltration studies at a hillside in Hubei Province, China." *Eng. Geol. (Amsterdam)*, 57, 31–38.

Schmertmann's Swell Sensitivity - Revisited

Alan J. Lutenegger, P.E., Member, ASCE

Professor of Civil & Environmental Engineering, University of Massachusetts, 27 Marston Hall, Amherst, Ma. 01003; lutenegg@ecs.umass.edu

ABSTRACT: In perhaps one of his less famous papers, Schmertmann (1969) introduced the concept of Swell Sensitivity as the ratio of the swelling behavior of a remoulded soil to the swelling behavior of the same soil in the undisturbed state. In effect, this Technical Note, which appeared in Geotechnique, essentially established the importance of inherent soil structure and natural bonding on the swelling potential of fine-grained soils. Observations made by Schmertmann and presented in the Note showed that some remoulded soils which have lost their natural structural bonds through simple mechanical remoulding have a higher swelling potential than the undisturbed soil. In the author's opinion, this paper is an important but lesser appreciated paper, in part because of the success of Schmertmann's other more popular papers. Swell Sensitivity has great practical importance to practicing geotechnical engineers, one of the main themes of all of Schmertmann's work. In the current paper, the author reexamines Swell Sensitivity as defined and described by Schmertmann (1969) and illustrates the behavior for a number of expansive soils from around the U.S.

KEY WORDS: clay, swelling, swell tests, shrink tests

INTRODUCTION

The swelling characteristics of some soils may be substantially altered by disturbance and remoulding. Generally, the swelling potential of an undisturbed soil is less that the swelling potential of the same soil in a remoulded state. Schmertmann (1969) described the effect of mechanical remoulding and the resulting increase in swelling as "Swell Sensitivity" and quantified this value as the ratio of the swelling potential of a remoulded soil to the swelling potential of an undisturbed soil. This short Note illustrates one of the recurring themes in Schmertmann's technical publications, recognizing the practical approach to recognizing practical problems of importance to the practicing geotechnical engineer. The swelling behavior of remoulded soils has considerable practical implications especially related to soils under floor slab and footings or placed against structures such as basement walls, retaining walls and pipes. Some potentially expansive soils may be inadvertently disturbed and remoulded by the contractor while performing site work. This can lead to more problems than might otherwise have been encountered.

Schmertmann (1969) provided some useful hints for the engineer as to which kinds of clays may show high Swell Sensitivity noting that "it seems reasonable to suspect any active clay with a low liquidity index and with breakable bonds will exhibit significant swell sensitivity. The engineer should suspect any old, highly overconsolidated clay as yet unweathered by natural processes. On the other hand relatively young and normally consolidated clays will probably have little bond formation and therefore little or no swell sensitivity". Three specific points are noted; 1). the clay must contain active (expansive) clay minerals, 2). the clay must be aged and highly overconsolidated, and 3). the clay should have a low liquidity index. All three of these characteristics are relatively simple for the engineer to determine. Intuitively, most highly overconsolidated clays have a low liquidity index since they exist at a natural water content well below the liquid limit and closer to the plastic limit.

In this paper, laboratory test results are presented by the author to illustrate the difference in swelling characteristics between undisturbed soil and the same soil prepared after simple mechanical remoulding. In most (but not all) cases the swelling characteristics of remoulded soils are much greater that the swelling of undisturbed soil. The difference is related to the initial structure of the soil and the subsequent breakdown of structure that occurs during remoulding.

SCHMERTMANN'S SWELL SENSITIVITY

Schmertmann (1969) defined Swell Sensitivity as "the ratio of the swelling index remoulded/swelling index undisturbed, or C_{SR}/C_{SU}". Using test results from an expansive clay from the Gainesville, Florida area, he showed that the Swell Sensitivity could be as high as 20, although most tests showed a value of around 4. The swelling index was obtained from the unloading curve of oedometer samples. An important test procedural point was noted by Schmertmann (1969) "when determining swell sensitivity it appears necessary to measure undisturbed swell behavior before imposing any large strains". This implies that large compression strains tend to destroy the internal structure and bonds so that in effect the undisturbed sample becomes destructured sufficiently to exhibit swelling behavior more closely to the remoulded soil. This might suggest that a "free swell" oedometer test in which the specimen is allowed to swell with either zero or a small nominal vertical surcharge might be an appropriate alternative test procedure.

Schmertmann (1969) also used the volume change results obtained from both "free swell tests" in which a specimen was placed in a dish of distilled water and allowed to swell by increasing water content and oedometer swell tests to illustrate Swell Sensitivity". Both procedures gave nearly the same results for Swell Sensitivity.

It is well known that remoulding or disturbance of clays can significantly influence the volume change characteristics, and many studies have shown that the consolidation behavior can be altered by this destructuring. Other investigations to directly determine the influence of remoulding on the swell behavior of clays are relatively limited. One could likely search the literature and find isolated examples of unloading oedometer curves on both undistrubred and remoulded clays at their natural water content, but most of these examples have been concerned largely with the influence of remoulding on the compression and not the swelling behavior.

DEFINING SWELL SENSITIVITY

The concept of Swell Sensitivity as introduced by Schmertmann (1969) was to provide a basis for comparing the swelling behavior of a remoulded soil with the same soil in an undisturbed state. Since there may be a number of ways to make this comparison, three different approaches might be considered by the engineer depending on the specific needs of a project.

Swell Sensitivity from Oedometer Free Swell

One of the primary objectives of the tests performed in this study was to evaluate the swelling behavior under zero surcharge or overburden stress. This is termed "Free Swell" and was accomplished by measuring the vertical expansion or 1D heave of specimens in the oedometer cell with no applied load. Under these conditions, Swell Sensitivity is defined as:

$$SS_{FS} = (\%\ \text{Swell (remoulded)})/(\%\ \text{Swell (undisturbed)}) \quad (1)$$

SS_{FS} is a numerical comparison of the amount of expansion or heave experienced by the two samples. The remoulded behavior is used in the numerator of Eq. 1 since in most cases the value of SS_{FS} will have a value greater than 1.0.

Since Free Swell conditions represent the maximum possible heave that an individual soil sample can experience; i.e., surcharge will allow less swelling, it might be expected that SS_{FS} represents an approximate maximum value that might be expected for a given soil subject to surcharge. Since the actual % volume change is measured in the test, this procedure provides a comparison of the potential volume change characteristics of the undisturbed and remoulded soil. Ratios of remoulded to undisturbed "free swell" tests presented by Schmertmann (1969) for five sets of tests on Gainesville clay ranged from about 3 to 4. These values illustrate how the remoulding or disturbance can dramatically affect the swelling behavior.

Swell Sensitivity from Oedometer Reloading Swell Pressure

The swelling pressure generated by a soil may be estimated by either applying load quickly as a specimen takes on water and begins to swell in an oedoemeter cell or allowing the soil to Free Swell in the oedometer and then apply incremental loading to bring the soil back to its original height. Applying loads quickly is sometime difficult to do accurately to just maintain the original specimen height while allowing the specimen to first swell and then reloading to the original height allows the engineer to obtain both the Free Swell amount and the swell pressure. Using the swell pressure from reloading, the Swell Sensitivity may be defined as:

SS_{SP} = (Swell Pressure (remoulded))/Swell Pressure (undisturbed)) .... (2)

In contrast to the Free Swell test procedure previously described, this method provides an estimate of the surcharge pressure required to return the sample to its initial void ratio. This is more-or-less an indirect measurement of the swelling pressure. As such, the value of SS_{SP} defines the ratio of swelling pressures instead of the ratio of volume change. One issue with allowing an undisturbed specimen to initially free swell is that some bonds may be broken during swelling. Schmertmann (1969) actually presented swell pressure results for two sets of tests on Gainesville clay obtaining the swell pressures as "Equilibrium vertical pressure to maintain natural void ratio when specimen allowed free access to distilled water". Results of two tests given by Schmertmann (1969) give ratios of this definition of remoulded/undisturbed swell pressure of about 2.0 and 6.7.

SOILS TESTED

Results from free swell Oedometer and Oedometer Swell Pressure Tests obtained on seven clays are presented. The soils represent five very stiff highly overconsolidated clays of moderate to high activity and two soft, near normally consolidated clays of medium to moderate activity. The five stiff overconsolidated clays are of low liquidity index and are representative of the type of clays that should exhibit significant Swell Sensitivity as described by Schmertmann (1969). A summary of the typical properties of the seven clays is given in Table 1.

TABLE 1. Summary of Typical Properties of Soils Tested.

Soil	L.L. (%)	P.L. (%)	L.I.	Clay (%<0.002 mm)	Activity
Houston	32	12	0.60	21	0.9
Centralia	45	18	0.10	35	0.8
Leon	55	17	0.10	48	0.8
ABE	60	25	0.10	46	1.0
Yazoo	100	52	-0.20	80	0.6
Ouachita	66	29	0.20	62	0.6
Omaha	-	-	-	63	-

SWELL TEST PROCEDURES

Oedometer Free Swell Tests

Free swell oedometer tests were performed on samples obtained from 76 mm diameter Shelby Tubes. Specimens were trimmed from the undisturbed samples into a 63.5 mm diameter stainless steel oedometer ring with an initial height of 19.0 mm. Remoulded specimens were prepared by taking adjacent soil from the Shelby Tube sample and mechanically remoulding the soil into the ring to achieve the same water content and dry density as the undisturbed specimen. The mechanical remoulding simply consisted of kneading the soil by hand into the oedometer ring. Porous stones were placed on the bottom and top of the specimen. After placing the ring in the oedometer cell, a dial gauge/LVDT was placed in contact with the top cap, an initial reading was obtained and water was added to the cell. Swelling was allowed for 24 hours.

Some soils will achieve full swelling within the 24 hour time period, while others will require much longer periods of swell time. In general, this is related to the specific composition of the specimen, such as clay content, mineralogy, etc., the initial water content and density of the specimen, and the initial hydraulic conductivity. A 24 hour time period was selected for the tests presented in this paper to provide a constant swell period for all soils and a reasonable period of time that consultants might use.

At the end of the 24 hour swell period, the specimen was removed and placed in a convection oven to obtain the final water content. Swelling behavior was evaluated by allowing samples to swell in an oedometer cell after being inundated with tap water. For remoulded tests, specimens were initially trimmed from undisturbed samples adjacent to the specimens trimmed for undisturbed testing. Remoulding was accomplished by initially cutting the trimmed specimens and then remoulding them back into the oedometer ring by hand to give density and initial water content as close as possible to the undisturbed samples. Since swelling behavior of remoulded soils is known to depend on aging, no aging was permitted in these tests and swell tests were performed immediately after remolding.

Oedometer Swell Pressure Tests

Oedometer swell pressure tests were also performed on selected samples by incrementally loading specimens following the 24 hour free swell procedure to determine the vertical stress required to return the specimen to the original height. These tests were performed on a separate set of specimens than was used for the Free Swell tests.

RESULTS

Results of Oedometer Free Swell Tests

Houston, Texas

Stiff Beaumont Clay from the National Geotechnical Experimentation Site at the University of Houston in Houston Texas was tested. Samples were obtained using 76 mm thin-walled Shelby tubes. Table 2 presents results of Free Swell tests performed on samples from three different depths. In two of the three cases the free swell of the remoulded specimens is higher that that of the undisturbed specimens and the clay shows very high Swell Sensitivity. The middle specimen showed the highest swell potential, several times higher than specimens from the other two depths and only gave a Swell Sensitivity of about 1.

TABLE 2. Oedometer Free Swell Test Results from Houston, Texas.

Specimen	Initial Water Content (%)	Final Water Content (%)	Water Content Change (%)	24 hr. $\Delta H/H_o$ (%)	R/U
14-16U	35.5	38.3	3.3	0.33	10.4
14-16R	33.6	42.3	8.7	3.44	
24-26U	23.8	35.3	11.5	11.19	0.9
24-26R	24.1	32.0	7.9	10.57	
31-33U	14.9	16.5	1.6	0.11	25.5
31-33R	14.2	19.2	5.0	2.80	

Centralia, Missouri

Samples of clay were obtained from Centralia, Missouri where a number of residential structures show signs of severe cracking and seasonal movement from the shrink/swell behaviour of the clay. The clay is a very stiff, highly overconsolidated and highly weathered Pleistocene glacial clay paleosol deposit. The clay generally has

TABLE 3. Oedometer Free Swell Test Results from Centralia, Missouri.

Specimen	Initial Water Content (%)	Final Water Content (%)	Water Content Change (%)	24 hr. $\Delta H/H_o$ (%)	R/U
2-2U	19.2	23.5	4.3	3.37	2.4
2-2R	19.0	29.3	10.3	8.20	
2-4U	24.2	28.3	4.1	1.69	2.4
2-4R	23.7	27.5	3.8	4.00	
2-6U	15.4	23.0	7.6	11.51	1.1
2-6R	16.2	25.0	8.8	12.55	
2-8U	19.7	23.3	3.6	6.01	1.2
2-8R	18.0	23.9	5.9	7.29	

low water content and typically shows values of Liquidity index around zero or even slightly negative, i.e., the natural water content is sometimes just at or slightly below the plastic limit. Table 3 gives results of Free Swell tests indicating Swell Sensitivity on the order of 1 to 2.4.

Leon, Iowa

Shelby tube samples of a high plasticity overconsolidated glacial clay paleosol from south central Iowa were obtained using 76 mm Shelby tubes. This clay is highly weathered and has moderate Activity. The clay is known to be composed of montmorillonite as the dominant clay mineral. Table 4 gives results of Free Swell tests which indicate Swell Sensitivity of about 1.5.

TABLE 4. Oedometer Free Swell Test Results from Leon, Iowa.

Specimen	Initial Water Content (%)	Final Water Content (%)	Water Content Change (%)	24 hr. $\Delta H/H_o$ (%)	R/U
3-1U	23.4	24.4	1.0	1.59	1.4
3-1R	21.2	23.7	2.5	2.29	
4-1U	21.6	24.8	3.2	2.65	1.5
4-1R	19.5	24.0	3.5	3.91	
7-1U	20.0	29.2	9.2	7.06	1.5
7-1R	21.6	32.2	10.6	10.80	

ABE, Iowa

Samples were obtained from another clay paleosol site in southern Iowa. The clay is very similar to the clay paleosol from the Leon site, but has slightly higher plasticity and higher Activity. The Free Swell test results given in Table 5 indicate higher Swell Sensitivity ranging from about 2.5 to 7.5.

TABLE 5. Oedometer Free Swell Test Results from ABE, Iowa.

Specimen	Initial Water Content (%)	Final Water Content (%)	Water Content Change (%)	24 hr. $\Delta H/H_o$ (%)	R/U
1-7U	26.3	28.3	2.0	0.92	4.4
1-7R	26.0	30.8	4.8	4.07	
1-8U	23.2	25.8	2.6	2.33	2.4
1-8R	24.2	28.0	3.8	5.52	
2-7U	28.4	29.8	1.4	1.13	2.5
2-7R	27.6	31.1	3.5	2.83	
6-5U	20.7	24.3	3.6	2.51	2.9
6-5R	21.0	26.7	5.7	7.17	
9-7U	28.2	31.1	2.9	0.59	7.5
9-7R	28.0	33.9	5.9	4.41	
9-8U	24.3	26.1	1.8	1.05	6.8
9-8R	24.6	29.5	4.9	7.12	

Yazoo, Mississippi

Samples of Yazoo clay were obtained from west central Mississippi. This stiff clay has the highest plasticity and highest Activity of all the clays tested. In comparison to the other clays tested, this clay shows the highest swelling potential as given by the Free Swell test result in Table 6, but only shows moderate Swell Sensitivity. The absolute values of 24 hr. swelling are exceptionally high, but Swell Sensitivity is only on the order of about 1 to 3. In this case it appears that the influence of remoulding may be partially masked by the high swell potential.

TABLE 6. Oedometer Free Swell Test Results from Yazoo, Mississippi.

Specimen	Initial Water Content (%)	Final Water Content (%)	Water Content Change (%)	24 hr. $\Delta H/H_o$ (%)	R/U
14-15U	38.2	60.1	21.9	27.13	1.1
14-15R	36.8	62.0	25.2	28.77	
24-25U	33.6	63.6	30.3	40.96	0.9
24-25R	36.3	64.7	28.4	38.68	
29-30U	43.9	59.7	15.8	5.95	3.2
29-30R	44.3	61.7	17.4	19.00	
39-40U	44.8	60.8	16.0	12.91	1.4
39-40R	43.7	68.2	24.5	17.96	

Ouachita, Louisiana

Samples of alluvial clay were obtained from Ouachita, Louisiana. The clay has a high plasticity but a higher liquidity index than the stiff clays previously described. Because of this, there is a tendency for the clay to take on less water and therefore swell less. The results given in Table 7 show lower swell amounts than most of the other clays but still show a Swell Sensitivity of about 1.5 to 2.

TABLE 7. Oedometer Free Swell Test Results from Ouachita, Louisiana.

Specimen	Initial Water Content (%)	Final Water Content (%)	Water Content Change (%)	24 hr. $\Delta H/H_o$ (%)	R/U
SPL-1U	36.8	37.8	1.0	1.40	1.7
SPL-1R	36.4	39.4	3.0	2.41	
SPL-2U	35.1	36.8	1.7	2.64	1.4
SPL-2R	39.3	40.3	1.0	3.60	
SPL-3U	34.0	36.9	2.9	2.65	2.0
SPL-3R	34.3	39.4	5.3	5.28	

Omaha, Nebraska

Alluvial clay was also obtained from near Omaha, Nebraska. This clay is very similar to the clay from Ouachita and overall showed the lowest swell potential of any of the soils tested as indicated in Table 8. Swell Sensitivity was similar to the Ouachita samples.

TABLE 8. Oedometer Free Swell Test Results from Omaha, Nebraska.

Specimen	Initial Water Content (%)	Final Water Content (%)	Water Content Change (%)	24 hr. $\Delta H/H_o$ (%)	R/U
5-6U	33.2	37.0	3.8	1.77	2.3
5-6R	32.0	38.6	6.6	4.09	
6-7U	46.2	53.1	6.9	2.67	1.0
6-7R	46.3	50.4	4.1	2.67	
7-8U	50.0	52.7	2.7	1.28	1.6
7-8R	48.6	51.5	2.9	2.08	
8-9U	55.4	57.4	2.0	0.81	1.8
8-9R	52.9	55.1	2.2	1.45	
9-10U	51.8	52.0	0.2	0.88	1.1
9-10R	51.9	52.2	0.3	0.93	
10-11U	52.7	52.9	0.2	0.52	1.2
10-11R	48.8	51.3	2.5	0.65	

Overall, it does appear that the combined use of Liquidity Index and Activity may give some important clues as to those soils that may exhibit high Swell Sensitivity. It may be important to make a more detailed investigation at these and other soil characteristics such as Cation Exchange Capacity and Specific Surface to help identify these problematic soils and the specific mechanics resulting in high Swell Sensitivity.

Results of Oedometer Swell Pressure Tests

Oedometer swell pressure tests were performed on selected samples from Leon, Iowa and Ouachita, Louisiana. Results of these tests are given in Tables 9 and 10. Recall that swell pressure was measured by reloading to the original void ratio after first allowing the specimen to free swell for 24 hrs. The results given in Tables 9 and 10 give Swell Sensitivity in terms of both the free swell and the swell pressure and suggest that for the samples tested there is little to no difference in the Swell Sensitivity from either test procedure, i.e., free swell vs. swell pressure, although the absolute values of Swell Sensitivity are low. Clays with higher Swell Sensitivity might show different results but unfortunately were not available for testing.

TABLE 9. Oedometer Swell Pressure Test Results from Leon, Iowa.

Specimen	Initial Water Content (%)	24 hr. $\Delta H/H_o$ (%)	Swell Pressure (kPa)	R/U ($\Delta H/H_o$)	SS_{SP}
3-1U	26.2	3.08	119.6	1.4	1.4
3-1R	28.5	4.43	167.5		
4-1U	23.0	3.41	83.7	0.6	1.0
4-1R	22.9	2.12	83.7		

TABLE 10. Oedometer Swell Pressure Test Results from Ouachita, Louisiana.

Specimen	Initial Water Content (%)	24 hr. $\Delta H/H_o$ (%)	Swell Pressure (kPa)	R/U ($\Delta H/H_o$)	SS_{SP}
SPL-1U	40.6	2.63	71.8	1.8	1.2
SPL-1R	39.4	4.65	83.7		
SPL-2U	32.9	1.60	83.6	1.6	0.9
SPL-2R	33.1	2.53	76.6		
SPL-3U	30.8	2.56	107.7	1.3	0.7
SPL-3R	30.3	3.39	76.5		

Influence of Swelling Time

The engineer may be curious as to whether the testing time may affect measured Swell Sensitivity. The author selected a 24 hour swelling period for convenience and to expedite the testing; Schmertmann (1969) used a period of 3 days in determining Swell Sensitivity but actually tested some samples for 14 days. The amount of time required for a clay to fully expand when allowed free access to water will vary with the clay. Some clays may require long periods of time swell to their maximum amount while others may fully swell in a few hours. This depends on the composition of the clay, the initial water content and dry density and the initial permeability.

Table 11 gives results of both short term (1 day) swelling and long term (30 day) swell tests on the same clay. One of the samples (2-7) shows very little difference while the other sample (9-8) shows a significant difference in Swell Sensitivity, although the long term tests on the sample had lower initial water content. The engineer needs results quickly and but may wish to choose an appropriate testing time for a particular clay.

TABLE 11. Short Term (1 day) and Long Term (30 day) Oedometer Free Swell Test Results from ABE, Iowa.

Specimen	Initial Water Content (%)	Final Water Content (%)	Water Content Change (%)	$\Delta H/H_o$ (%)	R/U
2-7U-1	28.4	29.8	1.4	1.13	2.5
2-7R-1	27.6	31.1	3.5	2.83	
2-7U-30	26.2	31.0	4.8	4.97	2.1
2-7R-30	25.4	34.1	8.7	10.24	
9-8U-1	24.3	26.1	1.8	1.05	6.8
9-8R-1	24.6	29.5	4.9	7.12	
9-8U-30	19.7	25.1	5.4	6.72	2.2
9-8R-30	19.8	30.6	10.8	14.45	

Influence of Initial Conditions

Groundwater conditions can change as a result of seasonal variations in moisture regime at a site either increasing or decreasing the water content in a soil. These changes can produce swelling or shrinking in the soil. The engineer may wish to know if swelling of the soil is influenced by the initial water content or initial dry density. It is expected that if an in situ soil were allowed to take on water and swell to the maximum amount that some of the internal bonds would be lost. In natural ground this would rarely occur but the soil may undergo small changes in water content.

Table 12 gives results of two sets of oedometer free swell tests performed on the same clay with differing initial water content. In addition to testing specimens at the natural (as sampled) water content, a second set of specimens was prepared after first allowing the samples to air dry a small amount over several days leading to a lower water content and higher initial dry density. While the actual swell amount is much higher in the dried set of specimens as expected, the Swell Sensitivity is actually lower in the dry set. This may suggest that the same clay may exhibit different Swell Sensitivity depending on its initial state. Clearly this needs to be examined further.

TABLE 12. Oedometer Free Swell Tests with Different Initial Conditions from Leon, Iowa.

Specimen	Initial Water Content (%)	Final Water Content (%)	Water Content Change (%)	24 hr. $\Delta H/H_o$ (%)	R/U
1-7U	26.3	28.3	2.0	0.92	4.4
1-7R	26.0	30.8	4.8	4.07	
1-7U	21.5	28.1	6.6	9.76	1.8
1-7R	19.8	28.7	8.9	11.61	

SHRINK SENSITIVITY

The loss of internal bonds that allow a mechanically remoulded soil to swell more than the same soil in the undisturbed state should have a similar effect on the *shrinkage* characteristics of the soil. That is, the loss of bonds from mechanical remoulding should allow the soil to achieve constant volume at a lower water content at the lower limit of shrinkage as compared to the undisturbed soil. In order to investigate this, shrink tests were performed using the procedure described by Briaud et al (2003). Specimens of both undisturbed and mechanically remoulded soil were allowed to shrink slowly in air over several days. Periodically, the diameter and height of the specimen were determined using a set of digital calipers in order to calculate the volume. At the same time, the total mass of the specimen was obtained using an electronic balance so that the water content could be calculated. At the end of the drying period, when no further significant change in volume had occurred, the specimen was placed in a convection oven so that the oven dried weight and oven dried volume could be obtained.

Figure 1 shows undisturbed and remoulded shrinkage curves obtained for two soils; 1). a sample of clay from Gainesville, Florida (provided to the author by Dr. J. Schmertmann), and 2) a sample of clay from Leon, Iowa. The ***Shrink Sensitivity*** is defined herein as the ratio of the water content at the lower limit of shrinkage of the remolded specimen to the water content at the lower limit of shrinkage of the undisturbed specimen. The water content at the lower limit of shrinkage may be interpreted as the intersection of two straight lines, one drawn along the drying curve and the other drawn vertically at the final volume.

As can be seen from Figure 1 the remoulded specimens shrink to constant volume at a lower water content as compared to the undisturbed specimens. Shrink Sensitivity of both of these two clays is on the order of 1.3. A reference or "Intrinsic" value of the lower limit of shrinkage for a fully remoulded soil, such as one that is initially reconstituted at a water content above the liquid limit as from the Shrinkage Limit Tests, as opposed to one remoulded at natural field water content, gives an even lower water content. As an example, Freeman (1936) presented results between the Shrinkage Limit obtained from undisturbed samples of Flushing Meadow Clay and the Shrinkage Limit obtained from conventional Shrinkage Limit Tests. For 20 pairs of tests, the average ratio of the Undisturbed Shrinkage Limit to Reconstituted Shrinkage Limit was 1.13. Again, these data illustrate the influence that initial soil structure can have on the volume change characteristics.

CONCLUSIONS

The results presented in this paper reinforce the concept of Swell Sensitivity introduced by Schmertmann nearly 40 years ago. And further illustrate that the swelling behavior of mechanically remoulded soils can be significantly greater than that of the same soil in the undisturbed state. Swell Sensitivity of the soils tested by the author ranged from about 1 to 25. Higher values of Swell Sensitivity were generally observed in stiffer highly overconsolidated clays but even some of these clays showed low values of Swell Sensitivity. Further detailed characterization is needed to help understand the specifics that differentiate Swell Sensitivity of different clays. While the fundamental mechanism is loss of bonds through remolding, the engineer might wish to identify a soil with significant swell sensitivity from basic soil tests. Soils that exhibit Swell Sensitivity will also likely exhibit Shrink Sensitivity.

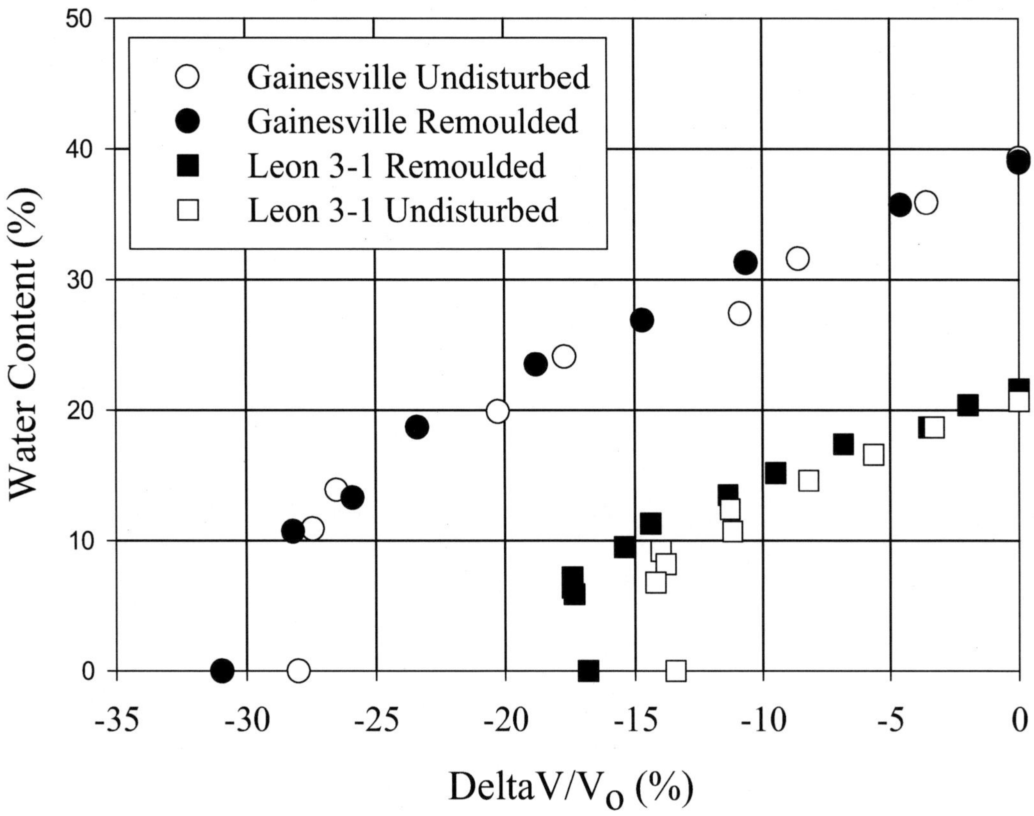

FIG. 1. Undisturbed and Remoulded Shrinkage Curves for Two Clays.

REFERENCES

Briaud, J.L Xiong, Z. and Moon, S., 2003. Shrink Test-Water Content Method for Shrink and Swell Predictions. *Journal of Geotechnical and Geoenvironmental Engineering*, ASCE, Vol. 129, No. 7, pp. 590-597.

Freeman, G.L., 1936. Soil Survey of the Flushing Meadow Park Site, Long Island, New York. Proceedings of the 1st. International Conference on Soil Mechanics, Vol. l, pp. 25-30.

Schmertmann, J.H., 1969. Swell Sensitivity. *Geotechnique*, Vol., 19, No. 1, pp. 530-533.

Physical Modeling of Raft on Settlement Reducing Piles

Vincenzo Fioravante[1], Daniela Giretti[2] and Michele Jamiolkowski[3]

[1]Professor of Geotechnical Engineering, University of Ferrara; vincenzo.fioravante@unife.it
[2]Ph.D. Student, University of Ferrara; daniela.giretti@unife.it
[3]Professor of Geotechnical Engineering, Technical University of Torino; michele.jamiolkowski@polito.it

ABSTRACT: The paper presents the results of extensive centrifuge tests modeling rigid circular piled rafts laying on a bed of loose very fine silica sand. The tests were aimed at investigating the behavior of rafts on settlement reducing piles. The testing program included: an unpiled raft, rafts on 1, 3, 7 and 13 piles. In each test, some model piles were instrumented with load cells to determine the distribution of load along the shaft. Beneath the rafts, two types of model piles, close-ended and free headed were installed: quasi displacement (QD) and quasi-non displacement (QND) piles. The obtained results permitted figuring out the role of piles in terms of their effectiveness as settlement reducers and to quantify the load sharing mechanism between piles and the raft-soil contact. The tests were aimed at investigating the load transfer mechanisms adopted in the design approach, and in particular at validating a numerical code which can be used in engineering practice. The paper describes the details of experiments undertaken, the adopted procedures and some of the results; where not specified, all the experimental data given in this paper referred to model scale. The adopted numerical code is briefly described and its validation, via test results on quasi-non displacement piles, is also reported.

KEY WORDS: centrifuge modeling, displacement and non displacement piles, piled raft, settlement reducing piles

INTRODUCTION

The everyday design of piled rafts is based on the assumption that the total weight of the structure is supported by piles. An innovative approach involving the use of piles as settlement reducers was postulated by Padfield and Sharrock (1983), Hansbo and Jedenby (1983), Hansbo (1984), Cooke (1986).

The basic concept is that the load from the structure can be transmitted, via the raft, partly on the piles and partly on the foundation soil, as confirmed by many researchers [Randolph (1994), Poulos (1994), Horikoshi and Randolph (1998), Viggiani (1998, 2001), Mandolini et al (2005)].

This approach allows optimizing the piled rafts design and significantly reducing the number of piles with a modest increase in settlement compared to the conventional design.

The use in practice of the settlement reducing piles can be focused on two objectives: in case of rigid foundations at decreasing the total settlement and when dealing with flexible rafts, at reducing both the total and the differential settlements. The latter is achieved by focusing the design on the piles distribution in plane and on the spacing able to minimize the raft deflections. See [Sommer et al. (1991), Randolph and Clancy (1994), Burland (1995), El-Mossallamy (1996), Russo and Viggiani (1998), Viggiani (2001), Poulos (2001), Randolph (2003), Mandolini (2003), Randolph et al. (2004), Russo et al. (2004), Hanish et al. (2002)].

Within the above frame, this paper represents a continuation of the multi-g physical modeling of piled rafts [Fioravante (1997), Fioravante and Jamiolkowski (2005)] aimed at exploring the basic aspects of the problem, and possibly giving information functional to the foundations design using settlement reducing piles.

The similarity relationships between model and prototype are summarized in Table 1; in a model geometrically scaled N and subjected to a centrifuge acceleration of Ng, both stresses and strains are scaled at 1:1, displacements are reduced in the model by factor of N and forces are reduced by factor of N^2.

TABLE 1. Scaling factors

QUANTITY	PROTOTYPE	MODEL	QUANTITY	PROTOTYPE	MODEL
Length	N	1	Stress	1	1
Area	N^2	1	Strain	1	1
Volume	N^3	1	Mass Density	1	1
Velocity (projectile)	1	1	Energy density	1	1
Velocity (undrained conditions)	1	N	Time (Dynamic)	N	1
Acceleration	1	N	Time (Diffusion)	N^2	1
Mass	N^3	1	Time (Creep)	1	1
Force	N^2	1	Frequency	1	N
Energy	N^3	1			

EXPERIMENTAL SET-UP

a. The ISMGEO Geotechnical Centrifuge (IGC)

The model tests were performed in the IGC, which is made of a symmetrical rotating arm 6 m in diameter, 2 m in height and 1 m in width giving a nominal radius of 2 m. The arm holds two swinging platforms carrying the model container and the counterweight; during the test the platforms lock horizontally to the arm to prevent transmitting the working loads to the basket suspensions. An outer fairing covers the arm and they concurrently rotate to reduce air resistance and perturbation during in flight.

The arm's unusual shape offers the following advantages:

- A small distortion of the centrifugal field in the model since its main dimension is parallel to the rotation axis.

- An easy location of instruments close to the rotation axis because of the absence of a central shaft across the arm.
- A dynamic excitation of the model in a direction parallel to the rotation axis, so that the problem of Corioli's acceleration is removed.

The centrifuge has the potential of reaching an acceleration of 600g at a payload of 400 kg. The model maximum dimensions are: length = 1 m, height = 0.6 m, with = 0.5 m.

Fig.1 shows a scheme of the IGC; further details can be found in Baldi et al. (1988).

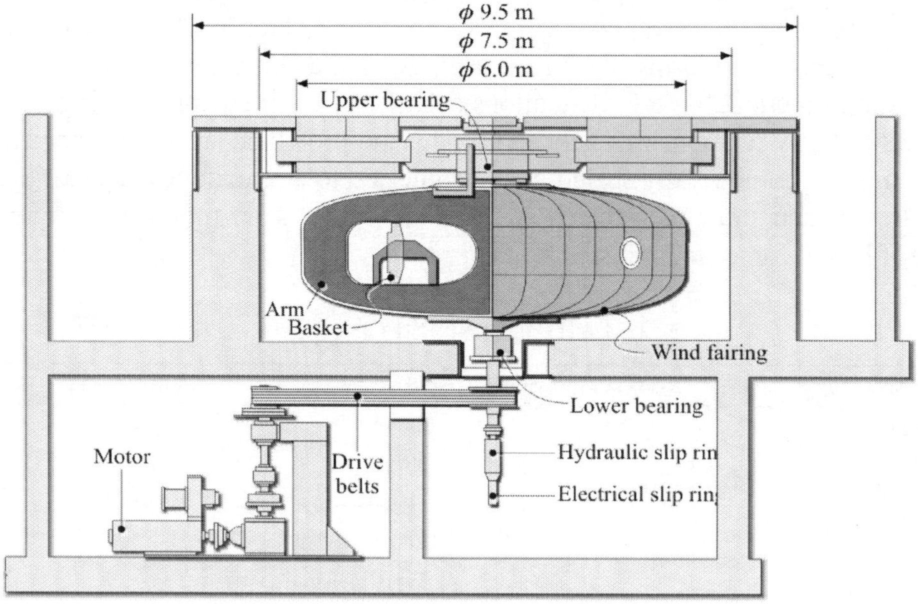

FIG. 1: The ISMGEO Geotechnical Centrifuge scheme.

b. Test sand, model raft and model piles

Centrifuge tests were performed using saturated fine silica sand, called FF sand (FFS), which is a very fine silica powder derived by grinding and sieving pit rocks. It is a uniform fine sand consisting of sub angular particles and made of 98.2% quartz, 1.3% feldspar and 0.5% mica.

It is characterized by:

- $\gamma_{min} = 11.58$ kN/m^3;
- $\gamma_{max} = 14.78$ kN/m^3;
- $\phi'_{cv} = 33°$;
- $G_s = 2.61$;
- $D_{50} = 0.093$ mm;
- $C_U = 1.89$.

The model raft was a 88 mm in diameter (d_r) and 15 mm in thickness (t_r) steel disc characterized by a modulus of elasticity E_r=2.1*10^5 MPa and a moment of inertia referred to the 1g prototype J_r=294.37 m^4. The very high raft-soil stiffness ratio, defined as:

$$K_{rs} = \frac{E_r}{E_s}\frac{1-\upsilon^2_s}{1-\upsilon^2_r}\left(\frac{t_r}{d_r/2}\right)^3 \text{ [-]} \qquad (1)$$

allows considering the raft as infinitely rigid [Gorbunov-Posadov et al. (1984), Horikoshi and Randolph (1997)].

The model piles employed in centrifuge tests were close-ended and free headed with a diameter D_p of 8 mm and a length L_{ap} of 160 mm (slenderness ratio of 20); they were made from an aluminium alloy pipe having the same geometrical properties and roughness, obtained by mechanical turning and measured by a micrometrical profilometer, whose average value was $R_t = 300$ μm.

It resulted a relative roughness $R_n=R_t/D_{50} = 3$, which produced an angle of interface friction δ' equal to the angle of soil resistance at critical state, ϕ'_{cv} [Yoshimi and Kishida (1981), Kishida and Uesugi (1987), Jardine et al. (1993), Foray et al. (1995), Garnier and Konig (1998)].

The model piles diameter with respect to particle size dimension, $D_p/D_{50} = 86$, allowed to minimize the scale effects [Garnier and Konig (1998), Foray et al. (1998), Fioravante (2002), Garnier (2002)].

Some of the model piles were manufactured in segments to allow accommodating load cells measuring the loads at various positions along the pile. Given the load cells design, the fairly small diameter of the model piles allows incorporating a maximum of two load cells per pile.

The instrumented piles were, therefore, assembled in two configurations: the upper load cell was placed in the pile head in both cases (hereafter referred to as position A) while the lower load cell was placed at the pile base (position B) or at mid-pile (position C). A model piles sketch is shown in Fig 2.

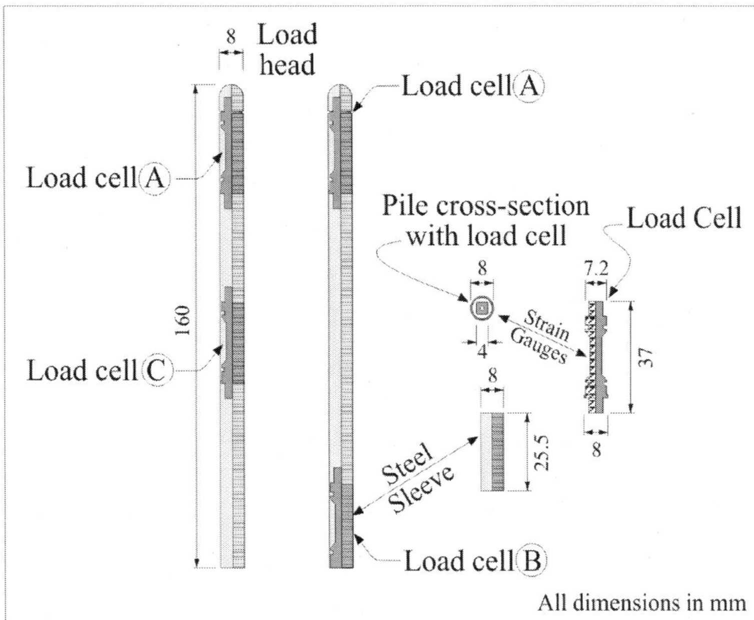

FIG. 2. Instrumented model piles.

The miniaturised load cells, made from stainless steel, had a square cross section as outlined in Fig. 2; two active strain gauges were coated with an acrylic moisture barrier on each of the four flat surfaces, one measuring pile-longitudinal strains, the other pile-transversal strains. The strain gauges were connected in a Wheatstone bridge, summing the pile-parallel strains while subtracting the pile-transversal strain on opposite faces, to permit cancelling bending and temperature effects in the readings, and giving a very accurate measurement of the axial load. The instrumented cross-section was covered by a protective stainless steel jacket avoiding soil radial stress acting on the gauges and was sealed with O-rings as a second moisture barrier.

Capillary wires connecting the load cell to the data acquisition system were guided through the centre cavity of the model piles. Where feasible, the drift of the offset and zero loads were carefully detected and minimised. The output signals from the load cells were amplified by a factor of 100.

To minimise the differential settlement between pile and soil during the consolidation, the weight of the model pile embedded was close to that of the substituted soil.

The adopted instrumentation allowed measuring the axial loads transmitted from the raft to the instrumented pile and estimating the average load transmitted through the pile shaft. Due to the axis-symmetry of the geometry and the loads, the forces measured on instrumented piles placed at the same radius were considered as belonging to the same equivalent pile; by combining the measures of the load cells, the local shaft friction transfer curves were evaluated.

c. Model preparation

The geometrical scaling factor of the models was N=100; all models were tested under an acceleration field of 100g, imposed in correspondence of the soil surface, the distortion of the acceleration field was accounted in the computation of the stress distribution with depth.

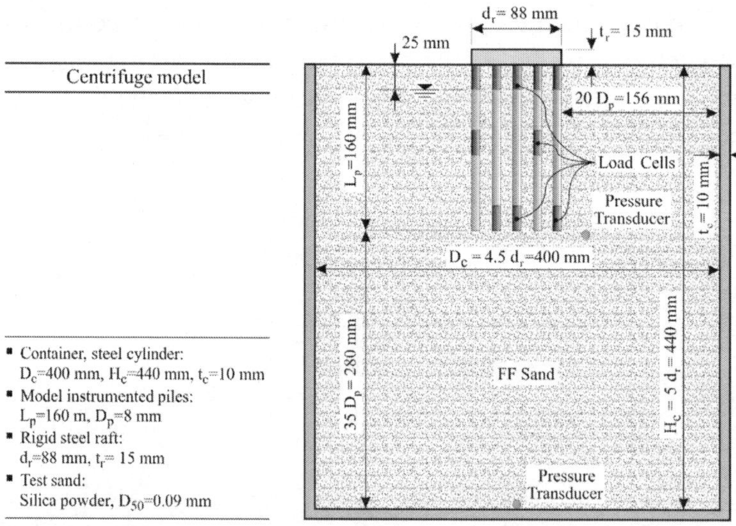

FIG. 3. Centrifuge model of piled raft

The soil model was reconstituted at low density by pluviating in air the dry sand into a cylindrical container at a very small constant height of fall. The adopted cylindrical container had rigid walls to avoid lateral displacements of the soil; its dimensions were 400 mm inside diameter and 440 mm height. A scheme of the model is shown in Fig. 3.

After deposition, tap water was flowed into the sand from the bottom of the container up to the soil surface at a small hydraulic gradient. Two miniaturized pore pressure transducers, inserted into soil mass, allowed monitoring water level during the tests.

The presented data referred to the average soil density measured at the end of the in flight consolidation; it was always about the same, i.e. $D_R \approx 30\%$ and it was assumed constant with depth.

The soil models uniformity and reproducibility were verified in flight by means of static cone penetration tests, using a miniaturized electrical piezocone (diameter d_c=11.3 mm and apex angle of 60°), penetrating at a rate of 2 mm/s.

The obtained cone resistance q_c profiles were compared with those computed using the empirical correlation (Fig.4) established on the basis of CPT's performed in a large calibration chamber in four silica sands (Jamiolkowski et al. 2003); the results are shown in Fig. 4, as referred to the prototype scale.

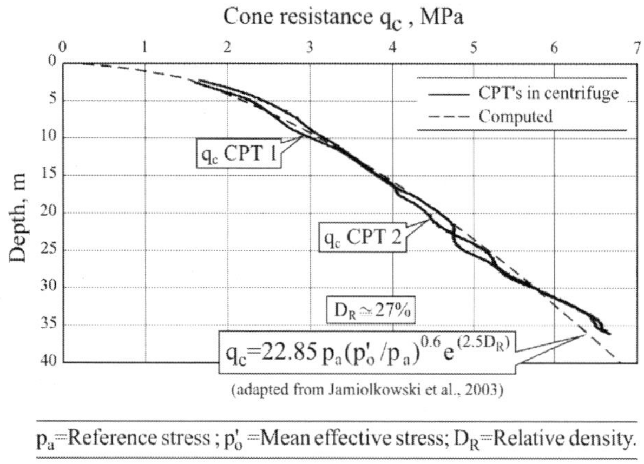

FIG. 4. CPT's carried out in the centrifuge, prototype scale.

A very rigid frame held two vertical displacement transducers to monitor the raft displacement; an external load cell and a hydraulic actuator were mounted on the container top; the settlement of the sand surface was measured at a distance of 50 mm from the container walls.

The model piles were installed following two different procedures. In the first, the piles were embedded in the soil specimen in pre-bored holes of about 7.8 mm in diameter for their full length and then the container was mounted into the centrifuge and accelerated to 75g. Since the soil surface settled 1-1.5 mm more than the piles, these were slightly pushed down by the raft plate to be aligned to the soil surface; then the raft plate was lifted and the centrifuge speed increased to the target value, thus a final consolidation occurred at 100g. Such "atypical" procedure was adopted to minimise the unavoidable problem of the different alignment of pile heads with respect to the soil surface. The piles installed like so, which simulate ideally bored prototype piles, were called "quasi non-displacement piles" (hereafter simply referred as QND).

The second procedure consisted in embedding the piles, to some extent, into pre-bored holes of about 7.8 mm in diameter for a length of 130 mm at 1g; then the container was subject to a centrifuge acceleration of 100g, then, after consolidation, piles group jacking took place simultaneously for 30-32 mm of displacement by means of the raft plate. Drained penetration conditions were ensured by a low displacement rate, as monitored by the pore pressure transducer placed near piles tips (Fig. 3). This procedure, called "quasi displacement" (henceforth QD), was assumed adequate to allow the fully mobilisation of the shearing and bearing capacity, but it produced a minor densification of the soil around the pile than a full length of piles complete penetration; however it was necessary to jack the piles group.

The raft plate, lifted after piles jacking, was then slowly lowered until pile head contact was achieved again, and then the loading tests were performed at a constant rate of load of 50 N/min: the compression testing load, applied by a hydraulic piston, was measured by external load cell. External and internal forces as well as raft and sand settlements were continuously recorded (acquisition frequency f = 0.1 Hz).

The water level was kept constant at 25 mm of depth from soil surface, as monitored by the pore pressure transducers, whose results showed constant values during the tests.

The two pile installation procedures adopted should approximate field conditions (bored vs. displacement piles); the gathered results were compared to analyze the piles effectiveness as settlement reducers with different initial conditions.

d. Testing Programme

Eight different models were tested and twenty loading tests were carried out as summarised in Table 2, with reference to Fig. 5:

- unpiled raft (tests n. 1);
- isolated single pile (tests n. 2, 3, 4, 5, 10,11);
- 1 piled raft (tests n. 6, 12);
- 3 piled raft, placed at the spacing i = 8.66 D_p (tests n. 7,8,13,14);
- 7 piled raft, placed at the spacing i = 5 D_p (tests n. 9, 15,16);
- 13 piled raft (tests n. 17,18);
- 3 piled raft, placed at the spacing i = 4.6 D_p (test n.19);
- 7 piled raft, placed at the spacing i = 2.5 D_p (test n.20);

TABLE 2: Testing programme.

Test No.	Test Name	Description	No. of piles	Test Layout	D_R (%)	Test No.	Test Name	Description	No. of piles	Test Layout	D_R (%)
1	URLT0	Unpiled raft			26	11	QND-PLT6	Isolated single pile	1		30
2	QD-PLT1	Isolated single pile	1		34	12	QND-PRLT5	Piled raft	1	L1	30
3	QD-PLT2	Isolated single pile	1		28	13	QND-PRLT6	Piled raft	3	L2	29
4	QD-PLT3	Isolated single pile	1		31	14	QND-PRLT7	Piled raft	3	L2	30
5	QD-PLT4	Isolated single pile	1		29	15	QND-PRLT8	Piled raft	7	L3	32
6	QD-PRLT1	Piled raft	1	L1	29	16	QND-PRLT9	Piled raft	7	L4	30
7	QD-PRLT2	Piled raft	3	L2	27	17	QND-PRLT10	Piled raft	13	L5	36
8	QD-PRLT3	Piled raft	3	L2	27	18	QND-PRLT11	Piled raft	13	L6	30
9	QD-PRLT4	Piled raft	7	L3	30	19	QND-PRLT12	Piled raft	3	L7	26
10	QND-PLT5	Isolated single pile	1		31	20	QND-PRLT13	Piled raft	7	L8	26

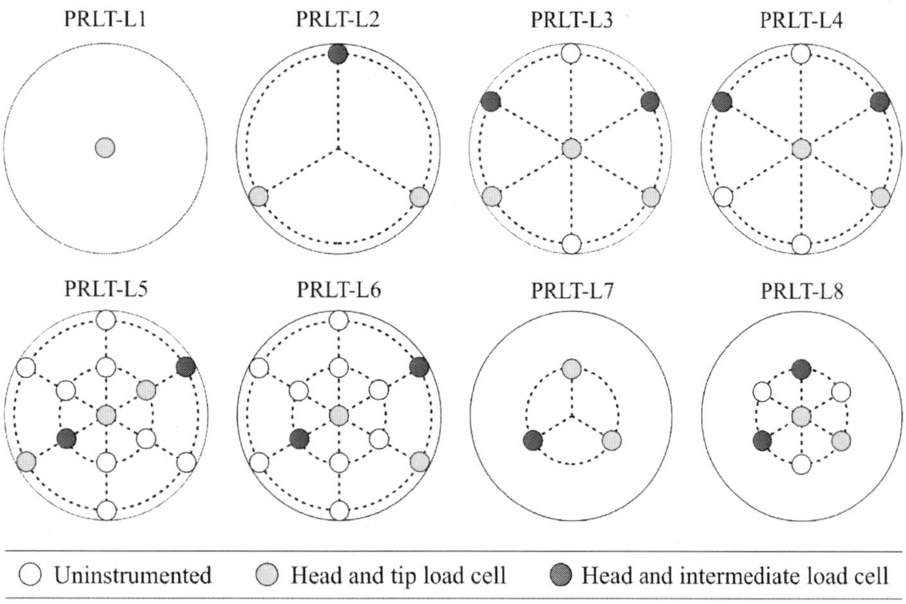

FIG. 5. Layout of the tested piled rafts.

TESTS RESULTS AND INTERPRETATION

a. Load tests on isolated pile

A series of six tests were carried out on isolated QD and QND model piles. The total applied load, Q_t and the base resistance, Q_b were directly measured during the test, while the load transferred through the shaft, Q_s was computed by means of the equilibrium equation:

$$Q_s(s) = Q_t(s) + W_p - Q_b(s) \quad [F] \tag{2}$$

where:

W_p = pile weight,
s = pile settlement; due to the inherent rigidity of the model pile, the settlement of the pile tip was assumed to be the one at pile head.

The results of two tests are reported in Fig. 6, where the total base and shaft loads transmitted through a QD pile as function of s are compared with analogous forces transmitted to the soil by a QND pile; all values being referred to the model scale.

As to the total load carried out by the QD pile, Q_t (s) increased to a ultimate value reached at s=0.77 mm, then reduced towards a constant post peak steady value, while the Q_t(s) carried by the QND pile showed a progressive mobilisation with settlements and its value increased toward that observed for the QD pile at a s ~ 10 mm.

As expected, the ultimate base resistance, Q_{bu} of the QD pile was almost fully mobilised at a settlement of 1 mm, while a value larger than 10 mm, was required to mobilize Q_{bu} for QND pile.

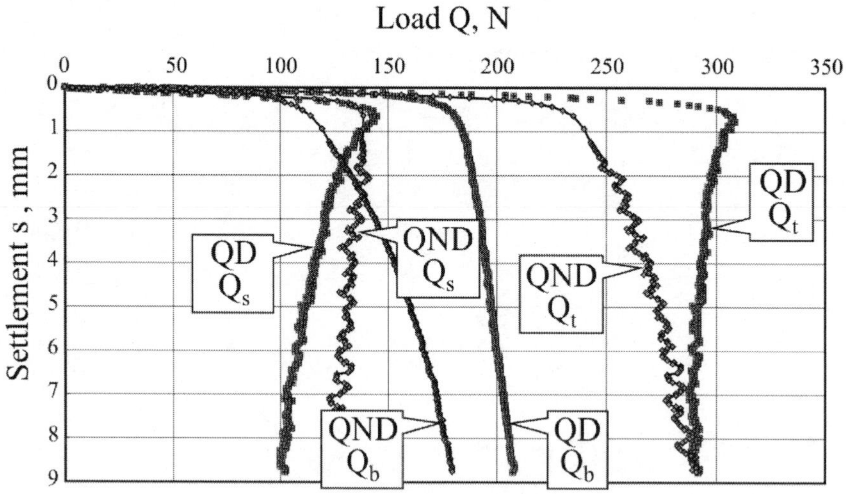

FIG. 6. Load tests on isolated piles.

As to the ultimate shaft resistance, Q_{su} it was achieved at almost the same settlement in both tests, i.e. 0.65 mm, however the peak value obtained for the QD pile was slightly larger than that achieved by the QND pile, subsequently Q_s decreased to a steady value as the settlement increased.

Sand dilatancy occurred in the interface zone, should cause the observed behaviour [Wernick (1978), Boulon and Foray (1986), Boulon (1988, 1989)]; dilatancy effects were more pronounced in QD piles due to the sand densification around piles shaft.

The residual forces, caused by the pile installation, were taken into account in the tests interpretation. The negative values of Q_s (at $Q_t = 0$) were originated by the relative settlement between soil and QND pile during the in flight consolidation, the negative skin friction and the pile self weight were equilibrated by the base resistance ($Q_b > 0$) measured for unloaded pile. As to the QD pile, upon the removal of the load required to obtain its desired penetration, it exhibited an elastic extension causing the mobilization of the negative friction along the upper segment of the shaft.

b. Load-settlement behaviour of QD piled rafts

Fig. 7 shows the relationship between the applied stress (q_r) and the relative settlement (s_r/d_r) as resulted from the tests on the rafts on QD piles. The same figure compares the test results obtained for an unpiled raft. As expected, the latter exhibits a much larger settlement at the same applied load. Comparing the settlements at q_r=550 kPa, the value of s_{ur} of the unpiled raft is 10 mm, while on 7 piles is of 0.7mm, being referred to model scale, likewise all the experimental data given in this paper.

Based on the above results, the settlement reduction factor ρ was computed as a ratio between the displacements of the piled raft, s_r, normalised with respect to that of the unpiled raft, s_{ur}, both measured at the same vertical applied loads:

$$\rho = s_r/s_{ur} \quad [-] \tag{3}$$

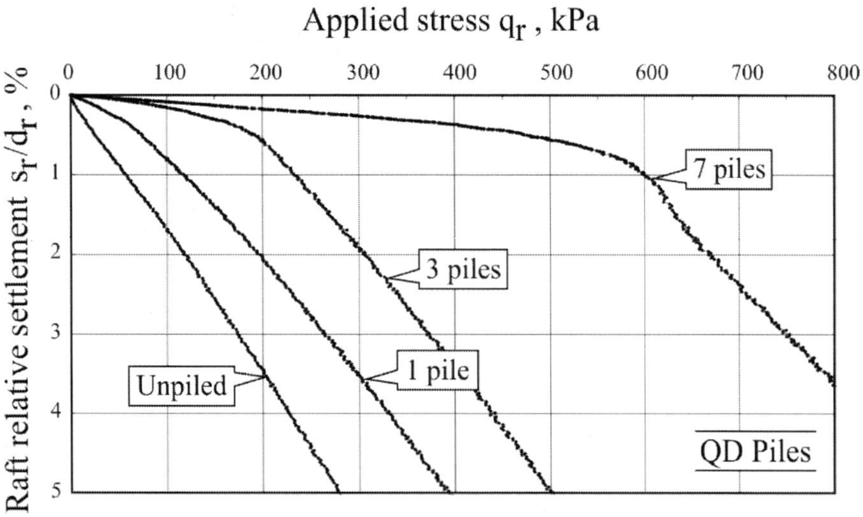

FIG. 7. QD piled rafts - Stress-settlement relationships.

Fig. 8 reports the ρ values, computed for q_r values equal to 300, 400, 600 and 800 kPa, plotted versus the number of piles underneath the raft. It is evident the benefit that even a few piles determine in terms of settlement reduction. For example a raft on 3 piles under q_r= 300 kPa settles only one third of that without piles.

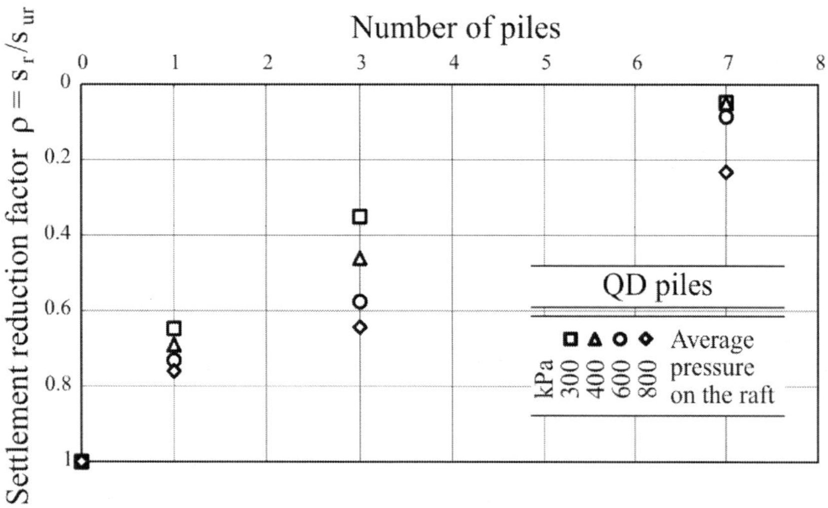

FIG. 8. QD piled rafts - Settlement reduction factor.

The values of the total applied load Q_t, the load acting on the piles Q_p and that transmitted directly by the raft to the soil Q_r, for the raft with 7 QD piles are reported in Fig. 9 as function of s_r/d_r.

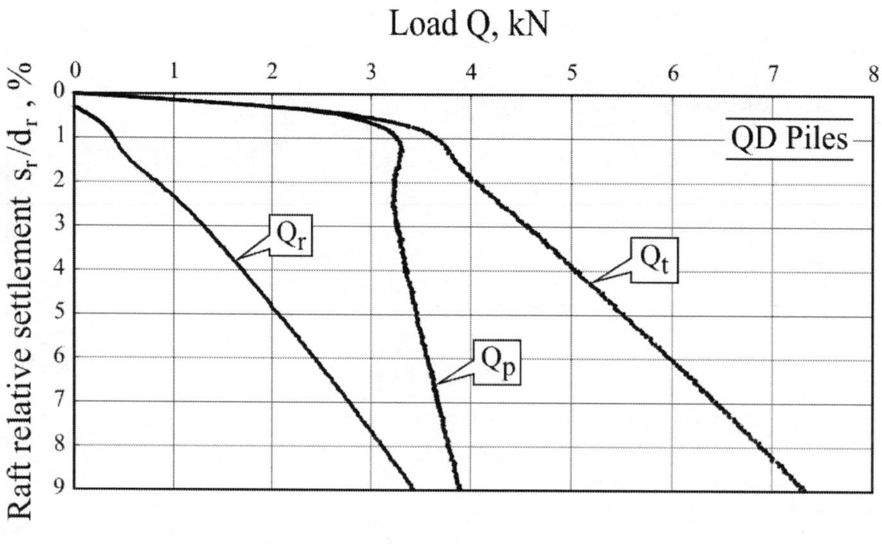

FIG. 9. QD 7 piled raft - Load sharing mechanism.

Due to the higher stiffness of piles and also because of a modest misaligning of their heads with respect to the sand surface, at small s_r/d_r, the load is entirely carried by piles. The load sharing between piles and soil at the direct contact with the raft is activated at s_r/d_r= 0.31%. From Fig. 9 it can be assumed that the ultimate piles load capacity Q_{pu}= 3.32 kN of the pile group is attained at s_r/d_r = 1.3%. As the settlement increases, the additional load mainly contributes to raise the Q_r value.

The load sharing mechanism between raft and piles for the rafts with 1, 3 and 7 QD piles, as deduced from centrifuge tests is illustrated in Fig. 10, where the ratio, Q_r/Q_t is plotted versus s_r/d_r. The figure illustrates how the load sharing mechanism between raft and piles, expressed via the ratio of Q_r/Q_t, decreases as the number of piles increases. For instance, taking as reference s_r/d_r=1% which indeed can be considered in the examined case, as settlement corresponding to the serviceability limit state, the measured values of Q_r/Q_t are 57, 30 and 12 percent, for 1-pile, 3-pile and 7-pile rafts.

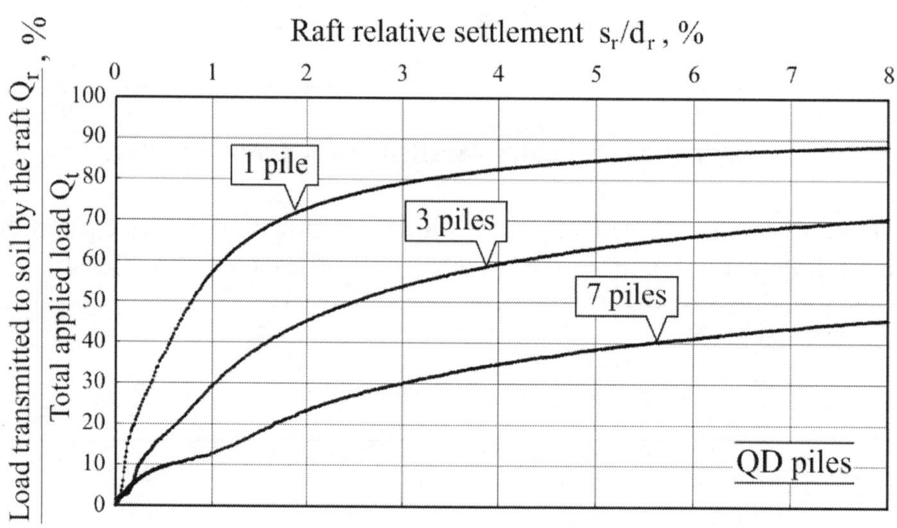

FIG. 10. QD piled rafts - Load sharing mechanism.

c. Load-settlement behaviour of QND piled rafts

Fig. 11 shows the stress-settlement relationships of QND piled raft tests; the stress applied on the raft q_r is plotted versus the s_r/d_r gathered from model tests on rafts with 1, 3, 7 and 13 QND piles. As a general trend, comparing Figs. 7 and 11, at a given value of q_r, the rafts on QND piles exhibit a larger relative settlement than those on QD piles, the trend becoming more marked as the number of piles increases.

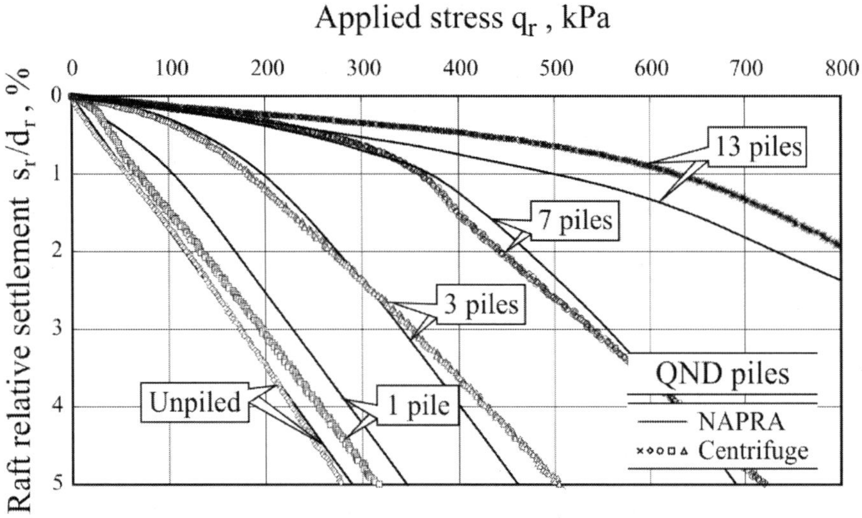

FIG. 11. QND piled rafts - Stress-settlement relationships. Centrifuge tests vs. NAPRA predicted values.

The settlements reduction factor ρ for QND piled rafts, for values of q_r equal to: 300, 400, 600 and 800 kPa, is plotted in Fig. 12 as function of the number of piles. Comparing the values of ρ for the same number of piles, the QND piles are less effective as settlement reducers with respect to the QD piles. As an example, in order to achieve $0.15 < \rho < 0.20$, it is required the raft on 7 QD piles (Fig. 8), as compared to raft on 13 QND piles (Fig. 12).

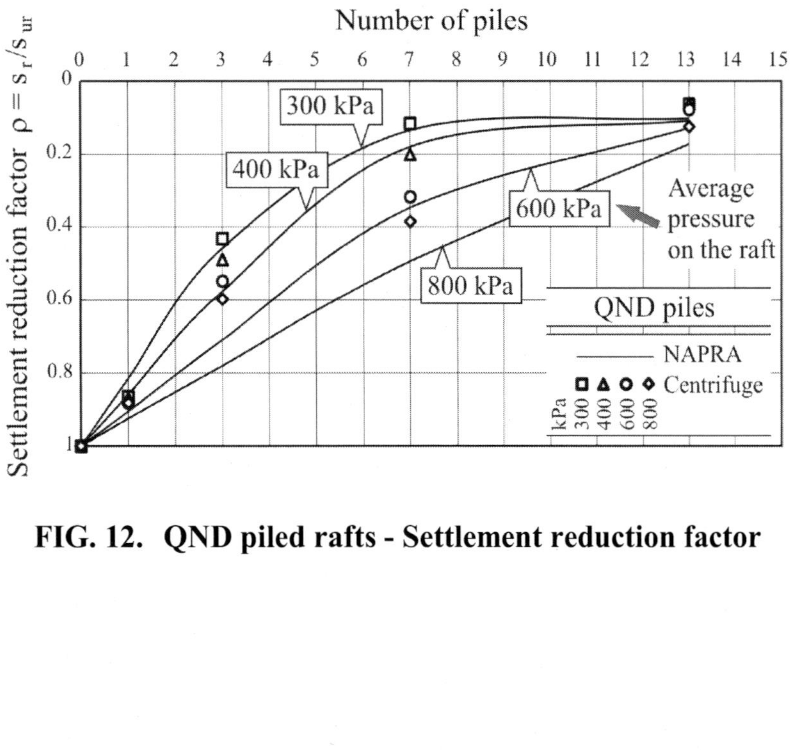

FIG. 12. QND piled rafts - Settlement reduction factor

Centrifuge tests vs. NAPRA predicted values.

The total applied load, Q_t, subdivided in components Q_r and Q_p as mobilised by the raft on 7 QND piles versus the s_r/d_r, is reported in Fig. 13. The ultimate value of Q_{pu} = 2.15 kN mobilized at s_r/d_r = 1.44% is significantly lower than that for the equivalent QD-piled raft, shown in Fig. 9.

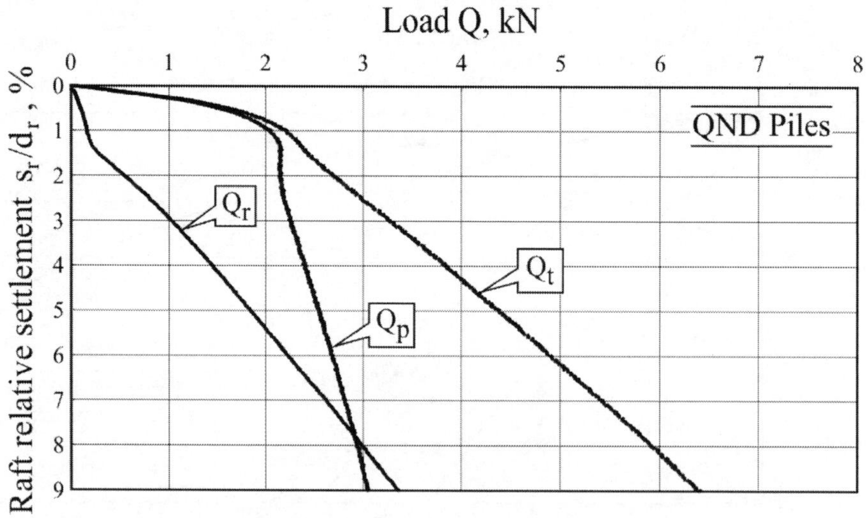

FIG. 13. QND 7 piled raft - Load sharing mechanism.

The load sharing mechanism between raft and piles, as deduced from the centrifuge tests carried out on QND piled rafts with 1, 3, 7 and 13 piles, is illustrated in Fig. 14, showing the ratio Q_r/Q_t versus s_r/d_r. Similarly to what has been observed for QD piles, as the number of piles increases, Q_r decays. At values s_r/d_r < 1.5%, of practical interest in design, the magnitude of Q_r becomes modest or even negligible in case of 13 and 7 piled rafts.

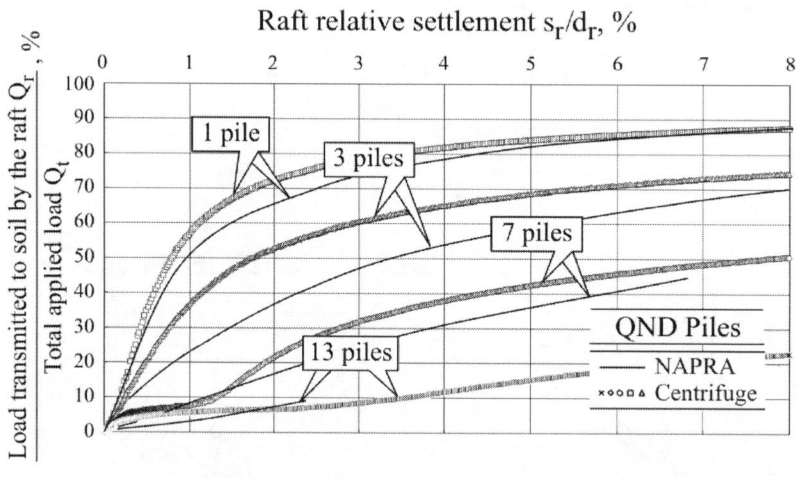

FIG. 14. QND piled rafts - Load sharing mechanism.
Centrifuge tests vs. NAPRA predicted values.

d. Load-transfer mechanism of the central pile in QD and QND piled rafts

The loads transferred from the rigid plate to the central pile of the 1-pile and 7-pile rafts, for both QD and QND piles, as deduced from centrifuge tests, are reported respectively in Figs. 15 and 16.

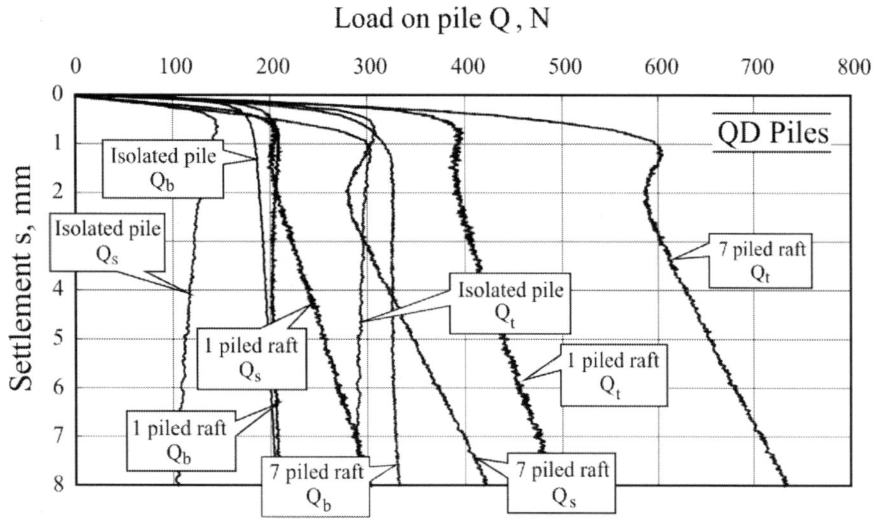

FIG. 15. QD piled rafts - Load-settlement relationships of the central pile vs. isolated pile.

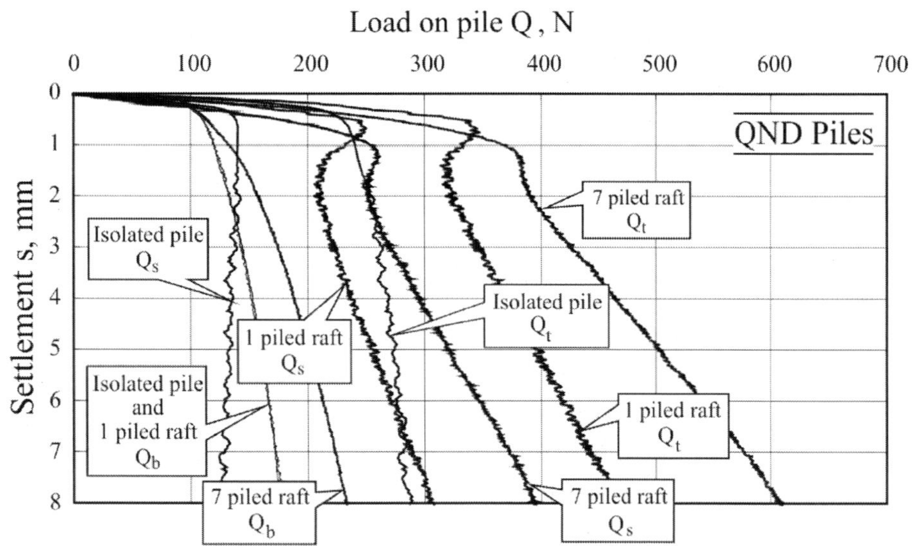

FIG. 16. QND piled rafts - Load-settlement relationships of the central pile vs. isolated pile.

They show the mobilized shaft load capacity (Q_s), the base load capacity (Q_b) and the total load acting on pile (Q_t) as function of the pile settlement (s). The relevant results are summarized in Table 3 and their qualitative analysis highlights the great influence of the interaction of the system raft-piles-soil on the mechanism of the pile capacity mobilization as compared to the response observed for an isolated pile.

TABLE 3: Load transfer mechanism - Central pile vs. isolated pile (all data referred to the model scale).

	QD-piles						QND-piles					
	Isolated pile		1-piled raft		7-piled raft		Isolated pile		1-piled raft		7-piled raft	
	Q_s N	s (mm)	Q_s N	s (mm)	Q_s N	s (mm)	Q_s N	s (mm)	Q_s N	s (mm)	Q_s N	s (mm)
At peak	146	0.64	208	0.77	302	0.75	140	0.65	249	0.70	260	1.27
Post peak	-	-	200	1.22	280	2.00	-	-	212	1.60	250	1.90
at large settlement	105	8.00	300	8.00	420	8.00	130	8.00	305	8.00	397	8.00
	Q_b N	s (mm)	Q_b N	s (mm)	Q_b N	s (mm)	Q_b N	s (mm)	Q_b N	s (mm)	Q_b N	s (mm)
Ultimate base capacity	206	8	210	8	332	8	180	8	180	8	233	8

Q_s = shaft load capacity; Q_b = ultimate base load capacity; s = pile settlement; QD = quasi displacement piles; QND = quasi non-displacement piles

The interpretation of the data presented in Figs. 15, 16 and Table 3 can be tentatively summarized as follows:

- The ultimate value of Q_{su} for both QD and QND isolated piles was mobilized at a value of s around 0.65 mm, referred to the model scale. For the piles under the raft the corresponding ultimate shaft capacity was mobilized at s values which are up to twice higher, with a trend to increase with the number of piles.
- The values of Q_{su} of the central pile within both 1-pile and 7-pile rafts were substantially larger than those observed for the isolated piles. The general trend for both QD and QND piled rafts was that Q_{su} increased with increasing piles number.
- After the peak had been attained, the Q_s of the isolated pile decreased with increasing the pile settlement. A more complex behaviour was observed for the central piles under the raft. Once the peak was mobilized, the Q_s exhibited a partial tendency to slightly decrease in the s range between 1.5 and 2.0 times that corresponding to the attainment of the peak. At larger settlements, as can be inferred from Table 3 and Figs. 15 and 16, the Q_s exhibits a steadily trend to increase, reaching at s = 8 mm values 50% higher than those at peak.
- The base load capacity Q_b mobilized for isolated piles was comparable to the 1-pile raft and was significantly lower than what observed for the central pile of the 7-pile raft. Contrarily to what was observed with Q_s, the magnitude of the load transmitted by the raft directly to the soil Q_r seemed to play a negligible influence on the values of Q_b mobilized for the central pile under the raft.

A comprehensive understanding of the differences in the mechanisms of the load capacity mobilization between isolated piles and those under the raft summarized above requires advanced numerical analysis and some additional centrifuge tests. At the present stage, the following tentative validations for the experimental data presented are submitted to the readers' attention.

The relevant differences between the shaft friction mobilization, as observed for the settlement reducing pile and the isolated pile, are mostly due to the load Q_r transmitted directly to the underlying soil by the raft. This load induces an increase of the stresses in the soil which, in turn, causes an increase of the shaft friction along the pile. Said effect of Q_r in case of 7-pile raft, is probably further enhanced by the "confining" effect exerted by peripheral piles, see the related rafts layouts in Fig. 5.

e. Tests results vs. NAPRA predicted behaviour

The centrifuge tests results have been compared against the results of numerical modelling of the piled rafts using the NAPRA (Non-linear Analysis of Piled Rafts) code.

This code, developed by Russo (1996, 1998) represents one of the most flexible tools for designing the piled rafts for which the load transmitted from the superstructure is partially conveyed on piles and partially on the soil, through the plate connecting the piles heads.

NAPRA considers a foundation plate of finite stiffness lying on piles heads connected with raft by hinges and is able to account for the following:

- Variable stiffness of the raft;
- Variable geometry and material properties of the piles;
- A stratified soil profile;
- Interaction at the tensionless contact between the raft and the underlying soil, considered as linear-elastic;
- Interaction at the contact between the raft and the piles, considered as non-linear elastic;
- Interaction between piles and surrounding soil assumed to be non-linear elastic.

The code analyzes the raft connecting the piles heads by F.E. method adopting quadrilateral four nodes, four-degree of freedom elements. The raft interacts with the underlying soil and piles under a system of vertical forces and moments (vertical and rotational springs) applied to the nodes of the F.E. mesh; if the node coincides with the head of a pile the corresponding spring behaviour will be non-linear elastic, while it will be linear elastic if it coincides with soil.

The load-settlement relationship of a single pile, that represents the input datum and the source of non-linearity in the system, is obtained computing its axial stiffness and taking into account the deformation characteristics of the relevant soil strata and the pile material properties. Combining the pile axial stiffness with an estimate of its ultimate load capacity, the NAPRA generates the non-linear load-settlement relationship for a single pile of hyperbolic shape.

The interaction between the raft and the piles-soil system is considered via the compliance matrix of the vertical displacements of the piles heads. The contribution to vertical displacements due to interaction among piles within the group, is computed via pile to pile interaction coefficients α_{ij} [Poulos (1994), Randolph (1994)] within the frame of the linear elasticity as originally adopted in the hybrid approach [Focht and Koch (1973), O'Neill (1983)], used to compute both horizontal and vertical displacements of piles groups.

In view of an earlier work carried out by the two senior writers [Fioravante and Jamiolkowski (2005)], the comparison of the centrifuge tests against the results of computations yielded by NAPRA are here limited to the rafts on quasi non displacement piles.

Fig.11 shows the relationship between the applied pressure and the relative raft settlement s_r/d_r as computed by NAPRA, compared against the centrifuge tests results. Despite a number of simplified assumptions for the soil behaviour and the pile to pile interaction, the computed q_r vs. s_r/d_r matched satisfactorily the observed behaviour of the modelled rafts in the entire range of the experiments carried out.

The role of piles as settlement reducing elements is highlighted in Fig.12 which shows the settlement reduction factor $\rho=s_r/s_{ur}$ as function of the piles number for the values q_r ranging between 300 and 800 kPa. The comparison between computed and measured values of ρ was acceptable for the range of q_r up to 600 kPa, while NAPRA underestimates them for the q_r =800 kPa.

The results yielded by NAPRA allow also quantifying the load carried through the raft-soil contact and that transmitted to the piles. This load sharing mechanism is illustrated in Fig. 14 displaying the ratio of the load transmitted directly from the raft to the soil Q_r to the total applied load Q_t.

The combined use of Figs such as 11 and 12 represents an efficient approach in the optimization of rafts design employing the settlement reducing piles.

SUMMARY AND CONCLUSIONS

The behavior of rafts on settlement reducing piles was investigated by means of extensive centrifuge tests on the model of rigid circular plate lying on a bed of loose very fine silica sand. Beneath the rafts, two types of model piles were installed: *quasi displacement* (QD) and *quasi non displacement* (QND) piles. The testing program included: unpiled raft, rafts on 1, 3, 7 and 13 piles. In each test a number of model piles were instrumented with load cells so that the load distribution along the shaft could be assumed. Load tests on isolated pile were also performed to establish the shaft and the base load carrying capacity as well as the load-settlement relationships of the two kinds of piles.

The obtained results demonstrate the piles effectiveness as settlement reducers and quantify the load sharing mechanism between piles and the raft-soil contact.

The data shown in Figs. 7 and 11 exemplify the trend of the raft relative settlement decreasing as the number of piles increases. At a serviceability pressure on the raft, the number of QD piles requested to reduce the settlement to an acceptable limit is lower than that of QND piles.

The efficiency of the piled raft quantified by means of the settlement reduction factor, $\rho = s_r/s_{ur}$, is presented in Figs. 8 and 12 from which it can be assumed that at a given stress level, for both QD and QND piles, ρ decreases as the number of piles increases; at the same time, the efficiency of QD piles is always larger than that of QND piles, i.e.: $(\rho)_{QD} < (\rho)_{QND}$. As an example, to obtain the same ρ value observed for 7 QD piles, the raft on 13 QND piles is required.

For the load sharing mechanism between piles and the raft-soil contact, see Figs.9 and 13; the gathered experimental results show that the phenomenon starts playing a relevant role when the load taken by piles approaches the ultimate carrying capacity of the group.

For example, referring to Fig. 14, it can be remarked that the contribution of the soil beneath the QND piled rafts becomes significant at s_r/d_r equal to 1.2% and 1.4% for 7 and 13 piled rafts respectively. This information should be accounted in design practice by optimizing the number of piles with respect to the constraints imposed by the allowable settlement of the constructed facilities.

The analysis of the gathered data allows the following considerations regarding the piled raft behavior and the related design issues:

An examination of the stress-settlement relationships, see Figs. 7 and 11, allows singling out a sharp change of their slope corresponding to a stress at which the ultimate load of the piles group is mobilized. As expected, for the same number of piles, the raft on QD piles exhibits a larger load carrying capacity than that on the QND piles.

As consequence, any rational design approach should account for the *variable stiffness* of the piled raft, controlled by the settlement level attained [Horikoshi and Randolph (1998)].

With this respect it is worthy to point out the following.

(a) At small settlements, the piled raft stiffness reflects almost entirely that of the pile group, which, in turn, depends on the number and type of piles adopted; as shown in Figs. 7 and 11, as the number of piles increases the piled raft stiffness increases. Moreover, these figures show that QD piled rafts exhibit a stiffer response if compared against that of QND piled rafts.

(b) The piled raft stiffness is further enhanced by the contribution of the dilatancy occurring in a thin shear band at the pile-soil interface during loading, which is responsible for the rise of the shaft load capacity at $s_r/d_r < 2\%$. [Wernick (1978), Boulon and Foray (1986), Boulon (1988, 1989)].

(c) At larger settlements, once the piles had reached their ultimate base and shaft capacity, the load transmitted via soil-raft contact Q_r, determined a remarkable increase in the radial stress on piles which, in turn, enhanced their shaft capacity. Accordingly, even at large values of s_r/d_r, the piled raft stiffness, although subject to degradation, remained well above the values observed for the unpiled raft at the corresponding level of s_r/d_r.

The above draws the attention on the fact that the rafts design on settlement reducing piles imposes a complex optimization process involving the magnitude of the allowable settlement, the decision about the requested piles load capacity and to focus on the definition of the crucial minimum number of piles.

This process requires the definition, for any raft configuration considered, of the relative settlement s_r/d_r and of the load sharing mechanism between piles and raft-soil contact represented by the Q_r/Q_t ratio. Such information can be computed by means of NAPRA code, which has been validated against the centrifuge tests results, reported in Figs. 11, 12 and 14 and referred respectively to the following relevant design information: stress applied on the raft (q_r) versus the relative settlement (s_r/d_r), settlement reduction factor (ρ) for different values of q_r as function of the piles number and quantification of the load-sharing mechanism between piles and raft soil contact via the Q_r/Q_t ratio. NAPRA results, despite some simplified assumptions incorporated, matched satisfactorily the results of the centrifuge tests carried out on rafts with QND piles.

Last but not least the writers recall the readers' attention to the fact that in all the tests presented herein the piles were always in the contact with the foundation structure. The recent experience with both centrifuge and advanced finite element modeling of the piled rafts, demonstrates that every time a granular layer is interposed between the raft and the piles head, the load sharing mechanism changes drastically with respect to what has been presented above.

This is illustrated in Fig. 17 comparing the response observed in centrifuge tests of a rigid rectangular raft on 9 piles with and without a 1 m thick interposed gravelly layer. As it can be seen from this figure, the piles of the raft with the gravelly layer beneath are exposed to the negative shaft friction involving half embedded pile length. Although, the analysis of the phenomenon is beyond the scope of this paper, its occurrence can be attributed to the circumstance that for the piles at the contact with the raft, the displacements compatibility at their interface is guaranteed. On the contrary, detaching the piles head from raft interface the displacements compatibility is no longer guaranteed.

In conclusion, if appropriately devised and interpreted, physical models, with the right awareness of scale and modeling technique effects, can be beneficially used to investigate the load transfer mechanisms in deep foundations. The results can also be employed as benchmarks in numerical modeling to ensure that the proper hypotheses are made in the computational model.

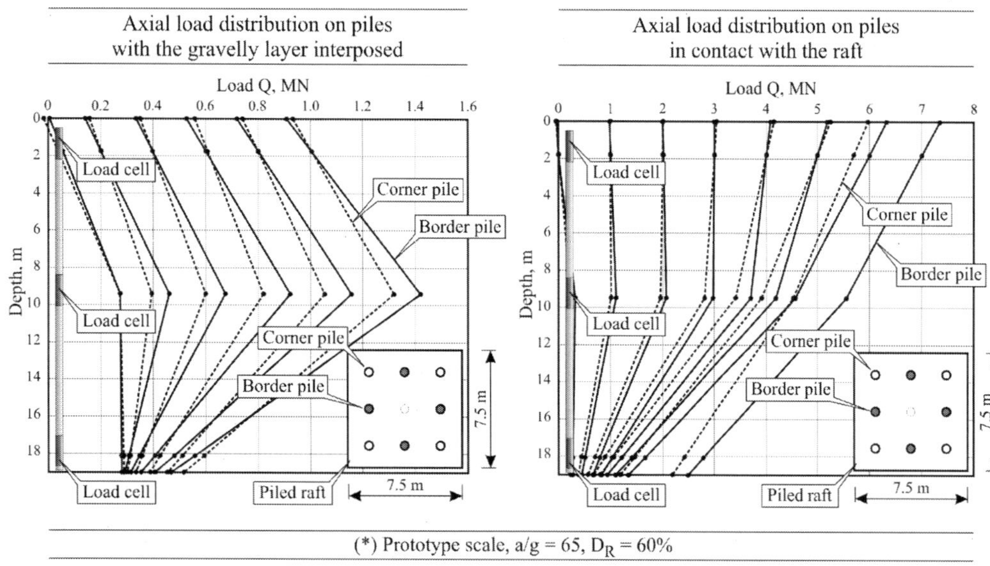

FIG. 17. Rigid raft with 9 QND piles: centrifuge test*

AKNOWLEDGMENTS

The authors appreciate the support of the technicians of the Istituto Sperimentale Modelli Geotecnici (ISMGeo), in particular Mr. Sergio Airoldi for the dedication during the experimental work. Last but not least, the writers express their appreciation to Mrs. Loredana Croce for her tireless effort in proof reading the text to make our English acceptable.

REFERENCES

Baldi, G., Belloni, G., Maggioni, W. (1988). *The ISMES Geotechnical Centrifuge.* Centrifuge 88, Paris, Corté J. F. Ed., Balkema, Rotterdam. Pp. 45-48.

Boulon, M., Foray, P. (1986). *Physical and Numerical Simulation of Lateral Shaft Friction along Offshore Piles in sand.* 3rd International Conference on Numerical Methods in Offshore Piling, Nantes. Editions Technip, Paris 127-147.

Boulon, M. (1988). *Numerical and Physical Modeling of Piles Behaviour under Monotonous and Cyclic Loading*. International Symposium on Modeling Soil-Water-Structure Interactions SOWAS '88, Delft, Kolkman,P.A. et al. Eds, Balkema, Rotterdam. Pp 285-293.

Boulon, M. (1989). *Basic Features of Soil-Structure Interface Behaviour.* Computers and Geotechnics vol. 7, No. 1-2, pp. 115-131.

Burland, J.B. (1995). *Piles as Settlement Reducers.* Proc. 19th Geotechnical National Congress. AGI. Pavia, Vol. 2, pp.21-34.

Cooke, R.W. (1986). *Piled Raft Foundations on Stiff Clays – A Contribution to Design Philosophy.* Geotechnique, Vol. 36, No.2, pp.169-203.

El-Mossallamy, Y.(1996) *"Ein Berechnungsmodel zum Tragverhalten der Kombinierten Phahl-Plattengrùndung".* Ph.D. Thesis ,Technical University of

Darmstadt,Germany.

Fioravante, V. (1997). *On the Physical Model Tests Capability to Reproduce some Basic Aspects of the Pile Foundation Behaviour.* 14th International Conference on Soil Mechanics and Foundation Engineering Hamburg. Balkema, Rotterdam, vol 4, pp 2209-2212.

Fioravante, V. (2002). *On the Shaft Friction Modelling of Non-Displacement Piles in Sand.* Soils and Foundations, Vol. 42, No. 2, pp. 23-33.

Fioravante, V. and Jamiolkowsi, M. (2005). *Physical Modelling of Piled Raft.* International Geotechnical Conference on Soil-Structure Interaction, Saint Petersburg. Ulitsky,V.M. Ed. ASV Publishers Saint Petersburg-Moscow, pp 89, 95.

Focht, J.A., and Koch, K.J. (1973). *Rational analysis of the lateral performance of offshore pile groups.* Proceeding of the 5th offshore technology conference 1973, Dallas, Texas, 1997, pp.701–8.

Foray, P., Balachowski, L., Labanieh, S. (1995). *Modélisation Physique des Ouvrages Géotechniques en Chambre D'étalonnage.* Colloque scientifique, Les modèles réduits en génie civil AUGC, Nantes. Sieffert, J.G. Ed., AUGC, pp. 55-68.

Foray, P., Balachowski, L., Raoult, G. (1998). *Scale Effect in Shaft Friction due to the Localisation of Deformations.* Centrifuge '98, Tokyo, Kimura, Kusakabe & Takemura Eds, Balkema, Rotterdam, pp. 211-216.

Garnier, J. and Konig, D. (1998). *Scale Effects in Piles and Nails Loading Tests in Sand.* Centrifuge '98, Tokyo, Kimura, Kusakabe & Takemura (ed.), Balkema, Rotterdam, pp. 205, 210.

Garnier, J. (2002). *Size Effects in Shear Interfaces.* Workshop on Constitutive and centrifuge modelling: two extremes, Monte Verità, Springman Ed., Swets & Zeitlinger Lisse, pp. 5-19.

Gorbunov-Posadov, M.I., Malikowa, T.A. and Solomin, W.I. (1984). *Analyses of Construction on Elastic Foundation.* Strojizdat, Moscow (in Russian).

Hanish, J., Katzenbach, R. and Kònig, G.(2002*)"Kombinierte Phahl-Plattengrùndungen".* Editor, Ernst & Sohn a Wiley Company.

Hansbo, S. (1984). *Foundation on Friction Creep Piles in Soft Clays.* International Conference on Case Histories in Geotechnical Engineering, St. Louis, Prakash, S. Ed. Vol.2, pp.913-922.

Hansbo, S. and Jendeby, L. (1983). *Case Study of Two Alternative Foundation Principle; Conventional Friction piling and Creep Piling.* Vag-och Vattenbyggaren, No.7-8, pp.29-31.

Horikoshi, K. and Randolph, M.F. (1997*). On the Definition of Raft-Soil Stiffness Ratio for Rectangular Raft.* Geotechnique, Vol 47, No.5, pp.1055-1061.

Horikoshi, K. and Randolph, M.F. (1998). *A Contribution to Optimum Design of Piled Raft.* Geotechnique, Vol 48, No.3, pp.301-317.

Kishida, H. and Uesugi, M. (1987). *Tests of the Interface between Sand and Steel in the Simple Shear Apparatus.* Geotechnique 37, No.1, pp. 45-52.

Jamiolkowski, M., Lo Presti, D.C.F., Manassero, M. (2003). *Evaluation of Relative Density and Shear Strength from CPT and DMT*. Soil Behavior and Soft Ground Construction Ladd Symposium MIT Cambridge Mass. GSP No. 119, Germaine J.T. et al Eds ASCE New York pp. 201-238.

Jardine, R.J., Lehane, B.M., Everton, S.J. (1993). *Friction Coefficients for Piles in Sands and Silts*. Conference on Offshore Site Investigation and Foundation Behaviour, Society for Underwater Technology, London 1992, Kluwer Academic Publishers Dordrecht, pp. 661-677.

Mandolini, A. (2003). *Design of Piled Raft Foundations: Practice and Development*. 4th International Geotechnical Seminar on Deep Foundations on Bored and Auger Piles, Ghent, Van Impe , W.F. Ed. Millpress, Rotterdam, pp. 59-80.

Mandolini, A., Russo, G. and Viggiani,C. (2005). Pile foundations: experimental investigation, analysis and design. 16th International Conference on Soil mechanics and Geotechnical Engineering. Osaka, Millpress Rotterdam, vol 1, pp. 177-213

O'Neil, M. W. (1983). *Group action in off-shore piles. Geotechnical practice in offshore engineering. Editor Stephen W. Wright. Geotechnical Engineering Division, ASCE, University of Texas, pp. 25-64*

Padfield, C.J. and Sharrock, M.J. (1983). *Settlement of Structures on Clay Soil*. Construction Industry Research and Information Institute, Special Publication 27, CIRIA London.

Poulos, H.G. (1994). *An Approximate Numerical Analysis of Piled-Raft Interaction*. International Journal for Numerical and Analytical Methods in Geomechanics, Vol.18, No.2, pp.73-92.

Poulos, H.G. (2001). *Piled raft foundations: design and applications*. Geotechnique No 2. Pp. 95-113.

Randolph, M.F. (1994). *Design Methods for Pile Groups and Piled Rafst*. 13th Intenational Conference on Soil Mechanics and Foundation Engineering, New Delhi, vol 5, pp.61-82.

Randolph, M.F. (2003). *Science and Empiricism in Pile Foundation Design*. 43rd Rankine Lecture. Geotechinque Vol. 53 No. 10, pp.847-875.

Randolph M.F., Jamiolkowski M.B. and Zdravkovic,L.(2004). Load carrying capacity of foundations. Advances in Geotechnical Engineering, The Skempton Conference, Institution of Civil Engineers London, Jardine, R. J., Potts, D.M. Higgins, K.G. Eds Thomas Telford London, vol. 1, pp. 207-240.

Randolph, M.F. and Clancy, P. (1994). *Design and Performance of a Piled Raft Foundation*. Settlements '94, *Austin Texas,* ASCE GSP 40, Yeung,A.T.; Felio,G.Y. Eds. pp.314-324.

Russo, G. (1996). *Interazione terreno struttura per piastre su pali*. Ph. D. Thesis. Department of Geotechical Engineering, University of Napoli Federico II.

Russo, G., (1998*). Numerical Analysis of Piled Raft. International Journal for Numerical and Analytical Methods in Geomechanics, Vol. 22, Issue 6 , pp 477 – 493.*

Russo, G. and Viggiani, C. (1998). *Factors Controlling Soil-Structure Interaction for*

Piled Rafts. International Conference on Soil-Structure Interaction in Urban Civil Engineering. Katzenbach, R., and Arslan,U. Eds Darmstadt. Vol 2 pp. 297-322.

Russo, G., Viggiani, C. and De Sanctis, L. (2004). *Piles as Settlement Reducers: a Case History.* Advances in Geotechnical Engineering, The Skempton Conference, Institution of Civil Engineers London, Jardine, R. J., Potts, D.M. Higgins, K.G. Eds Thomas Telford London vol 2 pp.1143-1154.

Sommer, H., Tamaro, G. and De Benedittis, C. (1991). *Messe Turm, Foundations for tallest building in Europe.* 4th International Conference on Piling and deep foundations.DFI, Stresa. Balkema Rotterdam, vol 1, pp. 139-145.

Viggiani, C. (1998). *Pile Groups and Piled Rafts Behaviour.* 3rd International Geotechnical Seminar on Deep Foundations on Bored and Auger Piles, Ghent, Van Impe, W.F. Ed. Millpress Rotterdam, pp 77-94.

Viggiani, C. (2001). *Analysis and Design of Piled Foundations.* 1st Arrigo Croce Lecture. Rivista Italiana di Geotecnica, Vol 35, No.1, pp. 47-75.

Wernick, E. (1978). *Skin Friction of Cylindrical Anchors in Non-Cohesive Soils.* Symposium on Soil Reinforcing and Stabilising Techniques, New South Wales Institute of Technology Sydney, pp 201-219.

Yoshimi, Y., and Kishida, T. (1981). Friction Between Sand and Metal Surface. 10th International Conference on Soil Mechanics and Foundation Engineering, Stockholm, vol 1, pp. 831-834.

NOTATION

C_U uniformity coefficient
D_{50} mean diameter of soil grains
D_c container diameter
D_p pile diameter
D_R relative density
E_r raft Young's modulus
E_s soil Young's modulus
FF-FFS sand
G_S specific gravity of soil grains
H_c container height
J_r raft moment of inertia
K_{rs} raft-soil stiffness ratio
L_p pile length
N geometrical scaling factor
QD quasi displacement pile
QND quasi non displacement pile
Q load
Q_b base resistance
Q_{bu} ultimate base resistance
Q_p load carried by the piles
Q_{pu} ultimate piles load capacity
Q_r load transmitted directly by the raft to the soil
Q_s shaft resistance
Q_{su} ultimate shaft resistance
Q_t total applied load
R_n normalized pile roughness
R_t pile roughness
W_p pile weight
d_c piezocone diameter
d_r raft diameter
f frequency
g gravitational acceleration
i piles spacing
p_a atmospheric pressure
p'_0 mean effective stress
q_r average stress on the raft
q_c cone resistance
s settlement
s_r piled raft settlement
s_{ur} unpiled raft settlement
s_r/d_r raft relative settlement
t_c container thickness
t_r raft thickness
α_{ij} pile interaction coefficients
γ_{max} maximum unit weight
γ_{min} minimum unit weight
δ' angle of interface friction
ν_r Poisson's ratio of raft material
ν_s Poisson's ratio of soil
ρ settlement reduction factor
ϕ'_{cv} angle of friction at critical state

Simple Shear Testing and Behavior of Norwegian Quick Clays – Revisited

Arvid Landva[1], Ph.D., P.Eng., FEIC

[1]Specialist Consultant, GEMTEC Ltd., 191 Doak Road, Fredericton, New Brunswick, Canada E3C 2E6; arvid.landva@gemtec.ca

ABSTRACT: The analysis of the 1959 landslide at Furre, Norway, had given an effective friction angle φ' of 7°, whereas large-scale in-situ shear box tests were interpreted to yield an angle of 9°. Triaxial tests had given values of φ' in the range 21-25°. It was concluded at the 1967 Oslo conference that the slide mechanism at work was that of retrogressive sliding, i.e. the slide went out piece by piece rather than moving out as a single flake. In the present paper it is postulated that the extremely low friction angles are not the real friction angles. The various failures, whether in the field or in the laboratory, whether drained or undrained and whether in the direction of initial shear or in the opposite direction, all occur when the real friction is fully mobilized along some plane in the failing clay. At this stage, if it is not physically possible for failure to take place along such a plane, i.e. if conventional sliding failure is hindered by boundary restraints, the clay will display a discontinuity appearing as a relatively sharp reduction of the gradient of the stress-strain curve.

KEY WORDS: Boundary restraints, direct simple shear, drained, failure envelopes, failure modes, field direct shear tests, Furre slide, interpretation, Mohr circles, simple shear, undrained tests

INTRODUCTION

Some 40 years ago, Bjerrum and Landva (1966) published the results of a preliminary series of direct simple shear tests on Norwegian quick clay. During the time period 1959-1963, the junior author of the 1966 paper had spent much time, under the expert guidance of Dr. Bjerrum, to develop equipment that was capable of transferring undisturbed samples of the quick clay without disturbance due to the laboratory testing equipment, and to develop methods of testing that would simulate as closely as possible the relevant shear stress-strain conditions in the field.

Hutchinson and Bjerrum (1962) had described a major landslide at Furre, Norway (Fig. 1), for which the angle of friction φ' had been determined to be about 7°, as calculated from the slide. Large-scale in-situ shear box tests (Hutchinson 1961) had suggested an effective angle of friction of about 9°. In comparison, triaxial tests had given values in the range 22°-25°.

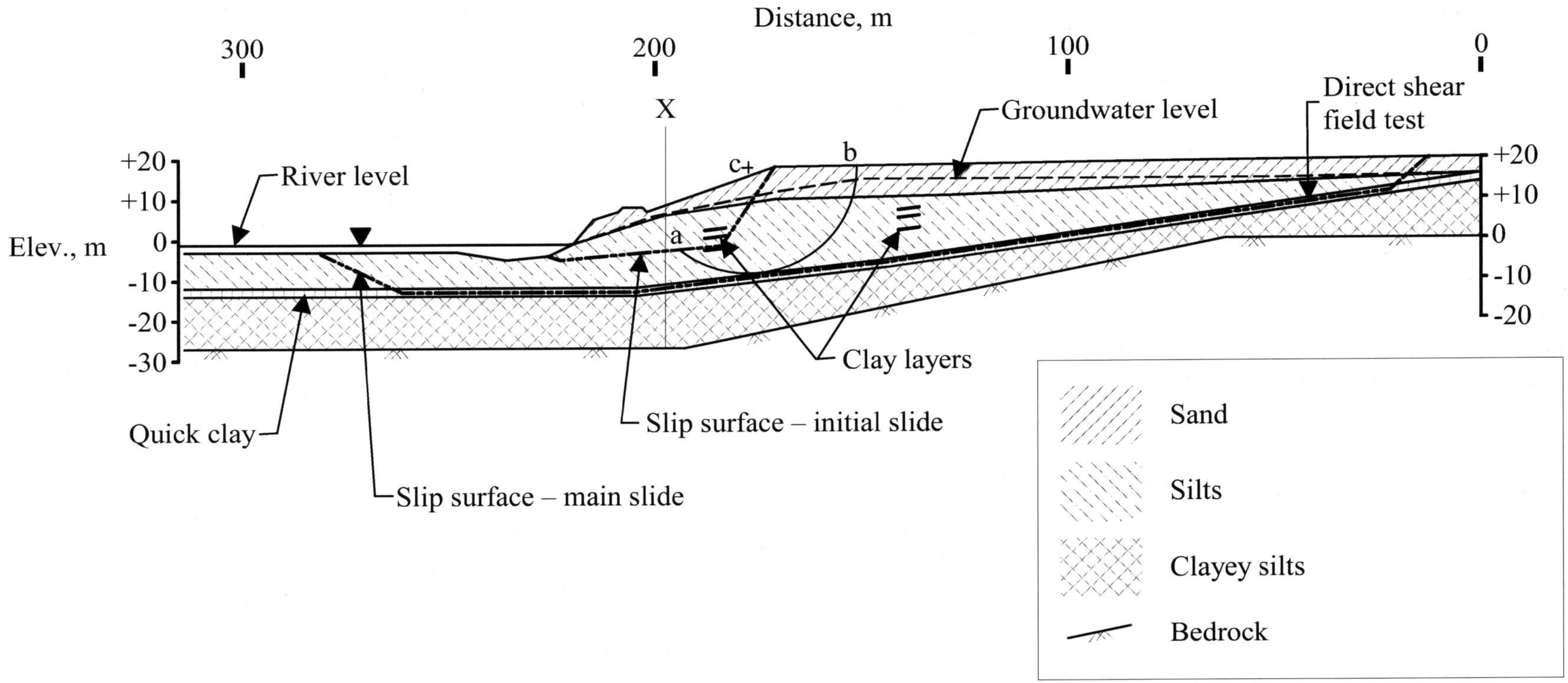

FIG 1. Non-distorted section through initial and main slides, 1959 Furre landslide

In the general report on the shear strength of soft clays at the Oslo Geotechnical Conference, Kenney (1967b) concluded, from a study of a group of slides in soft clay, that the mechanism at work was that of retrogressive sliding. Bjerrum and Kenney's (1968) research had indicated that quick clays exhibit a "structural rigidity" of a cohesive nature which at small strains and at a certain critical shear stress begins to break down. They reasoned that (i) the first part of the stress-strain curve is a measure of the resistance of the intact mineral structure against shear distortion, (ii) the strain is essentially the result of an elastic deflection of the structure, and (iii) in NC clays the geometric arrangement of the structure is governed by the condition that, immediately after consolidation, it would be just strong enough to carry the consolidation stresses, i.e. during consolidation the structural arrangement changed continuously, adapting itself to the gradually increasing effective stress.

The volume changes which occur when the structure collapses were no longer believed to be related to the macro-shear strain. The micro-shear failures causing the structure to collapse were believed to occur randomly in all directions, and the energy released by the volume change would be dissipated as heat and not have any effect whatsoever on the shear strength of the clay.

In the present paper it is postulated that the extremely low friction angles assigned to Norwegian quick clays on the basis of slide analyses and field and laboratory tests are not the fully mobilized friction angles, but the result of interpreting the various failure modes in ways that are not really compatible with the behavior of the clay. The various failures, whether drained or undrained or whether a result of shearing in the direction of the initial shear stresses or opposite to that direction, occur when the friction is fully mobilized (e.g. $\varphi' = 22° - 25°$, Hutchinson and Bjerrum 1962) along some plane in the failing clay. If actual sliding along such a plane is not physically possible, the clay will display a stress-strain discontinuity at this stage of the test. The state of stress at this point would correspond to a Mohr failure circle touching a conventional failure envelope (e.g. $\varphi' = 22° - 25°$), but the potential failure planes would be too steep to permit a sliding failure.

It is with some nostalgia that the writer recalls early-morning discussions decades ago with his mentor, the late Dr. Bjerrum, in the Norwegian Geotechnical Institute (NGI) laboratory, next to the direct simple shear equipment and various graphs showing the progress of the current test, about the strange behavior of the direct simple shear drained tests, i.e. the discontinuity and the temporary flattening of the stress-strain curve at low strain and what could possibly be the reason. The writer now argues that the sample had then in fact reached a state of failure, but that it could not physically fail until it was kinematically possible, i.e. until the direction of the failure planes was such that sliding failure became physically possible.

The writer also recalls with pleasure his many interesting discussions in 1962-63 with Dr. Schmertmann in the soils laboratory of the Norwegian Geotechnical Institute on many geotechnical topics and especially on the subject of direct simple shear (DSS) testing. In particular, Dr. Schmertmann's guidance with respect to the adaptation of the DSS test to his CFS/IDS tests, both drained and constant volume (undrained), was essential for such tests to be carried out with the DSS equipment. Contact has been maintained throughout the years since then, the writer having received much valuable advice from John Schmertmann both at his home, by electronic means and during visits and fishing trips in New Brunswick, Canada.

A NOTE ON APPARATUS DESIGN AND TEST INTERPRETATION

Prototype Design

Figure 2 shows a side view of the 1965-design direct simple shear apparatus (Bjerrum and Landva 1966). The design today is considerably modernized, but the general principle of applying normal and shear forces is the same.

The distributions of normal stresses are shown in Fig. 3 (Lucks et al. 1973). According to Lucks et al., the shear stress on the central horizontal plane is remarkably close to constant.

The direct simple shear apparatus was built to perform drained and constant volume (undrained) tests on specimens which were strained in simple shear and plane strain by having the specimen laterally confined. The original confinement was a reinforced rubber membrane. A later method of confinement used by the writer involves the type of design described by Kjellman (1951), who used of a number of thin brass rings to confine the specimen to truly K_0 conditions. Inside of the brass rings the specimen is protected from drying with a thin rubber membrane.

During the consolidation phase the major principal stress is equal to the vertical consolidation pressure σ_N' and the minor principal stress is the horizontal stress $K_0 \sigma_N'$.

Original Test Interpretation

In the case of **drained tests**, Bjerrum and Landva (1966) interpreted the discontinuity in the stress vs strain curve (such as *a* in Fig. 4) to be an indication that the structure of the clay was breaking down, i.e. the shear stress was exceeding a certain critical value, in the case of Fig. 4 corresponding to a shear stress ratio τ/σ_c value of about 0.20. No consideration was given to the possibility that failure might be hindered by boundary restraints.

The value of the shear stress ratio $\tau/\sigma_c = 0.20$ at the discontinuity was interpreted as corresponding to tan φ', i.e. $\varphi' \approx 11^\circ$. Thus it was assumed that failure would occur, whether in the field or in the laboratory, at a value of the effective friction angle very much lower than the values normally obtained from triaxial tests.

The interpretation of the drained tests was strongly influenced by an analysis of the 1959 Furre slide, which at the time of carrying out the simple shear drained tests was interpreted as being a flake slide failing under drained conditions. The Furre failure surface had been found to follow a thin quick clay layer dipping towards the river at a slope varying from 9° at the rear to 3° beneath the river. The effective angle of friction of the clay was considered to correspond to the average slope of the failure surface, which was about 7°.

Large-scale field shear box tests at Furre displayed the same type of discontinuity as that observed in the laboratory tests, the τ/σ_c ratio being 0.16, corresponding to an effective angle of friction of about 9° if this ratio is interpreted to be equal to tan φ'.

In the case of **undrained tests**, the maximum value of τ/σ_c (such as *b* in Fig. 5) was again interpreted to correspond to a break-down of the clay structure, and again no consideration was given to the possibility that boundary restraints might affect the true value of the shear strength. Since the maximum value of τ/σ_c was typically 0.19 and this was considered to correspond to tan φ', the value of the effective angle of friction would be about 11°.

Revised Test Interpretation

In the case of **drained tests**, it is assumed that the first part of the test, say up to a shear strain of about 5% (Duncan and Dunlop 1969), is one of a state close to that of pure shear. The Mohr stress circles for this part would therefore be concentric, or nearly so, and the end of this part of the test would yield a Mohr stress circle just touching the failure envelope. The apparent failure planes would be one set closely parallel to the failure envelope and one set vertical. However, physical failure would not be possible along either set of planes since the specimen is restrained by the top and bottom load plates and by the confining reinforced membrane or brass rings. In other words, failure is not possible until it becomes *kinematically* possible (i.e. not restrained by its boundaries). Straining of the sample beyond this state will cause the Mohr circles to move to the right (increasing stresses), and they are now constrained to continue touching the failure envelope, i.e. their points of tangency will move up the envelope away from the origin, as discussed later in connection with Fig. 7.

In the case of **undrained tests**, by the time the maximum applied shear stress on the horizontal plane is reached, the stress state for the planes closely parallel to the failure envelope (and also for the vertical planes) is one of failure, i.e. the Mohr stress circle for this state has just come into contact with the failure envelope (Wroth 1984 and 1987). However, as in the case of drained tests, physical failure is not possible along either set of planes. Straining of the sample beyond the state representing the first contact of the Mohr circles with the failure envelope will cause the Mohr circles to move to the left (decreasing stresses), being constrained to touch the failure envelope, its point of tangency moving down the envelope towards the origin (ibid.), as discussed later in connection with Fig. 8.

Current Relevance of Previous Data

The laboratory tests carried out at the Norwegian Geotechnical Institute and the field tests at Furre were carried out with the utmost care and by highly trained personnel. The *results* obtained with both the laboratory and the field equipment may therefore be considered to be as representative as practically possible. On the other hand, the *interpretation* of these results may have changed since the time of testing almost half a century ago.

The original interpretation suggests that the effective angle of friction of Norwegian quick clays is in the range 7 to11 degrees, as based on carefully conducted laboratory and field tests and detailed analyses. However, a reinterpretation taking into consideration the boundary conditions of the testing apparatus (section 2.3) has shown that the effective angle of friction of Norwegian quick clays is really in the range 22 to 25 degrees. It is obvious that the use of the lower range (7 to 11 degrees) would lead to highly uneconomical and probably impractical results.

The risk is always there that results applicable to one soil may be used for other, seemingly similar, soils. The writer therefore considered it important to point out that the original interpretation could not be correct for any soil.

Current Importance of Topic

The interpretation of direct simple shear tests offered in this paper is applicable not only to Norwegian quick clays, but to many other types of soil and to both direct simple shear and direct shear tests. The writer has found that the behavior of the soils in drained direct simple shear, which is the most common type of test, typically corresponds to that described above.

The topic of the behavior of soils during direct simple shear tests seems an important one considering the fact that many of these tests, small-scale and large-scale, are carried out in many laboratories and that it is important therefore to interpret them correctly. As demonstrated in this paper, and as indeed pointed out by one of the reviewers of the paper, the interpretation of direct simple shear tests is not a trivial matter.

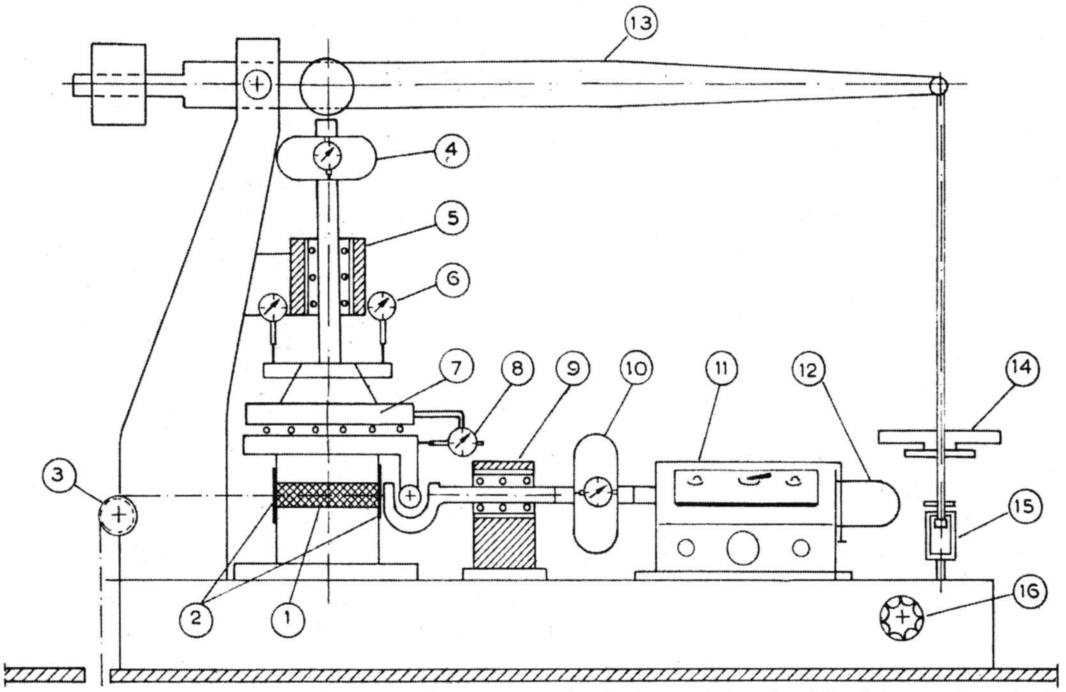

1 - Sample, 2 - Reinforced rubber membrane, 3 - Wheels for applying dead load (contr. Stress), 4 - Load gauge for vertical load, 5 - Ball bushing, 6 - Dial gauges for measurement of vertical deformation, 7 - Sliding box, 8 - Dial gauge for measurement of horizontal deformation, 9 - Ball bushing, 10 - Load gauge for horizontal force, 11 - Gear box, 12 - Exchangeable servogear motor, 13 - Lever arm, 14 – Weights, 15/16 - Clamping and adjusting mechanism used for constant-volume tests

FIG 2. General principle of direct simple-shear apparatus, design 1965

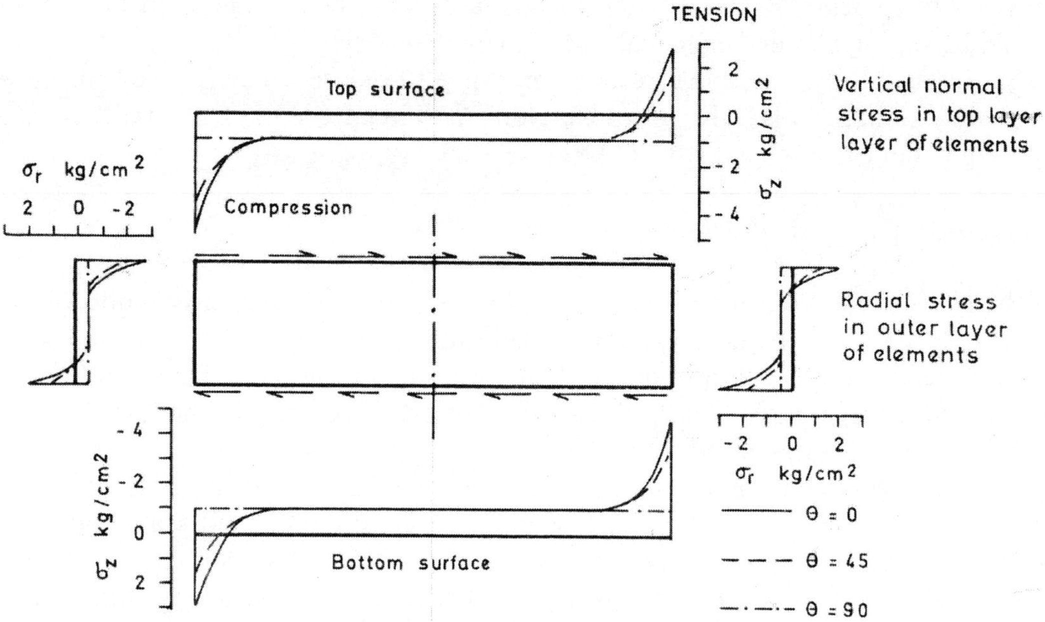

FIG 3. Normal stresses on NGI direct simple shear sample. θ = orientation of any radial plane (after Lucks et al. 1973).

EARLY DIRECT SIMPLE-SHEAR TESTS AT NGI

All the simple shear tests reported in this paper were carried out on samples prepared with equipment especially designed to prevent disturbance to the sample. Similar equipment prepared for triaxial testing is described and reported by Landva (1964).

Consolidated Drained Tests

The τ/σ_c vs ε_h and ε_v vs ε_h relationships shown shaded in Fig. 4 represent the envelopes of such relationships for six consolidated drained (CD) tests, all carried out on the Manglerud quick clay at a σ_o'/σ_c ratio of about one and all showing a discontinuity (then referred to as a critical stress) such as that at or near point *a*. Points *a* to *d* represent typical average values, used for the construction of the corresponding Mohr stress circles (Fig. 7).

It is postulated in this paper that the structure of the clay in fact starts to break down as soon as the shear stress is applied and that the said discontinuity occurs when the structure has broken down more or less completely, i.e. when failure has been reached. However, a sliding failure along slippage planes is hindered by the boundary conditions of the testing equipment. Thus, physical (sliding) failure in a simple shear sample with a height-diameter ratio of 1:8 cannot take place until the failure plane inclination is as flat as 8H:1V or about 7°.

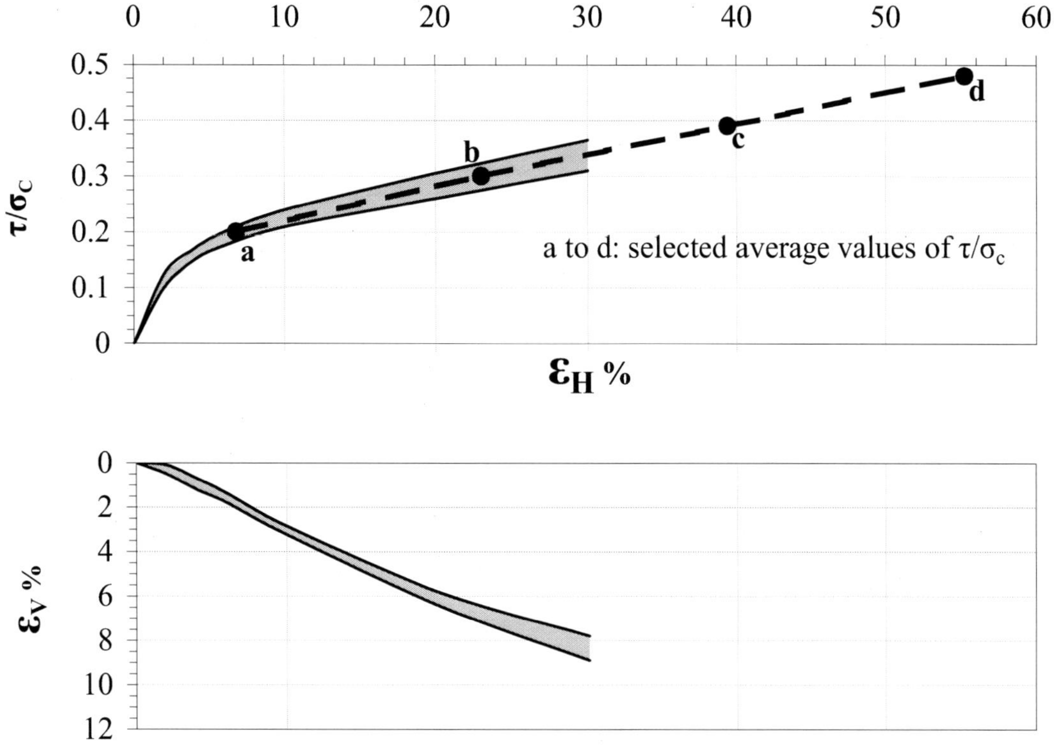

FIG 4. DSS CD tests on Manglerud quick clay. Tests reported by Bjerrum and Landva (1966) as showing a critical stress.

Consolidated Constant-Volume Tests

The τ/σ_c vs ε_h and σ_N'/σ_c vs ε_h relationships shown in Fig. 5 represent the envelopes of thirteen consolidated constant-volume (CCV) tests, all carried out on the Manglerud quick clay at a σ_o'/σ_c ratio of about one. As stated above, the reduction of the normal stress is taken to be equivalent to the induced pore pressure Δu (Dyvik et al. 1986). The τ/σ_c ratio at failure (point *b*), which typically occurs at a shear strain ε_h of 5 to 8 %, is in the range 0.175 to 0.225. The equivalent induced pore pressure ratio $(\sigma_c- \sigma_N')/\sigma_c$ is in the range 0.3 to 0.5. A single truly undrained simple shear test carried out with equipment designed to measure pore pressure directly gave a pore pressure ratio at failure of 0.33 and a τ/σ_c ratio at failure of 0.18.

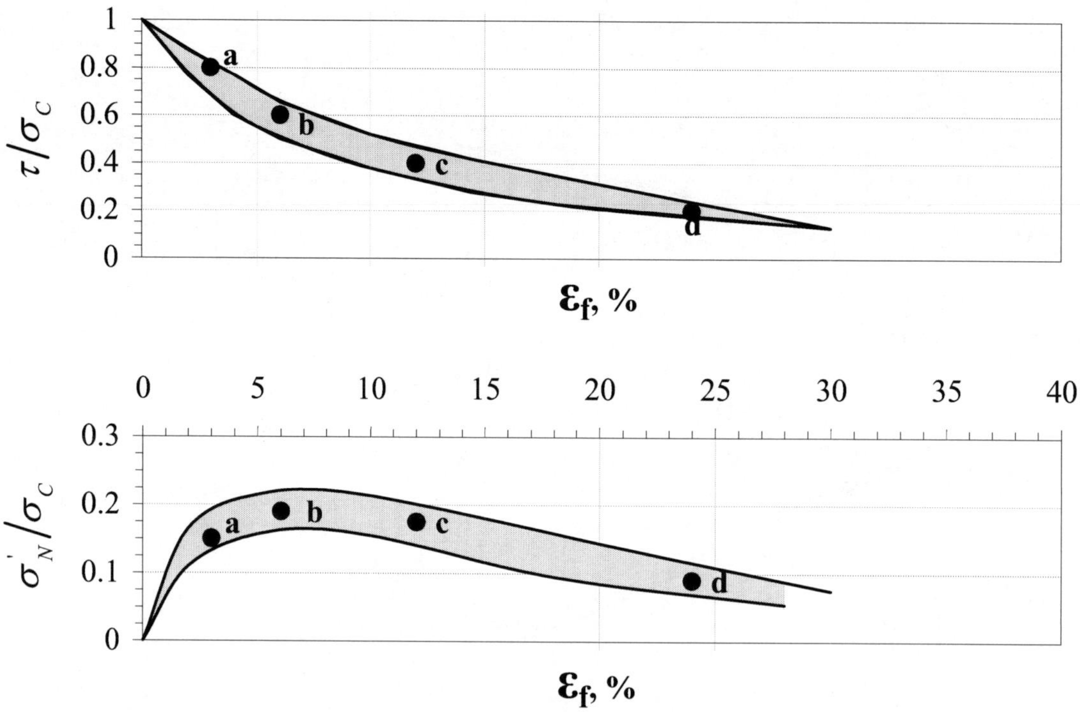

FIG 5. DSS CCV tests on Manglerud quick clay. Tests reported by Bjerrum and Landva (1966) for normally consolidated samples.

MOHR CIRCLES FROM INITIAL TO FINAL STATES OF STRESS

Failure Envelope from Triaxial Tests

The points shown in Fig. 6 represent samples of the Manglerud quick clay anisotropically consolidated in a triaxial cell to a vertical (major principal) stress equal to the effective overburden pressure (Landva 1962). The minor principal stress σ_3 (i.e. the cell pressure) was applied in steps of about 20 kPa, and for each step the major principal stress σ_1 was increased gradually up to a value corresponding to $\sigma_{c1} = \sigma_{c3}/K_o$. The criterion used to ensure K_o consolidation was that the volumetric compression of the sample was maintained equal to the volume of water squeezed out. The value of K_o was thus determined to be in the range 0.43 – 0.54. The gradient of the $\frac{1}{2}(\sigma_1' + \sigma_3')$ vs $\frac{1}{2}(\sigma_1 - \sigma_3)$ line is 23.4°. The $\frac{1}{2}(\sigma_1 - \sigma_3)$ intercept is assumed to be zero, giving a value of $\varphi' = 25.6°$.

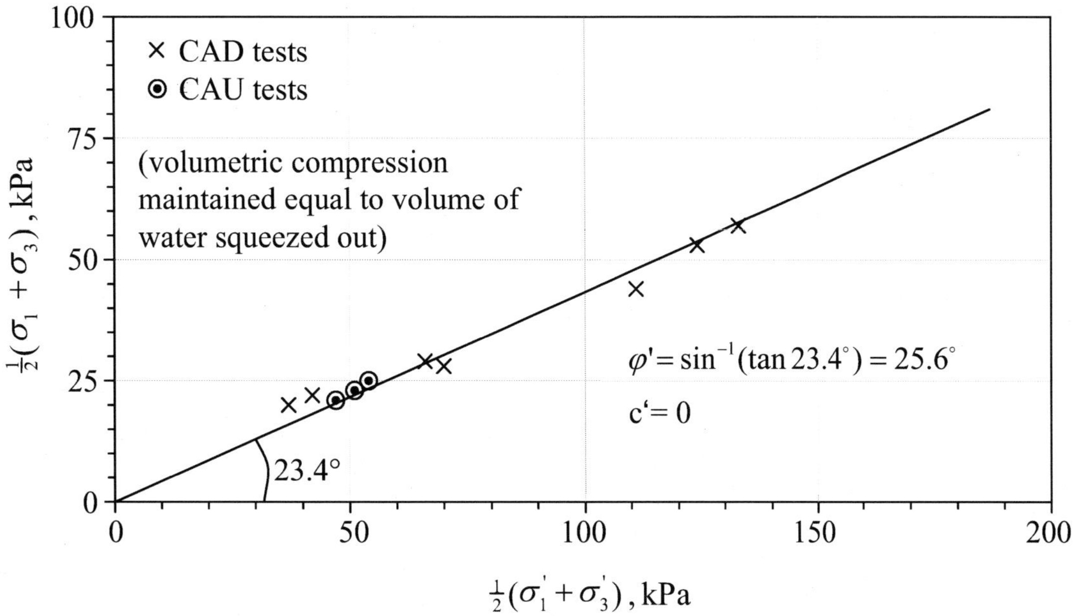

FIG 6. Results of triaxial tests on Manglerud quick clay, anisotropically consolidated.

Consolidated Drained Direct Simple Shear Tests

Circle C in Fig. 7 represents the state of consolidation for a value of K_0 = 0.53. This value of K_0 was chosen as a representative value for the Manglerud clay, based on the range 0.43 – 0.54 as determined above, 0.51 quoted by Bjerrum and Landva (1966) and 0.60 quoted by Kenney (1967a). The ordinate $\tau/\sigma_c = 0.21$ at $\sigma_n'/\sigma_c = 1.0$ (point *a*) corresponds to the discontinuity of the τ/σ_c vs ε_h relationship in Fig. 4 (point *a*).

The Mohr stress circle through point *a* is indeterminate, but can be approximated by assuming that either (i) a state of pure shear exists until the shear stress reaches point *a*, or (ii) point *a* represents failure. In case (i) the stress circle through *a* is concentric with the consolidation circle (see e.g. de Josselin de Jong 1988, page 545). In case (ii) the stress circle goes through *a* and is tangent to the failure envelope $c'/\sigma_c = 0$, $\varphi' = 25.6°$, as shown in Fig.7. If the structure of the clay has broken down when point *a* has been reached, as postulated above, circle I in Fig. 7 becomes the operative stress circle. However, it is not physically possible for the sample to fail at this stage, since the direction of the potential failure planes (f_I and the vertical planes) does not permit any sliding shear because of the physical constraints on the sample.

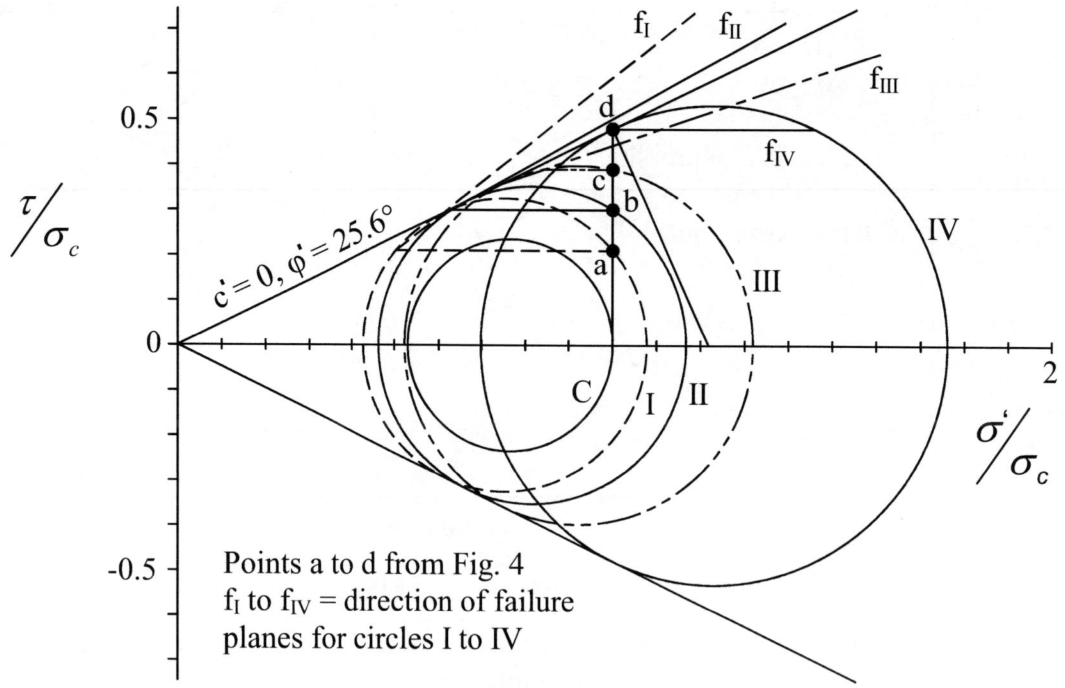

FIG 7. Mohr diagram of average DSS CD shown in Figure 4. $K_0 = 0.53$, Manglerud quick clay.

The sample remains in a state of failure as the shear strain is increased. Points *b* and *c* in Fig. 7 represent increasing shear strains, the corresponding Mohr circles (II and III) all being tangent to the failure envelope, the direction of the potential failure planes (f_{II} and f_{III}) still not permitting any sliding shear. Circle IV would represent the maximum possible shear stress, at which sliding shear could occur parallel to the direction of the applied shear stress, i.e. in this case horizontally. This is the Mohr circle conventionally plotted for direct shear tests. However, it can be seen from Fig. 4 that this state of shear, at $\tau/\sigma_c \approx 0.5$, would correspond to a shear strain close to 60 %. A simple shear configuration corresponding to such excessive strain would not, of course, be realistic.

Consolidated Undrained (Constant-Volume) Tests

Figure 8 is a Mohr diagram representing the average of eight consolidated constant-volume direct simple shear tests on Manglerud clay, as shown in Fig. 5. The failure envelope is as shown in Fig. 6, i.e. $c'/\sigma_c = 0$, $\varphi' = 25.6°$. Circle C again represents the state of consolidation for a value of $K_0 = 0.53$. Circle I is an intermediate circle through $\tau/\sigma_c = 0.15$ and $\sigma_N'/\sigma_c = 0.8$ as the sample is approaching $(\tau/\sigma_c)_{max}$.

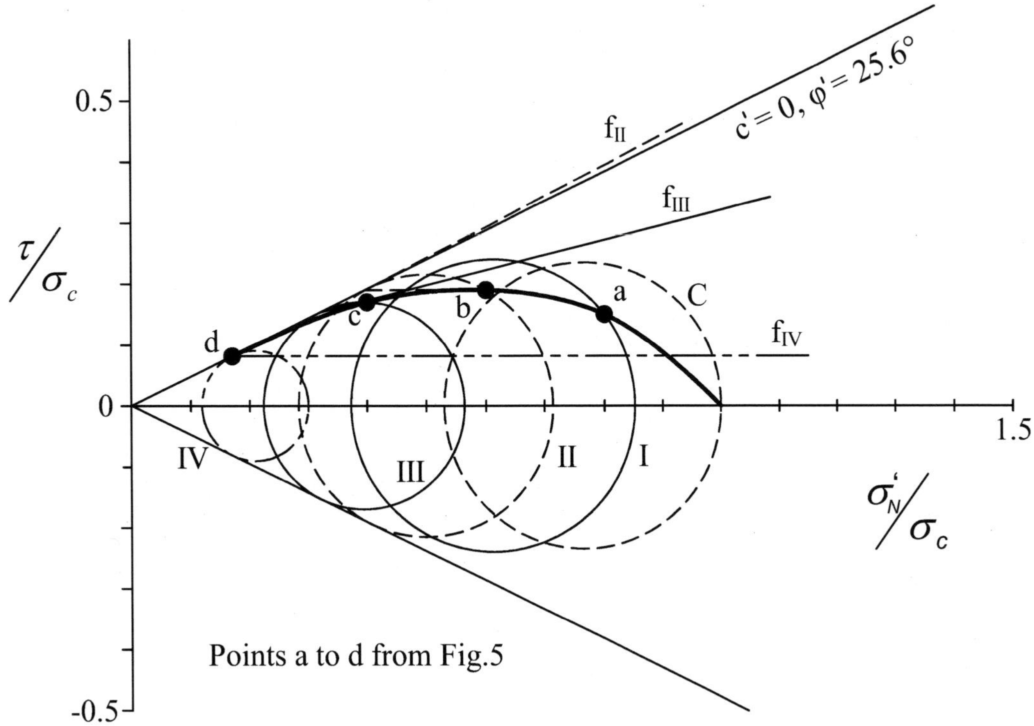

FIG 8. Mohr diagram of average DSS CCV shown in Figure 5. K_0 = 0.53. Manglerud quick clay. $c'/\sigma_c = 0$, $\varphi' = 25.6°$.

The value of $(\tau/\sigma_c)_{max}$ is 0.19 (point *b*). This is reached at $\sigma_N'/\sigma_c = 0.60$, which corresponds to an induced pore pressure ratio of $\Delta u/\sigma_c = 0.40$. Circles III and IV correspond to σ_N'/σ_c ratios of 0.40 and 0.20 and τ/σ_c values of 0.175 and 0.09, respectively. All circles to the left of (and including) circle II represent a state of failure under constant volume, i.e. under undrained conditions. However, as in the case of drained tests, it is not physically possible for the sample to fail, since the directions of the potential failure planes (f_{II} and f_{III}) do not permit any sliding shear because of the physical constraints of the sample.

Figure 9 is a Mohr diagram based on test results shown for Lierstranda clay by Kenney (1968, his figure 7). The failure envelope has been assumed to be the same as that for Manglerud clay, considering the similarity of the two clays (Table 1). The two samples were consolidated under a vertical pressure corresponding to a σ_o'/σ_c ratio of one and a shear stress ratio on the horizontal plane of 0.245. On the basis of the DSS CD test shown in Fig. 7, it may be assumed that the Lierstranda samples, after application of the consolidation shear stress, had already been brought to a state of failure during consolidation. In fact, the "consolidation" was really a DSS CD test where the shear stress was applied under conditions of controlled stress. The results of a typical controlled-stress DSS test are shown in Fig. 10. The usual discontinuity around 0.17 to 0.20 is again apparent. It had been expected, therefore, that failure would readily occur if a significant and sudden increase in shear stress were to be applied after the discontinuity (*a*) had been reached. However, such was not the case, despite a sudden increase in τ/σ_c of no less than 45%. Again, the physical constraints of the sample prevented outright failure.

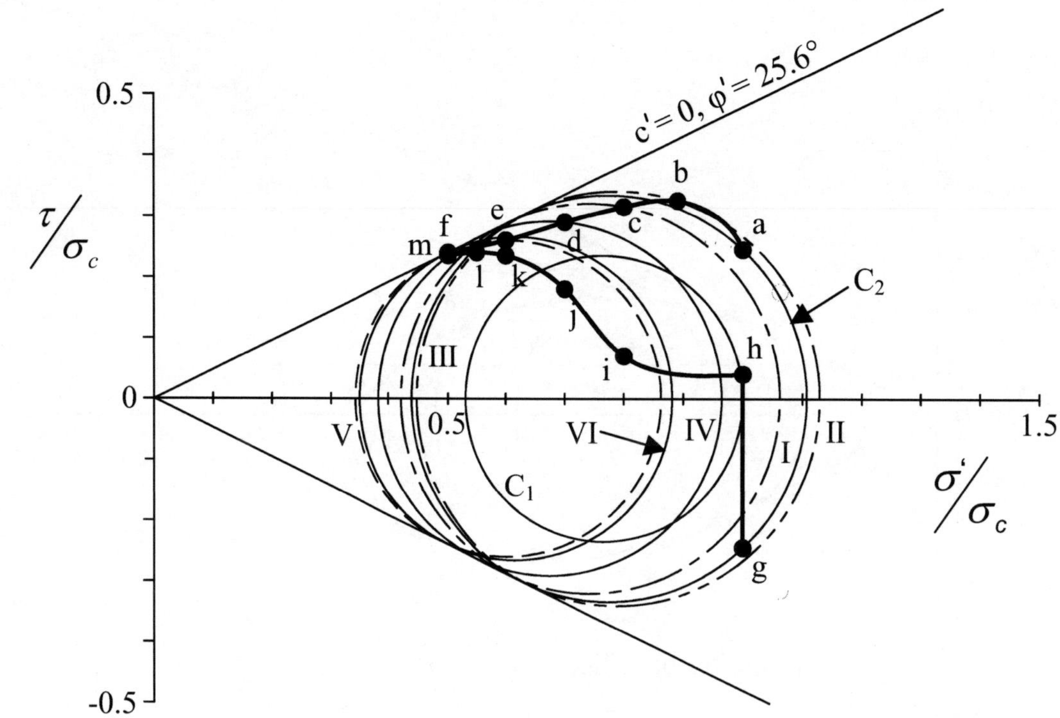

FIG 9. Mohr diagram for DSS CCV tests on Lierstranda quick clay (after Kenney 1968). $K_0 = 0.53$.

C_1 = consolidation before application of consolidation shear
C_2 = consolidation under $\sigma'/\sigma_c = 1$, $\tau/\sigma_c = 0.245$
a to e: selected from τ/σ_c and σ_N'/σ_c vs ε_H, test 4, Kenney 1968
f to k: selected from τ/σ_c and σ_N'/σ_c vs ε_H, test 2, Kenney 1968

Returning now to Fig. 9, after consolidation the samples were brought to failure under undrained conditions, one sample being deformed in the same direction as the consolidation shear stress (path *abcdef*, test 4, Kenney 1968) and the other deformed in the opposite direction (path *ghijklm*, test 2, ibid.). The lower peak value of $(\tau/\sigma_c)_{max}$ at *l*, as compared with *b*, was interpreted by Kenney as being a result of some destruction of the interparticle bonds, while the higher value at *b* was thought to be the result of <u>not</u> destroying these bonds. However, in the writer's opinion the two tests gave the same results, the failure circle being that through *f* and *m*, tangent to the failure envelope $c'/\sigma_c = 0$, $\varphi' = 25.6°$, and the failure surface being nearly horizontal. It is only at this stage that it becomes physically possible for the sample to fail. Said in other words, the τ/σ_c values at *b* and *l* are inflated because of the physical constraints of the sample. The true value of τ/σ_c is seen to be about 0.23.

The results of two special DSS tests are shown in Figs. 10 and 11. The controlled-stress test shown in Fig. 10 was described above. The CD/CCV test results shown in Fig. 11 were discussed by Bjerrum and Landva (1966), who suggested that the type of behavior displayed by the quick clay during this test might contribute to a better understanding of the landslide at Furre (where it had seemed that a large flake of soil resting on a thin layer of quick clay slid out as a consequence of a very small sudden increase in shear stress resulting from a relatively small failure of the river bank). However, the sample had already reached failure at point *e* (τ/σ_c = 0.225) and remained in a state of failure for the remainder of the test, regardless of whether it was sheared under drained or undrained conditions. This is illustrated in Fig. 12, where the Mohr circles through the various stages are all tangent to the failure envelope.

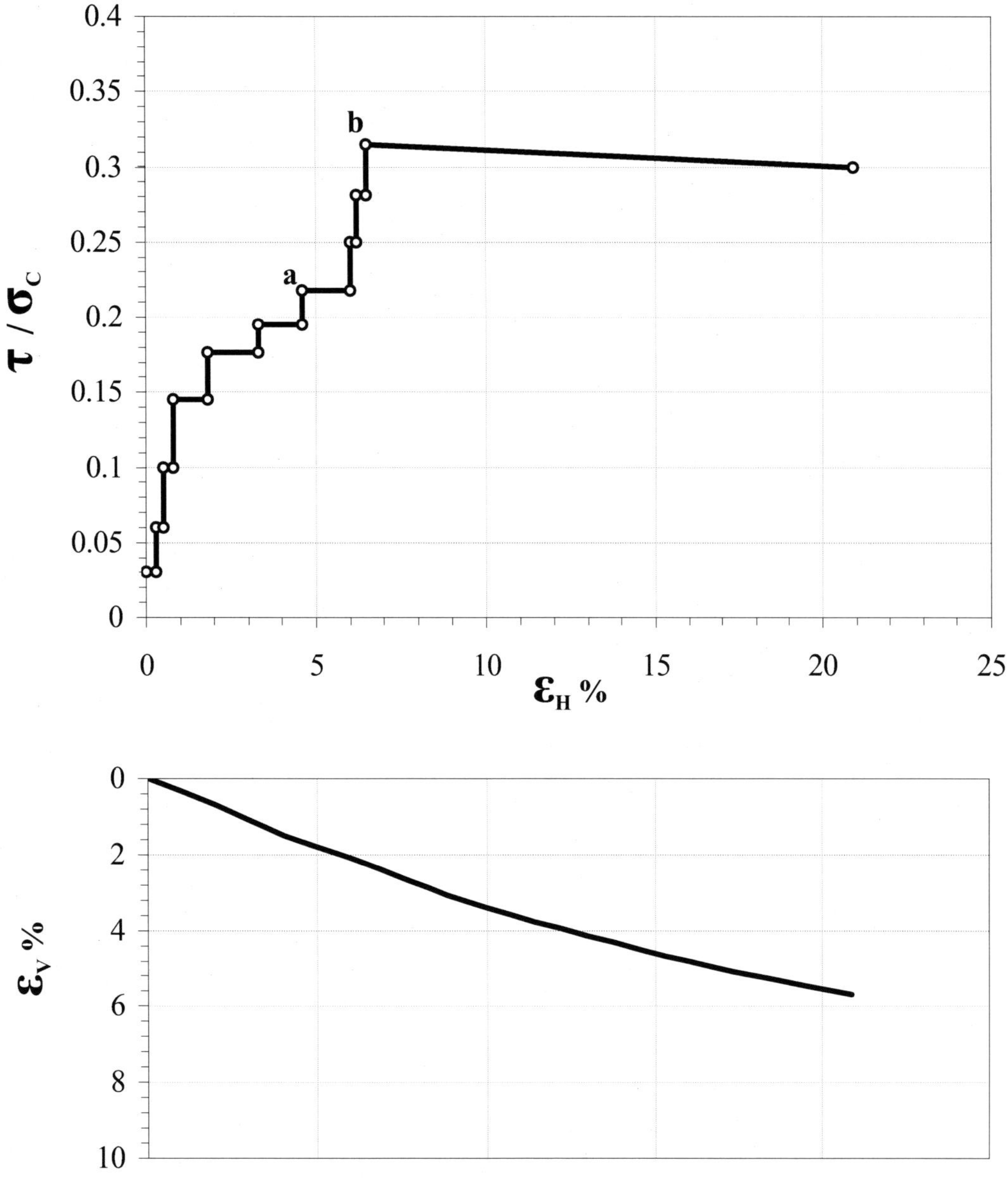

FIG 10. DSS CD test on Manglerud quick clay. Controlled stress.

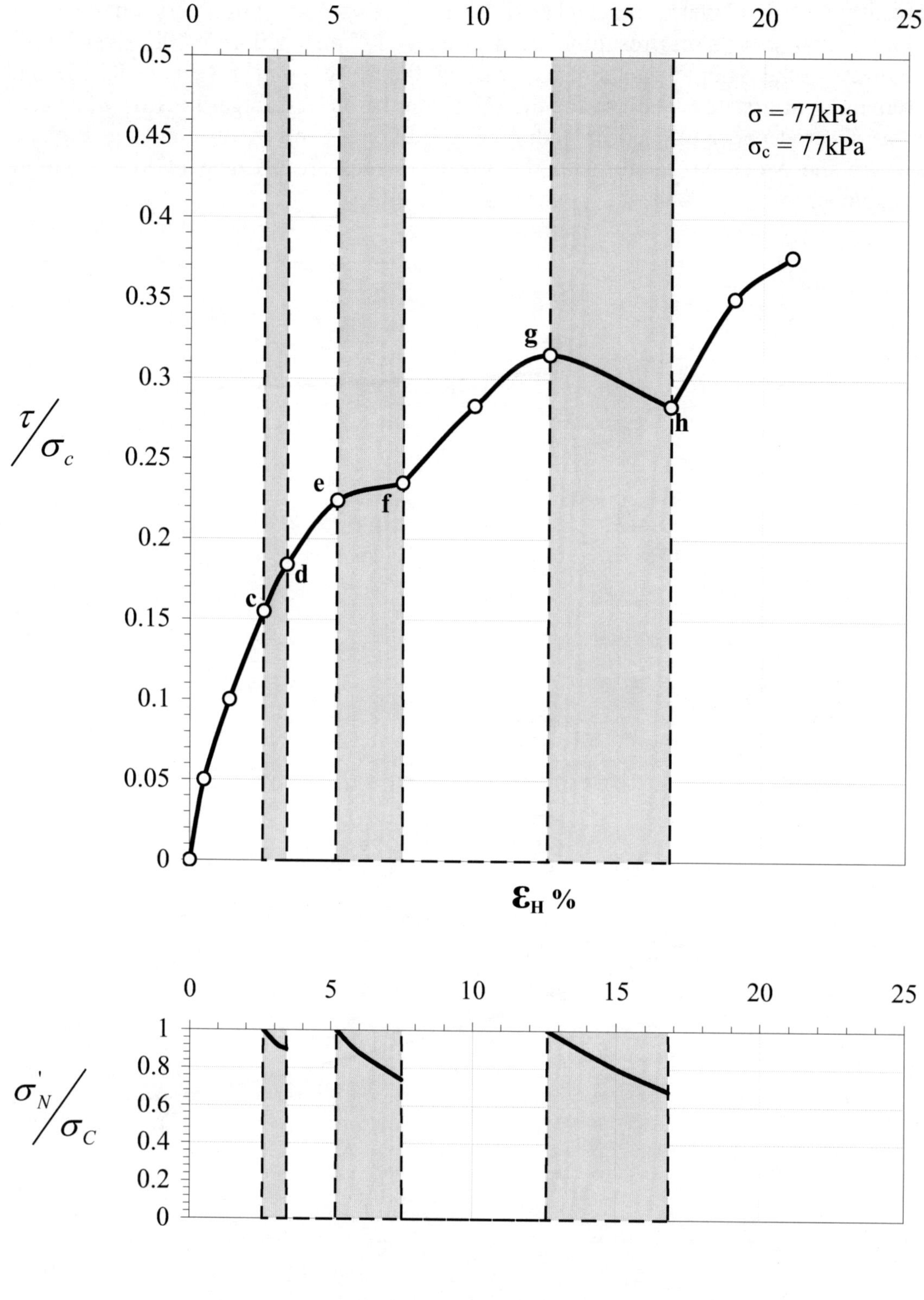

FIG 11. DSS CD/CCV test on Manglerud quick clay. Shaded portions (cd, ef, gh) represent CCV (undrained) shear.

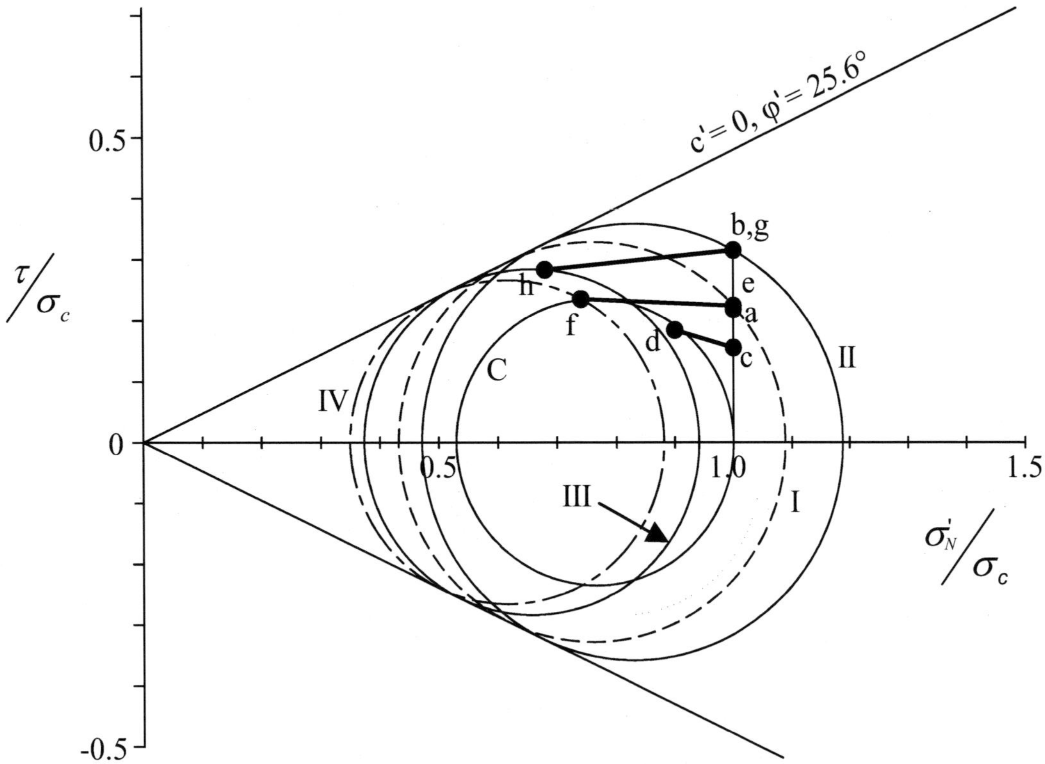

FIG 12. Mohr diagram for tests shown in Figures 10 and 11.

C = consolidation ($K_0 = 0.53$)
a,b from Fig. 10
c,d: before failure, Fig. 11
e-h: state of failure, Fig. 11
Circles I-IV: all tangent to failure envelope

TABLE 1. Manglerud, Lierstranda and Furre quick-clay properties

	Manglerud	Lierstranda	Furre
w %	35 - 40	31 - 40	37 - 48
w_p %	18 - 20	17 - 20	17 - 24
w_L %	23 - 25	25 - 32	22 - 39
γ kN/m^3	18.5	18.2	17.2 - 18.2
s_V kPa	5 - 12	10 - 20	8 - 13
Sensitivity	50 - 170	100 - 60	>100

THE 1959 FURRE LANDSLIDE INVESTIGATION

The Mohr diagram in Fig. 13 is based on the descriptions of the Furre slide by Hutchinson (1961), Hutchinson and Bjerrum (1962) and Kenney (1967b). Details of the testing arrangements in the field were described in an internal report prepared by Hutchinson and Rolfsen (1960). The author's copy of that report is a preliminary version, dated 25 March 1960, NGI file No. F.169-0, and marked "orettet" (not corrected) in pencil by Hutchinson.

The failure envelope in Fig. 13 corresponds to c' = 0, φ' = 24°. This is based on values quoted by Hutchinson (1961) of φ' = 25°, by Hutchinson and Bjerrum (1962) of φ' = 22° - 25° and by Hutchinson and Rolfsen (1960) of 20.9°.

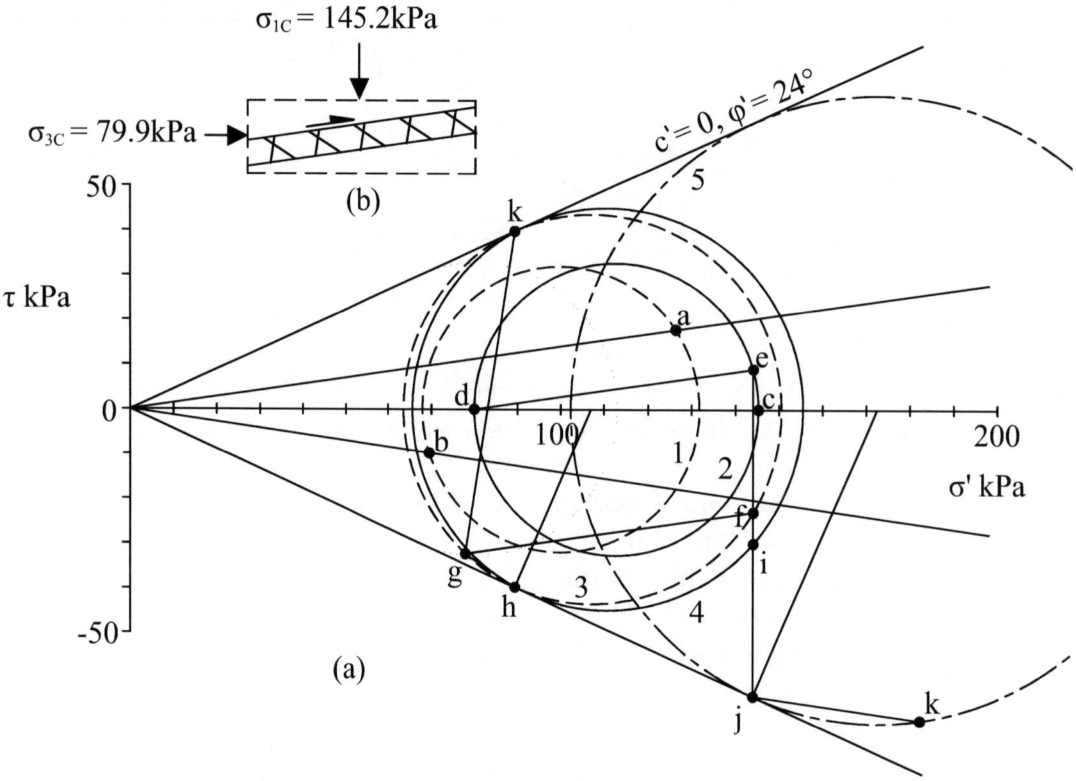

FIG 13. Mohr stress diagram for large Furre shear box test No. 3 (circles 2 to 5)

(a) Circle 1: In-situ before slide (σ_V=127.5 kPa, K_0=0.55)
Circle 2: Consolidation in Furre shear box 3 (145.2 kPa, K_0=0.55)
Circle 3: Failure in Furre shear box 3 (Hutchinson 1961)
Circle 4: Max. applied shear in Furre shear box 3
Circle 5: Theoretical max. possible shear stress in Furre shear box No. 3
(b) Directions of failure planes = directions of *gh* and *gk*

Circle 1 in Fig. 13(a) represents the conditions at Furre in situ before the slide. The inclination of the sliding surface is taken to be 8°. The distance *0a* corresponds to an effective overburden stress of 127.5 kPa (1.3 kg/cm^2, Hutchinson 1961). The distance *0b* corresponds to a stress of K_o×127.5 kPa, where K_o = 0.55 was assumed.

The stresses represented by points *a* and *b* form a pair of conjugate stresses, one being the resultant stress on the planes inclined at 8° to the horizontal and the other the resultant stress on the vertical.

The field shear boxes were installed to such a depth that the shear surface would be located in the middle of the apparently undisturbed quick clay left intact immediately below the thin sliding surface.

Circle 2 in Fig. 13(a) represents the stress conditions during consolidation in the in-situ shear box Furre No. 3 (ibid), which consisted of an upper and a lower frame separated by a continuous joint inclined at 8° to the horizontal. The vertical stress applied to the sample (on the horizontal plane) was 145.2 kPa, being the major principal stress (point *c*). The minor principal stress (on the vertical plane) would be $K_o \times 145.2$ kPa, which is 79.9 kPa (point *d*).

The next stage of the in-situ shear box No. 3 test was to unbolt the upper frame from the lower frame. The intersection of the line *de*, which is parallel to *0a*, with circle 2 shows the shear stress on the 8° joint plane to be about 9 kPa (at *e*). A shearing force was next applied to the upper frame in an up-slope direction parallel to the plane of the shear joint. When this load became equal to the down-slope component of the consolidation load, the shear stress on planes parallel to the joint was reduced to zero, i.e. these planes temporarily became the major principal planes.

The plot of shear stress vs. horizontal displacement (Hutchinson 1961, his figure 18) for the Furre in-situ shear box No. 3 test shows a very pronounced discontinuity at 23 kPa, which is point *f* in Fig. 13(a) (the same type of discontinuity as that shown by point *a* in Fig. 4 of this paper). The corresponding failure circle for the Furre test is circle 3 (dashed). The direction of the failure planes would be parallel to *gh* and *gk*, as shown in Fig. 13 (b). However, because of the very limited height of the failure zone (about 1 – 1.5 mm at failure, decreasing to close to zero at the end of the test), no sliding failure could take place, this being prevented by the physical constraints of the equipment. The test was terminated at a shear stress of about 30 kPa (point *i*, circle 4).

Theoretically, the shear stress at failure could be as high as that indicated by circle 5. However, as this would no doubt have required an excessive shear displacement and thus an unacceptable distortion of the failure zone, this state of stress is of little practical interest.

The initial slide surface in the river bank at Furre was located roughly parallel to and about nine metres above the main slide surface (Fig. 1)

The initial slide is described (Hutchinson 1961) as occurring "without warning, to the accompaniment of a loud roar". Hutchinson and Bjerrum (1962) describe it as starting with a "sudden and explosive slide in the river bank furthest downstream. During the following few seconds the main slide started to move outwards, coming to rest after 1 - 2 minutes." The extremely rapid occurrence of the initial slide caused a sudden unloading under the middle of the approximately 1000 m^2 slide area of about 180 to 235 kPa, assuming the thickness of the sliding mass to be in the range 10 to 13 m and its unit weight to be 18 kN/m^3. The porewater pressure would have been in the range 60 to 100 kPa before the slide, assuming the ground water level to be as shown in Fig. 1. It would thus seem possible that the rapid unloading of the clay in the failure zone of the initial slide may have produced spontaneous cavitation in the clay and that this may have been the cause of the loud roar and the sudden and explosive nature of the river bank slide.

The initial (river bank) slide resulted in an approximately 18 m high slope. A conventional 2-D undrained analysis of this slope (Fig.1, circular slide surface *ab*, center at *c*) indicates that an undrained strength of about 65 kPa would be required for a factor of safety of 1.0. In comparison, the measured undrained shear strength in the silt above the quick clay was in the range 35 to 50 kPa while that in the four quick clay layers was approximately in the range 10 to 20 kPa..

There seems to be little doubt that the retrogressive nature of the failure in the main slide (Kenney 1967b) was the result of a complete and rapid remolding, i.e. a reduction of its shear strength to practically zero, as the quick clay was rapidly remolded towards the rear of the slide. This domino effect would have been started by the initial river bank slide, which was then followed by a slide under undrained conditions of the outer portion of the main slide (*ab*, Fig. 1). Kenney (1967b) describes retrogressive sliding as "a slide going out piece by piece", and this seems to be an apt description of a very rapid succession of undrained slides, giving the impression of a flake type of slide.

The sliding masses above the slide must have exerted a tremendous lateral pressure on the lower portion of the sliding mass. The continuation of the slip surface out under the river, as observed, was therefore to be expected.

SUMMARY AND CONCLUSIONS

Drained direct simple shear tests carried out on samples of Manglerud quick clay showed a discontinuity in stress strain relationship at a relatively low strain, followed by a final stage of low but steady increase in shear stress with strain. No sliding failure took place, since such failure is hindered by the physical constraints of the equipment.

Undrained direct simple shear tests carried out on the gave an average $(\tau/\sigma_c)_{max}$ value of 0.19. Plotting these average values in a Mohr stress diagram shows that the $(\tau/\sigma_c)_{max}$ value is an inflated value, the magnitude of which is governed by the physical restraints of the sample. The value of τ/σ_c at the actual (sliding) failure stage is about 0.13.

In tests on a pair of samples of quick clay similar to the Manglerud clay, the samples were brought to failure under undrained conditions, one sample being sheared in the same direction as the consolidation shear stress and the other sheared in the opposite direction. The former gave considerably higher values of τ/σ_c than the latter, which at the time was interpreted to be the result of some destruction of the interparticle bonds, while the higher value was thought to be the result of not destroying these bonds. However, in the writer's opinion, the two tests gave the same results, both being in the same state of stress when the stress circles became tangent to the failure envelope. It is only at this stage that it becomes physically possible for the sample to fail

A review of the 1959 Furre slide investigations confirmed that the in-situ drained shear box tests displayed the same behavior as that of the direct simple shear tests on the Manglerud clay, i.e. exactly the same type of sharp bend in the stress-strain curves. The original interpretation of the sharp bend was that the corresponding τ/σ_c ratio represents tan φ', which would correspond to a friction angle as extremely low as $\varphi' = 9^o$. However, the writer has concluded that, while the value of τ/σ_c at the sharp bend certainly represents failure, the Mohr stress failure circle is tangent to a failure envelope described by $c' = 0$, $\varphi' = 24^o$. The original interpretation of the slide as a monolithic mass moving down along a continuous sliding surface inclined at about 7° would thus not be realistic. The conclusion at the 1967 Oslo conference that the main slide went out piece by piece as a rapid succession of failures towards the rear of the slide area seems quite compatible with a failure envelope of $c' = 0$, $\varphi' = 24^o$ and relatively high values of porewater pressure induced by the rapid failures.

ACKNOWLEDGEMENTS

The author acknowledges the profound influence of the late Dr. Laurits Bjerrum on his engineering training. The development of the direct simple shear equipment and the research results achieved with it stemmed from Dr. Bjerrum's phenomenal insight into soils testing and his constant motivation and encouragement.

The author owes a debt of gratitude to Dr. T.C Kenney for numerous discussions and recommendations during the course of the author's research work at the Norwegian Geotechnical Institute and The University of Trondheim. He also wishes to point out that the Landva 1964 paper listed in the references was in fact both initiated and prepared by Dr. Kenney.

Mr. Erik Heier, an accomplished and enthusiastic amateur machinist in Oslo, built most of the special equipment for the two direct simple shear prototypes and also offered much valuable advice for its design and production.

The author is indebted to GEMTEC LIMITED for their technical and financial support and wishes to thank especially Mr. Andrew DeMerchant for his patient and expert work on the figures and the editing of the paper.

REFERENCES

Bjerrum, L. and Landva, A.O. (1966). "Direct simple-shear tests on a Norwegian quick clay". *Géotechnique* 16 (1): 1-20.

Bjerrum, L. and Kenney. T.C. (1968). "Effect of structure on the shear behaviour of normally consolidated quick clays". *Proc. Geotechnical Conference Oslo 1967,* II: 3-11.

de Josselin de Jong, G. (1988). "Elasto-plastic version of the double sliding model in undrained simple shear tests". *Géotechnique* 38(4):533-555.

Duncan, J.M. and Dunlop, P. (1969). "Behavior of soils in simple shear tests". *Proc. 7th Int. Conf. on Soil Mech and Found. Eng..* Mexico, pp. 101-109.

Dyvik, R., Berre, T., Lacasse, S. and Raadim, B. (1987). "Comparison of truly undrained and constant volume direct simple shear tests". *Géotechnique* 37(1): 3-10. Also *Norwegian Geotechnical Institute Publ. No. 170 (1988).*

Hutchinson, J.N. and Rolfsen, E.N. (1960). "Drenerte skjærboksforsøk ved Furre i Namdalen" ("Drained shear box tests at Furre, Namdalen"). *Norwegian Geotechnical Institute,* internal report F.169-0, dated 25 March 1960.

Hutchinson, J.N. (1961). "A landslide on a thin layer of quick clay at Furre, Central Norway". *Géotechnique*, 11(2): 69-94.

Hutchinson, J.N. and Bjerrum, L. (1962). "Skredet ved Furre i Namdalen" (The landslide at Furre, Namdalen). *Norwegian Geotechnical Institute Publ. No. 49.*

Kenney, T.C. (1967a). "Shear strength of soft clay". *Proc. Geotechnical Conference Oslo 1967,* II: 49-55.

Kenney, T.C. (1967b). General report (panel discussion) on "Shear strength of soft clay". *Proc. Geotechnical Conference, Oslo 1967,* II: 107-110.

Kenney, T.C. (1968). "A review of recent research on strength and consolidation of soft sensitive clays". *Can. Geotech. J.* 5, 97-119.

Kjellman, W. (1951). "Testing the shear strength of clay in Sweden". *Géotechnique* 2, 225-232.

Landva, A.O. (1962). "An experimental investigation of the shear strength of normally consolidated clays". Dr. ing. thesis, The Norwegian Institute of Technology (previously Norges Tekniske Høgskole), Norway (in Norwegian, English summary).

Landva, A.O. (1964). "Equipment for cutting and mounting undisturbed specimens of clay in testing devices". *Norwegian Geotechnical Institute Publ. No. 56.*

Lucks, A.S., Christian, J.T., Brandow, G.E. and Høeg, K.. (1973). "Stress conditions in NGI simple shear test". *J. Soil Mech. and Found. Eng., Proc. ASCE, vol. 98, SM1.* Also *Norwegian Geotechnical Institute Publ. No. 96.*

Wroth, C.P. (1984). "The interpretation of in situ soil tests". *Géotechnique* 34(4): 449-489.

Wroth, C.P. (1987). "The behaviour of normally consolidated clay as observed in undrained direct shear tests". *Géotechnique* 37(1): 37-43.

NOTATION

CAU	Anisotropically consolidated undrained triaxial tests
CAD	Anisotropically consolidated drained triaxial tests
DSS	Direct simple shear
CD	Consolidated drained direct simple shear tests
CCV	Consolidated constant-volume (undrained) direct simple shear tests
φ'	Effective angle of internal friction
c'	Effective cohesion intercept
K_0	Coefficient of lateral pressure at rest
σ_N'	Effective normal stress
τ	Shear stress
σ_c	Consolidation pressure
ε_H	Horizontal strain
ε_V	Vertical strain
σ'_o	Effective overburden stress
Δu	Excess (induced) porewater pressure
σ_1'	Effective major principal stress
σ_3'	Effective minor principal stress

"Liquefaction": Hydraulic Compaction of Soil Near Rammed Piers

Richard L. Handy[1], Member, ASCE

[1] Distinguished Professor Emeritus, Iowa State University, Department of Civil, Construction and Environmental Engineering, Ames, Iowa, 50010 - 1502 270th St., Madrid, IA, 50156. rlhandy@iowatelecom.net.

ABSTRACT: Although normally associated with ground-shaking events such as earthquakes, liquefaction also is indicated to occur during dynamic installation of Rammed Aggregate Pier™ foundation elements in susceptible soils. Such piers nevertheless give a satisfactory performance, suggesting likelihood that drainage is contemporaneous with ramming. This was investigated by measuring lateral in-situ stresses at various depths and distances from the piers that are made by ramming graded aggregate in layers into open borings. Liquefaction is indicated by an undiminished transfer of radial stress for a distance outward into the soil. Total stress in the liquefied zone then is reduced by drainage that is accelerated by the hydraulic gradient caused by ramming. Simultaneous seepage forces act to compress and compact the soil; hence the coined term *"liquepaction."* Radial and tangential *in situ* stress measurements indicate that radial tension cracks occur in a peripheral elastic zone regardless of whether the soil is saturated or liquefied. In saturated soil the cracks act as a drainage gallery that may be extended by hydraulic fracturing. Stress measurements, which were made with the K_o Stepped Blade, also indicate that the radial cracks are propped open by soil invading from the liquefied zone.

KEY WORDS: hydraulic gradients, liquefaction, seepage forces, soil compaction, soil reinforcement

INTRODUCTION

Piers employed in the Geopier® method of soil reinforcement present a model that is relatively easy to study because of their large diameter, nominally 0.76 m (30 in.) before ramming, and the uniform high lateral contact stress created against the surrounding soil. Installation methods intentionally create high lateral stresses in soil around and between the piers.

Pier construction begins with boring a hole to the full depth of the pier, dumping a measured amount of coarse aggregate into the boring, and ramming the aggregate into a compact layer. The edge of the rammer is beveled at a 45° angle so a substantial part of the compactive effort is directed outward. Compaction of the first layer into an expanding hemispherical or bowl-shaped cavity gives way to an expanding cylindrical shape as successive layers of aggregate are added and compacted and pier construction continues upward.

Temporary liquefaction and the implications of liquefaction were not anticipated at the beginning of this investigation, which was initiated to simply measure lateral stresses created in the surrounding soil by ramming of aggregate to form a pier.

METHODOLOGY

Lateral stresses were measured *in situ* at various depths and distances from piers installed in soil ranging from loessial silt to silty sand and overconsolidated plastic clay. Measurements were made with the K_o Stepped Blade, which uses an extrapolation procedure to obtain hypothetical stresses on a zero-thickness blade. This removes effects of stress concentration that occur with conventional pressure cells. Excess pore water pressure also is accounted for in the extrapolation process, so measured stresses represent effective stresses plus static groundwater pressures. Coincidentally, the name for a prototype instrument was suggested in 1985 by the honoree of this volume, Dr. John Schmertmann.

EVIDENCE FOR LIQUEFACTION

Several different soil behavioral zones were identified from radial stress measurements. These include a zone showing a history of liquefaction of susceptible soils (Handy and White 2006a). The evidence for this is the undiminished transfer of radial stress from *a* to *b* in Fig. 1, across a radial distance of approximately 1.6 m (5 ft) from the pier surface. This pattern was repeated in other test series, and because of the increase in circumferential area is possible only if the material in this zone temporarily behaves as a liquid.

It may be conjectured that compression liquefaction (as distinguished from that caused by jetting) also may occur during dynamic installation of other foundation elements such as driven pile, where subsequent solidification may help to explain the rapid set-up or "freeze" that can occur if driving is interrupted.

After liquefaction starts, it becomes self-sustaining as ramming continues and lateral pressure is transferred farther outward through the growing mass of liquefied soil. In practice a standardized ramming time is used for each nominal 0.3 m (1 ft) lift of aggregate. The rate of compression diminishes as each lift expands and firms up as aggregate compacts.

When ramming stops, total stresses in the zone defined by *a* and *b* relax with drainage and become effective stresses that were measured along the line from *a* to *c*. Plastic theory applied to the slope on a log scale indicates an increase in friction angle from 25° to about 40°, indicative of compaction. This increase was confirmed with in-situ Borehole Shear Tests (Handy and White 2001a).

Liquefaction is prevented if the soil is not saturated or if, as in the case of overconsolidated expansive clay, the radial component of ramming pressure does not exceed the soil compressive strength sufficiently to break down the soil structure.

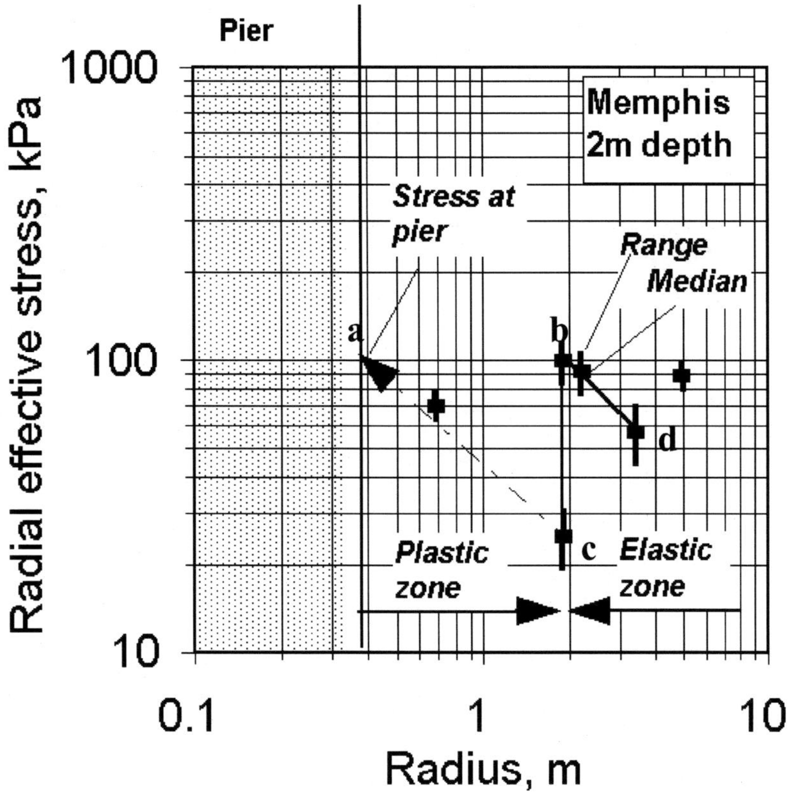

FIG. 1. Representative pattern of radial stresses measured *in situ* near a Rammed Aggregate Pier element in saturated loess soil. From Handy and White (2006a). The stress transfer from *a* to *b* is attributed to temporary liquefaction, and from *b* to *d* to elastic behavior with shrinkage cracking. The point to the right of *d* represents the pre-existing lateral K_o stress. With permission from ASCE.

CAVITY EXPANSION AND RADIAL CRACKING

Cavity expansion in an elastic medium theoretically causes an increase in radial stress and a simultaneous decrease in tangential stress, Fig. 2(a). The stress path on a Mohr diagram therefore is vertical, such that shearing will initiate at the failure envelope. The recipe for radial tension cracking is illustrated in Fig. 2(b), where a lower pre-existing lateral stress becomes tensile before shearing can occur. If the tensile stress exceeds the tensile strength of the soil, which typically is low and may be zero, radial cracking should initiate. This relieves tangential restraint in the elastic zone and results in a 1:1 slope such as from *b* to *d* in Fig. 1. This pattern is confirmed in other test sequences (Handy and White 2006b).

For radial cracking to occur, the tensile stress from cavity expansion also must exceed the pre-existing K_o stress, which increases with depth and with overconsolidation. In Fig. 1 the K_o stress is shown at a distance of 5 m, and was exceeded by the peak pressure at *b*, as required for liquefaction.

Radial tension cracks reduce the tangential minor principle stress to zero, so another control on liquefaction is the unconfined compressive strength of the soil. If that strength is not exceeded, the soil will retain its structure and continue to exhibit elastic behavior.

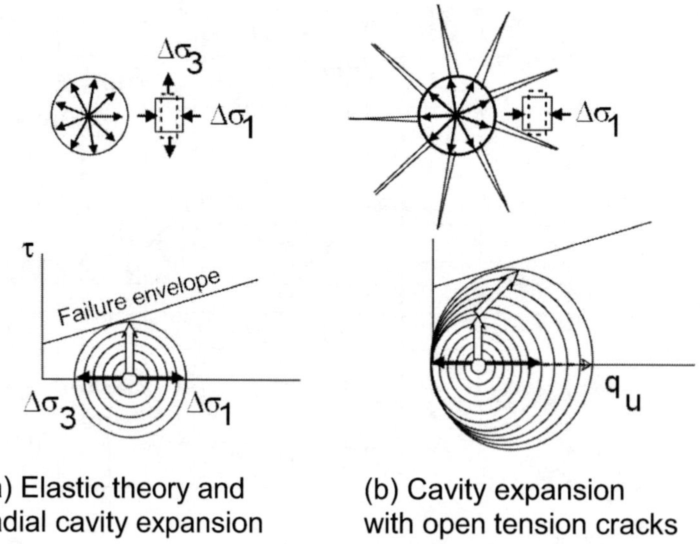

FIG. 2 Liquefaction can proceed in a saturated soil if the radial stress exceeds the soil unconfined compressive strength. Open arrows indicate stress paths.

DRAINAGE AND WEDGING

An obvious question is where does the water go as the liquefied soil drains and solidifies? The pier aggregate is well-graded "road stone" that when compacted has a void ratio that is too low to accommodate water expelled from the surrounding soil.

The open tension cracks should act as conduits for water, and water has been observed geysering upward from a nearby piezometer boring synchronous with blows of the rammer. Cracks therefore may be extended by hydraulic fracturing. On the other hand, stress measurements indicate that radial tension cracks also occur near Rammed Aggregate Pier elements in unsaturated soil, so this is an elastic response mechanism. Cracks also are observed extending to the ground surface where they have not been obliterated by development of a passive zone in the soil.

An open crack adjacent to liquefied soil should readily be penetrated by the soil. Evidence for this is the retention of stress at *b* in Fig. 1. Otherwise in elastic medium that stress should be reduced when the supporting pressure decreases as the liquefied soil drains and consolidates. By preventing the cracks from closing, soil wedges can be expected to preserve a tangential hoop stress that can support the radial stress at *b* through arching action.

Occasionally the peak stress measured at *b* is slightly higher than the lateral ramming stress at *a,* which could reflect the pulsing transfer of energy and capture of peak stresses by wedging.

COMPACTION OF A FILTER CAKE

Filters are commonly used in processing industries to remove solid particles from suspension. Particles accumulate as a filter cake that gradually compresses under the influence of seepage forces within the cake, until eventually the reduction in drainage rate is such that the cake is removed and the process is renewed. A similar densification and reduction in seepage rate occurs in weak sands during pumping from wells. Similarly, formation of a filter cake from the outward pressure of drilling mud creates a seepage-force holding pressure that helps to retain the exposed face of the soil.

Sridharan and Prakash (1999) propose a seepage-force-induced consolidation test for very soft sediments, and report good agreement with results from conventional odometer tests. Reversing the flow direction has the opposite effect, as Liu et al. (2006) report that upward acting seepage forces developed during load-induced consolidation can significantly decrease the consolidation rate below that which is predicted by Terzaghi's theory. A similar compression of sewage sludge by seepage forces sometimes is referred to as "shrinkage."

The rate of water flow through a porous filter with incompressible particles is proportional to the drop in fluid pressure across the filter. The hydraulic pressure transmitted as a result of liquefaction is the same as that at the pier surface, *a* in Fig. 1. The value of approximately 100 kPa (2100 lb/ft^2) is consistent with other data sets, and is the equivalent of hydrostatic pressure from a 10 m (33 ft) column of water. As a reference for comparison, if that hydrostatic pressure were directed upward it could create as much as 10 m of quicksand.

As the drainage rate is inverse to hydraulic gradient, the rate also is influenced by the thickness of the filter. As the pier is being constructed as a succession of lifts, this distance may remain relatively constant, shown schematically by the open arrows in Fig. 3. The main control on drainage rate may be the soil wedges deposited in the radial cracks, in which case the rate will depend on the conductivity, spacing, and width of the wedges, which is not known.

It was discovered through experience that the first lift at the bottom of each pier must consist of an enlarged bulb of open-graded aggregate that can accept water.

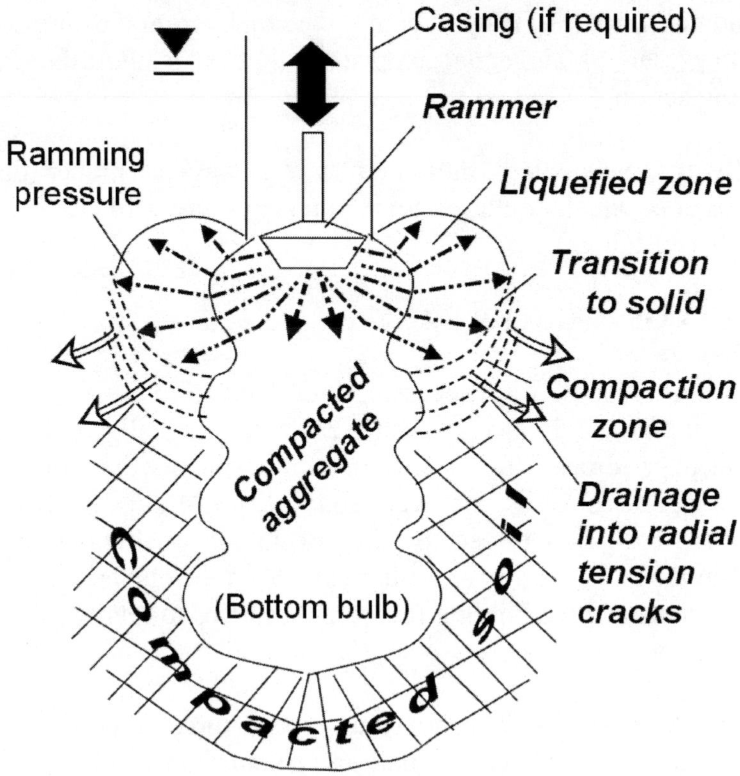

FIG. 3. Hypothetical model showing hydraulic compaction of successive lifts in a pier. Compaction is attributed to seepage forces caused by a high hydraulic gradient related to the ramming pressure. With permission of the Geopier Foundation Co., Inc.

SOIL ZONES AND CONTROL OF SETTLEMENT

A major application of the Geopier method for soil reinforcement is to control and reduce foundation settlement. One mechanism appears to be through increasing lateral stress in the matrix soil, which according to a Mohr circle representation should have the effect of increasing the preconsolidation pressure (Handy 2001). The distribution of radial stress in an elastic medium containing radial tension cracks is illustrated in Fig. 4(a). Liquefaction and the transfer of radial stress move this zone outward, Fig. 4(b). The mechanisms of settlement reduction in the liquefied zone therefore may include both a preconsolidation effect from the imposed lateral stress and a substantial densification, with the result that the Geopier method appears to be equally effective in soft, saturated soils as in harder soils that may or may not be saturated.

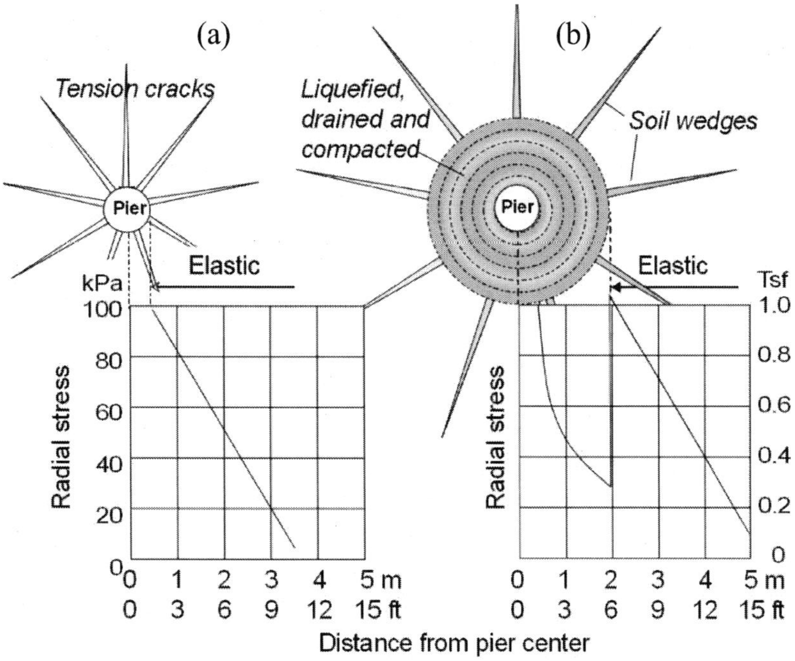

FIG. 4. Stress zones (a) without and (b) with temporary liquefaction. The graph at the right presents the same information as Fig. 1 with linear instead of logarithmic scales.

CONCLUSIONS

With regard to temporary liquefaction

1. Ramming of aggregate piers can trigger liquefaction of susceptible soils close to the piers.

2. A soil must be saturated and have an unconfined compressive strength that is lower than the horizontal component of the ramming pressure for liquefaction to occur.

3. Because liquefaction transfers radial stress outward, the surrounding zone of elastic compression also moves outward, thereby expanding the volume of soil that is exposed to a stress-induced increase in the preconsolidation pressure.

With regard to radial tension cracking

4. Radial pressure from ramming of aggregate piers causes radial compressive and tangential tension stresses in a surrounding quasi-elastic medium. Tension cracking occurs regardless of whether the soil is saturated, and will be prevented at depth by a high K_o stress.

5. Radial cracks create a drainage gallery for water expelled from the liquefied zone. Cracking is aided by hydraulic fracturing, but this is not a prerequisite for cracking to occur.

6. Temporarily liquefied soil invades adjacent tension cracks and creates soil wedges that preserve radial stress in the elastic zone by developing tangential hoop stresses.

With regard to drainage and compaction

7. The high hydraulic gradient caused by ramming hastens drainage of the liquefied soil.

8. As the liquefied soil drains it is compacted by seepage forces, analogous to compaction of a filter cake.

9. A consequence of liquefaction is to allow the Rammed Aggregate Pier or Geopier method of soil reinforcement to be as or more effective in soft, saturated soils as in stiffer, well drained soils.

10. Several factors can contribute to a decrease in foundation settlement. These include: (1) compaction of soil adjacent to the piers, discussed in this paper, (2) an increase in lateral confining stress acting on soil between the piers, and (3) stiffness of the compacted aggregate pier itself as a consequence of compaction and lateral confinement.

REFERENCES

Handy, R. L. (2001). "Does lateral stress really influence settlement?" *J. Geotech. Geoenviron. Eng.,* 127(7), 623-626.

Handy, R. L., and White, D.J. (2006a). "Stress zones near displacement piers: I. Plastic and liquefied behavior." *J. Geotech. Geoenviron. Eng.,* 132(1), 54-62.

Handy, R. L., and White, D. J. (2006b). "Stress zones near displacement piers: II. Radial cracking and wedging." *J. Geotech. Geoenviron. Eng.,* 132(1), 63-71.

Liu, J. C., Zhao, W. B. and Zai, J. M. (2006) "Research on Seepage Force Influence on One-Dimensional Consolidation." *Proc., GeoShanghai 2006, Am. Soc. Civ. Eng.* 203-209.

Sridharan, K and Prakash, A. (1999). "Simplified seepage consolidation test for soft sediments." *Geotechnical Testing Journal* 22:3:235-244.

ACKNOWLEDGEMENTS

This paper is partly the result of consulting research activities by the author for the Geopier Foundation Co. The author expresses his sincere thanks to Dr. Kord Wissmann and Dr. Nathaniel Fox of Geopier, Dr. David White, Iowa State University, Dr. John Schmertmann, and reviewers of this paper for their helpful questions and comments.

From Theory to Implementation of a CPT-Based Probabilistic and Fuzzy Soil Classification

Mehmet T. Tumay[1], Ph.D., P.E. , Fellow, ASCE; Murad Y. Abu-Farsakh[2], Ph.D., P.E., Member, ASCE and Zhongjie Zhang[3], Ph.D., P.E.

[1]Georgia Gulf Distinguished Professor Emeritus, Department of Civil & Environmental Engineering, Louisiana State University, Baton Rouge, LA 70803; and Visiting Professor, Dept. of Civil Engineering, Bogazici University, Bebek, Istanbul, Turkey; E-mail: mtumay@eng.lsu.edu

[2]Research Assistant Professor, Louisiana Transportation Research Center, Louisiana State University, Baton Rouge, LA 70808; and Adjunct Professor, Dept. of Civil Engineering, Louisiana State University, Baton Rouge, LA; E-mail: cefars@lsu.edu

[3]Geotechnical and Pavement Administrator, Louisiana Transportation Research Center, Baton Rouge, LA 70808; E-Mail: DocZhang@dotd.la.go

ABSTRACT: This paper discusses the development of an up-to-date computerized CPT (Cone Penetration Test) based soil engineering classification system to provide geotechnical engineers with a handy tool for their daily design activities. Five CPT soil engineering classification systems are incorporated in this effort. They include the probabilistic region estimation and fuzzy classification methods, both developed by Zhang and Tumay, the Schmertmann, the Douglas and Olsen, and the Robertson et al. methods. In the probabilistic region estimation method, a conformal transformation is used to determine the soil classification index, U, from CPT cone tip resistance and friction ratio. A statistical correlation is established between U and the compositional soil type given by the Unified Soil Classification System (USCS). The soil classification index, U, provides a soil profile over depth with the probability of belonging to different soil types, which more realistically and continuously reflects the in-situ soil characterization, which includes the spatial variation of soil types. The CPT fuzzy classification on the other hand emphasizes the certainty of soil behavior. The advantage of combining these two classification methods is realized through implementing them into visual basic software with three other CPT soil classification methods for friendly use by geotechnical engineers. Three sites in Louisiana were selected for this study. For each site, CPT tests and the corresponding soil boring results were correlated. The soil classification results obtained using the probabilistic region estimation and fuzzy classification methods are cross-correlated with conventional soil classification from borings logs and three other established CPT soil classification methods.

KEY WORDS: comparisons, CPT, fuzzy, probabilistic, soil classification

INTRODUCTION

During the past two decades, the cone penetration test (CPT) has gained wide popularity and acknowledgement as a preferred in situ tool for subsurface investigation and soil characterization. The CPT is a robust, simple, fast, reliable, and economical test that provides continuous sounding of subsurface sediments. The CPT test is basically conducted by advancing a cylindrical rod with a cone tip down into the soil. During penetration, the cone penetrometer simultaneously measures the cone tip resistance, q_c, and sleeve friction, f_s. When the piezocone penetration test (PCPT or CPTu) is used, the pore pressures generated during penetration can also be measured, depending on the location of the pressure transducer (at the cone face, u_1, behind the cone base, u_2, or behind the friction sleeve, u_3). The CPT/PCPT measurements can be effectively used in many geotechnical engineering applications, such as soil stratification and identification, and to evaluate different soil properties such as the strength and consolidation characteristics of the geomedia. This makes the CPT/PCPT technology valuable for a wide range of geotechnical engineering applications.

Due to the geometric design of the piezocone, pore water pressures generated behind the cone base (u_2) can influence the total stress measured by the cone tip. Therefore, the measured cone tip resistance (q_c) may have to be corrected for certain cone configurations to account for the effect of this pore water pressure developing behind the cone tip. Theoretically, the corrected cone tip resistance (q_t) is given by:

$$q_t = q_c + u_2\,(1\text{-a}) \qquad (1)$$

where a = A_n/A_c is the effective area ratio of the cone, A_n = cross-sectional area of the load cell, and A_c = projected area of the cone.

However, the authors' experience in primarily Louisiana soil-types and other similar soil-types outside of Louisiana showed that using either q_c or q_t does not appreciably change the CPT-based soil classification results by utilizing methodologies depending on tip resistance and friction ratio. Therefore, tip resistance q_c has been used throughout this study for soil classification purposes.

One important application of the CPT is its use in soil type identification and classification profiling. Several charts were proposed in the literature to classify the soil from the CPT (using q_c) or the PCPT (using q_t) data (e.g., Schmertmann 1978; Douglas and Olsen, 1981; Robertson et al., 1986; Robertson, 1990; Olsen and Mitchell, 1995). These charts were developed based on comparison/correlation between CPT/PCPT profiles and soil type data bases collected/evaluated from extensive soil borings. Thus the CPT soil classification depends on the physical response of the soil during cone penetration, which is directly related to the mechanical properties of the tested soils. According to Douglas and Olsen (1981), the CPT classification charts can not provide accurate prediction of soil type based on soil composition, but rather serve as a guide to the soil behavior type. The correlation between soil composition and mechanical properties is not simple, especially in transition zones of soil types, leading to probability of mis-classifying the soil type using the current CPT classification charts. To account for such probability of mis-classifying the soil, Zhang and Tumay (1999, 2000, 2003) developed a statistical-based probabilistic region estimation method to classify the soil from CPT data that

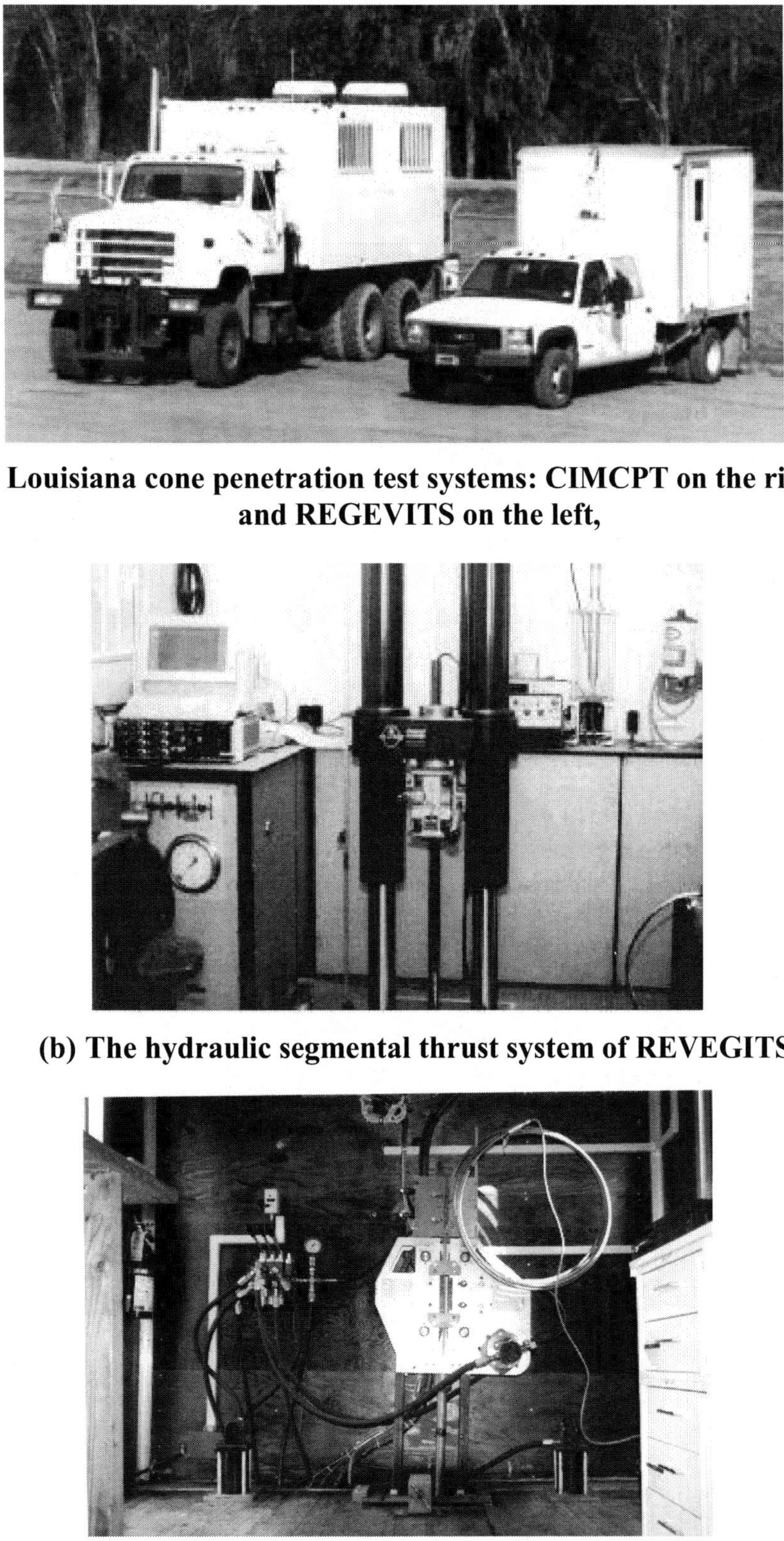

(a) Louisiana cone penetration test systems: CIMCPT on the right and REGEVITS on the left,

(b) The hydraulic segmental thrust system of REVEGITS,

(c) Continuous miniature cone/piezocone intrusion system of CIMCPT

FIG. 1: CPT systems managed by LTRC

involves uncertainty in the correlation between soil composition and soil mechanical behavior. This method provides a profile of the probability or the chance of having each soil type (clayey, silty, and sandy) with depth.

Due to the soft nature of soil deposits in Louisiana, the CPT is considered a preferred tool for site characterization. The Louisiana Department of Transportation and Development (LADOTD) operates three CPT systems. These systems are Louisiana Electric Cone Penetration System, LECOPS (Tumay, 1994), Research Vehicle for Geotechnical In-situ Testing and Support, REVEGITS (Tumay, 1996, 1998), and Continuous Intrusion Miniature Cone Penetration Test system, CIMCPT, (Tumay et.al, 1998). Currently, the CIMCPT and REVEGITS are managed by the Louisiana Transportation Research Center (LTRC). Figure 1 depicts a photograph of the CIMCPT system and REVEGITS. In order to facilitate the use of CPT technology for soil classification, a Visual Basic MS-Windows program was developed in which five CPT classification methods were implemented. These methods include the probabilistic region estimation and fuzzy classification methods, both developed by Zhang and Tumay (1999), the Schmertmann (1978), the Douglas and Olsen (1981), and the Robertson et al. (1986) methods. The program (www.ltrc.lsu.edu/downloads.html) performs the analyses on the CPT soundings using the selected CPT classification method and provides the geotechnical engineers with soil classification profile with depth.

SOIL CLASSIFICATION BY CPT

Soil identification and classification of soil stratigraphy can be achieved by analyzing the CPT data. The trends in CPT soil classification indicated that sandy soils usually have high cone tip resistance and low friction ratio; soft clay soils show low cone tip resistance and high friction ratio; organic soils such as peat have very low cone tip resistance and very high friction ratio; whereas overconsolidated soils tend to produce higher cone tip resistance and higher friction ratio.

Traditional CPT classification methods provide two-dimensional charts for soil classification based either on cone tip resistance (q_c or q_t), friction ratio (R_f), and pore pressure (u), or their normalization with respect to vertical overburden stress (σ_{vo}). These charts were developed through direct correlation between the CPT data (q_c, q_t, R_f, u) and the corresponding soil type determined from soil borings of the collected database. Several CPT charts have been proposed by investigators to classify the soil utilizing the CPT data (e.g., Schmertmann 1978; Douglas and Olsen, 1981; Robertson et al., 1986; Robertson, 1990; Olsen and Mitchell, 1995). While almost all the CPT methods (basically charts) give a specific classification to each soil layer along the penetrated depth; the probabilistic region estimation method proposed by Zhang and Tumay (1999) is unique in addressing the uncertainty in mis-classifying the soil. This statistical based method provides a profile of the probability or the chance of having each soil type (clayey, silty, and sandy) with depth. This method is similar to the classic soil classification methods which are based on soil composition. The following sections will summarize the CPT methods implemented and upgraded in the new Visual Basic software.

Probabilistic Region Estimation Method

The probability of incorrectly identifying soil type using the tradition CPT classification charts, especially in transition zones, motivated the development of the probabilistic region estimation method. This CPT classification method addresses the uncertainty of correlation between the soil composition and soil mechanical behavior.

In this method, conformal mapping was performed on the Douglas and Olsen (1981) chart to transfer the chart axis from the CPT data (q_c, R_f) to the soil classification index (U). The conformal transformation is accomplished using the following equations:

$$x = 0.1539\, R_f + 0.8870 \log q_c - 3.35 \qquad (2)$$

$$y = -0.2957\, R_f + 0.4617 \log q_c - 0.37 \qquad (3)$$

The soil classification index (U) is given as:

$$U = \frac{(a_1x - a_2y + b_1)(c_1x - c_2y + d_1)}{(c_1x - c_2y + d_1)^2 + (c_2x + c_1y + d_2)^2} - \frac{(a_2x + a_1y + b_2)(c_2x + c_1y + d_2)}{(c_1x - c_2y + d_1)^2 + (c_2x + c_1y + d_2)^2} \qquad (4)$$

The coefficients in equation 4 are: a_1 = -11.345, a_2 = -3.795, b_1 = 15.202, b_2 = 5.085, c_1 = -0.296, c_2 = -0.759, d_1 = 2.960 and d_2 = 2.477.

A statistical correlation was then established between the U index and the compositional soil type given by the Unified Soil Classification System (USCS). A normal distribution of U was established for each reference USCS soil type (GP, SP, SM, SC, ML, CL, and CH). Each U value corresponds to several soil types with different probabilities. Boundary values were used to divide the U axis into seven regions as described in Figure 2a. Soil types were further rearranged into three groups: sandy and gravelly soils (GP, SP, and SM), silty soils (SC and ML) and clayey soils (CL and CH). Figure 2a also gives the probability of having each soil group within each region. The original method gives constant probability of each soil type (represented by the step lines) regardless of the U value within the same region (R_1 to R_7 in Figure 2a). This will allow for the sudden drop in the probabilities as the U value crosses the border from one region to another. This method was further modified from the origin to allow smooth transition of probability (curved lines) with U values, and hence to provide a continuous profile of the probability of soil constituents with depth. An example of a U profile is presented in Figure 3, which is compared with q_c and R_f profiles and the corresponding probabilistic region estimation on data obtained from Manwell Bridge located in Evangeline, LA is presented in Figure 3.

Fuzzy Classification Method

Most of existing CPT soil classification methods are based on statistical correlation between the CPT profile data and the USCS soil classification; hence leading to soil identification according to their mechanical behavior. In contrary to other methods, the CPT fuzzy soil classification approach is fundamentally different in releasing the constraint of soil composition, and instead is based on the certainty of soil behavior (i.e., cone tip resistance and local friction).

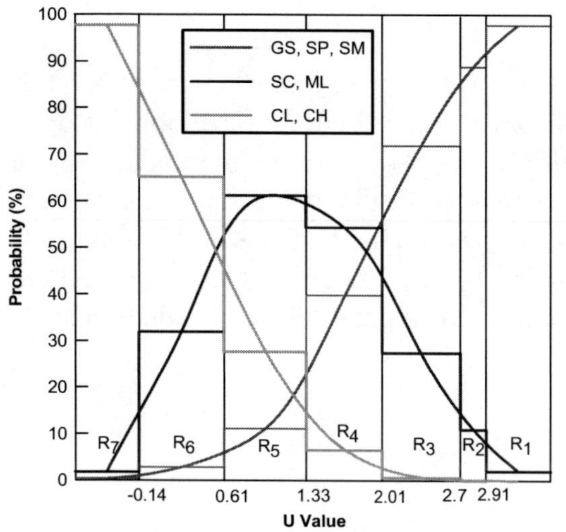

FIG. 2a: Regions' boundaries and the corresponding probabilities of each soil group

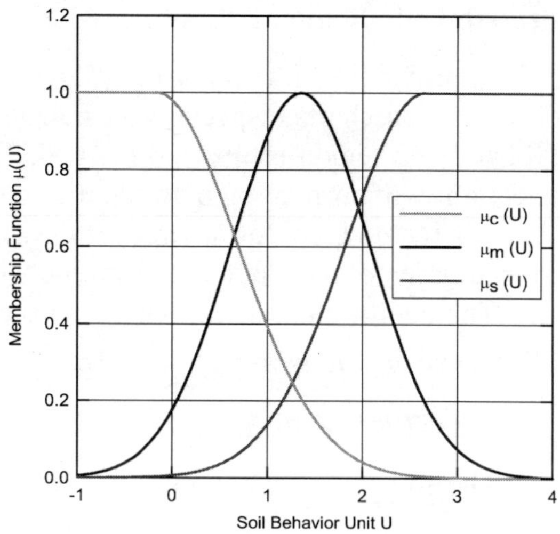

FIG. 2b: CPT fuzzy soil classification chart

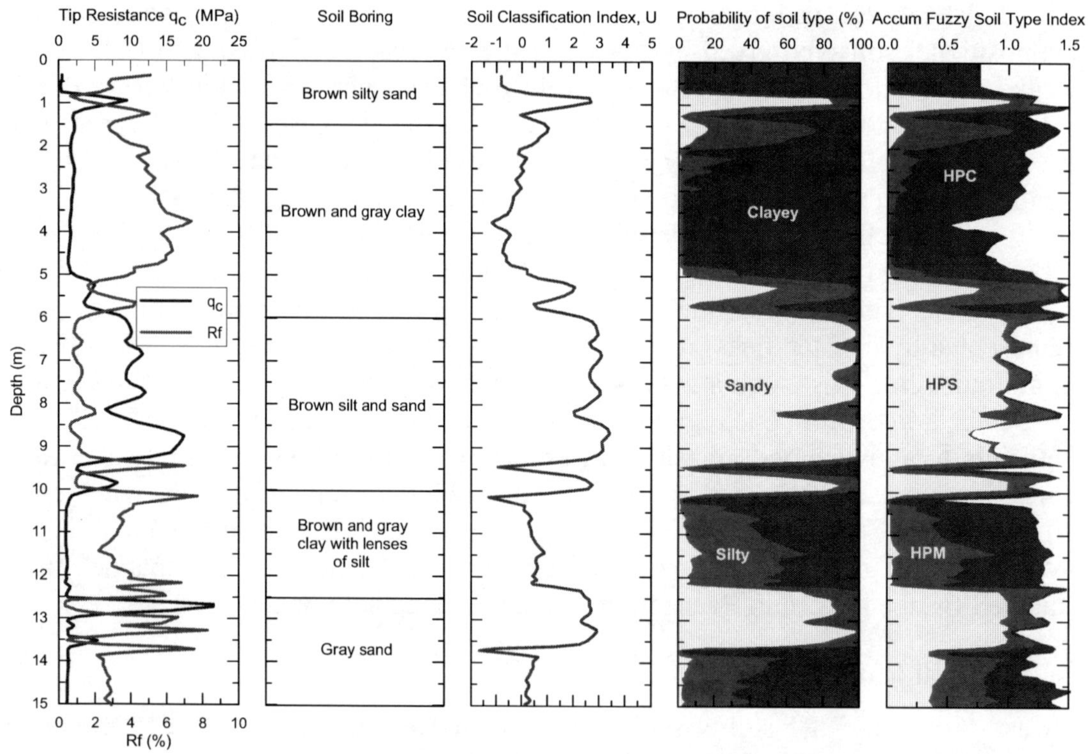

FIG. 3: Probability and fuzzy soil type index profiles for Manwell Bridge, Evangeline

In CPT fuzzy soil classification, three soil types are defined: Highly Probable Clayey soil (HPC), Highly Probable Mixed soil (HPM), and Highly Probable Sandy soil (HPS). The corresponding fuzzy membership functions of HPC, HPM, and HPS are given as (Zhang and Tumay, 1999):

$$\mu_c(U) = \begin{cases} \exp\left(-\frac{1}{2}\left(\frac{U+0.1775}{0.86332}\right)^2\right) & U \geq -0.1775 \\ 1.0 & U < -0.1775 \end{cases} \tag{5}$$

$$\mu_m(U) = \exp\left(-\frac{1}{2}\left(\frac{U-1.35}{0.724307}\right)^2\right) \qquad -\infty < U < \infty \tag{6}$$

$$\mu_s(U) = \begin{cases} 1.0 & U > 2.6575 \\ \exp\left(-\frac{1}{2}\left(\frac{U-2.6575}{0.834586}\right)^2\right) & U \leq 2.6575 \end{cases} \tag{7}$$

These empirical functions represent either "S" curve or "bell" curve types with a maximum membership value of one (1.0) for each soil type as depicted in Figure 2b. However, as seen in the figure, it is unlikely for all three membership values to have maximum values simultaneously, and that the accumulated sum depends on the U value. These empirical functions approximately relate the quantity change to quality change in soil composition and properties, reflecting an overall perspective of soil properties. The change is gradual from one soil type to another. The profile of fuzzy functions as compared to U profile and q_c and R_f profiles for Manwell Bridge, Evangeline Louisiana, are also shown in Figure 3.

Schmertmann Classification Method

The original mechanical CPT soil classification chart proposed by Schmertmann (1978, p.7) is shown in Figure 4a. Based upon CPT data taken from different sites in Louisiana, as well as CPT data taken from California, Oklahoma, Utah, Arizona and Nevada, as reported by Douglas and Olsen (1981), and comparison with soil borings, the original Schmertmann chart was modified by the first author as shown Figure 4b (Tumay, 1985). The chart depicts four distinct regions as identified by Douglas and Olsen (1981). Each region is further divided into sub-regions sorted out using Schmertmann classification modified slightly to reflect Louisiana research experience.

Douglas and Olsen Classification Method

Douglas and Olsen (1981) conducted comprehensive work correlating between the USCS soil classification and CPT data to develop a CPT-soil behavior type classification method. The development of this method was based on extensive data collected from sites in the western USA. The classification chart for the Douglas and Olsen method uses the cone tip resistance (q_c) and friction ratio (R_f) input parameters as shown in Figure 5. The chart shows the soil classification change (diagonally) from SP to SM to ML to CL to CH as the cone tip resistance decreases and friction ratio increases. Douglas and Olsen (1981) method demonstrates that the CPT classification charts can not provide an accurate prediction of soil type based on soil composition, but rather serve as a guide to soil behavior type (Lunne et al., 1997).

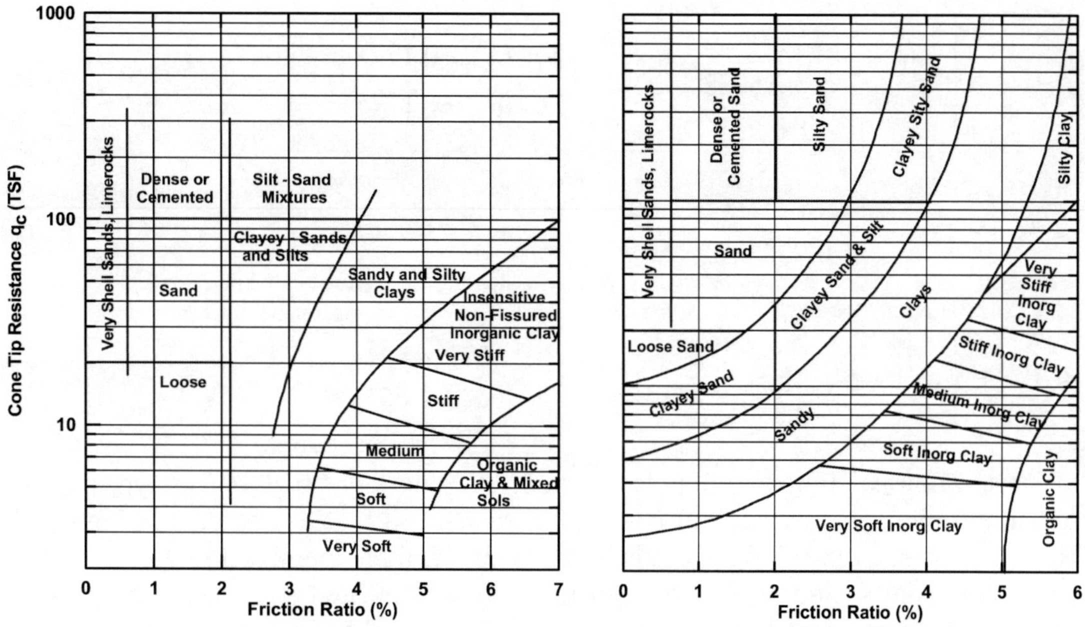

(a) Original **(b) Modified**

FIG. 4: Original and modified Schmertmann classification charts

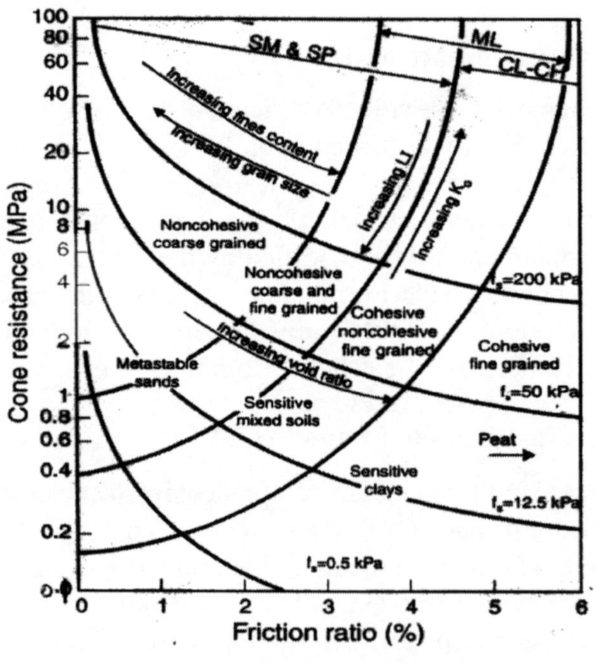

FIG. 5: Douglas and Olsen (1981) soil classification chart

Robertson et al. Classification Method

Robertson et al. (1986) developed a soil behavior type classification method derived from PCPT data (q_c, f_s, u). They proposed two charts, one chart uses corrected tip resistance (q_t) and friction ratio (R_f) as input data; while the other chart uses q_t and pore pressure parameter ($B_q = (u_2 - u_o)/(q_t - \sigma_{vo})$) as input data. They identified twelve different soil behavior types as shown in Figure 6. In case a soil falls within two different zones in respective charts, engineering judgment is required to classify the soil behavior correctly. Only the second chart was implemented in the visual basic soil classification software developed in this study.

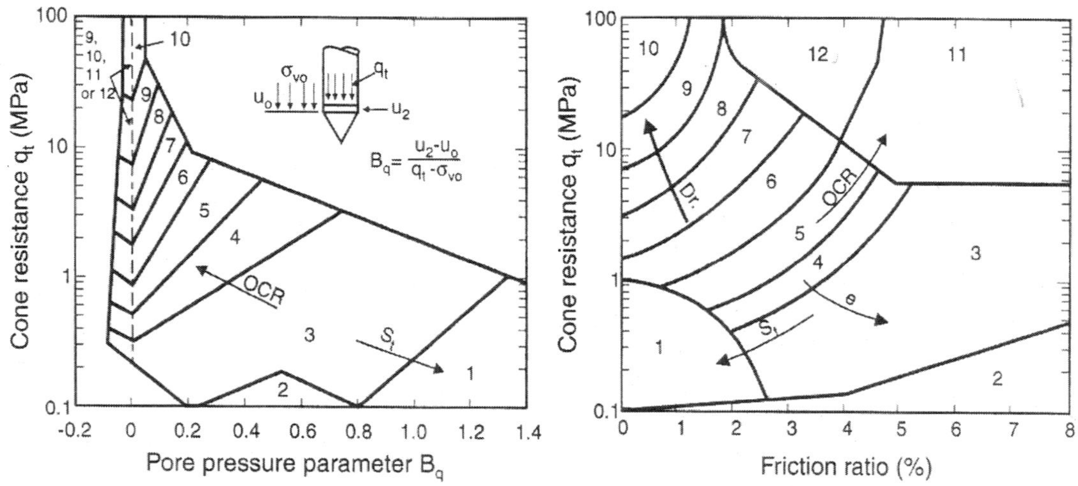

1. Sensitive fine grained, 2. Organic material, 3. Clay, 4. Silty clay to clay, 5. Clayey silt to silty clay, 6. Sandy silt to clayey silt, 7. Silty sand to sandy silt, 8. Sand to silty sand, 9. Sand, 10. Gravelly sand to sand, 11. Very stiff fine grained, 12. Sand to clayey sand.

FIG. 6: Robertson et al. (1986) soil classification charts

DEVELOPMENT OF SOIL CLASSIFICATION SOFTWARE

A visual basic soil engineering classification program, Louisiana Soil Classification by Cone Penetration Test Program (Figure 7), was developed utilizing the CPT data to provide geotechnical engineers with a user friendly methodology. Five CPT soil classification methods were implemented in this program. These include the probabilistic region estimation method and fuzzy classification method, both developed by Zhang and Tumay (1999), the Schmertmann (1978) method, the Douglas and Olsen (1981) method, and the Robertson et al. (1986) classification method. These methods use the cone tip resistance (q_c or q_t) and friction ratio (R_f) as input parameters.

The program is capable of reading CPT input data files of different units (SI, English, or millivolts raw data). Before running the program, the user can view the data file. The first step for the user is to input the project information. The program then plots the profiles of cone tip resistance, sleeve friction and friction ratio with depth. The user has the option to select the classification method and the corresponding display charts for output (graph and/or text). If the user selected a text chart for soil profile, the user can always change the layers manually. The program is available for free download from the LTRC Web site (www.ltrc.lsu.edu/downloads.html). Figure 7 describes the general features of the soil classification program.

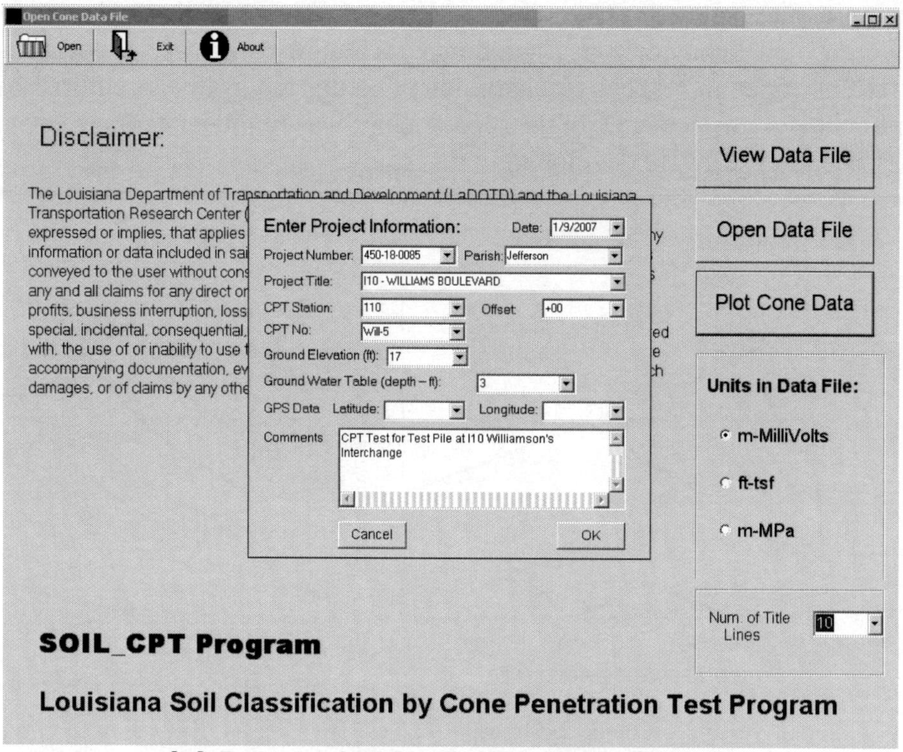

(a) Data and Information Input Screen

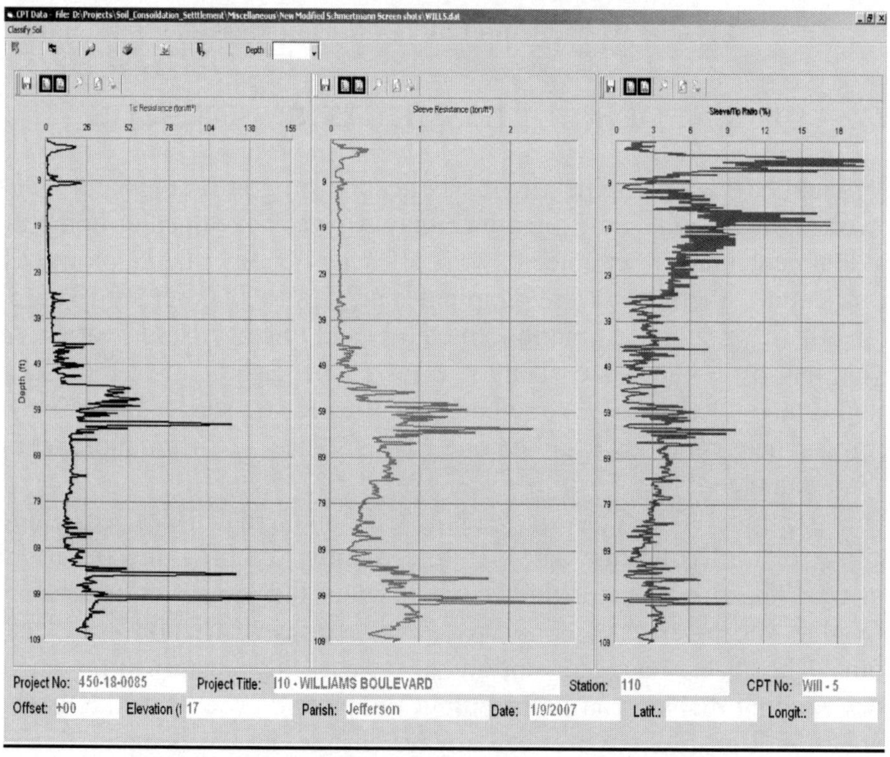

(b) CPT Profiles and Main Menu Screen

FIG. 7: Features of Louisiana Soil Classification by Cone Penetration Test Program (www.ltrc.lsu.edu/downloads.html)

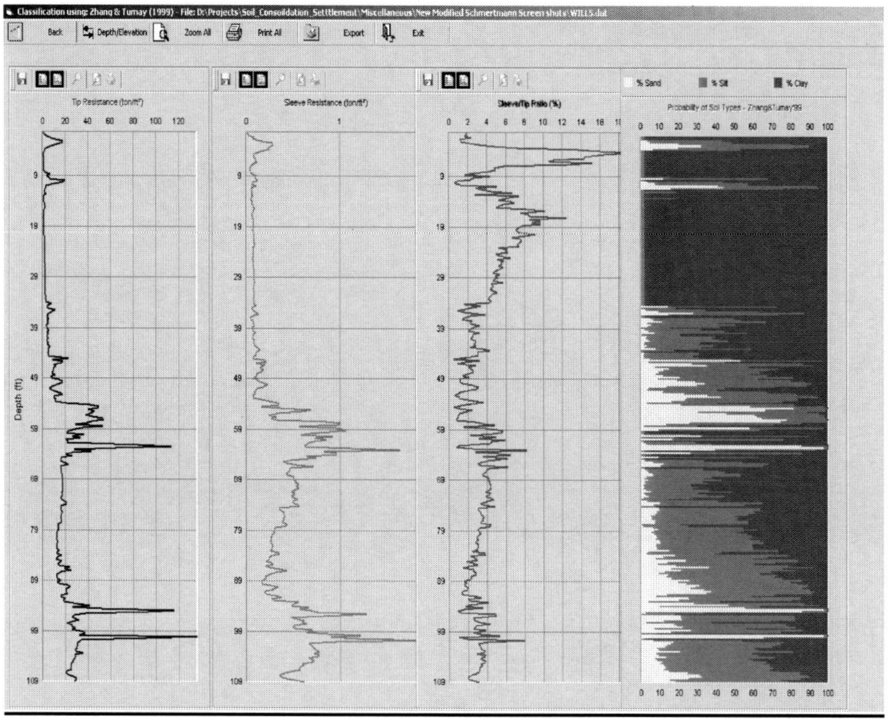

(c) Probabilistic Region Estimation Classification Method

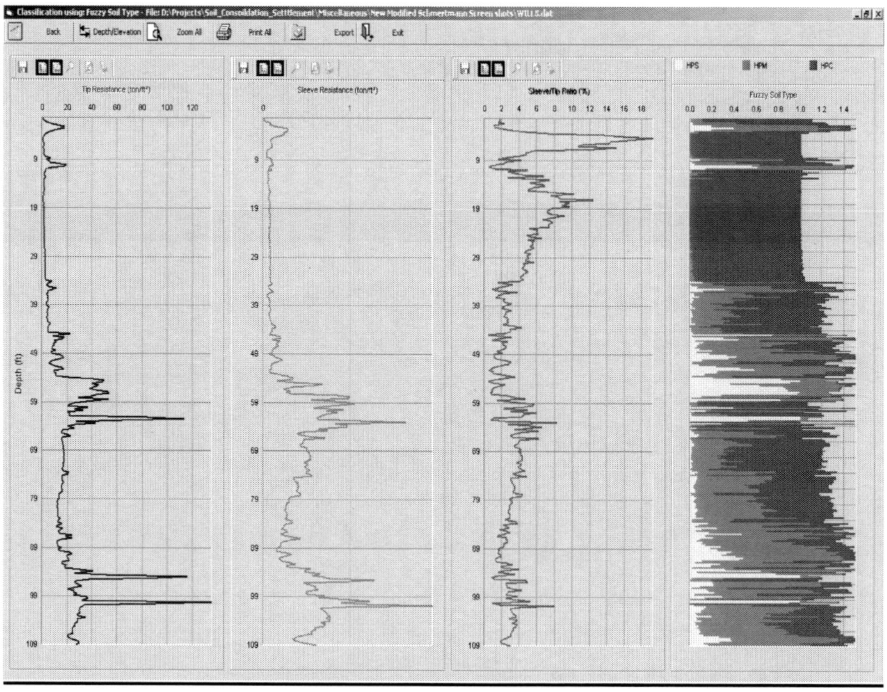

(d) Fuzzy Classification Method

FIG. 7: Features of Louisiana Soil Classification by Cone Penetration Test Program (cont.) (www.ltrc.lsu.edu/downloads.html)

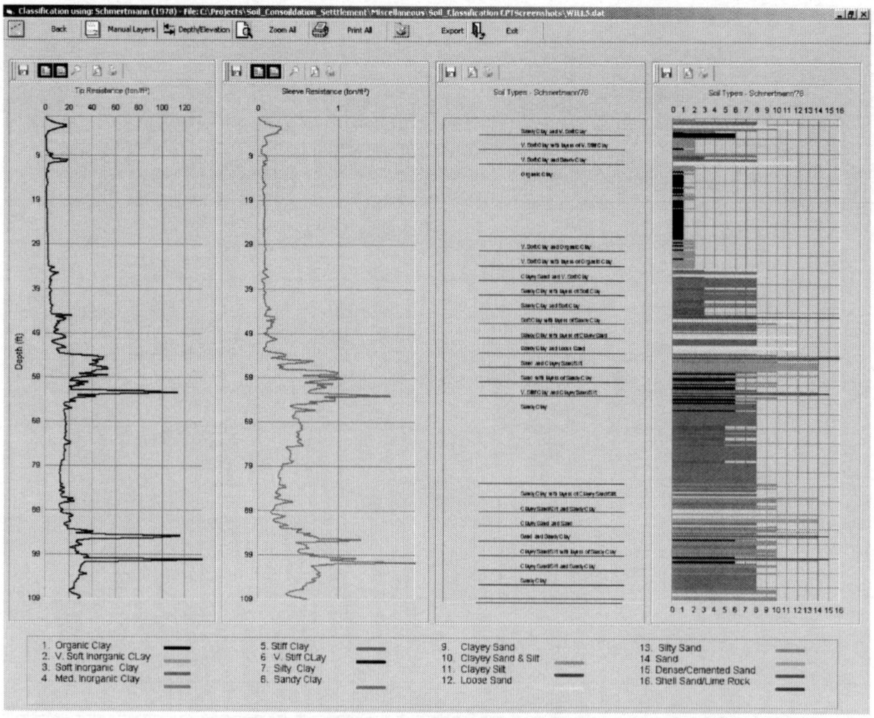

(e) Schmertmann Classification Method

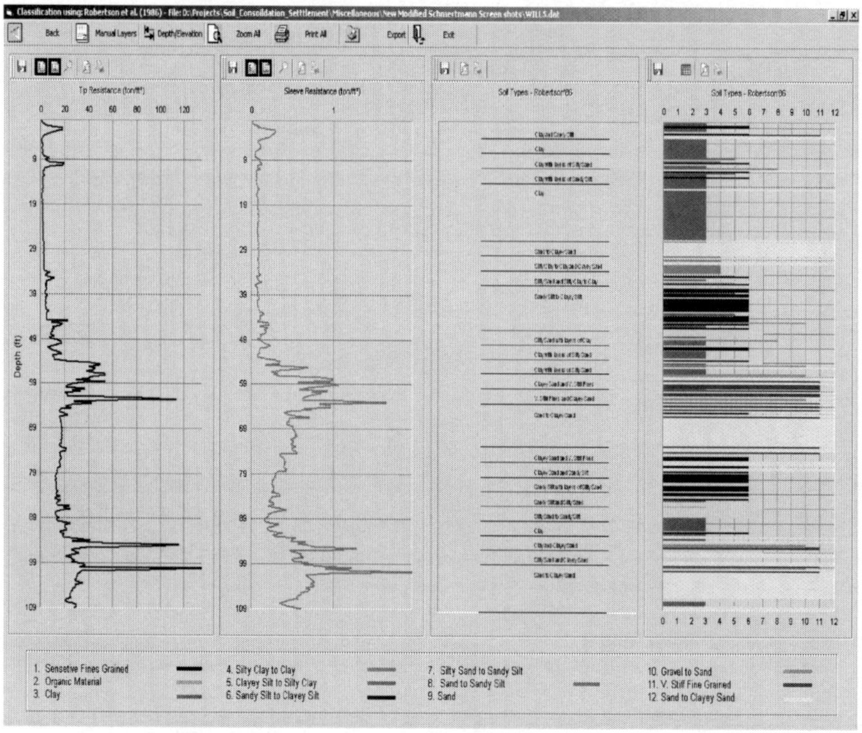

(f) Robertson et al. Classification Method

FIG. 7: Features of Louisiana Soil Classification by Cone Penetration Test Program (cont.) (www.ltrc.lsu.edu/downloads.html)

COMPARISON OF CPT CLASSIFICATION METHODS

Three sites in Louisiana were selected to compare the soil classification obtained using the probabilistic region estimation method and the fuzzy classification method with those obtained from other established CPT classification charts. In each site, cone penetration tests (CPT) were performed around drilled boreholes using a 10 cm^2 cone penetrometer. The borings in the Jennings site showed that the soil profile consists of about 20 feet of brown to gray medium to stiff clay and silty clay soil, followed by brown fine silty sand from 20 feet to 38 feet. Underneath, there is a layer of gray soft to medium silty clay extended from 38 feet to about 90 feet followed by a gray medium sand to the maximum depth of boring. Figure 8 presents the CPT results and the comparison of soil classifications obtained using the different CPT classification methods. The soil profile at the New Iberia site consists of soft to medium tan and gray silty clay soils down to 52 feet, silt and clayey silt from 52 feet to 62 feet, and silty clay soil interbedded with terraces of silts and sand from 62 feet to the maximum depth of boring. The CPT results and the corresponding soil classifications obtained using the different CPT classification methods for New Iberia site are presented in Figure 9. The soil boring at the Jefferson site indicates that the soil deposit consists of silty clay soils down to 50 ft, silty sand and sandy soils from 50 feet to 65 feet interbedded with lenses of silty clay soil, and medium to stiff silty clay soil interbedded with thin layers of silts and sands from 65 feet to the maximum depth of boring. Figure 10 presents the CPT results and compares the soil classifications for the Jefferson site obtained using the different CPT classification methods. The comparisons demonstrate that the probabilistic region estimation method and the fuzzy CPT classification method are superior in predicting, with "continuous and detailed" accuracy, the soil-type profile with depth.

SUMMARY AND CONCLUSIONS

This paper describes the implementation and general features of five CPT soil engineering classification methods in a visual basic program (Louisiana Soil Classification by Cone Penetration Test, LSC-CPT) for friendly use by geotechnical engineers in their daily activities. These include the probabilistic region estimation method and fuzzy classification method, both developed by Zhang and Tumay (1999), the Schmertmann (1978) method, the Douglas and Olsen (1981) method, and the Robertson et al. (1986) classification method. In Zhang and Tumay's two methods, a soil classification index, U, is determined and used to provide a continuous soil classification profile with gradual changes from one contiguous subsurface layer to another. The advantage of these two classification methods is demonstrated through comparison with soil borings in conjunction with three other established CPT classification charts. Three sites in Louisiana were selected for this comparison, which showed that the probabilistic region estimation and fuzzy CPT classification methods are capable of predicting, with good accuracy, continuous soil classification profile, including information on the probability of soil constituents in the layers encountered.

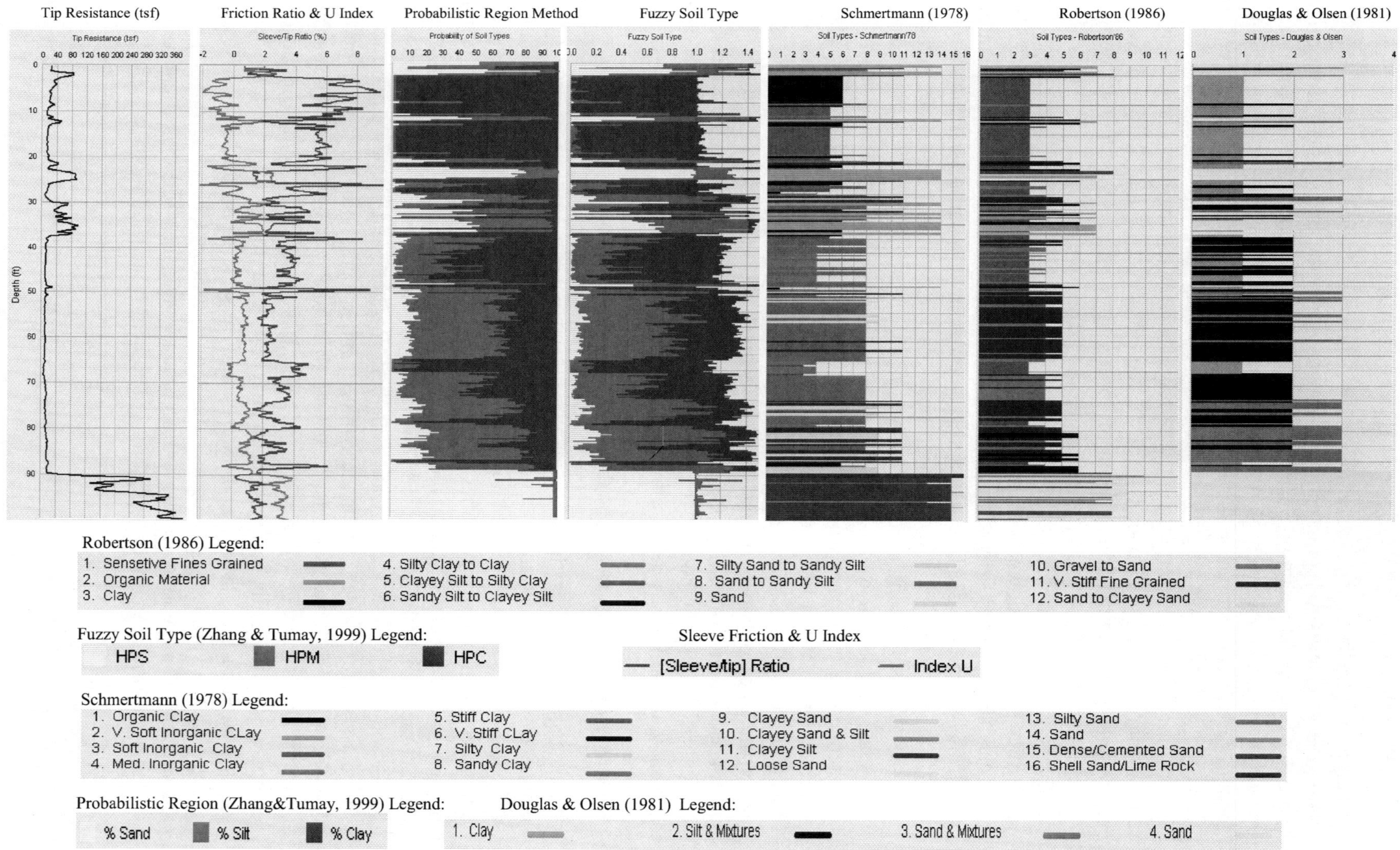

FIG. 8: CPT data and corresponding CPT soil classification for Jennings Site

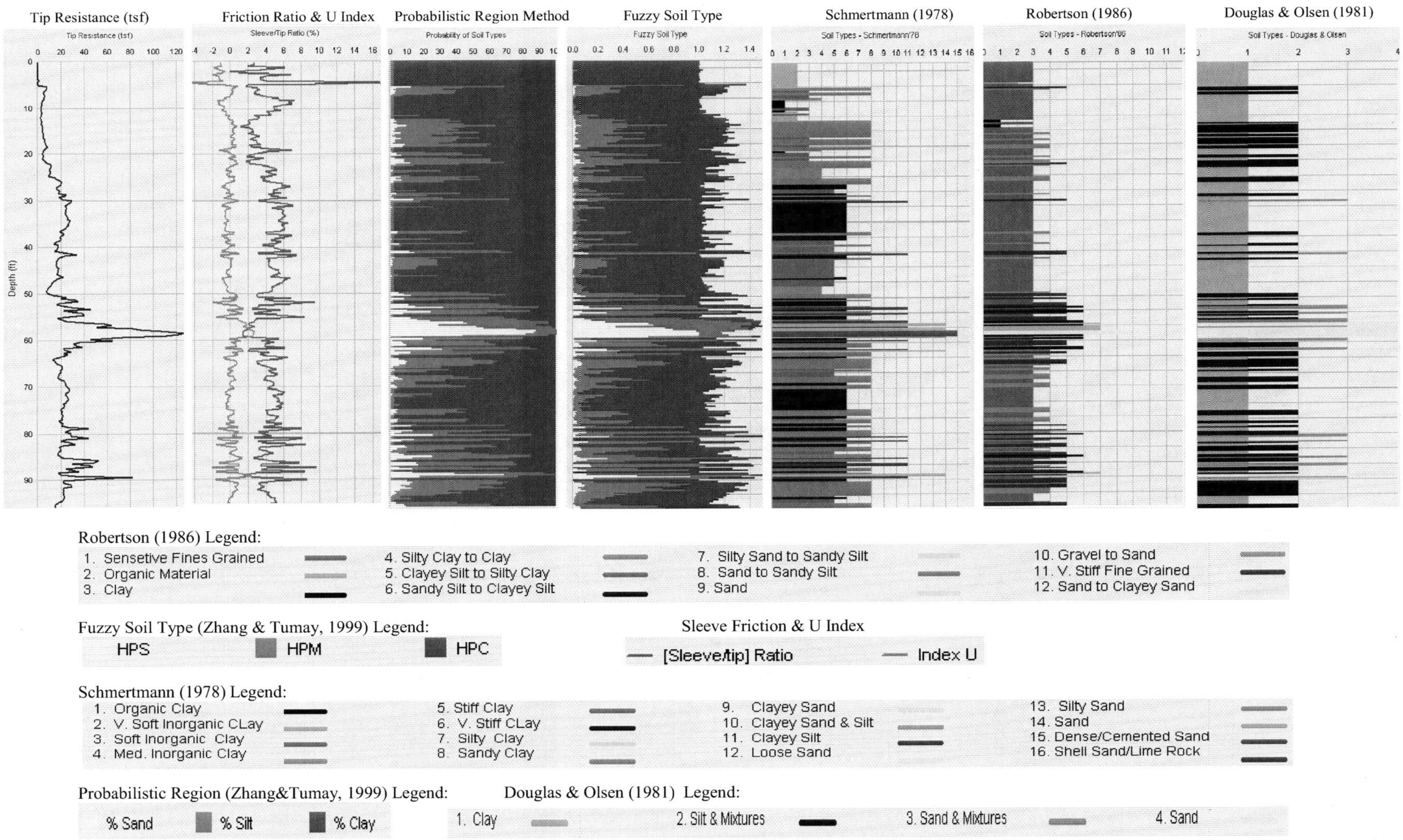

FIG. 9: CPT data and corresponding CPT soil classification for New Iberia Site

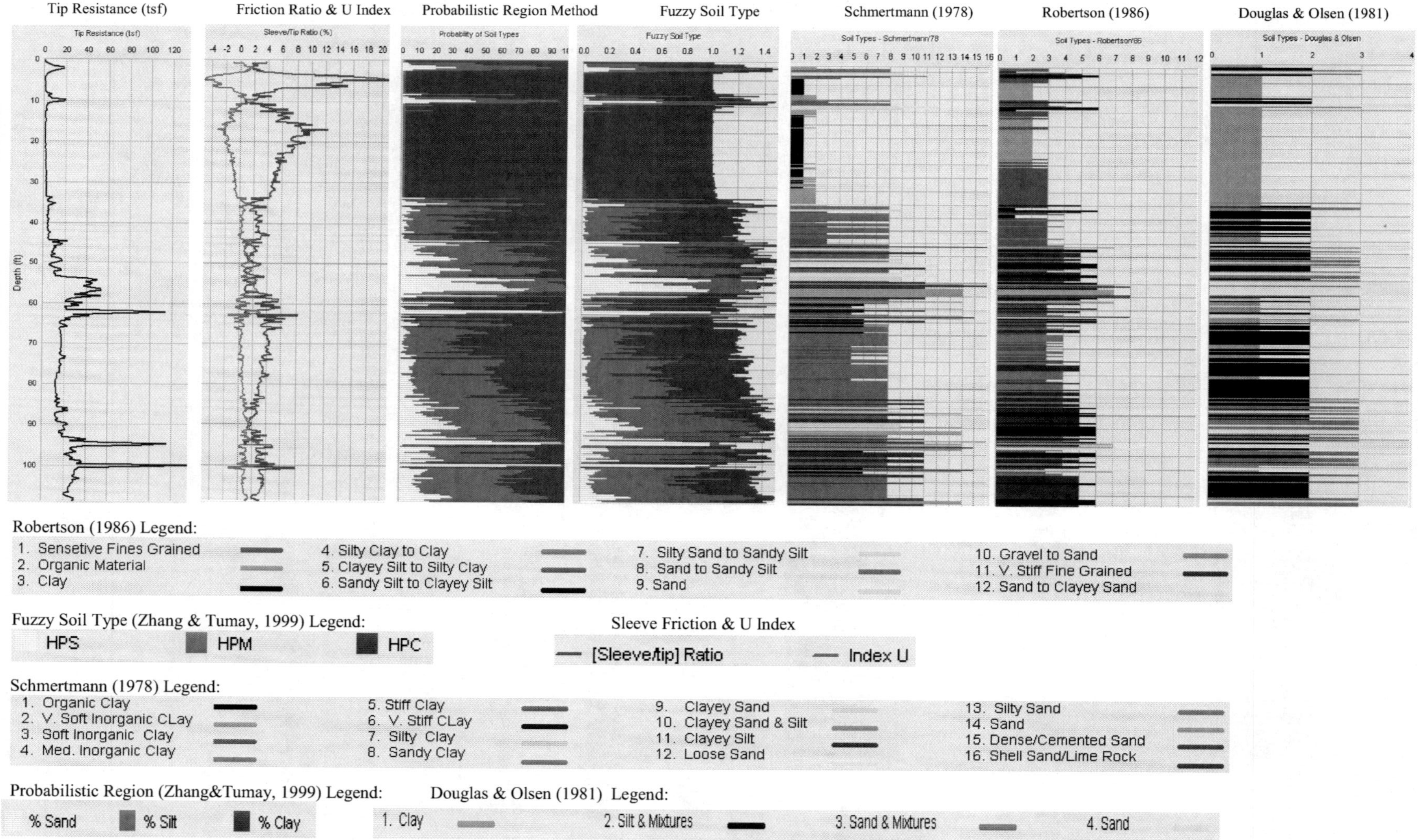

FIG. 10: CPT data and corresponding CPT soil classification for Williams-Jefferson Site

ACKNOWLEDGEMENTS

The authors of this paper (primarily the first author) were fortunate to have precious access to John H. Schmertmann's professional wisdom, technical expertise, scholarly efforts, wise council, and friendly advice. The geotechnical community is looking forward to seeing John around in the professional circles and benefiting from his treasure of knowledge, experience, and company for many more years. No doubt, he will keep busy and continue to be a credit to our esteemed profession in many, many ways.

The research effort that culminated the results summarized in this paper have been made possible over the years by various grants and research contracts from NSF, FHWA – Priority Technologies Program, and LADOTD. This support is hereby gratefully acknowledged. LSC-CPT was developed at the LSU/LTRC. The contents of this paper reflect the views of the authors, who are responsible for the facts and the accuracy of the data presented herein, and do not necessarily reflect official views or policies of the funding agencies.

This paper can best be viewed in color. For a color pdf copy, please send an e-mail request to the first author.

REFERENCES

Douglas, J. B. and Olsen, R. S. (1981). "Soil Classification using Electric Cone Penetrometer," Symposium on Cone Penetration Testing and Experience, Geotechnical Engineering Division, ASCE, St. Louis, pp. 209-227.

Lunne, T., Robertson, P.K., and Powell, J.J. M. (1997). *Cone Penetration Testing in Geotechnical Practice*, Blackie Academic and Professional, London, pp 312.

Olsen, R. S. and Mitchell, J. K. (1995). "CPT Stress Normalization and Prediction of Soil Classification," *Proceedings of the International Symposium on Cone Penetration Testing*, CPT'95, Linköping, Sweden, Vol. 2, pp. 257-262.

Robertson, P. K., Campanella, R. G., Gillespie, D., and Greig, J. (1986). "Use of Piezometer Cone Data," *Proceedings of the ASCE Specialty Conference on In Situ'86: Use of In Situ Tests in Geotechnical Engineering,* Blacksburg, Virginia, pp. 1263-1280.

Robertson, P. K. (1990). "Soil Classification using the Cone Penetration Test," *Canadian Geotechnical Journal*, Vol. 1, No. 27, pp. 151-158.

Schmertmann, J.H. (1978). *Guidelines for Cone Penetration Test, Performance and Design*, Report No. FHWA-TS-78-209, U.S. Department of Transportation, Washington, D.C., pp. 145.

Tumay, M.T. (1985). "Field Calibration of Electric Cone Penetrometers in Soft Soils – Executive Summary," Report No.FHWA/LA/LSU-GE-85/2, U.S. Department of Transportation, Federal Highway Administration, 37 pp.

Tumay, M.T. (1994). *Implementation of Louisiana Electric Cone Penetrometer System (LECOPS) for Design of Transportation Facilities, Executive Summary*, Report No. FHWA/LA_94/280 A&B, LTRC, Baton Rouge, LA.

Tumay, M.T., Kurup, P., and Boggess, R.L. (1998). "A Continuous Intrusion Electronic Miniature Cone Penetration Test System for Site Characterization," *Proceedings,*

International Conference on Site Characterization '98, Vol. 2. , Atlanta, GA, April 22-25, pp. 1183-1188.

Tumay, M.T. (1996, 1998). *In Situ Testing at the National Geotechnical Experimentation Sites – Phase1 & 2*, Contracts DTFH61-96-C-00017 & DTFH61-97-P-00161 Final Reports Phase 1 & 2, U.S. Department of Transportation, Federal Highway Administration; 101 pp + 4 x 1.44 MB disks and 154 pp + CD-ROM.

Zhang, Z., and Tumay, M.T. (1999). "Statistical to Fuzzy Approach toward CPT Soil Classification," *ASCE Journal of Geotechnical and Geoenvironmental Engineering*, Vol. 125, No. 3, pp. 179-186.

Zhang, Z, and Tumay, M.T. (2000), Closure to "Statistical to Fuzzy Approach Toward CPT Soil Classification," *ASCE Journal of the Geotechnical and Geoenvironmental Engineering*, Vol. 126, No. 6, June 2000, pp. 579-580.

Zhang, Z. and Tumay, M.T. (2003), "Nontraditional Approaches in Soil Classification Derived from the Cone Penetration Test," *ASCE Special Publication No. 121 on Probabilistic Site Characterization at the National Geotechnical Experimentation Sites*, ISBN 0-7844-06693, pp. 101-149.

Experimental and Analytical Investigation of Geogrid MSE Walls

D.V. Reddy, Ph.D.[1], P.E., M, ASCE, and Fernando Navarrete[2], Ph.D., M, ASCE

1Center for Marine Structures and Geotechnique, Department of Civil Engineering, Florida Atlantic University, PO Box 3091, Boca Raton, FL 33431.
Dvreddy@civil.fau.edu
[2]URS Corporation, 7800 Congress Avenue Suite 200, Boca Raton, FL, 33487

ABSTRACT: This study addresses the small-scale testing of Mechanically Stabilized Earth (MSE) walls reinforced with geogrids. Two High Density Polyethylene (HDPE) and two Polyethylene Terephthalate (PET) geogrid reinforced small-scale (1:5.5) MSE walls were tested. Additionally, a prototype full-scale MSE wall was tested, with one half of the wall reinforced with HDPE geogrids, and the other half with PET geogrids. Panel movements, strains in the reinforcement, and settlements were measured. The analytical work consisted of finite difference modeling of the small-scale MSE walls with Fast Lagrangian Analysis of Continua (FLAC) software.

The results of both the HDPE and PET walls testing showed that the deformations and strains encountered in the geogrids were very small, even at loads above normal working conditions. It was also observed that there were large differences in the values between the small-scale testing, and both the FLAC modeling and the full-scale MSE wall prototype testing. This is a consequence of the material property discrepancies and geogrid distortion for the FLAC modeling, and the lack of gravity simulation of full scale wall behavior. Therefore, although small scale testing can be useful for obtaining preliminary results, computer modeling and/or full prototype testing are still necessary to accurately determine the behavior of MSE walls.

KEY WORDS: experiment, FLAC, full scale, model, MSE wall, sand

INTRODUCTION

Mechanically Stabilized Earth (MSE) walls are composed of the backfill, reinforcing elements, and facings. The most common nonmetallic reinforcements are geogrids, which are designed specifically to provide soil reinforcement. The three-dimensional open structure of geogrids, interlocks with the surrounding soil, creating a positive connection between the two components of the MSE structure. Some of the advantages of MSE walls with geogrid reinforcement are their durability, simplicity, and rapidity of construction.

Due to the relatively limited experience with geogrids, there are uncertainties regarding the retention of their design properties after being subjected to construction stresses, construction damage, and exposure to in-soil environments over the expected design life. Testing of full size MSE walls presents great challenges due to their large size, associated self-weight, and the large loads they accommodate. Small-scale testing can help overcome some of these challenges; however, it presents limitations related to the scaling of gravity.

The purpose of this paper is to present the results of the small scale testing conducted as part of a large research project. The purpose of the larger research project was to use rational methods to enable the determination of applicable factors for use in LRFD design. An important objective of the small scale testing was to determine the viability of small scale testing of MSE walls. The small scale testing allowed the application of loads that otherwise will be extremely difficult to apply to a full scale MSE wall.

SMALL-SCALE TESTING

Four geogrid reinforced small-scale (1:5.5) MSE walls were tested. Two reinforced with High Density Polyethylene (HDPE) and two with Polyethylene Terephthalate (PET).

Materials

- HDPE Geogrid: UX-1400 SB
- PET Geogrid: Miragrid 3XT
- Facings: Clear Lexan 1/2" thick x 11" high x 10.5" wide, E = 2.38x109 Pa.
- Soil: SP (poorly-graded sands and gravely sands, little or no fines) Class II (coarse-grained one, clean) [ASTM D2321 and D2487].
  Sieve analysis: Based on the results of the sieve analysis the coefficients of uniformity and curvature were: C_u=3.75, and C_c=0.82.
  Soil compaction: A dry unit weight of 16.51 kN/m^3 (105 lb/ft^3) with optimum moisture content of 10.5% was obtained through laboratory Standard Proctor Tests (ASTM D698).
  Angle of friction: An angle of friction of 36 ° was obtained with direct shear testing.
  Quality Control: The soil was place on a 8.9 cm layers of sand/limerock, which was compacted with a hand compactor to 95-100% maximum relative compaction (Modified Procotor AASHTO T-180) within the range of optimum moisture content. The number of blows required to obtained the 95-100% compaction was determine prior to the testing and was applied to each layer.

Test Setup and Instrumentation

A loading frame to accommodate two 1:5.5 small-scale walls (separated by a vertical partition at the mid-location) was fabricated. Plane strain conditions were ensured with the stiffening of the sidewalls and a lubricated plastic placed in front of a sheet of clear Lexan.

The following instrumentation was used for the small-scale testing:

- 40 Channel Data Acquisition System,
- Two Digital Strain Indicators, and Switch and Balance Units
- Strain Gages-YFLA-2-3L
- Twenty-four Dial Gages with a measuring range of 2", and a least count of 0.001"

A schematic showing the location of the instrumentation is presented in Figure 1.

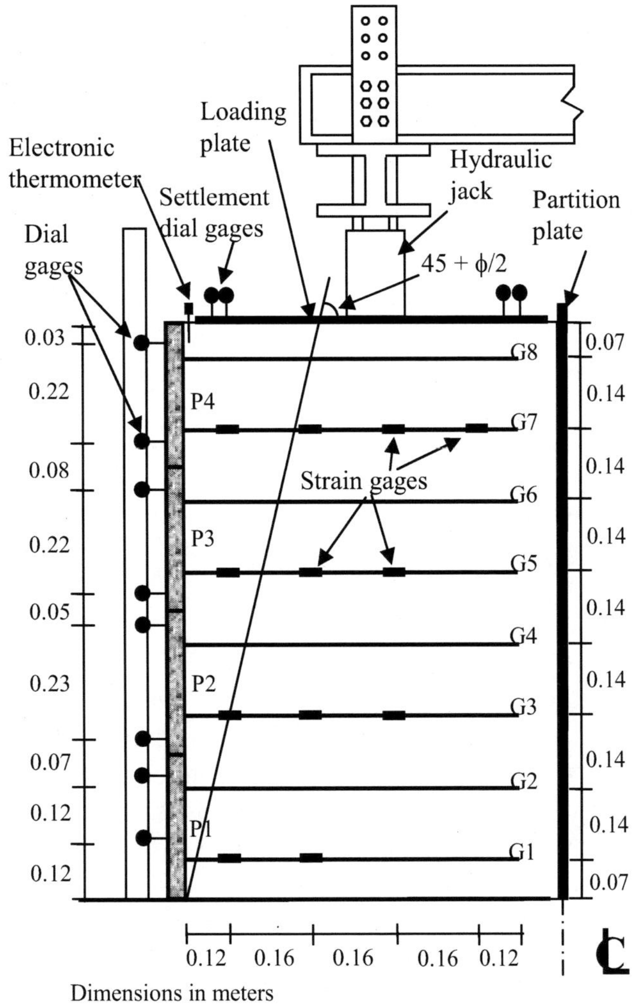

FIG. 1 Small-scale MSE wall instrumentation

Wall Design

The small-scale wall was constructed to simulate the prototype wall built at Florida Atlantic University in Boca Raton, Florida, with the following design parameters:

a) Wall height H = 6m, and length L= no restriction
b) Face angle = 90°
c) External Loads
    - Dead load = 10 kPa, to simulate a concrete pavement slab, 0.3m thick.
    - Live load = 10 kPa, to simulate the equivalent height of soil according to AASHTO LRFD specs. (FHWA H1-98-032).
    - No seismic loading
d) Facing: Segmental facing with concrete panels.
e) Construction requirements: 95% AASHTO T-99 Standard Proctor Compaction
f) Design / service life: short (<10 years)

Scale Effects

A complete analysis of the scale effects was carried out by using the Buckingham Pi theorem (Kundu and Cohen, 1990). Since this was a 1:5.5 model, the strength and stiffness of the geogrid had to be scaled down by 30.25. The surcharge needed to be scaled down by 5.5.

Geogrid Scaling

Tests were conducted to find the right geometry to obtain a 1 to 5.5 ratio for the pullout force for the geogrid. This was done by cutting some of the longitudinal ribs and all transverse ribs of the geogrids. For the PET geogrid, it was also necessary to reduce the length of the geogrids to achieve the 1 to 5.5 ratio for the pullout force. On the other hand, it was not possible to reduce the modulus of elasticity of the geogrids as weaker geogrids are not fabricated. This created a distortion in the geogrid scaling.

Test Procedure

At the end of construction of the small-scale MSE wall, the strain gages were read. Once the formwork, used for the construction was removed, the dial gages were installed, and readings of both the strain and dial gages were recorded; the wall was left unloaded for 24 hours and readings taken. Next, a loading plate with a load equivalent to a surcharge of 0.8 kPa, was placed on top of the wall. This was equivalent to 1/5.5 of the surcharge loading used in the prototype wall of 4.4 kPa, simulating twice the AASHTO HS-20 Highway Truck Loading (40 tons). This loading was maintained for 48 hours, after which readings were taken and a loading jack was placed at mid-location on the loading plate of each wall to induce a total surcharge of 3.64 kPa, which is 1/5.5 of the design surcharge of 20 kPa. This surcharge was applied for 72 hours.

The next surcharge level after 72 hours was 6.4 kPa to double the last surcharge increase to induce deformations in the walls; this was then increased with increments of 5.13 kPa to induce further deformations. Each surcharge level was maintained for a period of 72 hours. Readings were taken every day except on a couple of occasions due to rain. Once the surcharge of 21.8 kPa was reached, the increments were changed to 10.26 kPa up to 83.4 kPa, at which stage the experiment was stopped due to the large deformations encountered. Figure 2 shows a photograph of the small-scale test in progress.

Test Results

The results are presented for strains in the first, third, fifth and seventh geogrid layers (G1, G3, G5, G7), connected to the second column of panels from the left to the right. This is to avoid any influence of the boundary with the side walls. The spacing of the strain gages was 0.16m, and lateral displacements were recorded for the same second column of panels, with two dial gages per panel as shown in Figure 1. The settlements were measured with four dial gages mounted on the sides of the test frame to monitor the movements of the four corners of the loading plate. These measurements were averaged to obtain the overall settlement.

Strains in the Reinforcement

The strains were the averages of those measured in the two strain gages, located on the upper and lower faces of the geogrid. This was done to prevent possible distortion of the readings due to bending of the geogrid. As seen in Figure 1, the location of strain gages was selected taking into account the Rankine plane of failure.

FIG. 2 Finished Small-Scale MSE wall with PET geogrids

It was observed that for surcharges of up to 16.7 kPa, most of the strains due to creep were observed during the first 24 hours. For larger surcharge levels, most of the strains were observed in the first 72 hours; therefore, the loading intervals were adequate. At the same time, after the surcharge loading of 21.8 kPa, nonlinear behavior was observed. Very good agreement was observed in the deformation behavior for the two HDPE and two PET walls. The maximum strains were obtained in the first two strain gages in layer G3 and the three strain gages in layer G5.

Panel Displacements

For both the HDPE and the PET walls, the maximum displacements were found in the second and third panels (P2 & P3) and the bottom of the fourth panel (P4); this agreed with the strain results. There was good agreement between the two walls tested for both the HDPE and PET walls up to a surcharge of 21.8 kPa. For both HDPE and PET walls, large displacements were observed at the bottom of the four panel (P4), with small displacements at the top. This indicates that the displacements were primarily due to rotations because of relatively small load dispersion at the top of the wall.

Plate Settlement

The HDPE and PET walls showed good agreement for the two walls tested of each type for surcharges up to 21.7 kPa. The settlements were very similar for both the HDPE and PET walls, although the PET walls showed less settlement than the HDPE at the rear of the wall.

MSE WALL ANALYSIS

The small-scale MSE wall was modeled with FLAC (Fast Lagrangian Analysis of Continua) software version 3.3 from Itasca Consulting Group. The geogrids, soil, facings, and surcharges were modeled using the properties of the materials used in the small-scale MSE wall testing see Table 1. Although the material properties in the model were set to simulate the material properties of the small-scale test, some small differences were anticipated.

A finite difference grid of 21x40 was used, with Mohr-Coulomb Modeling, for all of the zones between the geogrid layers as shown in Figure 3. To generate the interfaces between the soil and the geogrid in both directions (top and bottom), the model "null" command was used to remove the rows representing the eight layers of geogrid. Then, the boundaries of the model were defined using the "generate" command.

The layers of geogrids were modeled with a flexible beam structure. The nodes and the structure characteristics of the geogrid elements were defined along the entire width of the model, and the interfaces between soil-geogrid/geogrid-soil were generated with Coulomb shear-strength material properties. The facing system was modeled by simulating four separate regions in the wall to correspond to the four different panels. Two nodes were defined in the locations, where the facings connect. The facing-soil interfaces were defined as elastically bonded. The geogrids and the facings were connected by attaching the first node of each geogrid layer to the respective facing node.

TABLE 1 Material Properties

DATA	STATUS	VALUE	FLAC VALUE	UNITS
Soil				
1.Density=	Given	1682.00	1682.00	kg/m3
2. Modulus of elasticity =	Assumed	1.32E+06	7.27E+06	N/m^2--N/m
3. Poisson's ratio =	Assumed	0.30	0.3	-
4. Angle of Friction =	Given	36.00	36	degree
5. Cohesion =	Assumed	2.00E+02	2.00E+02	Pa
6. Shear modulus (G)=	Calculated	5.09E+05	2.80E+06	Pa
7. Bulk modulus (K)=	Calculated	1.10E+06	6.06E+06	Pa
HDPE Geogrid				
1. Tensile modulus =	Given	7.29E+05	7.29E+05	N/m^2--N/m
2. Thickness	Given	0.001	0.001	m
3. Modulus of elasticity (E)=	Calculated	7.29E+08	7.29E+08	Pa
4. Inertia =	Calculated	8.33333E-11	8.33333E-11	m^4/m
5.Area (A= Thickness * Unit length)=	Given	0.001	0.001	m^2
PET Geogrid				
1. Tensile modulus=	Given	3.08E+05	3.08E+05	N/m^2--N/m
2. Thickness=	Given	0.001	0.001	mt
3. Modulus of elasticity (E)=	Calculated	3.08E+08	3.08E+08	Pa
4. Inertia=	Calculated	8.33333E-11	8.33333E-11	m^4/m
5.Area (A=Thickness*Unit length)=	Given	0.001	0.001	m^2
Facings				
1. Modulus of elasticity (E)=	Suggested	2.38E+09	2.38E+09	Pa
2. Thickness=	Suggested	0.013	0.013	m
3. Inertia=	Calculated	1.71E-07	1.71E-07	m^4/m
4. Height		0.92	0.92	ft
5. Width		0.88	0.88	ft
6. Thickness		0.50	0.50	in
4.Area (A=Thickness*Unit length)=	Calculated	0.013	0.013	m^2
Geogrid / Soil Interface				
1. Height of wall=	Given	1.09	1.09	m
2. Number of effective y-zones=	FLAC input file	32	32	grids
3. Angle of friction soil-geogrid=	*	36	36	degree
4. Kn=10*max[(K+ (4/3)G)/Dz]=	Calculated	5.22E+08	2.87E+09	N/m
5. Ks=	Calculated	5.22E+08	2.87E+09	N/m
Facing / Soil				
1. Length of the wall=	*Given	0.84	0.84	m
2. Number of effective y-zones=	FLAC input file	20	20	grids
3. Kn=10*max[(K+ (4/3)G)/Dz]=	Calculated	4.26E+08	2.34E+09	N/m
4. Ks=	Calculated	4.26E+08	2.34E+09	N/m

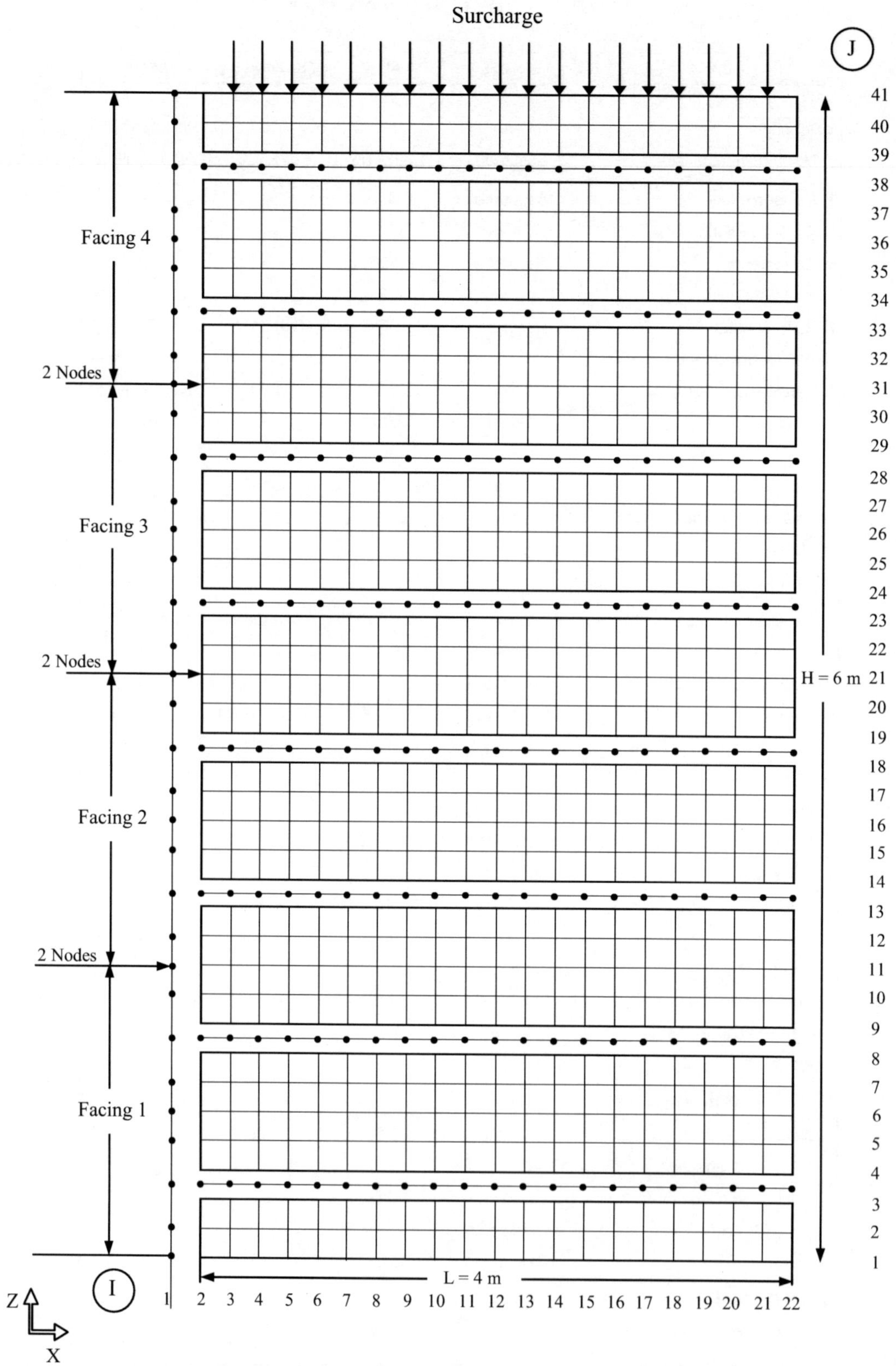

FIG. 3 Grid of the MSE wall modeled in FLAC software

The boundary conditions were defined as follows: no horizontal rotation was allowed at the middle of each facing, the right end of the geogrid was fixed in the x-direction, and the bottom boundary was fixed in both the x and y directions.

Multiple-stage construction was used to simulate the sequential construction process in the field. The initial gravitational stresses were obtained by applying the "solve"in the FLAC command to the system before applying the normal pressure. Equilibrium was established for each soil layer, and the surcharge applied after all layers were placed.

Comparison of the Small-Scale MSE Wall Test and FLAC Modeling

The comparison of the test and computer based values was carried out for the horizontal displacements of the facings, strains in the reinforcement, and settlements.

Horizontal Displacements of Facing Panels

The HDPE small-scale wall testing and the FLAC modeling results showed similar trends for panel displacements for all the different surcharge levels. The FLAC modeling values were larger in all the cases, but very similar for the fourth panel (P4), and the top of the third panel (P3) for the surcharge levels of 11.5 and 16.7 kPa. It was also observed that the differences between model and test values were larger for the PET MSE walls than for the HDPE ones.

Strains in the Reinforcement

It was observed that the HDPE and PET small-scale wall testing and the FLAC modeling results showed similar trends for strains for all the different surcharge levels. For the HDPE the values of the FLAC modeling were larger than the test values for the 0.0, 0.8, 3.6 and 6.4 kPa surcharges for all the geogrid layers. For the 11.5 and 16.7 kPa surcharges, the values of the FLAC modeling were larger for layers G5 and G7 and smaller than G1 and G3. For the PET MSE wall, the values of the FLAC modeling were larger in all cases.

Wall Settlements

It was observed that the forward tilting of the loading plates in the results agreed very well with the FLAC analysis for all the different surcharge levels. The settlements values in the FLAC modeling were larger in all the cases, but very similar, except for the 6.4 kPa surcharge for the HDPE wall, for which the model showed smaller settlements than those in the test.

Calibration of the Small Scale Test

The discrepancies in results between the small scale testing and the FLAC modeling were attributed to distortion due to the inadequate scaling of the geogrids and differences between the real material properties values and the ones used for the FLAC modeling. A multiplication factor ("a") was introduced to calibrate the small-scale test results to the FLAC modeling results. Table 2 shows the values of the factor "a" for the HDPE and PET MSE walls.

TABLE 2 Correction Factors for the HDPE and PET MSE Walls

Surcharge level (kPa)	Correction factor "a" HDPE MSE Walls	Correction factor "a" PET MSE Walls
0.0	0.13	0.07
0.8	0.18	0.14
3.6	0.18	0.18
6.4	0.63	0.35
11.5	0.86	0.39
16.7	0.86	0.7

Figures 4 and 5 show two plots for the horizontal panel displacements of the HDPE and PET small-scale MSE walls respectively. These plots present the results for the two small scale walls tested, the average of the two walls multiplied by factor "a" and the FLAC modeling results.

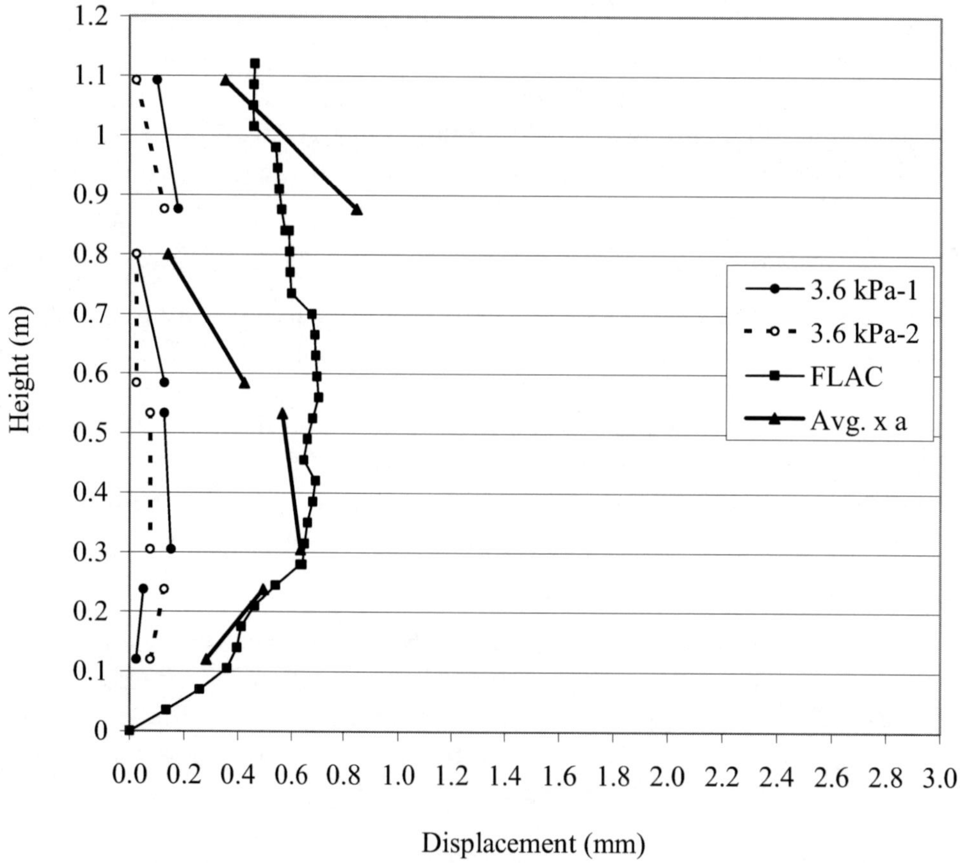

FIG. 4 HDPE MSE wall displacement, surcharge = 3.6 kPa

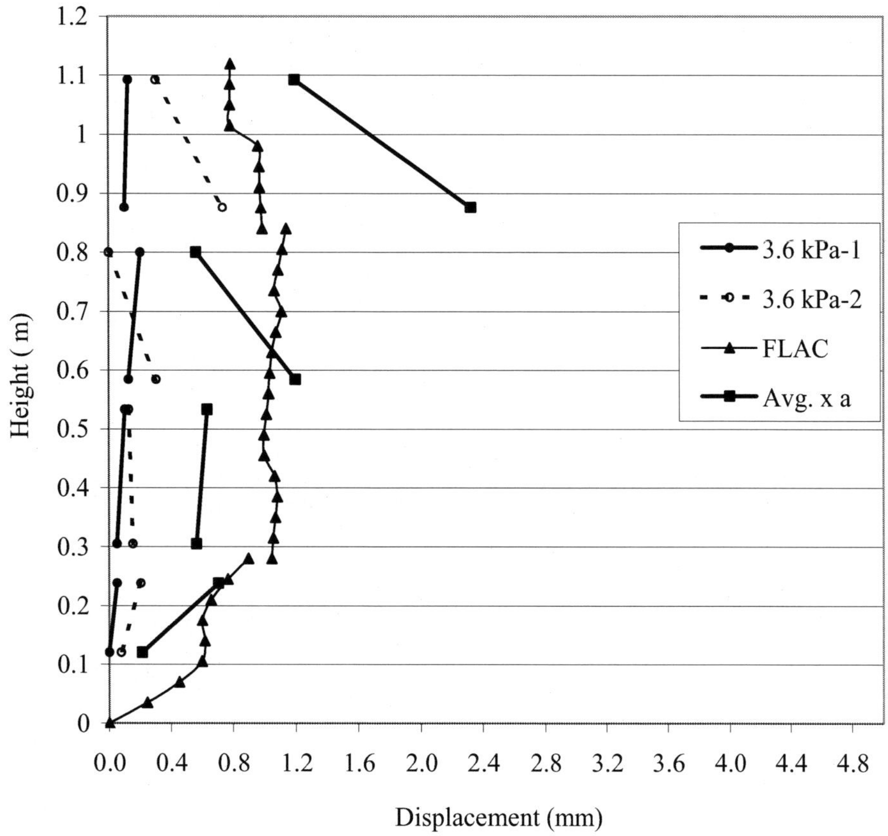

FIG. 5 PET MSE wall displacement, surcharge = 3.6 kPa

FULL SCALE TESTING

Additionally, a prototype full scale MSE wall was tested at FAU as shown in Figures 6 and 7. A preliminary comparison of the small-scale and the prototype MSE wall behavior, with results at 72 hours of testing for the prototype, indicated discrepancies due to scaling distortion related to the lack of gravity simulation and reduction of the geogrids properties for the small scale testing. The results should have been 5.5 times larger for the prototype if scaling the gravity was taken into account. In reality, the difference is much larger due to the lack of gravity and geogrids scaling as shown in Figure 8. The loading applied for the small scale testing presented was 0.8 kPa, compared to the large scale testing with 4.4 kPa, due to the scale effects based on the Buckingham Pi theorem (Kundu and Cohen, 1990). Figure 8 presents the results for the two small scale walls tested, and the results for the prototype.

Dancygier (1995) showed that when a centrifuge is not available, it is not possible to scale the gravitational effect. The loading in the wall is composed of two components: a) gravitational and b) non-gravitational. The gravitational component is the self-weight of the soil and the non-gravitational component refers to the external loadings applied to the top of the wall.

FIG. 6 Full Scale Wall Construction

FIG. 7 Full Scale Wall Testing

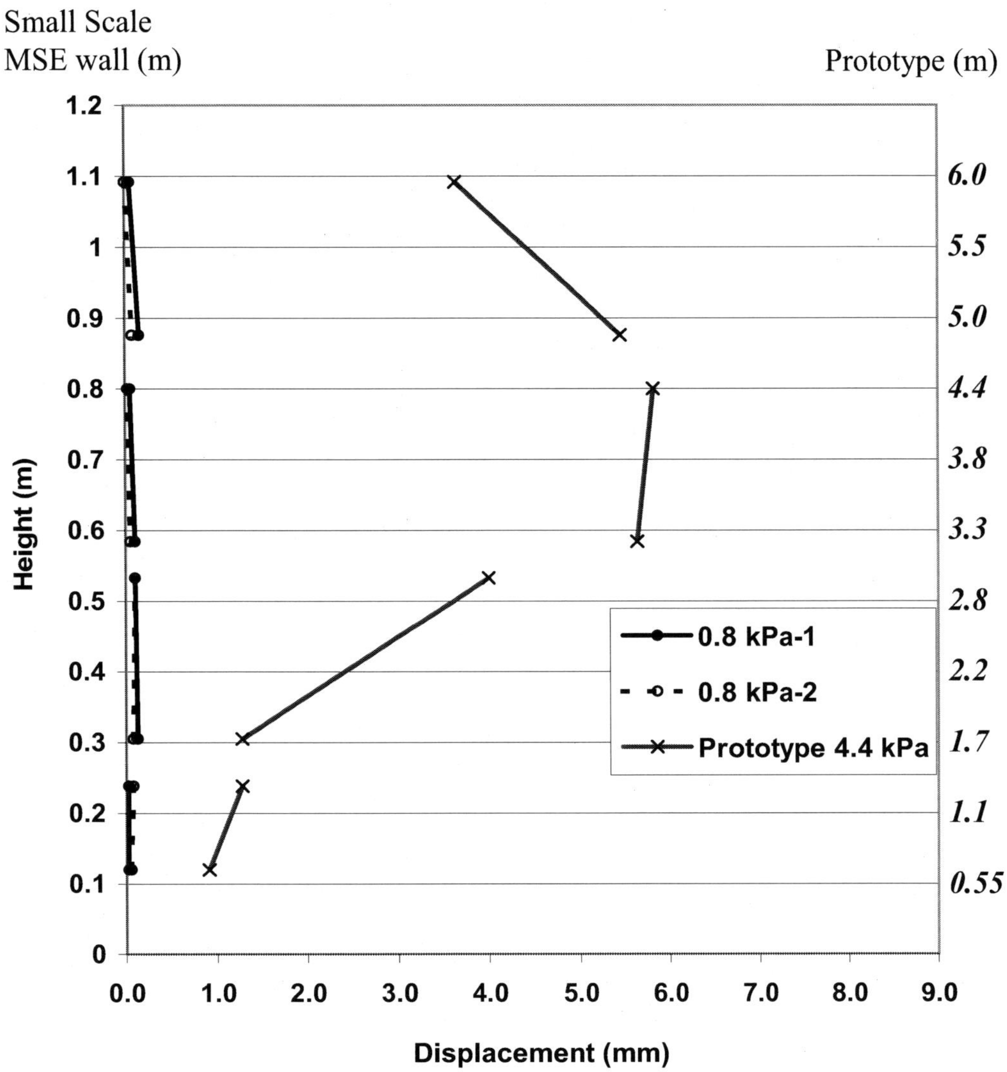

FIG. 8 Comparison between large and small scale MSE walls

CONCLUSIONS

The purpose of this paper is to present the small scale testing results. The small scale testing was part of a larger research project to use rational methods to enable the determination of applicable factors for use in LRFD design.

Results of the small-scale test indicate that, the lack of geogrid scaling affected both HDPE and PET geogrid reinforced MSE walls. The comparison between the small-scale MSE wall test results and the FLAC modeling showed the same trends, but the small scale testing values were smaller than those obtained by the model.

The differences between the test and the FLAC model values can be attributed mainly to the distortion created by the limitation in scaling the geogrid and to possible differences between the real values of the material properties, and the ones used for the FLAC modeling. This can be observed, as strains encountered in the geogrid were very small, even at loads above normal working conditions.

As presented in Figures 4 and 5 the average of the panel displacements for the two small scale walls tested for each type of geogrid was multiplied by factor "a". This was conducted to gain a better understanding of the distortion due to the lack of geogrid scaling. From this comparison we can observe that the general trend is similar. Panels 1, and 2 closely approach the result of the FLAC modeling once multiplied by factor "a". Panel 3 had smaller deformation while panel 4 showed deformation larger than any of the other panels.

Also, for both HDPE and PET walls, large displacements were observed at the bottom of the four panels (P4), with smaller displacements at the top. This indicates that the displacements were primarily due to rotations because of possible relatively small load dispersion at the top of the small scale wall, when applying large loads.

From Table 1 it can also be observed that the scaling distortion between the small scale test and the FLAC modeling had a larger effect on the small scale test when the surcharge load was closer to zero. As the load on the top of the wall increased, the distortion was reduced.

Additionally, comparison between the prototype MSE wall results at 72 hours and the small scale test should have shown results of the prototype being 5.5 times larger than the small-scale test results. However, the differences were much larger. This is a consequence of the lack of gravity simulation in the small-scale testing, as the loading in the wall is composed of two components: a) gravitational and b) non-gravitational.

Although small scale testing can be useful for obtaining preliminary tendency results, computer modeling and/or full prototype testing are still necessary to accurately determine the behavior of MSE walls. This will account for material property discrepancies and geogrid distortion for the computer modeling, and gravity simulation of full scale wall behavior. The limitation to accurately scale the geogrids due to the limited commercially available types of geogrids will also have considerable effects on the results of the small scale testing.

ACKNOWLEDGEMENTS

The authors gratefully acknowledge the Florida Department of Transportation (contract monitor: Mr. Peter Lai, Assistant State Geotechnical Engineer) for funding the project. The authors thank Dr. Alaa Ashmawy, formerly Associate Professor of Civil Engineering University of South Florida, now Dean of Engineering at American University of Dubai, and Ms. Aidee Cira, former Research Assistant, University of South Florida, for their valuable input with the FLAC modeling. Thanks are also due to the Tensar Earth Technologies, Inc. and the Reinforced Earth Company, Inc. for providing materials for the test program. Thanks are also due to the Consejo Nacional de Ciencia y Tecnologia (CONACYT) for its financial support to Dr. Fernando Navarrete.

REFERENCES

Dancygier A. N. (1995). "Quantitative Evaluation of Effect of Gravity on Small-scale Modeling, *Journal of Engineering Mechanics"*, Vol 121, No 7, pp 773-778.

Kundu P. K. and Cohen I. M., (1990). *"Fluid Mechanics*, Academic Press", Philadelphia, Pennsylvania.

In Situ Tests by Seismic Dilatometer (SDMT)

Silvano Marchetti[1], Paola Monaco[2], Gianfranco Totani[3] and Diego Marchetti[4]

[1]Professor, University of L'Aquila, Faculty of Engineering, DISAT, 67040 Monteluco di Roio, L'Aquila, Italy; marchetti@flashnet.it
[2]Assistant Professor, University of L'Aquila, Faculty of Engineering, DISAT, 67040 Monteluco di Roio, L'Aquila, Italy; p.monaco@ing.univaq.it
[3]Associate Professor, University of L'Aquila, Faculty of Engineering, DISAT, 67040 Monteluco di Roio, L'Aquila, Italy; totani@ing.univaq.it
[4]Studio Prof. Marchetti, Via Bracciano 38, 00189 Roma, Italy; diego@marchetti-dmt.it

ABSTRACT: The seismic dilatometer (SDMT) is the combination of the standard flat dilatometer (DMT) with a seismic module for measuring the shear wave velocity V_S. This paper summarizes the experience gained from SDMT tests performed at 34 sites. In particular the paper presents an overview of the SDMT equipment, comparisons of V_S measured by SDMT and by other methods, interrelationships G_0, E_D, M_{DMT} based on 2000 data points and a selection of significant SDMT results and related comments. The paper also illustrates the major issues of present research on use and applications of the SDMT, mostly focused on the development of methods for deriving the in situ G-γ decay curves and for evaluating the liquefaction resistance of sands based on SDMT results.

KEYWORDS: Seismic Dilatometer SDMT, Flat Dilatometer DMT, Shear Wave Velocity V_S, G-γ Curves, Working Strain Modulus, Liquefaction

INTRODUCTION

The seismic dilatometer (SDMT) is the combination of the traditional "mechanical" Flat Dilatometer (DMT) introduced by Marchetti (1980) with a seismic module placed above the DMT blade. The SDMT module is a probe outfitted with two receivers, spaced 0.5 m, for measuring the shear wave velocity V_S. From V_S the maximum shear modulus G_0 may be determined using the theory of elasticity.

Motivations of the combined probe:

- V_S (and G_0) are at the base of any seismic analysis.
- The G-γ decay curves are an increasingly requested input in seismic analyses and in general in non linear analyses.
- Increasing demand for liquefiability evaluations.
- Seismic site classification using directly V_S rather than N_{SPT} or c_u.

– Availability of the usual DMT results (e.g. constrained modulus M_{DMT}) for current design applications (e.g. conventional settlement predictions).

This paper comments on the most significant SDMT results obtained in the period 2004-2007 at 34 sites.

Information on the mechanical DMT, not described in this paper, can be found in the comprehensive report by the ISSMGE Technical Committee TC16 (2001).

THE SEISMIC DILATOMETER (SDMT)

The seismic dilatometer (SDMT) is the combination of the standard DMT equipment with a seismic module for measuring the shear wave velocity V_S.

The test is conceptually similar to the seismic cone SCPT. First introduced by Hepton (1988), the SDMT was subsequently improved at Georgia Tech, Atlanta, USA (Martin and Mayne 1997, 1998; Mayne et al. 1999). A new SDMT system (Fig. 1) has been recently developed in Italy. The seismic module (Fig. 1a) is a cylindrical element placed above the DMT blade, equipped with two receivers, spaced 0.5 m. The signal is amplified and digitized at depth. The *true-interval* test configuration with two receivers avoids possible inaccuracy in the determination of the "zero time" at the hammer impact, sometimes observed in the *pseudo-interval* one-receiver configuration. Moreover, the couple of seismograms recorded by the two receivers at a given test depth corresponds to the same hammer blow and not to different blows in sequence, which are not necessarily identical. Hence the repeatability of V_S measurements is considerably improved (observed V_S repeatability ≈ 1-2 %). V_S is obtained (Fig. 1b) as the ratio between the difference in distance between the source and the two receivers (S_2 - S_1) and the delay of the arrival of the impulse from the first to the second receiver (Δt). V_S measurements are obtained every 0.5 m of depth.

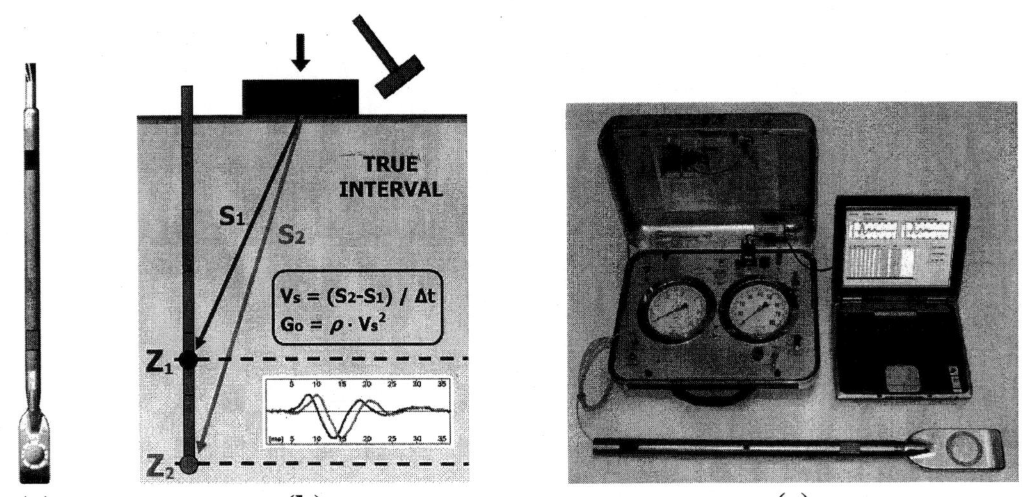

(a) (b) (c)

FIG. 1. Seismic Dilatometer: (a) DMT blade and seismic module. (b) Schematic layout of the seismic dilatometer test. (c) Seismic dilatometer equipment.

FIG. 2. **Shear wave source at the surface**

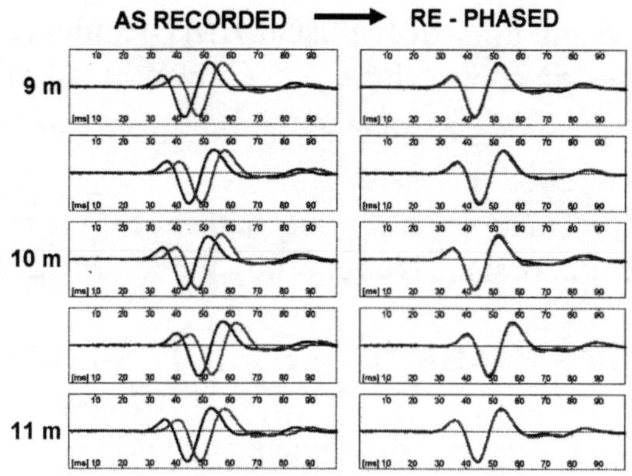

FIG. 3. Example of seismograms obtained by SDMT at the site of Fucino (Italy)

The shear wave source at the surface (Fig. 2) is a pendulum hammer (≈ 10 kg) which hits horizontally a steel rectangular base pressed vertically against the soil (by the weight of the truck) and oriented with its long axis parallel to the axis of the receivers, so that they can offer the highest sensitivity to the generated shear wave.

The determination of the delay from SDMT seismograms, normally carried out using the cross-correlation algorithm, is generally well conditioned being based on the two seismograms – in particular the initial waves – rather than being based on the first arrival time or specific marker points in the seismogram.

Fig. 3 shows an example of seismograms obtained by SDMT at various test depths at the site of Fucino (it is a good practice to plot side-by-side the seismograms as recorded and re-phased according to the calculated delay).

Fig. 4 (Fiumicino) is an example of the typical graphical format of the SDMT output. Such output displays the profile of V_S as well as the profiles of four basic

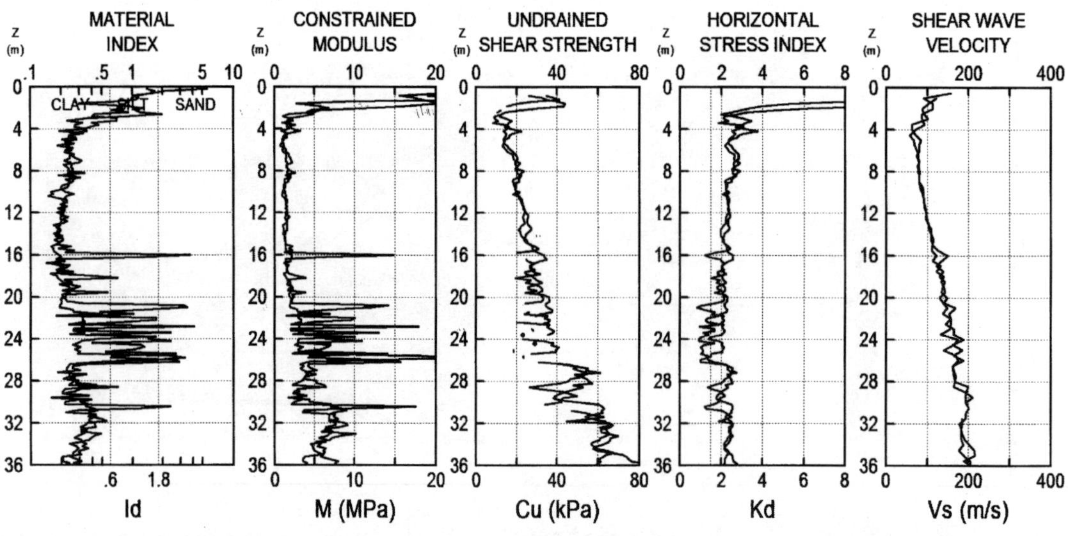

FIG. 4. SDMT profiles at the site of Fiumicino (Italy)

TABLE 1. Example of repeatability of V_S measurements by SDMT (Zelazny Most Tailing Dam, Poland)

Z [m]	V_S [m/s]	V_S values [m/s] corresponding to different hammer blows at each depth Z	Coefficient of variation [%]
7.00	179	178,178,180,180,180,179,179,180,180,180	0.50
7.50	231	234,232,232,230,229,231,232,229,230	0.68
8.00	225	227,225,224,225,225,225,226,226,225,224,224	0.40
8.50	276	276,276,280,273,275,273,271,273,287,281	1.68
9.00	296	291,286,301,292,296,288,301,300,304,303	2.09
9.50	248	244,251,250,247,250,249,250,249,242,248	1.11
10.00	292	292,289,290,293,289,292,289,292,296,295,293	0.79
10.50	320	321,323,320,325,323,325,316,314,308,321	1.61
11.00	291	293,291,293,291,291,290,290,291,290,290	0.38
11.50	321	324,320,320,322,320,322,319,319,320,320	0.48
12.00	309	311,307,311,309,309,311,309,309,307,311	0.50
12.50	286	287,285,285,285,287,285,285,287,287,287	0.35
13.00	265	265,265,265,264,265,265,265,266,265,266,264	0.24
13.50	280	287,276,279,276,276,276,294,275,278,279	2.08
14.00	312	313,312,312,322,310,312,310,310,310,312	1.10
14.50	298	301,298,299,299,298,296,299,298,299,298	0.44
15.00	309	307,309,307,309,309,309,309,309,309,309	0.29

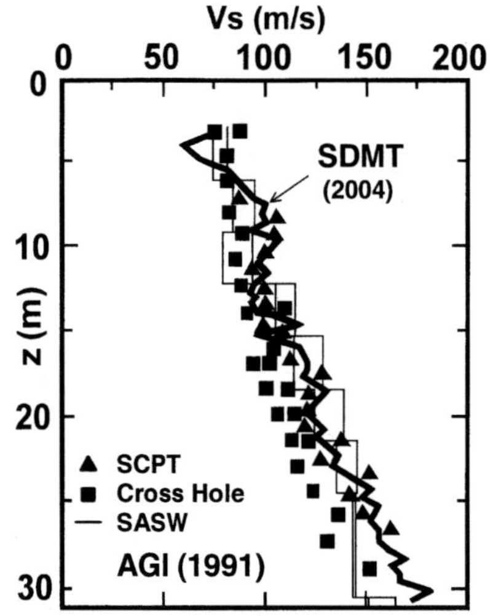

FIG. 5. Comparison of V_S profiles obtained by SDMT and by SCPT, Cross-Hole and SASW (AGI 1991) at the research site of Fucino (Italy)

DMT parameters – the material index I_D (soil type), the constrained modulus M, the undrained shear strength c_u and the horizontal stress index K_D (related to OCR) – obtained using current DMT correlations. It may be noted from Fig. 4 that the repeatability of the V_S profile is very high, similar to the repeatability of the other DMT parameters, if not better.

Table 1 shows an example of another kind of repeatability of V_S by SDMT (Zelazny Most Tailing Dam, Poland). Each V_S value at a given test depth corresponds to a different hammer blow. The coefficient of variation of V_S is in the range 1-2 %.

Such repeatability is more than adequate for normal engineering needs. However, especially in earthquake geotechnical engineering, where the design is increasingly based on the acceptability of the permanent displacements, accurate V_S (and G_0) estimates are a necessary prerequisite for correctly predicting such displacements.

V_S measurements by SDMT have been validated by comparison with V_S measurements obtained by other in situ seismic tests at various research sites. As an example Fig. 5 shows V_S comparisons at the research site of Fucino, Italy (NC cemented clay), extensively investigated at the end of the '80s. The profile of V_S obtained by SDMT in 2004 (Fig. 5) is in quite good agreement with V_S profiles obtained by SCPT, Cross-Hole and SASW in previous investigations (AGI 1991). Similar favourable comparisons are reported by various Authors, e.g. by Hepton (1988), McGillivray and Mayne (2004) and Młynarek et al. (2006).

INTERRELATIONSHIPS BETWEEN EXPERIMENTAL G_0, E_D, M_{DMT}

The experimental diagrams presented in this section have been constructed using same-depth G_0, E_D, M_{DMT} values determined by SDMT at 34 different sites, in a variety of soil types. The majority of the sites are in Italy, others are in Spain, Poland, Belgium and USA.

SDMT generates plentiful data points because each sounding routinely provides profiles of G_0, E_D, M_{DMT}. Of the over 2000 data points available, only 800 high quality data points have been considered, relative to "uniform" one-m soil intervals where log I_D, K_D, E_D, M_{DMT}, V_S all differ less than 30 % from their average – used then to plot the data points – to insure a proper match of the data.

The DMT parameters have been calculated with the usual DMT interpretation formulae as in Marchetti (1980) or Table 1 in TC16 (2001).

Diagrams of the Ratio G_0 / E_D

The ratio G_0/E_D are plotted versus K_D (Fig. 6a) and I_D (Fig. 6b). It can be seen that data points tend to group according to their I_D and K_D. From Fig. 6, if I_D and K_D are available, rough estimates of G_0 can be made from E_D, in absence of direct G_0 measurements.

Recognizable trends in Fig. 6 are: G_0/E_D is mostly in the range 1.5 to 3 in sand, 2.5 to 13 in silt, 3 to 25 in clay. The widest range and the maximum variability of G_0/E_D are found in clay. For all soils G_0/E_D decreases as K_D (related to OCR) increases.

Diagrams of the Ratio G_0 / M_{DMT}

The ratio G_0 / M_{DMT} are plotted versus K_D (Fig. 7a) and I_D (Fig. 7b). The diagrams indicate a wide range of the ratio G_0 / M_{DMT} ($\approx$ 0.5 to 25 for all soils), hence the unfeasibility of estimating the operative modulus M from G_0 by dividing G_0 for a fixed number. Again the data points tend to group according to their I_D and K_D. It can be seen that G_0 / M_{DMT} is strongly dependent on (at least) both soil type and stress history. Hence the use of only one parameter (e.g. c_u in cohesive soils) as an estimate for V_S (or G_0) for the seismic soil classification appears problematic.

Recognizable trends in Fig. 7 are: G_0/M_{DMT} is mostly in the range 0.5 to 3 in sand, 1 to 10 in silt, 1 to 20 in clay. The widest range and the maximum variability of G_0 / M_{DMT} are found in clay. For all soils G_0 / M_{DMT} decreases as K_D (related to OCR) increases.

If I_D and K_D are available, Fig. 7, where the dispersion is slightly less than in Fig. 6, is to be preferred to Fig. 6 to obtain rough estimates of the ratio G_0/M, i.e. G_0 from M or M from G_0, when only one of them is available.

Diagrams of the Ratio G_{DMT} / G_0

Fig. 8 shows the same experimental information as in Figures 6 and 7, but involves the additional modulus G_{DMT} derived from M_{DMT} using the formula of linear elasticity:

$$G = M / [2\,(1\text{-}\nu) / (1\text{-}2\nu)] \qquad (1)$$

For ν = 0.15, 0.20 and 0.25, the denominator in Eq. 1 would be 2.43, 2.67 and 3 respectively. For the conversion, the denominator 2.67 has been retained, i.e.:

$$G_{DMT} = M_{DMT} / 2.67 \qquad (2)$$

All the G_{DMT} have been derived from M_{DMT} using Eq. 2, then the ratios G_{DMT}/G_0 have been calculated too and plotted versus K_D (Fig. 8a) and I_D (Fig. 8b).

The reason of constructing Fig. 8 is the following. The ratio G/G_0 is the usual ordinate of the normalized G-γ decay curve and has the meaning of a strain decay factor. Since M_{DMT} is a *working strain modulus* one might hypothesize that G_{DMT} is a *working strain shear modulus* too, in which case G_{DMT}/G_0 could be regarded as the shear modulus decay factor at *working strains*.

It is emphasized that, at this stage, the legitimacy of using linear elasticity for deriving G_{DMT} from M_{DMT} (Eq. 2) and the assumption that G_{DMT} is a *working strain shear modulus* are only working hypotheses, likely more difficult to investigate than verifying that M_{DMT} is a *working strain constrained modulus* (the matter is discussed later in the paper). The very designation *working strain shear modulus* (or operative shear modulus) requires clarification. Anyway, if the above hypotheses were acceptable, Fig. 8 could provide, if I_D and K_D are known, rough estimates of the decay factor at *working strains*. If complete SDMT are available, then said rough estimates could be skipped and the decay factor could be obtained directly as the ratio between G_{DMT} from Eq. 2 and G_0.

Trends emerging from Fig. 8 are: (a) The G decay in sands is much less than in silts and clays. (b) The silt and clay decay curves are very similar. (c) For all soils the decay is maximum in the NC or lightly OC region (low K_D).

Fig. 7a and Fig. 8a are reproduced in a more readable format in Fig. 9 and Fig. 10 respectively, where the data points relative to clay, silt and sand have been plotted separately. Best fit equations are indicated for each of the six diagrams.

IN SITU G-γ DECAY CURVES BY SDMT

SDMT provides routinely at each depth, besides a *small strain modulus* (G_0 from V_S), also a *working strain modulus* (M_{DMT}). These two moduli could possibly be of help when selecting the G-γ curves. Such potentiality is heavily founded on the basic premise that M_{DMT} is a reasonable estimate of the operative *working strain modulus*. It is therefore considered appropriate to recall here the presently available evidence.

M_{DMT} as an Operative or Working Strain Modulus

The terms Operative Modulus and Working Strain Modulus are considered synonyms in this paper and are used interchangeably. They are defined as those moduli that, introduced into the linear elasticity formulae, provide reasonable estimates of the settlements under a shallow foundation (say for a safety factor Fs ≈ 2.5 to 3.5).

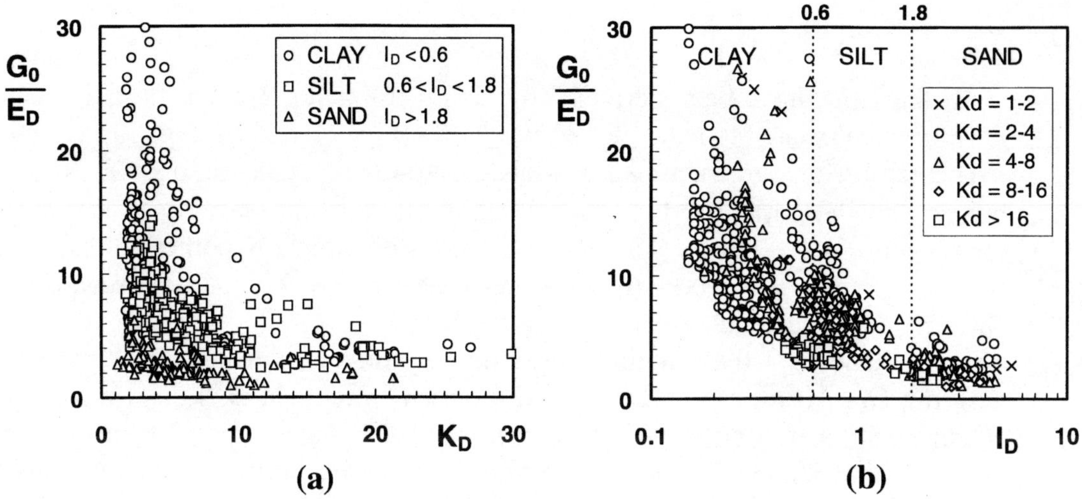

FIG. 6. Ratio G_0/E_D vs. K_D and I_D

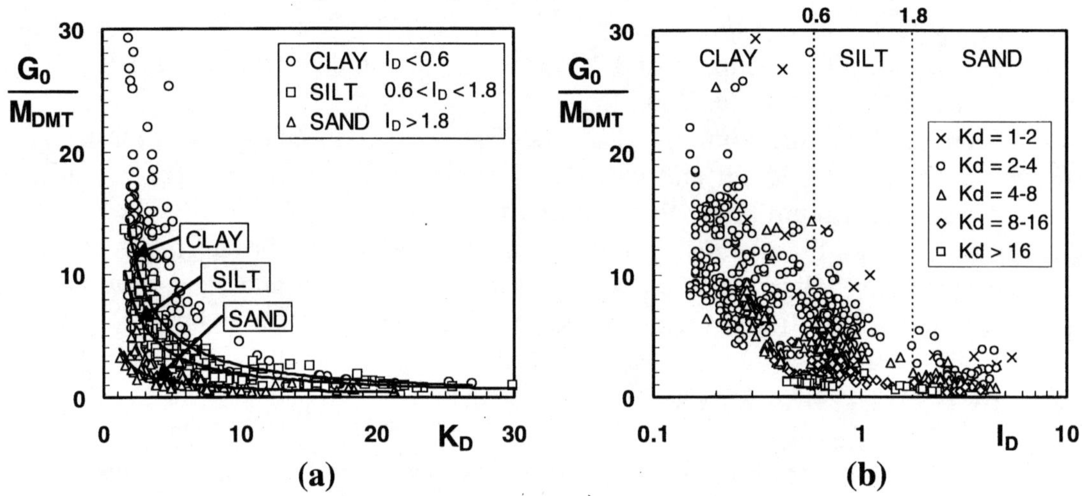

FIG. 7. Ratio G_0/M_{DMT} vs. K_D and I_D

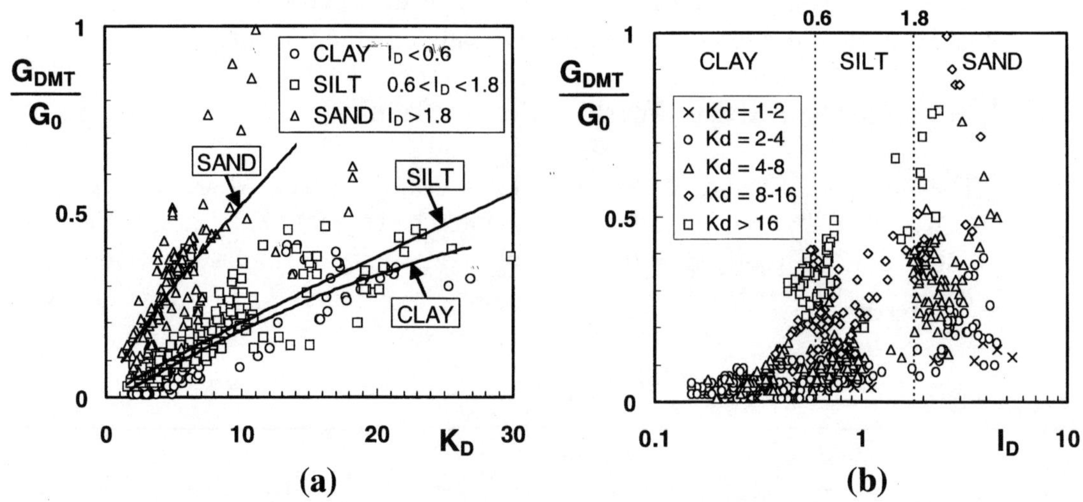

FIG. 8. Decay ratio G_{DMT}/G_0 vs. K_D and I_D

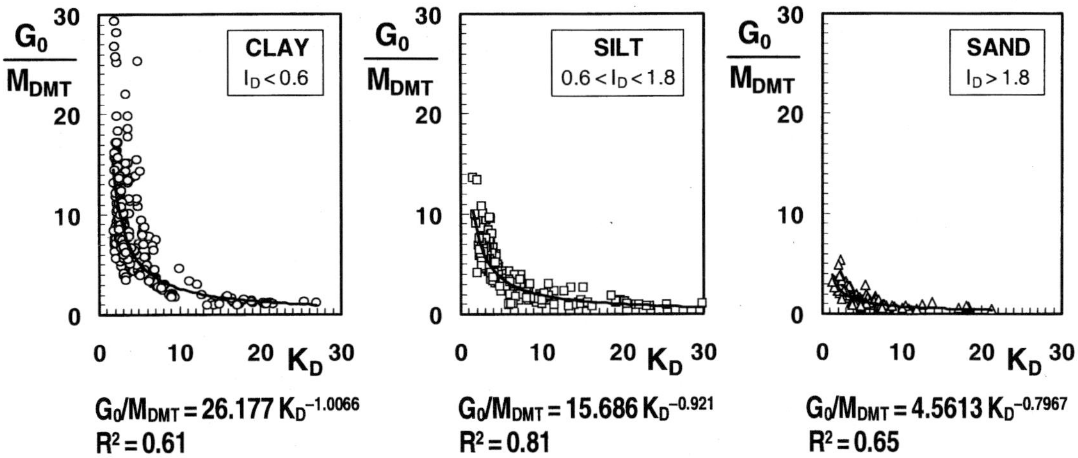

FIG. 9. Ratio G_0/M_{DMT} vs. K_D for clay, silt and sand

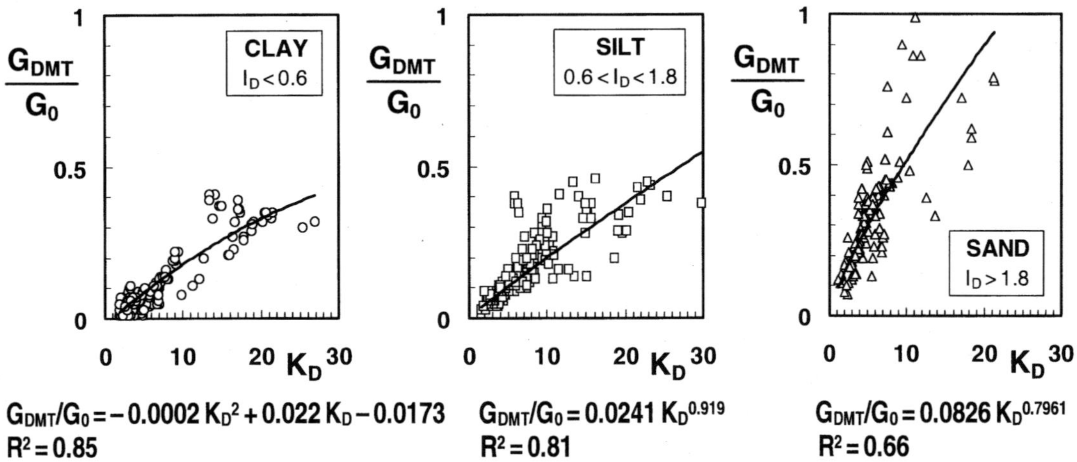

FIG. 10. Decay ratio G_{DMT}/G_0 vs. K_D for clay, silt and sand

Comparisons of Surface Settlements

Schmertmann (1986) reported 16 case histories at various locations and for various soil types, with measured settlements ranging from 3 to 2850 mm. In most cases settlements from DMT were calculated using the Ordinary 1-D Method. The average ratio DMT-calculated/observed settlement was 1.18, with the value of the ratio mostly in the range 0.7 to 1.3 and a standard deviation of 0.38.

Monaco et al. (2006) reviewed numerous other real-life well documented comparisons of DMT-predicted versus measured settlements. The average ratio DMT-calculated/observed settlement for all cases reviewed by Monaco et al. (2006) is ≈ 1.3, with an observed settlement within ± 50 % from the DMT-predicted settlement.

The above settlements comparisons appear to support the assumption that M_{DMT} is a reasonable estimate of the constrained *working strain modulus*.

Comparisons of Moduli

Even more direct, but rarely available, are data comparing M_{DMT} with moduli back-figured from local vertical strain measurements.

In 2002 a major research project, funded by the Italian Ministry of University and Scientific Research and by Consorzio Venezia Nuova, was undertaken by a consortium of three Italian Universities (Padova, Bologna and L'Aquila).

A full-scale cylindrical heavily instrumented test embankment (40 m diameter, 6.7 m height, applied load 104 kPa – Fig. 11a) was constructed at the site of Venezia-Treporti, typical of the highly stratified, predominantly silty deposits of the Venezia lagoon (Fig. 11b). The loading history, the progression of the settlements and the drainage conditions – practically fully drained – are shown in Fig. 11c.

A specific aim of the research was to obtain a profile of the observed 1-D operative modulus M under the center of the embankment. For this purpose a high precision sliding micrometer was used to accurately measure the *local* vertical strain ε_v at 1 m depth intervals.

M values were back-calculated from local vertical strains ε_v in each 1 m soil layer as $M=\Delta\sigma_v/\varepsilon_v$, with vertical stress increments $\Delta\sigma_v$ calculated at the mid-height of each layer by linear elasticity formulae (approximation considered acceptable in view of the very low ε_h).

Fig. 11d, which is believed to be one of the *most important results* of the Venezia-Treporti research, shows an overall satisfactory agreement between M_{DMT} and moduli back-figured from the test embankment performance, also considering the marked soil heterogeneity. Fig. 11e compares the observed versus DMT-predicted settlements at each depth. Again the agreement is rather satisfactory, considering that the DMT predicted settlements were calculated using the simple linear 1-D conventional approach $s=\Sigma(\Delta\sigma_v/M_{DMT})\Delta H$, where $\Delta\sigma_v$ is calculated by Boussinesq linear elasticity formulae.

As to the surface settlements, the total settlement measured under the center of the embankment at the end of construction (180 days) was $\approx$ 36 cm (Fig. 11c). The settlement predicted by M_{DMT} using the 1-D approach (before knowing the results) was 29 cm. Hence the 29 cm predicted by DMT (which does not include secondary) are in good agreement with the 36 cm observed settlement (which includes some secondary during construction).

(Cunha (2004) wondered if it is right to attribute the 7 cm settlement difference (20 %) to secondary or to other causes. The question is legitimate. But in practical circles settlement predictions within a factor of 1.5-2 are often considered satisfactory for normal design. Mayne (2005) reports an SPT-DMT settlement comparison where the SPT prediction was wrong by a factor of 5 (the DMT prediction "was in line with the observed performance"). Factors of 3 or 5 are not unusual. In general, engineering decisions on the foundation type are based on knowing if the foundation will settle 1 mm or 10 mm or 100 mm. While it is desirable to reduce the error factor below 5 or 3, once it is reduced to 1.5 to 1.2, additional efforts may not be worthwhile or even inherently unfruitful due the many uncertainties involved in the definition of the settlement components and in their precise measurement).

(a) Test embankment. Penetrometer truck for testing after construction.

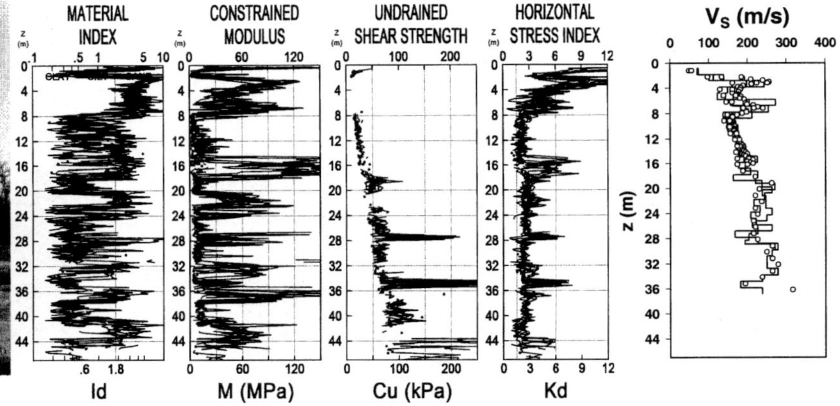

(b) Superimposed profiles of all SDMT data

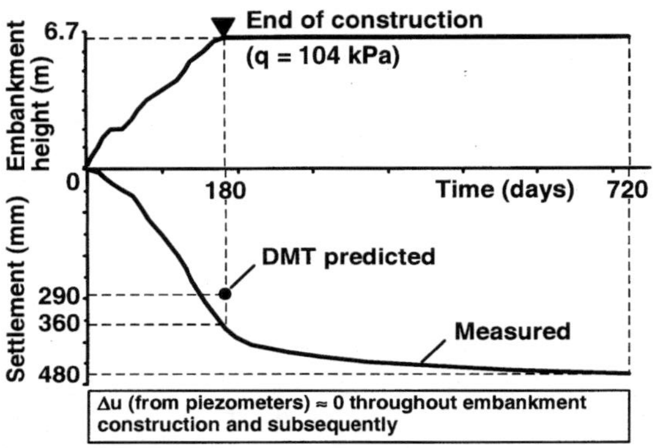

(c) Settlement vs. time at the center of the embankment and comparison of measured vs. DMT-predicted settlements at the end of construction

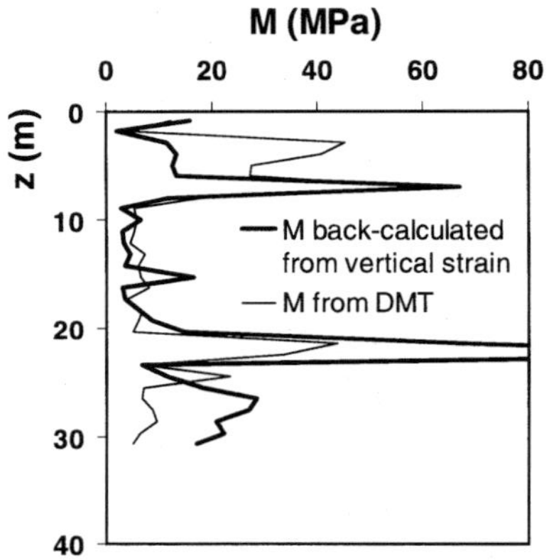

(d) M_{DMT} vs. M back-calculated from local ε_v measured at 1 m depth intervals under the center at the end of construction

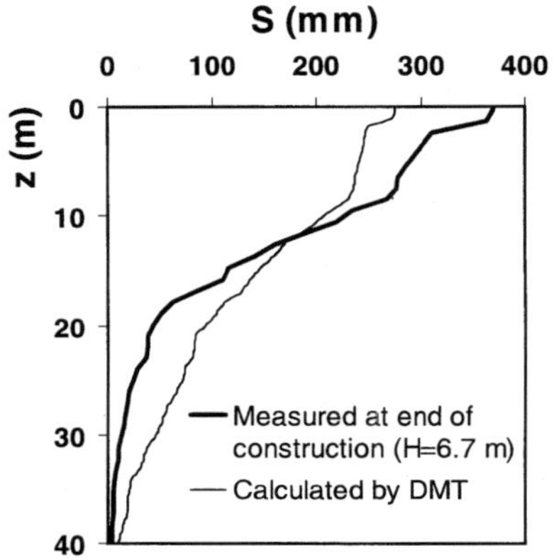

(e) Observed vs. DMT-predicted settlement under the center at the end of construction

FIG. 11. Venezia-Treporti Research Embankment. SDMT profiles. Predicted vs. observed moduli and settlements (Marchetti et al. 2006).

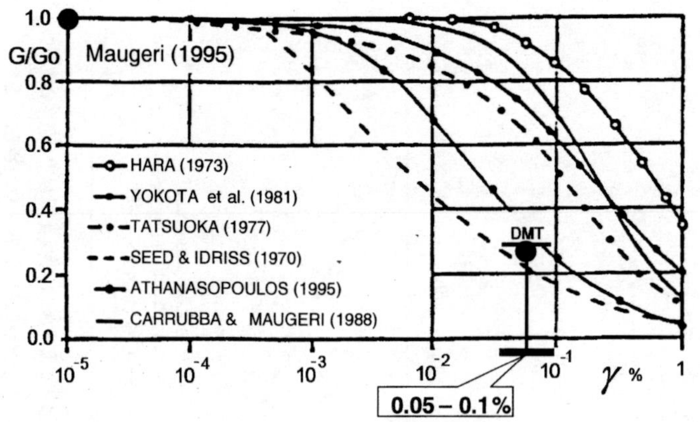

FIG. 12. Tentative method for deriving G-γ curves from SDMT

More details on the Venezia-Treporti research can be found in Marchetti et al. (2006), also containing numerous additional bibliographic references.

In conclusion also the Venezia-Treporti case-history supported the assumption that M_{DMT} is a reasonable estimate of the constrained *working strain modulus*.

A note of caution. In OC clays oedometer moduli, even from good quality samples, are generally too low to be taken as operative moduli in situ. As noted by Ladd (1971), quoting Terzaghi-Peck (1964), the compressibility of even good oedometer samples of OC clay may be 2 to 5 times larger than the in situ compressibility. Indeed many engineers feel that the Skempton-Bjerrum reduction factor for settlements in OC clays was prompted by the fact that oedometer-based settlement predictions in OC London clay were systematically too high. Moduli back-figured from local strain measurements are far more realistic and are preferable for calibration or comparison purposes.

Construction of the G-γ Curves

Such curves could tentatively be constructed by fitting "reference typical-shape" laboratory curves (see Fig. 12, where G is normalized to G_0) through two points, both obtained by SDMT: (1) the *initial modulus* G_0 from V_S, and (2) a *working strain modulus* G_{DMT} (Eq. 2). To locate the second point it is also necessary to know, at least approximately, the shear strain corresponding to G_{DMT}. Indications by Mayne (2001) locate the DMT moduli at an intermediate level of strain ($\gamma \approx 0.05$-$0.1\,\%$) along the G-γ curve. Similarly Ishihara (2001) classified the DMT within the group of methods of measurement of soil deformation characteristics involving an intermediate level of strain (0.01-1 %). The above indications, to be supplemented by further investigations, could possibly help develop methods for deriving in situ G-γ curves from SDMT.

Lines of research currently under investigation are:

(a) Enter the G_{DMT}/G_0 ratios of Fig. 8 in the vertical axis of "reference typical-shape" G-γ curves recommended in the literature for the corresponding soil type. The range of abscissas of the intersection points with the G-γ curves could possibly help to better define the shear strain corresponding to G_{DMT}.

(b) Develop a procedure for selecting the G-γ curve, among the typical curves recommended in the literature, making use of I_D for choosing the band of curves recommended for the soil type (sand or silt or clay), and K_D (possibly G_0/M_{DMT} too) for selecting one curve in the band.

(c) Evaluate for each of the 800 data points in Figures 6-7-8 the settlement under a simple loading scheme using the simple linear analysis with input M_{DMT} (operation equivalent to converting a DMT investigation into a "virtual" load test). Then calculate the settlement by non linear analyses with G-γ curves having variable rates of decay as input. By trial and error identify the G-γ curve (originating in G_0) producing agreement between the two predicted settlements. Consider such G-γ curve reasonably correct and use it in the development of procedures for selecting the G-γ curves from SDMT data.

DERIVABILITY OF THE OPERATIVE MODULUS M FROM G_0

This section discusses the possibility of deriving the operative modulus M by dividing G_0 by a constant.

Fig. 13 (Barcelona airport site) shows that, while the modulus M_{DMT} exhibits a drastic drop at ≈ 12 m depth, at the transition from an upper stiff sand layer to a lower very soft clay layer, V_S shows only a minor decrease. Hence $G_0 = \rho\ V_S{}^2$, even considering the power 2, is far from being proportional to the *working strain modulus* M. Similar lack of proportionality, with variations of the ratio G_0/M_{DMT} often of one order of magnitude, has been observed at many sites (including Venezia, Fig. 11d), suggesting that it is next to impossible (at least without local layer-specific correlations) to derive the *working strain modulus* by simply reducing the *small strain modulus* by a fixed percent factor (e.g. 50 %, Simpson 1999).

On the other hand the poor correlability was expected, since at low strains the soil tendency to dilate or contract is not active yet. Such tendency substantially affects the operative modulus M, but does not affect G_0. Said in a different way, M includes some stress history information, G_0 does not (Powell and Butcher 2004).

The high variability of the ratio G_0/M was already noted earlier in Fig. 7. That figure explains that part of the difference in the amount of decrease in the Barcelona M and V_S profiles is due to the transition from sand (where G_0/M_{DMT} is in the range 0.5-3) to clay (where G_0/M_{DMT} is in the range 1-20). Fig. 7 (or the equivalent Fig. 9), entered with I_D and K_D, can provide rough estimates of the ratio, hence M from G_0. However the direct measurements of both M and G_0 are preferable if accurate estimates of these parameters are required.

ESTIMATING OCR IN SAND FOR ADVANCED PREDICTIVE MODELS

Computer programs based on advanced soil models are increasingly used not only by researchers but also by practitioners for everyday design. Hence an important aim of in situ tests is to possibly provide input parameters for such models.

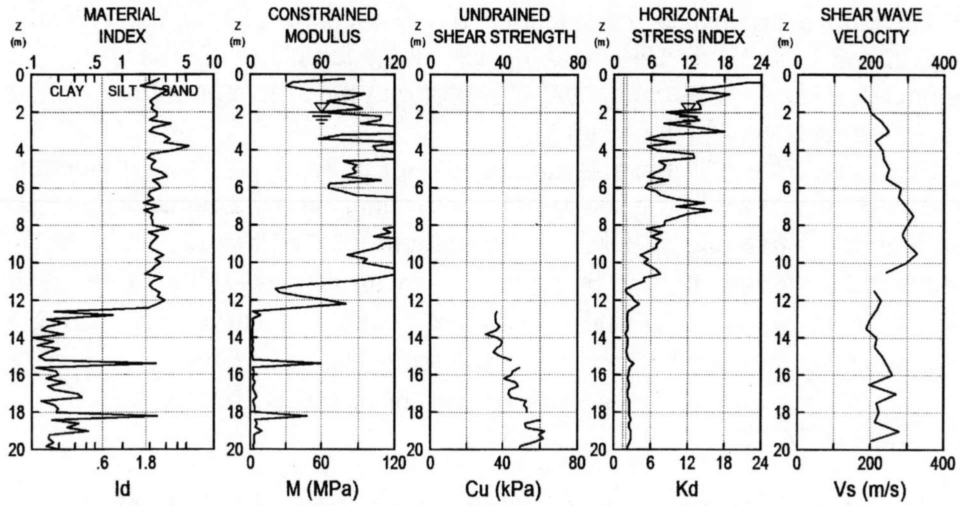

FIG. 13. SDMT profiles at the site of Barcelona – El Prat Airport (Spain)

An input parameter required by many models is OCR. In clays reasonable OCR estimates can be obtained by laboratory samples, or in situ sometimes by the DMT OCR-K_D correlations. In *sands*, however, OCR evaluations are problematic. As today, to the authors' knowledge, one of the least imprecise methods for evaluating OCR in sand by in situ tests is the one suggested by TC16 (2001). Such method consists in evaluating OCR based on the ratio M_{DMT}/q_c (q_c from CPT) using as a guide: $M_{DMT}/q_c = 5\text{-}10$ in NC sands, $M_{DMT}/q_c = 12\text{-}24$ in OC sands. The basis of such ratios is explained in TC16 (2001). In essence: when imparting stress history to the sand – for instance by compaction – M_{DMT} increases considerably faster than q_c (see e.g. Schmertmann et al. 1986, Jendeby 1992), hence M_{DMT}/q_c increases with OCR.

The above guide provides very rough estimates of OCR. The uncertainty however may not be due necessarily to the approximate nature of the correlations. Many sands could simply not have a well defined break in their insitu *e*-log *p* curve (Fig. 14) due to the complexity of the OCR sources (cementation, desiccation, water level fluctuations, etc.). Besides, layers of sands are often stratified, as the sand layers in Fig. 11b, with OCR presumably seesawing, so that the *break* in the *e*-log *p* of the composite layer will be rounded, and assigning a unique value to OCR of the layer could be inappropriate.

Uncertainty in the values of OCR represents a problem for many models requiring a precise value of OCR in *sand* because their predictions are generally critically sensitive to OCR. For instance in the case of Venezia-Treporti mentioned before, using advanced models with OCR = 1.2 or OCR = 1.6 leads to completely different settlement predictions, while the conventional analysis using the tangent modulus M_{DMT} produced a stable and satisfactory prediction (Fig. 11c).

The cause of the advanced model sensitivity can be understood even with the simple oedometer scheme. If a sand has an in situ *e*-log *p* curve of the type in Fig. 14b, and is modeled as in Fig. 14a, the predicted settlements will be highly dependent on the choice of the abscissa of the (in reality non existing) break.

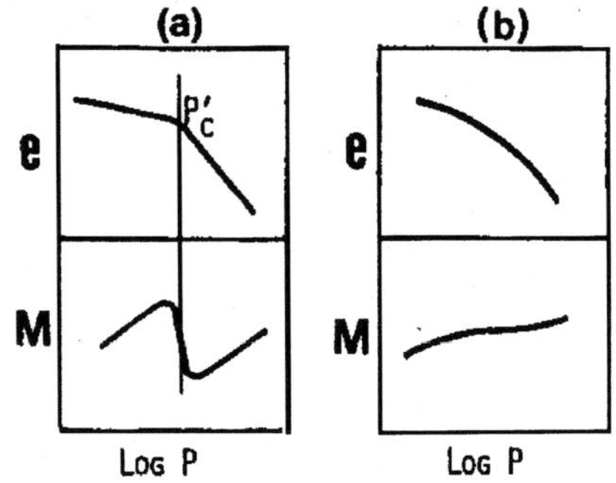

FIG. 14. Variation of e and M with p in presence/absence of a well defined break

Therefore *sands* whose in situ e-log p curve are of the type in Fig. 14b should probably be treated with models *attenuating* the stiffness discontinuity across the preconsolidation stress.

In short, the problem could be not so much the determination of the exact OCR in sand, but the possibly incorrect assumption of the model that the sand has a sharp break.

It is generally recommended to run both advanced and traditional analyses to avoid gross errors that are likely if one is not familiar with the details of the program. This recommendation appears appropriate in this case, because, at least in the described Venezia-Treporti case, the simplified analysis using M produced a stable and satisfactory prediction, while the "advanced" analysis was found to be critically dependent on the possibly elusive value of OCR.

USE OF SDMT FOR LIQUEFACTION

SDMT routinely provides, among other measurements, pairs of profiles of K_D and V_S – both correlated with the liquefaction resistance of sands. Hence SDMT permits to obtain two parallel independent estimates of liquefaction resistance CRR, one from K_D and one from V_S, using CRR-K_D and CRR-V_S correlations – where CRR is the cyclic resistance ratio, a basic input in the commonly used Seed and Idriss (1971) simplified procedure.

The use of V_S for evaluating CRR is well known. The most popular CRR-V_S correlation (Fig. 15) is the one proposed by Andrus and Stokoe (2000), modified by Andrus et al. (2004). CRR is obtained as a function of $V_{S1} = V_S\,(p_a/\sigma'_{v0})^{0.25}$, shear wave velocity corrected for the overburden stress σ'_{v0} (p_a = atmospheric pressure). The CRR-V_{S1} curves in Fig. 15 are for magnitude $M_w = 7.5$ earthquakes (magnitude scaling factors should be applied for different magnitudes).

Correlations CRR-K_D have been developed in the last two decades, stimulated by the recognized sensitivity of K_D to a number of factors which are known to increase liquefaction resistance, which are difficult to sense by other tests, such as stress history, prestraining, cementation, structure, and by the relationship of K_D to relative density and state parameter.

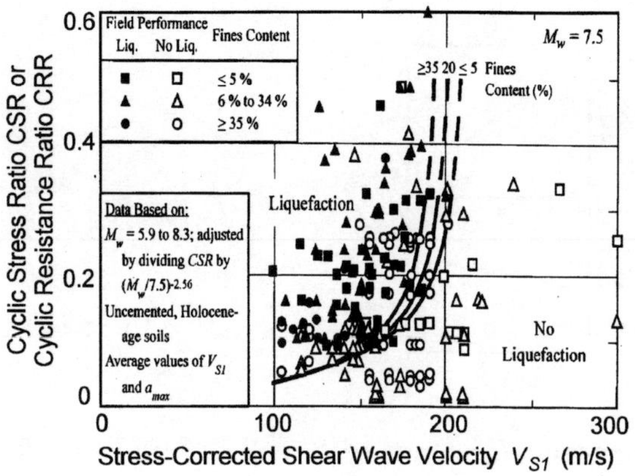

FIG. 15. Curves for evaluating CRR from V_S for clean uncemented soils (Andrus and Stokoe 2000)

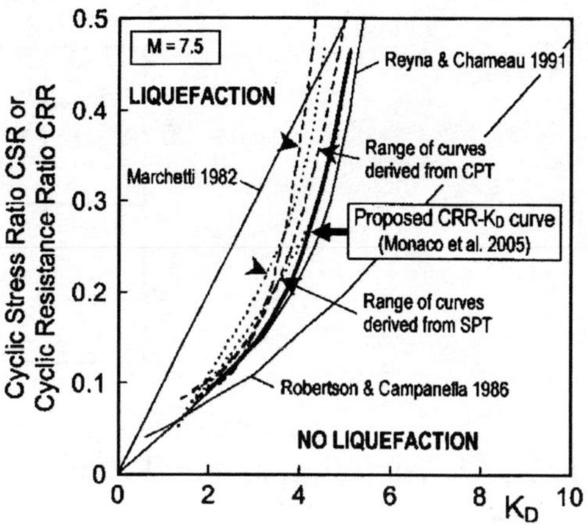

FIG. 16. Curves for evaluating CRR from K_D (Monaco et al. 2005)

A key element of the correlation CRR-K_D (Monaco and Schmertmann 2007, Monaco and Marchetti 2007) is the ability of K_D to reflect *aging* in sands. Calibration chamber data (Jamiolkowski and Lo Presti 1998) suggest that sensitivity to aging of sand is 3 to 7 times higher for K_D than for penetration resistance (see Fig. 6 in Monaco and Marchetti 2007). *Aging* in sand is a factor having a first order of magnitude influence on liquefaction behavior, as pointed out e.g. by Leon et al. (2006).

Fig. 16 summarizes the various correlations developed to estimate CRR from K_D (for magnitude M = 7.5 and clean sand) – to be used according to "simplified procedure" – including the latest CRR-K_D correlation (Monaco et al. 2005), based on all previous data.

Comparisons based on parallel measurements of K_D and V_S by SDMT at several sandy sites (Maugeri and Monaco 2006) have indicated that methods based on K_D and V_S often provide, at the same site, substantially different estimates of CRR. Generally CRR from V_S was found to be "more optimistic".

This finding opens the question "which CRR should be given greater weight", which is further discussed in the next section.

SDMT RESULTS AT VARIOUS TEST SITES

This section presents three examples of SDMT results considered of some interest.

OCR and K_D Crusts in Sand

"Crust-like" K_D profiles, very similar to the typical K_D profiles found in OC desiccation crusts in clay, have been found at the top of many sand deposits. Various indications (Maugeri and Monaco 2006) suggest that "K_D crusts" in sands reflect stress history (OCR, cementation, aging and/or other effects), rather than higher relative density. In the case shown in Fig. 17 (Catania), as in many other cases, the existence of a shallow *"stress history crust"* (believed by far not liquefiable) is clearly highlighted by the K_D profile, but almost "unfelt" by the V_S profile. This suggests a lesser ability of V_S to profile liquefiability.

Role of the Interparticle Bonding

The SDMT profiles in Fig. 18 (Cassino) show relatively high V_S values coexisting with very low values of K_D and moduli M. A possible explanation: the shear wave travels fast due to the interparticle bonding (typical of many volcanic sands in this area), preserved at small strains. By contrast K_D is "low" because it reflects a different material, where the interparticle bonding has been at least partly destroyed by the DMT blade penetration. As noted by Andrus and Stokoe (2000), weak interparticle bonding can increase V_S (measured at small strains), while not necessarily increasing resistance to liquefaction, a phenomenon occurring at medium to high strains (range of K_D measurement). Thus, for liquefiability, the K_D predictions could be possibly more fitting in case of strong earthquakes. Very light earthquakes, however, may not destroy bonding, then CRR evaluated by V_S may be appropriate in this case.

Limiting "No Liquefaction" Values of V_{S1} and K_D

The vertical asymptotes of the curves CRR-V_{S1} (Fig. 15) and CRR-K_D (Fig. 16) identify limiting values of V_{S1} and K_D (roughly, for clean sand, $V^*_{S1} = 215$ m/s and $K^*_D = 5.5$) for which liquefaction can be definitely excluded for any earthquake.

At the site of the Zelazny Most dam (Fig. 19) the indications derivable from such vertical asymptotes put in evidence a clear contradiction. While the values of $V_{S1} > 215$ m/s suggest "no liquefaction" even in case of strong earthquakes, the values of $K_D \approx 1.5$-2 indicate that liquefaction may occur above a certain seismic stress level (high cyclic stress ratio CSR). Even in this case, as noted before, CRR from V_S is "more optimistic" – one reason being that it partly relies on bonding or equivalent phenomena.

In the case of the Zelazny Most dam the contradiction has no practical effects, since the region is non seismic and both Fig. 15 and Fig. 16 indicate no liquefaction. But for high seismicity cases the question "which CRR should be given greater weight" remains open.

Figures 17-18-19 are also examples of a commonly noted feature: the much smoother shape of the V_S profiles compared with the M or K_D profiles.

SDMT INSIDE BACKFILLED BOREHOLES

In cases where the soil is too hard to penetrate (or even in rock), SDMT can be carried out inside a borehole backfilled with sand (only V_S, no DMT measurements).

The good agreement observed between V_S profiles obtained by parallel SDMT soundings carried out, at the same site, in the natural soil and in a backfilled borehole (Fig. 20) supports the reliability of V_S values obtained by this procedure.

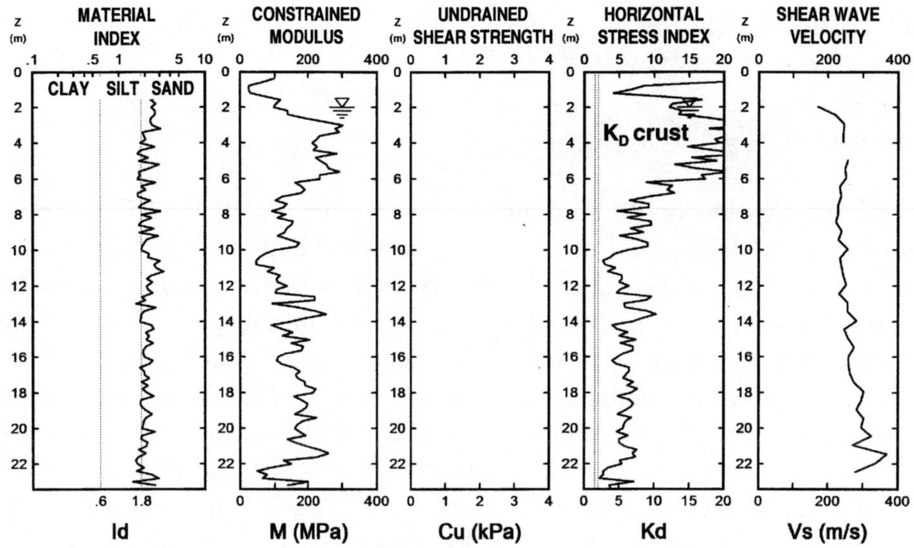

FIG. 17. SDMT profiles at the site of Catania – San Giuseppe La Rena (Italy)

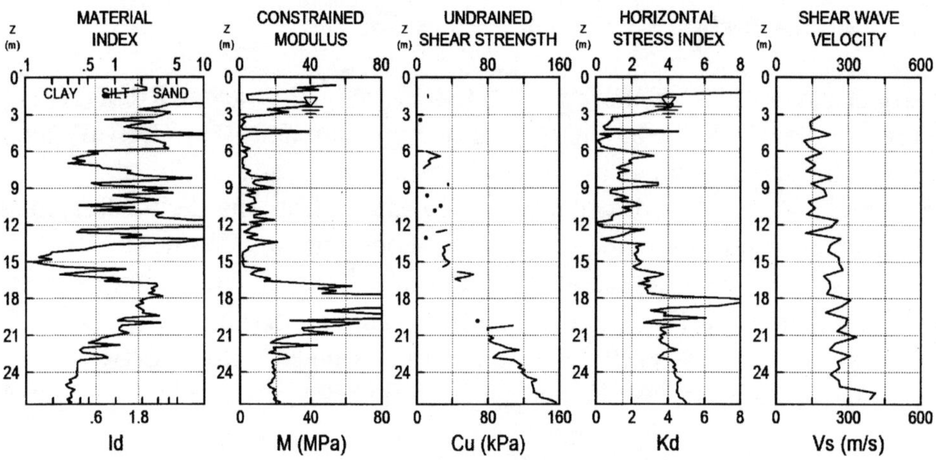

FIG. 18. SDMT profiles at the site of Cassino (Italy)

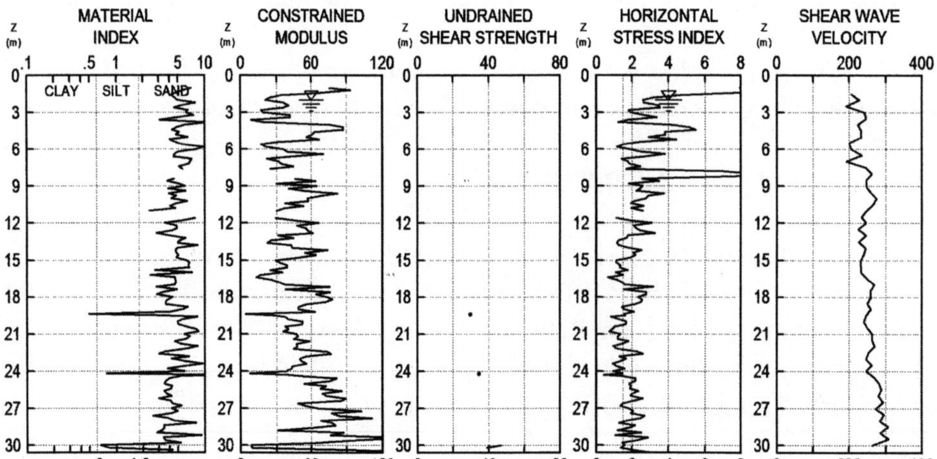

FIG. 19. SDMT profiles at the site of the Zelazny Most Tailing Dam (Poland)

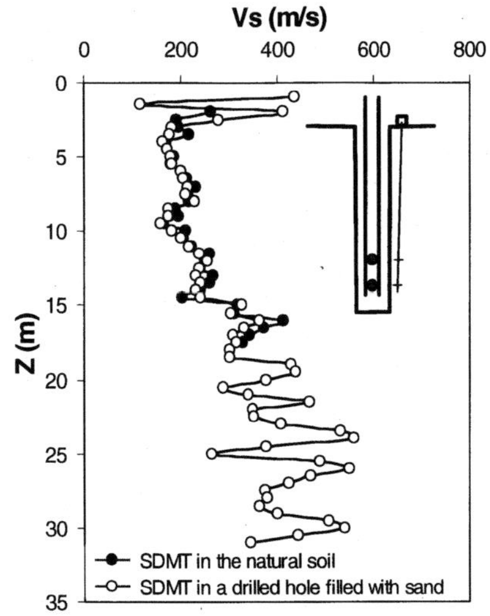

FIG. 20. Comparison of V_S profiles obtained by SDMT in the natural soil and in a backfilled borehole at the site of Montescaglioso – Ginosa (Matera), Italy

CONCLUSIONS

The seismic dilatometer (SDMT) provides accurate and highly reproducible measurements of the shear wave velocity V_S – a basic input parameter for seismic analyses. Besides V_S, SDMT provides the usual DMT results (e.g. constrained modulus M_{DMT}) for current design applications.

Based on a large number of results by SDMT, diagrams showing experimental interrelationships G_0, E_D, M_{DMT} have been constructed. In particular Fig. 9 and Fig. 10 illustrate the most significant observed trends.

Recent experience indicates that SDMT investigations can be performed with good results also in unusual conditions, e.g. *offshore* or in non penetrable soils (V_S-only measurements in backfilled boreholes).

Current research investigates the possible use of the SDMT for deriving "in situ" decay curves of soil stiffness with strain level, by fitting "reference G-γ curves" through two points provided by SDMT at different strain levels: the *small strain* shear modulus G_0 (from V_S) and a *working strain* modulus corresponding to M_{DMT}.

Deriving the operative modulus M for settlement predictions from G_0 appears arduous. Often to drastic variations in the M profile correspond barely visible variations in the G_0 profile. The ratio G_0/M varies in the wide range 0.5 to 20 (Fig. 7), hence it is far from being a constant, especially in clays and silts. Its value is strongly dependent on multiple information, e.g. soil type and stress history. Hence the use of only one information (e.g. c_u in cohesive soils) as a proxy of V_S (or G_0) for the seismic soil classification appears problematic.

If only mechanical DMT results are available rough estimates of G_0 from M can be obtained from Fig. 9.

The SDMT provides two parallel independent evaluations of the liquefaction resistance CRR from V_S and from K_D (horizontal stress index) by means of correlations CRR-V_S (Fig. 15) and CRR-K_D (Fig. 16), to be used in the framework of the Seed and Idriss (1971) simplified procedure. Preliminary studies indicate that methods based on K_D and V_S often provide substantially different estimates of CRR. In principle, the authors would propose to give greater weight to CRR by K_D for various reasons – above all the higher sensitivity of K_D to *stress history* and *aging*, factors which greatly increase liquefaction resistance. Very light earthquakes, however, may not destroy bonding, and in that case CRR evaluated by V_S may be more appropriate. The above obviously deserves additional verification, supported by real-life liquefaction case histories.

REFERENCES

AGI (1991). "Geotechnical Characterization of Fucino Clay." *Proc. X ECSMFE,* Firenze, 1, 27-40.

Andrus, R.D. and Stokoe, K.H., II. (2000). "Liquefaction resistance of soils from shear-wave velocity." *J. Geotech. Geoenv. Engrg.,* ASCE, 126(11), 1015-1025.

Andrus, R.D., Stokoe, K.H., II and Juang, C.H. (2004). "Guide for Shear-Wave-Based Liquefaction Potential Evaluation." *Earthquake Spectra,* 20(2), 285-305.

Cunha, R.P. (2004). "Discussion Report: Applications to Geotechnical Structures." *Proc. 2nd Int. Conf. on Site Characterization,* Porto, 2, 1391-1395.

Hepton, P. (1988). "Shear wave velocity measurements during penetration testing." *Proc. Penetration Testing in the UK,* ICE, 275-278.

Ishihara, K. (2001). "Estimate of relative density from in-situ penetration tests." *Proc. Int. Conf. on In Situ Measurement of Soil Properties and Case Histories,* Bali, 17-26.

Jamiolkowski, M. and Lo Presti, D.C.F. (1998). "DMT research in sand. What can be learned from calibration chamber tests." *1st Int. Conf. on Site Characterization,* Atlanta. Oral presentation.

Jendeby, L. (1992). "Deep Compaction by Vibrowing." *Proc. Nordic Geotechnical Meeting NGM-92,* 1, 19-24.

Ladd, C.C. (1971). "Settlement analyses for cohesive soils." *Special Summer Program on Soft Ground Construction,* Report R71-2, Soils Publication 272.

Leon, E., Gassman, S.L. and Talwani, P. (2006). "Accounting for Soil Aging When Assessing Liquefaction Potential." *J. Geotech. Geoenv. Engrg.,* ASCE, 132(3), 363-377.

Marchetti, S. (1980). "In Situ Tests by Flat Dilatometer." *J. Geotech. Engrg. Div.,* ASCE, 106(GT3), 299-321.

Marchetti, S., Monaco, P., Calabrese, M. and Totani, G. (2006). "Comparison of moduli determined by DMT and backfigured from local strain measurements under a 40 m diameter circular test load in the Venice area." *Proc. 2nd Int. Conf. on the Flat Dilatometer,* Washington D.C., 220-230.

Martin, G.K. and Mayne, P.W. (1997). "Seismic Flat Dilatometer Tests in Connecticut Valley Varved Clay." *ASTM Geotech. Testing J.,* 20(3), 357-361.

Martin, G.K. and Mayne, P.W. (1998). "Seismic flat dilatometer in Piedmont residual soils." *Proc. 1st Int. Conf. on Site Characterization,* Atlanta, 2, 837-843.

Maugeri, M. and Monaco, P. (2006). "Liquefaction Potential Evaluation by SDMT." *Proc. 2nd Int. Conf. on the Flat Dilatometer,* Washington D.C., 295-305.

Mayne, P.W. (2001). "Stress-strain-strength-flow parameters from enhanced in-situ tests." *Proc. Int. Conf. on In Situ Measurement of Soil Properties and Case Histories,* Bali, 27-47.

Mayne, P.W. (2005). "Unexpected but foreseeable mat settlements of Piedmont residuum." *Int. J. of Geoengineering Case Histories,* Vol. 1, Issue 1, 5-17.

Mayne, P.W., Schneider, J.A. and Martin, G.K. (1999). "Small- and large-strain soil properties from seismic flat dilatometer tests." *Proc. 2nd Int. Symp. on Pre-Failure Deformation Characteristics of Geomaterials,* Torino, 1, 419-427.

McGillivray, A. and Mayne, P.W. (2004). "Seismic piezocone and seismic flat dilatometer tests at Treporti." *Proc. 2nd Int. Conf. on Site Characterization,* Porto, 2, 1695-1700.

Młynarek, Z., Gogolik, S. and Marchetti, D. (2006). "Suitability of the SDMT method to assess geotechnical parameters of post-flotation sediments." *Proc. 2nd Int. Conf. on the Flat Dilatometer,* Washington D.C., 148-153.

Monaco, P. and Marchetti, S. (2007). "Evaluating liquefaction potential by seismic dilatometer (SDMT) accounting for aging/stress history." *Proc. 4th Int. Conf. on Earthquake Geotechnical Engineering,* Thessaloniki.

Monaco, P., Marchetti, S., Totani, G. and Calabrese, M. (2005). "Sand liquefiability assessment by Flat Dilatometer Test (DMT)." *Proc. XVI ICSMGE,* Osaka, 4, 2693-2697.

Monaco, P. and Schmertmann, J.H. (2007). Discussion of "Accounting for Soil Aging When Assessing Liquefaction Potential" by Leon, E. et al. (in *J. Geotech. Geoenv. Engrg.*, ASCE, 2006, 132(3), 363-377). *J. Geotech. Geoenv. Engrg.*, ASCE, 133(9), 1177-1179.

Monaco, P., Totani, G. and Calabrese, M. (2006). "DMT-predicted vs observed settlements: a review of the available experience." *Proc. 2nd Int. Conf. on the Flat Dilatometer,* Washington D.C., 244-252.

Powell, J.J.M. and Butcher, A.P. (2004). "Small Strain Stiffness Assessments from in Situ Tests." *Proc. 2nd Int. Conf. on Site Characterization,* Porto, 2, 1717-1722.

Schmertmann, J.H. (1986). "Dilatometer to compute Foundation Settlement." *Proc. ASCE Spec. Conf. on Use of In Situ Tests in Geotechnical Engineering In Situ '86,* Virginia Tech, Blacksburg. ASCE Geotech. Spec. Publ. No. 6, 303-321.

Schmertmann, J.H., Baker, W., Gupta, R. and Kessler, K. (1986). "CPT/DMT Quality Control of Ground Modification at a Power Plant." *Proc. ASCE Spec. Conf. on Use of In Situ Tests in Geotechnical Engineering In Situ '86,* Virginia Tech, Blacksburg. ASCE Geotech. Spec. Publ. No. 6, 985-1001.

Seed, H.B. and Idriss, I.M. (1971). "Simplified procedure for evaluating soil liquefaction potential." *J. Geotech. Engrg. Div.,* ASCE, 97(9), 1249-1273.

Simpson, B. (1999). "Engineering needs." *Proc. 2nd Int. Symp. Pre-Failure Deformation Characteristics of Geomaterials,* Torino.

TC16 (2001). "The Flat Dilatometer Test (DMT) in Soil Investigations - A Report by the ISSMGE Committee TC16." May 2001, 41 pp. Reprinted in *Proc. 2nd Int. Conf. on the Flat Dilatometer,* Washington D.C., 7-48.

Terzaghi, K. and Peck, R.B. (1964). "Soil Mechanics in Engineering Practice." John Wiley & Sons.

Analysis of Post-Installation Dynamic Load Test Data for Capacity Evaluation of Deep Foundations

Frank Rausche[1], P.E., M. ASCE, Garland Likins[2], P.E., M.ASCE, and Mohamad H. Hussein[3] , P.E., M. ASCE

[1]President, GRL Engineers, Inc., 4535 Renaissance Parkway, Cleveland, Ohio 44128, Frausche@pile.com

[2]President, Pile Dynamics, Inc., 4535 Renaissance Parkway, Cleveland, Ohio 44128, Garland@pile.com

[3]Vice President, GRL Engineers, Inc., 8000 South Orange Avenue, Orlando, Florida 32809, MHGRL@pile.com

ABSTRACT: Construction of deep foundations causes changes to the original geotechnical condition of the surrounding soils and rock. The nature, extent, and effects of these alterations, and subsequent time-dependent modifications, on the long-term load bearing capacity of the foundation elements are mainly functions of the characteristics of the supporting geo-materials and type of installation method. Post-installation Dynamic Load Testing (DLT) is commonly used for capacity evaluation of driven and drilled deep foundation elements and is performed under impacts of the pile driving hammer or a special large drop weight loading device. Use of a large drop weight loading device is usually appropriate for cast-in-place shafts and driven piles in high-setup soils to activate the full bearing capacity. The DLT procedure consists of applying a specific number of blows (typically 2 to 10 blows) and measuring pile/shaft strain and motion time-records, and the resulting displacement. This paper presents discussions on the analysis of post-installation DLT data for capacity evaluation of driven and drilled deep foundations. Technical aspects related to the proper test record selection for analysis considering pile/shaft set per blow, changes of capacity from blow to blow with subsequent impacts, energy levels, and total penetration during the test are delineated. Analytical procedure and engineering evaluation of results are discussed with recommendations for practical use. Test records from actual case histories are utilized to demonstrate data characteristics and illustrate proposed numerical methods and recommended analytical procedures for rational evaluation of DLT results for engineering use.

KEY WORDS: Bearing capacity , CAPWAP, drop weight, dynamic testing, end bearing, hammer, Pile Driving Analyzer, piles, residual stresses , setup, shafts, skin friction, soil resistance.

INTRODUCTION

Deep foundations are generally categorized into driven piles, augercast piles, and drilled shafts (also called bored piles, or caissons), each with distinctive features and characteristics. Driven piles are manufactured elements usually made of wood, concrete, or steel, uniform or tapered profile, uniform or composite material, low or high soil displacement, and are installed with either impact or vibratory hammers. Augercast piles are constructed in-place by pressure grouting utilizing a hollow stem continuous-flight auger or by a displacement method whereby no soil is removed from the ground. Drilled shafts may be constructed in a dry hole or by using slurry, may utilize temporary or permanent steel casing, or no casing at all. Both augered and drilled foundations usually incorporate a steel reinforcement cage or individual bars.

Construction of deep foundations alters the original geotechnical condition of the surrounding soils and rock. The nature, extent, and effects of these (temporary, or permanent) alterations and subsequent time-dependent modifications on the "long-term" load bearing capacity of the foundation elements are mainly functions of the geotechnical properties of the supporting geo-materials, foundation type's physical characteristics and installation mechanism (Coduto, 2001). Soil aging also has an effect (Schmertmann, 1991).

Construction of cast-in-place piles and drilled shafts may in some cases produce "soft-toe" effects due to soil loosening or unclean bottom resulting from sedimentation of suspended materials. Post-installation grouting may be used in special cases to enhance the end bearing resistance. The development of shaft resistance in skin friction along the length of augercast piles is influenced by soil type, the method of installation (i.e., soil displacement or non-displacement method, degree of over-rotation which can "mine" soils and reduce effective stresses, etc.), grout pressure and actual grout volume used for each depth increment.

Impact pile driving generally has desirable densification effects in non-cohesive soils resulting in enhanced pile capacity. Pile driving effects on cohesive soils are mainly remolding and development of excess pore water pressure, which result in a temporary change in the state of stress around the pile. Vibratory hammers can generate sometimes unpredicted reduced or increased soil density and strength.

Time-dependent changes of geotechnical conditions often have beneficial effects by increasing pile capacity over time; this phenomenon is commonly called "soil setup" or "pile freeze". Modern progressive engineering practice takes setup into account to optimize pile design and minimize foundation costs (Komurka, 2004). Pile driving into calcareous soils or stiff clays may produce permanent negative effects on the soils and pile-soil interaction that may not be recovered with time. Piles driven into weathered shale or saturated dense fine-grained soils may lose end-bearing capacity over time; this unfavorable effect is commonly called "relaxation".

The performance of a deep foundation is a function of its structural strength and integrity, geotechnical strength and deformation properties of the supporting soil or rock, pile/shaft-soil/rock interaction characteristics, and nature and magnitude of loads. Field assessment of a deep foundation element's response to applied load and evaluation of its ultimate load bearing capacity is typically made with static or dynamic loading tests. Static load testing is performed using conventional cumbersome loading methods, or by utilizing modern innovative procedures (Schmertmann, 1993). Dynamic Load Testing (DLT) is conveniently performed under the impacts of the pile driving hammer or a special-purpose large drop weight system. Static or dynamic load testing of cast in-situ shafts require that the grout or concrete has gained sufficient strength (typically at least five days after casting) to allow the geotechnical capacity of the foundation element to be evaluated. Dynamic load testing of driven piles may be performed during the pile driving process, and/or during restrike some time following initial installation (possibly ranging from a few hours for piles driven into sands to days or weeks for piles driven into clays or shale) for assessment of "long-term" pile load bearing capacity. However, project construction schedule requirements may restrain the maximum amount of waiting time allowed for post-installation testing. In all cases, both static and dynamic testing procedures provide information regarding the pile or shaft load bearing capacity at the time of testing.

The post-installation dynamic load testing procedure is typically performed under impacts of the pile driving hammer or a special-purpose large drop weight for the dynamic loading device. The latter is always needed for cast in-situ shafts or for driven piles in high-setup soils which sometimes require more energy to activate the full bearing capacity than available from the original pile driver. Testing typically consists of applying a specific number of blows (typically 2 to 10 blows) and measuring pile/shaft strain and motion time-records, and the resulting permanent set.

DYNAMIC LOAD TESTING

Modern dynamic testing methods of deep foundations are based on one-dimensional stresswave propagation theory and the measurement of force and velocity records at the pile/shaft top under impacts of the pile driving hammer, or a special-purpose drop weight.

Testing is routinely performed on driven piles and cast-in-place shafts (Hussein and Likins, 1995). For post-construction testing, the installation pile-driving hammer may be used during the restrike, provided it can mobilize sufficient capacity at the time of restrike. If not, such as in high setup situations which then result in "refusal" conditions, a larger hammer should be used to activate the added resistance. For testing of cast-in-place shafts, a drop hammer having a minimum weight equal to 1 to 2% of the ultimate pile capacity (or the desired proof load) will be needed for the test. Typically, two to four each reusable strain transducers and accelerometers are affixed around the perimeter of the pile or shaft at a location of approximately one to two diameters below its top. Field data acquisition and processing are made with specialty equipment such as the Pile Driving Analyzer® (PDA) system according the Case Method (Rausche et al., 1985). Post-installation restrike testing consists of dynamic measurements under typically 2 to 10 blows. The pile or shaft net permanent set under each blow is also independently measured. Dynamic test records are analyzed according to the CAPWAP® Method which is a computerized system identification numerical process employing signal matching based on stresswave propagation principles for determining soil/rock resistance parameters (magnitude and stiffness) acting along the shaft length in skin friction and under the toe in end bearing (Rausche et al., 1972). The analysis results include a load-movement graph simulating a static loading test. Comparisons between CAPWAP analyses and full-scale static load test results have shown excellent correlations (Likins and Rausche, 2004).

Since testing is done under a number of hammer blows, the data consists of as many test records. In many cases, the data include consistent signal features and one CAPWAP analysis is performed with records obtained under a typical blow representative of the whole test sequence (Seidel and Rausche, 1984). However, there are situations where analyzing only one test record would not be sufficient to fully represent the entire testing process, or the whole geotechnical circumstances related to the pile or shaft's ultimate load bearing capacity. For example, restrike testing using a diesel hammer may produce inconsistent records due to the variability of hammer stroke heights, and therefore energy levels, from blow to blow at the beginning of the restrike process. Sometimes DLT is performed by purposely varying the hammer energy by systematically increasing the drop height of the hammer to obtain dynamic data and set under successive test blows (Beim and Aoki, 1996).

The level of energy delivered to the pile or shaft top, along with set per blow, are important parameters that must be considered for the proper evaluation of the DLT results. A commonly quoted rule of thumb states that a test blow, producing a permanent pile set less than 2.5 mm may not measure the ultimate capacity, but only a maximum mobilized resistance or a lower bound pile bearing capacity. Such a test would still have value as a proof test if the mobilized capacity exceeds the required value. The 2.5 mm set value corresponds to a blow count of 400 blows/m (or 120 blows/ft) which is frequently quoted as a refusal blow count condition. Thus, at that driving resistance level uncertainty increases about complete resistance activation.

Incomplete resistance activation during a test blow with insufficient energy creates a problem when several low energy impacts cause soil setup resistance to be partially lost. In this case no single record indicates the total pile capacity. Superposition of results obtained from analyzing data under individual blows of a restrike test, or from an end of driving and restrike tests combination, attempts to correct this problem (Stevens, 2000; Hussein et al., 2002). Combining the results from multiple tests and/or blows has also been used to predict the long-term pile capacity in soils with time-dependent characteristics (Preim et al., 1989; Hussein et al., 2004; Rausche et al., 1996; Hussein et al. 2006). Superposition, however, is not a simple matter and, in some cases, the potential for overprediction exists. The present paper suggests guidance on test interpretation that may reduce the need for superposition.

THEORY OF RESISTANCE ACTIVATION

Static loading of a pile first causes relatively high displacements and full resistance activation near the pile top. With continued loading, resistance progressively activates along the pile. Once all shaft resistance and end bearing has been activated, both top and toe experience the same incremental displacements. In sensitive soils, after peak resistance has been reached, large pile displacements reduce resistance values along the pile so that the total peak resistance from all points along the pile is never simultaneously activated. This effect is sometimes called a progressive failure.

Dynamic loading first causes high downward motions and resistance development near the pile top, and short time periods later downward motions at lower and lower pile sections. The time between impact and the beginning of the pile toe movement can be determined by the wave travel; it depends on the speed of the stresswave which is a function of only the pile's elastic modulus and specific density. The downward motions generate pile displacement that activates static resistance forces. The soil also reacts with dynamic resistance forces which depend on the velocity of the pile particles and soil type.

Full static resistance activation requires that the pile displaces a distance necessary to reach soil plastification. Under ideal conditions this is an elastic deformation which is often called the "quake" in the pile dynamic literature (Smith, 1960). Figure 1 shows a typical two-load-cycle static test result together with the simulated load-movement plot from a dynamic test. This example was taken from a test series conducted at Apalachicola River and Bay Bridge in Florida in 1986. The dynamic restrike test on this voided 760 mm square precast, prestressed concrete pile followed the static test after a 5 day waiting period. The restrike permanent set per blow was 1.25 mm. Using the radiation damping soil model option, the CAPWAP program produced a good capacity match although it indicated a somewhat higher stiffness which could, at least in part, be due to the difference between the static and dynamic concrete elastic modulus.

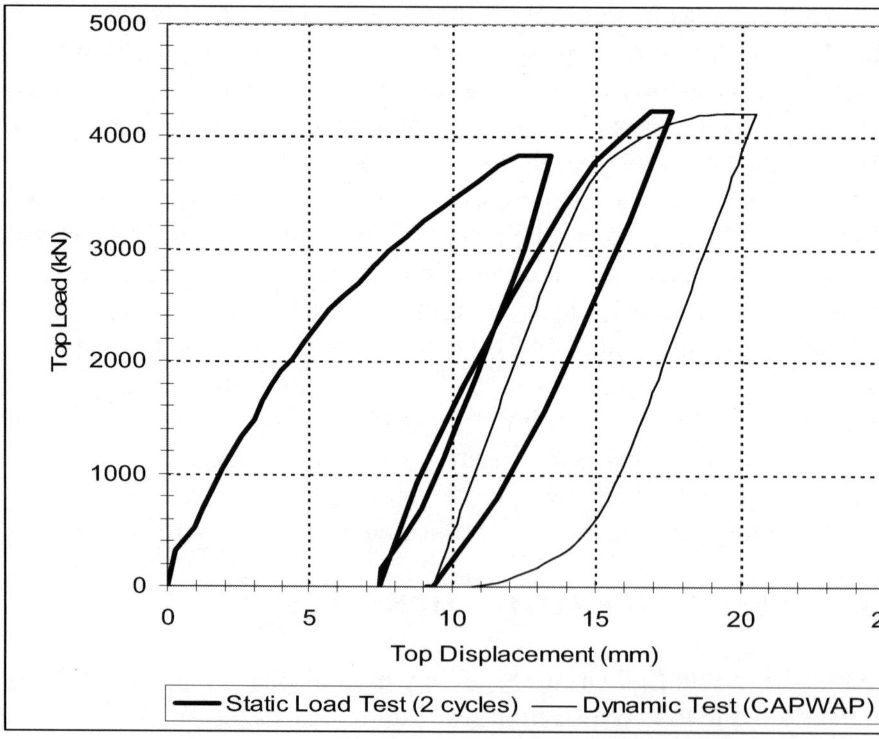

FIG. 1: Example of a two-cycle static test with subsequent dynamic

As is frequently observed, Figure 1 shows a first load cycle stiffness which is lower than that of the second load cycle. A similar behavior can be observed when applying several dynamic loads with successive hammer impacts. In either the static or dynamic case, a stiffened response can normally be observed under subsequent load cycles. Possible explanations include (a) residual stresses, (b) soil densification, (c) the gain of strength associated with increased effective stresses, and (d) possibly others, some of which are subsequently discussed.

Residual Stress Effects

Consider for the purpose of illustration the following case, observed by the authors. The first hammer blow of a dynamic restrike test series on long, flexible piles with high shaft resistance produced a 5 mm set while subsequent impacts only produced 1 or 2 mm sets. In this case soil setup generated most of the shaft resistance with time after pile installation. Thus, the piles experienced only negligible residual stresses during the installation process and practically were at a near zero residual stress state in the beginning of the restrike test. The first impact of the restrike test compressed the pile elastically; during rebound, the shaft resistance prevented the pile from fully expanding to a zero stress condition which created the high permanent top set. In this case it is possible and likely that the first impact did not move the pile toe, because the energy was used for pile compression. Later load cycles expended less energy on pile compression and therefore produced more permanent pile sets at both top and bottom. Clearly, residual stresses of a pile that was not subjected to high residual stresses during installation achieves a much larger top set under the first load cycle or restrike blows than later load cycles or impacts. The permanent pile set of the first blow is then not an indication that failure has been achieved in the test. Piles with high shaft resistance present already during installation would have a stiffer behavior during the load test because the pile would be precompressed by the residual stress effects. Residual stresses in the pile complicate the determination of the resistance distribution in either static or dynamic test; however, the total capacity activation is achieved under similar displacement conditions in either case.

Soil Densification

Once the pile toe begins to move, depending on the soil type and the type of construction, increasing toe displacements cause densification and thus increased capacities. In a driven pile this densification often has already taken place due to the pile installation and possibly due to the subsequent installation of neighboring piles. In a drilled shaft the soil at the bottom may be in a relaxed state during and after construction and large displacements may be required before the full capacity of the shaft can be activated. The allowable pile toe penetration, and therefore densification, under a static loading is a matter of judgment. Clearly, there is a vast difference between a driven displacement pile and a drilled shaft with the former displaying a much stiffer behavior. Both static and dynamic tests exhibit the same effects. Densification together with residual stress effects often makes the comparison between tests difficult: regardless of the order of static and dynamic tests for correlation; the earlier test densifies and prestresses the soil yielding a stiffer response for the later test (or later impact response).

Pile Toe Resistance

In granular soils, the pile displacements during static loading increase the effective stresses between the soil particles and thus allow for increased toe resistance forces. As a result, failure at the toe may not be clearly pronounced and, in fact, the toe resistance may increase significantly with increasingly large toe displacements during a static load test. During Dynamic Load Testing the toe displacements for each impact are generally relatively limited and the cumulative effect of several blows is necessary to duplicate this effect of increased toe pressures. However, repeated dynamic loadings can only produce these similar toe resistance gain effects if sufficient shaft resistance maintains high toe pressures between hammer blows thereby allowing for a build up of a high residual toe resistance. If the pile toe is allowed to unload (rebound) to low resistance values between hammer blows, pressures at the pile toe will not show the same increase that either static tests or dynamic tests with residual stresses can generate.

Time Effects

It is well known that soil setup (Komurka, 2004) and relaxation (Morgano and White, 2004; Thompson and Thompson, 1983) effects cause time dependent changes of static soil resistance for driven piles. Load testing, either dynamic or static, therefore has to be performed after a sufficiently long waiting period following pile installation. Multiple tests with various wait times can establish the capacity change trend with time; dynamic tests on production piles after only short waits can be used to confirm pile acceptance if they follow similar trends. Load testing can reverse some of the time dependent capacity gains or losses, particularly if the tests cause large deformations and associated pore water changes. In relatively sensitive soils, setup gains may be lost and cause progressive failure (as discussed above) such that the full setup capacity is never apparent in either the static or dynamic test. On the other hand, pile toe relaxation may be reversed (resistance increases again) if testing produces sufficiently large sets.

Creep

Under sustained loads near failure, pile settlements often increase because of creep effects in the soil. Dynamic testing obviously cannot determine creep and even most static load tests are similarly limited. In special cases, the authors have observed a constant rate of creep under maintained static loading for an entire month. Historically, creep effects occurring during a pile's service life have been limited by maintaining working loads at levels below one half of the failure load.

FAILURE CRITERIA

Static load test capacity can be calculated from pile top load-movement curves according to a variety of criteria (Fellenius, 1975; Fellenius, 2001) to arrive at a clear interpretation of the "ultimate capacity" of the deep foundation element. One generally accepted method is referred to as the "Offset" or "Davisson" criterion. It suggests that failure of a pile occurs when the pile toe has achieved a permanent set which is 4 mm plus the pile toe diameter or width, D, divided by 120. The corresponding pile top set is calculated as though the pile would support all load in end bearing. The Federal Highway Administration recommends application of the Offset Criterion (which is among the more conservative definitions) for driven piles of less than 900 mm diameter and an offset of D/30 for larger sizes (Hannigan et al., 2006). For drilled shafts and augercast piles, the D/30 offset criterion is also reasonable. Other criteria merely consider the elastic plus plastic pile top settlement. Duzceer and Saglamer (2002) demonstrated that the definition of failure by various methods can result in different failure loads for the same static test by a factor of two.

Proposed Failure Criterion for Dynamic Tests

The foregoing remarks show that failure is not easily defined (neither for the static nor the dynamic tests cases) because of the variety of possible pile stress states and soil failure modes. In particular, possible underprediction due to both lack of capacity activation and loss of soil setup, and potential overprediction due to relaxation, have to be considered. The following failure criterion for dynamic tests is recommended to address these concerns:

1. Most importantly, the pile should be tested after a sufficient waiting time after construction for time-dependent soil strength changes to take effect.
2. During the dynamic test, the pile toe should achieve a total (elastic plus cumulative permanent) displacement corresponding to a reasonable failure criterion. The set should not be too large or soil setup would be lost and it should not be too small for eliminating sensitive capacity gains that would be quickly lost by progressive failure. Also the toe set should be large enough for soil densification and increased stress effects to take place. The authors recommend a total toe displacement of D/60 for failure which is consistent with the toe quake recommendation of the GRLWEAP wave equation analysis program (PDI, 2005). Further discussions will follow in the case studies below.
3. To be comparable to service piles, a similar "virgin" pile (e.g., "sister pile") should be tested and evaluated so that capacities should not be affected by some preloading by either a previous static or dynamic load test that would generate residual stresses or densification and stiffen the pile response, particularly for drilled or augered piles.
4. For reasons of conservatism and because of the generally unknown residual stress state, the pile bearing capacity usually should not be based on the first

restrike test record. The first initial loading may include resistance components which could easily be lost when high static or dynamic deflections are applied, or cause changes in the residual stress state. Instead, we recommend applying a first blow with a relatively low energy which in all likelihood will not activate the full resistance. This is very important for piles that have an unexpectedly low capacity; the initial low impact will then result in a large permanent set and already reveal that low capacity with greater reliability than the higher energy impacts (which then need not be applied).

5. The results from the first dynamic test impact should be included in the calculated load-set curve, particularly for drilled shaft tests.
6. Where the geotechnical capacity exceeds the structural strength of the pile (e.g., piles founded in rock), the pile should only be loaded to a level sufficient to demonstrate that the required capacity has been activated to protect the pile integrity

When calculating the maximum toe movement in a restrike test, the permanent set of preceding impact loadings should be added. Dynamic test records of pile top force, F, and velocity, v, provide the necessary information for calculating the pile toe displacement, u_{toe}, and thus its maximum value from top force and velocity measurements. This can be done in closed form as follows:

$$u_{toe} = \int \{(W_{d1} - W_{u2}) / Z\} \, dt \qquad (1)$$

where $W_{d1} = ½ (F(t_1) + v(t_1)Z)$ and $W_{u2} = ½ (F(t_2) - v(t_2)Z)$ are the downward wave at time 1 and upward wave at time 2, respectively. Time t_2 occurs 2L/c after t_1 with 2L/c being equal to twice the pile length divided by the pile material's stresswave speed. The pile impedance, Z, is the product of pile cross sectional area, wave speed and material density. Alternatively, instead of calculating in closed form, CAPWAP analysis may be used to determine the maximum pile toe displacement (see Figure 2 in Case 1 below).

The dynamic pile toe displacement may or may not be associated with a particular permanent set. Thus a soil that allows for a large elastic set, i.e., quake, at the bottom could reach the failure displacement prior to achieving a permanent set. When considering the pile top deflection, the elastic pile compression would be additional to the pile toe displacement. Considering the completely different construction and loading history of driven piles and drilled shafts (the former with residual stresses and the latter with a relaxed soil at the toe), it would be reasonable to apply different failure criteria to both pile types. However, the proposed criterion and test evaluation procedure (i.e., items 1 to 6 above) appear to be reasonable for both types of pile as demonstrated by several case studies discussed below.

CASE STUDIES

Case 1: Steel H-Pile A

An H-pile 356x109 (14x73) was driven into silts and clays with N values typically less than 15 at depths less than 30 m, underlain by clays and silts with N values of 40 to 100 to the pile termination depth of 45 m. The end of initial driving blow count was 125 blows per meter. Following a waiting period of 21 days, the pile was tested during restrike using a Delmag D30-32 impact hammer with strokes as high as 2.8 m for an early blow at 5 mm set per blow, decreasing to 2.2 m stroke for later blows with up to 15 mm set per blow. A static load test (SLT) was performed 48 days after pile installation. CAPWAP analyses were performed with dynamic records obtained under several blows (i.e., blows 1 through 4, and representative later blows 8 and 25). The static load test results predicted by CAPWAP, including both top and bottom loads vs. set, are summarized sequentially in Figure 2. Blows 1 through 4 are shown cyclically, offset by the preceding cumulative sets, while blows 8 and 25 follow with gaps representing the intervening blows' corresponding sets. The CAPWAP produced pile-top load-movement corresponds very well to the measured static load test curve. The toe loading was calculated as a standard part of the CAPWAP analysis. Obviously the toe loading for this H pile is relatively low and the pile is basically a friction pile. The failure toe displacement for a 356 mm pile would be 6 mm which is exceeded by the toe maximum displacement of the second impact. The second impact therefore yields the dynamically calculated capacity (2051 kN) which correlates well with the static test result.

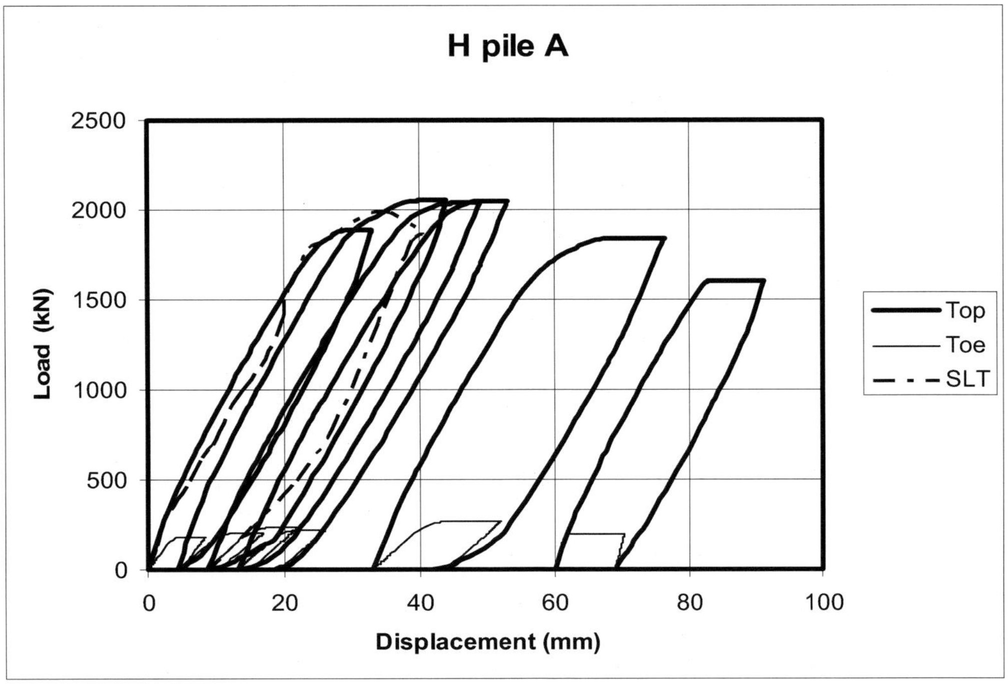

FIG. 2: Static load test results for H-pile A from CAPWAP and SLT

The shaft resistance distributions determined by CAPWAP analyses for several blows are shown in Figure 3. They clearly indicate the higher resistance in the lower, stiffer materials below 30 m. and the reductions of shaft resistance for later blows. Clearly, analysis of one of the later blows would under predict the shaft resistance. It is reasonable to expect that, after completion of the restrike, the capacity will again increase as soil setup regenerates the long-term service strength of the soil.

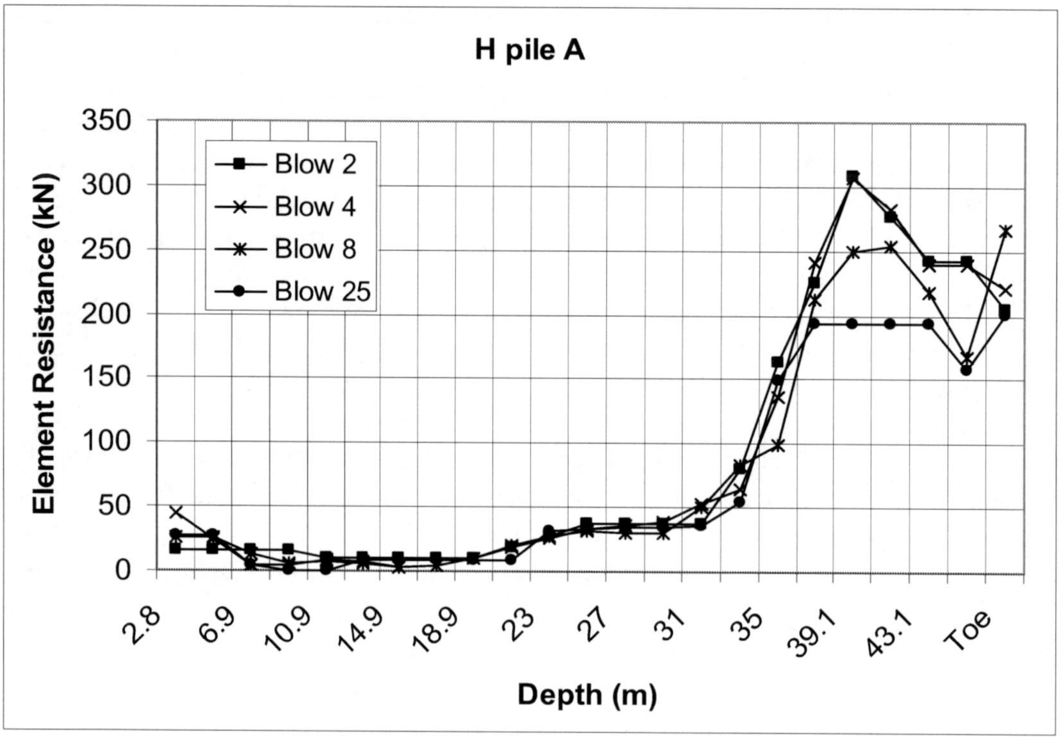

FIG. 3: Calculated resistance distributions for H- pile A

Case 2: H-Pile B

An H-pile 309 x 79 (12 x 53) was restruck with an APE D19-42 hammer for 10 blows. The first 5 blows resulted in 25 mm set. The next 3 blows produced an additional 25 mm set, and the final 2 blows also had 25 mm set. Soils were generally clayey fine sands (SC) or clay (CH) with SPT N values of 10 or less along the shaft. The toe was driven into Limestone (N-values of 50/5"). Figure 4 shows the CAPWAP results cyclically presented for the first 5 blows plus later blows 7 and 10. The shaft resistance decreases significantly between blow 1 and 2, with continued further decreases with subsequent blows. The end bearing gradually increased during the restrike. The resistance distribution curves calculated from several blows are shown in Figure 5. The results of the first blow include a very high shaft resistance and an almost completely activated toe resistance. The D/60 failure toe displacement, D/60, would be 5 mm and that has been achieved with the first restrike blow. According to rule 4 of the proposed failure criteria above, the second blow would therefore present the, admittedly somewhat conservative, bearing capacity result. This result is reasonable because it would not include soil setup effects that could be lost under a sudden (dynamic) or excessive load due to progressive failure.

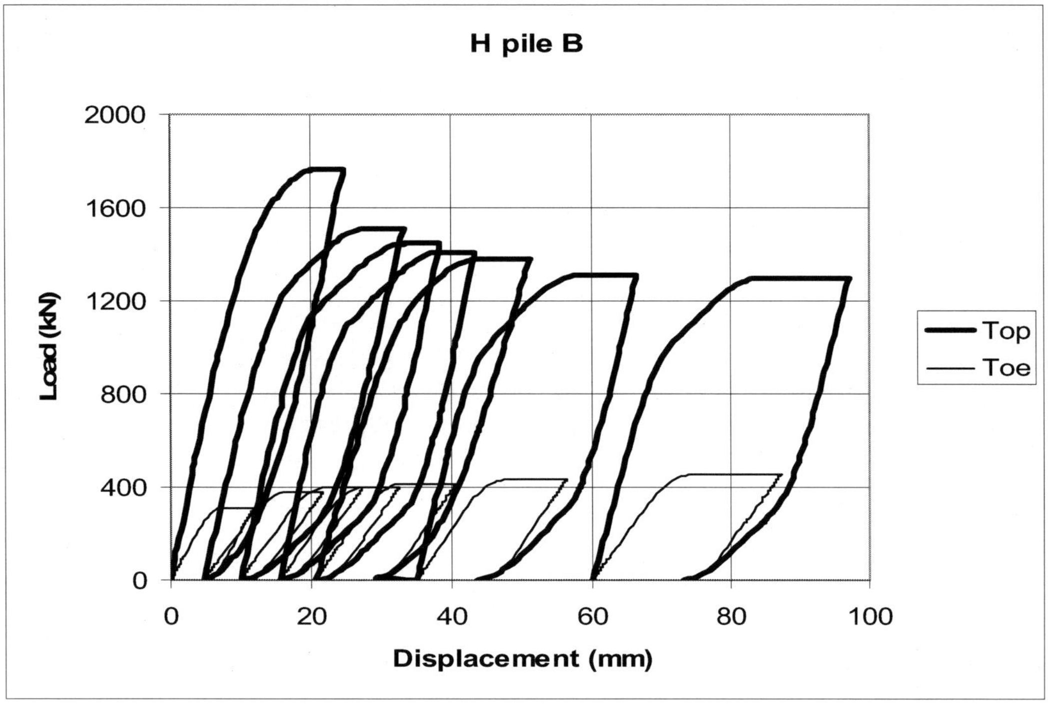

FIG. 4: H- pile B

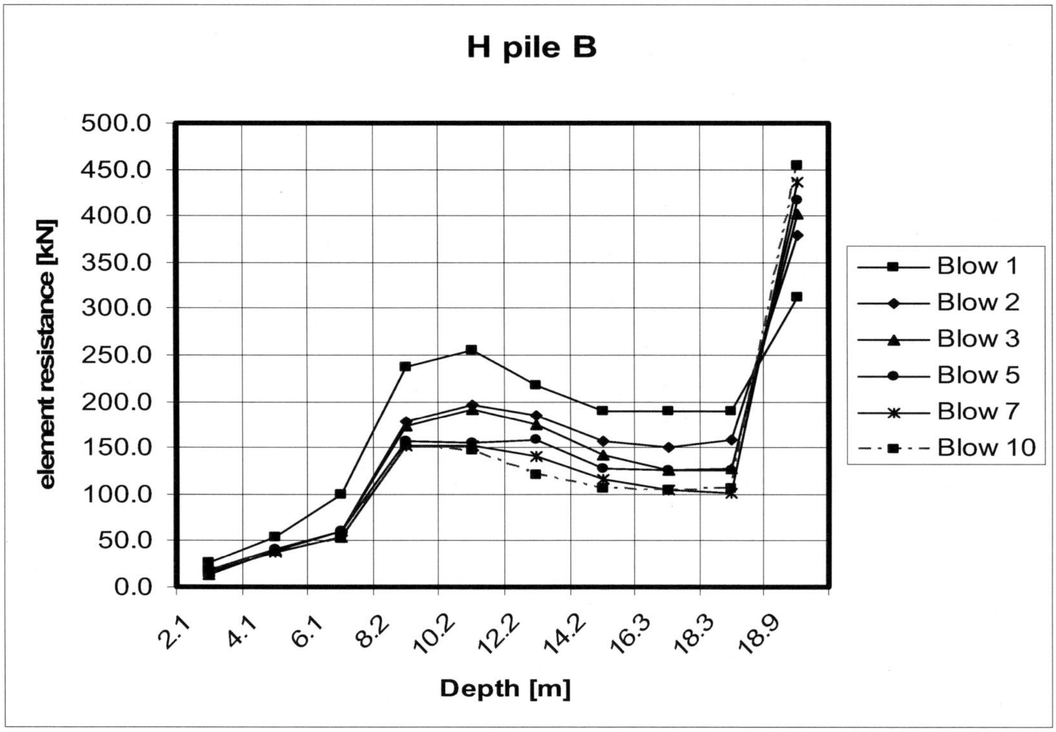

FIG. 5: H-pile B resistance distribution

Case 3: Prestressed Concrete Pier A

A 610 mm (24-inch) square prestressed concrete (PSC) pile was driven at a pier location for a large bridge to a depth of 25 m into saturated sand (fine and dense) with silts and clays by an APE D62-22 diesel hammer to a blow count of 270 blows per meter (3.7 mm set per blow) at a stroke of 2.7 m. After a waiting period of one week, the restrike of 10 blows resulted in a total net displacement of 30 mm (average 3.0 mm set per blow) at a hammer stroke of 2.7 m.

Figure 6 shows the CAPWAP load-set curves calculated from the records of the first 5 blows and blow 10. While there is relatively little change in the total capacity from blow to blow, the end bearing significantly increases and the shaft resistance significantly decreases from blow to blow as seen in Figure 6. This behavior is consistent with the soil type. During initial driving, and again upon redriving like in this sequence of blows, negative pore pressures increase producing dilation in the silt or fine sand and, therefore, high negative pore water pressures. These pressures, in turn, cause temporarily larger effective stresses and thus more end bearing resistance.

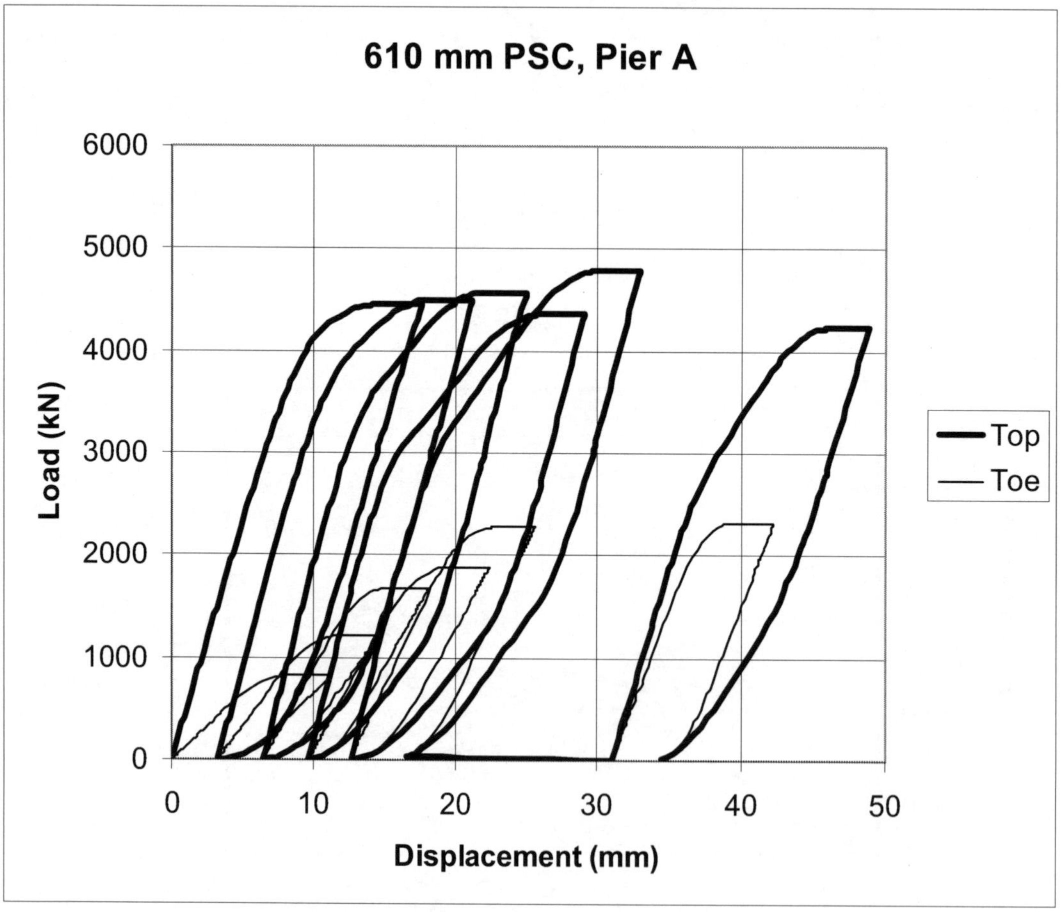

FIG. 6: Prestressed concrete pile, Pier A - CAPWAP results

With time, hydrostatic pore pressures are re-established and the toe then experiences relaxation effects. It is perhaps coincidental that the shaft reduction (e.g., loss of skin friction setup) approximately balanced the toe increase during redriving. Other observed cases with toe in dense saturated fine sands and silts have shown an increase in total capacity from blow to blow during restrike, reflecting the temporary change in pore pressure condition during driving. Although the total capacity was relatively insensitive during this test series, it would be important in most cases to carefully choose an early blow in restrike to properly account for the shaft setup and toe relaxation effects.

In this case, the failure toe total displacement would be 10.2 mm (610/60) and this displacement has been reached or exceeded already during the first restrike blow. Again from rule 4, the result of the second blow should be used for capacity determination and it would yield approximately 4500 kN capacity like the first restrike hammer blow.

Case 4: Prestressed Concrete Pile at Pier B

At Pier B of the same project as Case 3, the same D62-22 hammer drove and tested another 610 mm PSC pile, this time to only 19 m penetration depth. Restrike testing resulted in 45 mm set for 10 blows (4.5 mm set per blow, on average). With less penetration, and considering that soil characteristics vary across larger sites such as this one, the apparent behavior has both similarities and differences.

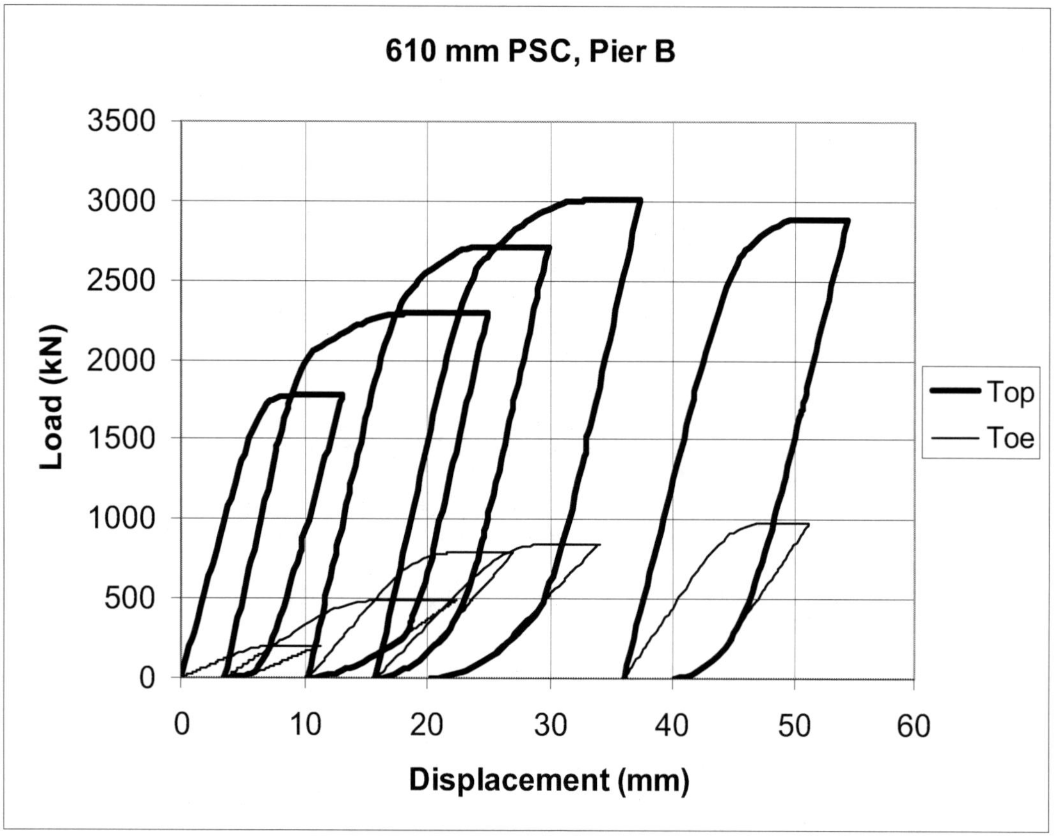

FIG. 7: Prestressed concrete pile, Pier B – CAPWAP results

It is clear from Figures 6 and 7 that the end bearing increase is somewhat similar to that observed at pier A. However, the shaft resistance remains fairly constant through this restrike. Thus the total capacity increases rather substantially during this 10 blow restrike, and selecting an early blow in the restrike is critical so that the in-place pile capacity is not overpredicted. Blow 1 transferred to the pile only 1/3 the energy of blow 2, and thus blow 1 may have not fully mobilized the capacity. Again, the record of the first blow should not be used for capacity assessment although it is valuable to judge the activation of both shaft resistance and end bearing. Also, since the first impact generated just barely the required failure toe displacement of 10.2 mm, the second hammer blow would be acceptable as the indicator of full capacity. The ultimate capacity of this pile is therefore likely to be in the range of 2200 kN, or less than half that at pier A owing to a lower end bearing, and with less depth of penetration, a lower shaft resistance.

Case 5: Drilled Shaft

This case study presents results obtained from dynamically testing an 1.8 m diameter drilled shaft with an embedment depth of approximately 20 m. Dynamic Load Testing consisted of applying four hammer impacts with a large drop weight, which produced a total cumulative shaft net penetration of about 14 mm. The CAPWAP calculated load-movement curves shown in Figure 8 indicate that the maximum top and toe displacements reached 31 and 27 mm, respectively.

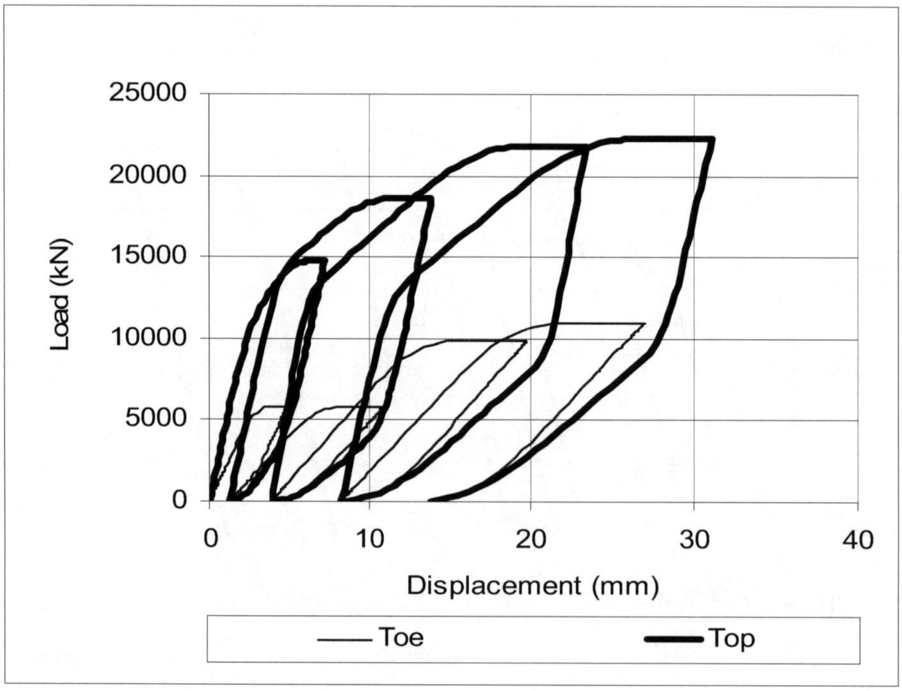

FIG. 8: CAPWAP calculated load cycles for 1.8 m diameter drilled shaft

The first two blows partially activated the available end bearing while the shaft resistance was activated under the second impact. The results of the third blow shows a higher capacity (as compared to the second impact) primarily from the increase in end bearing; the rate of end bearing increase diminished with the fourth blow, and comes close to achieving the D/60 toe displacement criteria; it is likely that the next impact would have achieved the D/60 criteria at a load not much different than the result of this fourth blow. Note that the presentation of successive load-set curves allows for a reasonable means to evaluate the in-place activated capacity and the increase in resistance due to the development of additional end bearing resistance with further penetration. It should be added that for a drilled shaft, relaxation effects at the toe are typically not of concern since the geomaterial below the pile toe is in a disturbed state due to the mechanics of the installation method.

Case 6: Augercast Pile

An augercast (also referred to as augered cast-in-place or ACIP) pile of diameter 510 mm and length 21.2 m was installed into soils consisting of very soft clay for 11 m underlain by very dense sand. After allowing the pile to harden for a few weeks, a 60 kN drop weight applied 3 blows with drop heights 0.25, 0.50 and 0.75 m respectively, resulting in permanent set per blow of 0.9, 2.1 and 3.3 mm, respectively. The dynamic test results are presented in Figure 9. The failure toe displacement required for D/60 is 9 mm, and the third blow satisfies the criteria. The predicted capacity for the third blow, which is 2500 kN, was submitted to the client before the static test was performed, and it correlates almost exactly with the static load test performed about two weeks later.

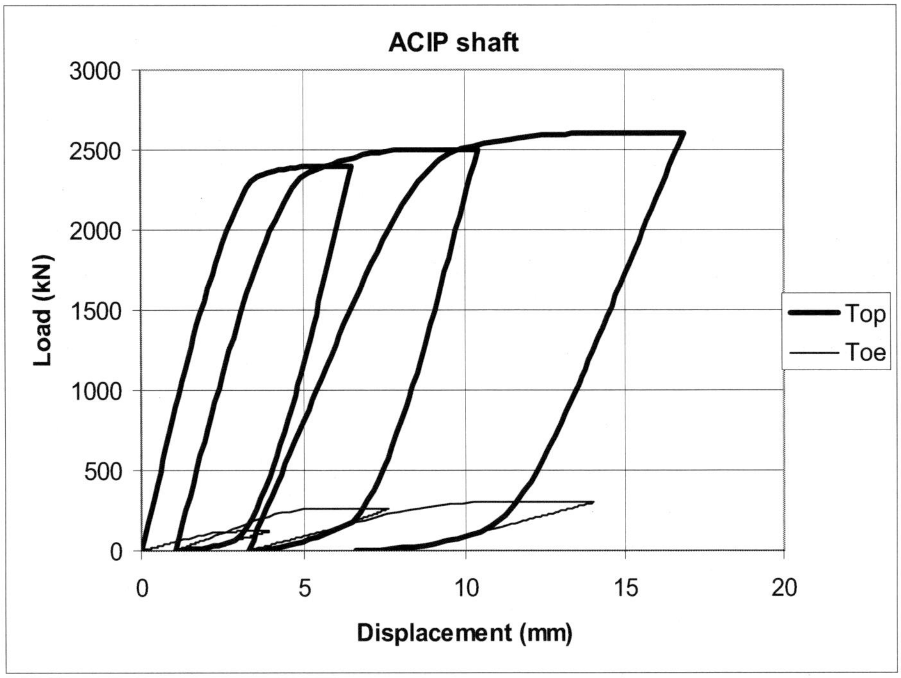

FIG. 9: ACIP CAPWAP results

SUMMARY AND CONCLUSIONS

This paper proposes a standardized procedure in which Dynamic Load Test records are selected and evaluated when assessing the post-installation load bearing capacity of driven piles, augercast piles, and drilled shafts. The method is closely related to the Offset or Davisson Failure Criterion that is frequently used for static load test evaluations. Different from current practice, the new criterion does not require that a specific minimum permanent set per blow be achieved but rather that the pile or shaft achieves a cumulative toe displacement of D/60 or more, where D is the pile or shaft diameter. In addition, a first load cycle (impact) of a lesser energy should be applied but its results should not be used to establish pile or drilled shaft capacity although it may be considered for the evaluation of resistance development trends. CAPWAP analysis of the dynamic records obtained under the first blow can also serve to complement the calculated load-set relationships of the second and later impacts. It is also sometimes helpful to perform CAPWAP analyses with data obtained under several blows as necessary and plot their load-set curves in a sequential presentation for assessment of setup and relaxation effects. The example records discussed in this paper demonstrated a variety of situations which can be clearly identified by the proposed analysis method. The method incorporates aspects that warrant further study and consideration for practical application and engineering use.

ACKNOWLEDGMENTS

The authors would like to acknowledge the thorough efforts of the reviewers and their constructive comments in enhancing and improving this paper. Most importantly, the pioneering work of Dr. John H. Schmertmann, P.E. in the area of dynamic testing of the SPT and his early consulting work involving dynamic pile testing through Schmertmann and Crapps, Inc. are recognized and acknowledged.

REFERENCES

Beim, J and Aoki, N. (1996), "Dynamic load test method with variable energy", *Proceedings of the 5th international Conference on the Application of Stresswave Theory to Piles*, edited by F. Townsend, M. Hussein, and M. McVay, The University of Florida, pp. 274-281.

Coduto, D.P. (2001), "Foundation design – principles and practices", *2nd Edition*, Prentice-Hall, Inc. Publishers, www.prenhall.com, 883 p.

Duzceer, R. and Saglamer, A., (2002), "Evaluation of load test results", *Proc. of the Ninth Int'l Conf. on Piling and Deep Foundation*, Nice France, Deep Foundations Institute.

Fellenius, B., (1975), "Test loading of piles. Methods, interpretation and new proof testing procedure", *ASCE GT9*, (Vol 101).

Fellenius, B. H., (2001). "What capacity value to choose from the results of a static loading test". *Fulcrum Winter 2001 issue*, Deep Foundations Institute

Hannigan, P.J., Goble, G.G., Likins, G.E., and Rausche, F., 2006. Design and Construction of Driven Pile Foundations (two volumes), Report # FHWA-NHI-05-042, National Highway Institute, Washington D.C.

Hussein, M. and Likins, G.E. (1995), "High-strain dynamic testing of drilled and cast-in-place piles", *Proceedings of the Deep Foundations Institute's 20th Annual Meeting*, Charleston, S.C., www.dfi.org, pp. 129-148.

Hussein, M. Sharp, M., and Knight, W. (2002), "The Use of superposition for evaluating pile capacity", *ASCE Geotechnical Special Publication No. 116: Deep Foundations 2002, An International Perspective on Theory, Design, Construction, and Performance* – edited by M.W. O'Neill and F.C. Townsend, pp. 6-21.

Hussein, M., Mondello, B., and Alvarez, C. (2004), "Simulated pile load-movement incorporating anticipated soil set-up", *ASCE Geotechnical Special Publication No. 126: Geotechnical Engineering for Transportation Projects* – edited by M.K. Yegian and E. Kavazanjian, pp. 1153-1160.

Hussein, M., Woerner, W., Sharp, M., and Hwang, C. (2006), "Pile driveability and bearing capacity in high-rebound soils", *Proceedings if the ASCE Geo-Institute's Geo-Congress on February 2 –March 11, 2006*, Atlanta, Georgia, CD.

Komurka, V. E. (2004), "Incorporating set-up and support cost distributions into driven pile design", *ASCE Geotechnical Special Publication No. 125: Current Practices and Future Trends in Deep Foundations* – edited by J.A. DiMaggio and M.H. Hussein, pp. 16-49.

Likins, G. and Rausche, F. (2004), "Correlation of CAPWAP with static load tests", *Proceedings of the 7th International Conference on the Application of Stresswave Theory to Piles,* The Institution of Engineers Malaysia, pp. 153-161.

Morgano, C.M. and White, B.A. (2004), "Identifying soil relaxation from dynamic testing", *Proceedings of the 7th International Conference on the Application of stresswave Theory to Piles*, The Institution of Engineers Malaysia, pp. 415-422.

Pile Dynamics, Inc. (2005), "GRLWEAP help manual", Cleveland, OH.

Preim, M., March, R., and Hussein, M. (1989), "Bearing capacity of piles in soils with time dependent characteristics", *Proceedings of the International Conference on Piling and Deep Foundations* – edited by J.B. Burland and J.M. Mitchell, held in London on May 15-18, 1989, A.A. Balkema Publishers, Rotterdam, Netherlands, pp. 363-370.

Rausche, F., Moses, F. and Goble, G.G. (1972), "Soil resistance perditions from pile dynamics", *ASCE Journal of the Soil Mechanics and Foundations Division*, (Vol. 98, No. SM9), pp.917-937.

Rausche, F., Goble, G.G., and Likins, G. E. (1985), "Dynamic determination of pile capacity", *ASCE Journal of Geotechnical Engineering*, (Vol. 111, No. 3), pp.367-383.

Rausche, F., Richardson, B., and Likins, G. (1996), "Multiple blow CAPWAP analysis of pile dynamic records", *Proceedings of the 5th international Conference on the Application of Stresswave Theory to Piles*, edited by F. Townsend, M. Hussein, and M. McVay, The University of Florida, pp. 435-446.

Schmertmann, J. H. (1991), "The mechanical aging of soils", *The Twenty-Fifth Terzaghi Lecture, ASCE Journal of Geotechnical Engineering*, (Vol. 117, No. 9), pp. 1288-1330.

Schmertmann, J.H (1993), "The bottom-up, Osterberg cell method for static testing of shafts and piles", *ASCE, Progress in Geotechnical Engineering Practice Proceedings, Central Pennsylvania ASCE*, Hershey, PA, 7p.

Seidel, J. and Rausche, F. (1984), "Correlation of static and dynamic tests on large diameter drilled shafts", *Proceedings of the 2nd International Conference on the Application of Stresswave Theory to Piles*, edited by G. Holms, H. Bredenberg, and C.J. Grävare, A.A. Balkema Publishers, Rotterdam, Netherlands, pp. 313-318.

Smith, E.,A., L. (1960), "Pile driving analysis by the wave equation", *Journal of Soil Mechanics and Foundations, ASCE*, August

Stevens, R. F. (2000), "Pile acceptance based on combined CAPWAP analyses", *Proceedings of the 6th International Conference on the Application of Stresswave Theory to Piles*, edited by S. Niyama and J. Beim, A.A. Balkema Publishers, Rotterdam, Netherlands, pp. 17-27.

Thompson, C. D. and Thompson, D.E. (1983), "Real and apparent relaxation of driven piles", *ASCE Proceedings of Symposium 6 at the 1983 Spring Convention in Philadelphia, PA*, edited by G.G. Goble, pp.103-114.

The Role of Ram Momentum in Pile Driving
Part IV: Applications

David K. Crapps[1], PE, Member, ASCE

[1]Principal Engineer, David K. Crapps, Inc., dcrapps@attglobal.net

ABSTRACT: The author explored the importance of momentum in pile driving in a paper published in GSP No. 125 (see Crapps, 2004). Parts I through III of the GSP No. 125 paper presented the pertinent theory and discussed the role of momentum in pile force, penetration and pile specifications. The entire hammer-pile-soil system must be considered for successful pile installation. This paper presents relationships to estimate the pile force and ram momentum required for pile penetration for specification and hammer selection purposes. These relationships and principles presented herein may be used to assist with pile design, to evaluate the effects of site variability, to assist with hammer selection, to evaluate the effects of hammer performance variability and to reduce and/or settle pile claims.

KEY WORDS: force, impedance, impulse, momentum, momentum/impulse ratio, pile driving, prediction equations, specifications

INTRODUCTION

The force and momentum required to drive a pile depends upon the entire hammer-pile-soil system. Problems associated with pile installation often occur in the early stages of construction resulting in a delay of all subsequent construction activities. Delay claims of this type are often expensive to settle.

The Owner provides the site and usually selects the Designer and sometimes selects the Contractor. The Owner often expects the Designer to develop contract documents that pass all the risks to the Contractor. The Designer selects the pile type and size and usually includes allowable stresses in the specifications. The Contractor selects the hammer to drive the pile while maintaining the stresses within the specified allowable limits. The Contractor normally has only a few weeks to develop his bid while the Designer often has months or even a year or more for design. Therefore, the Designer and Owner should provide as much information as practical for the Contractor to evaluate in preparing his bid. The Designer, with input from an experienced geotechnical engineer, should include sufficient information (including borings, elevations and water table information) for the Contractor to evaluate the site conditions, evaluate site variability and to understand the risks involved. The goal of this paper is to provide a better understanding of the role of momentum in pile driving to assist Designers with pile design and Contractors in hammer selection which will hopefully reduce the cost to the Owner from claims.

This paper concentrates on impact hammers used to drive piles. The theory applies to all types of impact hammers for which the impact velocity may be estimated accurately. The equations apply primarily to gravity hammers, single acting air/steam hammers and hydraulic hammers. The results are less accurate for diesel hammers, especially at small strokes (say less than about 60% of the maximum stroke) and for double acting air/steam hammers. (The theory herein does not apply to vibratory hammers.)

The equations for force apply to piles with a constant cross section and account for different pile materials (wood, concrete and steel). However, the major emphasis in this paper is directed toward concrete piles because they have more constraints than other piles because they are generally weaker in tension than compression. The square wave approximations given herein apply to impedance ratios normally used to drive concrete piles and are more accurate for cases for which all the penetration occurs in one wave cycle. The equations become less accurate for cases where the time to maximum incident force is less than the wave length generated in the pile by the hammer.

HAMMER FACTS

The author has made a long search for one number that will uniquely define a hammer for specification purposes. Energy has traditionally been used for this purpose. However, energy has many limitations because it is proportional to impact velocity squared. The author has advocated the use of momentum primarily because momentum is proportional to impact velocity to the first power, also true for other important pile driving variables including hammer damping (see Pile Dynamics (1997), pile force, incident velocity, pile damping, side damping, point damping, etc. The author recognizes that momentum alone also has limitations because the entire hammer-pile-soil system must be considered.

Lowery, Hirsch, et al (1969) presented a solution to the wave equation for a single degree of freedom hammer (one mass and one spring) for a long slender elastic pile. A study of the these equations and subsequent research by the author has revealed a number of facts about hammers that contribute to a better understanding of the role of the pile hammer in the impact driving of piles. Some of these facts are listed below.

1. Impedance methods accurately estimate the maximum pile head stress (or force) in the pile due to impact.

2. The maximum input pile stress (or force) is directly proportional to the impact velocity. Impact velocity is the most important variable for control of forces for a given impedance ratio.
3. The impedance ratio and impact velocity fully define the maximum impact force for a given pile material. (Note that three items are required - the hammer impedance, the pile impedance and the impact velocity or some variation of these three; for example, impedance ratio, momentum and hammer frequency. (See Crapps (2004) and Notation herein for definitions).
4. The time to a given force (for example, the time to the maximum force) is inversely proportional to hammer frequency for a given impedance ratio. Therefore, the family of force-time curves for the infinite number of frequencies available for a given impedance ratio, may be reduced to one curve by presenting the curve as force versus frequency multiplied by time (F vs $\omega_h t$) curve. Also note that one may present a dimensionless curve by normalizing the results by dividing force by the maximum force (F/F_{max} vs $\omega_h t$)
5. The area beneath the force-time curve (pile impulse) is inversely proportional to the hammer frequency.
6. Multiple degree of freedom hammers may be approximated by a single degree of freedom hammer, both with the same mass and impact velocity – therefore, the same momentum at impact. (See Crapps (2004) and Appendix A for details.)
7. Other "hammer facts" will become apparent later in this paper.

Theoretically, there are an infinite number of hammers due to the infinite combinations of ram masses and cushion stiffnesses. However, hammer manufacturers offer a finite number of hammers and practical limits are placed by available and relatively durable pile cushion materials. The following discussions will set practical bounds to facilitate hammer selection.

THE ROLE OF ALLOWABLE STRESSES & FORCES

All pile materials have limitations and possible pile damage results when these limits are exceeded. Therefore, pile specifications should include allowable stresses. These allowable limits should consider the ultimate strength of the pile material(s), effect of prestressing, as well as high stress fatigue of the pile material(s). (See Crapps (1977) and Schmertmann & Crapps (1977) for a discussion of allowable stresses. Also see Florida Department of Transportation (2006) for allowable stresses used successfully by the FDOT since the early 1980's. AASHTO (2004), FHWA (2006) and PDCA (2007) provide additional references for allowable driving stress limits.)

The maximum allowable stress defines a maximum allowable force for the pile. (In the following discussion, one may substitute allowable stress for allowable force). Boundary conditions, wave shape and wave length must be considered to keep the pile forces within allowable limits. The maximum compression force in the pile will be between the maximum input force at the pile head and twice the maximum input force (when driving to a rigid boundary). One can see that boundary conditions at the pile tip may require limiting the input force into the pile to 0.5 times the allowable compressive force at final driving. Fortunately, a pile will withstand a few blows above most allowable forces. Therefore, a practical refusal limit will help to prevent damage when a rigid boundary is contacted.

The highest value of tension in the pile occurs with little or no resistance at the pile tip. In this case, the reflected force may equal the value of the incident compression force. One can see that boundary conditions at the pile tip may require limiting the input force into the pile to 1.0 times the allowable tension force. The allowable tension force for a wood or steel pile is approximately the same as the allowable compression force (different sign; but, same absolute value) as the allowable compressive stresses. The maximum input force for steel and wood piles lies between 0.5 and 1.0 times the maximum allowable compression force.

The allowable pile tension force for concrete piles is much lower than the allowable compression force. For example, the ratio of allowable tension to allowable compression may range from about 0.05 for reinforced concrete piles when tension cracking is not allowed to about 0.35 for prestressed concrete piles. Therefore, the maximum input force for concrete piles must be between 0.05 and 1.0 times the maximum allowable compressive stress for reinforced concrete piles and between 0.35 and 1.0 times the allowable compressive stress times the pile area for prestressed concrete piles.

From a practical standpoint, the most effective stress/force control results through control of the impact velocity. However, cushion selection is also important for stress/force control, especially when driving concrete piles. The following discussion helps to quantify the role of allowable force/stress in pile driving including hammer and cushion selection.

Crapps (1977) and Schmertmann & Crapps (1977) defined the "characteristic impedance" of material as ρc (the product of the mass density and the wave speed). The characteristic impedance is unique to a given material. Table 1 summarizes values of the characteristic impedance for selected wood, concrete and steel piles.

Crapps (2004) presented Equation (1) as a method to estimate the incident maximum force in a pile:

$$F = \frac{C_f Z_h Z_p V_0}{Z_h + Z_p} = C_f M_r V_0 \left(\frac{\omega_h}{1 + Z_R}\right) \tag{1}$$

(See the "Notations" section in this paper, Crapps (2004), Crapps (1977) and Schmertmann & Crapps (1977) for definitions of the variables for force. Also see Kummel (1984) for a method proposed to determine force for Delmag diesel hammers.) For a given pile, there is a limiting or "allowable" particle velocity for compression, V_{apc} and a limiting or "allowable" particle velocity for tension, V_{apt} to maintain the stresses within allowable limits as shown in Equations (2a) and (2b).

TABLE 1 - Selected Theoretical Pile Material Properties

Material	f'_c MPa	E MPa	γ kN/m^3	c m/s	$Z_o = \rho c$ kN·s/m^3
Wood:					
1.	13.79	8,970	6.60	3,650	2,460
Concrete:					
2a.	34.47	27,790	23.56	3,400	8,170
2b.	41.37	30,440	23.56	3,560	8,550
2c.	48.26	32,880	23.56	3,700	8,890
2d.	55.16	35,150	23.56	3,820	9,190
Steel:					
3a.	248.21	206,840	77.28	5,120	40,370
3b.	344.74	206,840	77.28	5,120	40,370
3c.	413.69	206,840	77.28	5,120	40,370

(see **NOTATION** section for unit abbreviation definitions and SI to English unit conversions)

$$V_{apc} = \frac{\sigma_{apc}}{\rho c} = \frac{F_{apc}}{Z_p} \tag{2a}$$

$$V_{apt} = \frac{\sigma_{apt}}{\rho c} = \frac{F_{apt}}{Z_p} \tag{2b}$$

Theoretically, one can use a rigid hammer without a pile cushion ($Z_h = \infty$), to drive piles provided the impact velocity, V_0, is less than or equal the allowable particle velocity. However, impact is not always perfect and cushioning is required to account for hammer-pile misalignment. Many hammers offer only one or two selections of strokes and therefore impact velocities. From a practical standpoint, one must use a pile cushion when driving concrete piles to reduce the particle velocity to acceptable values. The cushion design involves selection of a cushion stiffness to reduce the hammer impedance, Z_h, to the level required to maintain the stresses within allowable limits for a given impact velocity. One may picture the role of the cushion as a mechanism to reduce the particle velocity in the pile from the value of the impact velocity of the hammer to the allowable particle velocity used to protect the pile from damage.

Table 2 contains allowable stresses for selected materials and allowable particle velocities for compression, V_{apc}, and tension, V_{apt}. The allowable stresses are based upon Florida Department of Transportation (2006) specifications.

TABLE 2 - Allowable Stress & Particle Velocity

Material	f'_c MPa	f_{pe} MPa	σ_{apc} MPa	V_{apc} m/s	σ_{apt} MPa	V_{apt} m/s
Wood						
1.	13.79	-	14.48	5.90	14.48	5.90
Concrete						
2a.	34.47	0.00	24.13	2.95	1.58	0.19
2b.	34.47	6.89	18.96	2.32	8.82	1.08
2c.	41.37	6.89	23.79	2.78	8.98	1.05
2d.	48.26	6.89	28.61	3.22	9.11	1.02
2e.	55.16	6.89	33.44	3.64	9.24	1.01
Steel						
3a.	248.21	-	223.39	5.53	223.39	5.53
3b.	344.74	-	310.26	7.68	310.26	7.68
3c	413.69	-	372.32	9.22	372.32	9.22

Note that the Table 2 allowable stresses are based upon FDOT specs (see FDOT, 2006) with the allowable tension based on piles over 60 ft. (18.3 m) long

Equation (3a) defines the stress ratio.

$$\sigma_{Rac} = \frac{\sigma_{apc}}{(\rho c)V_0} = \frac{\sigma_{apc}}{Z_0 V_0} \tag{3a}$$

Multiplying the numerator and denominator of Equation (3a) by the area of the pile, A_p, gives Equation (3b), which defines the force ratio. Also note from Equation (2a) that F_{apc}/Z_p equals V_{apc}. Also define the velocity ratio, V_{Rac} as V_{apc}/V_0 and one may see that the force ratio and velocity ratio have the same value as shown in Equation (3b).

$$F_{Rac} = \frac{\sigma_{apc} A_p}{(\rho c A_p)V_0} = \frac{F_{apc}}{Z_p V_0} = \frac{V_{apc}}{V_0} = V_{Rac} \tag{3b}$$

Also note that the stress ratio, force ratio and velocity ratio have the same value as shown in Equation (3c).

$$\sigma_{Rac} = F_{Rac} = V_{Rac} = \frac{C_f Z_R}{(Z_R + 1)} \tag{3c}$$

It follows that equation (4a) may be used to estimate the required impedance ratio for a given pile problem.

$$Z_R = \frac{F_{Rac}}{C_f - F_{Rac}} = \frac{\sigma_{Rac}}{C_f - \sigma_{Rac}} = \frac{V_{Rac}}{C_f - V_{Rac}} \tag{4a}$$

Equation (4b) substitutes the appropriate valve of tension for the compression value and be used to determine the value of the impedance ratio for control of tension stresses.

$$Z_R = \frac{F_{Rat}}{C_f - F_{Rat}} = \frac{\sigma_{Rat}}{C_f - \sigma_{Rat}} = \frac{V_{Rat}}{C_f - V_{Rat}} \tag{4b}$$

Note that calculated zero or negative value of the denominator in Equations (4a) and (4b) indicates that an infinite impedance ratio is acceptable. (A value of impedance ratio of 100 may be used as a close approximation of the effects of a hammer with an impedance ratio of infinity.) The value of C_f given in Crapps (2004) was 1.22 for the WEAP program. Note in the above equations that ρc and Z_p are related to the ability of the pile material to generate stress or force for a given particle velocity while σ_{apc} and F_{apc} are a related to limitations of the pile material. Designers should recognize that an increase in the pile area will increase the allowable dynamic force just as it increases the allowable static force. In some cases an increase in pile area may make pile installation much easier to accomplish without pile damage.

PRACTICAL RANGE OF IMPEDANCE RATIOS

Theoretically one can select a value of Z_R to always keep the stresses within allowable limits. However, there are practical limits.

Wood Piles: Typically, impedance ratios between about 1.5 and 10 are appropriate for the driving of wood piles depending upon impact velocity.

Steel Piles: Typically, impedance ratios between 2.0 and 100 (which behave closely to an infinite impedance ratio) are appropriate for the driving of steel piles depending upon the strength of the steel and impact velocity.

Concrete Piles: Typically, impedance ratios between about 0.35 and 1.75 are appropriate for the driving of concrete piles depending upon the strength of the concrete and impact velocity. Note that there are practical limits for the thickness of the pile cushion. For example, the pile helmet must have sufficient depth to accommodate the pile cushion thickness. The author is aware of contactors using pine plywood pile cushions between 25 mm (1 inch") and 690 mm (27"). However, the practical limits are between about 75 mm (3 inches) and 380 mm (15 inches).

For a simple hammer (single degree of freedom), equation (5) may be used to estimate the cushion thickness.

$$t_{pc} = \frac{M_r A_{pc} E_{pc}}{(Z_R Z_p)^2} \tag{5}$$

(See Crapps (2004) and Appendix A for details of how to develop a single degree of freedom approximation for more complicated hammers.) Equation (A5) in the Appendix provides the equivalent single degree of freedom hammer stiffness.

FORCE REQUIRED FOR PENETRATION

The force required for penetration depends upon the entire hammer-pile-soil system. Theoretically, the incident force must be greater than 0.5 times the resistance for an end bearing pile. Soil damping increases the required force. However, a higher incident force results in higher damping and only a fraction of the additional force is effective in increasing the pile set per blow. An increase in force is required to overcome an increase in soil quake and to provide a greater set per blow. Pile damping also increases the required force in a minor way and will be ignored herein.

One may correctly reason that there is no unique force required for penetration because penetration depends upon the entire hammer-pile-soil system. However, one may estimate the required force for a selected hammer for a given pile-soil system and a desired penetration per blow.

Fischer (1960) used the "graphodynamical method to study the pile driving process and noted that "...the penetration is mainly determined by the area of the pulse above half the ground resistance..." Crapps (2004) presented results of a pile study showing that the area above the resistance in the double wave down curve controls the piles set. This is consistent with the findings of Fischer because the incident force is 0.5 of the wave doubled down force for the ideal case of a long pile. We know that the actual required force for penetration will be greater than O.5R due to damping and the effects of boundary conditions.

Crapps (2004, Fig. 3) presented an idealized impulse wave showing how the impulse is distributed for a point bearing pile. (Note that the discussions below apply to a point bearing pile.) The area of the impulse, A_s, required for pile set and A_Q, required for the quake may be estimated using equations (6a) and (6b).

$$A_s = S(Z_p + JR) \tag{6a}$$

$$A_Q = Q(Z_p + 0.5JR) \tag{6b}$$

These equations are valid for any wave shape. Review of Fig. 3 from Crapps (2004) shows that part of the area required for the quake and all of the area required for pile set are above the resistance in this idealized impulse. Equations (6a) and (6b) reveal information concerning the area. However, one must know, or assume, the shape of the wave and the wave length to estimate the peak force required for penetration. The problem was simplified by limiting consideration to the penetration for one wave cycle and keeping the penetration constant (1.27 mm or 0.05 inches per blow). (Note that 20 blows per inch (0.05 inches per blow = 20 blows per 25.4 mm) is often specified as "practical refusal".)

The author found that a square wave provides reasonable estimates of force over the practical range of impedance values (0.35 to 1.75) used for the driving of concrete piles. Figure 1 demonstrates how the wave area may be apportioned. The variables shown in Figure 1 are defined below and apply to a point bearing pile:

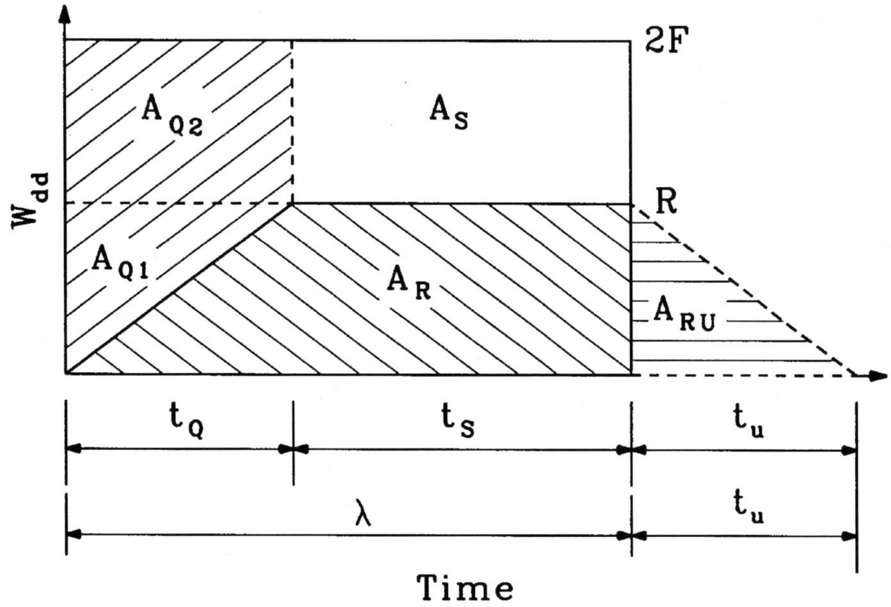

FIG. 1 Simplified Wave Down, Double Impulse Wave

The sum of A_{Q1} plus A_{Q2} equals the total impulse area, A_Q, required for the quake. Note that A_{Q1} lies below the point resistance, R, while A_{Q2} lies above R. The quake is reached at time, t_Q. The impulse area required for pile set is designated as A_S. The time to reach the pile set is designated as t_S. The impulse area below the resistance up to the effective wave length, λ, is designated as A_R. The unload area is designated as A_{RU} which occurs over the unload time, $t_{U,}$ and after the time to reach the pile set.

Crapps (2004) defined Q_{max} as the maximum quake which occurs when the pile set, S, equals zero and S_{mas} as the maximum pile set which occurs when the quake, Q, equals zero. Equations (7a) and (7b) provide the values of Q_{max} and S_{mas} for a square wave.

$$Q_{\max} = \frac{(2F - 0.5R)\lambda}{(Z + 0.5JR)} \tag{7a}$$

$$S_{\max} = \frac{(2F - R)\lambda}{(Z + JR)} \tag{7b}$$

Equations (8a), (8b) and (8c) define b and $b'$and λ:

$$b = \frac{A_Q}{A_{Q\max}} = \frac{Q}{Q_{\max}} = \frac{t_Q}{\lambda} \tag{8a}$$

$$b' = \frac{S_{max}}{Q_{max}} = b\left(\frac{Z + 0.5JR}{Z + JR}\right) \tag{8b}$$

$$\lambda = \frac{(Z_R + 1)}{\omega} = \text{effective wave length approximation} \tag{8c}$$

A square wave approximation was tested for 29 of the 31 wave equation cases presented in Crapps (2004). (Two cases with impedance values much greater then 2.0 were eliminated because they are outside the limits normally used for the driving of prestressed concrete piles.) These 29 cases had Z_R values ranging from 0.08 to 1.99 which more than covers the estimated impedance range between 0.35 and 1.75 estimated by the author as the practical range for driving prestressed concrete piles. Equation (9) provides an estimate of the required force to drive a pile.

$$F = 0.64\left[R + \left(\frac{A_S + bA_Q}{\lambda}\right)\right] = 0.64\left[R + \left(\frac{(S + b'Q)(Z + JR)\omega}{\lambda}\right)\right] \tag{9}$$

The force in equation (9) is the maximum incident force. Note that the force (F) occurs on both sides of the equation (on the left and in b and b'). However, the square wave was used to provide an estimate of b or b' and the calculated force was correlated with the maximum incident force from the wave equation. Equation (9) has a correlation coefficient of 0.967, a bias of 0.999 and a coefficient of variation of about 10.6% for the 29 cases analyzed.

MOMENTUM REQUIRED FOR PENETRATION

Equation (9) provides an estimate of the maximum incident force required for penetration for a given hammer-pile-soil system. Note that Figure 1 approximates the area as a square wave up through completion of the pile set. Thereafter, the unload portion of the curve was approximated as a triangle. For regression purposes, the unloading area of the impulse, A_{RU} was assumed equal to the area beneath the resistance curve to the time the quake is reached at t_Q.

Equation (10) was used to determine the value of t_Q for the square wave approximation.

$$t_Q = \frac{A_Q}{(2F - 0.5R)} \tag{10}$$

Equation (11) provides an estimate of the ram momentum at impact for the 29 cases mentioned above which covers the likely range of impedance ratios used for the driving of prestressed concrete piles.

$$M_R V_0 = 0.41\left[2F\lambda + \frac{Rt_Q}{2}\right] \tag{11}$$

Use the values of force, F, from equations (9) in equations (10) and (11). Note that Equation (11) is valid for the same range of impedance ratios valid for equation (9). This equation has a correlation coefficient of 0.987 , the bias was 0.98 and the coefficient of variation was about 8.8% for the 29 cases considered.

SITE VARIABILITY

Site variability may include variability in soil resistance, soil damping and soil quake. The equations presented in this section may be used to approximate the effects of site variability.

Variation in Resistance: Static analyses of soils data at the site may be used to estimate the variability of resistance at the site.

Change in Soil Quake: This paper uses a pile with all the resistance at the tip to develop relationships, which are normally conservative, to demonstrate principals involved in pile driving. The primary effect of an increase in soil quake is that more impulse is required to maintain the same pile set per blow or the set is reduced for the same impulse. Crapps (2004) introduced the concepts of S_{max} and Q_{max}. A straight line approximation of a plot between pile set and quake for a given hammer-pile-soil system with quake and set as the only variables may be extrapolated to provide the values of S_{max} and Q_{max}. The slope of the line is S_{max}/Q_{max} which has been defined herein as b'. Equation (12) defines a straight line approximation between pile set, S, and Quake.

$$S_{\max} = S + b'Q \tag{12}$$

Dividing Equation (12) by S_{max} provides Equation (13).

$$\frac{S}{S_{\max}} + \frac{Q}{Q_{\max}} = 1 \tag{13}$$

Equation (13) may be used to estimate the effect of a change of quake upon pile set. For best results, actual wave shapes should be used. However, a square wave approximation can be useful for variation studies.

Change in Soil Damping: A change in soil damping changes the value of S_{max} and Q_{max}. The value of b' will also be different. Equations (14a), (14b) and (14c) may be used to estimate the effect upon each of these variables.

$$S_{\max 2} = S_{\max 1}\left(\frac{Z + J_1 R}{Z + J_2 R}\right) \tag{14a}$$

$$Q_{\max 2} = Q_{\max 1}\left(\frac{Z + 0.5 J_1 R}{Z + 0.5 J_2 R}\right) \tag{14b}$$

$$b_2^{'} = \frac{S_{\max 2}}{Q_{\max 2}} \tag{14c}$$

If the soil Quake, Q is maintained constant, and the damping is increased, the pile set will decrease. Equation (15) may be used to estimate the new pile set, S_2.

$$S_2 = S_{\max 2} - b_2^{'} Q \tag{15}$$

SPECIFICATIONS AND SPECIAL PROVISIONS

This section provides suggestions to incorporate the momentum impulse concept into specifications. (See Part III of Crapps (2004) for additional details.) The author recommends including the following in pile specifications or special provisions:

1. **Allowable driving stresses:** Maintaining driving stresses below allowable levels are a must to prevent pile damage during driving. The Florida Department of Transportation Standard Specifications include allowable stresses for the driving of prestressed concrete piles. They have been used successfully for many years and the author recommends the allowable stresses included therein.
2. **Practical Refusal:** Specifying practical refusal protects the hammer and pile from damage. The Florida Department of Transportation Standard Specifications include a practical refusal value of 20 blows per 25 mm (20 blows per inch). This value has been used successfully for many years and the author recommends its use. Note that reaching practical refusal signals that the Contractor must take appropriate steps to insure pile installation without damage. Reaching practical refusal usually indicates the need for a penetration aid (jetting, predrilled pile hole, etc). However, in some cases, practical refusal may signal the need to change the cushion or even the hammer.
3. **Penetration Aids:** The need for penetration aids and the method of payment should be addressed in the pile specifications or special provisions to prevent or minimize construction claims.
4. **Variable Stroke Hammer:** The author recommends a variable stroke hammer for the driving of all piles. However, a variable stroke hammer is a must for the driving of concrete piles to maintain tensile stresses within allowable limits.

5. **Hammer Reserve:** A hammer reserve factor, H_{RF}, should be included in the specifications to allow extra hammer capacity to drive though layers with an ultimate resistance higher than that required to provide the safe resistance time the factor of safety (S.F.) or the factored load divided by the resistance factor (Φ) for Load and Resistance Factor Design (see (AASHTO, 2004) and (Likins, 2004) for further information). The hammer reserve factor also helps to insure there is reserve hammer capacity to guard against thc cffccts of overestimates of hammer efficiency and other factors unknown at the time the specifications are developed. The author recommends a hammer reserve factor of 1.5 for most projects. (When a FS of 2 is specified the ultimate driving resistance, R, would be 3 times the safe resistance, $R = H_{RF}(S.F.)R'$ (= 1.5 x 2.0 x R'). The absolute minimum hammer reserve factor should be 1.25 for a uniform site with soils typically providing an increase in resistance with depth.
6. **Momentum to Impulse Ratio:** The momentum to impulse ratio, M_{IR}, is a new item which is the author's attempt at one number to specify an adequate hammer. Note that the major variables for the hammer-pile-soil system are included. The equation for specification purposes for concrete piles would be as follows:

$$F = 0.64\left[H_{RF}(S.F.)R' + \left(\frac{(S + b'Q)(Z + JH_{RF}(S.F.)R')\omega_h}{(Z_R + 1)} \right) \right] \tag{16}$$

$$t_Q = \frac{A_Q}{(2F - 0.5H_{RF}(S.F.)R')} \tag{17}$$

$$M_{IR} = \frac{M_R V_0}{0.45[2F\lambda + 0.5H_{RF}(S.F.)R't_Q]} \tag{18}$$

$$M_{IR} \geq 1.0$$

Where:

$R' = R/(S.F.)$ or

$R' = \Phi R$

$S.F.$ = factor of safety

Φ = soil resistance factor

Note that the coefficient (0.41) in the denominator of Equation (11) was raised to 0.45 (about one standard deviation) in Equation (18) to help guard against under predictions of the required momentum. Field trial may show that this coefficient will require adjustment.

7. **Expected Range of Damping and Quake:** The engineer normally has much more time to evaluate a site than the Contractor has to prepare his bid. Many claims could be avoided if Engineers provide estimates of the soil damping and quake in the Plans or Special Provisions. In the event additional construction costs are encountered because the actual quake and damping

values are outside the expected range, all parties could reasonably agree that there is a "changed condition" unexpected by the engineer. Requiring inclusion of expected values of damping and quake would encourage a design phase test program to provide reasonable values. A design phase test programs is usually cost effective and often pays many times over by eliminating or minimizing construction claims.

See Appendix B for an example using the key equations in the preceding sections.

SUMMARY & CONCLUSIONS

1. Momentum concepts were used to develop a method for hammer and cushion selection based upon allowable stresses which were translated to allowable forces and allowable particle velocities and finally maximum appropriate impedance values.
2. A momentum to impulse ratio was developed which considers the major hammer-pile-soil variables. The author proposes using this new variable for hammer selection and specification purposes. Like all new concepts, the momentum to impulse ratio should receive "field trial" and refined as necessary before final adoption.

ACKNOWLEDGEMENTS

Dr. Schmertmann has been a mentor, partner and friend for over 34 years beginning in 1973 as Chairman of the author's Ph.D. Committee. The author appreciates and gratefully acknowledges the interaction and constructive criticism on this and many other projects through the years.

The results described herein are from research beginning in 1973 and from subsequent pile related projects sponsored by the Florida Department of Transportation which the author acknowledges with appreciation. Additional knowledge has been gained from continuing in-house research.

The author also owes much to his wife, Lee, for her patience while he pursued his interest in the pile driving process in school and the many years of in-house research since.

NOTATION

The following symbols and abbreviations are used in this paper.

A = area (usually pile area)

A_p = area of pile

A_{pc} = area of pile cushion (normally assumed equal to pile area)

A_Q = impulse area required for the quake

A_{Q1} = area for quake when the wave down, doubled is < the point resistance, R

A_{Q2} = area for quake when the wave down doubled is ≥ the point resistance, R

A_Q = $A_{Q1} + A_{Q2}$ = total impulse area required for the quake

A_R = impulse area required for the soil resistance, R

A_{RU} = impulse area required for the soil resistance, *R*, unload
A_S = impulse area required for the pile set
b = coefficient = A_Q/A_{Qmax}
b' = coefficient = S_{max}/Q_{max}
C_f = force coefficient (= 1.22 for WEAP Program)
c = wave speed of pile material
E = Young's modulus (usually pile Young's modulus)
E_{pc} = Young's modulus for pile cushion
F = force in pile, normally incident force (also see W_{dd})
F_{apc} = allowable compression force
F_{apt} = allowable tension force
F_{Rac} = force ratio for compression
F_{Rat} = force ratio for tension
fps = feet per second = 0.3048 m/s
ft = foot = 0.3048 m
g = gravitational constant = 9.807 m/s^2 = 32.17 ft/sec^2
H_e = "equivalent stoke" = rated kinetic energy divided by the ram weight
H_{RF} = hammer reserve factor
K_1 = stiffness of hammer element 1
K_2 = stiffness of hammer element 2
K_n = stiffness of hammer element n
K_h = stiffness of hammer
kcf = kips per cubic foot (kcf x 157.1 = kN/m^3)
kip = 1,000 lbs = 4.4482 kN
ksi = kips per square inch (ksi x 6.895 = MPa)
ks/cf = kip seconds per cubic foot (ks/cf x 157.1 = kN·s/m^3)
J = Smith damping factor (for end bearing in this paper)
lb = pound = 4.44822 Newtons
M_1 = mass of hammer element 1, usually the mass of the ram
M_1 = mass of hammer element 1
M_2 = mass of hammer element 2
M_n = mass of hammer element n
M_r = mass of ram = W_r/g
m = meter = 3.2808 ft
N = Newton = (1/4.4482) lbs
Q = soil quake (point quake in this paper)
Q_{max} = the maximum quake for a given incident wave (causes a zero pile set)
Q_{max1} = Q_{max} for case 1
Q_{max2} = Q_{max} for case 2
R = ultimate soil resistance (ultimate point resistance in this paper)
R' = $R/(S.F)$ or ΦR = safe or factored resistance, respectively
S = pile set = penetration per blow
$S.F.$ = safety factor
S_{max} = the maximum pile set for a given incident wave (when quake = zero)
S_{max1} = S_{max} for case 1
S_{max2} = S_{max} for case 2

T	= hammer time = time for wave to travel through hammer
t_{pc}	= thickness of pile cushion
t_Q	= time to reach soil quake
t_S	= time to reach pile set
t_U	= time to unload soil quake
V_0	= impact velocity = $(2g\eta He)^{0.5}$
V_{apc}	= allowable particle velocity for compression
V_{apt}	= allowable particle velocity for tension
V_{Rac}	= velocity ratio for compression
V_{Rat}	= velocity ratio for tension
W	= weight of hammer element
W_1	= weight of hammer element 1, usually the ram weight
W_2	= weight of hammer element 2
W_n	= weight of hammer element n
W_r	= weight of ram
W_{dd}	= wave down, doubled = $2F$ for "long" piles (sufficiently long for no interference at the pile head from the returning wave).
Z_o	= characteristic impedance of material = ρc
Z_h	= hammer impedance
Z_p	= pile impedance
Z_R	= the impedance ratio = the ratio of the hammer impedance to the pile impedance
Φ	= resistance factor
δ	= total static deflection of hammer elements
δ_1	= static deflection of hammer element 1
δ_2	= static deflection of hammer element 2
δ_n	= static deflection of hammer element n
γ	= unit weight of pile material
λ	= effective wave length (for pile quake and pile set)
ρ	= mass density of pile material = γ/g
η_h	= hammer efficiency
σ_{apc}	= allowable pile compression stress
σ_{apt}	= allowable pile tension stress
σ_{Rac}	= stress ratio for compression
σ_{Rat}	= stress ratio for tension
ω	= frequency (usually hammer frequency)
ω_1	= frequency of hammer element 1
ω_2	= frequency of hammer element 2
ω_n	= frequency of hammer element n
ω_h	= hammer frequency

REFERENCES

American Association of State Highway and Transportation Officials (2004) "AASHTO LRFD Bridge Design Specifications", 3rd Edition, Washington, DC.

Crapps, David K. (1977), *Wave Equation Study and Recommended Specifications for the Impact Driving of Prestressed Concrete Piles*, Ph.D. Dissertation, University of Florida, Gainesville, Florida.

Crapps, David K. (1997), Project R850 unpublished notes, Schmertmann & Crapps, Inc., Gainesville, FL.

Crapps, David K. (2004), "THE ROLE OF MOMENTUM IN PILE DRIVING", *CURRENT PRACTICES AND FUTURE TRENDS IN DEEP FOUNDATIONS*, DiMaggio, Jerry A. and Hussein, Mohamad H., Editors, ASCE, *Geotechnical Special Publication No. 125*, pp. 276-309.

Federal Highway Administration (2006), "*FHWA Manual on Design and Construction of Driven Pile Foundations*" Vol. 1, FHWA-NHI-05-042 and Vol. 2, FHWA-NHI-05-043, Washington, DC.

Fischer, H.C. (1960), "On Longitudinal Impact IV, New Graphodynamical Pulse Method of Computing Pile Driving Processes", *Applied Science Research*, Section A, Vol. 9, pp. 93-138.

Florida Department of Transportation (1996), *Standard Specifications for Road and Bridge Construction*, Edition of 1996, Tallahassee, Florida.

GRL (1995), "GRLWEAP, Wave Equation Analysis of Pile Driving", Goble, Rausche Likins and Associates, Inc., Cleveland, Ohio.

Kummel, F., (1984) "The Kummel Method for Calculation of Impact Forces in Piles", *The Proceedings of the Second International Conference on the Application of Stress-Wave Theory on Piles*, Stockholm, 27-30 May, pp. 273-281.

Likins, Garland, "PILE TESTING – SELECTION AND ECONOMY OF SAFETY FACTORS", *CURRENT PRACTICES AND FUTURE TRENDS IN DEEP FOUNDATIONS,* DiMaggio, Jerry A. and Hussein, Mohamad H., Editors, ASCE, *Geotechnical Special Publication No. 125*, pp. 239-252.

Lowery, Lee Leon, Jr.; Hirsch, T. J.; Edwards, Thomas C.; Coyle, Harry M. & Samson, Charles H. (1969). "Development of Equations for Impact Stresses in a Long, Slender Pile", Appendix A, "Pile Driving Analysis State of the Art", *Research Report 33-13 (Final)*, Texas Transportation Institute, Texas A&M University, College Station, Texas, January.

Pile Driving Contractors Association (2007), "INSTALLATION SPECIFICATION FOR DRIVEN PILES", *PDCA Specification 102-07*

Pile Dynamics, Inc. (1997), "Wave Equation Analysis Program", Cleveland, Ohio.

Schmertmann, John H. & Crapps, David K. (1977), *Wave Equation Study and Recommended Specifications for the Impact Driving of Prestressed Concrete Piles*, Final Report Number D657 for Florida Department of Transportation, College of Engineering, Department of Civil Engineering, University of Florida, Gainesville, Florida, June 1977.

APPENDIX A: EQUIVALENT SINGLE DEGREE HAMMER EQUATIONS

For convenience, the equations to approximate the equivalent single degree of freedom hammer for a more complex multiple degree of freedom hammer are included in Appendix A herein (see Crapps (2004) for additional details).

The ram mass and velocity (therefore, momentum) are the same for the actual and the equivalent single degree of freedom hammer approximation. The stiffness and hammer frequency are modified according to the following relationships. See Crapps (2004) and the Notation section above for additional information.

$$\omega_1 = \sqrt{\frac{K_1}{M_1}} \tag{A1a}$$

$$\omega_2 = \sqrt{\frac{K_2}{M_2}} \tag{A1b}$$

$$\omega_n = \sqrt{\frac{K_n}{M_n}} \tag{A1c}$$

$$T = \frac{1}{\omega_1} + \frac{1}{\omega_2} + \ldots + \frac{1}{\omega_n} \tag{A2}$$

$$\delta_1 = \frac{W_1}{K_1} \tag{A3a}$$

$$\delta_2 = \frac{W_1 + W_2}{K_2} \tag{A3b}$$

$$\delta_n = \frac{W_1 + W_2 + \ldots W_n}{K_n} \tag{A3c}$$

$$\delta = \delta_1 + \delta_2 + \ldots + \delta_n \tag{A3d}$$

$$Z_h = \frac{W_r T^{0.5}}{\delta^{0.25} g^{0.75}} \tag{A4}$$

$$K_h = \frac{Z_h^2}{M_r} \tag{A5}$$

$$\omega_h = \frac{Z_h}{M_r} \tag{A6}$$

APPENDIX B: EXAMPLE

This Appendix provides an example using the equations herein. A contractor is bidding on a project and the Florida Department of Transportation specifications will be used for allowable stresses. The piles are 610 mm (24 inches) square prestressed concrete piles, 30.5 m long (100 feet) which have a design load of 1,334 kN (300 kips). The pile concrete strength will be 41.37 MPa (6.0 ksi) and the effective prestress level in the piles is 6.89 MN (1.0 ksi). A factor of safety of 2.0 has been specified and the contractor must provide a hammer reserve factor of no less than 1.5. The contract documents include borings which indicate that the piles must be driven through weak soils and tensile stresses are a concern. The specifications require a variable stroke hammer and define Practical Refusal as 20 blows per 25 mm (20 blows/inch). The contract documents also tell the contractor that the expected point quake will be 7.62 mm (0.3 inch) and the expected point damping will be 0.492 s/m (0.15 s/ft).

Table 2 provides the allowable stresses: Concrete, 2c provides 23.79 MPa (3.45 ksi) allowable compression and 8.98 MPa (1.30 ksi) allowable tension. Allowable particle velocities are also provided in Table 2: 2.781 m/s (9.125 fps) for the allowable particle velocity for compression and 1.049 m/s for the allowable particle velocity for tension. Note that the value of the allowable particle velocities could also have been calculated using the allowable stresses and the value of Z_0 from Table 1 (8550 kN·s/m^3 = 54.44 kip·sec/ft^3) in Equations (2a) and (2b).

Analyses Step 1

The first step is to choose the type of hammer to drive the piles. This choice will provide an estimate of the impact velocity. The contractor in this example prefers air/steam hammers with a long stroke of 0.914 m (3.0 feet) and a short stroke of 0.457 m (1.5 feet). The efficiency factor for this hammer is 0.67 which provides an estimated impact velocity of 3.466 m/s (11.37 fps) at long stroke and 2.451 m/s (8.04 fps) at short stroke. Equation (3a) provides the stress ratio (23.79 x 1000)/(8550 x 3.466) = 0.803 for compression and (8.98 x 1000)/ (8550 x 2.451) = 0.428 for tension. Equation (4b) provides the impedance ratio for the hammer. The maximum impedance ratio for compression is 1.926 [=(0.803/(1.22 – 0.803)] and 0.540 [= (0.428/(1.22 -0.428)] for tension. Tension controls in this example. Note that impact velocities and allowable stresses control the hammer selection process to this point.

Analyses Step 2

The second step is to select a trial hammer which provides the pile force and impact momentum. The contractor has an air/steam hammer with a ram weight of 133.45 kN (30 kips) which he would like to use. The mass of the ram equals 13.61 kN·s^2 (= 133.45/9.807). He now needs to determine the pile cushion he will likely use. The pile impedance, Z_p equals the pile area multiplied by Z_0. The value of Z_0 from Table 1 was 8,550 kN·s/m^3 which provides a pile impedance value of 3,181 kN·s/m. The hammer impedance will equal the pile impedance, Z_p, times the impedance ratio, Z_R (= 3,181 x 0.540 = 1,718 kN·s/m). The contractor will use a pine plywood cushion with a modulus of 206.84 MPa (30 ksi). Equation (5) provides an estimate of the pile cushion thickness required to keep the tension stresses within allowable limits (13.61 x 0.61^2 x 206.84 x 1000/(0.54 x 3,181)^2 = 0.355 m (14.0 inches) if the hammer is a single degree of freedom hammer. From Equation (A6), the hammer frequency, ω_h, equals 126.2 s^{-1} (= 1717.7/13.61).

Analyses Step 3

The third step is to compare the force provided by the hammer to the estimated force required. Equation (1) estimates the force provided by the hammer. Note that the contractor will use the hammer at full stroke to provide the final required ultimate resistance. The pile force equals = 1.22 x 13.61 x 3.466 x 126.2/(1 + 0.54) = 4,712 kN.

The ultimate resistance equals the hammer reserve factor, H_{RF}, multiplied by the safety factor, S.F. times the design load, R', which equals 4002 kN (= 1.5 x 2.0 x 1,334). The value of the wave length, λ, from Equation (8c) equals 0.0122 s [=(0.540 + 1)/126.2]. The value of A_S from Equation (6a) equals 6.540 kN·s [= (1.27/1000) x (3,181 + 0.492 x 4,002)]. The value of A_Q from Equation (6b) equals 31.741 kN·s [= (7.62/1000) x (3,181 + 0.5 x 0.492 x 4,002)]. The value of Q_{max} equals 0.0217 m [= (2 x 4,712 – 0.5 x 4,002) 0.0122/(3,181 + 0.5 x 0.492 x 4002)] The value of b from Equation (8a) equals 0.351 [= (7.62/1000)/0.0217)]. Equation (9) provides an estimate of the required force which equals 3,489 kN {= 0.64[4,002 + (6.540 + 0.351 x 31.741)/0.0122]}. Note that the force provided by the hammer (4,712 kN) is greater than the required force (3,489 kN) to drive the pile.

Analyses Step 4

The fourth step is to check to make sure the hammer provides adequate momentum to meet the specification requirements. Equation (17) provides the quake time, t_Q for the required force which equals 0.00638 s [= 31.741/(2 x 3,489 – 0.5 x 1.5 x 2.0 x 1,334))] The required momentum ratio from Equation (18) equals 1.065 {= [(13.61 x 3.446)/(0.45 x (2 x 3,489 x 0.0122 + 0.5 x 1.5 x 2.0 x 1,334 x 0.00638)]} The momentum ratio of 1.065 is greater than 1.000 and the hammer meets the specification requirements.

Design of Foundations for a Wind Turbine Employing Modern Principles

Lymon C. Reese[1], Honorary Member, ASCE, and Shin-Tower Wang[2], Member, ASCE

[1] Professor Emeritus, Dept. of Civil Engineering, The Univ. of Texas, Austin, Texas, PH (512) 244-6464; email: *lymonreese@aol.com*
[2] Project Manager, Ensoft, Inc. 3003 W. Howard Lane, Austin, Texas 78728, PH (512) 244-6464; email: *ensoft@ensoftinc.com*

ABSTRACT: A design of the foundation for a wind turbine is presented employing a single pile and a companion design employs a pile group. Modern principles, using concepts of soil-structure interaction, are recommended in the designs. An important criterion in design is to limit the rotation at the foundation to within a tolerance of 0.001 radians in order to restrict the movement at the turbine, which is typically 50 m to 80 m above the ground. The method of design, based on the use of *p-y* curves and *t-z* curves for taking soil-structure interaction into account, is rational and practical for wind-turbine foundations.

KEY WORDS: design, foundations, lateral loads, pile groups, piled rafts, piles, rotation, soil-structure interaction, wind turbines

INTRODUCTION

The installation of wind farms in the United States and abroad has become increasingly popular and the popularity of these environmentally favorable systems is expected to continue. The obvious advantages are the decreased dependence on petroleum and the cleaner atmosphere. The disadvantages are the vagaries of the wind, the first cost of the installations, and the possible destruction of wildfowl.

Texas has seen in the last ten years the birth of a growing market for wind energy, with an explosion of development. Energy from wind totals more than 3,000 megawatts (MW) and includes some of the world's largest wind farms. For example, the Desert Sky Wind Farm (CPS Corporation, *www.desertskywind.com*), completed in 2001, is a 160.5 MW wind-power-generation facility located near the far West Texas town of Iraan, Pecos County. The site includes 107 turbines, each rated at 1.5 MW, spread over a 15-square-mile area on Indian Mesa.

A wind turbine, that is a component in a wind farm, is presented in a photograph, Fig. 1. The height of the tower can range from 22 to 120 m with the rotor diameter up to 76 m. The largest turbine can generate power up to 3.6 MW. The stability of the foundation and keeping the rotation of the wind turbine at the top of the tower within operational limits are important subjects for the design of wind-farm foundations. The aim of this paper is to discuss the major issues that need to be considered when designing foundations for wind turbines.

FIG. 1 View of a Wind Turbine in Texas with Oil Pump in Background (photo, courtesy of CPS Energy Corporation, Desert Sky Wind Farm, *www.desertskywindfarm.com*, 2004)

DIMENSIONS OF A TYPICAL WIND TURBINE

The size of a wind turbine and the nature of the loading present problems not commonly encountered by many foundation engineers. Based on information provided by the Desert Sky Wind Farm (2004), the following data are for a particular turbine in a wind farm and may be instructive to many engineers.

1. Height of tower is approximately 200 feet (equal to a 20-story building).
2. Diameter of rotor-assembly (sweep of blades) is 231 feet (greater than the wingspan of a Boeing 747-400).
3. Total height of tower and blades is 328 feet (taller than the Statue of Liberty).
4. Length of each blade is 112 feet.
5. Weight of nacelle (module that houses generator at top of tower) is 56.2 tons.
6. Operating wind speeds ranges from 8 to 56 mph.
7. Revolutions per minute during operation ranges from 11 to 20 (one revolution every 3 to 5 seconds).
8. A weather station accompanies each turbine.
9. Automatic "yaw" control is included to keep turbine facing into wind.
10. The weight of rotor assembly is 36.3 tons.
11. The weight of the entire turbine is163.3 tons.
12. The concrete foundations are designed specifically for the turbine described and for a particular soil and are 14 feet in diameter and 18 to 26 feet deep.

LOCATION OF SYSTEM

A number of factors are considered in locating a wind farm but the primary factor is the character of the wind. Meteorological data are usually available for a historical period that will guide the selection of a favorable site. Other factors to be considered are attitude of local community, possible interference with transportation facilities, and transmission of the power. A number of wind farms have been located at sites that are near-shore or offshore. While the subsurface should be important consideration, the assumption is usually made that a successful foundation can be constructed at almost any site.

Prudent and detailed studies are needed to ensure that the wind-generation facility will not adversely impact the natural environment in the area. For examples, areas that are commonly used by endangered species of birds should be regarded as unsuitable for wind development. The wind industry is working with environmental groups, federal regulators, and other interested parties to develop methods of measuring and mitigating effect on birds of operating wind turbines. An archaeological study is also recommended for site selection.

TYPES OF TOWERS

The towers may be tubular steel, tubular concrete, lattice, three-legged, or a guy-wired pole. The size of the turbine, along with possible concern of the local community about esthetics, will dictate in most instances the selection of the design. The use of guy lines is not feasible when the diameter of the turbine is large with respect to the height of the tower.

LOADS ON STRUCTURE

The loads on the tower consist of those generated by the wind and gravity but gravity is usually modest in comparison to the loads from the wind. The principal characteristic of the wind that governs the design of a turbine is represented by the 10-minute wind speed and the standard deviation of the wind speed. The turbulence, the wake, the wind shear, and extreme winds must also be taken into account. Other factors to be considered are hurricanes, earthquakes, rain, snow, and hail. Under certain conditions, consideration should be given to the possibility that ice may form on the turbine. The horizontal loads on the tower and turbine must be evaluated in consideration of the range of events that may occur over the life of the structure and may require a special effort (Saranyasoontorn and Manual, 2006).

The nature of the maximum turbine-induced loads at the base of the foundation are commonly the coupled bending moment and shear force and are shown in Fig. 2, along with a possible soil profile. In most cases, rocking vibration dominates the foundation response. The rotor frequency is usually less than 1 Hz and within the low-frequency range.

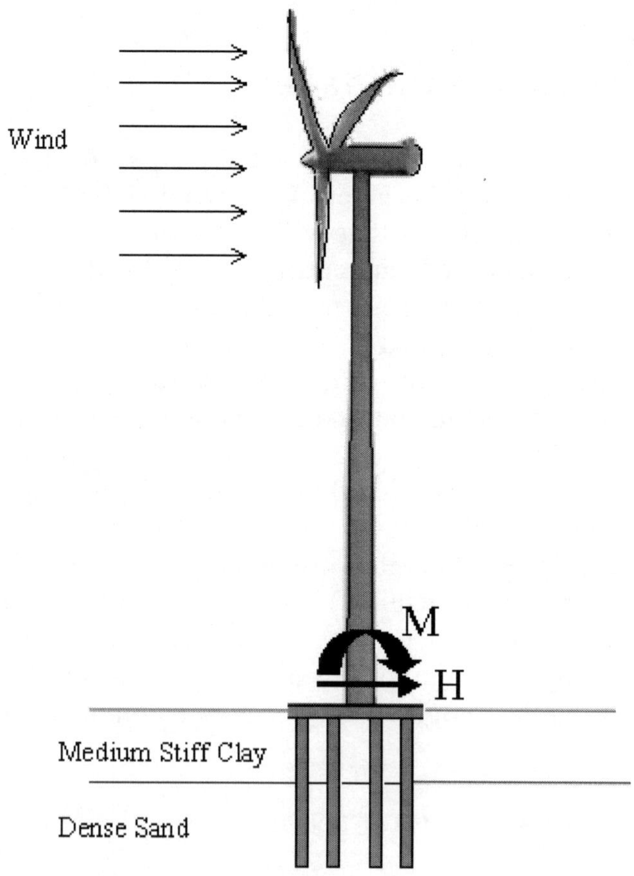

FIG. 2 Dominant Lateral Loads on Foundations

TYPES OF FOUNDATIONS

Large, square, shallow foundations were frequently used for wind farms in Europe in the past (Bu, 2005). However, the nature of the loading and increase of the height of towers, where a shear and a relatively large moment occur at the ground surface, usually dictate a deep foundation in the United States.

A single pile, probably with a relatively large diameter, may support a wind turbine. In many locations, the horizontal load may be relatively constant but extreme winds and other variations in the weather must be addressed in design. If a single pile is employed, ground-surface deflection will be caused by both the shear and the moment. A large ground-surface deflection should be avoided to eliminate locked-in deflection or loss of resistance of the near-surface soils. A group of piles in concept is a more favorable foundation because axial resistance along the piles can support the gravity load and the moment, and the shear can be distributed to the piles in the group. The use of a pile group can increase the natural frequency of the system and decrease the amplitude of vibration.

METHODS OF FOUNDATION ANALYSIS

The interaction between the structure and the soil around the foundation must be considered carefully in view of the relatively large values of ground-line shear and overturning moment. The analysis must fundamentally demonstrate the foundation is reasonably safe against complete upset by the overturning moment and shear force, taking time-related factors into account. The analysis also needs to meet the serviceability requirement for the wind turbine at the top of the tower with serviceability being typically controlled by the deflection and rotational tilt. Rotation stiffness is, in general, expressed in units of GN-m/radian as defined by moment (M) divided by rotation (θ). Rotational stiffness plays an important role in the foundation performance with a value sufficiently large so as not to change significantly the total system of the wind-turbine structure. Typically, the guidelines for acceptable performance of the foundation under maximum operating loading are listed in Table 1.

TABLE 1 - Typical Design Criteria for Wind-Turbine Foundations

Item	Acceptable tolerance under maximum operating loading condition
Maximum groundline deflection	5 mm
Rotation tilt at the foundation	0.001 radians
Rotation stiffness	25 GN-m/radians

With regard to deep foundations, a single pile or a pile group may be used. When a pile group is used, the interactions between pile to pile and pile to soil must be considered. According to Bu (2005) dynamic pile-group effects under dynamic loads differ considerably from pile group effects for static loads. Because the dynamic response of a pile group is very complex and the natural frequency range of wind turbine foundations is in the lower range, simplified engineering approximations based on the static condition are accepted in practice.

Ultimate-Limit-State Analysis

The ultimate capacity of the foundation can be obtained by employing the fundamental principles of mechanics to compute the resistance of the foundation when the soil and rock are in the fully plastic state. There are several methods to computing the ultimate lateral resistance of the soil and, hence, computing the factor of safety against overturning. These methods include those of Broms (1964) and Hansen (1961). In each method the capacity of the foundation to sustain moment may be computed by employing the ultimate force from the soil or rock above and below a point of rotation. The methods are fundamentally deficient because of failure to compute deflection of the foundation, a factor of great importance in the design of foundations for wind turbines, and in other designs as well.

Dynamic Foundation Analysis

The elastic-half-space (EHS) method assumes an embedded rigid block within an elastic half space and a solution is obtained from solving closed-form equations for horizontal stiffness and rocking stiffness. This method is used commonly for design of machine foundations and for spread footings for wind turbines. With the advance of the finite-element method and boundary-element method, computer software was developed, assuming linear elasticity, to solve more complicated cases such as the foundations embedded in layered soils and pile-supported foundations. The advantages of using the dynamic method are:

1. The foundation and the surrounding soil in the system are bounded together as an elastic continuum. Therefore, the stress-strain at every location (or sub-domain) inside the system will be included in the analysis, taking into account the tip resistance due to lateral deflection.
2. The solutions are well established for design of machine foundations where computed deformations are small.

Some problems to be addressed in employing the dynamic method are listed here.

1. The solution assumes that a perfect bonding is present between the soil and the foundation, an assumption that is never true in reality. The computed value of rotation stiffness based on this assumption will yield a higher value than will be computed if some slip occurs between the soil and foundation.
2. The foundation is assumed to be perfectly rigid, an assumption which is not quite true for a deep foundation with a depth to width ratio greater than 2.
3. The deformation of machine foundations, as noted above, is expected to be relatively small, but foundations of wind turbines, operating at relatively low frequencies, are expected to experience large deformation. The soil cannot be classified as a linear-elastic material for most of the loadings of wind turbines. The selections of the proper values of the shear modulus and other elastic parameters are difficult for the range of deformations experienced by foundations for wind turbines, especially for the soil near the ground surface.

Analysis by Soil-Structure Interaction

The alternative method for computing the rotational stiffness of the foundations for wind turbines is to use the nonlinear load-transfer curves. Recommendations for computing load-transfer curves are in the technical literature, derived from experiment and analysis. The model for lateral loading (Fig. 3) employs *p*-y curves that have been developed by analyzing the results from many loading tests of instrumented piles. Engineers in numerous countries understand the theory for soil-structure interaction and the computational procedures used in the nonlinear model. The load-transfer model for axial loading (Fig. 4) employs *t-z* curves and *Q-w* curves to represent the response of the soil. Of special interest is to note that Det Norske Veritas and Risø National Laboratory (DNV/Risø) presented a summary of concepts of *p-y* and *t-z* curves in the section on foundations in their book: *Guidelines for Design of Wind Turbines* (DNV/Risø, 2001). While recommendations for developing load-transfer curves have been given in the literature for many types of soil and for types of loading, more data on field performance will enhance to value of soil-structure interaction.

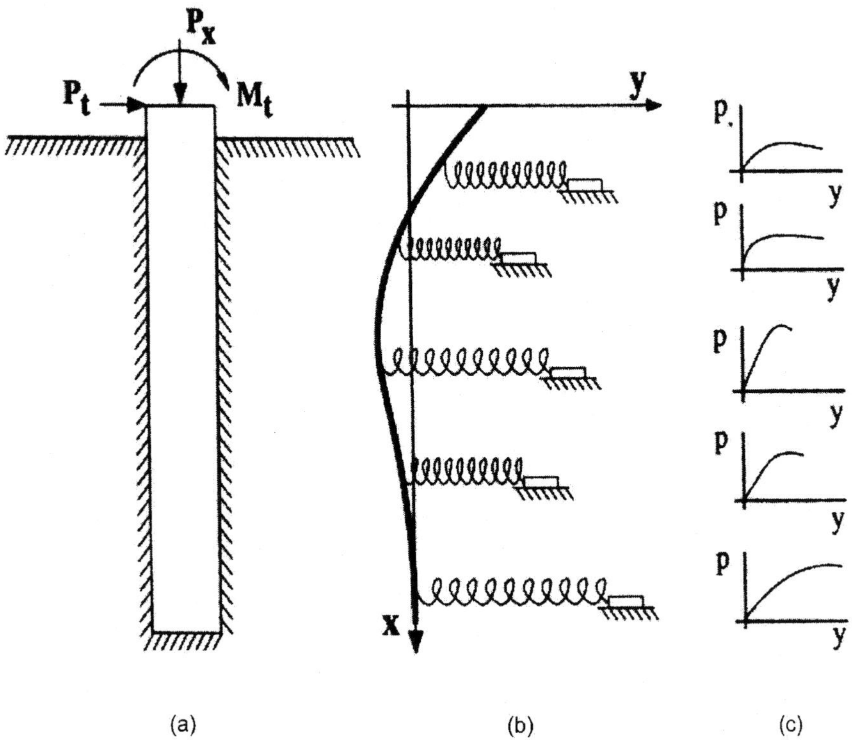

FIG. 3 Model of laterally loaded pile

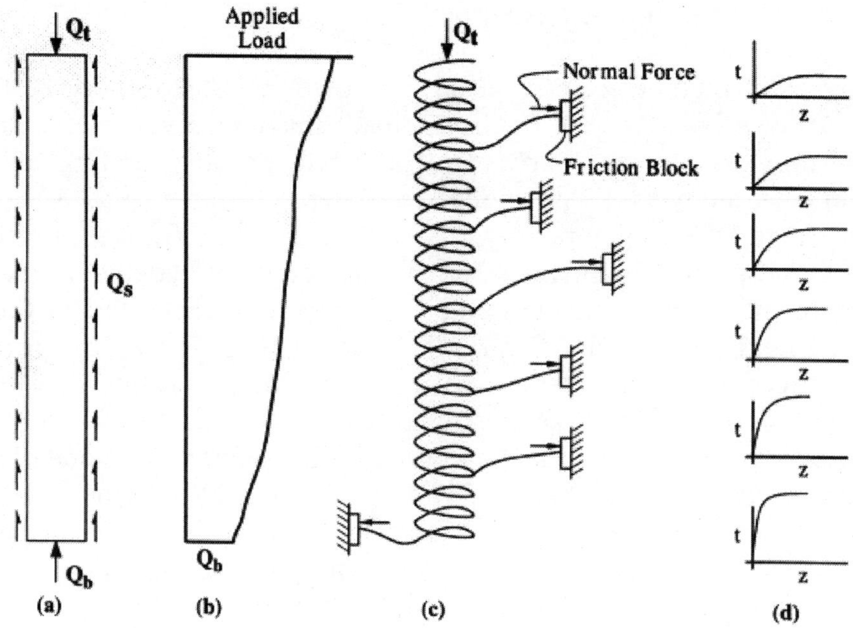

FIG. 4 Model of axially loaded pile

Two factors that have a significant effect on the rotational stiffness at the pile head are: 1. the deformation of surrounding soil; and 2. the flexibility of the structural member. If a deep foundation is long and flexible, the *p-y* method for a single pile subjected to lateral loads likely provides excellent accuracy for rotation stiffness at the pile head, provided the pile-head deflections are moderate and in the range where the *p-y* curves are well defined. The *p-y* method is based on a solution of the governing equations of mechanics for a structural member used to model the behavior of the pile and considers explicitly both the nonlinear soil behavior in terms of *p-y* curves and the nonlinear bending stiffness (*EI*) of the pile or drilled shaft.

If a deep foundation is short and rigid, the rotational stiffness is dominated by the rigid-body rotation of the foundation. For the two-dimensional case, rigid-body rotation is computed by representing numerically the deformation of the soil near the ground surface, the deformation of soil near the tip, and the deformation of the soil beneath of the tip.

An expression for foundation stiffness, commonly adopted in structural analysis for the two-dimensional case where shear and moment are applied at the pile head, is given with a full 3 by 3 stiffness matrix as:

$$\begin{bmatrix} V \\ H \\ M \end{bmatrix} = \begin{bmatrix} K_{11} & 0 & 0 \\ 0 & K_{22} & K_{23} \\ 0 & K_{32} & K_{33} \end{bmatrix} \begin{bmatrix} \delta \\ y \\ \theta \end{bmatrix}$$

Where V = axial load at the pile head,
H = lateral (shear) load at the pile head,
M = bending moment at the pile head,
K_{11} = axial stiffness of the pile,
K_{22} = lateral stiffness of the pile,
K_{33} = rotational (rocking) stiffness,
K_{23} , K_{32} = coupling stiffness between shear and moment,
δ = axial displacement at the pile head,
y = lateral displacement at the pile head, and
θ = rotation (slope) at the pile head,

Typically, the off-diagonal terms (K_{23}, K_{32}) are equal for a linearly elastic system. However, when system is nonlinear, expressing the off-diagonal terms (coupling terms) is inconvenient. For a nonlinear system the foundation stiffness is more conveniently expressed as:

$$\begin{bmatrix} V \\ H \\ M \end{bmatrix} = \begin{bmatrix} K_{11} & 0 & 0 \\ 0 & K_{22} & 0 \\ 0 & 0 & K_{33} \end{bmatrix} \begin{bmatrix} \delta \\ y \\ \theta \end{bmatrix}$$

where K_{22} and K_{33} are the equivalent shear stiffness and rotational stiffness, respectively.

The equivalent stiffness of K_{22} is derived on the basis of the magnitude of the applied moment. A family of curves of shear-vs-deflection, under differing magnitudes of bending moment, can be obtained as shown in Fig. 5.

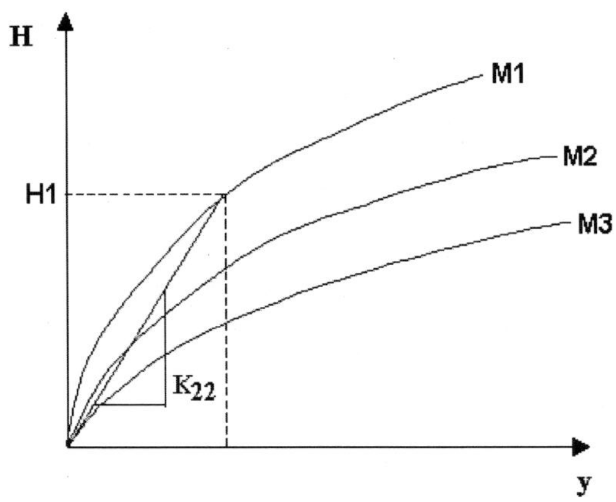

FIG. 5 Equivalent stiffness K_{22} for translation.

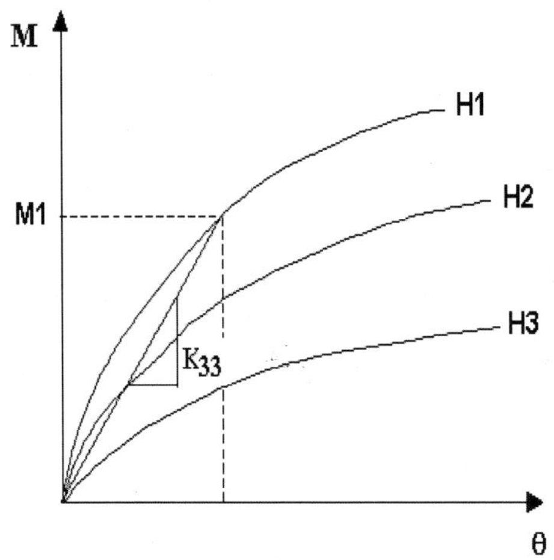

FIG. 6 Equivalent rotational stiffness K_{33}

Similarly, a family of curves of bending moment vs rotation, under differing magnitudes of shear, can be obtained as shown in Fig. 6.

The analysis of a pier of large size, as from an open-ended tube under overturning moment, should include in the model the components of axial load-transfer, as shown in Fig. 7. The side friction and tip resistance are important components that resist the overturning moment on the foundation. Such a model (referred to as the Group model) is available and has been applied in the design of suction piles for the offshore industry (Wang and Arréllaga 1999).

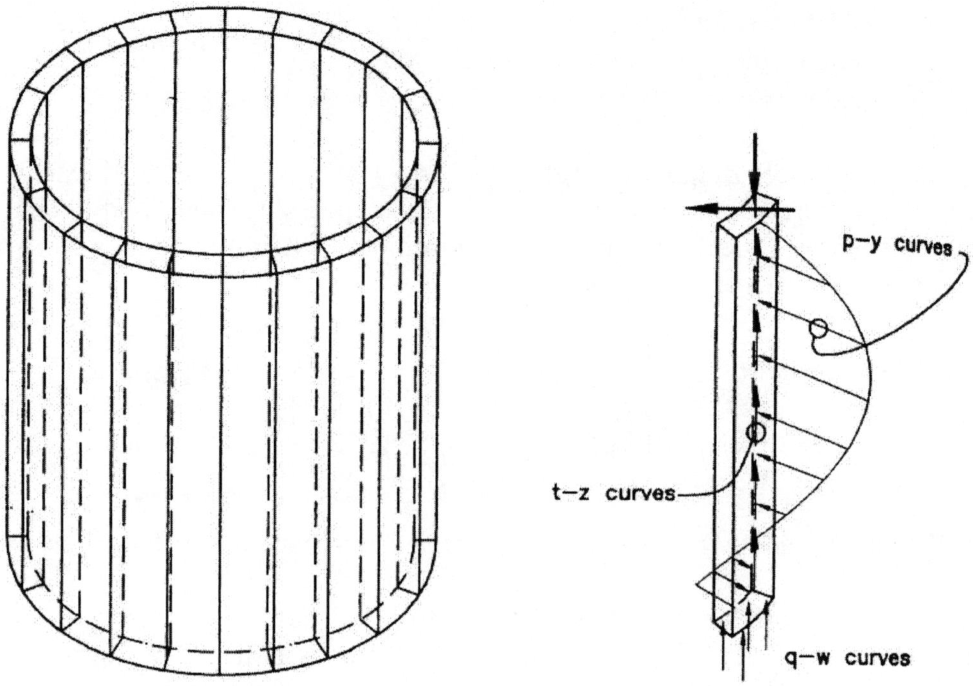

FIG. 7 Soil-Resistance Response for a Circular-Ring Foundation of Large Size

The Group model employs well-accepted soil-resistance curves; *p-y* curves for lateral soil resistance, *t-z* curves for axial side resistance, and a *Q-w* curve for the axial tip resistance. Numerical values can be ascribed to the load-transfer curves for these resistances by using recommendations from technical literature and data from the local subsurface conditions. The assumption of a rigid cap on top of the pile-group foundation, which is common practice, allows the behavior of the foundation to be analyzed. Calibration can be accomplished by use of data from field measurements.

EXAMPLES OF FOUNDATION ANALYSIS

For these examples, the results of a subsurface investigation are presented along with the loadings from wind. Seismic loading is assumed to be unimportant. A tubular-steel pipe is selected for the tower. The first example presents the design of a single pile for the foundation and a companion example presents the design of a pile group. Modern principles, employing concepts of soil-structure interaction, are implemented in both designs. An important criterion in design is to limit the rotation at the foundation to a value of 0.001 radians or less in order to limit the movement at the turbine, approximately 80 m above the ground in the examples. A comparison of the two methods of support is given, and the advantages and disadvantages of each method are presented below.

Subsurface Conditions At Site

The subsurface conditions for the example designs consists of 2 m of clayey fill at the top underlain by 20 m of sandy lean clay. The soil properties and strength parameters are listed in Table 2.

TABLE 2 - Soil Layers and Soil Properties

Layer, m	Description	Unit weight, kN/m^3	Cohesive strength, kPa	ε_{50}
0 – 2.3	Clayey fill	18.7	86.2	0.005
2.3 – 3.8	Sandy lean clay	18.7	105	0.005
3.8 – 5.3	Sandy lean clay	18.7	144	0.005
5.3 – 8.5	Sandy lean clay	18.7	129	0.005
8.5 – 20	Sandy lean clay	18.7	160	0.005

Loading Information on Foundation

The height of the turbine tower is approximately 80 m (260 ft). The information on loading used in the examples is provided in Table 3.

TABLE 3 - Loading Conditions

Loading Item	English Units	S.I. Units
Horizontal Load	65 kips	289 kN
Axial Dead Load	282 kips	1256 kN
Bending Moment at Base	15,000 ft-kips	20,250 kN-m

Single Drilled Shaft with Diameter of 2.4 m (8 ft)

The largest size of drilled shafts commonly built in the United States is 2.4 m (8 ft) in diameter. A single drilled shaft with a diameter of 2.4 m (8 ft), in mostly stiff clay, is selected for the first example for the foundation. The steel ratio for the proposed 2.4 m OD drilled shaft was selected as 3% and a concrete strength of 27,500 kPa (4,000 psi) was selected. A penetration of 15 m was used in the analysis. Although the bending stiffness (*EI*) of the 2.4 m shaft has a significant value (30,000,000 m^2-kN), the drilled shaft still behaves as a flexible structure under such a large overturning moment, as shown in Fig. 8.

The deflection at the top of the shaft was computed to be 24 mm, a value over the tolerance of 5 mm shown in Table 1. The rotational stiffness is approximately 12 GN-m/radians which is also less than the required 25 GN-m/radians in the specification. The computations clearly indicate that the single drilled shaft with a diameter of 2.4 m (8-ft) will not meet the design criteria for this particular wind turbine.

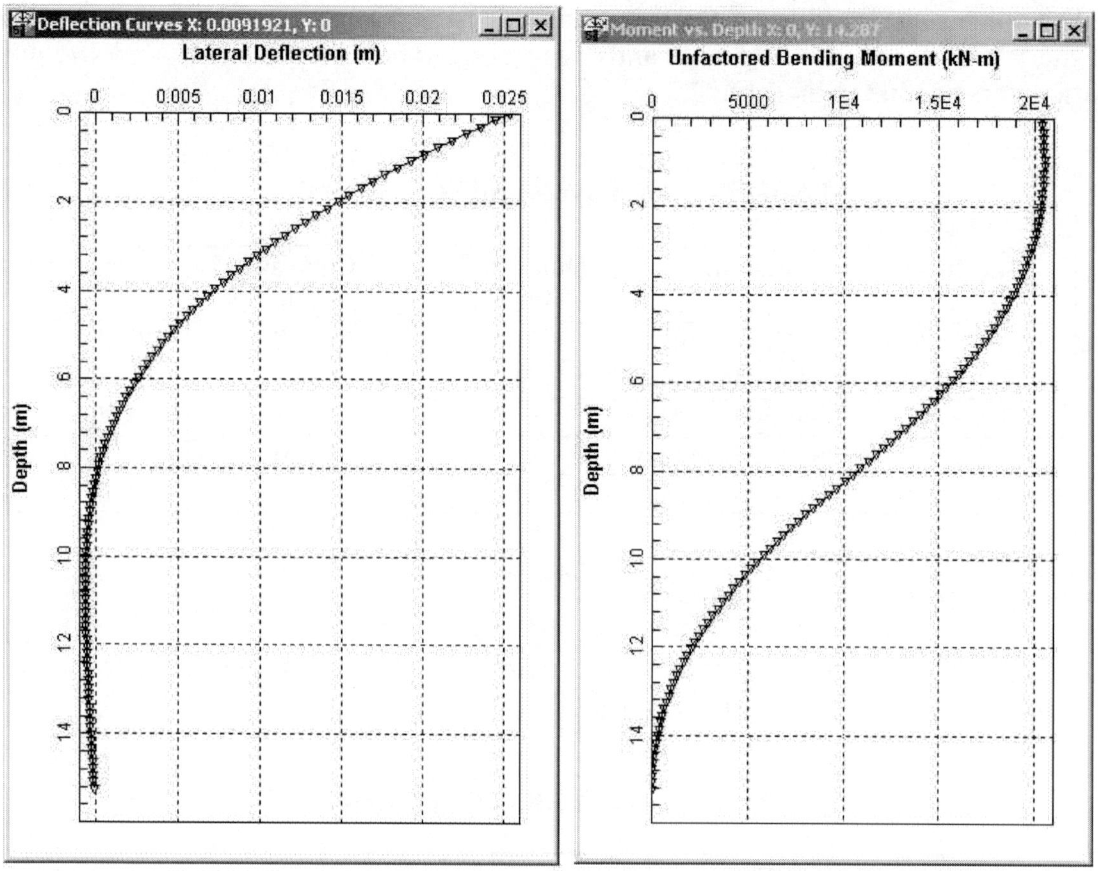

FIG. 8 Computed values of deflection and bending moment for a single drilled shaft with a diameter of 2.4 m (8 ft)

Pile Group with 12 Drilled Shafts in a Circular Arrangement

Because the single shaft cannot be used to meet the design criteria, an alternate design is a group of piles that provide the stability and rotational stiffness. A total of twelve drilled shafts with a diameter of 0.76 m were arranged in a circular pattern, as shown in Figure 9. The diameter of the circle formed by the drilled shaft is approximately 4.5 m and the center-to-center spacing between two adjacent drilled shafts is approximately 1.18 m. Each drilled shafts has a steel ratio of 2% and strength of concrete of 27,500 kPa (4,000 psi). A penetration of 15 m was used in the analysis.

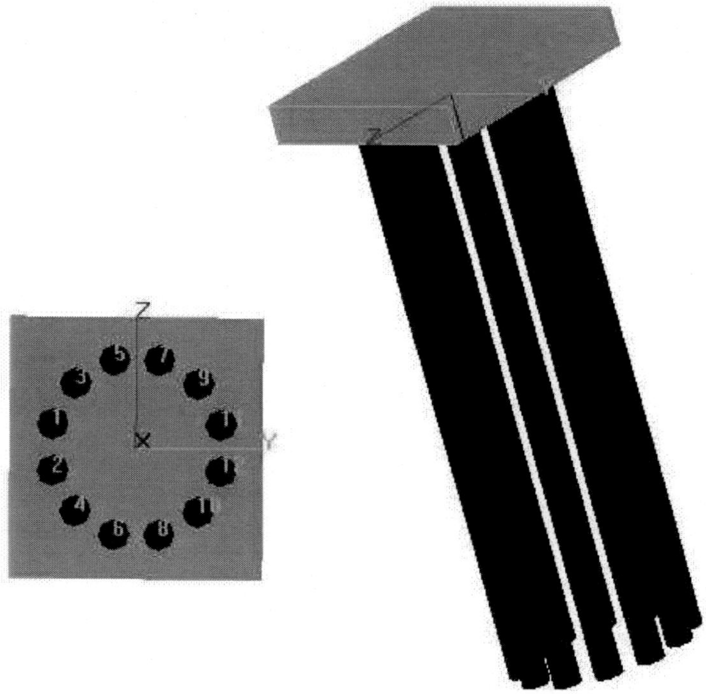

FIG. 9 Configuration of 12 piles in a group

Computations of the deflection of the group and the bending moment of each pile in the group can be accomplished by of Computer Program GROUP (Ensoft, Inc., 2003). Program GROUP can be used to analyze the combined effects of axial and lateral loading by computing the mobilized soil reactions and the displacement of the foundation. Bogard and Matlock (1983) presented the results of the loading of a closely spaced group of piles in a circular configuration and gave an analytical procedure for taking into account the effect of close spacing. When the piles are close together to form a circle, the soil contained within the circle moves with the piles. A group reduction factor of 0.35 was calculated under lateral loading, based on the recommendations of Bogard and Matlock.

Figure 10 shows the computed values for the group of deflection and bending moment as a function of depth. The deflection at the top of group was computed to be 1.2 mm, within the tolerance of 5 mm shown in Table 1. The maximum bending moment developed in each pile was approximately 105 m-kN, less than the computed value of ultimate bending moment of 550 m-kN. The rotational stiffness was computed to be 26 GN-m/radians, greater than the required 25 GN-m. The tilt at the top of the foundation was computed to be 0.0008 radians that also meets the requirement.

The computations clearly indicate that the pile group with 12 drilled shafts can meet the design criteria for this particular wind turbine. A representation of the group is shown in Fig. 11 where the color code indicates that the largest moment is developed near the top of each of the drilled shafts.

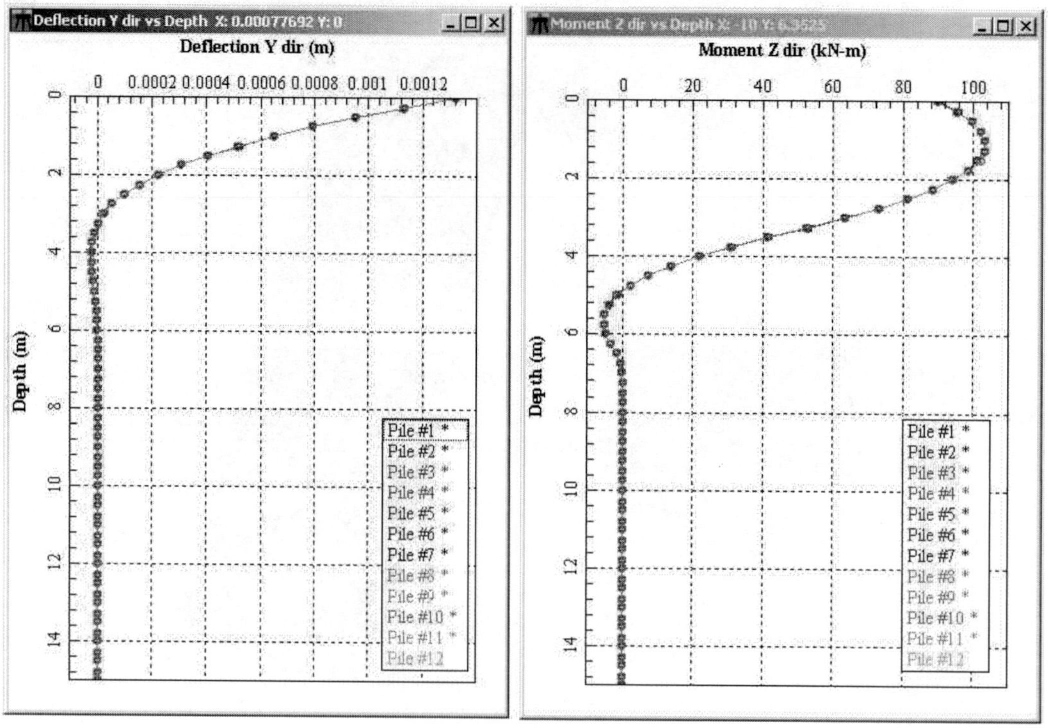

FIG. 10 Diagrams of deflection and bending moment for drilled shafts in the group

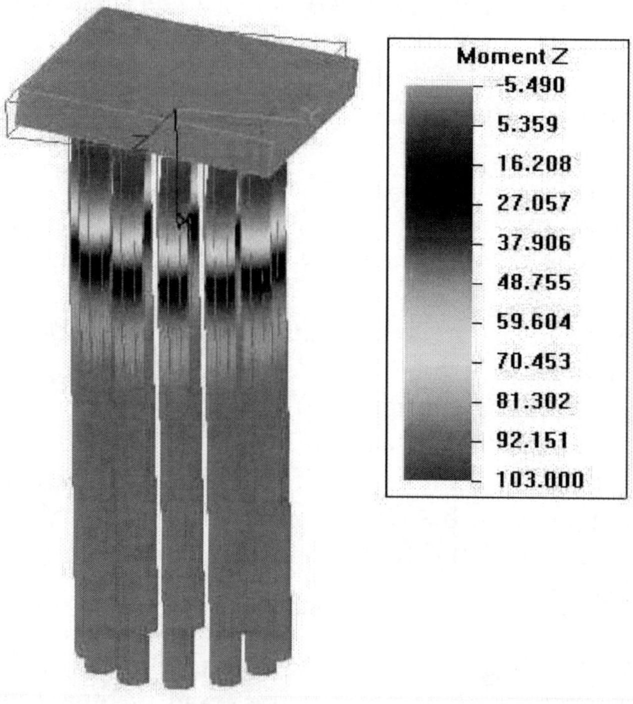

FIG. 11 Representation of group with color code

CONCLUSIONS

With the increase of the capacity and size of modern wind turbines, the design of foundation becomes a challenged task. Taking into account the nonlinear behavior of the soil and rock, the analysis and design employing concepts of soil-structure interaction are highly recommended. For a foundation with a group of piles, *p-y* curves and *t-z* curves are rational soil models for analyzing the behavior of the foundation because those soil-resistance curves, presented in technical literature, were derived on the basis of many field loading tests. Improvements on soil models are still needed, and the establishment of a monitoring system by the wind-power industry for foundations in the field can be very helpful. Engineers can have improved tools to achieve safe designs in a difficult environment and will be better able to meet the challenges of the future.

REFERENCES

Bogard, D. and Matlock, H., "Procedures for Analysis of laterally Loaded Pile Groups in Soft Clay," Proceedings, Specialty Conference on Geotechnical Engineering in Offshore Practice, ASCE, April 1983.

Broms, B. B., "Design of Laterally Loaded Piles," *Proceedings*, American Society of Civil Engineers, Vol. 91, No. SM3, May, 1965, pp. 77-99.

Bu, D. S., "Engineering Considerations for Design of Wnd Farm Foundations," Technical Note, Ground Engineering, December 2005.

CPS Energy Corporation, Desert Sky Wind Farm, *www.desertskywindfarm.com*, 2004

Det Norske Veritas and Risø National Laboratory, *Guidelines for Design of Wind Tunnels,* 2001,253 pages.

Ensoft (2003). Computer *Program GROUP V.7 – Analysis of a Group of Piles Subjected to Axial and Lateral Loading*, Austin, Texas.

Hansen, J. B., "The Ultimate Resistance of Rigid Piles Against Transversal Forces," *Bulletin 12*, Danish Geotechnical Institute, 1961.

Saranyasoontorn, K., and Manual, L., "Design Loads for Wind Turbines Using the Environmental Contour Method," *Transactions*, American Society of Mechanical Engineers, November, 2006, pp. 554-561.

Wang, S. T. and Arréllaga, J. "Study of the Behavior of a Suction Pile With a Simplified Numerical Model," Proceedings of the OTRC-99 Conference, Austin, Texas, April 29-30, 1999, Published by ASCE, Geotechnical Special Publication, pp. 234-246.

Axial Pile Capacity of Large Diameter Cylinder Piles

Peter Lai[1], Michael McVay[2], David Bloomquist[3], and Dhuruva Badri[4]

[1] Assistant State Geotechnical Engineer, Florida Department of Transportation, email: Peter.Lai@dot.state.fl.us
[2] Professor, Dept. of Civil & Coastal Eng., University of Florida, email: mcm@ce.ufl.edu
[3]Associate Professor, Dept. of Civil & Coastal Eng., Fellow, ASCE, University of Florida, email: dave@ce.ufl.edu
[4] Graduate Student, Dept. of Civil & Coastal Eng., University of Florida

ABSTRACT: This paper reports on the behavior of large diameter cylinder piles under both driving and static conditions. Of special interest is the understanding of the tip resistance and the associated soil column within the pile under static and dynamic conditions. An analysis of driving forces, shows that inertia on the soil plug rapidly exceeds the static skin friction between the pile and soil on the inside of the cylinder. The analysis of the soil column for static conditions was undertaken with an Eulerian finite element viscous analysis, ADINA-F. The static analysis revealed that even when the cylinder "cookie cut" under driving it behaved "plugged" under static conditions for a multitude of soil strengths and wall thicknesses. Next, based on a database of 35 load tests (22 concrete & 13 steel) unit side shear and end bearing for cylinder piles as a function of pile material (i.e., steel and concrete) and insitu SPT N were developed and compared successfully to Schmertmann's original values for small diameter piles. Finally, LRFD resistance factors, ϕ, for the proposed cylinder pile design were determined.

KEY WORDS: Cylinder piles, dynamic resistance, plugging, static resistance

INTRODUCTION

Prestressed/post-tensioned concrete cylinder piles were initially developed in 1949 by the now defunct Raymond International, Inc. They are used throughout the U.S. and other countries to support navigable waterway bridges because of their large diameters and corresponding large lateral resistance, i.e., moment capacities. Generally, large diameter cylinder piles are designed similar to smaller diameter open pipe piles. Axial capacity is determined from the sum of side friction for individual soil layers along the pile plus the tip resistance at the bottom of the pile. The tip resistance is the product of the mobilized tip stress times either the total cross-sectional area of the pile or the ring area of the pile's cylinder, depending on whether the soil column within the pile is moving with, or detached from, the pile.

During the initial driving, soil enters the inside of the cylinder pile as a column at a rate equal to the rate of pile penetration. Generally, with pile penetration, the soil column develops friction along the inside wall of the pile. If the side friction between the soil column and the cylinder wall is large enough, the soil column will attach itself to the pile and both will move down as one unit. Herein, this phenomenon will be referred to as "plugged", whereas "unplugged" is the case in which the soil column moves separate from the pile.

Paikowsky (1990) was the first to extend Hvorslev's (1949) research on soil arching in soil samplers to the soil column in a cylinder pile during driving and subsequent static loading. Specifically, the soil column shown in Figure 1, has a convex (upward) curvature arch in the upper zone, attributed to a bulb of reduced stresses during the pile's initial penetration, and a concave (downward) curvature arch at the tip of the plug, attributed to a bulb of increased vertical stresses due to penetration in a plugged or semi-plugged mode.

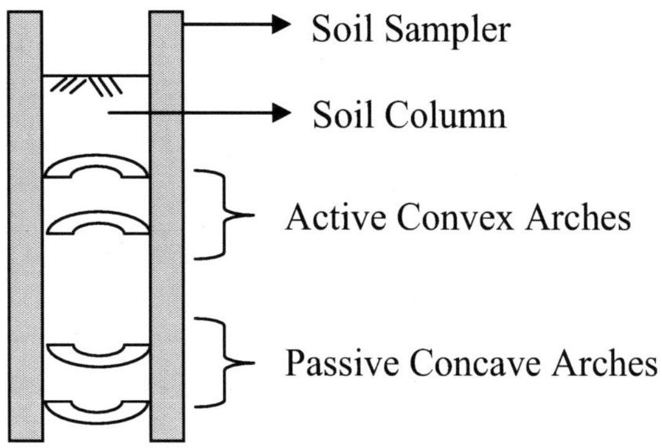

FIG. 1. Vertical stress isobars within the soil column, Hvorslev (1949)

When subjected to loading (static or dynamic), load is transferred through the soil mass to the inner walls of the pile. When the load exceeds the arch capacity, dilation and shear take place along the arch; allowing penetration of additional soil until a new stable arch forms, signifying that the arch resistance exceeds the upward pushing forces (tip capacity). This process accounts for zones of varying soil densities within the pile cylinder or soil column. For plugged piles, the densest soil zone occurs approximately one quarter of the cylinder's diameter (1/4 D) above the tip. In case of unplugged piles, the densest zone exists at the pile tip with diminishing density upwards in the soil column. Unfortunately, Paikowsky (1990) does not identify under what conditions a cylinder pile will plug, especially as a function of pile diameter, wall thickness, soil properties or dynamic conditions (i.e., hammer energies, velocities, etc.).

FORCES ON SOIL PLUG DURING DRIVING

During initial driving, the pile tip penetrates into soil and the soil enters the inside of the pile cylinder from the tip. As shown in Fig 2(A), as the pile penetrates, the height of the soil column, L, increases and the side friction, F_s, acting between the soil and the inside of the pile's wall increases. Acting against the soil-wall friction is the weight of the soil column and the vertical upward force at the bottom of the soil column from the underlying soil. If the side friction increases to a value greater than the end resistance, Q_t, the soil column tends to attach itself to the inside of the pile (plugged) and subsequently moves with the pile. With succeeding hammer blows, the soil column will experience a downward motion (i.e., moving with the pile), and the soil column will develop an additional inertia force, **I**, equal to the mass of the soil plug times the pile's downward acceleration, Fig. 2(B). The inertia force on the soil column acts upward in opposite direction of the pile motion (i.e., downward).

If F_s is of sufficient magnitude to overcome inertia, **I**, and tip resistance, Q_t ($q_{tip} \cdot A_{cross}$: total cross-sectional pile area), *then the pile remains plugged during driving* (i.e., $F_s + W > Q_t + I$). However, if the pile were to remain unplugged during driving, it still may act plugged (i.e., $F_s + W > Q_t$) during static loading. In this case the end bearing calculation, Q_t, would remain the same, i.e., unit end bearing (stress) would be multiplied by the total pile cross-sectional area - however q_{tip} may be less than that of the plugged value. If the pile remains unplugged during static loading (i.e., $Fs < Q_t$), then Q_t should be computed using the unit end bearing (q_t: stress) times the cross-sectional area of the pile material alone [i.e., $\pi\ (D_o^2 - D_i^2)/4$] and not the total pile cross-sectional area [$\pi\ (D_o^2)/4$] as with the plugged scenario.

For small diameter pipe piles, e.g., 0.3 m (12") to 0.46 m (18"), the inertia forces on the soil column are generally small, which can be ignored, thus only F_s and Q_t are used for assessing the plugged or unplugged status. However, as the diameter of cylinder piles have become larger, the latter is no longer true. Specifically, the inertia force, **I**, will become the dominant force that F_s must overcome as the cylinder diameter increases.

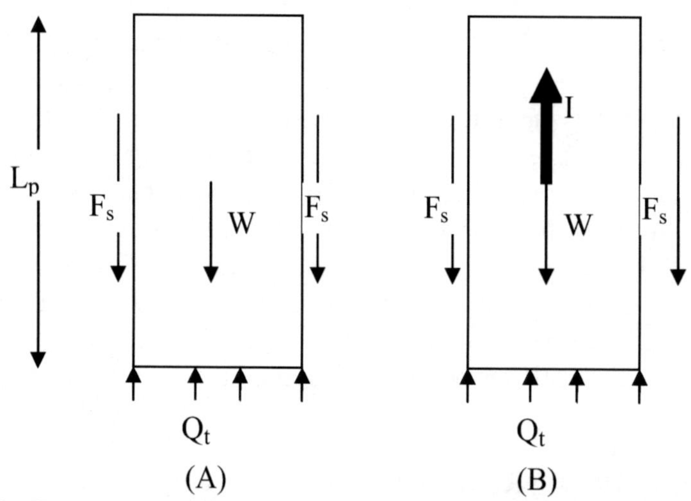

FIG. 2 Free body diagram of soil column within the pile during driving.

The latter is shown above through the calculation of soil plug height, L_p (Fig. 2), required for F_s to overcome Q_t and **I**, for a given plug acceleration (i.e., pile acceleration during driving).

To show a comparison of the identified forces, the calculations with representative values are provided below:

a) **Unplugged state during driving,** i.e., Fig. 2(A), behaves plugged during static loading - vertical force equilibrium gives:

$$F_s + W = Q_t \tag{1}$$

Substituting in unit skin friction, f_s, and unit tip resistance, q_t, and soil column areas (i.e., surface and cross-sectional) gives:

$$(\pi D_i f_s)L_p + (\frac{\pi D_i^2 \gamma}{4})L_p = (\frac{\pi D_i^2}{4})q_t \tag{2}$$

Simplifying,

$$L_p(f_s + \frac{\gamma D_i}{4}) = \frac{q_t D_i}{4} \tag{3}$$

and solving for L_p, the required soil column height for equilibrium between side friction, F_s, and end bearing, Q_t (using total cross-sectional area) is

$$L_p = \frac{q_t D_i}{4 f_s + \gamma D_i} \tag{4}$$

where:

L_p	=	Length of the soil column (m).
D_i	=	Diameter of soil column/ inner diameter of the pile (m).
γ	=	Total unit weight of soil (kN/m^3).
f_s	=	Unit skin friction between the soil column and pile (kPa).
F_s	=	Total skin frictional force (kN).
q_t	=	Unit end bearing at pile toe (kPa).
Q_t	=	Total tip resistance on the soil column (kN).
W	=	Weight of the soil column (kN).

b) **Plugged conditions during pile driving,** Fig. 2(B) - vertical force equilibrium gives:

$$F_s + W = Q_t + \mathrm{I} \tag{5}$$

Substituting in unit skin friction, f_s, unit tip resistance, q_t, mass of soil column (W/g = (Volume x unit wt)/g), and soil column areas (i.e., surface, and cross-sectional) gives:

$$(\pi D_i f_s)L_p + (\frac{\pi D_i^2 \gamma}{4})L_p = (\frac{\pi D_i^2}{4})q_t + (\frac{\pi D_i^2 \gamma}{4})\frac{L_p a_p g}{g} \tag{6}$$

Simplifying,

$$L_p(f_s + \frac{\gamma D_i}{4} - \frac{\gamma D_i a_p}{4}) = \frac{q_t D_i}{4} \tag{7}$$

And solving for L_p, yields,

$$L_p = \frac{q_t D_i}{4 f_s + \gamma D_i (1 - a_p)} \tag{8}$$

where, $\mathbf{a_p} \geq 0$, is a dimensionless number identifying the soil column acceleration in terms of g_s, i.e., $a_p \cdot g$ (m/sec^2).

Eqs. 4 and 8 are quite similar with the exception of the average soil column acceleration, a_p, in the denominator of Eq. 8. The latter will result in a negative term when a_p>1which is magnified by the inner pile diameter, D_i, and will result in a significant increase in soil column height for equilibrium conditions.

A parametric study was carried out on piles, during driving, with six different diameters, i.e., 1.372 m, 0.965 m, 0.61 m, 0.508 m, 0.406 m, 0.305 m (54", 38", 24", 20", 16", and 12"), one soil type (silt), a pile length of 24.4 m (80 ft) (adapted from the St. Georges, FL project) with soil column g-forces varying from 0-g to 45-g. Table 1 tabulates the various parameters and their values used in the study.

TABLE 1. Fixed Parameters and Values

Parameter	Value
Soil Type	Silt
Pile Material	Concrete
SPT N Value for Skin	25-blows, f_s = 56.1 kPa (0.586 tsf)
SPT N Value for Tip	30-blows, q_t =1178 kPa (12.3 tsf)
Pile Length	24.4m (80 ft)

Fig. 3, shows the computed soil column heights for soil diameters (inner diameter of the pile) of the 1.372 m, 0.965 m, 0.61 m, 0.508 m, 0.406 m, 0.305 m (54", 38", 24", 20", 16", and 12") piles using Eqs. 4 & 8. Consider the 1.72 m (54") concrete pile for example; an average pile-soil acceleration of 10 g, a soil column height of approximately 12.2 m (40 ft) (Eq. 8) is required to develop sufficient skin friction, Q_s (f_s = 56.1 kPa (0.586 tsf) x surface area), to overcome end bearing, Q_t =863 kN (96.9 tons), and inertia force, **I**. If the acceleration of the soil column increases to 15 g, a soil column height equal to the total pile length, 24.4 m (80 ft) is required, i.e., the soil column will move separately from the pile. A line labeled "critical g-force value" in Fig. 3 identifies if a particular size pile will plug. At g-force values above the critical g-force value there is no "plug", or the pile will "cookie cut" into the soil mass. At soil column acceleration levels below the critical g-force value, the soil within the pile "plugs," i.e., attaches to the inside of the pile and both move as a single mass. The length of the latter soil-column is directly proportional to the g-force during the driving process (Fig. 3).

Table 2 summarizes the critical soil-column accelerations, for various pile-cylinder diameters, at which the soil will remain "unplugged" during driving.

TABLE 2. Critical g-force Values for Various Soil Column Diameters

Soil Column Diameter	Critical g-force Value
1.371 m (54")	15-g
0.965 m (38")	24-g
0.609 m (24")	42-g
<0.508 m (20")	>45-g

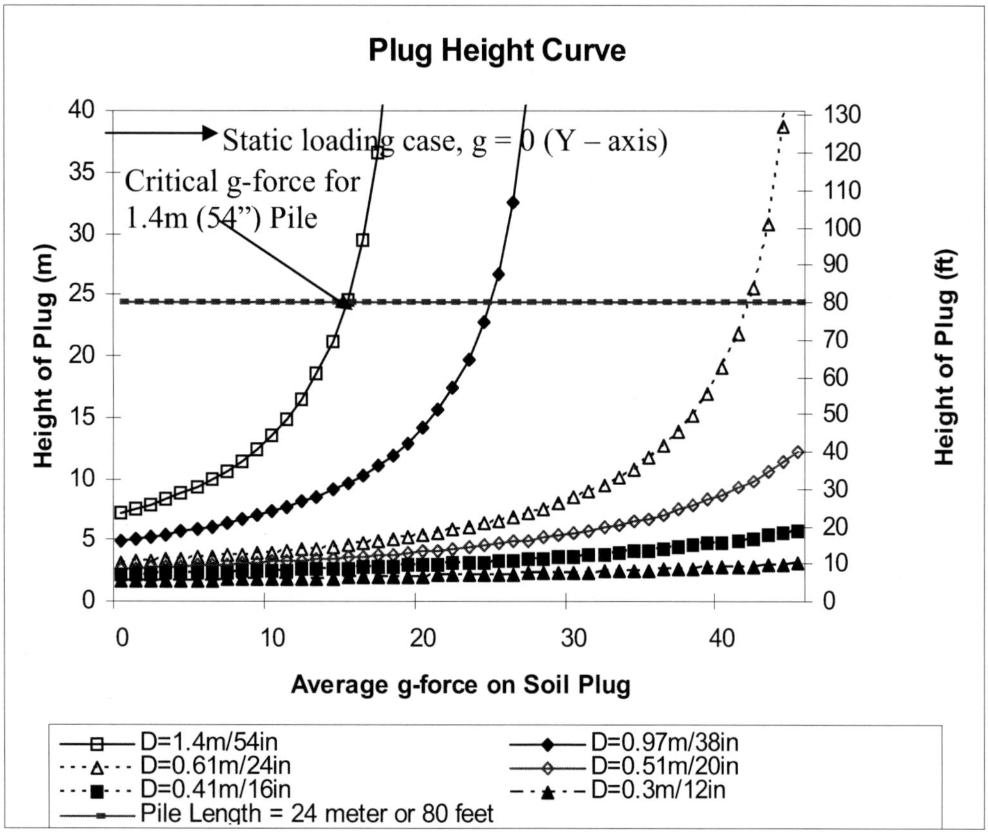

FIG. 3 Variation of soil column height with soil column diameter and g-forces.

Evident from the Figure and Table, significant soil friction, $\mathbf{F_s}$ (or soil column height), is required to overcome the inertia effects, **I**, for large diameter cylinders. Also, the relationship between pile acceleration levels for plugging is not linear with pile diameter. For example, the required soil column accelerations for "plugging" of a 0.965 m (38") and a 0.609 m (24") diameter pile is below 24 g and 42 g, respectively. The latter may be attributed to a linear relationship between skin friction and pile diameter, but a quadratic relationship between Inertia and pile diameter.

Of special interest is the typical acceleration levels that the "plugged" soil mass may experience during pile driving. The latter is very much controlled by the type of pile being driven, i.e., concrete and steel. Both the allowable stresses and the particle velocities for each are quite different, with concrete being much lower than steel. For this research, a number of diesel and air steam hammers were investigated along with different pile sizes, cushions, and soil types. Generally, for prestressed concrete piles, average pile acceleration values varied from 40 to 60 g using Florida Department of Transportation specifications. However, in the case of steel piles, much higher g levels, i.e., over 100 were seen. The latter is in agreement with reviewed literature data (Stevens, 1988), which reports measured pile accelerations ranging from 160 g to 215 g during driving of large diameter offshore steel pipe piles. Evident, large diameter steel pipe piles are far less likely to plug than similar diameter concrete piles during driving even when driven into the same material.

Also as shown in Fig. 3, [Static Load case, g=0] is the required soil column heights to mobilize the full static tip resistance, i.e., $Q_t = q_{tip} A_{cross}$, from internal side friction. The 1.371 m (54") diameter pile had the longest soil column height at 7.62 m (25 ft), but less than 24.4 m (80 ft), suggesting that even though it didn't plug during driving, it still may act plugged under static load conditions.

Numerical (ADINA-F) Simulation of Static Load Testing To shed light on the potential for a soil "plugged" condition under static loading, (but "unplugged" during driving), a FEM study of a pile "slowly" penetrating a viscous material (i.e., soil) was undertaken (McVay et. al, 2004). The problem was characterized as a viscous axisymmetric two-dimensional steady state fluid flow problem, where the soil particles move along streamlines around the pile as suggested by Azzouz and Baligh (1990).

Figs. 4 and 5 show typical soil velocity profiles for unplugged and plugged conditions, respectively. Fig. 4 shows an axisymmetric pile section, with the soil mass directly adjacent to the pile having the highest velocities (viscous shear), with only a small radial zone of soil directly adjacent to the pile wall (black area) moving down at higher velocities. The core of the soil column within the pile is moving with a much smaller velocity. The large differential relative velocity between the soil column and the pile signify that the pile is not plugged. In Fig. 5, the soil inside and outside the pile has almost reached a uniform downward velocity, i.e., the relative velocity of the pile and the soil column is almost zero, indicating formation of soil plug (pile and soil column moving together). The small non-uniform velocity zones within the soil plug in Fig. 5 may support the arching phenomenon (Fig. 1) put forth by Paikowsky (1990).

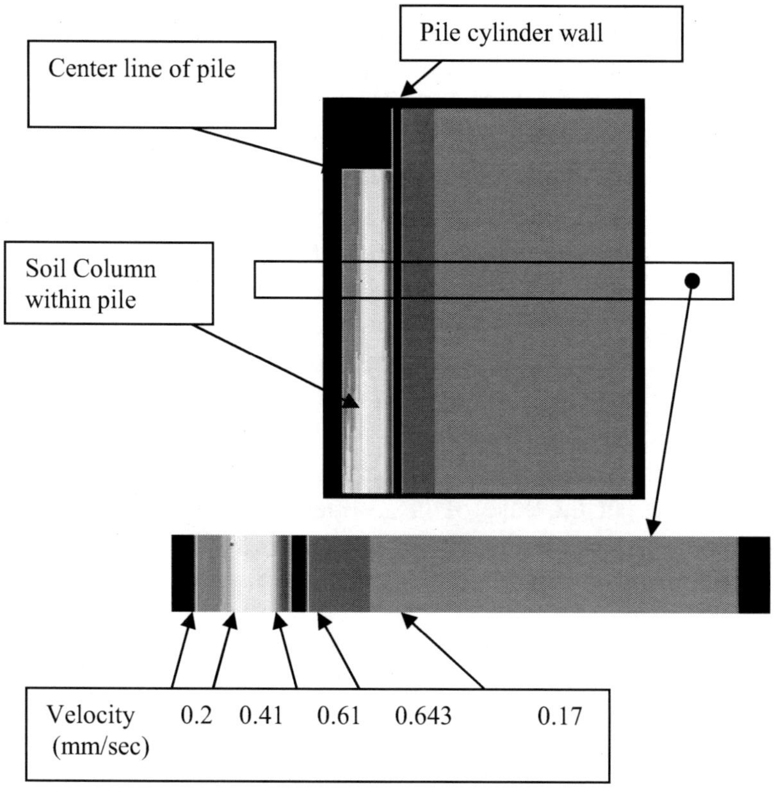

FIG. 4 Velocity profile of soil around cylinder pile - unplugged condition

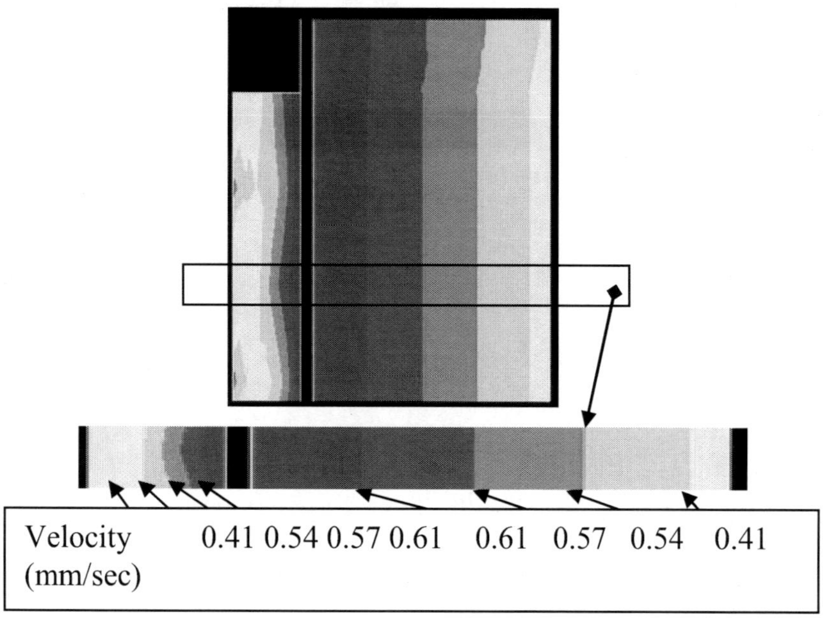

FIG. 5 Velocity profile of soil around cylinder pile - plugged condition

Using the viscous axisymmetric two-dimensional steady state fluid flow model, the following soil/pile parameters were varied: 1) pile wall thickness, and 2) shear strength of soil. All variations were applied to a 1.371 m (54") OD cylinder pile, 203 mm (8") wall thickness, and 24.4 m (80 ft) length, representative of St. George Island, Florida bridge installation. The soil, modeled as a fluid, was considered either compressible (e.g. loose sand) or incompressible (e.g. dense sand – no volume change). A discuss of the results of the parameter variation follows.

Pile Shell Thickness As suggested by Paikowsky et al. 1990, the soil column (i.e., plug) movements tend to stabilize faster with increased pile wall thickness, as shown in Fig. 6. This phenomenon can be explained by the larger soil mass displaced by the pile wall which must pass either within or around the pile, and to the possibility of soil "plugging." In addition, both loose and dense sands exhibit similar response to wall thickness variations. Note, in the case of large open steel pipes, e.g., 1.371m (54") with small wall thickness – 25.4 mm (1"), the possibility of soil "plugging" under static conditions is doubtful.

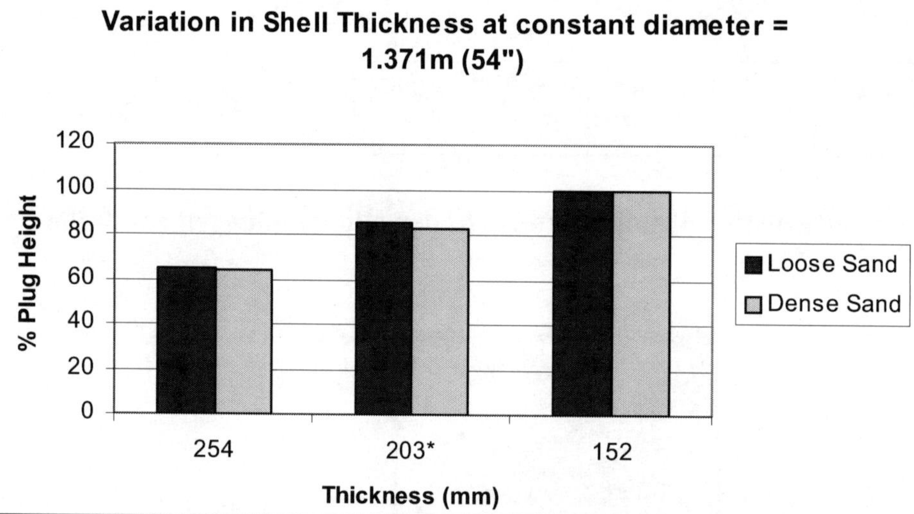

* Indicates standard pile used for St. George Island, FL. Project for comparison. Outer Diameter = 1.371m (54"), Shell Thickness =203mm (8"), Unit Weight = 157.1 N/m^3 (100 pcf) Dynamic Shear Strength/Friction angle = 1000 Pa-sec/30 deg, Rate of Penetration = 12.7mm/sec (0.5inch/sec).

FIG. 6. Variations in shell thickness with height of soil plug.

Variations in Soil Shear Strength For the numerical simulation, the soil was modeled as a viscous fluid with dynamic shear strengths ranging from 500 Pa-sec to 2000 Pa-sec. According to Chow (1981), the latter corresponds to Mohr-Coulomb angles of internal friction between 27.5 degrees and 35 degrees. Again, the pile analyzed was the St Georges Island's 24.4 m (80 ft) pile with a 203 mm (8") wall thickness. Fig. 7 shows the influence of soil shear strength on plug height. For instance, the soil column heights increased 40% when the friction angle was reduced from 35° to 27.5°.

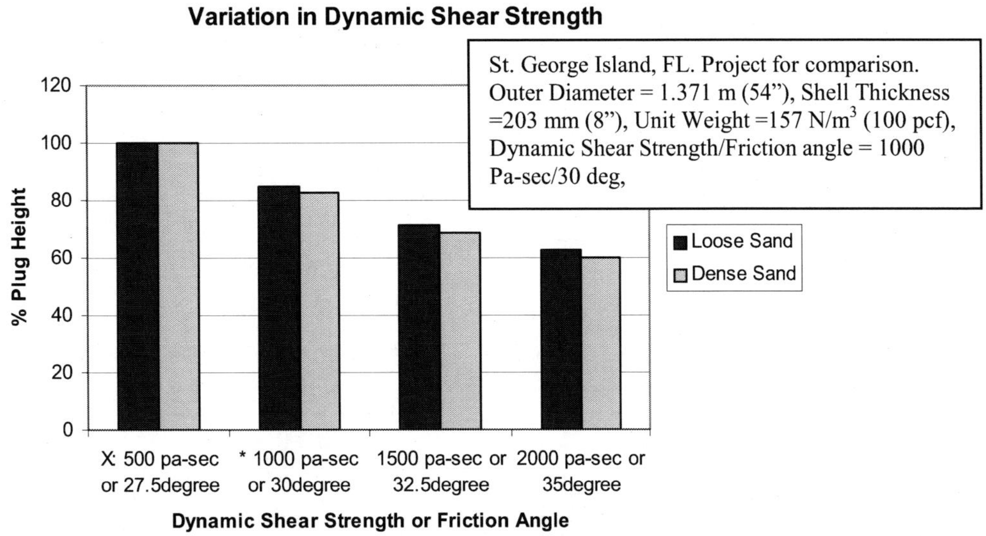

FIG 7. Variations in friction angle or dynamic Shear strength

Based on the numerical analyses, cylinder piles, especially concrete, appear to behave "plugged", under static loading, suggesting that tip resistance may be computed from unit end bearing times total cross-sectional area. The analyses will consider both i.e., total cross-sectional, and ring area of the cylinder.

Database & Unit Side Shear and End Bearing for Cylinder Piles In order to develop design procedure for large diameter cylinder piles, unit side shear and end bearing had to be established based on soil properties or insitu testing (i.e., SPT, CPT, etc.) for sites which had cylinder piles installed and load tested. Over eighty geotechnical consulting companies, federal and state department transportation (DOTs) agencies were contacted throughout the U.S. and around the world. A total of 35 load tests were collected (22 concrete & 13 steel). Of the 35 load tests with associated load vs. settlement data, 21 of the piles reached the modified Davisson failure (elastic shortening with offset of 0.15 + D/30) used by a number of DOTs for large diameter piles.

Shown in Table 3 is a description of the piles (left side number of piles with specific OD, right side, number of piles with given wall thickness). Presented in Table 4 is a description of soil types adjacent to the piles. The latter was used in establishing unit skin friction and end bearing vs. SPT N value for specific soil types. For instance, shown in Fig. 8 is the measured unit skin vs. SPT N values for concrete cylinder piles founded in clay along with the best fit trend line, $f_s = 48.84 \text{ Ln}(N) - 60.71$ [$f_s = 0.51 \text{ Ln}(N) - 0.634$].

TABLE 3. Cylinder Pile Sizes and Wall Thickness

Outer Diameter	Number of Piles	Shell Thickness	Number of Piles
2.13m (84")	1	203mm (8-inch)	4
1.83m (72")	2	175mm (6.9-inch)	2
1.68m (66")	3	152mm (6-inch)	7
1.37m (54")	18	127mm (5-inch)	9
1.07m (42")	10	25.4mm to 51mm (1-inch to 2-inch)	5
0.91m (36")	1	<25.4mm (1-inch)	8

TABLE 4. Soil Type and Description

Soil Type	Soil Type Around the Pile	Number of piles
1	Plastic Clay	13
2	Clay-silt-sand mixtures, very silty sand, silts and marls	11
3	Clean Sands	18
4	Soft Limestone, very shelly sands	2

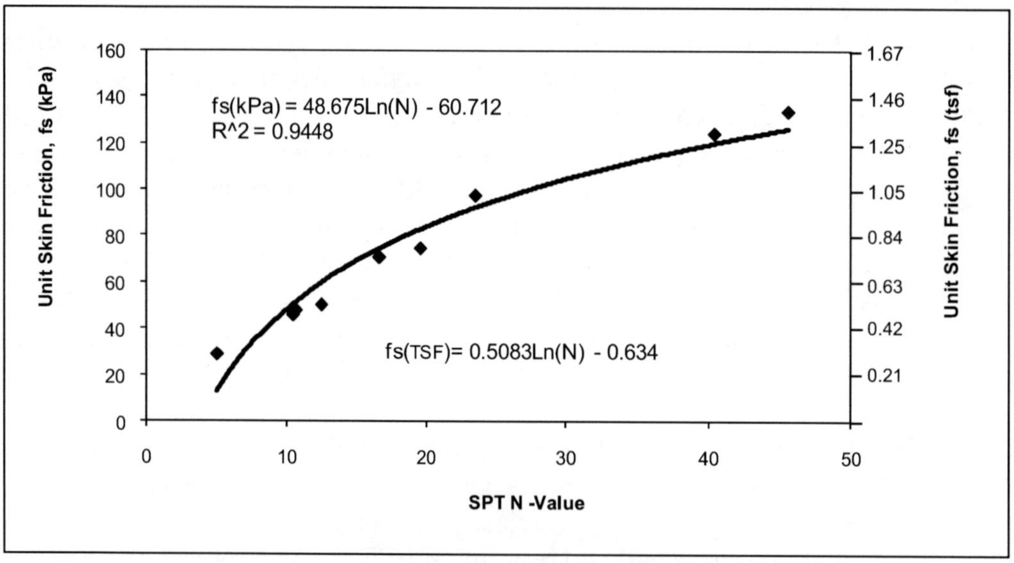

FIG. 8 Skin friction vs. SPT N-Value for concrete cylinder piles in clay.

For each of the 21 load tests, the total resistance was separated into unit skin friction and end bearing, and plotted against the uncorrected blow counts (N), as a function of soil type and pile material. Subsequently, best-fit (regression) skin friction and end bearing curves vs. SPT N values for different pile material type were constructed. Tables 5 and 6 present the back computed unit skin and end bearing for both steel and concrete cylinder piles based on SPT N values.

TABLE 5. Unit Skin Friction and End Bearing for Steel Cylinder Piles

Soil Description	Unit Skin friction, f_s (kPa) [f_s (tsf)]
Plastic Cay	$f_s = 40.56 * Ln(N) - 51.75$ [$f_s = 0.4236 * Ln(N) - 0.5404$]
Clay-Silt-Sand Mixtures	$f_s = 2.68 + 4.31 * N - 0.086 * N^2 + 6.11E-4 * N^3$ [$f_s = 0.029 + 0.045 * N - 8.98E-4 * N^2 + 6.371E-6 * N^3$]
Clean Sands	$f_s = 19.42 * Ln(N) - 25.34$ [$f_s = 0.2028 * Ln(N) - 0.2646$]
Soil Description	**Unit End Bearing, q_t (kPa) [q_{tip} (tsf)]**
Plastic Cay	$q_t = 21.32 N$ [$q_t = 0.2226 N$]
Clay-Silt-Sand Mixtures	$q_t = 39.27 N$ [$q_t = 0.4101 N$]
Clean Sands	$q_t = 54.35 N$ [$q_t = 0.5676 N$]

TABLE 6. Unit Skin Friction and End Bearing for Concrete Cylinder Piles

Soil Description	Unit Skin friction, f_s (tsf)	Unit End Bearing, q_{tip} (tsf)
Plastic Cay	$f_s = 48.67 * Ln(N) - 60.71$ [$f_s = 0.5083 * Ln(N) - 0.634$]	$q_t = 21.32 N$ [$q_t = 0.2226 N$]
Clay-Silt-Sand Mixtures	$f_s = 31.27 * Ln(N) - 26.05$ [$f_s = 0.3265 * Ln(N) - 0.272$]	$q_t = 39.27 N$ [$q_t = 0.4101 N$]
Clean Sands	$f_s = 1.8 N - 2.83$ [$f_s = 0.0188 N - 0.0296$]	$q_t = 54.35 N$ [$q_t = 0.5676 N$]

Of interest is a comparison of unit skin friction and end bearing for both small width/diameter concrete or steel piles with the large cylinder pile values.

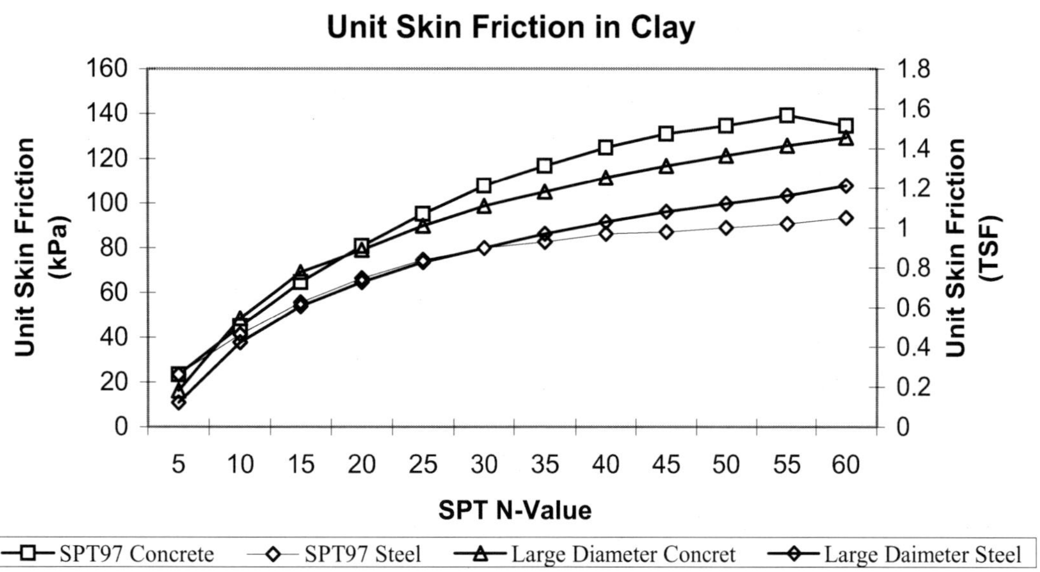

FIG. 9 Comparison of Unit Skin Friction for Small & Large Diameter Piles

For instance, Fig. 9 shows a comparison of unit skin friction for both small and large width/diameter piles driven into clay. The small width/diameter ratio piles, identified with SPT97 (an FDOT software program) are based on FDOT Research Bulletin 121 by Schmertmann (1967) which separated piles based on soil type (plastic clay, sand-clay-silt mix, clean sand, and limestone) and pile type (concrete & steel). Evidently, the unit skin frictions vs. SPT N values are very similar for small and large width/diameter piles, with greatest variation being pile material type (steel & concrete). Similar observations were observed for the other soil types (Tables 5 & 6).

Recommended LRFD Resistance Factors, ϕ, for Cylinder Pile Design The Load and Resistance Factored Design (LRFD) specifications as approved by AASHTO recommend the use of load factors to account for uncertainty in the loads, and resistance factors to account for the uncertainty in the materials. This criterion can be written as:

$$\phi R_n \geq \eta \sum \gamma_i Q_i \qquad (9)$$

where: R_n = Nominal resistance (predicted pile capacity),
η = Load modifier to account for effects of ductility, redundancy and operational importance. The value of η usually ranges from 0.95 to 1.00. Herein, η = 1.00,
Q_i = Load effect,
γ_i = Load factor. Based on current AASHTO recommendation, the following factors are used: γ_D = 1.25 for dead load, γ_L = 1.75 for live load,
ϕ = Resistance factor (Typically ranges from 0.3 to 0.8).

The ϕ factor can be viewed as the reciprocal of factor of safety (FS) in Allowable stress design (ASD). The LRFD approach is viewed as having the following advantages:

- It accounts for variability in both resistance and load. For instance, in allowable stress design (ASD), no consideration is given to the fact that different loads have different levels of uncertainty. For example, the dead load can be estimated with a higher degree of accuracy vs. the live load.

- It achieves relatively uniform levels of safety based on the strength of soil and rock for different limit states and foundation types based on risk assessment.

- It has the potential to provide more consistent levels of safety in the superstructure and substructure as both are designed using the same loads for known probabilities of failure. In ASD, selection of a factor of safety is subjective, and does not provide a measure of reliability in terms of probability of failure.

There are generally three levels of probabilistic design that can be used to estimate resistance factors, ϕ for deep foundations. The one used by AASHTO and FHWA is referred to as the first order second moment method (FOSM). According to Barker, et al, 1991 and Withiam, et al, 1997, using an assumed lognormal distribution function for resistance (R_n), and bias factors (λ_R , λ_{QD}, λ_{QL}), the resistance factor, ϕ, can be obtained from:

$$\phi = \frac{\lambda_R(\gamma_D \frac{Q_D}{Q_L} + \gamma_L)\sqrt{\frac{1+COV_{QD}^2 + COV_{QL}^2}{1+COV_R^2}}}{(\lambda_{QD} \frac{Q_D}{Q_L} + \lambda_{QL})\exp\left(\beta_T\sqrt{\ln[(1+COV_R^2)(1+COV_{QD}^2 + COV_{QL}^2)]}\right)} \tag{10}$$

where: γ_D = Dead load factor (1.25, recommended by AAHSTO (*1*)),
γ_L = Live load factor (1.75, recommended by AAHSTO 1996/2000 (*1*)),
Q_D/Q_L = Dead/live load ratio, varies from 1.0 - 3.0 (spans, L = 17- 52 m)
λ_R = Resistance bias factor,
COV_R = Resistance coefficient of variation,
λ_{QD}, λ_{QL} = Dead load and live load bias factors,
λ_{QD} = 1.08, λ_{QL} = 1.15 (recommended by AASHTO 1996/2000),
COV_{QD}, COV_{QL} = Dead load and live load coefficients of variation,
COV_{QD} = 0.128, COV_{QL} = 0.18 (recommended by AASHTO (*1996*))
β_T = Target reliability index; due to the redundancy of pile groups, AASHTO and FHWA recommend using β from 2.5 to 3.0 for pile foundations (Barker, 1991).

Of major importance in estimating, ϕ, from Eq. 10 are resistance bias factor (λ_R), and the resistance's coefficient of variation (COV_R). Both are found from: 1) the predicted pile capacities, R_{ni}, and 2) the measured (Davisson) capacities, R_{mi}, for each pile in the database, from which the following database statistics are computed:

Mean: $$\lambda_R = \frac{\sum \lambda_{Ri}}{N} \tag{11}$$

Standard deviation: $$\sigma_R = \sqrt{\frac{\sum(\lambda_{Ri} - \lambda_R)^2}{N-1}} \tag{12}$$

Coefficient of variation: $$COV_R = \frac{\sigma_R}{\lambda_R} \tag{13}$$

where: $$\lambda_{Ri} = \frac{R_{mi}}{R_{ni}} \tag{14}$$

R_{mi} = Measured resistance (Davisson Capacity) for a pile case.
R_{ni} = Predicted (nominal) resistance for an individual pile case.

Using the soil descriptions, SPT N values, the unit skin friction, end bearing (Tables 5 & 6) and total capacity (R_{ni}) were predicted for the 21 large diameter cylinder piles, along with their bias ($\lambda_{Ri} = R_{mi} / R_{ni}$), as shown in Table 7. In estimating end bearing, both total cross-sectional pile area (Full C/s – Table 7), and ring area only of the pile cylinder wall were considered. Shown in Figure 10 is a comparison of measured and predicted capacities for the 21 piles using the full cross-sectional area given in Table 7.

TABLE 7. Measured vs. Predicted Pile Capacities

#	Project	Measured*	Ring Area **	λri **	Full C/s ***	λri ***
1	St. Georges Island Project	1050	803.92	1.31	1146.31	0.92
2	St. Georges Island Project	1400	1208.45	1.15	1604.15	0.87
3	St. Georges Island Project	1275	1081.69	1.17	1499.86	0.85
4	St. Georges Island Project	1425	834.08	1.71	1237.07	1.15
5	Chesapeake Bay Bridge LT-4	700	477.75	1.47	770.94	0.91
6	Chesapeake Bay Bridge LT-6	460	324.34	1.42	399.26	1.15
7	San Mateo Hayward Bridge	775	699.23	1.11	729.47	1.06
8	South Test-4	650	328.54	1.98	553.28	1.17
9	Woodrow Wilson Bridge	1475	1185.97	1.24	1600.36	0.92
10	Woodrow Wilson Bridge	1500	1230.65	1.22	1530.27	0.98
11	Woodrow Wilson Bridge	900	781.81	1.15	951.49	0.95
12	Salinas River Bridge	750	517.45	1.45	862.22	0.87
13	Port Of Oakland 27NC	625	462.39	1.35	544.57	1.15
14	Port Of Oakland 17NC1	512.5	439.08	1.17	592.09	0.87
15	Port Of Oakland 10NC1	450	400.47	1.12	598.50	0.75
16	Port Of Oakland 31NC	600	532.43	1.13	592.55	1.01
17	I-880 Oakland Site 3C	450	382.82	1.18	467.41	0.96
18	I-880 Oakland Site 3H	600	506.92	1.18	600.44	1.00
19	Santa Clara River Bridge 13	925	947.67	0.98	1205.60	0.77
20	Santa Clara River Bridge 7	1070	996.90	1.07	1436.45	0.74
21	Berenda Slough Br 4	800	702.77	1.14	798.97	1.00

* Measured capacity in Tons.
** Capacity prediction using end bearing over the ring area in Tons.
*** Capacity prediction using end bearing over the entire cross sectional area in Tons.

Next, using the individual pile biases (λ_{Ri} – Eq. 14, Table 7), the mean (Eq. 11), standard deviation (Eq. 12) and coefficient of variation (Eq. 13) were found for the entire database (21 piles, Table 8). Subsequently, the LRFD resistance factors, ϕ, for approach 1 (ring area) and approach 2 (total cross-sectional area) were assessed from Eq. 10 using the latter statistics with a target reliability, β_T, of 2.75. Note, a reliability, β_T, of 2.75 has a probability of failure of 3 out of 1000 which is the midpoint of AASHTO recommendations.

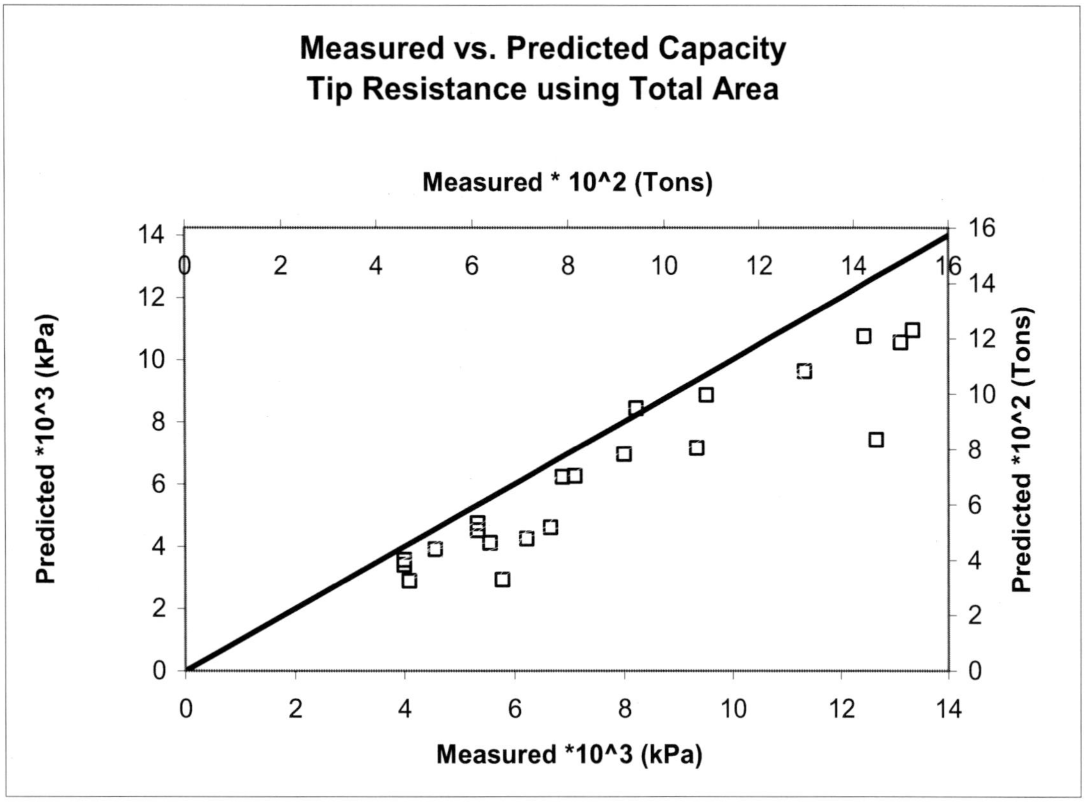

FIG. 10. Comparison of Measured vs. Predicted pile capacity

TABLE 8. Statistics and LRFD Resistance Factors, ϕ, for Predictive Methods

	Approach – 1 (Ring Area)	Approach – 2 (Full C/S)
λ_R	1.27	0.955
σ_R	.23	.13
COV_R	.18	.13
ϕ	.76	.61

Evident from Table 8, the use of ring area for end bearing calculations, results in a conservative prediction (i.e., λ_R = measured capacity / predicted capacity), as well higher standard deviation versus the full cross-sectional area prediction. However, because the ring area is more conservative (i.e., higher λ_R) and COV_R are similar for both methods, the LRFD resistance factor ϕ, (Eq. 10) is larger in approach-1. To judge the quality of the method, one should compute the percentage of Davisson Capacity that is available for design (i.e., ϕ / λ_R x 100%) which is 59.8% for Approach-1 vs. 63.9% for Approach-2, suggesting that even though large diameter cylinder piles will behave "unplugged" during driving, they behave as "plugged" under static load testing.

SUMMARY AND CONCLUSIONS

It is found that soil arching alone cannot explain the existence or lack of "soil plugging" inside cylinder piles during driving. The inertia force from the intruding soil column increases quadratically with the pile diameter whereas the skin friction on the column increases linearly with diameter. A simple parametric study found that soil plugging would not occur in piles larger than 24" in diameter. However, even though a cylinder pile may not plug during driving, it still could behave "plugged" under static load conditions due to a lack of inertia force. A FEM study characterizing the static load test as a viscous axisymmetric two-dimensional steady state fluid flow problem, revealed:

Shell Thickness - A 40% increase in soil column height (from 60% to 100%) for "plugging", if the shell thickness was reduced from 10" to 6". The increase was attributed to: 1) as the diameter of the soil column increases (i.e., pile wall decrease), the skin friction force increases linearly ($f_s \cdot A_{surf}$), but the tip resistance that it balances increases quadratically ($Q_t = q_t \cdot A_{cross}$); and 2) the thicker the pipe wall, the larger the volume of soil mass being displaced and the more likely the soil, i.e., sand, silt, etc. will dilate within the cylinder, which would develop high shear strengths at the pipe-soil interface, and plug.

Friction Angle - A 40% increase in the soil column's height for "plugging" if the friction angle of the soil was reduced from 35 degrees to 27.5 degrees. The reduction in the soil's angle of internal friction reduces the shear strength of soil and pile/soil friction along the inside wall of the pipe.

Outer Diameter - A 10% increase in the soil column height before plugging when the diameter increased from 48" to 54" (same shell thickness of 8"). This is in accordance with explanation (1), the large the diameter, the unit skin friction increases linearly, and the tip stress increasing quadratically, requiring longer lengths to plug.

Based on both the driving analysis, as well as the quasi-static finite element study, it was decided to consider both total cross-sectional and ring area in estimating a pile's tip resistance from insitu tests for pile design. Subsequently, a database of 35 cylinder piles with outside diameters ranging from 91.4 cm (36") to 213.4 cm (84") and wall thicknesses from 1" to 8" was collected along with their static load test and insitu data. Of the 35 tests, 21 reached the modified Davisson capacity (PL/AE + D/30) from which both unit skin friction and end bearing were assessed. Both, the skin friction and end bearing were separated by soil type and average SPT N value. Interestingly, both the unit skin friction and end bearing vs. SPT N values developed for small diameter/width piles, Schmertmann RB121 (1967) agreed closely with the large diameter cylinder piles. Subsequently, a first order second moment method (FOSM) assessment of LRFD resistance factors, ϕ, for the proposed design (i.e., unit skin & end bearing vs. SPT N) based on the 21 load tests and either total cross-sectional area or pile ring area for end bearing calculations were performed. Resistance factors, ϕ, of 0.76 and 0.61 were found for the recommended unit skin and tip resistance curves with the varied tip areas, i.e., ring area or full cross-sectional area for tip bearing calculations. The latter factors suggest that the ring area approach is very conservative; moreover, even though the total cross-sectional area approach had the lower resistance factor ϕ, it had the higher percentage of Davisson capacity (ϕ / λ) available for design.

REFERENCES

AASHTO, 1996/2000. LRFD Bridge Design Specification, 2nd Edition-2000 Interim Revisions. Washington, DC.

Azzouz, A.S., Baligh, M.M., and Whittle, A.J., (1990). "Shaft resistance of piles in clay." J. Geotech. Engrg., ASCE, 116(2), 205-221.

Barker, R. M, Duncan, J. M., Rojiani, K. B., Ooi, P. S. K., Tan, C. K., and Kim, S. G. (1991). *Manuals for the Design of Bridge Foundations.* NCHRP Report 343, Transportation Research Board, National Research Council, Washington D.C.

Chow, Y.K. (1981). "Dynamic Behavior of Piles". Ph.D. Thesis, University of Manchester.

Hvorslev, M.J., (1949) "Subsurface Exploration and Sampling of Soils for Civil Engineering Purposes," Report on a Research Project of the Committee on Sampling and Testing Soil Mechanics and Foundations, Division ASCE, Edited and Printed by Waterway Experiment Station, Vicksburg, Miss. 1949.

McVay, M.C., Badri, D. and Hu, Z. (2004) , Final report, Determination of Axial Pile Capacity of Prestressed Concrete Cylinder Piles, FDOT Contract No. BC 354 RPWO-60, 121 pages.

Paikowsky, S.G., and Whitman, R.V., (1990). "The Effects of Plugging on Pile Performance and Design," Canadian Geotechnical Journal, vol. 27, pg. 429-440.

Schmertmann, J. (1967), "Guidelines for use in the Soils Investigation and Design of Foundations for Bridge Structures in the State of Florida," FDOT Research Bulletin 121A, 87 pages.

Stevens, R.F. (1988),"The Effect of Soil Plug on Pile Drivability in Clay", Application of Stress-Wave theory to Piles, Third International Conference, May 25-27.

Withiam, J., Voytko, E., Barker, R., Duncan, M., Kelly, B., Musser, S., and Elias, V., (1998). Load and Resistance Factor Design (LRFD) of Highway Bridge Substructures. FHWA Publication No. HI-98-032, July, Washington, D.C.

Effective Stress Analysis and Set-up for Shaft Capacity of Piles in Clay

Bengt H. Fellenius, Dr.Tech., P.Eng., M.ASCE

1905 Alexander St. SE, Calgary, Alberta, T2G 4J3. <Bengt@Fellenius.net>

ABSTRACT: A case history of repeated dynamic and static loading tests in Alberta on two pipe piles during dissipation of driving-induced pore pressures is presented together with three reanalyzed published case histories involving similar records. The four case histories demonstrate that, for each case, the same effective-stress proportionality coefficients, beta-coefficients, fit the capacities at different degrees of dissipation of excess pore pressures. For two of the test sites, the beta-coefficients back-calculated from the tests differed considerable from the values determined from the soil plasticity relation, while for two, the agreement is good. For one case, the back-calculated shaft resistance agreed well with the values of vane shear strength, while a less good agreement was found for the other tests. Neither case showed good agreement was found for methods combining undrained shear strength and effective overburden stress. Capacity calculations for two cases employing methods based on CPT soundings gave excellent agreement with one test and a poor agreement with the other. The increase of capacity due to aging after dissipation of excess pore pressures did not agree with cited recommendation for calculations of aging effect.

KEY WORDS: analysis, clay, effective stress, examples, piles, setup

INTRODUCTION

It is well-accepted in engineering practice that load-transfer for piles in sand is proportional to effective overburden stress. Since more than 40 years, investigations on instrumented piles in clay (e.g., Bjerrum et al. 1965; 1969, Endo et al. 1969, Burland 1973, Konrad and Roy 1987, Bond and Jardine 1995) have established that this applies also to load transfer in clay. Yet, many practioners still consider the unit shaft resistance to be a function of the undrained shear strength of the clay and calculate the unit shaft resistance along a pile from the undrained shear strength values determined from in-situ tests such as vane-shear tests, converted from results of CPT-soundings, or obtained from laboratory tests on recovered samples using various types of undrained tests. The enduring popularity of total stress analysis is partially due to the ease and low cost of field-vane testing and of obtaining representative samples in cohesive soil for laboratory testing. When data are available that can be combined with various theoretical solutions for shear forces based on bearing capacity formulae or more elaborate approaches, it is understandably tempting to correlate these values to pile unit shaft resistance. Inasmuch as this approach is based on and adjusted to observations of actually measured pile shaft resistance, the results can often be used to predict closely the shaft resistance for similar piles in soils of similar geologic background and subjected to similar loading conditions—experience is the paramount tool for geotechnical engineers. However, in the author's opinion, such empirically correlated undrained values are little more than index quantities, useful for the designer working in a known soil, but the values do not correlate consistently to pile unit shaft resistance in general.

The universality of effective stress analysis in clays, as opposed to total stress analysis, is demonstrated in a case history from Alberta involving full-scale static and dynamic testing of two pipe piles, with correlation to field vane, CPTU soundings, and measured pore pressures. The results are put into context by results from reanalysis of three other case histories from full-scale tests in Bangkok Thailand, Vancouver, and Quebec. The case data results are first correlated to the effective overburden stress and then discussed in direct reference to undrained shear strength and to methods combining undrained shear strength and effective stress.

CASE HISTORIES

Bangkok, Thailand

Results of short-term measurements of load-transfer were presented by Indraratna et al. (1992) on a 20 m long, 400 mm diameter concrete pile driven in highly plastic, soft to firm, marine clay in Bangkok, Thailand. The water content of the clay is about 90 % and the plasticity index is about 50. The undrained shear strength determined by field vane tests ranges from about 10 KPa at 4 m depth increasing progressively to about 40 KPa at 18 m depth. The groundwater table lies at a depth of about 2 m and the pore pressure distribution is hydrostatic within the upper 20 m depth. Interrupting the pile installation of the pile, a series of uplift loading tests were performed one week after driving at depths of 8 m, 12 m, 16 m, and 20 m. The author has back-calculated the distribution using effective stress analysis to approximately match the pile capacities for each depth. For all four tests, a good fit was obtained for a beta-coefficient of 0.25, constant along the full pile lengths.

Vancouver, BC, Canada

Davies (1987) and Robertson et al. (1988) reported the results of a series of static loading tests performed in Fraser River deltaic soils at Lulu Island, Vancouver, BC. The soil profile consists of a surficial 2 m thick sand fill placed on 13 m of very soft clay with silt and sand lenses that overlies a 13 m thick layer of compact micaceous sand followed by soft clay and silty clay to large depth and containing numerous silt and sand lenses. Results from vane shear tests are not available. The water content of the clay at 2 m through 15 m depth is about 30 % and the plasticity index is about 15. The upper 2 m to 3 m and the lower about 2 m thick portions of the sand

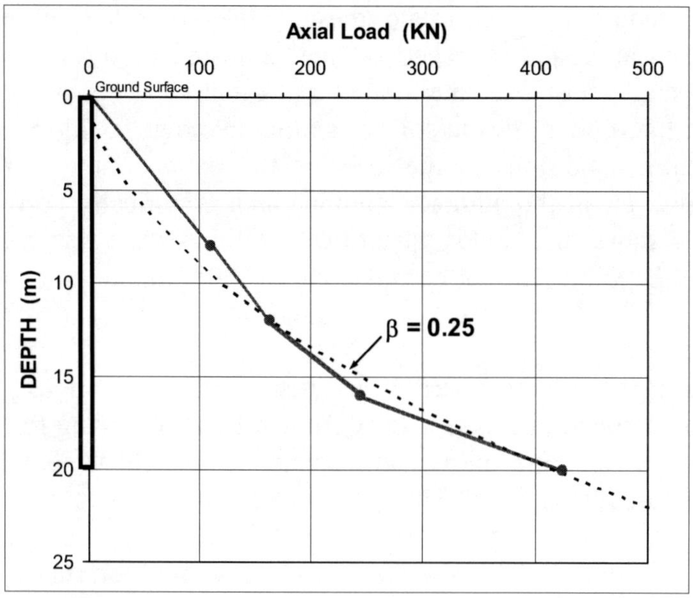

FIG. 1 Distribution of capacity in tension (uplift) tests at different depths (Data from Indraratna et al. 1992)

layer are interspersed with clay and silt lenses. The CPTU cone stress, q_t, ranges from about 8 MPa through about 10 MPa. The groundwater table lies at a depth of about 1 m and the pressure distribution is hydrostatic within the embedment depth of the piles. The soils are similar to those at the adjacent Annacis Island, described by Bazett and McCammon (1986).

Three 324-mm diameter closed-toe pipe piles driven to embedment depths of 13.7 m, 16.8 m, and 31.1 m were subjected to static loading tests at 197 days, 85 days, and 38 days after initial driving, respectively. The toe of Pile 13.7 m was about 1.3 m (about 4 pile diameters) above the sand. Pile 16.9 m was driven about 1.9 m into the weaker, clay and silt interspersed upper portion of sand layer. Pile 31.1 m was driven about 3 m below the bottom boundary of the sand layer, i.e., about 10 pile diameters below the sand/clay boundary.

The load-movement curves of the static loading tests are presented in Fig. 2. What capacity value to evaluate from the tests is a matter of judgment and personal preference. The sloping dashed line in the figure rising from the abscissa is the offset elastic line whose intersection with the curves defines the offset-limit load, commonly considered to be a lower-bound capacity value. The increase after the offset-limit load is considered to be mostly due to increase of toe resistance. While full dissipation of driving-induced pore pressures might not have occurred for the 38-day test on Pile 31.1 m, they have probably dissipated for the 85-day test on Pile 16.8 m and for the 197-day test on Pile 13.7 m.

Two methods for calculating the pile capacity were applied to the CPTU sounding: the CPT-based LCPC method (Bustamante and Gianeselli 1982) and the CPTU-based

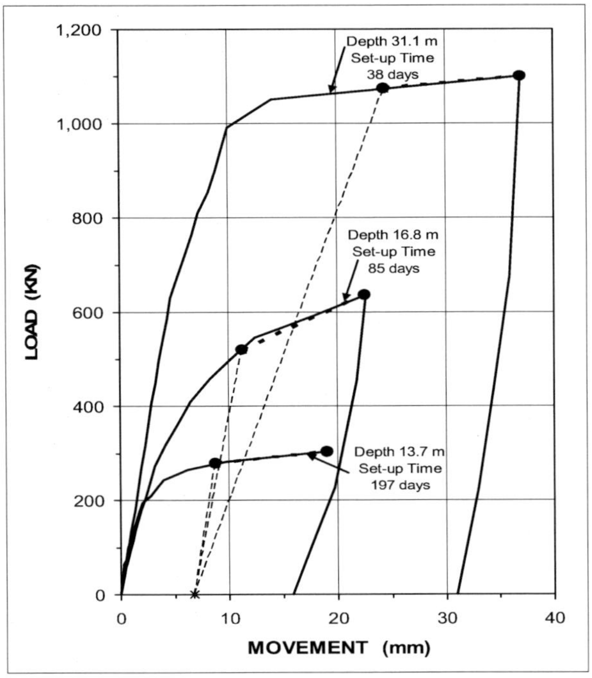

FIG. 2 Load-movement curves from static loading tests on three 324 mm diameter, closed-toe, driven pipe piles in Vancouver, BC. Data from Davies (1987).

Eslami-Fellenius method (Eslami and Fellenius 1997). Fig. 3A presents the distribution of the total shaft resistances according to the two methods. The shaft resistance distributions are very similar, which is a coincidence because the methods differ appreciably: The CPT-method uses the cone stress uncorrected for pore pressure on the cone shoulder, disregards the sleeve friction, and includes cut-off limits of shaft resistance based on ranges of cone stress and either of two main soil types, clay and sand, as input by the user. In contrast, the CPTU-method uses corrected cone stress values and determines the soil type directly from the sounding, separating the profile on six soil types for every measurement point. Sleeve friction is relied on for the determination of soil type and is referenced in determining the toe resistance.

The fit of the loading test results to an effective stress analysis was made by dividing the shaft resistance values determined by the CPTU-method by the effective overburden stresses at corresponding depths, resulting in equivalent values of ß-coefficients. The average values in the three main soil layers were 0.20, 0.30, 0.20, and 0.15, respectively. These values are low, which probably is due to the presence of mica flakes in the soil. As shown by Gilboy (1928), micaceous soils are more compressible and weaker as opposed to non-micaceous soils. The average shaft resistances were used to determine the distribution of total capacity The calculated resistance distributions are shown in Fig. 3B together with the resistances of each test shown as pairs of square dots, with the first indicating the estimated shaft resistance and the second the total resistance of the test. The average of the CPT and the CPTU

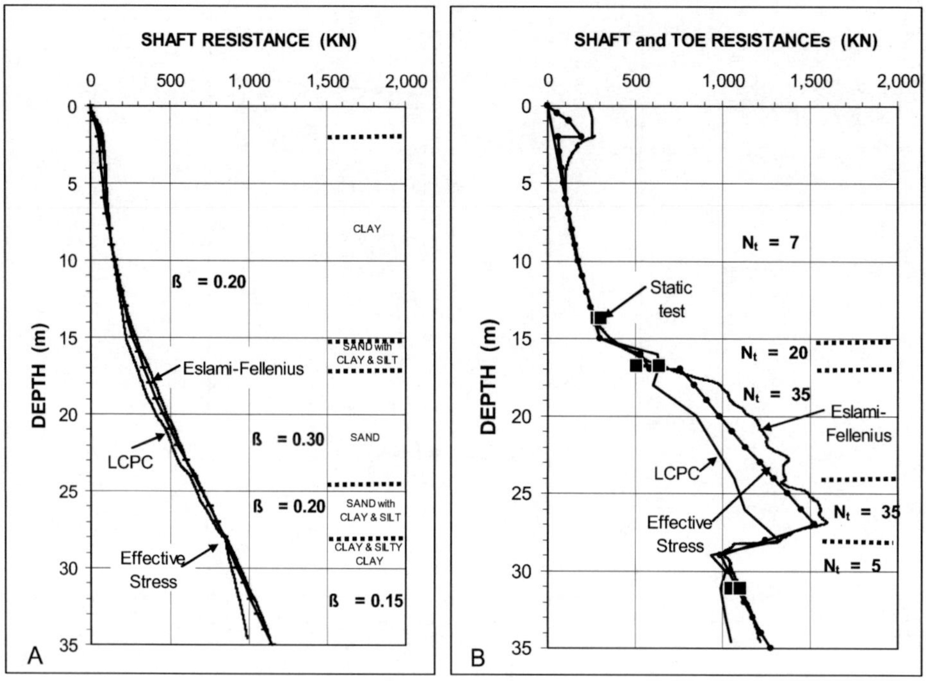

FIG. 3 Distributions of shaft resistance (A) and total resistance (B) from effective stress calculations fitted to the distributions calculated from the CPT-LCPC and the CPTU Eslami-Fellenius methods and to capacity from the static loading tests (Bustamante and Gianeselli 1982; Eslami and Fellenius 1997).

methods is indicated in the figure as "effective stress". The beta-coefficients and toe coefficients employed are also indicated in Fig. 3.

Quebec, Canada

Konrad and Roy (1987) and Roy et al. (1981) reported static loading tests on two 7.6 m long, 219 mm diameter closed-toe pipe piles jacked into Champlain clay at the St Alban test site, west of Quebec City. The soil profile consists of thin weathered clay crust on 8 m of soft silty marine clay. The plasticity index of the silty clay is about 20 and the water content is appreciably higher than the liquid limit. The pore pressure distribution is hydrostatic from a groundwater table located at a depth of 0.4 m. The deposit is overconsolidated with an OCR of about 2.2 developed due to secondary consolidation or aging of the clay. The undrained shear strength measured by in-situ vane tests varies linearly from 10 KPa at 2 m depth to a maximum of 30 KPa at the bottom of the soft silty clay layer. The values are the average determined in 17 soundings with the envelope of the low and high boundaries deviating about ±30 % from the average.

The piles were instrumented to separate shaft and toe resistances, and piezometers were installed at different depths next to the piles. The ultimate shaft resistances determined at the end of the jacking installation and at several static loading tests performed at 4 days, 8 days, 20 days, and two years after the installation are shown in Fig. 4. The values are shown as a ratio (%) between ultimate resistance at the specific test over the resistance (77 KN) of the two-year test, and ratio (%) of the excess pore pressures measured on the specific test occasion over the excess pore pressures immediately after the pile installation.

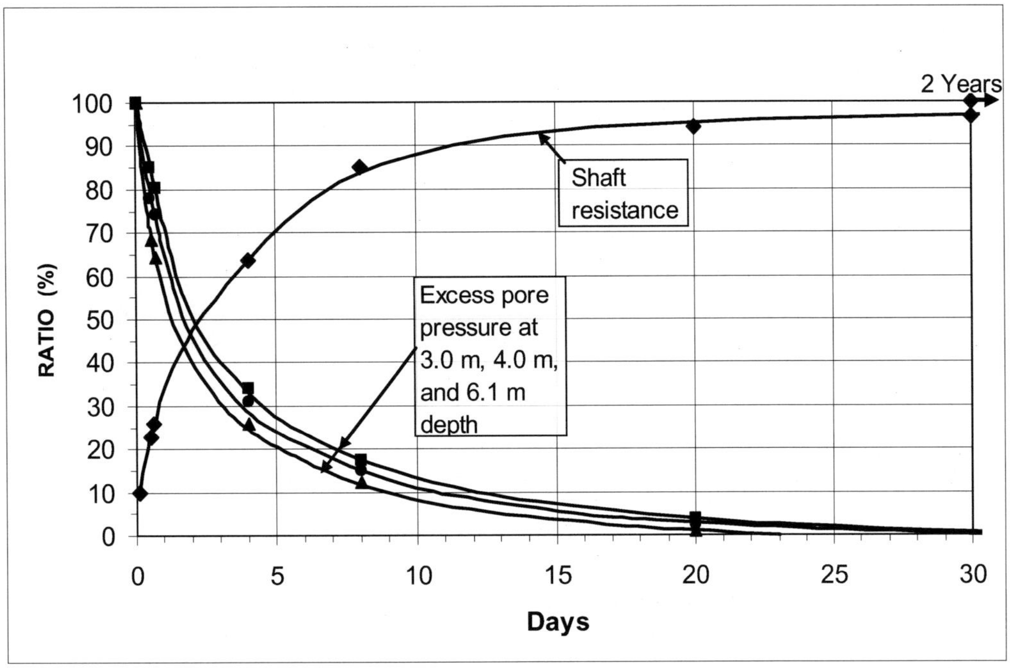

FIG. 4 Shaft resistance and excess pore pressure measurements at installation and at several static loading tests during up to two years after the pile was jacked into the soil at Quebec. (Data from Konrad and Roy 1987)

The two-year ultimate shaft resistance corresponds to a beta-coefficient of 0.6 (ratio of shaft shear to effective overburden stress). When the analysis is repeated for the effective stress adjusted for the excess pore pressures measured at 4 days, 8 days, and 20 days after the installation, a fit between the actual effective stresses and the measured ultimate resistances is achieved with beta-coefficients of 0.7, 0.6, and 0.6, respectively. That is, essentially the same beta-coefficient applies to the shaft resistances measured for all three of the mentioned testing occasions during the ongoing dissipation of the excess pore pressures.

North of Edmonton, Alberta, Canada

In 1990, Alberta Department of Transportation, undertook a long-term study of the development of capacity with time for two 324 mm diameter, closed-toe, concrete-filled, pipe piles at a highway bridge site located about 180 Km Northwest of Edmonton, Alberta, for which project the author served as a consultant. The site conditions are presented by Diyaljee and Pariti (2002).

The two test piles were driven on October 30, 1990, about 2.4 m apart in line with three previously installed anchor piles to depths of 16 m and 20 m. Static loading tests were performed 15 days, 30 days, and 4 years (1,485 days) after EOID. Restrike tests were carried out one day after end-of-driving and at the 30-day tests the day before the static loading test. The 30-day restrike consisted of 5 to 8 blows to a total penetration of about 50 mm.

The clay is an eroded, transported, and re-deposited glacial till clay interspersed at 5 m depth by an about 0.5 m thick layer of silty sand. The groundwater table lies at a depth of 1.5 m and the pore pressures are hydrostatically distributed. The soil profile consists of 7.5 m of silty clay with a water content of about 35 % through 70 %, a Liquid Limit of about 60 % through 70 %, a Plastic Limit of about 15 % through 40 %, and a Plasticity Index of about 25 %. Three field vane tests were performed at 8 m, 11 m, and 17 m depth, showing almost linearly increasing values from 45 KPa through 85 KPa. The Janbu modulus number ranges from 12 through 20 and the recompression modulus is about fifteen times larger. The OCR is about 3. Triaxial consolidated and undrained tests and direct shear testing on the clay indicated a strain-softening soil having a friction angle ranging from 21° through 25° with a residual (post peak) value of about 21°. A small cohesion intercept was found in the range of 10 KPa through 25 KPa.

At a depth of 7.5 m lies a second 0.5 m thick layer of silty sand. The sand is followed by soft silty sandy gravelly ablation clay till that continues to the end of the borehole at a depth of about 26 m. The ranges of water content and indices of the clay till are about the same as those of the upper clay layer. Consolidation tests on samples from the clay till show the Janbu modulus number of the clay to range from 19 through 21 and the recompression modulus to be about ten times larger. The OCR is about 2.

The results of a CPTU sounding at the pile testing site indicate a relatively homogeneous cone stress in the clay and the clay till with a cone stress, q_t, ranging from about 1 MPa through 2 MPa. According to Kulhawy and Mayne (1990), the undrained shear strength can be determined by subtracting the overburden total stress from the corrected cone stress and dividing the balance with a factor "N_{kt}" ranging from 10 through 20. A N_{kt}-factor of 10 makes a good fit between all three vane shear values and the q_t-values at corresponding depths.

Prior to the pile driving, six piezometers were installed on the center of the test pile group to depths of 3 m, 6 m, 9 m, 11 m, 16 m, and 18 m, respectively. As indicated in Fig. 5, the piezometer measurements show that the pile driving produced considerable excess pore pressures that dissipated with time. The actual time for full dissipation was not measured. However, the extrapolation in the logarithmic plot, Fig. 4B, indicates that the dissipation was completed 70 days after the driving.

In Fig. 6, the vertical distribution of pore pressures are shown as measured before the pile driving, 15-days, 30 days, and 4 years after the pile driving. (The 4-year measurements were obtained in a new set of piezometers also installed near the test piles and to the same depths as the original piezometers). The measurements show that significant excess pressures still remained at the time of the 30-day static loading tests.

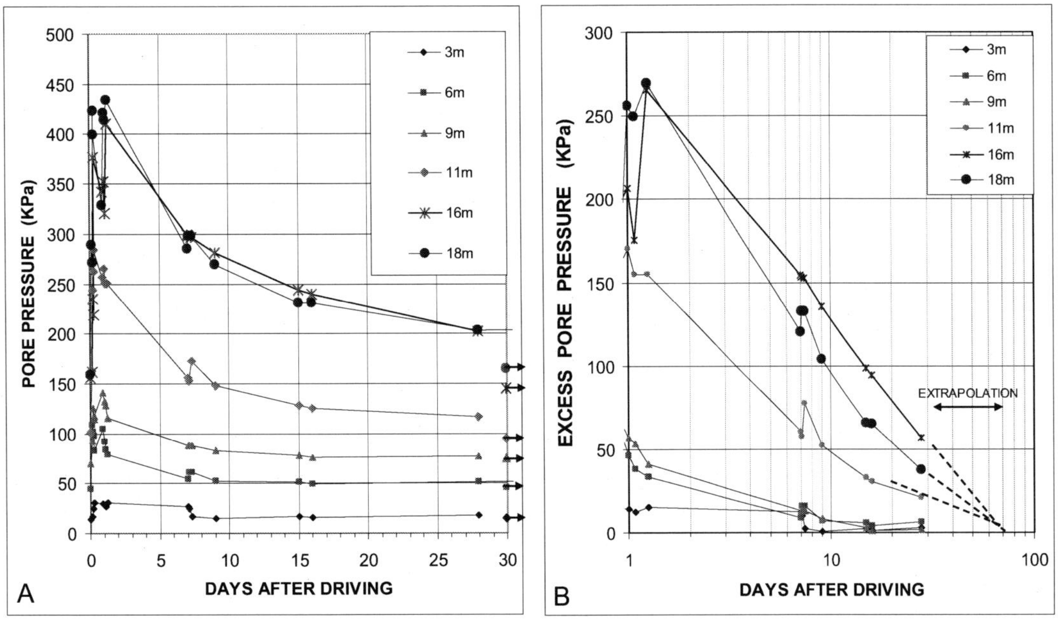

FIG. 5 Dissipation of measured pore water pressures with time (days) after driving. A. 0 - 30 days, Linear scale. B. 1 - 100 days, Logarithmic scale

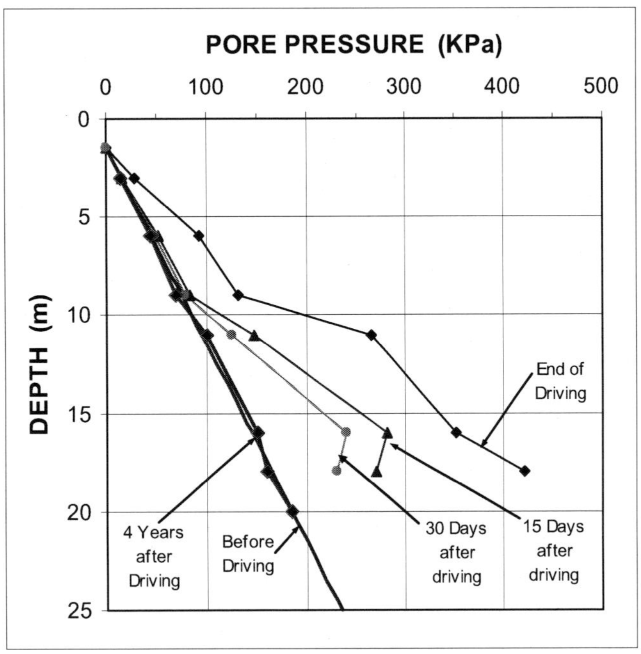

FIG. 6 Distribution of measured pore water pressures with depth.

Fig. 7 presents the load-movement curves of the static loading tests on Piles 16 m and 20 m, respectively. The 15-day and 30-day tests on Pile 16 m were carried to movements in excess of 100 mm. After a movement of about 12 mm, the applied load reached a peak, which is taken as the ultimate resistance (capacity) of the test. Thereafter, the load required to move the pile diminished. This is an indication that the toe resistance of both piles is very small for all tests. When evaluating change of pile capacity with time by means of repeated compression tests, so-called "stage-testing", it is essential that the toe resistance is small. Otherwise, the movement-dependent increase of toe resistance from one test to the next could falsely indicate a capacity increase with time (Fellenius 2002).

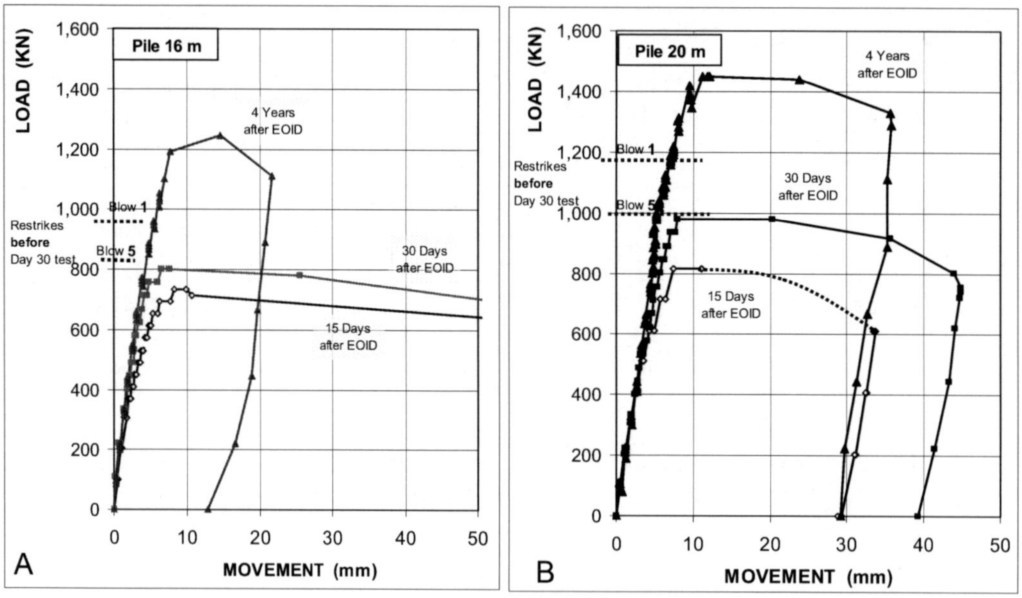

FIG. 7 Load-Movement curves from the tests of Pile 16 m. A. Pile 16 m, B Pile 20 m.

For both test piles, the capacity increased from one test to the next. This phenomenon is called "set-up". With regard to shaft resistance, set-up is usually credited to dissipation of excess pore pressures and reconsolidation of the disturbed clay. The piles were restruck the day before the 30-day static loading tests. Dynamic analysis (CAPWAP) on the restrike data show that the restrike reduced the capacity from blow to blow and, therefore, also the capacity in the subsequent static loading test. That is, the 30-day capacities would have been higher had the restrike not been performed.

The CPTU analyses of resistance distribution show a gradually increasing unit shaft resistance with depth and a small toe resistance, no more than about 50 KN to 100 KN. However, the CPT methods considerably underestimate the capacity of the test piles. Fig. 8 shows a comparison between the capacities determined in the test and calculated from five cone methods: Eslami and Fellenius (1997), Tumay (Tumay and Fakhroo 1981), Schmertmann (1975), Dutch (DeRuiter and Beringen

1979), and LCPC (Bustamante and Gianeselli 1982). The full height of the columns is the capacity and the lowest part of each column is the toe resistance. Moreover, unlike the Vancouver case, the Eslami-Fellenius and the LCPC methods give very different results. The reason lies in the somewhat arbitrary cut-off limits of shaft resistance imposed in the LCPC method and its use of uncorrected cone stress values. This reason explains both the agreement between the two methods found in the Vancouver case and the disagreement found in the Alberta case.

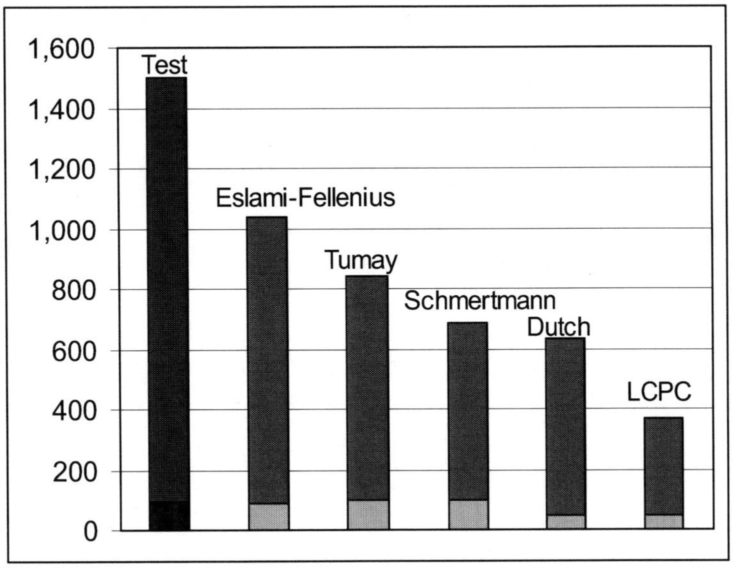

FIG. 8 Capacities determined in the static loading test on the 20 m long pile compared to capacities calculated by five CPT methods for the Alberta case.

The mentioned gradual increase of unit shaft resistance with depth supports that the resistance is controlled by effective stress. Although the pore pressure distribution is well established, the distribution of total stress along the piles to be less well defined than desirable: while the site is generally horizontal, some ten metre beyond the piles, the natural ground slopes gently down toward a river located at a distance of about 20 m away from the piles, and a few metre away from the test pile location is a two metre high road embankment. Calculations using upper bound ground surface contours (no embankment but slope toward the river and vice versa) versus horizontal ground surface show the uncertainty in the amount of effective stress to be about ±15 %. Therefore, an acceptably representative effective stress distribution for the long-term condition can be determined on the assumption of horizontal ground surface in the vicinity of the test piles. For this condition, a beta-coefficient of 0.5 and a toe-bearing coefficient (N_t) of 6, result in a fit to the capacity of the 4-year static loading tests for both piles.

The difference of effective stress for the 15-day and 30-day tests on the piles in relation to the 4-year tests is known because it is equal to the difference of pore pressure illustrated in Fig. 6. By multiplying the pore pressure differences to the hydrostatic line with the beta-coefficient and subtracting the values from the 4-year distributions of resistance, the resistance distributions are determined for the 15-day and 30-day tests. The so-determined distributions are shown in Fig. 9 for the three static loading tests on Piles 16 m and 20 m, respectively.

The calculated capacity for the 30-day test is larger than the capacity of the static loading test. As indicated in the foregoing, the 30-day static loading test is affected by the preceding restrike. The calculated 30-day capacities should indeed come out larger than the values found in the static loading tests. That the agreement between the calculated capacities is as good as shown in Figs. 9A and 9B is probably somewhat fortuitous. The main finding is that a reasonable agreement between the calculations for the three pairs of static loading tests, performed at different extents of pore pressure dissipation, is obtained by assigning the soil-pile response to be proportional to the effective overburden stress.

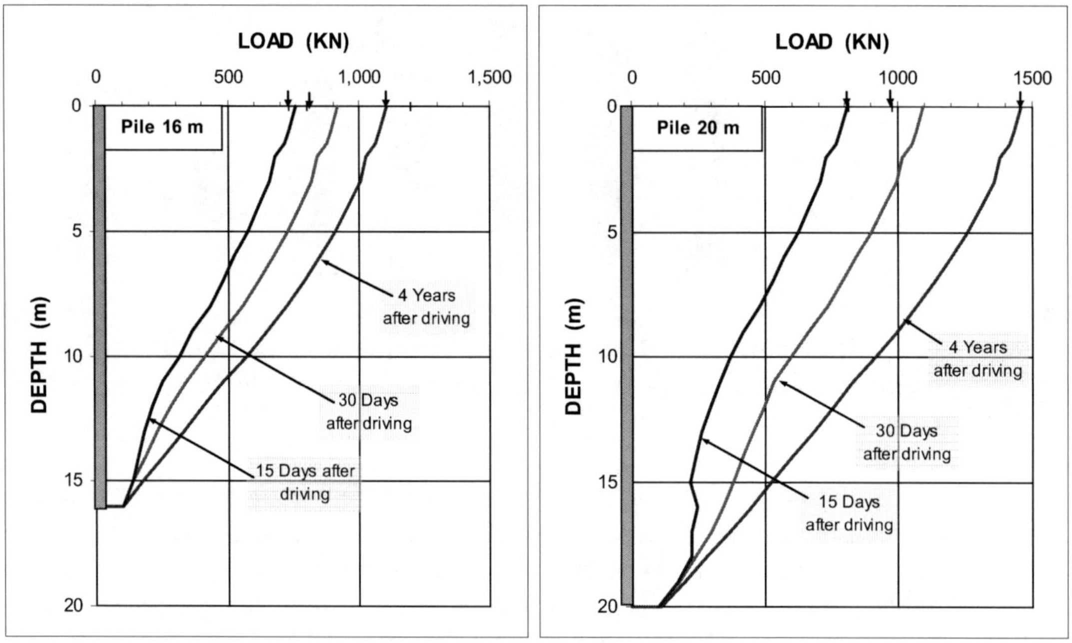

FIG. 9 Load distribution for the 15 day and 30 day loading test capacities calculated by the same coefficient values as used for the 4 year test and with adjustment to the measured excess pore pressures. The small arrows on the abscissa indicate capacities determined in the loading tests. A. Pile 16 m; B. Pile 20m.

DISCUSSION

Shaft Resistance

In principle, the shaft resistance is a friction coefficient times the normal effective stress acting against the pile shaft. The friction coefficient is usually considered to be equal to the tan δ', where δ' is the interface friction angle. Bozozuk (1972), incorporated a M-coefficient equal to tan δ'/tan φ' and suggested that this ratio is about 0.8. The coefficient, K_s, applied to the overburden stress, is usually set equal to the earth stress coefficient at rest, K_0, which for normally consolidated clays is taken to be equal to 1 - sin φ'. In overconsolidated clays, Burland (1973) suggested that the beta-coefficient for overconsolidated clays should increase in proportion to the square root of the OCR, also suggested by Esrig and Kirby (1979). Thus, the beta-coefficient is a function of several parameters: the ratio between tan δ'/tan φ', or tan φ' if the shearing is considered to occur a distance out from the pile shaft, the pile surface material and roughness, the earth stress coefficient, the OCR, as shown in Eq. 1.

$$\beta = M\ \tan\varphi'\,(1-\sin\varphi')(OCR)^{0.5} \tag{1}$$

where M = tan δ'/tan φ'
φ' - effective soil friction angle
δ' = effective interface friction angle; a function of pile surface texture and material
OCR = overconsolidation ratio

Jardine et al. (2005) proposed a relation for the ß-coefficient according to Eq. 2, obtained from fitting measured pile capacity to a series of soil parameters.

$$\beta = 0.8\{2.2 + 0.016(OCR) - 0.87\lg(S_t)\}(OCR)^{0.42}\,(2h/b)^{-0.2}\,\sigma'_z\,\tan\delta' \tag{2}$$

where OCR = apparent overconsolidation ratio
S_t = sensitivity, related to "relative void ratio index" from oedometer tests
h = height of point above pile toe; $h/b \leq .4$
b = pile diameter
δ' = interface friction angle

Regarding Eq. 1, not all of the three parameters are determined for a routine design case and, when they are, they vary considerably depending on natural variation and method of determining them. Also influencing is the method of pile installation, and the soil gradation, mineralogical composition, density, and depositional history, etc. In fact, once the potential ranges of each parameter are considered, for any specific case, the beta-coefficient calculated per Eq. 1 will range from the very small to the very large. In the author's opinion, chances are good that an experience-based judgment call for the beta-coefficient directly would be considerably closer to the actual value than any value determined from the mentioned parameters, however well researched in separate tests.

As to Eq. 2, fitting a series of precise two-digit constants and exponents to approximate indices, such as sensitivity (determined in an oedometer test, no less) and OCR values is hardly applicable to engineering design or analysis. The author finds it difficult to take such a relation seriously.

It should be recognized that the beta-method (as the effective stress method often is called) is a simplistic method. It disregards several factors, such as that the imposed movement of the pile surface in a static loading test causes rotation of principal stresses along the pile surface, can result in strain softening or strain hardening of the soil, and generates small additional pore pressures which reduce the effective stress against the pile surface. Still, as demonstrated in the above four case histories, actual shaft resistance is proportional to the effective overburden stress, and, therefore, the results of a test can be correlated through the beta-coefficient, and the test results be applied to the design of other similar piles at the particular site.

The question of importance to persons active in design of pile foundations is to what extent the beta-coefficient can be determined from soil parameters established in a routine site investigation at a site and applied to a design without first being "calibrated" in a loading test.

Karlsrud et al. (2005) compiled results from shaft resistances determined in tests on pile in clay and presented a relation between the beta-coefficient and the plasticity index, I_P, shown in Fig. 10. The beta-coefficients determined from the cases in Thailand, Vancouver, Quebec, and Alberta are added to the original data. Of the latter, the coefficients from Thailand and Vancouver cases lie close to the mean curve of the original plot. However, the coefficients from Alberta and Quebec deviate considerably from the Karlsrud et al. (2005) curve and data. That the first two cases are from normally consolidated clay, while the latter two are from overconsolidated clay may have had an influence. That is, according to Eq. 1, the beta-coefficients for the Alberta and Quebec would be about 50 % larger than for a normally consolidated clay. However, the values are larger than so and about twice to three times larger than indicated by the mean curve shown in Fig. 10. In the author's opinion, the scatter of data and the fact that two of the four cases deviate from the proposed relation make the plasticity index not directly useful for a specific design case.

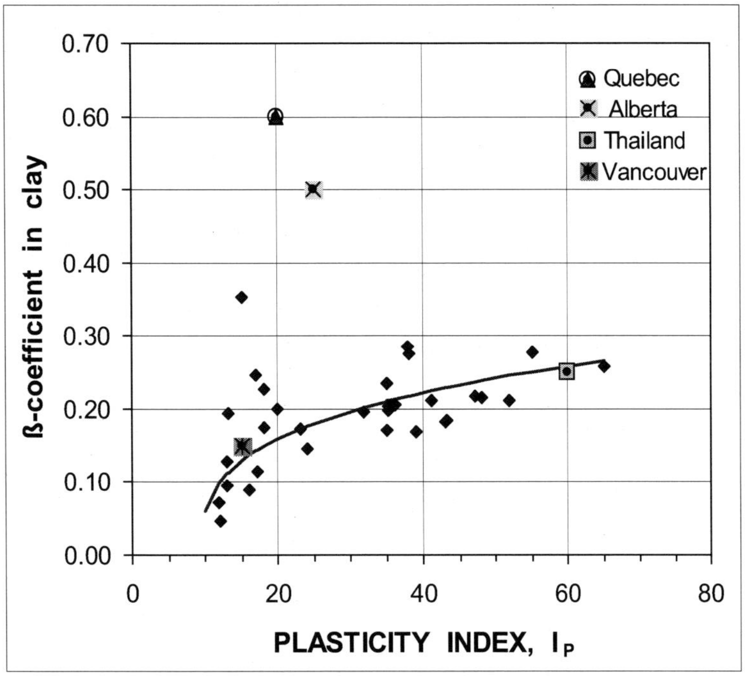

FIG. 10 Beta-coefficient versus average I_P for piles in clay (Data from Karlsrud et al., 2005). The coefficients evaluated for the cases in Quebec, Alberta, Thailand, and Vancouver are added.

The most common method for calculating shaft resistance is the so-called alpha method, which correlates the shaft resistance, r_s, to the undrained shear strength of the clay, τ_u, via a reduction coefficient, α (Eq. 3). For concrete and wood piles in soft clays (soft clay is defined as $\tau_u \leq 25$ KPa), α is equal to unity, while it is about 0.8 for steel piles (CFEM 1992). For undrained shear strengths of 50 KPa and 100 KPa, the α-coefficient is about 0.6 and 0.4, respectively.

$$r_s = \alpha \, \tau_u \tag{3}$$

where r_s = unit shaft resistance along the pile
α = alpha coefficient
τ_u = undrained shear strength

Vijayvergiya and Focht (1972) combined undrained shear strength and effective stress proposing a method which equals the shaft resistance to a coefficient, λ, times the sum of the effective stress and twice the undrained shear strength, as shown in Eq. 4. The lambda-coefficient ranges from 0.5 at the ground surface to 0.1 at a depth of 60 m. The lambda method is mostly applied to long, large diameter offshore piles in the Mexican Gulf.

$$r_m = \lambda(\sigma'_m + 2\tau_m) \tag{4}$$

where r_m = mean unit shaft resistance along the pile
λ = lambda coefficient
σ'_m = mean overburden effective stress
τ_m = mean undrained shear strength

Randolph (2003), quoting Randolph and Murphy (1985), proposed determining the unit shaft resistance from three alternative combinations of undrained shear strength and effective stress, as shown in Eqs. 5 - 7.

$$r_s = 0.5(\tau_u)^{0.50} (\sigma_z')^{0.50} \tag{5}$$

$$r_s = 0.5 (\tau_u)^{0.75} (\sigma_z')^{0.25} \tag{6}$$

$$r_s = 0.55(\tau_u)^{0.7} (\sigma_z')^{0.3} (\frac{40}{D/b})^{0.2} \tag{7}$$

where r_s = unit shaft resistance along the pile
τ_u = undrained shear strength
σ'_z = overburden effective stress
D = embedment depth
b = pile diameter

The case histories from Thailand, Quebec, and Alberta include records of undrained shear strength determined from field vane tests. The Thailand and Alberta only include a few values from one vane test, while the Quebec case includes a large number of vane tests. The vane shear values are used to calculate the unit shaft resistance according to Eq. 3 and Eqs. 5 - 7. The results are plotted in Figs. 11 - 13 versus the unit shaft resistance determined from the results of the static loading tests for the full dissipation of excess pore pressures. No reduction has been made to the undrained shear strength values, i.e., alpha is equal to unity and the B-figures show the actual distribution of the vane shear strength.

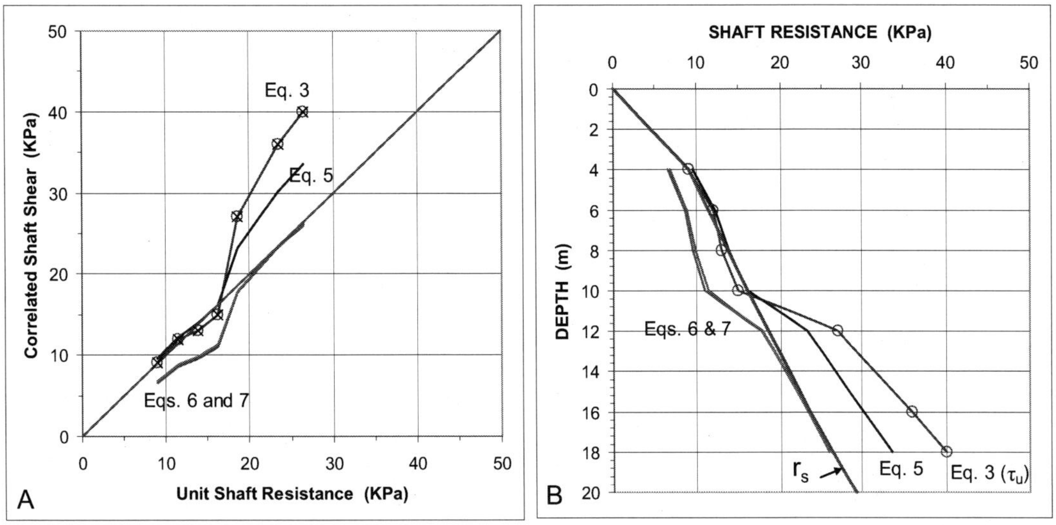

FIG. 11 Comparison between actual unit shaft resistance and shaft resistance calculated according to Eq. 3 and Eqs. 5 - 7 for the Thailand case. A. Direct correlation. B. Distribution with depth.

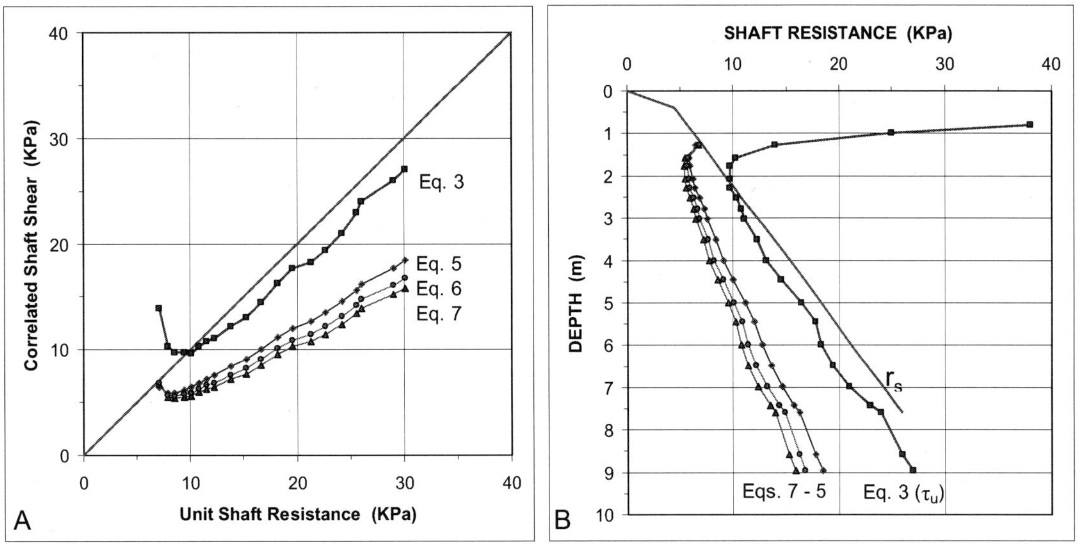

FIG. 12 Comparison between actual unit shaft resistance and shaft resistance calculated according to Eq. 3 and Eqs. 5 - 7 for the Quebec case. A. Direct correlation. B. Distribution with depth.

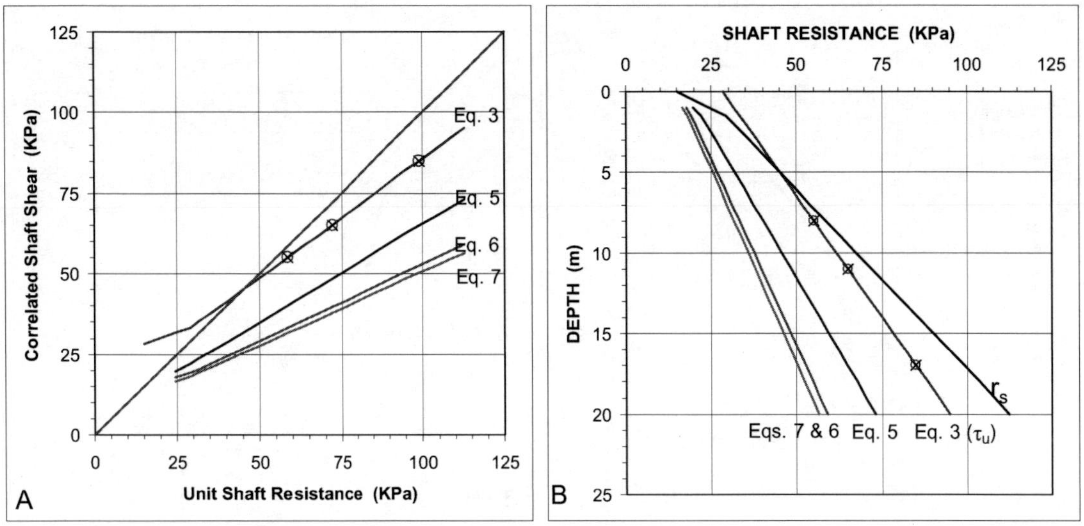

FIG. 13 Comparison between actual unit shaft resistance and shaft resistance calculated according to Eq. 3 and Eqs. 5 - 7 for the Alberta case. A. Direct correlation. B. Distribution with depth.

Of the four methods employing the undrained shear strength for determining shaft resistance, after full dissipation of excess pore pressure, the undrained shear strength values from Eq. 3 show the best correlation. For the Thailand case (Fig. 11), had the adjustment for shear strength exceeding 25 KPa been applied, the correlation would have been next to perfect. In contrast, the same adjustment for the Alberta case would instead have increased the difference considerably. The calculated values for unit shaft resistance according to Eqs. 3, 5 - 7 differ little from each other, but deviate considerably from the measured shaft resistance.

The Vancouver case does not include any records of undrained shear strength. However, it does include good CPTU records. The two CPT methods applied, the LCPC and the Eslami-Fellenius methods, give very similar distributions of shaft resistance that agree well with the resistances calculated from the static loading tests. In contrast, when the CPT sounding is applied to the Alberta case, the results by the two methods differ considerably from each other. Both underestimate the distribution determined from the static loading test. It is interesting to notice that the capacities calculated by E-F-method for the Alberta case agree quite well with the capacities determined in the 30-day test. The piezometer measurements indicate that only about two-thirds of the excess pore pressures had dissipated at that time. However, it is normally assumed that 30 days wait is sufficient time for the excess pore pressures to dissipate, as the Quebec case also indicates. In the absence of pore pressure data, it might have been thought that the 30-day test was after completed set-up, with the belief supported by that the 30-day capacity being in good agreement with the CPTU method capacity, and the increased capacity found in the two-year test be an effect of aging. When planning a test, it is obviously prudent to install a couple of piezometers to verify that the construction-induced excess pore pressures have dissipated, and, indeed, to determine the actual distribution of also the neutral pore pressures.

Set-up due to aging

Pile capacity will increase after completed pile installation. First, due to dissipation of pore pressure and, then, due to aging — marginally or appreciably. Bullock et al. (2005) indicated a relation for increase of capacity with elapsed time after end of driving. The relation would allow a test performed before full set-up and the final capacity to be calculated — extrapolated — from the relation shown in Eq. 8.

$$Q_t = Q_{REF}\,(1 + A\,\lg\frac{t}{t_{REF}}) \tag{8}$$

where Q_t = capacity at Time "t"
Q_{REF} = capacity at Time "t_{REF}"
A = a constant expressing ratio of capacity increase for a ten-fold increase of elapsed time (days) beyond t_{REF}

Bullock et al. (2005) adapted the capacity at the 1 day elapsed time after end of initial driving as the reference capacity. In contrast, Karlsrud et al. (2005) recommended that the reference capacity should be the capacity after full dissipation of pore pressures and suggested that the reference should be the capacity at 100 days after the driving, being the capacity after full dissipation of the excess pore pressures developed during the installation of the pile. The condition that the excess pore pressures have indeed dissipated during the first 100 days is neither true for all single piles nor for all pile groups. For example, Fellenius (2006) reported that the dissipation time developed from the driving of a single pile in soft marine clay was six months, and Bozozuk et al. (1978) reported a dissipation time of eight months for a pile group in Champlain clay.

Karlsrud et al. (2005) indicated that the A-value in Eq. 7 would follow Eq. 9.

$$A = 0.1 + 0.4(1 - \frac{I_P}{50})\,(OCR)^{-0.8} \tag{9}$$

where A = the A-constant of Eq. 8
I_P = plasticity index, $I_P \leq 50$
OCR = overconsolidation ratio

As the lower limit values of plasticity index and OCR are zero and unity, respectively, the A-constant according to Eq. 9 cannot become larger than 0.5. Moreover, a plasticity index larger than 50 is input as 50. Therefore, however large the overconsolidation; the A-constant cannot be smaller than 0.1, i.e., $0.1 < A < 0.5$.

Waiting 100 days after driving before testing the pile is normally not practical, however. Therefore, a practical solution to find the 100-day capacity could be by performing a few, minimum two, tests during the set-up period and plotting the pile capacities versus time in a semi-log plot, as illustrated in Fig. 14. The 100 day

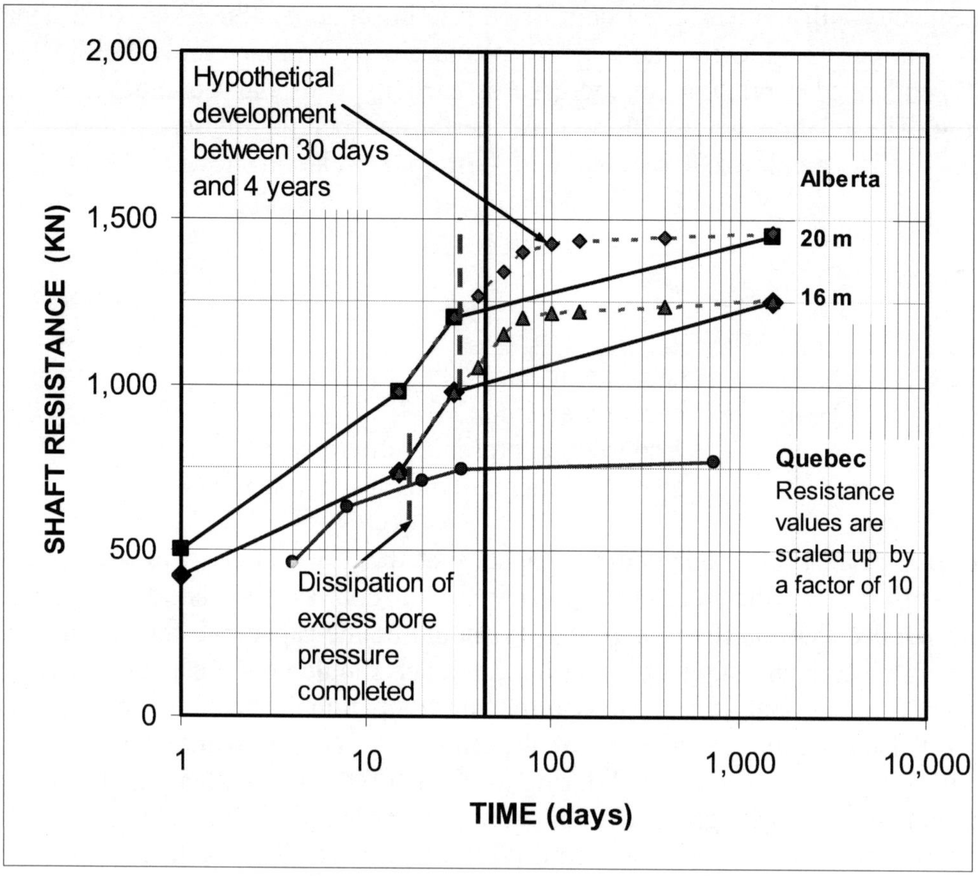

FIG. 14 Capacities versus days after initial driving.

capacity can then be determined from such a graph by extrapolation. Once the 100-day capacity is determined, Eqs. 8 and 9 could be used to indicate the increase of capacity due to aging beyond the 1 day and 100 day values, respectively.

However, extrapolation from capacities measured during the period of excess pore pressure to the capacity after the pore pressures have dissipated is rather inaccurate. Such hypothetical extrapolations are indicated in Fig. 13 for the Alberta case, taking the lead from the development for the Quebec case. It is only because of this lead, and because the actual four-year test results are known, that the Alberta extrapolations appear reasonable.

The Alberta case plasticity index and OCR values are about 25 and 3, respectively. Inserting these values into Eq. 9 returns A = 0.4. With this value for A, Eq. 8 indicates that, after twelve years, the capacity would increase by about two-thirds beyond that at 100 days. Similarly, the A-value for the Quebec case is 0.5, suggesting an 80 % capacity increase. Considering that, for both cases, capacity increase during the first four years and two years, respectively, is marginal, the calculated twelve-year capacity increases appear rather unrealistic.

There are very few tests performed many years after the piles were installed, when also the 100-day capacity is determined. However, several such long-term tests have been performed in connection of measurements of drag load due to settling soil (Fellenius 2006). These tests do not indicate increase of resistance once the ultimate shaft resistance has been mobilized in either the negative or positive shear directions. It would seem that the capacity increase due to aging (as opposed to dissipation of excess pore pressure) is rather small and requires very long time to reach appreciable values. "Appreciable" refers to values that can be put to use in a design.

Bullock et al. (2005) suggested that actual design of a piled foundation should make use of the set-up with time determined according to Eqs. 8 and 9 in order to obtain a reference pile capacity larger than tested values. The approach presupposes that the so-extrapolated capacity value be given the same assurance (i.e., same factor of safety) as that found in the test. The author considers this to be a debatable suggestion. The capacity used as a *tested capacity* in a design should not be larger than the maximum capacity actually found in a test. It is a different matter that Eqs. 8 and 9 can be used to indicate the potential success of investing time and money to carry out a repeated test to establish a larger capacity for the design. Moreover, a designer is free to carry out a *theoretical calculation of the capacity* and apply a larger factor of safety (or smaller resistance factor) to the results of that calculation.

Verifying capacity through testing

In addition to having the field investigation including CPTU soundings and vane shear tests, frequently, to optimize the use of the piles, testing for capacity is warranted. Note, a static loading test does not replace a thorough site investigation. Ideally, the test should not start until piezometer readings show that excess pore pressures developing from the construction have dissipated, i.e., after set-up is completed. Performing more than one test adds to the assurance of the design. However, static loading tests are expensive. The ideal approach is to combine static and dynamic testing, carrying out several dynamic tests for every static test. And, of course, correlating the test types to each other by carrying out a dynamic testing on the pile on which also a static test is performed. In this regard, the combination of static and dynamic tests on the Alberta piles show that to obtain a good correlation between the tests, the static test should be performed before the dynamic. This is not a trivial recommendation, because, in arranging for the static loading test, the pile head is usually cut close to the ground, which makes a subsequent dynamic test awkward. It is necessary, however, because the dynamic test involves several impacts, each of which move the pile more against the soil more than the movement imposed in the static loading test, causing a larger disturbance to the bond between the pile surface and the soil. Moreover, it is probable that many published correlations of capacity determined from static loading tests and dynamic tests, showing the dynamic test to "overestimate" the capacity (taking the subsequent static result as "true") involve this error. It is a different matter, and not any less important, that a reliable correlation must include the two tests carried within a very short time frame, if the tests are made while excess pore pressures still exist.

CONCLUSIONS

1. The increase of pile capacity demonstrated in the Quebec and Alberta cases is clearly a function of dissipation of excess pore pressures developed during the pile driving.
2. For each of the Quebec and Alberta cases, the same beta-coefficient fitted the back-calculated shaft resistance distribution for the tests performed at different times and at different level of dissipation of excess pore pressures.
3. For the Bangkok and Vancouver cases, the same beta-coefficient fitted the loading tests on the different length piles.
4. The correlation of beta coefficient to the plasticity index relation was close in the Vancouver and Bangkok cases, but very much off in the Quebec and Alberta cases.
5. To be reliable for a design case, analysis from CPTU tests requires calibration to each particular design project site.
6. For the Bangkok and Quebec cases, the undrained shear strength was close to the shaft resistance of the static loading tests after full dissipation of the pore pressures. This is also the finding for the Alberta case, provided unreduced vane shear values were used (alpha coefficient equal to unity).
7. The undrained shear strength cannot be used to estimate the shaft resistance before the excess pore pressures have dissipated.
8. Eqs. 5 - 7 for combining the undrained shear strength and the effective stress, applied to Bangkok, Quebec, and Alberta cases, did not provide a good agreement to the shaft resistance values of the static tests.
9. For the Alberta case, the increase of capacity with time after full dissipation of the excess pore pressures was small and did not agree with the aging relations, Eqs. 8 and 9.
10. The shaft resistance calculated from the CPTU soundings agreed very well with the static tests for the Vancouver case, but disagreed considerably for the Alberta case.
11. The pile capacity determined in the 30-day dynamic tests on the Alberta piles agreed very well with static tests, when the disturbance from one test to the following was considered. The results showed that when carrying out static and dynamic testing, it is important to perform the dynamic test after the static test.
12. The results show that in calculating pile capacity for a specific design case, the effective stress method is the most reliable approach. However, the beta-coefficients cannot be taken uncritically from published compilations of case histories, but must be taken from a well-designed test on a instrumented pile at

the considered site or from a case record of such a case at a representative site. Those same conditions apply also to calculations from CPTU soundings as well as to geotechnical parameters and indices used when calculating pile response to load.

REFERENCES

Bazett, D.J. and McCammon, M., 1986. "Foundations of the Annacis cable-stayed bridge". Canadian Geotechnical Journal 23(4) 458-527.

Bjerrum, L., Johannessen, I.J., and Eide, O., 1969. "Reduction of negative skin friction on steel piles to rock". Proc. 7th ICSMFE, Mexico City, August 25 - 29, Vol. 2, pp. 27 - 34.

Bond, A.J. and Jardine, R. J., 1995. "Shaft capacity of displacement piles in a high OCR clay". Geotechnique 45(1) 3-23.

Bozozuk, M. 1972. "Downdrag measurements on a 160-ft floating pipe pile in marine clay". Canadian Geotechnical Journal 9(4) 127 136.

Bozozuk, M., Fellenius, B. H. and Samson, L., 1978. "Soil disturbance from pile driving in sensitive clay". Canadian Geotechnical Journal 15(3) 346-361.

Bullock, P.J., Schmertmann, J.H., McVay, M.C., and Townsend, F.C., 2005. "Side shear set-up I: Test piles driven in Florida". ASCE Journal of Geotechnical and Environmental Engineering 131(3) 292-300.

Bustamante, M. and Gianeselli, L., 1982. "Pile bearing capacity predictions by means of static penetrometer CPT". Proc. of the 2nd European Symposium on Penetration Testing, ESOPT II, Amsterdam, May 24 - 27, A. A. Balkema, Vol. 2, pp. 493 - 500.

Burland, J. B., 1973. "Shaft friction of piles in clay". Ground Engineering, London, 6(3) 30-42.

Canadian Foundation Engineering Manual, CFEM, 1992. Third Edition. Canadian Geotechnical Society, BiTech Publishers, Vancouver, 512 p.

Davies, M., 1987. "Predicting axially and laterally loaded pile behavior using in-situ testing methods". M.A.Sc. Thesis, Univ. of BC, Vancouver, 187 p.

DeRuiter, J. and Beringen F.L., 1979. "Pile foundations for large North Sea structures". Marine Geotechnology 3(3) 267-314.

Diyaljee, V. and Pariti, M., 2002. "Capacity of pipe piles in cohesive ground". ASCE International Deep Foundation Congress. Geotechnical Special Publication No. 116, Edited by M.W. O'Neill, and F.C. Townsend, Orlando, Florida, February 14 - 16, 2002, Vol. 2, pp. 1318 - 1334.

Endo, M., Minou, A., Kawasaki T, and Shibata, T, 1969. "Negative skin friction acting on steel piles in clay". Proc. 8th ICSMFE, Mexico City, August 25 - 29, Vol. 2, pp. 85 - 92.

Eslami, A. and Fellenius, B.H., 1997. "Pile capacity by direct CPT and CPTu methods applied to 102 case histories". Canadian Geotechnical Journal 34(6) 886-904.

Esrig, M.E. and Kirby, R.C., 1979. "Advances in general effective stress method for the prediction of axial capacity for driven piles". 11th Annual Offshore Technology Conference, Houston, TX, 437-449.

Fellenius, B.H., 2002. "Determining the true distribution of load in piles". ASCE, International Deep Foundation Congress, An International Perspective on Theory, Design, Construction, and Performance, Geotechnical Special Publication No. 116, Edited by M.W. O'Neill, and F.C. Townsend, Orlando, Florida, February 14 - 16, 2002, Vol. 2, pp. 1455 - 1470.

Fellenius, B.H., 2006. "Results from long-term measurement in piles of drag load and downdrag". Canadian Geotechnical Journal 43(4) 409-430.

Fellenius, B.H. and Eslami, A., 2000. "Soil profile interpreted from CPTu data". Proceedings of Year 2000 Geotechnics Conference, Southeast Asian Geotechnical Society, Asian Institute of Technology, Bangkok, Thailand, November 27 - 30, 2000, Editors Balasubramaniam, A.S., Bergado, D.T., Der Gyey, L., Seah, T.H., Miura, K., Phien wej, N., and Nutalaya, P., Vol. 1, pp. 163 - 171.

Gilboy, G., 1928. "The compressibility of sand-mica mixtures". Proceedings of American Society of Civil Engineers, ASCE, 54(6) 555-568.

Indraratna, B., Balasubramaniam, A.S., Phamvan P., and Wong, Y.K., 1992. "Development of negative skin friction on driven piles in soft Bangkok Clay". Canadian Geotechnical Journal 29(3) 393-404.

Jardine, R., Chow, F., Overy, R., and Standing, J., 2005. "ICP Design methods for driven pile in sands and clays". Thomas Telford Publishing, London, 105 p.

Karlsrud, K., Clausen, C.J.F., and Aas, P.M., 2005. "Bearing capacity of driven piles in clay, the NGI approach". Proceedings of International Symposium on Frontiers in Offshore Geotechnics, Perth Sept. 2005, A.A. Balkema Publishers, pp. 775 - 782.

Konrad, J-M and Roy, M. 1987. "Bearing capacity of friction piles in marine clay". Geotechnique 37(2) 163-175.

Kulhawy, F.H. and Mayne, P.H., 1990. "Manual on estimating soil properties for foundation design". Electric Power Research Institute, EPRI, 250 p.

Randolph, M.F., 2003. "Science and empiricism in pile foundation design". 43rd Rankine Lecture. Geotechnique 53 (10) 847-875

Randolph, M. F. and Murphy, B. S., 1985. "Shaft capacity of driven piles in clay". Proc. 17th Annual Offshore Technology Conference, Houston, Vol. 1, pp. 371 - 378.

Robertson, P.K, Campanella, R.G., Davies, M.P., and Sy, A., 1988. "Axial capacity of driven piles in deltaic soils using CPT". Proceedings of the First International Symposium on Penetration Testing, March 20 - 24, Orlando, 10 p.

Roy, M., Blanchet, R., Tavenas, F., and Larochelle, P., 1981. "Behavior of a sensitive clay during pile driving". Canadian Geotechnical Journal 18(1) 67-85.

Schmertmann, J. H., 1978. "Guidelines for cone test, performance, and design". Federal Highway Administration, Report FHWA-TS-78209, Washington, 145 p.

Tumay, M.T., and Fakhroo, M., 1981. "Pile capacity in soft clays using electric QCPT data". ASCE, Proceedings of Conference on Cone Penetration Testing and Experience, St. Louis, October 26 - 30, pp. 434 - 455.

Vijayvergiya, V. N. and Focht, J. A. Jr., 1972. "A new way to predict the capacity of piles in clay". 4th Annual Offshore Technology Conf., Houston, Vol. 2, pp. 865 -- 874.

Wave Forces on Bridge Decks during Hurricanes and Impact on the Foundation

Jean-Louis Briaud[1] and Hamn-Ching Chen[2]

[1]Professor and Holder of the Buchanan Chair, Zachry Dpt. of Civil Engineering, Texas A&M University, College Station, Texas 77843-3136, USA (briaud@tamu.edu)
[2]Professor, Zachry Dpt. of Civil Engineering, Texas A&M University, College Station, Texas 77843-3136, USA

ABSTRACT: This paper addresses the problem of wave forces on bridge decks created by water flowing against and over the decks such as in the case of surges due to hurricanes. Such waves develop horizontal and vertical loads on the bridge decks. These decks in turn may fail by horizontal sliding since they are not tied to the piers. One solution is to tie the decks down to the piers but then the foundation has to resist the additional load. This paper takes a first look at the magnitude of such forces and at the impact they are likely to create on the foundation of the piers.

KEY WORDS: bridges, deck, failure, forces, foundations, hurricane, models, storm surge

INTRODUCTION:

Hurricanes create storm surges which raise the level of the water near coastal bridges. This increase in water level and the accompanying wind is associated with high waves which break against the side of bridge decks. The reduced weight of the bridge deck due to buoyancy combined with the cyclic horizontal push from the waves has knocked bridge decks off their supports in recent hurricanes in Mississippi, Florida and in Louisiana. This is due in part to the fact that bridge decks are typically not tied down to the piers. There are talks about tying the bridge decks down as a remedy for this problem. If such a solution is selected for many of the coastal bridges, the bridge deck will not move away but the foundation load will have to resist the additional load created by the waves.

HURRICANES

Major hurricanes are very large and can be 250 miles in diameter. They move relatively slowly at speeds of the order of 25 mph. Therefore they pass on top of a bridge for about 10 hours. During these 10 hours, they create waves which vary in height and in length. A wave length of 60 m, a wave height of 6 m and a period of 6 seconds are not uncommon for such waves. Therefore a hurricane can generate 6000 waves on a bridge during its passage over it.

A hurricane is a large rotating mass of moisture with wind speeds being the lowest (calm) in the eye of the hurricane and the highest at the wall of the eye. The winds then decrease as a function of the distance from the center of the rotating mass. The barometric pressure is also lowest in the eye but increases radially away from the center.

BRIDGE DAMAGE

By friction between the air and the ocean surface, the hurricane pushes the water forward ahead of itself. This phenomenon generates a surge of water in the form of a wave which can be 4 to 8 m in height and many kilometers long (Fig.1). During Hurricane Katrina, the surge reached 3 m at Lake Pontchartrain, 4.6 m at Lake

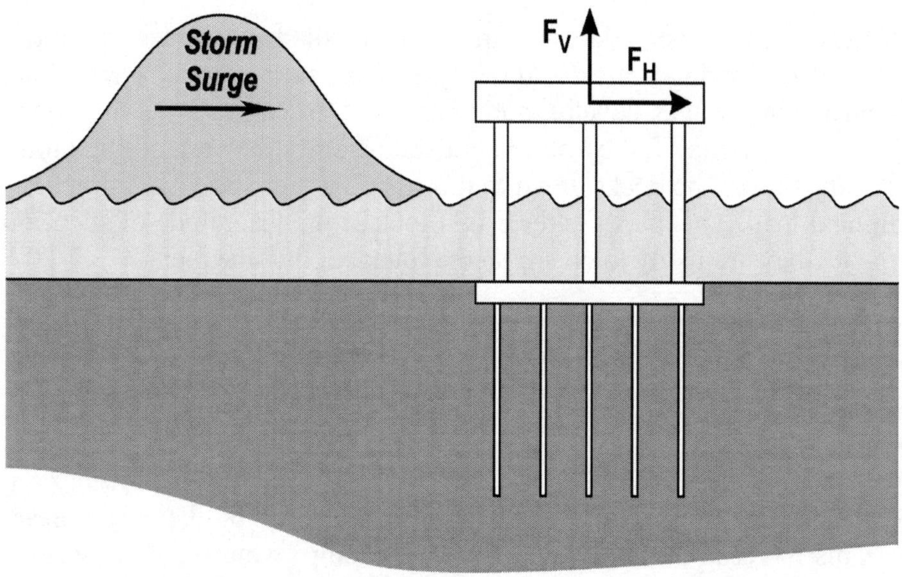

FIG. 1. Hurricane driven storm surge approaching a bridge.

Borgne, and 8.5 m at Bay of Saint Louis. Such water heights are sufficient to reach the height of the bridge deck for many coastal bridges.

In 2005 during Hurricane Katrina in Louisiana and in 2004 during Hurricane Ivan in Florida, many bridges were overtopped by the surge. During that overtopping, the side of the bridge deck was repeatedly pounded by waves and then overtopped. The cyclic horizontal forces ratcheted the bridge decks horizontally out of their support and after enough cycles, some of the decks fell into the water.

QUALITATIVE EXPERIMENT

An experiment was performed in an apparatus called the EFA or Erosion Function Apparatus (Briaud et al. 2001). This apparatus is designed for different types of erosion experiments. The bridge deck was made of a steel plate resting on two supports (Fig. 2). The water flow was set at 0.5 m/s and the water level was gradually increased in the conduit. The deck did not move until the water level reached half way up the thickness of the simulated deck. It was noted that the water level was lower under the deck than upstream and that bubbles could be trapped under the deck as the water rose. The movement of the deck was very slow and gradual for a few minutes until the deck had moved enough to fall off the supports. This qualitative experiment demonstrated that the failure of the bridge deck is a gradual ratcheting process and not one single violent event as could have been the case.

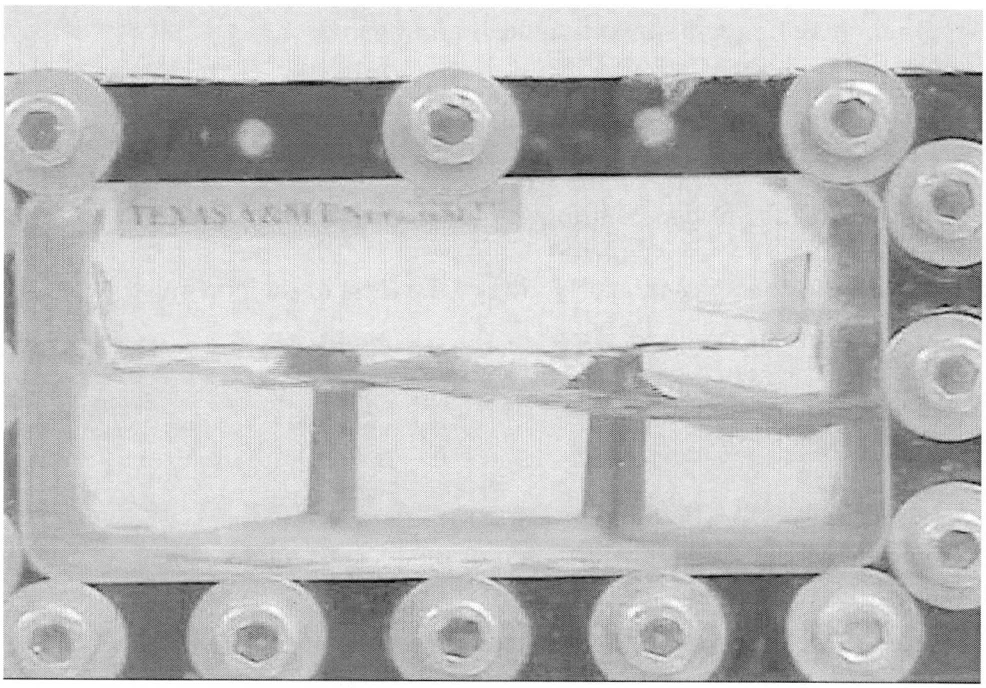

FIG. 2. Simulated bridge deck in a water flow

WAVE FORCES ON BRIDGE DECKS

The literature is limited on the topic of wave forces on bridge decks. The following authors have contributed to the state of the art. They include Bea et al. (1999), Tirindelli et al. (2002), and McConnell et al. (2003 and 2004). A good summary of the existing knowledge can be found in the work of Douglas et al. (2006). Douglas et al. also recommend the use of the following equations as interim guidance and provide examples calculations in their report to FHWA. The forces on the bridge deck are decomposed into the vertical and the horizontal components. Each component is made of two parts, one dealing with the slow varying forces and one dealing with the impact forces. The slow varying forces are

$$F_v = c_{v\text{-}va} F_v^* \qquad (1)$$

$$F_h = (1 + c_r (N-1))\, c_{h\text{-}va} F_h^* \qquad (2)$$

where:
F_v = estimated, vertical, wave-induced load component
F_h = estimated, horizontal, wave-induced load component
F_v^* = "reference" vertical load defined by Eq. 3
F_h^* = "reference" horizontal load defined by Eq. 4
$c_{v\text{-}va}$ = empirical coefficient for vertical "varying" load (recommended value = 1)
$c_{h\text{-}va}$ = empirical coefficient for horizontal "varying" load (recommended value = 1)
c_r = reduction coefficient for reduced horizontal load on the internal (i.e. not the wave ward-most) girders (recommend value, 0.4)
N = number of girders supporting the bridge span deck

$$F_v^* = \gamma\,(\Delta z_v)\, A_v \qquad (3)$$

where:
A_v = the area the bridge contributing to vertical uplift, i.e. the projection of the bridge deck onto the horizontal plane
Δz_v = difference between the elevation of the crest of the maximum wave and the elevation of the underside of the bridge deck
γ = unit weight of water (64 lb/ft^3 for salt water)

$$F_h^* = \gamma\,(\Delta z_h)\, A_h \qquad (4)$$

where:
A_h = the area the projection of the bridge deck onto the vertical plane
Δz_v = difference between the elevation of the crest of the maximum wave and the elevation of the centroid of A_h
γ = unit weight of water (64 lb/ft^3 for salt water)

The impact forces are

$$F_v = c_{v\text{-}im} F_v^* \qquad (5)$$

$$F_h = c_{h\text{-}im} F_h^* \qquad (6)$$

$c_{v\text{-}im}$ = empirical coefficient for the vertical "impact" load (recommended value = 3)
$c_{h\text{-}im}$ = empirical coefficient for the horizontal "impact" load (recommended value = 6)

Douglas et al. explain that the slow varying forces and the impact forces do not necessarily peak at the same time. They also recommend a factor of safety of two.

NUMERICAL SIMULATION: AN EXAMPLE

A numerical simulation was performed to obtain some idea of the magnitude of the forces generated during a hurricane surge. The program used was CHEN3D which has been well documented before (Chen, 2002). The thickness of the bridge deck was 2 m. The wave length was 60 m, the wave period 7 s, the wave height 5.2 m, and the wave speed 10 m/s. Fig. 3 shows the water surface and the velocity vectors.

Fig. 4 shows a plot of the vertical force per unit length of bridge as a function of time while the wave is passing by the bridge. Note that there are two peaks because the passage of two waves was simulated. The vertical force was taken as the integration of the pressures over the under side of the deck.

Fig. 5 shows a plot of the horizontal force per unit length of bridge as a function of time while the wave is passing by the bridge. Note that there are two peaks because the passage of two waves was simulated. The horizontal force was taken as the integration of the pressures over the vertical side of the deck.

The weight of the bridge deck was calculated to be 1470 kN/m. As can be seen from Fig. 4 the highest uplift force is close to the weight of the bridge deck. The maximum horizontal force is 80 kN/m or a mean pressure on the side of the bridge deck of 40 kN/m2. If a coefficient of friction of 0.5 is used, the maximum vertical force which could be overcome to push the deck would be 160 kN/m. This is very small compared to the dead weight of the deck. Therefore this numerical simulation indicates that it is the combination of the horizontal load and the decrease in vertical load which allow the horizontal ratcheting to take place.

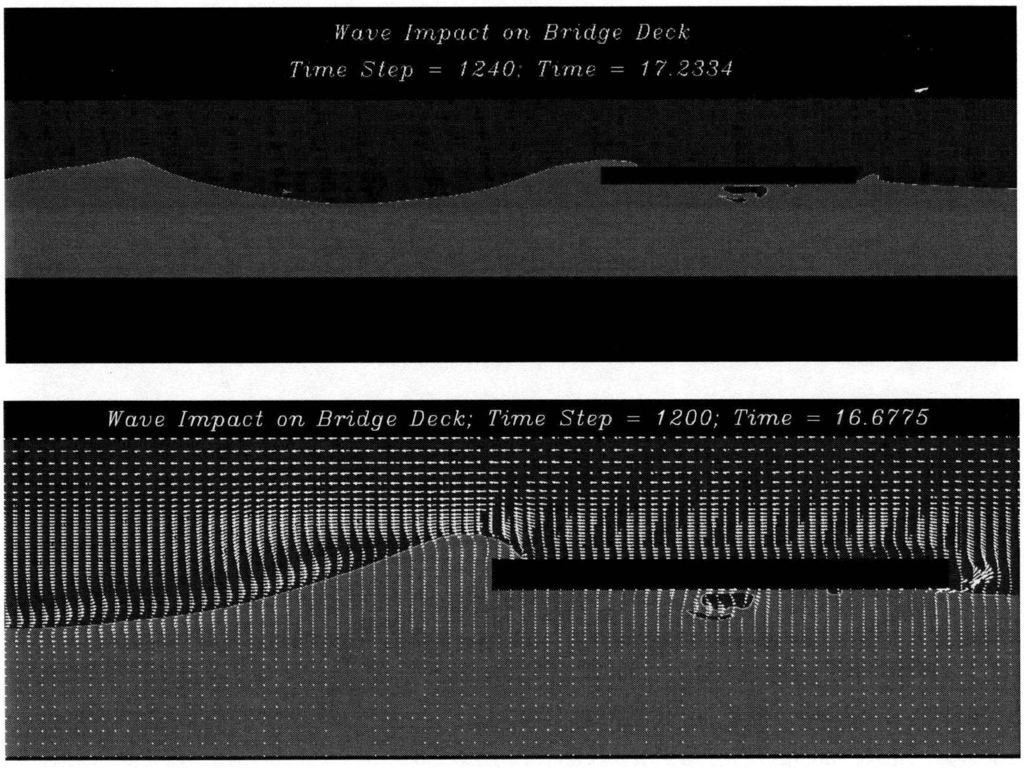

FIG. 3. Water surface and velocity vectors from the numerical simulation

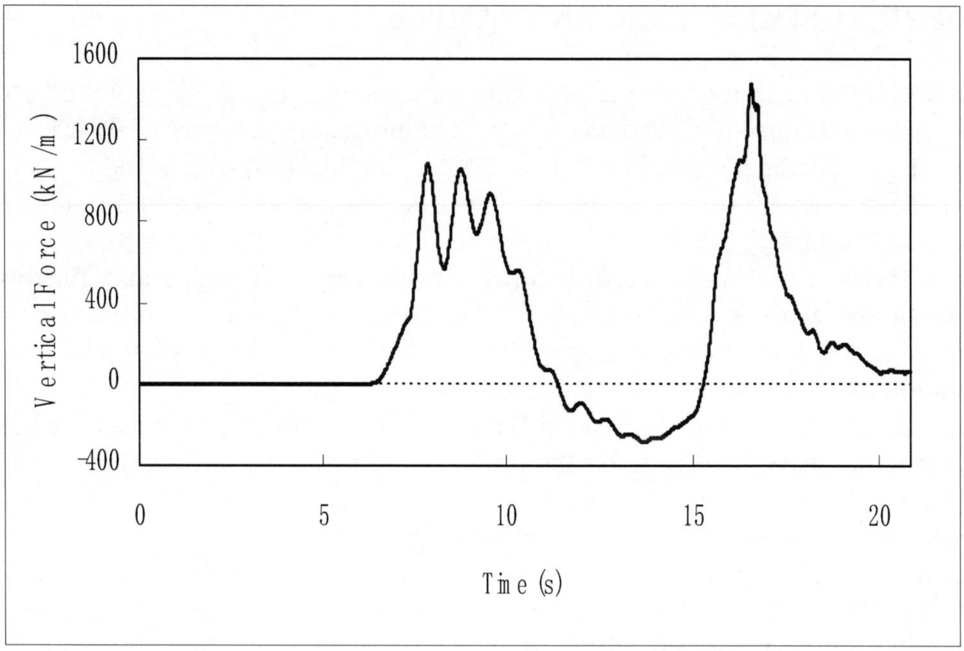

FIG. 4. Vertical force as a function of time during the passage of the wave

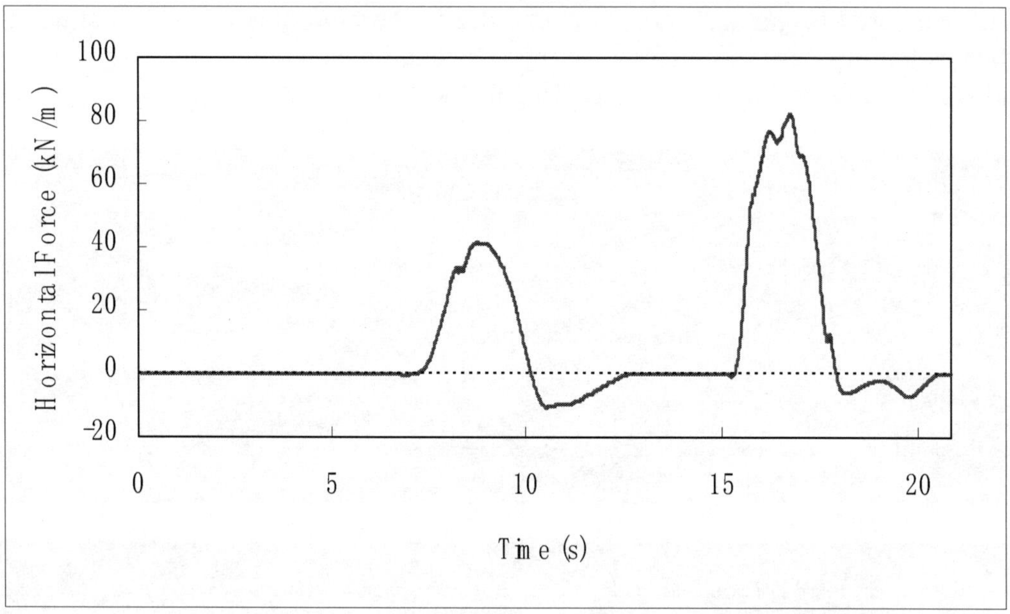

FIG. 5. Horizontal force as a function of time during the passage of the wave

VERTICAL LOAD IMPACT ON THE FOUNDATION: AN EXAMPLE

Considering the undesirable movement of bridge decks during storm events, it is considered to tie the deck down to the pier on coastal bridges. This creates additional loads on the existing foundation. The example studied in the numerical simulation indicates that the connection deck-pier should be designed for vertical tensile forces which are very small since the uplift force is of the order of the weight of the bridge deck and for horizontal (shear) forces calculated for a horizontal pressure of about 40 kN/m^2 (80 kN/m line load divided by the 2 m thick deck).

These loads must be resisted by the existing foundation in addition to the other loads the foundation was designed for. The loads created by the waves and to be resisted at the foundation level are an uplift vertical load, a horizontal load, and an overturning moment. The existing foundation should be able to resist the vertical uplift load relatively easily; indeed the net uplift load is near zero since the uplift load is of the order of magnitude of the weight of the deck.

The overturning moment generated is resisted in pull out of the piles on one side and in compression on the other (Fig. 6 and 7). A 3 by 7 pile group consisting of 1 m diameter pipe piles was assumed in this case with the dimensions shown in Fig. 6. The moment to be resisted is 80,000 kN.m (80kN/mx50mx20m). Such a moment amounts to an uplift for the outer piles of 572 kN/pile (80,000kN.m / (3x2x15m + 3x2x6.67m + 3x2x1.67m)). Considering that the bridge deck weighs 75000 kN (50m x 30m x 2m x 25kN/m3) and that there are 21 piles, the design load per pile is 3571 kN. The uplift load is the combination of the uplift due to the pressure under the deck and the uplift force due to the overturning moment. If the uplift pressure under the deck cancels the weight of the deck, then the uplift load of 572 kN must be resisted in tension by the outer piles. Since the piles are 47 m long and 1 m in diameter, the pile-soil interface shear stress required to resist such uplift load is 3.9 kN/m^2. Such friction value is very low and the existing foundation is very likely to be able to resist such an additional load without any problem.

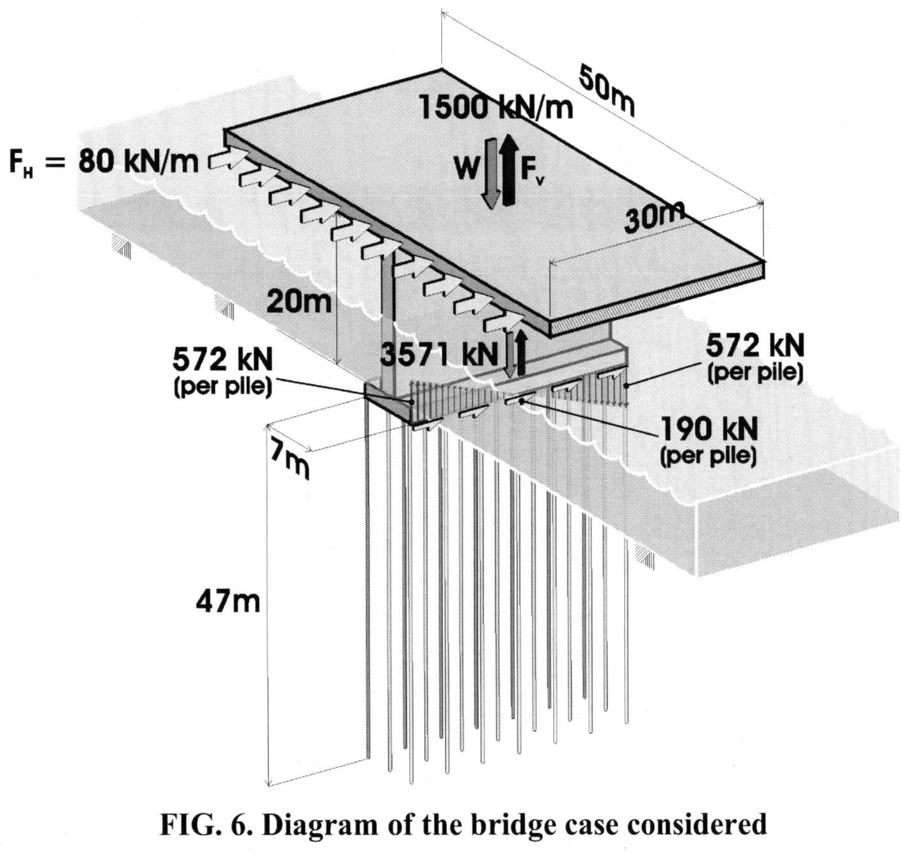

FIG. 6. Diagram of the bridge case considered

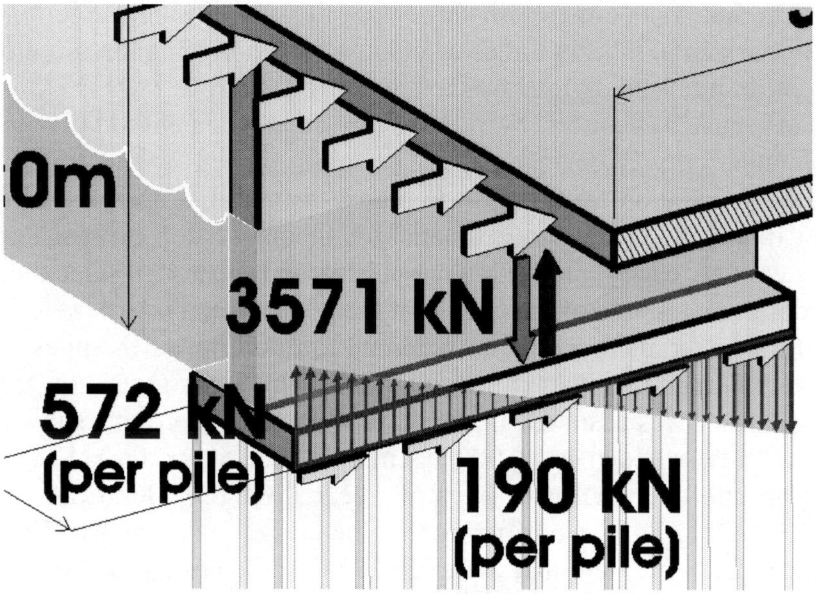

FIG. 7. Loading on the bridge deck and on the foundation

If the bridge foundation experiences significant scour during the hurricane, this conclusion may not be applicable and checking uplift resistance after scour should be performed.

On the compression side, the additional load per pile is also quite low since it is 572 kN compared to the design load of 3571 kN. The existing foundation should be able to resist this 16% increase in compression load without problem if the design factor of safety is sufficient.

HORIZONTAL LOAD IMPACT ON THE FOUNDATION: AN EXAMPLE

The existing foundation also has to resists the additional shear load at the pile head. This shear load is 190 kN/pile (80kN/mx50m/21piles). The ultimate horizontal load H_{ou} that a pile can resist (soil failure) can be calculated using the SALLOP method proposed by Briaud (1997). The equation is:

$$H_{ou} = 0.75\, p_L\, B\, D_v \tag{7}$$

Where p_L = limit pressure from the pressuremeter (PMT)

B = projected pile width

$D_v = (\pi/4)\, l_o$ with $l_o = (4EI / K)^{1/4}$ for $L > 3\, l_o$

$D_v = L/3$ for $L < l_o$

E = modulus of pile material

I = moment of inertia of pile

$K = 2.3\, E_o$

E_o = pressuremeter (PMT) first load modulus of soil
L = pile length

For the case at hand the following numbers were retained: EI = 161000 kN.m^2, E_o = 1000 kN/m^2, p_L = 100 kN/m^2. Close to the sea bottom, river bed, or lake bed, the soil is likely to be relatively weak and a limit pressure of 100 kN/m^2 was considered. Such a limit pressure might be found for a clay with an undrained shear strength of about 15 kN/m^2 or a sand with an SPT blow count of 2 blows per foot. The above numbers lead to the following calculated quantities: l_o = 4.1 m; D_v = 3.2 m; H_{ou} = 241 kN. Since the imposed load is 190 kN, this shows that the existing foundation may not be able to resist the horizontal load applied during the storm without some major movement.

CONCLUSIONS

The case which is presented in this article is purely fictitious. The numbers used for the parameters involved are reasonable but in no mean representing the complete range of parameters for coastal bridges. If bridge decks are going to be tied down to the bridge piers, new forces will have to be resisted by the foundation in addition to the design forces originally considered. The example presented indicates that the additional uplift and compression loads induced by the wave forces may not lead to a need for additional foundation piles. However the example shows that the horizontal shear load applied at the mud-line cannot be readily absorb by the existing foundation without additional piles. The numerical simulation which was performed indicates that the waves created an average pressure on the side of the bridge deck equal to 40 kN/m^2.

REFERENCES

Bea, R. G., Xu, T., Stear, J., and Ramos, R. (1999) Wave Forces on Decks of Offshore Platforms, Journal of Waterway, Port, Coastal and Ocean Engineering, American Society of Civil Engineers, May/June, vol. 125, no. 3, p. 136-144.

Briaud, J.L. (1997), “SALLOP: Simple Approach for Lateral Loads on Piles,” Journal of Geotechnical and Geoenvironmental Engineering, Vol. 123, No. 10, pp. 958-964, ASCE, New York.

Briaud J.-L., Ting F., Chen H.C., Cao Y., Han S.-W., Kwak K., (2001), “Erosion Function Apparatus for Scour Rate Predictions”, Journal of Geotechnical and Geoenvironmental Engineering, Vol. 127, No.2, pp. 105-113, Feb. 2001, ASCE, Reston, Virginia.

Chen H.-C (2002), “Numerical Simulation of Scour Around Complex Piers in Cohesive Soils”, Proceedings of the First international Conference on Scour of Foundations, Vol. 1, p. 14-33, Texas A&M University, College Station, USA (briaud@tamu.edu).

Douglas S.L., Chen Q.J., Olsen J.M., Edge B.L., Brown D., (2006), "Wave Forces on Bridge Decks", Report to the Federal Highway Administration, Washington D.C. (http://www.southalabama.edu/usacterec/waveforces.pdf).

McConnell, K. J., Allsop, W., Cuomo, G., and Cruickshank, I. C. (2003) New Guidance for Wave Forces on Jetties in Exposed Locations. Proceedings of the Sixth Conference on Coastal and Port Engineering Developing Countries (COPEDEC VI), Colombo, Sri Lanka, 20 pp.

McConnell, K., Allsop, W., and Cruickshank, I. (2004) Piers, jetties, and related structures exposed to waves: Guidelines for hydraulic loadings. Thomas Telford Press, London . 148 pp.

Tirindelli, M., Cuomo, G., Allsop, W., McConnell, K. (2002) "Exposed Jetties: Inconsistencies and Gaps in Design Methods for Wave-Induced Forces." Proceedings of the International Conference on Coastal Engineering, American Society of Civil Engineers, 13 pp.

Vibration & Shear Wave Velocity Measurements at the I-10 Blackwater River Bridge

Woods, Richard D.[1] P.E., F.ASCE and Crapps, David K.[2] P.E., M.ASCE

[1]Professor Emeritus, University of Michigan, rdw@umich.edu
[2] Principal Engineer, David K. Crapps, Inc., dcrapps@attglobl.net

ABSTRACT The I-10 Blackwater Bridge (parallel bridges) experienced settlements since its completion in 1968 of sufficient magnitude to create a rough ride. Florida Department of Transportation maintenance crews releveled the bridge a number of times by jacking and shimming the superstructure. In 1995 the FDOT noted additional settlements after Hurricanes Erin and Opal. Total settlements had reached up to about 100 mm (4 inches) at that time. The FDOT requested and was granted emergency funds from the Federal Highway Administration to replace the Blackwater Bridge.

Vibrations, especially from heavy truck traffic, were suspected as the likely cause of the settlements. Driving piles for the replacement bridge would create additional vibrations of a magnitude that would likely induce settlement. This paper describes the study made at the Blackwater Bridge site using test piles at two locations to document the magnitude of settlement induced in the existing bridge by shear waves from pile driving. Crosshole shear wave velocities were made at each test site location and down-hole measurements were made of the soil vibrations induced by pile driving. No settlement was detected in the nearby piers. This paper considers possible explanations of the no-settlement including "soil aging", a favorite topic of Dr. Schmertmann (see Schmertmann (1991)). This paper also discusses the complication of detecting shear wave signatures from pile driving. Our field exercise provided confirmation that the vertical angle between the source and the receiver has a major impact on the interpretation of shear wave arrivals and amplitudes.

KEY WORDS: mechanical aging of soils, method of multiple working hypotheses, observational approach, settlement, shear waves, vibrations

INTRODUCTION

The Blackwater River I-10 Bridge (hereafter referred to as the Blackwater Bridge), located east of Pensacola and downstream of Milton, Florida, was completed in 1968. The bridge was 893.4 meters (2931 feet) long consisting of 40 spans. The pile bent supported spans were typically 16.8 meters (55 feet) and the pier supported spans were typically 28.0 meters (92 feet). The channel crossing was a three-span continuous unit consisting of a 36.6 meter span (120 feet), a 51.8 meter (170 feet) across the channel followed by a second 36.6 meter (120 feet) span. Figure 1 shows a view of the bridge looking to the west and Figure 2 a view looking east at the channel crossing. The eastbound bridge is on the left and the westbound bridge on the right in Figure 1 and the opposite in Figure 2. Prestresssed concrete piles (610 mm = 24 inch square) provided foundation support.

FIG. 1 View to West of Blackwater Bridge River

FIG. 2 View to East at Blackwater Bridge Channel

The Bridge(s) settled enough that that Florida Department of Transportation Maintenance crews found it necessary to jack and shim several locations to provide a smoother ride for the traveling public and especially the heavy truck traffic on this major east-west freight route serving Florida. Some of the pier footings, including the channel pier shown in the left foreground of Figure 2, were badly cracked.

The bridge was inspected in 1995 after Hurricanes Erin and Opal and additional settlements up to about 25 mm (1 inch) were noted. By this time there was over 100 mm (4 inches) of settlement. The FDOT requested and was granted emergency funds from the Federal Highway Administration to replace the Blackwater Bridge. The project was considered an emergency replacement and the FDOT elected to use the Design-Build approach. The Traylor Brothers, Inc. TEAM was selected by FDOT to design and build the new bridge. Finley McNary Engineers, Inc. was the Designer.

Schmertmann & Crapps, Inc. (S&CI) was engaged by the FDOT to provide an investigation of the likely causes of the bridge settlement and to provide recommendations to prevent a reoccurrence in the new bridge (see Crapps & Schmertmann (1995) for a report of this project). Some of the proposed testing was shifted to the construction contract for the new bridge. Therefore, the FDOT made S&CI available as a member of the winning Design Build Team to provide vibration related consulting services during design and construction of the replacement bridge (see Crapps, Schmertmann and Pang (1996) for a report of this project).

The first author of this paper was a consultant to S&CI throughout both assignments and the second author was in responsible charge of the work. The authors worked with Dr. John H. Schmertmann on this project to evaluate the causes of settlement and to provide recommendations to the FDOT to prevent a reoccurrence in the new bridge. Dr. Schmertmann was actively involved in the project, especially during the investigative work. We applied two techniques taught and advocated by Dr. Schmertmann throughout his career. The first was the "method of multiple working hypotheses" and the second was the "observational approach". Vibration induced settlement was one of the hypotheses. The observational approach was used during the test program described herein and during construction to minimize the effects of construction on the traveling public in the event there was damaging settlement due to pile driving vibrations.

Measurements were made of vibrations from traffic and were considered by the authors of sufficient magnitude from truck traffic to induce settlement. One of the goals of the S&CI study was to duplicate and document settlement from vibrations applied to the bridge either by truck traffic and/or that may occur during pile driving for construction of the new bridge. Two test sites were selected for study. Shear wave velocity profiles were made at each test site using crosshole methods. Three test piles were driven at each test site while vibration measurements were made in the bridge and down hole measurements were made of soil vibrations. No settlement was detected during driving of the test piles with vibration levels normally expected to cause settlement. One of the explanations was another favorite topic of Dr. Schmertmann – aging of soils (see Schmertmann (1991)).

METHOD OF MULTIPLE WORKING HYPOTHESES

The method of multiple working hypotheses was first proposed in 1890 by T. C. Chamberlin (see Chamberlin (1890)). This technique has been a favorite of Dr. Schmertmann's for use with complex problems. The method consists of an initial list of hypotheses providing an answer to the problem at hand. Each hypothesis is tested as additional information is gained. Each hypothesis is either rejected or retained. Multiple cycles of this process are repeated until one hypothesis remains. The remaining hypothesis is likely the best answer.

Chapter 3 of the Crapps and Schmertmann (1995) report listed the following possible causes (hypotheses) of settlement of the existing bridge:

1. Traffic Vibrations
2. Settlement below pile tips
3. Negative Skin Friction Induced Settlements
4. Long Term Creep
5. Inadequate Driving Resistance
6. Consolidation of clay or peat layers below the pile tips
7. Scour
8. Effects of cofferdams
9. Combinations

There were available data to support each of these hypotheses. However, vibrations appeared to be the most likely cause of settlement. Additional information was needed to fully apply the method of multiple working hypotheses. The observational approach was used to assist in providing additional evidence from a test program.

SURVEYS TO DETERMINE THE AMOUNT OF SETTLEMENT

Determining the magnitude of settlement was not as easy as one may imagine. The original plans provided plan elevations for the entire structure (footings, caps and deck were of most interest). The FDOT made available the results of a deck surveys made in 1992 and again in 1995. As part of our study, a survey was made of the existing top of footing elevations. Figure 3 shows Dr. Schmertmann performing a survey of the pier footings to provide reference elevations for the observational approach during the test program and during construction of the new westbound bridge. The height of grout and steel shims were measured. All of the data were considered to provide the best estimates of settlement of the entire bridge. The data showed that 48% of the bents and piers had experienced settlements greater than or equal to 25 mm (1 inch) with maximum settlement up to about 100 mm (4 inches).

FIG. 3 Dr. Schmertmann performing settlement survey.

OBSERVATIONAL APPROACH

Borings were made at all the proposed new pier locations and they showed loose sand layers located below the elevation of the existing bent/pier piles. The design loads are much larger today than in 1968 when the existing bridge was built and Contractors use larger hammers now than then. Analyses estimated vibrations from pile driving would produce higher levels of vibration in the existing bridge than traffic had imposed. Analyses showed that vibrations from the driving of the piles for the replacement bridge could cause additional settlements up to 150 mm (6 inches) if a 162,700 Nm (120,000 lb-ft hammer was used to drive the piles and the new bridge was in close proximity to the old bridge.

The projected pile driving vibrations warranted further study. Therefore, S&CI proposed two test sites, each with three piles to gain additional on-site information. This test program was included in the Traylor Brothers, Inc. Design Build Team work. East and westbound traffic was diverted onto the eastbound bridge on which the tests were conducted in the early stages of the project.

Figure 4 shows the Test Site 1 layout near existing pier 14L. The layout for Test Site 2 was similar and was located near existing Pier 20L. The "150 feet" (45.7 m) dimension shown is the width of the Right of Way and the northern R/W is at the top of the figure. Note that the existing bridge was located as near as practical to the existing northern R/W line to minimize the effects of vibration to the extent practical.

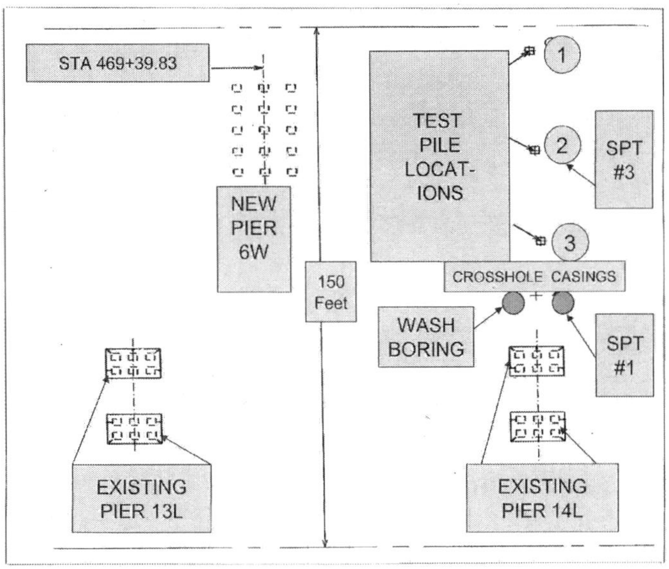

FIG. 4 Plan View of Test Site #1 (TS-1)

Pioneering research at the University of Florida, under the direction of Dr. John H. Schmertmann, showed that SPT blow counts (up to at least N equal 50) are inversely proportional to the enthru, the energy in the rods (see Schmertmann and Palacios (1979)). Dr. Schmertmann energy calibrated all of the SPT drill rigs used at the Blackwater Bridge site (see Schmertmann (1996)). All of the SPT borings, at both test sites, used the same Failing 1500 drill rig, equipped with a safety hammer and the energy calibration showed that 72% of the theoretical energy entered the rods (enthru) during routine sampling

Two casings were grouted in place at each of the test sites for Crosshole shear wave measurements. Sensors were placed in the same casings during the driving of the three instrumented test piles to measure *in situ* soil vibrations. The adjacent existing pier footings were monitored for settlement. No settlement was detected (within about ± 0.5 mm (0.02 inch)) even though vibration levels were reached which were expected to induce settlement.

The pile installation method for production pile driving was selected to minimize vibration transmitted to the existing bridge and included jetting piles to the maximum practical depth before driving. This initial jetting generally placed the new pile tips below the tip elevation of the existing adjacent pier/bent locations.

The existing parallel bridges remained in service during construction of the new westbound bridge. However, precautions were taken to protect the public in the event time was needed to make repairs or to re-level the bridge due to storms or construction related activities. Detours were constructed at the ends of the parallel bridges so that all traffic could be quickly routed over either of the bridges or could proceed as originally planned using both bridges. At critical times, all of the traffic was placed on the original eastbound bridge so that it would be removed as far as practical from pile driving. (Upon completion of the new westbound bridge, all traffic was placed on it, the two existing parallel bridges were removed and then the new eastbound bridge was constructed.)

The observational approach was closely followed throughout the entire pile driving operations for the new westbound bridge to prevent unpleasant surprises. Vibrations reaching the four closest piers or bents were monitored for vibration and the closest Pier/Bent was monitored for settlement during driving. Settlements were small and the pile driving operations proceeded as planned.

APPROACH TO MEASUREMENTS AND PREDICTIONS

Soil borings logs, shown in Fig. 5 for Test Site 1 (TS-1), were available for pre-design and showed weak, loose, noncohesive soils below the tips of the existing foundation piles which were driven to a depth of about 8.5 m (28 ft) into the river bed. Logs for Test Site 2 (TS-2) were similar. These short piles were believed to be susceptible to vibration induced settlement during driving of the piles for the replacement bridge as well as from ongoing truck traffic on the bridge.

It was anticipated that settlement of the existing bridge piers due to driving of the replacement bridge piers could be predicted based on the amplitude of vibrations experienced by the soil supporting the existing bridge piles. To estimate this settlement, the volumetric strain occurring in the soil in the -40 ft to -90 ft (-12.2 m to -27.4 m) elevation range would be summed for all the blows required to drive the piles for the new bridge. Volumetric strain ($\Delta H/H$) would be assumed equal to the shearing strain, γ_s, and calculated from the relationship:

$$\gamma_s = \frac{V_v}{V_s} \quad (1)$$

Where γ_s is shearing strain,
Vv is p-p particle velocity, and
Vs is shear wave velocity.

An empirical equation of the standard form (function of SPT blow count and confining pressure) for shear wave velocity would be developed specifically for this site from measurements of shear wave velocity in crosshole seismic tests;

$$Vs = a + bN^n (\sigma_o)^m \quad (2)$$

Where: Vs is shear wave velocity,
N is SPT Blow Count (uncorrected N values used in this study),
σ_o is effective octahedral confining pressure, and

a,b,n & m are empirical constants.

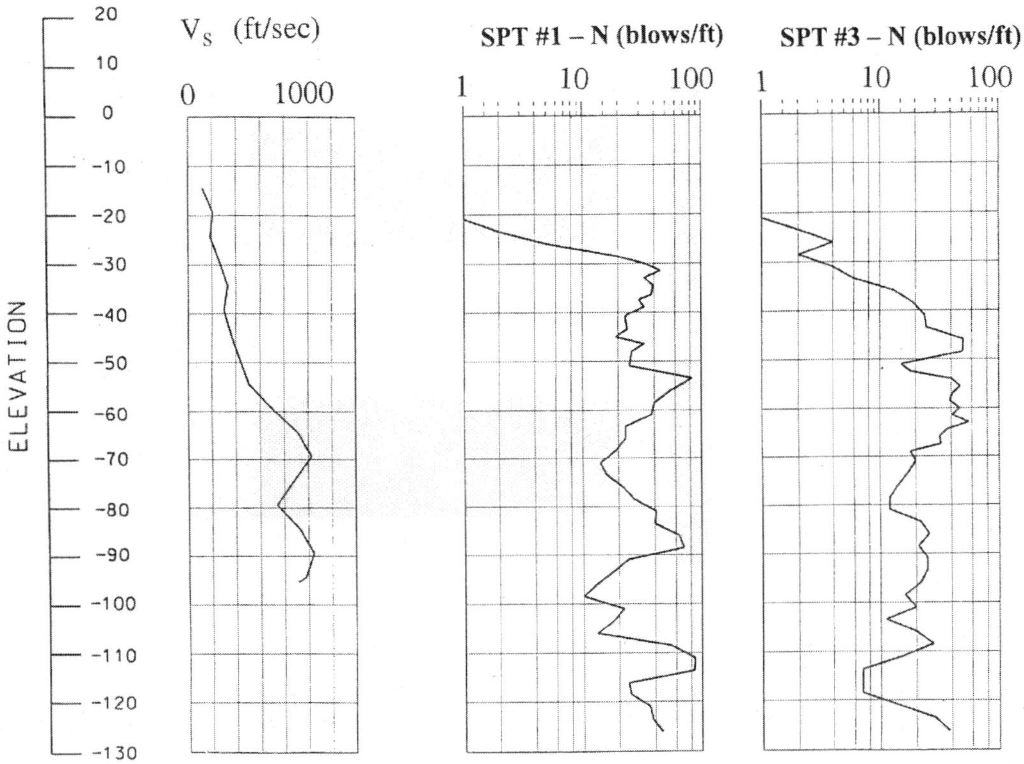

FIG. 5 . Shear wave velocity and SPT profiles at TS-1.
(1 ft = 0.305 m and 1 ft/sec = 0.305 m/s)

Attempts would be made during the crosshole tests to obtain tomographic type data for waves traveling at angles across horizontally bedded soil strata as well as waves traveling horizontally from borehole to borehole.

Particle velocity would be predicted by again developing a site specific empirical equation from measured energy required to drive the replacement bridge foundation piles, and the distance to the existing bridge pile supported piers. This empirical equation would take the common form (Woods & Jedele, 1985) of:

$$Vv = a\,(\text{Scaled distance})^{-n} \tag{3}$$

Where: Vv is resultant p-p particle velocity,
scaled distance is (actual distance/$\sqrt{}$energy), and
a & n = empirical constants.

FIELD MEASUREMENTS

Crosshole seismic tests were performed at both test sites and a shear wave velocity profiles developed to a depth of about 30.5 m (100 ft) below mud line at each site. Vertically polarized shear waves were generated in the crosshole source borehole with a hydraulic packer hammer, shown in Fig. 6a & 6b, and the arrivals of the shear waves at transducers in a receiver borehole were detected with triaxial arrays of geophones coupled to the casing with pneumatic expanders. Shear wave velocities were chosen to characterize pile support for both the existing and replacement bridge

(a)

(b)

FIG. 6 . (a) Hydraulic SV borehole shear source with platen expanded, (b) Dr. Woods guiding hammer down borehole.

because both foundations were in saturated, noncohesive soils. Primary waves (P-waves) would only reveal the compression wave velocity in the void space water of the saturated soils and not of the skeleton of the soil. Furthermore, shear wave velocity could be used in conjunction with ground motion measurements to calculate shear strain in a process of estimating shake-down settlement. The use of shear waves and particularly the choice of using SV-waves led to one of the key conclusions/observations of this testing program as described in the next section.

Geophone packages at 3 elevations in one of the crosshole casings at each test site are shown in Fig. 7. These three component geophone packages recorded vertical and two orthogonal horizontal motions. During driving of the test piles, vibrations were recorded at these three elevations in the crosshole casing and at triaxial geophone arrays on the piers of the existing bridge. Figure 8 shows a work boat next to the two crosshole casings extending above water and Fig. 9 shows a geophone recorder on a pier of the existing bridge.

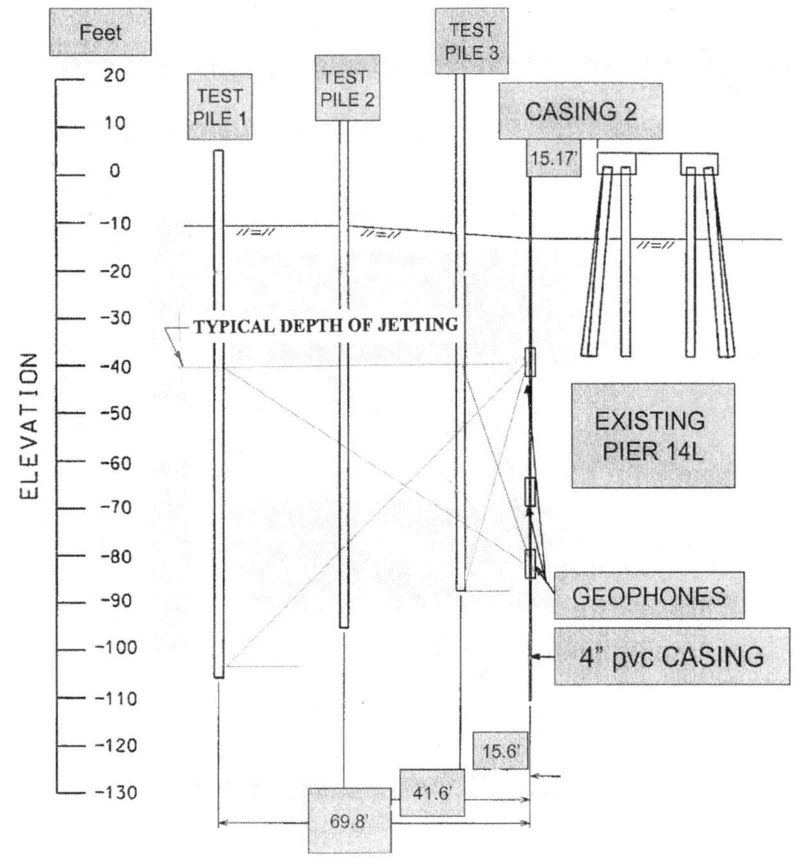

FIG. 7. Elevation view of Test Site #1

FIG. 8 Workboat next to crosshole recorder casings.

FIG. 9 Blastmate vibration on existing bridge pier.

Attempts were made to determine the SV- wave arrival time and amplitude from each blow of the pile driving hammer and these velocities and amplitudes could be used to estimate shear strain in the calculation of settlement. The tip elevation of the piles continually changed during driving so the angle from horizontal between the tip of the pile and any of the three geophone packers changed continuously. Figure 7 demonstrates the continually changing angle as the pile passes elevation -40 feet (-6.1 m) to the final tip elevations (for example, elevation -105 feet (32.0 m) for Pile 1 at TS-1). The geophones were placed such that, during driving, data were recorded with the tip of the piles approximately at the same elevation (0 degrees) as one of the geophones to approximately 74 degrees to one of the geophones.

Crosshole seismic test performed in boreholes 1 & 2 at TS-1 and TS-2 also employed attempts to detect shear waves in a pseudo-tomographic mode. Angles from the horizontal at the source in those cases for the SV-wave ranged from 0° to about +/-80°. For most of these angles the arrival of the vertical component of the seismic wave was impossible to detect. Likewise, shear wave amplitudes could not be measured. Failure to identify arrival times meant velocities could not be calculated and failure to determine shear wave amplitude meant that shear strains could not be calculated.

UNDERSTANDING SV-WAVES

At the time these measurements were made, it was well known that the amplitude of the SV-wave generated by a vertical shear source in a borehole produced a distribution of shear energy illustrated by Fig. 10.

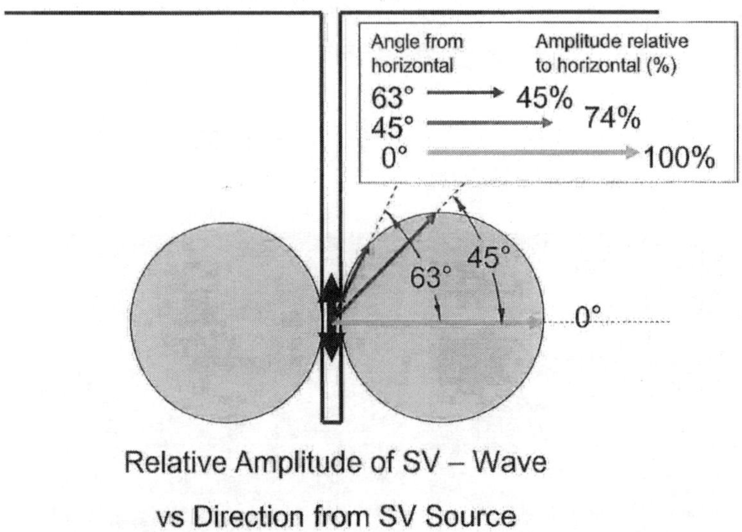

FIG. 10 Shear Wave Energy Distribution from Shear Source (after Mooney, 1974)

This distribution of energy is the same for both crosshole and up-hole seismic methods and demonstrates why the up-hole SV-wave method is not feasible and why steep-angle ray paths cannot be used in crosshole. With a SV source at depth in a borehole, the up-hole process requires identification of a ground motion at the surface near the borehole. However, the maximum SV energy travels away from the source in a horizontal direction and the amplitude at other directions from horizontal can be approximated by circles tangent to the borehole at the center of the source. A ray of energy traveling from the source to the surface would be created from only small amplitude energy, making identification of the vertically traveling wave difficult. From Fig. 10 it can be seen that for a ray of energy traveling 45° from the horizontal, the amplitude of the SV wave would be about 74% of the energy at emanating at 0° from the horizontal. For an SV wave traveling on a ray at 63° from the horizontal, the energy would be only 45% of that in the horizontal direction. With this distribution of energy, picking SV arrivals at large angles from horizontal would be difficult, but in most cases would be possible for angles up to at least 45°.

Down-hole seismic measurements do not suffer from this difficulty. In that method, energy is created in the ground at the surface and near to the cased borehole. That energy travels downward to a receiver in the borehole on a nearly vertical angle. The source energy is maximum in the downward direction making identification of seismic energy easy in the down-hole method.

In more recent work by Kim et al (1999) it is shown that not only does the SV amplitude change with angle from the horizontal, but the whole seismic wave signature changes. Those researchers developed a FEM for wave propagation in both isotropic media and cross-anisotropic layered media. Figure 11 shows the isotropic system where the shear wave velocity (Vs) is 183 m/s (600 fps), Poisson's ratio (ν) is 0.33, and the unit weight of the medium (γ) is 15.7 kN/m^3 (100 pcf). A

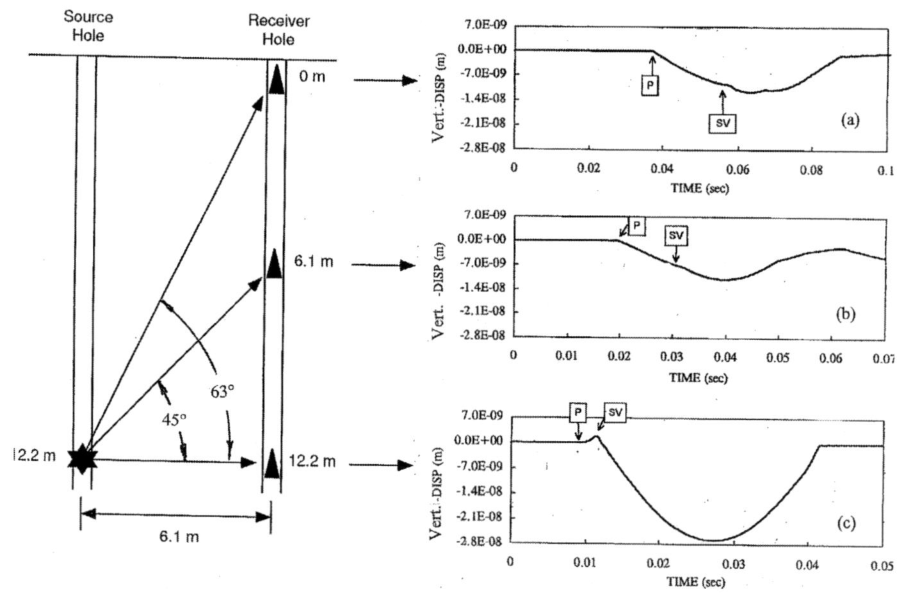

FIG. 11 Waveforms for SV-wave from an SV source in an isotropic medium: (a) at surface 12.2m above source level, (b) at 6.1m above source level, and (c) at source level. (after Kim et al, 1999)

frequency of 11.6 Hz yielding a wavelength of 5.49 m (18 ft) was also assumed. Kim et al used a vertical SV energy source which yielded seismic wave signatures shown in Fig. 11 for three elevations in an adjoining borehole. At 0° inclination from horizontal, the SV wave is clearly identifiable, but at 45° and 63° to the horizontal the SV-wave is not at all identifiable. The arrivals of the SV–wave in Figs. 11 (a) & (b) were calculated based on the shear wave velocity in the isotropic medium. Neither the arrival time of the SV wave nor the amplitude of the wave are identifiable.

Kim et al used a similar configuration with layers and cross-anisotropy to study the SV-wave propagation generated by a vertical shear source. Table 1 shows the properties of the anisotropic layers shown in Fig. 12. It is seen again that the 0° declination SV-wave is clear, but SV-waves traveling along other ray paths are even less clearly identifiable than in an isotropic medium, Fig. 11. The SV-wave arrivals in Figs. 12 (a) & (b) were calculated based on velocities in layers and refraction of waves between layers. It is likely that the alluvial soils in the stream bed of the Blackwater River may more closely resemble the cross-anisotropic model of Fig. 12 than Fig 11 but in either case it is clear why SV-wave velocity and amplitude could not be determined for almost any angle except 0° to the horizontal.

TABLE 1 – Cross-Anisotropic Shear Wave Velocities

Layers	V_{XY} m/s	V_{XZ} m/s	V_{XX} m/s	V_{ZZ} m/s
Layer I	122	134	201	226
Layer II	183	198	302	338
Layer III	244	268	405	451
Layer IV	305	332	503	564

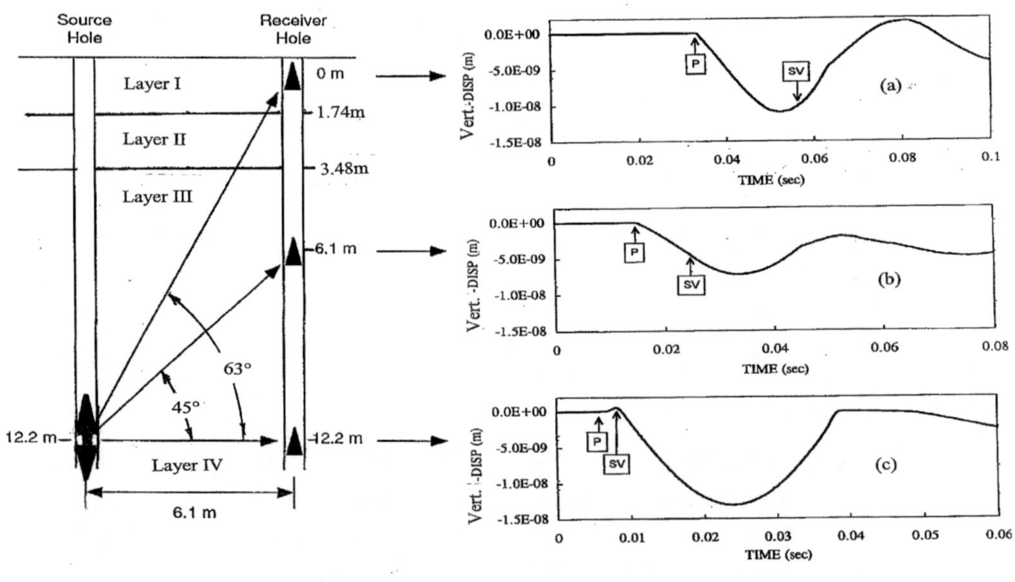

FIG. 12 Waveforms for SV-wave from SV excitation in anisotropic layered medium: (a) at surface 12.2m above source level, (b) at 6.1m above source level, and (c) at source level. (after Kim et al, 1999)

ESTIMATES OF VIBRATION SHAKE-DOWN SETTLEMENT

Even though many wave propagation measurements at this site were unproductive with regard to predicting settlement because of the above discovery, the majority of the settlement prediction process could be accomplished through other measurements. The shear wave velocity from the crosshole seismic tests was correlated with SPT blow count (N) for this site. As previously mentioned, the Failing 1500 SPT rig using a safety sampler was energy calibrated with 72% of the energy into the rods (enthru). The following site specific empirical Eq. (4) was developed:

$$Vs = a + b\,(N_{SPT})^{0.265}\,(\sigma_o)^{1.059} \tag{4}$$

Where: Vs = shear wave velocity (V_S = ft/sec when a = 146.7, b = 73.0, N_{SPT} in blows per foot, and σ_o in ksf) or Vs = m/s when a = 44.71, b = 0.37, N_{SPT} in blows/305 mm, and σ_o in kN/m^2),
N_{SPT} = SPT N value (enthru = 72% for drill rig used in this study), and
σ_o = effective octahedral stress, [K_o was assumed to be 0.5].

This empirical value of Vs was very good up to velocities of about 400 ft/sec (121.9 m/s) and approximate at higher velocities. Velocities higher than this value were generally at depths greater than elevation – 15.2 m (-50 ft), and soils at deeper elevations had higher N values and were less likely to undergo shake-down settlement.

To complete the calculation of strain by equation (1), the p-p particle velocity, Vv, was also needed. Initially it was expected that a distance/energy relationship could be developed based on energy imparted to the soil at the pile tip. Coupling of energy into the ground from a pile being driven into the ground is a complex phenomenon. The tip of the pile must create a bearing capacity failure in the ground causing plastic deformations. At some distance from the pile tip those deformations decay to the elastic range and seismic waves propagate outward from the pile tip. The sides of the pile near the tip also create shear waves during bearing capacity failure. Because of non-homogeneity of the soil and non-ideal pile behavior, P- & S-waves interact during driving to create derivative waves including SH-waves. Shear energy is also transferred from the shaft of the pile to the surrounding soil as the pile is driven, but the soil at those elevations has already been brutalized by the tip passing through. Consequently, transfer of energy from the shaft by shear is a small fraction of the total energy transferred. Therefore it was assumed that the enthru energy from the pile driving analysis was all delivered to the soil at the tip. Initially it was expected that energy transferred to the soil at the tip of the pile would be traced through vibration measurements in the cased borehole at three elevations, Fig. 7 for example, and vibration amplitude at the three levels of geophones would provide the p-p particle velocities need for equation (1).

Because it was impossible to identify SV-waves and estimate peak particle velocity from most of the ray paths from the tips of the test piles to the geophone locations, a different approach was adopted. Resultant peak particle velocities were determined from measurements of vibration at the three elevations in the cased hole, Fig. 7. The distance from the test pile to all the geophones in the cased holes was taken as the diagonal distance from the pile tip to the geophone location in the crosshole casing. From this data a correlation between pile energy, distance, and peak particle velocity was developed:

$$V_r = K\left(\frac{D}{\sqrt{EMX}}\right)^{-1.25} \tag{5}$$

Where V_r = p-p resultant particle velocity (V_r = in/sec with K = 0.11, D in ft and EMX in lb-ft; or V_r = mm/sec with K = 0.523, D in meters and EMX in Nm)

D = vector distance from tip of pile tip being driven to point of interest

EMX = maximum energy in the pile (Enthru)

Using (4) & (5), shear strain in the soil under the existing bridge piers was calculated for each foot of driving of the piles for the replacement bridge. Estimates of settlement then were made by calculating relative density of the soil from blow count, volumetric strain based on strain calculated above, and number of cycles of vibration from each blow on the piles. These calculations showed a considerable potential for vibration induced settlement of the existing bridge (up to 6 inches) during driving of piles for the replacement bridge.

A parallel study using precise leveling techniques was implemented to measure actual settlement. However, these measurements showed no detectable settlement of the existing bridge piers (within the survey accuracy of about 0.5 mm (0.02 inch) during the test program. Very small apparent settlements (an average of about 0.3 mm (0.012 inch) with a maximum of about 1.5 mm (0.06 inch) were measured in the existing bridge piers during the driving of the production piles for the new westbound bridge.

It is postulated that preconditioning from prior vibrations, including those from storms, higher than expected silt content of the soil profile and development of slight cementation because of aging of the noncohesive soils led to the over prediction of settlement. Silt content and cementation were not considered in the literature studies in which settlement at given relative densities and at given strain levels were predicted. Future studies to better quantify the threshold velocity required for settlement in aged soils and those with higher silt contents would likely improve estimates of vibration generated settlements.

CONCLUSIONS

1. Approximately half of the then existing Blackwater Bridge bent/pier locations settled 25 mm (1.0 inch) or more and maximum settlements of up to about 100 mm (4 inches) were reached between the time the bridge was completed in 1968 and 1995. Measurements made in 1995 after two close proximity hurricanes showed these events may have contributed up to 25 mm (1 inch) of the total measured settlements.

2. The Method of Multiple Working Hypotheses was used to evaluate the cause(s) of settlement of the existing bridge.
3. The cause of settlement was likely due to a combination of factors. The piles supporting the original I-10 Bridge(s) were relatively short. The soils supporting the original bridge were lightly cemented, had relatively light design loads and were driven with a relatively low energy hammer. Lateral loading from wind loads (especially from hurricanes) together with loss of upper capacity due to scour may have resulted in a shift of the center of resistance to a lower elevation closer to the pile tip resulting in settlement of layers below the pile tips.
4. The Observational Approach was used thought the investigation of the existing bridge and throughout pile driving for the new bridge to provide early warning of potential damaging vibrations and subsequent settlement.
5. The settlement of the existing bridges was negligible (within the survey accuracy estimated to be ± 0.5 mm (± 0.02 inch) during the test program. The settlement during production pile driving for the westbound bridge was small (maximum of 3 mm (0.12 inch) with an estimated survey accuracy of ± 1.5 mm (0.06 inch). The average apparent settlement during production pile driving for the new westbound bridge was about 0.3 mm (0.012 inch).
6. The small settlements from pile driving may have been due to preconditioning from prior vibrations including those from storms, combined with aging of the soils, and possible higher than anticipated silt content in the soils. These factors may have increased the "threshold" vibrations beyond which additional settlement would occur.
7. Vibrations from pile driving proved difficult to interpret. It was especially difficult to determine the arrival time for shear waves.
8. A pile penetrating into the ground generates a complicated wave pattern at its tip. Strong P-waves and S-waves are generated. S-waves are also generated along the sides of the pile.
9. The authors now recognize that the angle between the tip of the pile (the source) and the sensors is important because of the effect of energy loss, especially in an anisotropic soil. The authors recommend that the sensors follow the elevation of the source as the pile penetrates to make interpretation of shear waves possible.

REFERENCES

Chamberlin, T. C. (1890), *Science*, (old series), 15, 92, reprinted by University of Wisconsin, 2005 (see www.biotech,wisc. edu/chamberlin)

Crapps, David K. and Schmertmann, John H. (1995), BLACKWATER RIVER I-10 BRIDGE ESTIMATED SETTLEMENTS FROM TRAFFIC VIBRATIONS AND PILE DRIVING, Schmertmann & Crapps, Inc. Project No. 972, State Project N0. 58002-3417, report prepared for the Florida Department of Transportation, July.

Crapps, David K., Schmertmann, John H. and Pang, Shing, (1996), EFFETCS OF VIBRATIONS AT BLACKWATER RIVER I-10 BRIDGE (3 Volumes), Schmertmann & Crapps, Inc. Project No. 985, State Project N0. 58002-3449, report prepared for Finley McNary Engineers, Inc., TRAYLOR BROTHERS, INC. Design Build Team for the Florida Department of Transportation, August.

Kim, S-M, Liao, S-Y, and Roesset, J.M. (1999), "Simulation of the Crosshole Method in Isotropic and Anisotropic Media," *International Journal for Numerical and Analytical Methods in Geomechanics,* Vol. 23, pp 1101-1119.

Mooney, H. M. (1974), "Seismic Shear Waves in Engineering," *Journal of the Geotechnical Engineering Division,* ASCE, Vol. 100, No. GT8, Aug., pp.905-923.

Schmertmann, John H. (1991), "The Mechanical Aging of Soils", The Twenty-Fifth Karl Terzaghi Lecture, *Journal of Geotechnical Engineering*, Vol. 117, No. 9, September, pp. 1288-1330.

Schmertmann, John H. (1996), "Report on SPT Energy Calibrations of the Williams Earth Sciences SPT Drill Rigs # DR-2, DR-4, DR-6 and DR-7 While Working on the Blackwater Bridge Replacement Project", Report for Design Build Team, Project No. 985, Feb. 15.

Schmertmann, John H. and Palacios, Alejandro (1979), "Energy Dynamics of SPT", *Journal of Geotechnical Engineering*, Vol. 105, No. GT8, August, pp. 909-926.

Woods, R.D. and Jedele, L.P. (1985), "Energy-Attenuation Relationships from Construction Vibration," *Vibration Problems in Geotechnical Engineering*, Proceedings of a symposium sponsored by the Geotechnical Engineering Division ASCE, Detroit Annual Convention, Detroit, October 22, 1985, pp.229-246.

NOTATION

The following symbols and abbreviations are used in this paper.

a = empirical constant
b = empirical constant
D = distance from pile being driven to centerline of crosshole casing (ft)
EMX = maximum energy in the pile (Enthru) (ft-lb)
H = height of layer
K = empirical constant in Eq. (4)
K_o = at rest pressure (assumed to be 0.5 in this study)
n = empirical constant
N_{SPT} = blows/foot
V_r = p-p resultant particle velocity (in/sec)
V_s = shear wave velocity,
TS = test site
σ_o = effective octahedral stress (ksf)
ΔH = change in height of layer
$\Delta H/H$ = volumetric strain

Mitigation of Liquefaction Potential of Silty Sands

James K. Mitchell, P.E., Hon. M. ASCE[1]

[1]University Distinguished Professor, Emeritus, Department of Civil & Environmental Engineering, Virginia Tech, Blacksburg, VA 24061-0105, jkm@vt.edu

ABSTRACT: Mitigation of liquefaction potential of soils with fines contents greater than 15 to 20 percent usually requires special adaptation of the dynamic densification methods – Deep Dynamic Compaction, Explosive Compaction, Vibro-Compaction, Compaction Piles – that are ordinarily used to improve loose, clean cohesionless soils. However, these methods can be used to attain the needed improvement in many soils if closely spaced vertical sand, gravel, or prefabricated (wick) drains are installed prior to construction. Soil-cement columns or walls installed by deep soil mixing or jet grouting can be used to provide both reinforcement and containment of liquefied soil. Removal and replacement of liquefiable silty and clayey soils using compaction, admixture stabilization or substitution by another material; e.g., roller compacted concrete, is also possible; however, the cost may be high, and consideration must be given to dewatering needs and the stability of open excavations. Mitigation of the liquefaction potential of high fines content sandy and silty soils using permeation grouting is not feasible because the low hydraulic conductivity of such soils.

KEY WORDS: cost, improvement, liquefaction, methods, mitigation, sands, silty

INTRODUCTION

Whereas the liquefaction potential of clean cohesionless soils may be relatively easy to assess, and there are several well-proven ground improvement methods for its mitigation, the problem is complicated in soils containing fines (particles finer than 0.074 mm) in amounts greater than a few percent. When it has been determined that silty or clayey sand layers or deposits at a site are susceptible to liquefaction, the selection of the best method for mitigation of this liquefaction risk is often complicated by the fact that the most economical and effective ground improvement techniques for clean sands are not likely to be very effective if the fines content exceeds a certain amount, usually approximately 10 to 20 percent. In such cases other methods, or modifications of the traditional methods, must be called upon to attain the level of improvement that is needed. Attaining the necessary improvement of liquefiable soils may be both difficult and expensive.

The objectives this paper are to evaluate the suitability of different methods for improvement of liquefiable soils that contain more than about 10 to 20 percent fines, to note special considerations in their application, and draw some conclusions and guidelines for use in future projects.

LIQUEFACTION POTENTIAL OF SOILS CONTAINING FINES

According to Idriss and Boulanger (2007) the term "liquefaction" is used to describe the degradation and loss of strength under cyclic loading of cohesionless soils; i.e., gravels, sands, and very low plasticity silts; whereas, "cyclic failure" is used to describe the strength loss in clays and plastic silts. In this paper no clear distinction is made between liquefaction and cyclic failure, as the focus is on preventing soil structure breakdown and loss of strength and stiffness.

The physics and mechanics of liquefaction are detailed in Idriss and Boulanger (2007) and the numerous references cited therein. For the present study it is sufficient to focus on the impact of fines on liquefaction and cyclic load resistance, and this is illustrated by the well-known correlations between cyclic stress ratio required to cause liquefaction (CSR) and penetration resistance, normalized to an effective overburden stress of 1 atm and magnitude 7.5 earthquakes. Fig. 1 shows the relationship based on the corrected and stress-normalized value of $(N_1)_{60}$ derived from the results of Standard Penetration Tests (SPT), and Fig. 2 shows the relationship in terms of the stress-normalized value of tip resistance q_{c1N} deduced from Cone Penetration Tests (CPT).

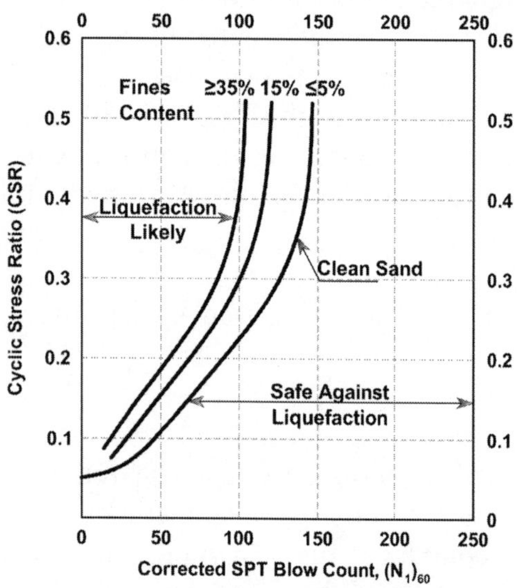

FIG. 1 Liquefaction triggering curves based on Standard Penetration Test Resistance for M7.5 EQ and 1 atm pressure (adapted from Youd, et al, 2001)

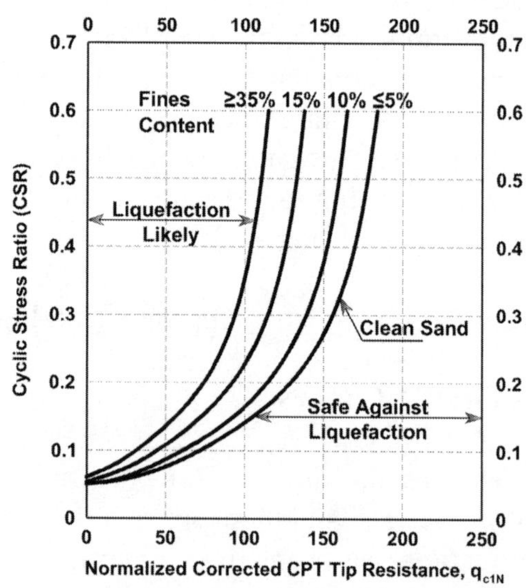

FIG. 2 Liquefaction triggering curves based on Cone Penetration Test Resistance for M7.5 EQ and 1 atm pressure (adapted from Idriss and Boulanger, 2007)

In each case it may be seen that the higher the fines content, for values up to 35 to 50 percent, the greater is the cyclic stress ratio (CSR) needed to cause liquefaction. Idriss and Boulanger (2007) state that a Plasticity Index (PI) of about 7 percent separates soils that are "sand-like" from soils that are "clay-like" in seismic behavior. Correlations such as those in Figs. 1 and 2 are strictly applicable only to sand-like soils. Nonetheless, clay-like soils can still undergo significant strength loss under seismic loading and require ground improvement to provide needed stability.

The inhibiting effect of increasing fines content on liquefaction can be attributed to the interference of the small particles with sand particle movement and adhesion imparted by silt and clay plasticity in comparison with clean sand. Further increase in resistance is not observed for fines contents greater than about 35 to 50 percent because these fines contents are high enough to provide a matrix that prevents direct interference between the coarser particles, and the strength is essentially equal to that of the matrix. These same factors influence the breakdown of soil structure and ease of densification of a sandy soil containing fines, processes that are essential for effective soil densification by dynamic methods. Plastic fines have a greater than proportional effect on resistance to liquefaction and ease of densification than do non-plastic fines.

GROUND IMPROVEMENT METHODS IN RELATION TO SOIL TYPES

Numerous methods for soil stabilization and ground improvement are available. All rely on one or more basic types of treatment, usually differing only in the materials and the specifics of equipment and procedures used.

The soil gradation ranges for which different methods of soil improvement have been used for mitigation of liquefaction risk are shown in Fig. 3. Each method may not be useful over the full range of gradations indicated, because specific soil and site constraints may impact its applicability considerably. This is illustrated for various methods of in-situ treatment by the information about their characteristics, effective depth of treatment, implementation, improvement level, and advantages and limitations given in Table 1. The methods can be grouped into the following categories relative to their applications for improvement of liquefiable soils:

1. Drainage – gravel drains, sometimes referred to as earthquake drains.
2. Void filling – penetration grouting.
3. Densification in-situ - deep dynamic compaction (DDC), explosive compaction (EC), vibrocompaction (VC), and compaction grouting.
4. Densification in-situ with accompanying installation of drainage and reinforcing columns – vibro-replacement stone columns, compaction piles.
5. In-situ columns and walls of cemented soil – Cement Deep Soil Mixing (CDSM) and Jet Grouting (JG).
6. Remove and replace, with or without admixtures.

Comprehensive information describing the technology of all of these methods is readily accessible; e.g., Mitchell (1981), USACE (1999), ASCE (1978, 1987, 1997), Moseley and Kirsch (2004), and literature provided by ground improvement specialty contractors.

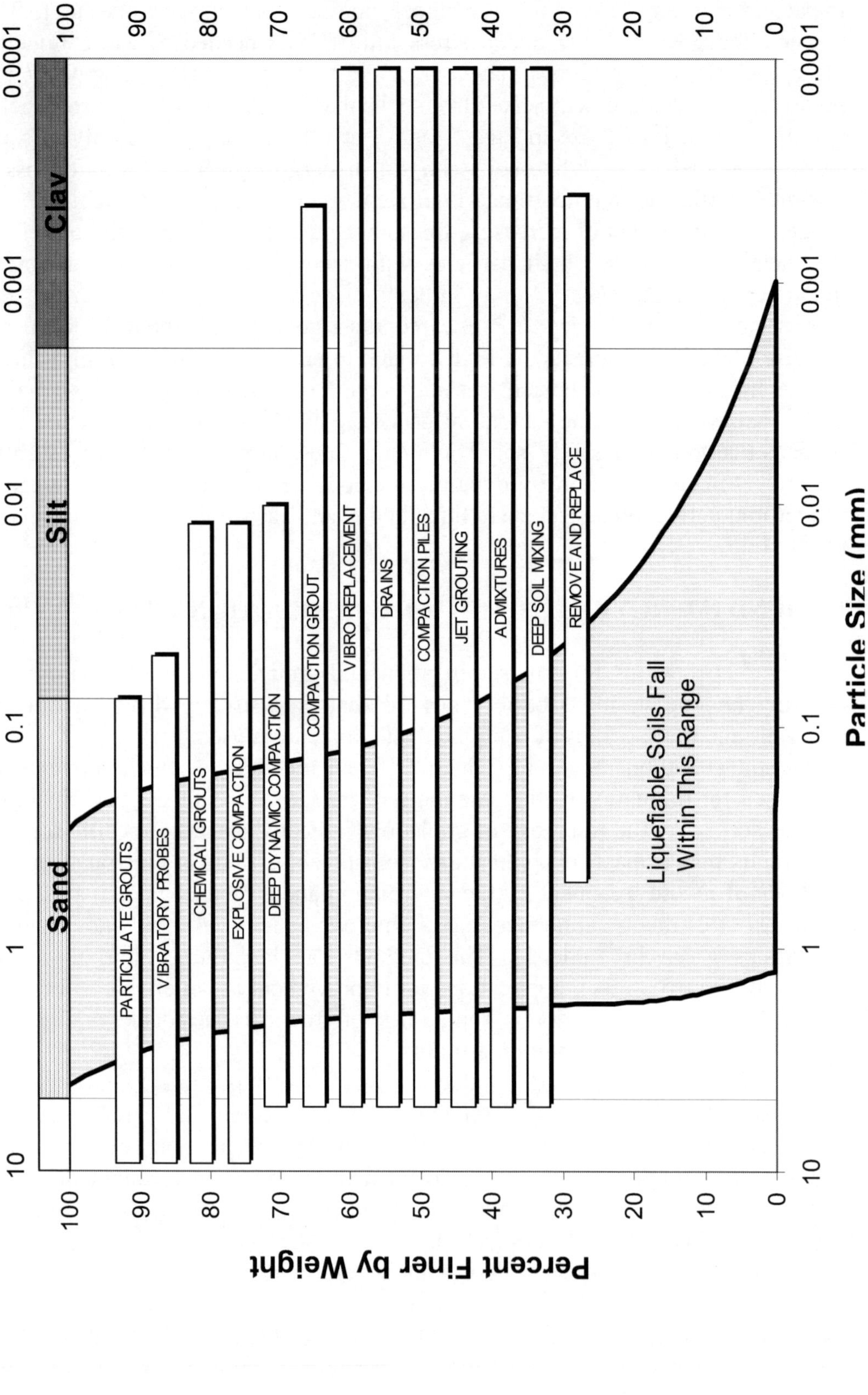

FIG. 3 Applicable soil gradation ranges for different methods of ground improvement

TABLE 1. Methods of Ground Improvement for Mitigation of Liquefaction Risk in Saturated Silty and Clayey Sands

Method	Most Suitable Soil Type	Effective Depth	Attainable Maximum Improvement	Advantages	Limitations
Gravel Drains	Sands, silty sands	20 m (?)	Reduce pore pressure buildup, Intercept pore pressure plumes	• Inexpensive • Full area treatment not required • Limited performance record	• May require very close spacing • EQ-induced settlement not prevented • Treated ground must have high hydraulic conductivity
Penetration Grouting	Sands and coarser materials	Unlimited	Void filling and solidification	• No excess pore pressure or liquefaction • Can localize treatment area	• High cost • Fines prevent use in many soils
Deep Dynamic Compaction (DDC)	Saturated sands and silty sands; partly saturated materials	Up to 10 m	D_r = 80 % $(N_1)_{60}$ = 25 q_{c1} = 10-15 MPa	• Low cost • Simple • Good for large areas	• Limited effective depth • Clearance required • High mobilization cost • Vibrations
Explosive Compaction	Saturated clean sands	>30 m	D_r = 75 % $(N_1)_{60}$ = 20-25 q_{c1} = 10-12 MPa	• Inexpensive • Simple technology • Can use at greater depths than other dynamic methods	• Vibrations • Safety issues • Psychological barriers

TABLE 1. Methods of Ground Improvement for Mitigation of Liquefaction Risk in Saturated Silty and Clayey Sands (Cont.)

Method	Most Suitable Soil Type	Effective Depth	Attainable Maximum Improvement	Advantages	Limitations
Vibrocompac-tion,	Sands, silty sands, gravelly sands < 20% fines	30 m	D_r = 80+ % $(N_1)_{60}$ = 25 q_{c1} = 10-15 MPa	• Uniformity with depth in a given soil type • Moderate cost	• Special equipment needed • Best in clean sand • Unsuitable in soils with cobbles and boulders • Backfill required in most cases
Compaction Grouting	Any rapidly consolidating, compressible soil	Unlimited	Up to D_r=80+% $(N_1)_{60}$ = 25 q_{c1} = 10-15 MPa (Soil type dependent)	• Controllable treatment zone • Useful in soils with fines	• High cost • Post-treatment loss of prestress • Slow
Vibro-replacement Stone Columns	Silty and clayey sands, silts, clayey silts	30 m	$(N_1)_{60}$ = 20 q_{c1} = 10-12 MPa	• Provides drainage and reinforcement • Uniformity with depth in a given soil • Bottom feed dry process puts fill where needed	• Special equipment needed • Unsuitable in soil with cobbles and boulders • Fines may intermix with and clog columns • Backfill may be costly • Difficult QA/QC
Sand and Gravel Compaction Piles	Can use in most soil types	20 m	Up to $(N_1)_{60}$ = 25-30, q_{c1} = 10-15 MPa, depending on soil type	• Proven effectiveness • Provides drainage and reinforcement • Uniformity with depth	• Special equipment needed • Slow • High cost

TABLE 1. Methods of Ground Improvement for Mitigation of Liquefaction Risk in Saturated Silty and Clayey Sands (Cont.)

Method	Most Suitable Soil Type	Effective Depth	Attainable Maximum Improvement	Advantages	Limitations
Cement Deep Soil Mixing	Most soil types	20 m	Depends on size, strength and configuration of DCSM elements	• Positive ground reinforcement • Contains liquefiable soil within high strength grid walls	• Requires special equipment • Brittle elements • Expensive
Jet Grouting	Any soil, difficult in highly plastic clays	Unlimited	Depends on size, strength and configuration of high strength jetted elements	• Controllable treatment zone • Useful in soils with fines • High strength columns	• High cost • Difficult to QA/QC
Remove and Replace, with or without admixtures	All soils	A few m	High density fills to strong, cemented materials	• Can design to desired improvement level • Easy to QA/QC	• May be expensive • May require dewatering • Excavations may impair stability of adjacent ground • Temporary support of existing structures may be needed

DISCUSSION AND ILLUSTRATION OF THE METHODS

Drainage

The concept and theory underlying the use of closely spaced vertical drains for liquefaction mitigation was developed by Seed and Booker (1977), and further analysis and refinement of the method has been made in the years since. The purpose of the drains is to dissipate excess pore pressure in a liquefiable soil concurrently with its development during earthquake shaking. In this way the strength reduction is limited and liquefaction is prevented. Sand, gravel and prefabricated vertical drains have been used. The ground disturbance and vibrations induced by the installation process are believed to cause some improvement through densification. Ground settlements caused by the 1993 Kushiro-Oki and 1995 Hyogoken-Nambu earthquakes in Japan were estimated to have been reduced by 50 percent or more in areas containing sand drains, with part of this reduction resulting from the improvement caused during their installation.

The hydraulic conductivity of the native soil is the most important parameter controlling the effectiveness of drainage elements in dissipating excess pore pressure generation during earthquake shaking. Even small amounts of fines cause large reductions in hydraulic conductivity, because the hydraulic conductivity of cohesionless soil varies approximately as the square of the 10 percent (or in some cases, the five percent) particle size. This means that even closely spaced drains by themselves cannot be relied upon as an effective means for preventing liquefaction in silty sands. Nonetheless, drains may be useful when used in conjunction with other methods that depend on densification, such as deep, dynamic compaction and explosive compaction, owing to their ability to remove water quickly during the densification process.

Void Filling

If the water in the pores of a loose soil can be replaced by a material of sufficient viscosity or has even only modest shear strength, then rapid collapse of the soil structure will not be possible, and liquefaction will be prevented. Thus, penetration grouting can be an effective, albeit often expensive, method for mitigation of liquefaction risk in loose clean sands. Both particulate, usually cement and mico-fine cement, and chemical, most-often silicate based, grouts are used. As may be seen in Fig. 3 and further illustrated in Fig. 4, the range of soil types in which penetration grouting can be used is limited. The presence of even a few percent fines is likely to render the process ineffective, so penetration grouting is seldom a viable option for mitigation of liquefaction risk in silty sands.

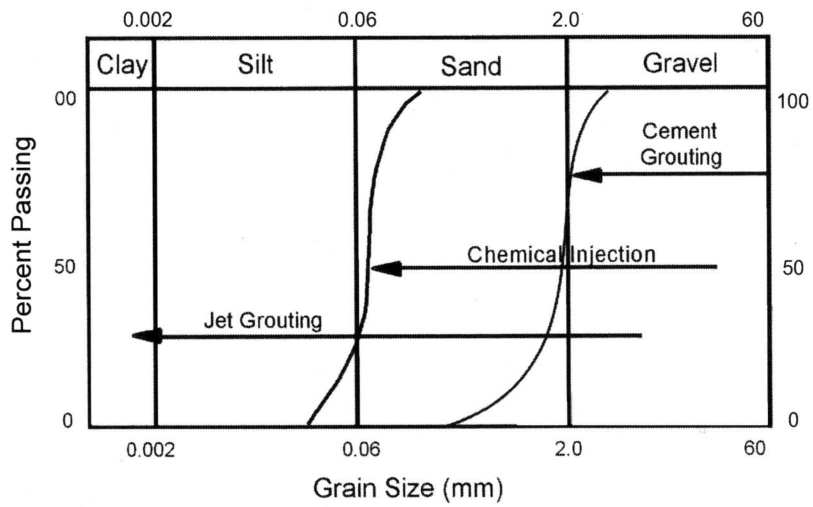

FIG. 4 Range of gradations suitable for permeation using cement and chemical grouts

In-situ Densification

Deep dynamic compaction (DDC) and Vibrocompaction (VC) are usually among the first methods to be considered for in-situ densification of cohesionless soils for mitigation of liquefaction risk, with explosive compaction (EC) being an economical alternative in some cases. The amount of improvement that can be achieved is greatest in clean sands and falls off rapidly as the fines content increases, so if the non-plastic fines content is 15 to 20 percent or more, little densification is possible by these methods alone. Clay fines are even more inhibiting, with a little as 3 percent finer than 2 μm rendering the methods ineffective. This behavior is illustrated by Fig. 5 which shows how the increase in penetration resistance that was obtained by vibratory compaction of a reclaimed sandy deposit in Japan decreased with increasing fines content. Measurements of the increase in SPT resistance as a function of distance from sand compaction piles showed significant increases for soil with fines content less than 20 percent, modest increases for fines content between 20 and 40 percent, and no increase for fines content greater than 40 percent (Kitazume, 2005). Another example is shown in Table 2 listing the increase in penetration resistance values caused by densification of liquefiable foundation layers at the site of the reconstructed Jackson Lake Dam in Wyoming using DDC. The magnitude of the increase in penetration resistance decreased with increasing fines content, as indicated by the Unified Soil Classifications from clean sand and gravel (GP and SP) to silty sand (SM) to low plasticity silt (ML).

An additional factor that contributes to the effectiveness of densification in increasing penetration resistance is that the process can result in significant increases in lateral stress in the ground. This may be especially important for compaction piles, vibro-compaction and vibro-replacement.

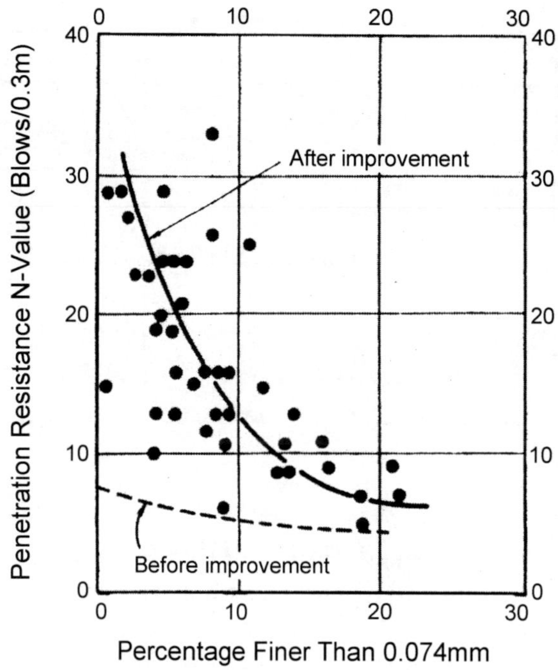

FIG. 5 – Effect of fines content on the increase in penetration resistance using vibratory compaction (redrawn from Saito, 1977)

TABLE 2. Effect of Soil Type on Increase in Penetration Resistance By Deep Dynamic Compaction at Jackson Lake Dam

Unified Soil Type	Number of Tests	ΔN (blows per ft.)
GP and SP	80	14.0
SM	53	10.4
ML	70	5.8
SM-ML	6	10.4
All	209	10.3

The ranges of soil gradations that can be most effectively improved by dynamic in-situ densification methods is reasonably defined by the curves shown in Fig. 6, which was developed originally for DDC by Lukas (1995). The effective range can be extended to higher fines contents using compaction grouting, as this method allows more time for consolidation under the pressure caused by the expanding grout bulb. Overall, however, most silty and clayey sands with fines contents in excess of 15 to 20 percent can not be improved to the level required using DDC, EC, or VC methods alone.

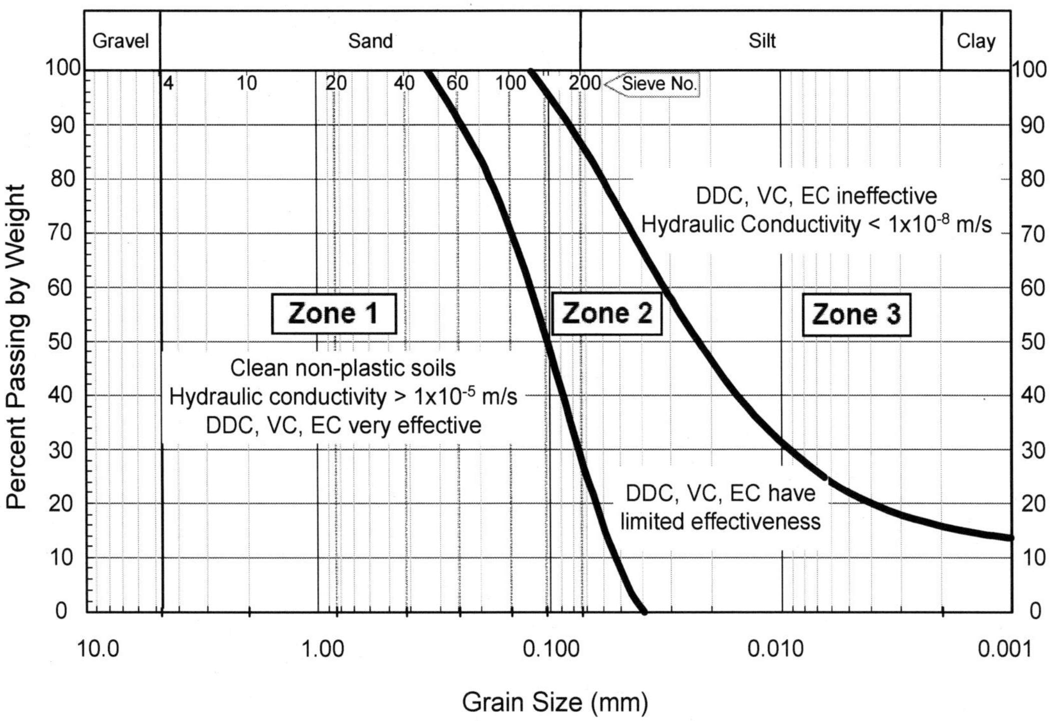

FIG. 6 Effective ranges of soil types for effective densification using dynamic densification methods. DDC is Deep Dynamic Compaction, VC is Vibro-compaction, EC is Explosive compaction (modified from Lukas, 1995)

Cost is always a consideration in the selection of a densification method. Although the actual cost depends on project size, location, depth of treatment, mobilization of equipment, availability of contractors, environmental considerations, and other factors, comparative costs for the dynamic densification methods discussed herein are given in Table 3.

TABLE 3. Comparative Costs of Vibro-Densification Method
(A comparative cost of 1 may be of the order of \$2 to \$3 per cu yd of improved ground in the USA in 2005)

Treatment Method	Relative Cost (per unit volume)
Explosive Compaction	0.25 - 1
Deep Dynamic Compaction	0.5 - 2
Vibratory Probe	~ 3
Vibro-Compaction	2 - 10
Vibro-Replacement	3 - 10

In-situ Densification with Accompanying Drainage and Reinforcement

Both Vibro-replacement stone columns (VR) and compaction piles[1] (CP), although differing in their methods of installation, can densify liquefiable soils with higher fines contents than can be improved using DDC, EC, or VC. In fact cases have been documented where loose silty soils with fines contents greater than 50 percent have been improved by these methods; e.g., Mitchell and Wentz (1991). The reasons for this are that the strong vibrations during installation of the SC and CP, along with the high radial lateral pressures generated during the cavity expansion created by the backfill densification both breakdown the native soil structure and generate excess pore pressure. If the hydraulic conductivity of the native soil is high enough, then the soil densifies during the VR and CP process. Since the time a given element of the ground is subjected to strong input energy is greater during stone column and compaction pile installation than it is during DDC or EC, there is more time for expulsion of water, and VR and CP methods can better densify soils with lower hydraulic conductivity; i.e., more fines, than can DDC and EC.

Stone columns and compaction piles provide two additional benefits; namely, vertical drainage through the column and reinforcement. Thus during a seismic event they can aid in partial dissipation of excess pore pressure, and their higher strength and stiffness relative to the native soil means that they will carry a proportionally higher shear stress than the soil, thereby reducing the cyclic stress ratio acting on the soil. At present there is no consensus as to how to account for these effects quantitatively, and some agencies do not allow for them in design.

The ability of VR and CP, and DDC, and perhaps EC, to densify soils with even higher fines contents can be enhanced by installation of drains to shorten the drainage paths for escape of water during the construction process. The successful improvement of an up to 18 m (60 ft) thick deposit of interbedded, generally cohesionless, fluvial-lacustrine sandy and silty foundation materials beneath Salmon Lake Dam in Washington State, Luehring, et al (2001), provides a good example. Both extensive SPT and CPT testing indicated that these materials were susceptible to liquefaction under the design earthquake. Fines contents ranged up to more than 60 percent.

Prefabricated vertical drains (wick drains) were installed on 3 or 6 ft spacing to the full 58 to 61 ft (17.7 to 18.6 m) target depths of the stone columns throughout the treatment area. The drains were located at the mid-points of the sides of the triangles formed by the stone columns, as shown in Fig. 7. Dry, bottom feed stone columns were spaced 6.0 ft on centers. Stone column area replacement ratios of 22 and 35 percent were used. The column diameters were 3.0 ft or 3.75 ft, with the larger diameter specified for the area underlain by the finer-grained soils.

---

[1] Several proprietary compaction pile systems; e.g., Franki Piles, Geopiers, Vibro Piers, are available for installation of compacted gravel, stone, or aggregate piers that can be used as foundation support elements for structures. Sand Compaction Piles, installed using special machines and methods, are used extensively in Japan (Kitazume, 2005). They may also improve the ground into which they are driven by vertical and horizontal vibrations, compression and cavity expansion in the horizontal direction, increase in lateral pressure and drainage.

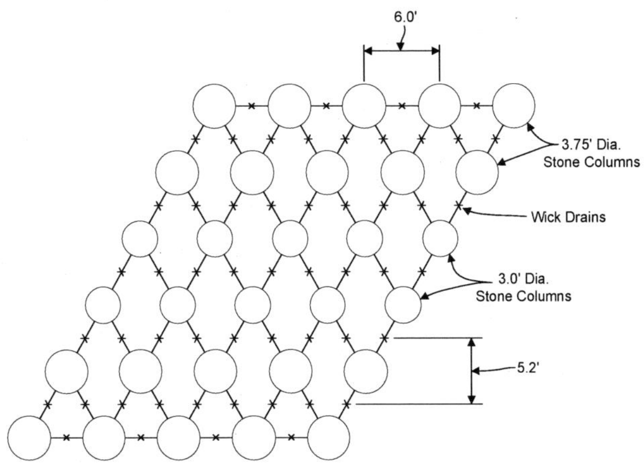

FIG. 7 Pattern of Stone Columns and Wick drains used for improvement of sandy and silty foundation soils at Salmon Lake Dam (adapted from Luehring, et al. (2001)

The vertical drains enabled rapid drainage of the ground during the stone column installation process, with accompanying densification and large increases in both the SPT and CPT penetrations resistances, as may be seen in Table 3 and Fig. 8, adapted from Luehring, et al. (2001). The values in Table 3 are also consistent with past findings that the maximum penetration resistance that can be attained by treatment decreases as the fines content increases.

Dise, et al., (1994) report the effective treatment of silty sands (SM) and sandy silts (ML) at three dams (Jackson Lake Dam, Mormon Island Auxiliary Dam, and Steinaker Dam) using methods to enhance the pore pressure dissipation. These included wick drains, drainage trenches, controlled sequencing patterns for the weight drops, increased time between passes, and reduced energy per drop in confined areas. As is the case for vibro-replacement stone columns, the wick drains should be installed to the full depth of ground requiring improvement.

TABLE 3. SPT Penetration Resistance Increases by Combined Stone Columns and Wick Drains in Salmon Lake Dam Foundation Materials (after Luehring, et al., 2001)

Soil Type	Average fines % <0.074 mm	Average percent clay <0.005mm	Average pre-treatment $(N_1)_{60}$	Average post-treatment $(N_1)_{60}$	Percent Increase
Silt	65	11	12	23	88
Silty sand	49	5	17	33	95
Poorly graded sand with silt	10	2	21	40	92
Silty gravel with sand	12	3	15	52	236

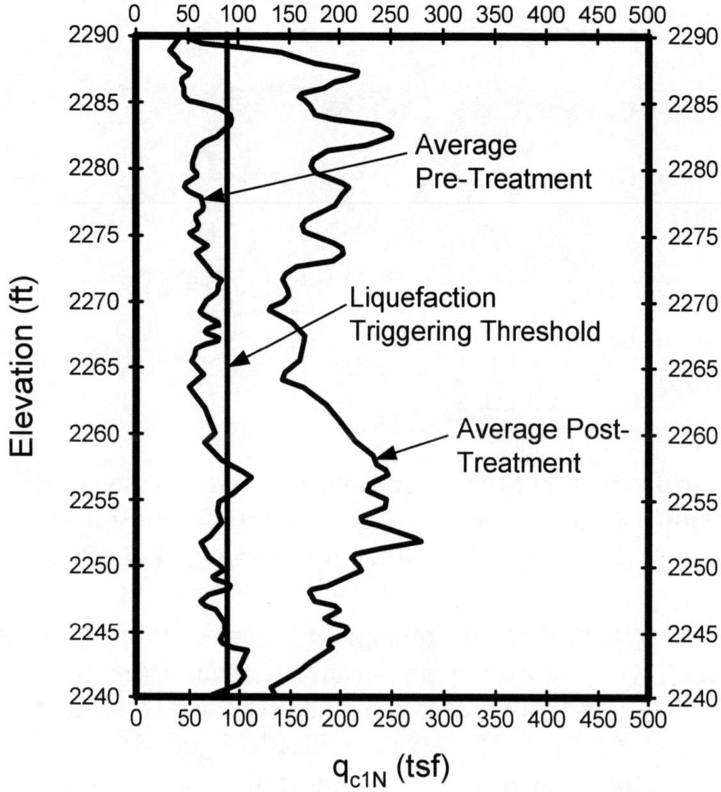

FIG. 8 Increase in penetration resistance from ground treatment at Salmon Lake Dam resulting from combined wick drains and vibro-replacement stone columns (adapted from Luehring, et al., 2001)

An analytical methodology for quantifying the pore pressure generation and dissipation and densification associated with stone column and deep dynamic compaction supplemented by wick drains in saturated sands and silty soils is described by Shenthen, et al., (2004). Given the complexity of the problem, several simplifying assumptions and approximations were required in order to develop a numerical solution. Nonetheless, simulation results agreed quite well with field observations of the influences of pre-densification relative density and area replacement ratio of post-treatment SPT penetration resistance. They showed that wick drains should enable densification of silty soils having hydraulic conductivities as low as 1×10^{-8} m/s if stone columns are spaced at 2 diameters or less, center-to-center, corresponding to an area replacement ratio of 20 percent or more. The analysis of two case histories where DDC was supplemented by vertical drains gave good agreement between the predicted and measured improvement in SPT penetration resistance. Shenthen, et al., (2004) concluded that soils with hydraulic conductivity greater than about 1×10^{-6} m/s can be densified by DDC or VR without supplemental drainage.

In-situ Columns and Walls

Both Cement Deep Soil Mixing (CDSM) and Jet Grouting (JG) have assumed an important role in mitigating liquefaction risk in recent years in addition to their past functions for foundation and excavation support and as cutoff walls. As indicated in Fig. 1 these methods can be used in a wide range of soil types and gradations, with high fines content liquefiable soils being very well-suited for treatment. Each can be used to depths of up to 30 m or more. Mobilization, site preparation, and demobilization costs may be well over $100,000, and treatment cost may exceed $100/ cu yd, with the cement and additives accounting for about half of this. Each requires special equipment, and each may produce significant spoils that must be dealt with in some manner. The design strength of properly constructed CDSM and JG elements is high; e.g., 150-200 psi, considerably greater than can be obtained by the other methods of soil improvement. Both methods are relatively difficult to QA/QC, with uniformity of CDSM better than can be obtained using JG.

Both wet and dry mixing processes are available for CDSM; however, the wet process is better suited for use in potentially liquefiable soils. Isolated columns and walls formed of overlapping columns can be formed using either CDSM or JG. Complete treatment of a liquefiable soil layer by either method is impractical and prohibitively expensive. Accordingly, isolated columns can be installed to provide the necessary vertical compressive reinforcement and lateral shear resistance in the event of soil liquefaction. Alternatively, walls, typically about 3 ft in thickness, can be used to form cells that both provide enhanced structural resistance and containment for liquefied soil. The Trench Cutting Remixing Deep Wall Method (TRD), recently developed in Japan and now available in the U.S involves a chain saw like process for wet cement mixing that produces continuous walls about 0.2 m thick and up to 30 m deep. This process tends to homogenize the soil vertically, as opposed to vertical axis processes used for CDSM and JG where more of the vertical stratification of the soil profile is retained.

Jet grout columns can be installed over specified depth ranges as opposed to DCSM columns that ordinarily must extend full depth. At the Carrefour Shopping Center site along Izmit Bay in Turkey, JG columns were used both to provide additional support to footings for a supermarket and to reduce liquefaction risk in a liquefiable layer at a depth of 6.5 to 9 m (Martin, et al., 2004). The soil in this layer, classified as SM with SP and SC lenses, contained 30 percent finer than No. 200 sieve and 14 percent clay size (<2μm). Without treatment the factor of safety against liquefaction was estimated to be about 0.6, with an earthquake-induced settlement of 60 to 100 mm.

Jet grout columns installed in a primary grid were 0.6 m in diameter, at a spacing of 4 m center-to-center and extended from the ground surface to a depth of 9 m, as shown in Fig. 9. The columns in a secondary grid, installed to reduce the liquefaction risk in the silty sand stratum, were only 2.5 m long and located at the intermediate points of the primary grid, as shown. The Kocaeli earthquake in 1999, magnitude 7.4, subjected the site to an estimated horizontal ground acceleration of 0.24g. No evidence of liquefaction, structural or ground damage, or settlement was observed.

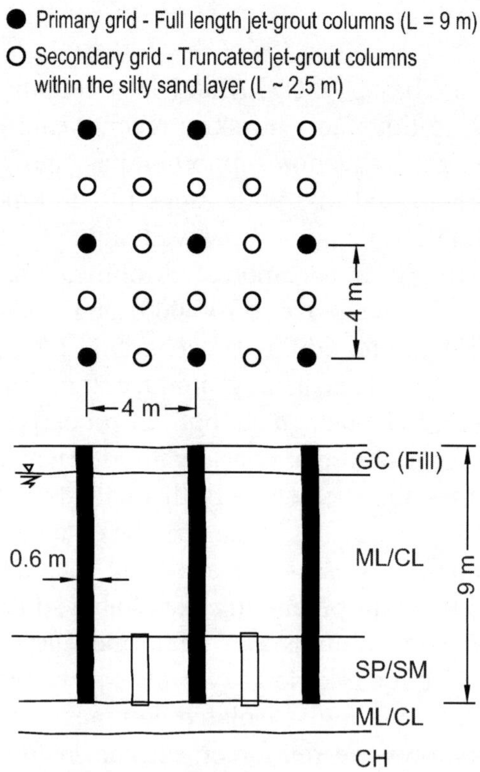

FIG. 9 Layout of the Jet Grout Columns at the Carrefour Shopping Center (from Martin, et al., 2004)

Other areas on the site, where there was no jet grouting, and adjacent warehouses and apartment buildings where no ground treatment was used experienced settlements of 50 to 100 mm.

Remove and Replace, With or Without Admixtures

The improvement of high fines content liquefaction-susceptible soil deposits by removal and replacement with the same soil either recompacted to an acceptable state, mixed with a stabilizing admixture and then recompacted, or replaced with a better material; e.g. Roller Compacted Concrete, provides the best assurance that there will be no future risk of liquefaction. The technologies and construction procedures are well developed and QA/QC is straightforward. However, remove and replace is usually not the method of choice, because of high cost, the need to handle large volumes of ground, dewatering requirements, environmental impacts, and risks due to stability concerns associated with temporary open excavations.

INNOVATIONS AND NEW POSSIBILITIES

The methods for mitigation of the liquefaction potential of cohesionless soils with high fines content described in the preceding paragraphs have been used successfully in the field. There are at least two other concepts that might be useful, although at this stage they have been demonstrated only in the laboratory.

Permeation Grouting Using Electro-Osmosis

Whereas permeation grouting is prohibitively expensive and time consuming for soils with hydraulic conductivity less than about $1 x 10^{-5}$ to $1 x 10^{-6}$ m/s, electro-osmosis can move fluids into and through silty soils with hydraulic conductivity down to $1 x 10^{-7}$ to $1 x 10^{-8}$ m/s. Thevanayagam and Jin (2003) describe experiments where electro-kinetic injection of colloidal silica and silicate grouts was effective in producing significant increases in the strength of a 40 percent sand – 60 percent silt mixture. It could be inferred that these strength increases would be accompanied by and increase in liquefaction resistance.

Partial Saturation

Both theoretical analyses (Martin, et al., 1975) and experimental measurements; e.g., Yoshimi, et al. (1989), Ishihara, et al. (2001), Yegian, et al. (2007) have shown that even small reductions in saturation can reduce the pore pressure generation under cyclic loading and increase resistance to liquefaction significantly. Accordingly, if gas bubbles could be generated and distributed uniformly within potentially liquefiable initially saturated silty sand, then the risk of liquefaction would be reduced. The challenge, of course will be to devise practical methods for doing this in the field.

Gases are generated during densification by explosive compaction. Changes in penetration resistance are used as the usual measure of the effectiveness of treatment; however, these data do not provide any direct measure of whether the liquefaction resistance has been increased beyond that commonly attributed to densification. Air bubble inclusion during vibrocompaction, vibro-replacement, deep dynamic compaction and jet grouting is also to be expected, and there appears to have been little if any study of the consequences. Yegian, et al (2007) generated oxygen and hydrogen gas by electrolysis within saturated loose sand and measured considerably reduced pore pressure ratios during cyclic loading. Rebata-Landa and Santamarina (2007) did laboratory experiments that showed how the generation of nitrogen gas bubbles in sands, silts, and clayey sands using biologically mediated processes will result in large reductions in P-wave velocity, corresponding to reductions in saturation. This, in turn, would indicate an increased resistance to liquefaction.

CONCLUSIONS

Removal and replacement of high fines content liquefiable soils by carefully controlled fills of recompacted soil, admixture stabilized soil, or roller compacted concrete provides the best insurance against liquefaction. However, the high cost, concerns about stability of excavated slopes, and dewatering requirements will often preclude use of excavate-replace methods in favor of either some form of dynamic densification, deep soil mixing or jet grouting.

Owing to the inhibiting effects of fines on both liquefaction and densification after liquefaction, mitigation of the liquefaction potential of soils with fines contents greater than 15 to 20 percent requires special adaptation of the dynamic densification methods – Deep Dynamic Compaction, Explosive Compaction, Vibro-Compaction, and Compaction Piles – that are ordinarily used to improve loose, clean cohesionless soils. Experience shows that installation of closely spaced gravel or prefabricated vertical drains can provide sufficient drainage and densification during DDC, VC and CP construction to make such soils safe against liquefaction. Drains are likely to be effective if the soils to be densified by dynamic compaction methods have a hydraulic conductivity less than about $1x10^{-8}$ m/s.

Columns and walls constructed in-situ using cement deep soil mixing and jet grouting can provide both reinforcement and encapsulation of liquefiable silty and clayey sands, thus protecting against settlements and lateral displacements in earthquakes. These methods are likely to be more costly and slower than dynamic densification methods; however, they can provide a higher level of improvement provided the columns and walls are properly constructed and adequately bonded to or keyed into non-liquefiable strata above and below the liquefiable layer,

Permeation grouting of liquefiable soils with high fines contents using traditional hydraulic methods is not a viable for liquefaction mitigation owing to the low permeability of these soils.

New methods for liquefaction mitigation in high fines content soils using electrokinetic injection and reduction in saturation by gas bubble generation using mechanical, electrolysis, or biological methods are candidates for further development.

REFERENCES

ASCE (1978). *Soil Improvement – History, Capabilities and Outlook,* Committee on Placement and Improvement of Soils, Geotechnical Engineering Division, 182 pp.

ASCE (1987). *Soil Improvement – A Ten Year Update*, J.P. Welsh, ed., Geotechnical Special Publication No. 12, 329 pp.

ASCE (1997). *Ground Improvement, Ground Reinforcement, Ground Treatment: Developments 1987 – 1997,* V. R. Schaefer, ed., GSP No. 69.

Dise, K., Stevens, M.,G., and Von Thun, J.L. (1994), "USBR Experiences Using DDC at 3 Dams for Mitigation of Liquefaction Risk, ASCE GSP No.45, 1-23.

Idriss, I. M. and Boulanger, R.W. (2007). *Soil Liquefaction During Earthquakes, Second Edition,* Earthquake Engineering Research Institute, in press.

Ishihara,K., Tsukamoto, Y., Nakazawa, H., Kamada, K., and Huang, Y. (2002). "Reisistance of Partly Saturated Sand to Liquefaction with Reference to Longitudinal and Shear Wave Velocities," *Soils and Foundations,* 42 (6), 93-105.

Kitazume, M. (2005). *The Sand Pile Compaction Method,* A.A. Balkema Publishers, 232 pp.

Luehring, R., Snortland, N., Stevens, M. and Mejia, L. (2001). Liquefaction Mitigation of a Silty Dam Foundation Using Vibro-Stone Columns and Drainage Wicks: A Case History at Salmon Lake Dam," Bureau of Reclamation *Water Operation and Maintenance Bulletin,* No. 198, 1-15. http://www.usbr.gov/infrastr/waterbull

Lukas, R.G. *Geotechnical Engineering Circular No. 1 – Dynamic Compaction.* Federal Highway Administration. Washington DC, 1995.

Martin, G.R., Seed, H.B., and Finn, W.D.L. (1975). "Fundamentals of Liquefaction under Cyclic Loading," *J. Geotech. Eng. Div.,* ASCE, 101 (5), 423-438.

Martin, J.R., Olgun, C.G., Mitchell, J.K., and Durgunoglu, H.T. (2004), *J. Geot. And Geoenv. Engrg.,* ASCE, (130), 6, 561-571.

Mitchell, J. K. (1981). "Soil Improvement: State of the Art," *Proc. Tenth Intl. Conf. on Soil Mechs. & Found. Engrg.*, Stockholm, Sweden, Vol. 4, 509-565.

Mitchell, J.K. and Wentz, F.J. (1991), "Performance of Improved ground During the Loma Prieta Earthquake," *Report No. UCB/EERC-91/12,* Earthquake Engineering Research Center, University of California, Berkeley, 93 pp.

Moseley, M.P. and Kirsch, K.(2004). *Ground Improvement.* 2nd ed. Spon Press. NY.

Rebata-Landa, V. and Santamarina, J.C. (2007). "Mechanical Effects of Biogenic Gas Bubbles in Soils (Nitrogen Gas)", *J. Geotech. And Geoenv. Eng.,* ASCE, in press.

Saito, A. (1977). "Characteristics of Penetration Resistance of a Reclaimed Sandy Deposit and Their Change through Vibratory Compaction," *Soils and Foundations,* 17 (4), 32-43.

Seed, H. B. and Booker, J. R. (1977). "Stabilization of Potentially Liquefiable Sand Deposits Using Gravel Drains," *J. Geotech. Eng. Div.,* ASCE, 103, (GT7), 757-768.

Shenthen, T., Nashed, R., Thevanayagam. S., and Martin, G.R. (2004). "Liquefaction Mitigation in Silty Soils Using Composite Stone Columns and Dynamic Compaction," *J. Earthquake Eng. And Eng. Vibrations,* 3(1), 39-50.

Thevanayagm, S. and Jin, W. (2003). "Electro-osmotic grouting for liquefaction mitigation in silty soils," *Proc. Third Int. Conf. on Grouting and ground Treatment,* ASCE GSP No. 128, (2) 1507-1517.

USACE (1999)."Engineering and Design Guidelines on Ground Improvement for Structures and Facilities," Publication No. ETL 1110-1-185, U.S. Army Corps of Engineers, Engineering Division, Directorate of Civil Works, Washington, DC. *http://www.usace.army.mil/inet/usace-docs/eng-tech-ltrs/etl1110-1-185/toc.htm*

Yegian, M.K., Eseller-Bayat, E., and Alshawabkeh, A. (2007). "Induced Partial Saturation for Liquefaction Mitigation: Experimental Investigation," *Journ. Geotech. and Geoenv. Eng.,* ASCE, 133(4), 372-380.

Yoshimi, Y., Yanaka, K., and Tokimatsu, K. (1989). "Liquefaction Resistance of a Partially Saturated Soil," *Soils and Foundations,* 29(2), 157-162.

Youd, T.L., et al. (2001). "Liquefaction Resistance of Soils: Summary Report from the 1996 NCEER and 1998 NCEER/NSF Workshops on Evaluation of Liquefaction Resistance of Soils," *Journ. Geotech. and Geoenvir. Eng,* ASCE 127(10), 817-833.

ACKNOWLEDGEMENTS

I am privileged and honored to be invited to contribute this special symposium and volume honoring my good friend and colleague, John H. Schmertmann. His contributions over many years have provided numerous important advances to the state-of-knowledge and the state-of-practice of Geotechnical Engineering from which we have all benefited.

I thank Crysta Highfield and Nestor Suarez for preparation of the figures.

Loading Tests: Interpretation and Prediction of their Results

Luciano Décourt[1]

[1]Luciano Décourt Engenheiros Consultores Ltda. Av. Brig. Faria Lima, 1616 / 2nd floor – 01451-001 São Paulo, SP. – Brazil, decourt@decourt.com.br

ABSTRACT: This paper aims at providing engineers with understandable load (stress) movement curves whenever load tests are available. The idea is to allow correct interpretations of test results. When they are not available, the idea is to predict what would have been their results on basis of conventional field tests, the Standard Penetration Test, SPT, the Cone Penetration Test, CPT or the Menard Pressuremeter, PMT.

KEY WORDS: loading tests, shaft friction capacity, stiffness, stiffness plot, toe capacity.

INTRODUCTION

Foundation design usually involves two steps: Initially, the bearing capacity or the failure load is determined. Then, a safety factor is adopted and an allowable load is determined by diving the capacity with the safety factor. In the majority of the cases, this is all that is done. Only special cases also include an evaluation of load movements of the foundations under their allowable loads.

The conventional way to compute bearing capacity makes use of Terzaghi type formulas, the so-called triple N formulas. When calculating settlement, the conventional way is to employ linear elastic theory or empirical formulas, all of which apply relationships between load (stress) and movement.

In order to check the validity of capacity prediction, only one method is fully reliable to carry out load tests on prototypes. However, theoretical bearing capacity relations, which are usually based on rigid plastic theories, yield ultimate values, regardless of movement. Load tests, while being the practical way to check computed values, on the other hand, do not define ultimate values. Only pairs of loads (or stresses) and movements are given. The ultimate value, i.e., the capacity, has to be defined from the test data rather arbitrarily, which requires the use of one of the numerous failure and/or extrapolation criteria that have been proposed.

THE CONVENTIONAL BEARING CAPACITY

Terzaghi (1942) proposed a definition of the bearing capacity of piles in load tests as the load $(Q_u)_c$ corresponding to a head pile deformation of 10% of the pile diameter.De Beer (1988) resumes the problem of establishing limit values for bored piles, defining physical $(Q_u)_u$ and conventional failure $(Q_u)_c$.

For piles in clays, the load corresponding to a movement of 10% of the pile diameter was also used by DeBeer (1988) to define the bearing capacity. For piles in sands, the capacity was defined as the load corresponding to a movement of 25% or 30% of the pile diameter.

EXTRAPOLATION OF LOAD TESTS RESULTS

It is not usual to carry out load tests up to deformations corresponding to, at least, 10% of the pile diameter (or 10% of the equivalent width of foundation in the case of footings). The most well known reference as far as methods for extrapolation of load test results are concerned is Fellenius (1980). An updated version of this paper, very import but nevertheless much less known, is Fellenius (2001). The methods mentioned in Fellenius's papers are presented below. (After Fellenius - 1980, 2001)

FELLENIUS (1980)	FELLENIUS (2001)
Davisson	Davisson
De Beer	De Beer
Brinch Hansen 80%	Brinch Hansen 80%
Brinch Hansen 90%	---
Chin	Chin-Kondner
Mazurkiewicz	---
Fuller & Roy	---
Butler & Roy	---
Van der Veen	---
---	Décourt

When comparing Fellenius (2001) with Fellenius (1980) one observes that five methods have been discarded and four maintained. Besides, a new method was included, Décourt (1996a,1999).

THE DÉCOURT STIFFNESS METHOD

Décourt (1996a,1999) proposed a method for extrapolation of load test results. One of the most important features of this method is the recognition that some foundations may fail but others will never fail. This procedure makes use of the so-called Stiffness Plot and is known in Brazil as the Stiffness Method. Using load test results, the stiffness is computed (stiffness is the load or stress divided by the corresponding movement) and plotted against the applied load. The physical failure $(Q_u)_u$ is defined as the load corresponding to zero stiffness. Zero stiffness corresponds to infinite deformation and it cannot be reached in practice. So-defined, it can be stated that physical failure has never been reached in a single load test, ever carried out in the world.

It was found that only two possibilities exist:

a) From some level of the test, the stiffness decreases linearly with increasing load, as indicated in Figure 1. The ultimate pile capacity $(Q_u)_u$ is defined as the load for which the extrapolated stiffness line intersects with the horizontal axis (abscissa). In this paper, for both sands and clays the conventional bearing capacity $(Q_u)_c$ is considered as being the load corresponding to a pile top movement, "s", of 10% of the pile diameter, "d".
b) From some level of the test, the stiffness decreases non-linearly with increasing load, approaching asymptotically a limit value as indicated in Figure 2.

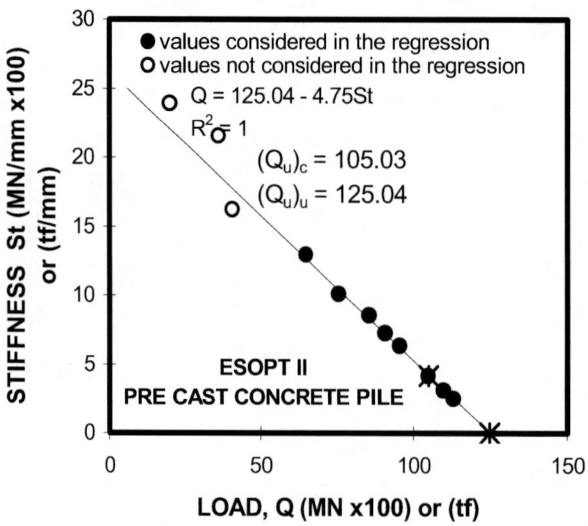

FIG. 1. Stiffness plot for a displacement pile. (After Décourt - 1999)

A footing foundation and a pile toe of a bored pile will never reach failure. Figure 2 shows a case of a footing foundation.

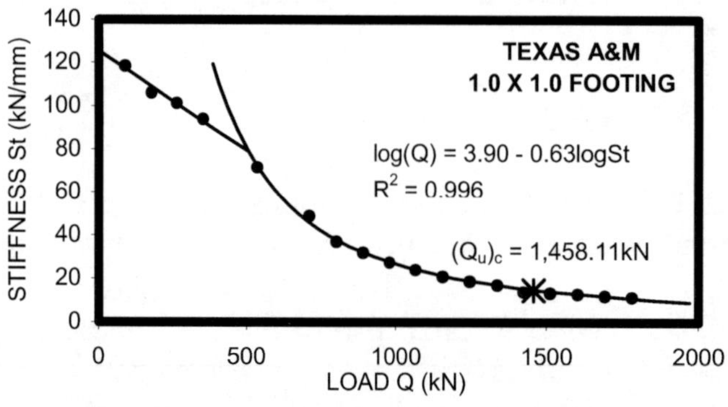

FIG. 2. Stiffness plot for a shallow foundation. (After Décourt - 1999)

When the increase of load with the decrease of stiffness is linear – Figure 1 - it is very easy and reliable to define both, the physical failure $(Q_u)_u$ (load for zero stiffness) and the conventional failure $(Q_u)_c$ (load for s = 0.1d), showing what may be called a failure behavior. When this increase is non-linear – Figure 2 - it is quite clear that only the conventional bearing capacity $(Q_u)_c$ might have some meaning, showing what may be called a non-failure behavior. The fundamental conclusion is that some load test may show the full capacity to have been neared – Figure 1 - but others will never reach the ultimate load – Figure 2.

Décourt (1999) upon analyzing results from more than 140 loading tests concluded that generally speaking the foundations that may fail are:

- Floating piles, which are piles that derive their capacity exclusively, or perhaps mostly from side friction (all types of piles).
- Displacement piles.

Figure 1 presents an example showing the stiffness curve from a load test on a reinforced concrete displacement pile, showing clearly both the conventional $(Q_u)_c$ and the ultimate $(Q_u)_u$ failure loads. Some very rare cases of foundations tested in cemented soils may eventually show a different behavior as exemplified in Figure 3. However, this almost perfect rigid-plastic behavior is rather an exception than a rule, even for cemented soils and soft rocks.

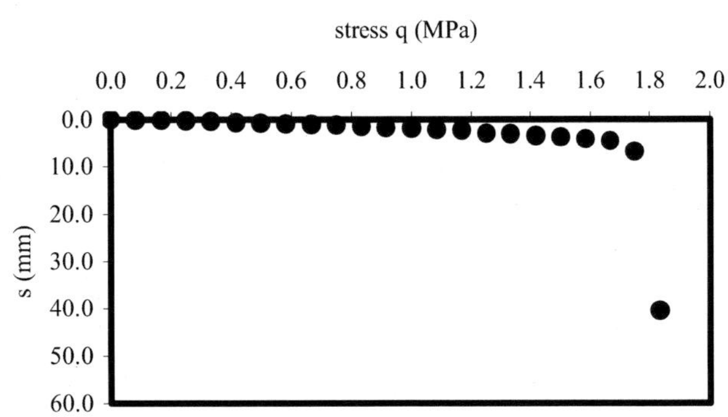

FIG. 3. Load-movement curve for a footing on a cemented (lateritic) soil. (After Décourt - 2003)

FUNDAMENTALS OF THE METHOD FOR LOAD TEST INTERPRETATION

Shallow Foundations and Bases of Bored Piles

There is no doubt that load tests represent the most reliable way for determining the load (stress)-settlement behavior of foundations. The results of a load test carried out on a square shallow foundation with a width of 2.50m will be used as an example of the application of the method, Décourt (1999).

Step I: The Load Test Data

The results of the load test are presented, as measured, in Figure 4, which also shows the extrapolation of the results according to the proposed method, Step II.

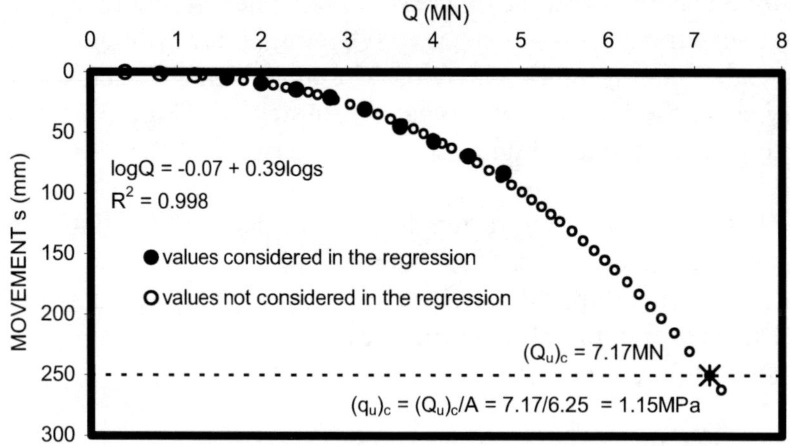

FIG. 4 - Load test data on a 2.50m x 2.50m shallow foundation and extrapolation of the results. (After Décourt - 1999)

Step II: The Basic Regression

The basic regression may be established for load (Q) or stress (q) versus movement (s). Generally speaking, it is postulated that the load or stress-movement curve may be divided in two parts. The initial one corresponds to either, the pre-stressing branch or in the case of cemented soils to stresses lower than the yield stress σ'_p. It is usually an approximately straight line. For higher stresses, the stress-movement behavior is represented by a parabolic curve. The latter turns out to be another straight line, provided both, loads (stresses) and movements are plotted in a log scale.

The proposed procedure starts with statistic regressions. Initially, only the last three points on the curve are considered.

Other regressions follow, considering successively the last four points, the last five points, and so on. Two criteria for judging the quality of the correlations are considered: the value of the coefficient of correlation R^2 and the number of points entering in the correlation. It is quite obvious that the higher the value of R^2 and the greater the number of data points entering in the correlation, the more reliable the statistically generated curve will be.

In many cases, however, these two conditions are fulfilled by different combinations of criteria. As of yet, no general criterion of preference has been established, the decision to adopt one or another being a matter of judgment of the engineer. Ideally, the value of R^2 should be higher than 0.999 and not lower than 0.99. Moreover, in most of cases, the separation between the straight line (if existent) and the main curve is easy to be visually established. The rebound curve also helps to separate these two behaviors. The Stiffness Plot however is the most important tool to help in identifying the limit of pre-stressing as will be explained later.

Once the equation of the curve is chosen, an estimate of the conventional failure load $(Q_u)_c$ or failure stress $(q_u)_c$ can easily be made, as indicated in Figure 4. It is enough to enter in the correlation with a settlement of 10% of the width of the footing, in the example, 250mm. If the footing has a shape other than square, an equivalent width (B_{eq}) is defined, $B_{eq} = \sqrt{A}$, "A" being its area. Values of $(Q_u)_c$ of 7.17MN and of $(q_u)_c$ of 1.15MPa are obtained, Figure 4.

Step III: The Normalized Plot

Once $(q_u)_c$ is determined, the variation of $q/(q_u)_c$ x s/B_{eq} is established, as indicated in Figure 5.

$$\log\ (q/(q_u)_c) = C' + C''\ \log\ (s/B_{eq}) \qquad (01)$$

Theorectically, C' should be equal to C". In most of the cases, they are not equal, but pretty close together. A value C is then introduced, which is the average of C' and C".

$$C = (C' + C'')/2 \qquad (02)$$

$$\log\ (q/(q_u)_c) = C + C\ \log\ (s/B_{eq}) \qquad (03)$$

C is called Coefficient of Intrinsic Compressibility. In the given example, as indicated in Figure 5, C = 0.39

$$\log\ (q/(q_u)_c) = 0.39 + 0.39\ \log\ (s/B_{eq}) \quad R^2 = 0.998 \qquad (04)$$

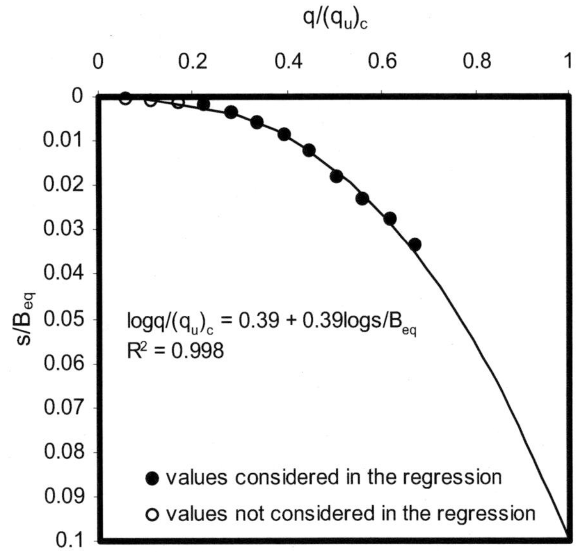

FIG. 5. Stresses x movements in a normalized plot. (After Décourt - 1999)

Step IV: The Stiffness Plot

Step four aims to check if the postulate of the non existence of geotechnical physical failure $(Q_u)_u$ (point of zero stiffness), is correct for the load test being analyzed. It also helps in identifying if the soil has been pre-stressed or if it is normally consolidated.

The conventional bearing capacity $(Q_u)_c$ is determined from the Stiffness Plot shown in Figure 6. In the example, $(Q_u)_c$ is 7.16MN and $(q_u)_c$ is 1.15MPa. It is clear from this figure that for loads lower than about 1.2MN, the soil behavior is different, indicating that up to this level the soil behaves as pre-stressed and/or cemented. It is also quite clear from this figure that the physical failure will never be reached.

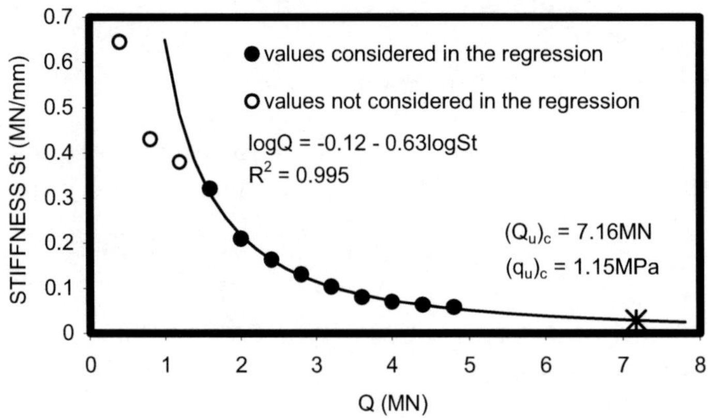

FIG. 6. Stiffness Plot for the load test presented in Figure 4. (After Décourt - 1999)

Shaft Friction Capacity

Bored piles, as any other type of pile, may show shaft friction failure. In conventional load tests (not instrumented) it is not an easy task to separate shaft friction from end bearing. However, there is a type of load test, developed independently more than 20 years ago, by da Silva (1983), in Brazil, and Osterberg (1984), in USA, which allows the development of pure side friction. A hydraulic jack called Expancell in Brazil and Osterberg cell, O-cell, in USA is placed close to the tip of the pile, and the jack pushed simultaneous upward to establish the side resistance and downward to establish the load-movement for the pile toe.

In most such load tests, carried out under the Author's supervision, a slightly belled base is made just below the cell level in order to ensure a complete elimination of the shaft resistance in the part of the pile below the cell. Some authors question the validity of the shaft bearing capacity as determined by this test, arguing that the loading is in the upward direction and the usual loading is in the downward direction. However, the differences, in the Author's opinion, are of second order of magnitude. Both tests are compression tests. Besides, the octahedral stresses, which are fundamental for the shaft bearing capacity, are rigorously equal, for both types of test, at least at the beginning of the loading.

Very recently the author carried out a research program analyzing a large number of such tests. The results were partially presented in Décourt (2006a). A very important load test will now be used to exemplify this method. It is the test, carried out by Loadtest International Inc. in South Korea which is presently a world record, Figures 7 and 8.

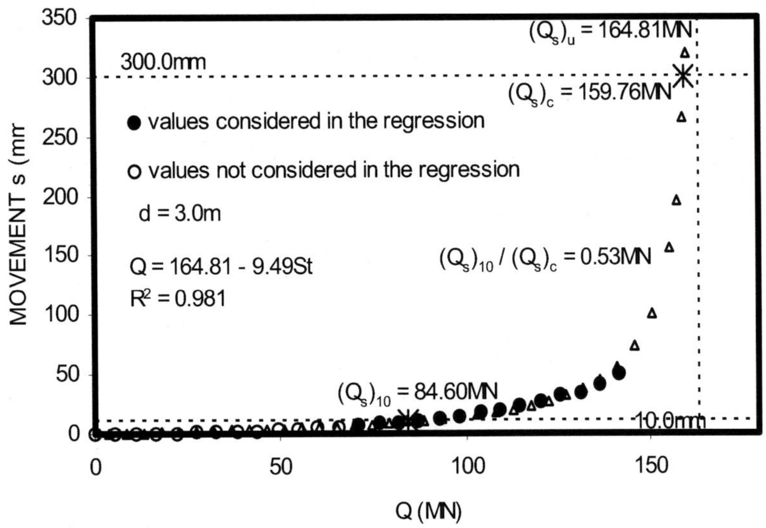

FIG. 7. Load test data and the extrapolated curve, statistically established.

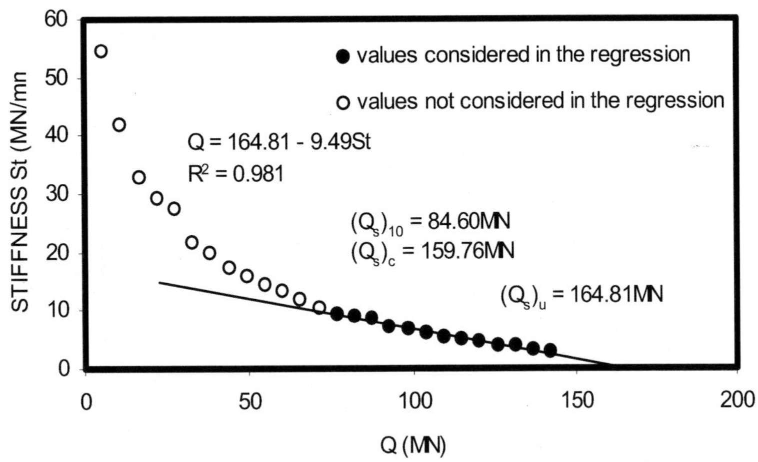

FIG. 8. Stiffness Plot for the load test shown in Figure 7.

The method allows the visualization of the failure process and a mathematically correct estimate of both, conventional - $(Q_s)_c$ and physical - $(Q_s)_u$ failure loads. Moreover, it also allows the identification of some problems that occurred during the

test. In the Author's opinion, the statistically determined correlation represents the load-settlement relationships even better than the measured values themselves.
In analyzing Figures 7 and 8, it becomes clear that for the three largest (last) loads the actual movements should have been larger than the measured values. For those familiar with carrying out loading tests, this is a quite common situation, because of the difficulties in maintaining the applied loads when the deformations are substantially increasing.

One important point to be emphasized is that the value of Q_s does not reach an ultimate value for displacements of about 10mm as assumed by many authors. On the contrary, the shaft friction load keeps on increasing with deformations. Besides, it is quite clear that the assessment of both, the conventional $(Q_s)_c$ and the ultimate $(Q_s)_u$ capacities can be easily made, even in cases where the loading has been prematurely interrupted. In the mentioned example, if the maximum test load had been 92.81MN with a corresponding movement of 12.70mm and provided the stiffness straight line could be properly defined, the linear regression would yield to Q = 146.96 – 7.26St, R^2 = 0.923. The computed values of $(Q_s)_c$ and $(Q_s)_u$, would have been 143.49 and 146.96 respectively a 10% difference from the actual values shown in Figures 7 and 8.

Defining α as the angle that the straight line makes with the load axis, and $(Q_s)_{10}$ as the shaft friction for a movement of 10,0mm an attempt was made to correlate tgα with the ratio r, $(Q_s)_{10}/(Q_s)_c$. The values of tgα and $(Q_s)_{10}/(Q_s)_c$ of 40 load tests were plotted and are shown in Figure 9. The conclusion is that the stiffer the pile-soil system is, (higher values of tgα) the higher this ratio, r, will be. The ratio r is therefore a measure of the error made when $(Q_s)_{10}$ is assumed to be the ultimate value. In another words, it may represent an additional shaft resistance as compared to that usually assumed, $(Q_s)_{10}$.

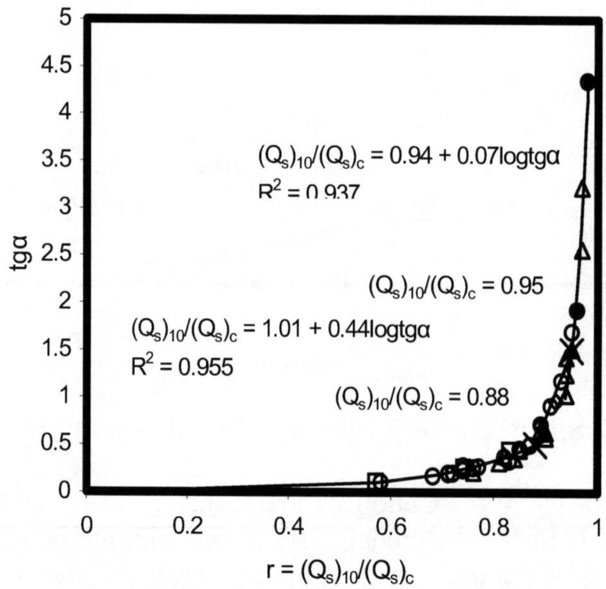

FIG. 9. $(Q_s)_{10}/(Q_s)_c$ as a function of tgα.

UNIQUENESS OF THE NORMALIZED LOAD (STRESS) MOVEMENT CURVE

For Gibson soils, the shear strength, i.e. the penetration resistance of the SPT (N_{SPT}) and the point resistance of the CPT (q_c) increase with octahedral stresses. Because for normally consolidated soils, the octahedral stresses increase with depth, it may be assumed that other characteristics also increase with depth.

The bearing capacity theories state that for sands, the capacity, $(q_u)_c$, is a function of the sand density, the dimension of the footings (B) and the vertical effective stress, $q = \gamma z$

$$(q_u)_c = \frac{1}{2}\gamma B N\gamma + qN_q \tag{05}$$

It is often said, which by the way is quite obvious, that $(q_u)_c$ increases with the density of the sand, the width of the foundation (B) and the vertical effective stress (q).

As already mentioned, to take into account size and depth effects, a representative shear strength, which in practice is assessed via penetration tests, SPT or CPT for example, should be considered. Normally not only one, but many SPTs or CPTs are carried out for a given job. The strength of the bearing layer is then assessed taking into account the results of as many tests as possible carried out nearby and establishing statistical regression of their results with depth. To reduce the effect of the natural scatter of the data, a suggestion is made that average values for each depth be considered.

A statistical regression of these average values with depth is then established. According to Décourt (1999), the characteristic value of the penetration resistance is the value corresponding to a depth of 0.7 B_{eq} below the level of the foundation. So, if a 2.50m width square footing is placed at a depth of 1.10m, the characteristic depth (z_c) will be:

$$z_c = 0.7 \times 2.50 + 1.10 = 2.85\text{m} \tag{06}$$

Décourt (1999) suggested that in a normalized plot, the stress-movement curve is unique, regardless of the density of the sand, the size and depth of the foundation. The stresses are normalized by the value corresponding to the conventional bearing capacity $(q_u)_c$ (q for s = 0,1d) and the movements by the width (or diameter) of the foundation. All analyses so far carried out by the author have confirmed this suggestion, Décourt (1999).

To check once more the validity of such proposal, a comparison is now made between conventional stress-settlement curves and normalized ones. Décourt (2006b) analyzed the data provided by Lee and Salgado (2005). In Figure 10 the load-settlement curves of the tests carried out on three footings with 1.0, 2.0 and 3.0m in diameter and relative densities of 50%, 70% and 90% are shown. It is observed the obvious increase of q with B and D_r.

In Figure 11 the same data are presented. However, in this case the stresses have been normalized by the values of $(q_u)_c$ and the settlements by the footing diameters. The results of the nine load tests merge approximately in a single curve.

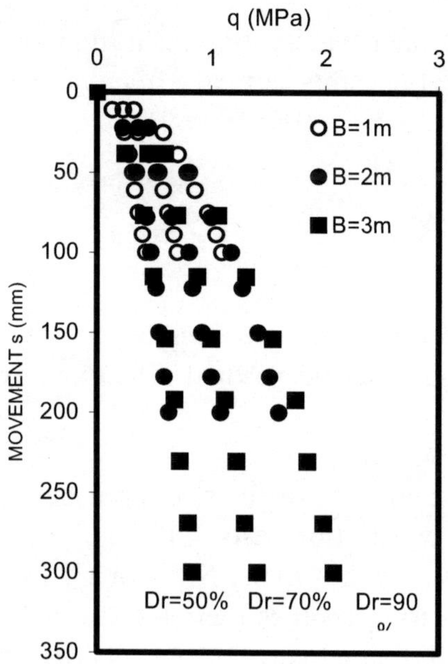

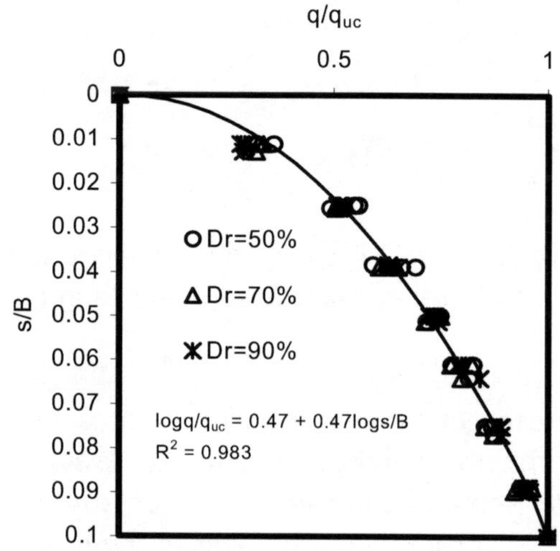

FIG.10. q (MPa) x s (mm). Footings with various diameters, B, on sands with various relative densities, D_r. (After Décourt – 2006b)

FIG. 11. Data from Figure 10 in a normalized plot. (After Décourt – 2006b)

INTERPRETATION OF CONVENTIONAL LOAD TESTS ON PILES

Conventional load test on piles are much more complex to interpret than those with Expancells (O-cells) because the load is transferred to the soil simultaneously by shaft friction and end bearing. During the first stages of the loading, the transfer is mostly by shaft friction. For large deformations the opposite happens.

For rigid piles, the transition between these two behaviors may be assumed to occur for settlements of about 2% of the pile diameter. Using the MBM as proposed by Décourt (1993), estimates of both, Q_s and $(Q_u)_c$ may be easily made.

Brierley et al (1979) proposed an interesting method for assessing side shear Q_s, which is based on the displacements of the toe of the pile. However, considering that these displacements frequently are not measured, and that the differences between toe and top displacements are expected to be very small for bored piles of common lengths (10.0/15.0m), it was postulated that the use of top displacements would also allow reasonable estimates of Q_s. This new version of the Brierley Method was designated the Modified Brierley Method (MBM), Décourt (2003).

Figure 12 presents the basis for the interpretation of such tests.

Following Décourt (1993) a conservative estimate (lower bound limit) of Q_s is 900.42kN, Figure 12. According to the method that now is being proposed, as shown in Figures 13 and 14, an upper limit of Q_s may be obtained using the Stiffness Plot. The straight line mainly represents the linear correlation obtained for small settlements (typically less than 2% of the pile diameter for rigid piles). $Q_s = 1{,}247.43 - 2.41St$ provides estimates of $(Q_s)_c$ and $(Q_s)_u$, as shown in Figures 13 and 14. The curve that follows the straight line mainly represents an end bearing behavior. An assessment of $(Q_u)_c$ may be obtained using log x log correlations obtained for settlements higher than approximately 2% of the pile diameter, $\log Q = 3.48 - 0.24 \log St$, Figures 13 and 15.

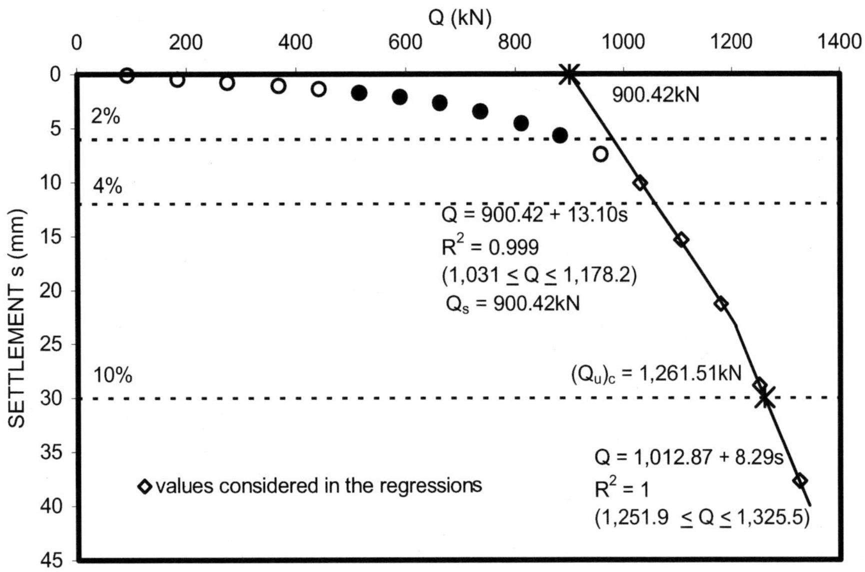

FIG. 12. Load test data and estimates of Q_s and $(Q_u)_c$ using the MBM, Décourt (1993). Also shown the six points used in the regression of Figure 13.

These two statistically determined equations may be used to represent the full load-settlement curve, as shown in Figure 15.

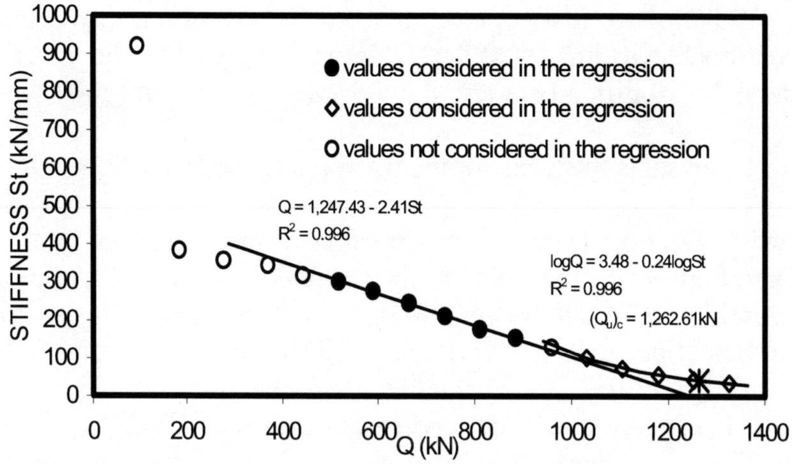

FIG. 13. Stiffness Plot and statistical regressions

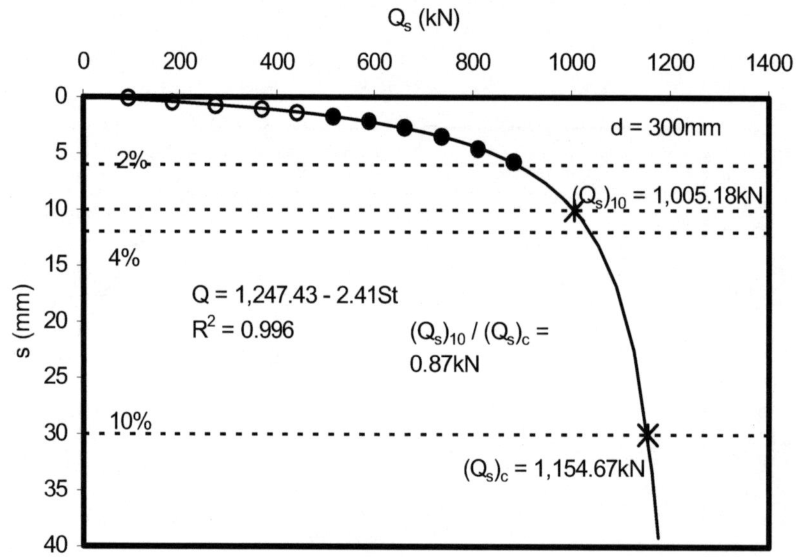

FIG. 14. Development of Q_s with movement(s)

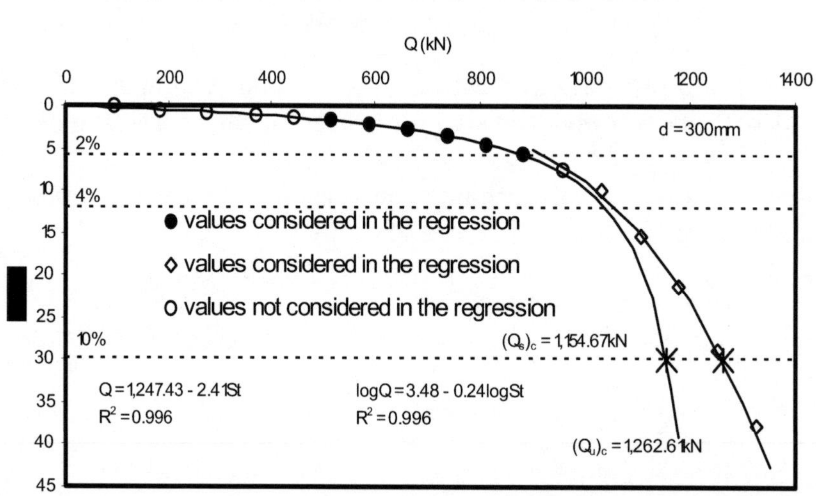

FIG. 15. Curves Q x s and Q_s x s put together with estimates of $(Q_u)_c$ and $(Q_s)_c$.

PREDICTION OF LOAD TEST RESULTS

Correlations of Failure Stresses with the Penetration Resistance of the SPT

The prediction of the load test curve depends on the knowledge of the failure stress in conventional load tests.

Considering that many different types of piles exist, the Author suggests computing the bearing capacity for a reference concrete displacement pile with a round section. For other types of pile, coefficients α and β are applied to adjust, respectively, the end bearing and the shaft friction. For the reference pile, $(q_u)_c$ is given by:

$$(q_u)_c \text{ (kPa)} = KN_{60} \qquad (07)$$

For other piles:

$$(q_u)_c \text{ (kPa)} = \alpha KN_{60} \qquad (08)$$

All information presented in Tables I, II and III were taken from the text book Foundations, Theory and Practice, Décourt (1996b).

The values of K are given in Table I, and the values of α in Table II.

TABLE I. Values of the coefficient K for the reference pile as a function of the soil type, $(q_u)_c$ (kPa) = KN_{60}

Soil Type	Values of K (kPa)
Sands	333
Intermediate Soils	175
Clays	100

It must be emphasized that, following De Beer (1988), for bored piles in sands, the failure stress, $(q_u)_c$ is supposed to correspond to a movement of 30% of the pile diameter.

TABLE II. Values of the coefficient α as a function of pile and soil types

Soil Type \ Pile Type	Bored	Bored with Bentonite Slurry	Continuous Flight Auger	Root	Injected under High Pressure
Sands	0.50	0.50	0.30*	0.50*	1.0*
Intermediate Soils	0.60	0.60	0.30*	0.60*	1.0*
Clays	0.85	0.85	0.30*	0.85*	1.0*

*values are no more than indications due to shortage of available data.

The unit shaft friction of the reference pile was found to be independent of the soil type.

$$(q_s)_c \text{ kN/m}^2 = 10\,(N_{60}/3.6 + 1) \tag{09}$$

For piles other than the reference pile, a correction factor β is introduced, Table III.

$$(q_s)_c = 10\beta(N_{60}/3.6 + 1) \tag{10}$$

TABLE III. Values of the coefficient β as a function of pile and soil types

Soil Type \ Pile Type	Bored	Bored with Bentonite Slurry	Continuous Flight Auger	Root	Injected under High Pressure
Sands	0.50	0.60*	1.0*	1.50*	3.0*
Intermediate Soils	0.65	0.75*	1.0*	1.50*	3.0*
Clays	0.80	0.90*	1.0*	1.50*	3.0*

*values are no more than indications due to shortage of available data.

For shallow foundation $(q_u)_c$ has been correlated with N_{60}, Décourt (1995), Table IV.

TABLE IV. Conventional bearing capacity of spread footings, $(q_u)_c$ (kPa) = KN_{60}

Soil Type	Values of K (kPa)
Sands	95
Intermediate Soils	80
Clays	65

These values might also be used to assess the end bearing capacity of bored piles. However, it is very important to recognize that these failure values correspond to movements of 10% of the pile diameter and not 30% as assumed in Table I.

Correlations of Failure Stresses with Other Field Test Results

Failure stresses may also be correlated to other in-situ tests such as the CPT and the PMT. Lee and Salgado (2005) proposed to correlate $(q_u)_c$ with the cone resistance, q_c, of the CPT, suggesting that this correlation would depend on the sand relative density (D_r) and the coefficient of earth pressure at rest, K_0. Disregarding extreme values, like relative densities, D_r of 30% and 90% and K_0 values lower than 0.45 and higher than 0.7, $(q_u)_c$ would be in the range 0.25/0.30 q_c, the smaller coefficients corresponding to higher values of both D_r and K_0:

$$(q_u)_c = 0.25 / 0.30\ q_c \qquad (11)$$

Combarieu (1997) suggested a correlation between $(q_u)_c$ and the limit pressure of the pressuremeter, P_L.

$$(q_u)_c \approx P_L \qquad (12)$$

Stress-Movement of Shallow Foundations and Bases of Bored Piles

The stress corresponding to the conventional failure is established on basis of penetration test results N (SPT/SPT-T), q_c (CPT) or P_L (PMT)). The other unknown parameter is the Coefficient of Intrinsic Compressibility, C, Décourt (1999).

C is a function of mineralogical and some other characteristics of soil grains and, therefore, it does not need to be determined for each specific case. Once load test results exist in a given area, and therefore the results of C are known, they may be used in other cases dealing with similar soils.

For most of the soils of São Paulo city, with quartz particles, for clays as for sands, sedimentary or residual, C varies between 0.35 and 0.45. A value of 0.42 has been successfully used by the author whenever no specific load test results are available.

For the dune sands from Labenne, France, C is higher, varying between 0.38 and 0.63, C = 0.55 fitting well most of the test results. For the much more compressible grains of calcareous sands from Kuwait and Australia, the values of C are much higher, varying between 0.6 and 0.83, C = 0.75 fitting well most of the load test results.

Once $(q_u)_c$ and C are determined, load-movement relationships for any stress level, may be easily determined:

$$\log (q/(q_u)_c) = C + C \log (s/B_{eq}) \quad (13)$$

The higher the adopted value of C, the more conservative the load-movement prediction will be.

Load-Movement of Piles

The available methods for assessing the bearing capacities of piles aim to calculate the pile toe capacity and the unit shaft friction. Considering that in almost all the cases it is not clearly specified which failure values have been considered by the authors, it is assumed that conventional bearing capacities have been considered (values of loads for settlements of 10% of the pile diameter).

Décourt (1993) proposed the already mentioned Modified Brierley Method, MBM, for determining the actual load-settlement curve in a schematic and conservative way.

However, the determination of a more-accurate curve is highly desirable. The author is presently working in this direction. The idea is to take into account the computed values of $(Q_s)_c$ and $(Q_u)_c$ and log x log correlations.

One of these curves would represent the branch of low settlements, that is mostly a shaft friction behavior. The other would represent the branch of high settlements, that is mostly an end bearing behavior. The values to be considered in the regressions are shown in Table V.

TABLE V. Predicted load-settlement curve

	Loads	Settlements (mm or %d)
I – Shaft Friction	$0.45 (Q_s)_c$	2.0
	$0.9 (Q_s)_c$	10.0
II – End Bearing	$0.9 (Q_s)_c$	10.0
	$(Q_u)_c$	0.1d

In Figure 16 a comparison is made between the predicted curve and the loadtest results in Fig. 12.

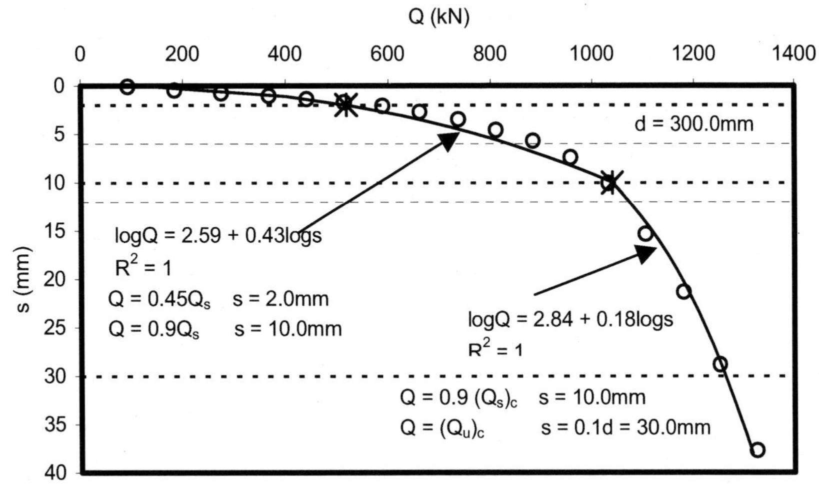

FIG. 16. Comparison between the predicted load-settlement curve and the actual data.

CONCLUSIONS

Methods are proposed for a more-correct interpretation of load test results. These methods allow conventional and ultimate failure values to be properly determined. Besides, in case of non-instrumented piles, procedures are suggested for obtaining the shaft friction and the toe loads. Considering that for routine jobs load tests are usually not available, methods have been proposed to predict what these curves would have been like. The basic parameters $(q_u)_c$ and $(q_s)_c$ are determined on basis of conventional field tests like the SPT (especially the SPT-T), the CPT and also the PMT (the limit pressure P_L).The movements of the foundations, for any load (stress) level, may be easily determined using the correlations here proposed.

REFERENCES

de Beer, E. E. (1988). "Different Behavior of Bored and Driven Piles". *Proceeding of the 1st International Geotechnical Seminar on Deep Foundation on Bored and Auger Piles*, pp. 47-78, Ghent, Belgium.

Brierley, G. S., Thompson, D. E. and Eller, C. W. (1979). "Interpreting end-bearing Pile Load Test Results". *ASTM Special Publication 670* – Behavior of Deep Foundations, Boston, pp. 181-198.

Combarieu, O. (1997). "Capacité Portante des Fondations Superficielles. Pressiomètre et Essais de Laboratoire". *Bulletin des Laboratoires des Ponts et Chaussées*, 211, pp. 53-72.

Décourt, L. (1982). "Prediction of the Bearing Capacity of Piles Based Exclusively on N Values of the SPT". *Proceedings of ESOPT II*, pp. 29-34, Amsterdam.

Décourt, L. (1993). "Predicted and Measured Behavior of Non Displacement Piles in Residual Soils". *Proceeding of the 2nd International Geotechnical Seminar on Deep Foundations on Bored and Auger Piles*, pp. 369-376. Ghent, Belgium.

Décourt, L. (1993/1995). "On the Load-Settlement Behavior of Piles". III Odair Grillo Lecture, given in 1993 and published in 1995. *Soils and Rocks*, vol. XVIII, no 2, pp. 93-112.

Décourt, L. (1995). "Prediction of Load Settlement Relationships for Foundations on the Basis of the SPT-T". "*Ciclo de Conferencias Internacionales Leonardo Zeevaert*", pp. 87-104, Mexico.

Décourt, L. (1996a). "Foundation Failures Assessed on Basis of the Concept of Stiffness" (in Portuguese). *SEFE III*, Vol. I, pp. 215-224, São Paulo.

Décourt, L. (1996b). "*Foundations, Theory and Practice*". Chapter 8.1 - Piles (in Portuguese), pp. 265- 301, edited by Waldemar Hachic, Frederico . Falconi, José Luiz Saes, Régis G. Q. Frota, Celso S. Carvalho and Sussumu Niyama, ABMS/ABEF.

Décourt, L. (1999). "Behavior of Foundations Under Working Load Conditions". *Proc. of XI Panamerican Conference on Soil Mechanics and Geotechnical Engineering*. Vol. 4, pp. 453-488, Foz do Iguaçu (Brazil).

Décourt, L. (2003). "Behaviour of a CFA Pile in a Lateric Clay". *4th International Geotechnical Seminar on Deep Foundations on Bored and Auger Piles*, pp. 301-308. Ghent, Belgium.

Décourt, L. (2006a). "Development of the Lateral Friction Between Pile and Soil on Basis of Load Tests Using Hidrodynamic Expansive Cells" (in Portuguese). *COBRAMSEG 2006*, Curitiba, PR. (Brazil).

Décourt, L. (2006b). "Discussion of "Estimation of Bearing Capacity of Circular Footings on Sands Based on Cone Penetration Test" by Junhwan Lee and Rodrigo Salgado. *ASCE, Journal of Geotechnical and Geoenvironmental Engineering*, Vol. 132, no. 11, pp. 1511-1513.

Décourt, L. and Quaresma, A. R. (1978). "Bearing Capacity of Piles on Basis of SPT" (in Portuguese). *Proceedings of VI COBRAMSEF*, Vol. I, pp. 45-53, Rio de Janeiro, RJ. (Brazil).

Fellenius, B.H. (1980). "The Analysis of Results from Routine Pile Loading Tests". *Ground Engineering*, pp. 19-31.

Fellenius, B. H. (2001). "What Capacity Value to Choose from the Results of a Static Loading Test". *Fulcrum. The Newsletter of the Deep Foundations Institute.*

Lee, J. H. and Salgado, R. (2005). "Estimation of Bearing Capacity of Circular Footings on Sands Based on Cone Penetration Test", *ASCE, Journal of Geotechnical and Geoenvironmental Engineering*, Vol. 131, no. 4, pp. 442-452

Osterberg, J. O. (1984). "A New Simplified Method for Load Testing Drilled Shafts". *Foundation Drilling*. Vol. XXIII, no. 6, ADSC, pp. 9, July/August, 1984.

da Silva, P.E.C.A.F. (1983). "*Hidrodynamic Expansive Cell. A New Way to Perform Loading Tests*". Belo Horizonte, MG. (Brazil): 106 p.

Terzaghi, K. (1942). "Discussion of Pile Driving Formulas". *Proceedings of ASCE*, 68 (2), pp. 311-323.

The Easy Button for Driven Pile Setup: Dynamic Testing

Paul J. Bullock[1], P.E., M. ASCE

[1]Senior Engineer, GRL Engineers, Inc., 3636 NW 23rd Place, Gainesville, FL 32605, pbullock@pile.com

ABSTRACT: Field inspectors often observe a dramatic increase in penetration resistance between initial driving and restrikes of driven piles, a result of additional static "setup" capacity. Designers often ignore setup when specifying pile lengths and blow count acceptance criteria, and pile capacity design methods, calibrated against static load tests, already incorporate some setup that occurs during the test preparation delay. Research indicates that setup results mostly from an increase in side shear and occurs in all soil types as a result of combined pore pressure dissipation, consolidation, and mechanical aging effects. Repeated static tests provide a reliable measurement of setup, but also delay construction. Penetration resistance (set per blow) does not reliably measure setup during isolated restrikes as it varies per blow due to inconsistent driving system performance and changing pile capacity. Instrumented dynamic tests, using strain gages and accelerometers, provide an "easy button" to verify hammer energy and pile capacity during all phases of installation. This paper reviews pile setup, its causes, and related properties. It presents a simple method, with examples, to measure and analyze setup from dynamic test results. An engineer may use these results to reduce the length, size, or number of production piles, and adjust the required set per blow. Caveats include staged testing effects and consideration of the setup contribution from different soil strata.

KEY WORDS: aging, driven pile, dynamic load test, end bearing, restrike, side shear, signal matching, time dependence, ultimate static capacity

INTRODUCTION

Field inspectors and engineers often observe a change from "easy" to "hard" driving during restrikes following the initial installation of driven piles, and possibly during brief driving interruptions for cushion changes, template removal, or equipment breakdowns. They may also measure unexpectedly high capacity during a static load test (typically 3 to 14 days after the end of driving) following low penetration resistance. Depending on site stratigraphy and pile type, this "setup" may increase the pile capacity by more than

100% over time. However, many engineers do not include setup in design calculations, or attempt to assess it during construction. Ignorance of time changes can also lead to foundation problems, as piles driven into some soil-rock profiles may actually decrease in capacity with elapsed time, or "relax". With the relative simplicity and widespread availability of dynamic testing methods (see equipment in Figure 1), engineers can now measure changes in static capacity through a series of restrikes specified in the project documents, benefiting the designer through capacity verification, the client through economy in the constructed foundation, and the contractor through reduced driving requirements and shorter pile lengths. This paper reviews when to expect setup (almost always), the causes of setup and relaxation, setup predictor tests, and staged testing effects. It presents a relatively simple method for measuring and analyzing side shear setup using dynamic tests, with numerical examples based on field data.

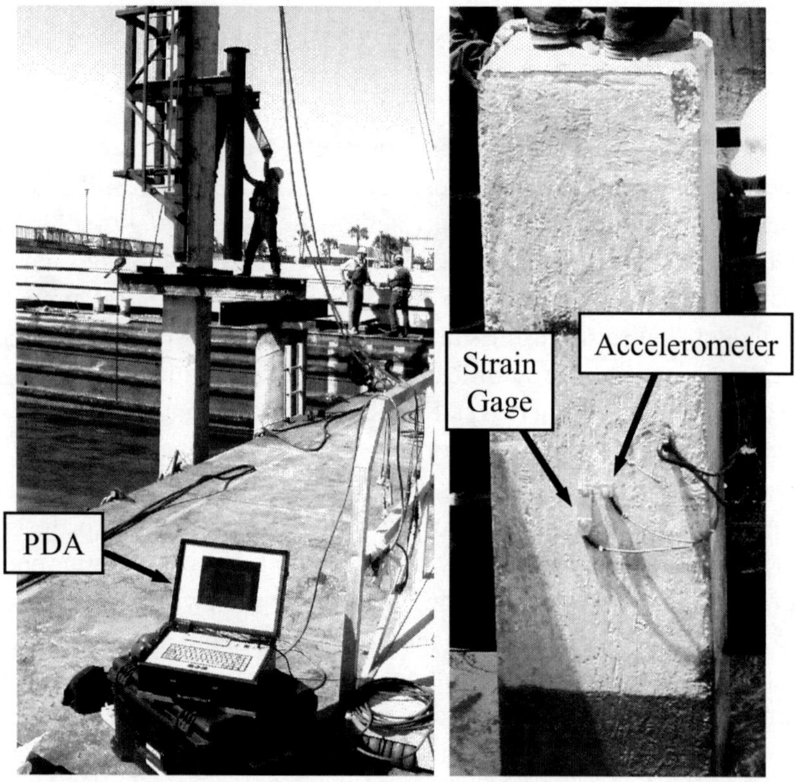

FIG. 1 Dynamic Test Equipment: Pile Driving Analyzer® (PDA) from Pile Dynamics, Inc.

SETUP OR RELAXATION?

Driven pile setup seems the rule rather than the exception, and it seems to occur in all soil types, including cohesionless soils. A few examples covering a variety of soil and pile types include (among many available):

- Bullock et al. (2005) - four Florida sites, ***sand, clay, mixed soils***, prestressed concrete piles, dynamic tests followed by three to six static tests over

16 to 1727 days elapsed time using Osterberg cell tests to separate side shear and end bearing, 12 to 32% side shear increase per log cycle of time.

- Fellenius et al. (1989) - Wisconsin, ***sandy clay and silty sand***, driven steel pipe and H-piles, static and dynamic tests, capacity increase averages 50% between 1 and 21 days.
- Karlsrud and Haugen (1985) - Norway, ***sensitive, overconsolidated clay***, jacked closed-end, pipe piles, static tests, side shear increase averages 30% between 6 and 30 days.
- Kehoe (1989) - two Florida sites, ***mixed cohesive soils***, driven square, prestressed concrete piles, static and dynamic tests, capacity increases average 58% and 200% within 11 days after driving (72% of increase within 100 min)
- Lukas and Bushell (1989) - five Illinois sites, ***soft to stiff clay***, driven steel pipe and H-piles, static tests, capacity increase averages 25% between 10 and 32 days in stiff clay and 50% between 10 and 82 days in soft clay.
- Seidel et al. (1988) - Australia, ***alluvial sands over limestone***, driven prestressed, concrete pile, static and dynamic tests, 80% capacity increase 535 days after initial driving (see Figure 3).
- Tavenas and Audy (1972) - Quebec, ***uniform medium sand***, hexagonal, reinforced concrete piles, static tests, average capacity increase of 70% between 0.5 and 15-20 days.

Foundation engineers also sometimes observe driven pile "relaxation", but fortunately much less often than setup (and than commonly feared). Some reports of this phenomenon, based solely on blow count, have less creditability due to changes in the driving system or inefficient hammer operation during restrikes. Relaxation may result from temporary confining effects, or from the creation of negative pore pressures that grip the pile during initial penetration and then later dissipate (see Chow et al., 1998). The following stratigraphy may raise these concerns:

- sands confined by a cofferdam, or closely spaced piles, in which lateral confining stresses may later relax
- strong soils that dilate during penetration (e.g. dense fine sands), creating negative pore pressures that later dissipate
- weak, foliated, sedimentary and metamorphic rock (e.g. shale), in which driving stresses may relax due to creep

Given the predominance of evidence suggesting setup (Bullock, et al., 2005, Chow, et al., 1998, etc.), the foundation engineer, after considering the above list, often may reasonably assume de facto setup for a given pile foundation. Of course, if lacking local experience, he or she will prudently require verification testing of this assumption via the relatively simple procedure provided below.

REVIEW OF SETUP BEHAVIOR

Side Shear Only: Bullock et al. (2005), Axelsson (1998a), and Chow et al. (1998) concluded that setup occurs primarily as a result of side shear increase, not end bearing. Penetration of the pile pushes soil outward and away from the pile, destructuring and shearing it to a greater extent adjacent to the side of the pile than at the pile tip, and thus reducing the side resistance during installation (and increased aging effects).

Conventional Pile/Soil Side Shear: Eq.1 provides a relatively simple and conventional estimate of side shear, which shows that setup could result from either, or both, an increase in the horizontal effective stress or an increase in the strength components.

$$f_s = \text{unit side shear} = c'_a + \sigma_h'(\tan \Phi') \quad \text{..........} (1)$$

where: c'_a = pile-soil adhesion
σ_h' = effective horizontal stress
Φ' = pile-soil drained friction angle

What Changes?: Bullock et al. (2005), Chow et al. (1998), and Axelsson (1998a), all concluded that the setup process initially increases the strength components of Eq.1 through consolidation and the effective stress as result of both consolidation and collapse of soil pushed away from the pile during installation. However, they postulate that later setup (after about 30 days) occurs as a result of mechanical aging increasing the friction angle, a process that Schmertmann (1991) identified as affecting soil behavior in many other situations as well. For cohesive soils, consolidation and aging effects seem to overlap, resulting in an apparently continuous setup process. However, pore pressures dissipate rapidly in well drained soils (sand), possibly resulting in a time gap between the two effects. Therefore, as found by Bullock et al. (2005), pile tests in clean sands often show no increase in side shear until aging effects begin, possibly one to two weeks after driving.

Continuing Process: The research literature indicates that side shear setup begins almost immediately after the end of driving (EOD), and continues for many years, though the rate of increase approaches zero over time. Initially the capacity increase may exceed 100% within the first 24 hrs after driving. A satisfactory mathematical model of setup should describe a continuing process. Of course at some point, typically in the range of 60 to 180 days after EOD, the foundation must bear a substantial portion of the design load (including construction loads), and later changes may not provide usable design capacity. Some engineers simply multiply the EOD capacity by a conservative constant, effectively ignoring later increases, e.g. Table 1 from Rausche et al. (1996). These factors generally underestimate the actual setup trend, but may prove useful for wave equation driveability studies to determine the reduced static resistance to driving from that calculated for various analytical methods (Hannigan, et al, 2006).

TABLE 1. Prescriptive Setup Factors from Rausche et al. (1996)

Soil Type Adjacent to Pile	Range of Setup Factor	Recommended Setup Factor
Clay	1.2 - 5.5	2.0
Silt - Clay	1.0 - 2.0	1.0
Silt	1.5 - 5.0	1.5
Sand - Clay	1.0 - 6.0	1.5
Sand - Silt	1.2 - 2.0	1.2
Fine Sand	1.2 - 2.0	1.2
Sand	0.8 - 2.0	1.0
Sand-Gravel	1.2 - 2.0	1.0

Whole-Pile Semilog-Linear Setup Factor, A: The total pile capacity (side shear plus end bearing) generally follows a linear increase versus the log of the time elapsed after the EOD. Figure 2 shows a best fit line through both static and dynamic test data from Seidel et al. (1988) illustrating this trend. Skov and Denver (1988) proposed the dimensionless setup factor, A, in Eq.2 to represent the relative increase in pile capacity per log cycle of elapsed time. The plot in Figure 3 shows the Figure 2 data with the resulting setup factor, $A = 0.10$ based on the best-fit line in Figure 2. Though sometimes useful, this approach incorrectly includes end bearing in the setup process and thus reduces the setup factor by increasing the reference capacity. Note that Eq. 2 proves less accurate early (<10 min) and late (> 3 years) in the setup process, but generally adequate within the range of practical application (15 minutes to 1 year).

$$\frac{Q}{Q_0} = A\log\left(\frac{t}{t_0}\right)+1=\left(\frac{m_Q}{Q_0}\right)\log\left(\frac{t}{t_0}\right)+1 \qquad (2)$$

where: A = Dimensionless setup factor
Q = Whole pile capacity at time t
Q_0 = Whole pile capacity at initial reference time t_0
t = Time elapsed since EOD
t_0 = Reference time at start of log-linear capacity increase
m_Q = Semilog-linear slope of Q vs. log t (see Figure 2)

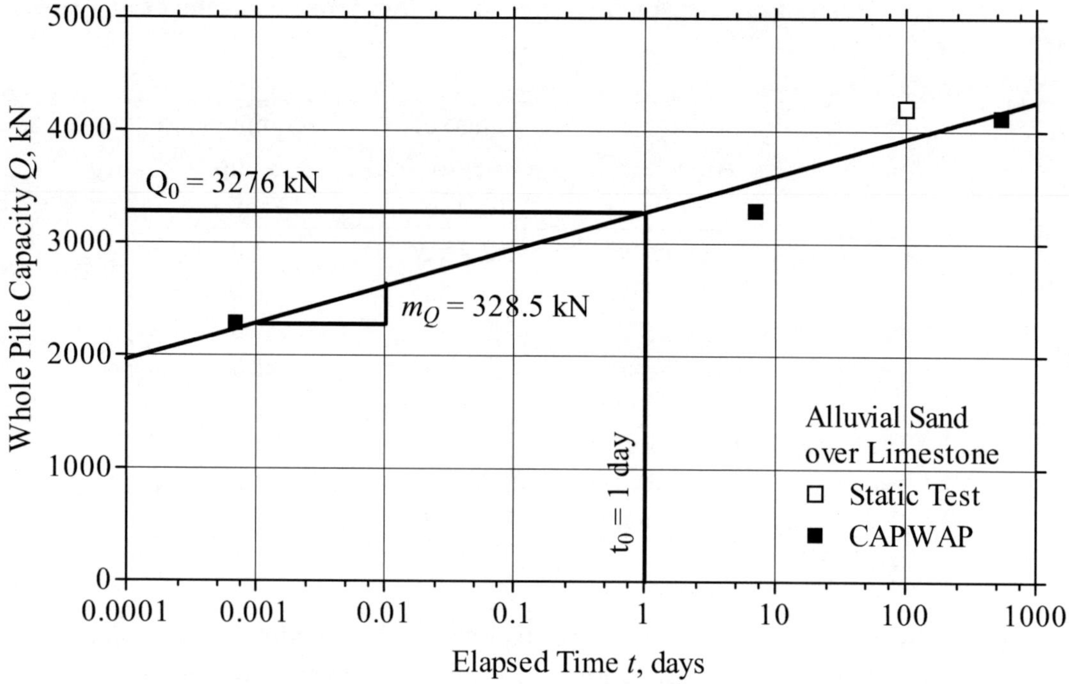

FIG. 2. Pile Capacity Data from Seidel et al. (1988)

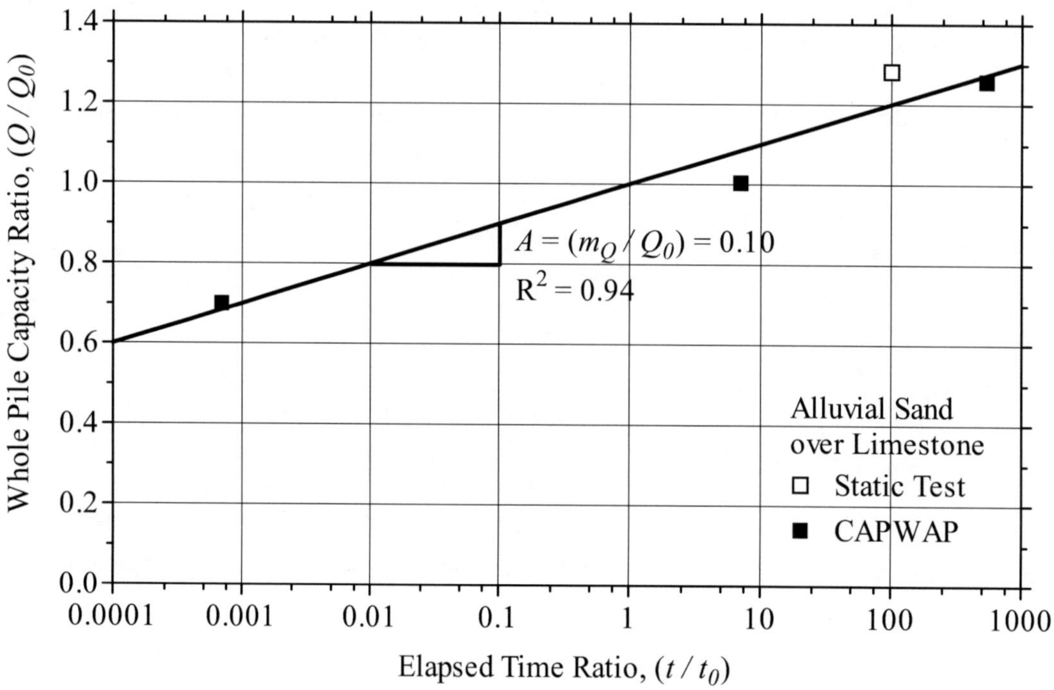

FIG. 3. Setup Factor for Data from Seidel et al. (1988)

Side Shear Semilog-Linear Setup Factor, A: Bullock et al. (2005) recommend three modifications of the analysis proposed by Eq.2:

- Use $t_0 = 1$ day as this removes the difficulty of finding the actual start of the semilog-linear setup process while providing a convenient global reference.

- Use A to describe the side shear component only, not the whole pile capacity, as the end bearing component may change in magnitude due to differing pile sizes or site variability and does not increase appreciably due to setup. Though the setup factor decreases when including end bearing, differences in bearing between piles may lead the engineer to over predict the total capacity using whole-pile setup factors.

- As an approximation based on experience, plot the EOD capacity (determined from dynamic testing) at 1 min elapsed time. Capacity measurements at fixed times should prove more reliable.

Setup factors based on side shear as shown in Eq.3 provide a more reliable comparison between test and production piles, for which the designer may expect $A = 0.1\text{-}0.8$ using $t_0 = 1$ day.

$$\frac{Q_S}{Q_{S0}} = \frac{f_S}{f_{S0}} = A\log\left(\frac{t}{t_0}\right)+1 = \left(\frac{m_S}{Q_{S0}}\right)\log\left(\frac{t}{t_0}\right)+1 \qquad (3)$$

where:

A	=	Dimensionless setup factor
Q_S	=	Side shear capacity at time t
Q_{S0}	=	Side shear capacity at initial reference time t_0
f_S	=	Unit side shear capacity at time t
f_{S0}	=	Unit side shear capacity at initial reference time t_0
t	=	Time elapsed since EOD, days
t_0	=	reference time, recommended to use 1 day
m_S	=	Semilog-linear slope of Q_S vs. log t

Pile Size: Note that Eq.3 works equally well with either force or unit resistance as the ratio (Q_S / Q_{S0}) cancels the side area of the pile (perimeter times length). Figure 4, from data presented by Bullock (1999), shows dynamic test results from the Seabreeze site in Daytona Beach, FL for restrikes performed at 15 min to 4 days after the EOD of eighteen, 610-mm-square, prestressed concrete piles. These test piles produced a similar side shear setup factor, A = 0.18, to that found from the combined dynamic and static tests of a 457-mm-square, prestressed concrete, research pile, A = 0.21. The scatter of the data for the 610 mm piles in Figure 4 likely results from site variability and estimation of the side shear using the Case method capacity (not signal matching). Bullock et al. (2005) found that reducing the reference time of 1 day in proportion to the ratio of the squared, effective pile diameter produced similar setup factors for this same

457-mm test pile compared with torque tests on a 51-mm-diameter, driven sampler (see below). However, this effect may prove less important when comparing full-size piles, especially for incremental differences in pile size.

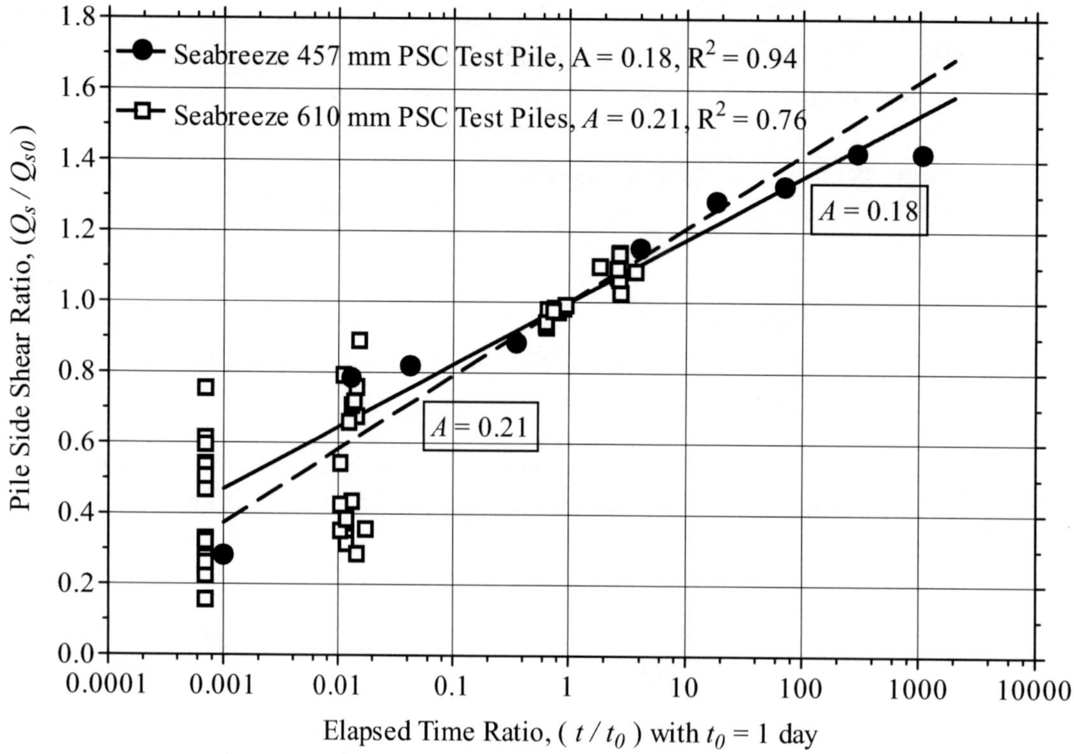

FIG. 4. Setup Factor from Bullock (1999) for Two Pile Sizes at FL Site

Setup Predictor Tests: While this paper focuses on dynamic tests of driven piles, predictor tests based on model piles may provide the designer an opportunity to assess pile setup from insitu tests such as: torque tests performed on the Standard Penetration Test (SPT) sampler by Bullock and Schmertmann (2003), uplift tests performed on the SPT sampler by Rausche et al (1996), or tests of driven rods performed by Axelsson (1998b).

Setup Reduction for Staged Testing: Engineers investigating setup routinely perform repeated, or "staged", tests of pile capacity on the same pile, typically resulting in small axial movements (< 50 mm) during each test (dynamic or static). The resulting pre-shearing effect of one test on the next does not remold the volume of soil that a larger penetration might affect, such as during initial driving. Staged testing apparently increases side shear at a quicker rate, and results in a higher setup factor than for piles tested only once, or "unstaged", at a specific elapsed time following the EOD. Research results reported by Karlsrud and Haugen (1985) for steel pipe piles, Miller (1994) for steel pipe piles, and Bullock and Schmertmann (2003) for torque tests of the SPT sampler show that staged tests significantly increase side shear capacity when compared to unstaged tests in clay soils. Although the range of soil types affected requires further research, Bullock et al (2005) recommend reducing the setup factor calculated from staged test results by 60%, or a factor of 0.4, when applying it to unstaged piles. This reduction factor does not apply to the staged test pile itself, as it develops and maintains the capacity actually measured.

DYNAMIC MEASUREMENT OF STATIC CAPACITY

Static load tests require time, money, manpower, and heavy equipment, all essential commodities on a construction site. Dynamic tests provide an attractive "easy button" alternative to static testing, mobilizing minimal test equipment and personnel on short notice with little impact to the construction schedule. Ground conditions around the test pile generally do not impede dynamic tests, and the engineer can test any accessible pile.

Dynamic measurements during driving require strain gages and accelerometers (see Figure 1) mounted near the top of the test pile to measure and record the instantaneous pile velocity and force generated in the pile by each hammer blow. Through advances in instrumentation, computer equipment, and user-friendly software, the effort required to obtain and analyze dynamic test data has decreased greatly during the last 25 years. Analysis of the test data, preferably with a signal-matching program such as CAPWAP®, separates dynamic and static resistance to estimate the overall static capacity, as well as the side shear and end bearing components. While estimation of the side shear and end bearing components has less reliability, analytical expertise and consideration of site stratigraphy can improve this.

For 303 comparisons up to a capacity of 36 MN, Likins et al. (2004) report 0.98 for the ratio of the CAPWAP® estimate to static load test measurement, with a 17% coefficient of variation. Site-specific calibration of dynamic test results using static tests may provide additional design confidence on sites with unfamiliar soil conditions, unusual piles, large projects, or heavy loads. Like a static load test terminated prior to failure, dynamic tests with less than about 1 to 2 mm penetration per blow may not fully mobilize, and therefore tend to under predict, the pile's static capacity. For best comparison, both tests should attain a failure load on the same pile, and the dynamic test should follow 12 to 24 hrs after the static test (not the reverse).

ASTM International provides a standard test method (D4945) for high strain dynamic testing, and Likins et al. (2000) indicate that many governmental and code organizations worldwide accept and encourage dynamic tests. Because of the wide acceptance and availability of dynamic testing, and the accuracy of static capacity estimates from signal-matching techniques, the foundation engineer has an ideal tool to quickly assess pile setup using restrike tests, preferably scheduled to minimize construction delays. Of course, for large projects, unusual soil profiles, and large setup factors, the prudent engineer will still require static tests to verify dynamic results.

WHEN TO RESTRIKE (EARLY AND OFTEN)

Possibly as a legacy from static tests that often require a week or more to prepare and perform, many engineers specify long restrike times to assess setup. Of course, when comparing dynamic test results to a static test, the engineer should request a restrike performed immediately following (not before) the static test. However, with the exception of piles driven in clean sand, restrike tests can provide considerable, and often dramatic, setup information as little as 15 min after the EOD.

Figure 5 shows a best-fit $A = 0.30$ with $R^2 = 0.99$ for the combined static and dynamic side shear capacity measurements from the Aucilla test pile reported by Bullock et al. (2005), with the EOD capacity plotted at 1 min elapsed time. The dynamic tests at the EOD, 15 min, and 60 min provided a best-fit $A = 0.28$ with $R^2 = 0.95$, and the static tests from 1 day up to 4.7 years later provided a similar best-fit $A = 0.32$ with $R^2 = 0.98$.

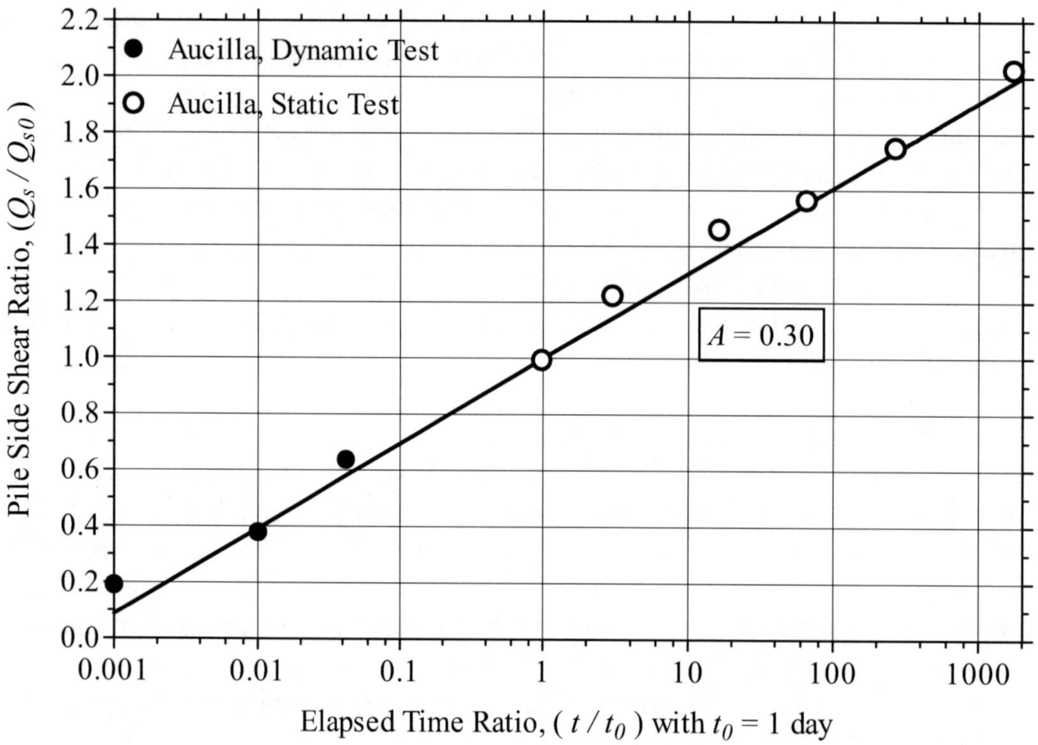

FIG. 5. Side Shear Setup for Florida Test Pile in Bullock et al. (2005)

The setup factor A describes the relative capacity increase with respect to the 1-day capacity, Q_{S0}, but a comparison with the EOD capacity proves much more remarkable. Table 2 shows the ratio of the setup capacity of the pile with both the EOD ($t = 1$ min) and 1-day reference. The results in Table 2 indicate that a restrike only 15 min after the EOD showed a dramatic increase of 97% of the EOD capacity, while restrikes at 1 hr and 24 hrs showed increases of 232% and 417% respectively.

TABLE 2. Side Shear Setup for Florida Pile (Aucilla) from Bullock et al. (2005)

Elapsed Time from EOD, t days	Time Ratio t/t_0 t_0 = 1day	Measured Side Shear Q_S kN	Q_S/Q_{S0} t_0 = 1day for Q_{S0} = 977 kN	Q_S/Q_{S1} t_1 = 1 min for Q_{S1} = 188 kN
0.0007	0.0007	188	0.19	1.00
0.010	0.010	370	0.38	1.97
0.042	0.042	624	0.64	3.32
0.98	0.98	972	1.00	5.17
2.97	2.97	1197	1.22	6.36
16.1	16.1	1427	1.46	7.59
65.1	65.1	1528	1.56	8.13
265	265	1712	1.75	9.11
1727	1727	1982	2.03	10.54

Therefore, while later restrikes provide greater reliability for setup trend analysis, early restrikes may provide adequate data without unnecessary delay to the project. For best results, obtain dynamic test results from the EOD and two or three restrike capacities at 15 min, 1 hour, and 1 to 3 days after the EOD. More data develops a better trend line, and the prudent designer will require that some restrikes verify the desired capacity. Note that if relaxation proves important, a 15 min restrike should quickly identify this problem as well. (Note that a geometric time progression for restrikes provides equally spaced data on a semi-logarithmic plot, though it may test a contractor's patience.)

ANALYSIS STEPS

The following examples illustrate the use of dynamic tests and restrikes to determine the setup factor and predict pile capacity. The basic steps include:

1. Perform dynamic tests (measure force and velocity in pile) during the EOD and a minimum of two restrikes, with accurate set per blow measurements (important for analyses).
2. Perform signal-matching analysis of an average blow at the EOD
3. Perform signal-matching analysis for an early blow of each restrike (typically second or third blow with adequate transferred energy to mobilize pile capacity)
4. End bearing typically does not change much during restrikes with only a few inches permanent set. Consider (don't force) this in signal-matching analyses.
5. Tabulate the estimated static capacity versus elapsed time since EOD for each blow analyzed. Use the time of first blow to calculate elapsed time from EOD for restrikes, and as a first approximation, use 1 min for the EOD capacity.
6. Plot side shear and total capacity versus the logarithm of elapsed time.

7. Find a best fit line through the side shear (slope m_S) and through the total capacity. They should have similar slopes with a relatively constant difference of the average end bearing. (If not, adjust as necessary, usually assuming the total capacity slope as more accurate unless the end bearing is suspected to increase.)
8. Using the adjusted best fit line (slope m_S), calculate the reference side shear capacity Q_{S0} at the reference time $t_0 = 1$ day.
9. Calculate the side shear setup factor $A = (m_S / Q_{S0})$
10. Calculate and plot the ratio of (Q_S / Q_{S0}) versus the time ratio (t / t_0). Add the setup line with slope A through the point $(Q_S / Q_{S0}) = 1$ and $(t / t_0) = 1$.
11. Use Eq.3 to estimate side shear at desired times. Add end bearing to obtain total pile capacity at these times. For reliability, perform at least some late restrikes to verify the desired capacity.

EXAMPLES FROM LOUISIANA AND THE RED SEA

The engineer can perform the above analyses quite easily in a spreadsheet program, such as Excel or Quattro. Figure 6 shows a spreadsheet analysis of static pile capacity (Class A prediction) for dynamic tests of a 762-mm-square, prestressed concrete pile driven 33 m into cohesive Louisiana soils at Rigolets Pass. The static test performed 14 days after the EOD agrees reasonably well with the Figure 6 trend line of predicted capacity based on restrikes at 15 min, 1 hr, and 1 day.

When possible, a comparison of setup results from multiple test piles at the same site provides additional confidence in the results. Figures 7 to 9 show setup results from six, 1372-mm-diameter, open-ended, steel pipe piles, driven near-shore on the Red Sea through 20 m of water and penetrating 31 to 40 m into calcareous sand, silty sand, and weak limestone to bear on competent limestone. Figures 7 and 8 respectively show the total capacity and side shear obtained from CAPWAP analyses. The setup factor for these piles ranged from 0.13 to 0.18 with an average of 0.15 as shown in Figure 9, or a side shear increase of 15% per log cycle of elapse time in days. This setup factor falls at the low end of the expected range, possibly as a result of the primarily calcareous soils and less soil displacement and destructuring caused by an open-ended pipe pile.

		Elapsed	**CAPWAP Capacity**				Capacity Estimate with Setup				
	Pile	**Time**	**Shear**	Ratio	**Toe**	**Total**	Setup	Shear	Ratio	Toe	Total
Blow	**Penetr.**	t	Q_S	Q_S/Q_{S0}	Q_B	Q	Time	Q_S	Q_S/Q_{S0}	Q_B	Q
Type	**m**	**days**	**kN**		**kN**	**kN**	days	kN		kN	kN
EOD	**13.46**	**0.0007**	**323**	0.164	**3156**	**3479**	0.0007	229	0.116	3172	3401
SC1	**13.46**	**0.0160**	**750**	0.380	**3281**	**4031**	0.0160	982	0.497	3172	4154
SC2	**13.58**	**0.0424**	**1301**	0.659	**3111**	**4412**	0.0424	1216	0.616	3172	4388
SC3	**13.69**	**0.924**	**2008**	1.017	**3166**	**5174**	0.924	1956	0.990	3172	5128
		(EOD plotted at 1 min)					2.0	2141	1.084	3172	5313
Average Q_B		3,172 kN		5463 kPa			4.0	2308	1.169	3172	5480
t_0		1.0 day					8.0	2474	1.253	3172	5646
Q_{S0}		1975 kN		(at t_0)			16.0	2641	1.337	3172	5813
m_S		552.8 kN		(slope of Q_S)			30.0	2792	1.413	3172	5964
A		0.280		(unitless setup factor)			**14.0**	**(Static Test)**			**6540**

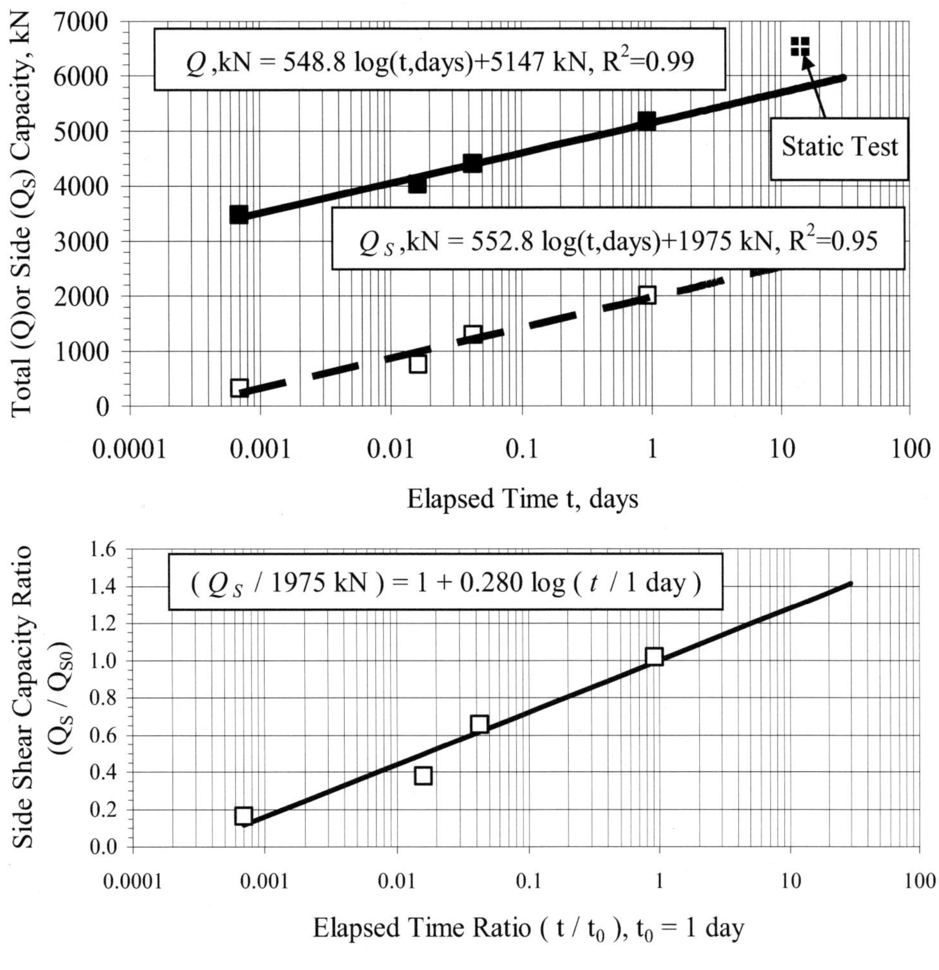

FIG. 6. Spreadsheet Analysis for Rigolets Pass Test Pile (required input in bold)

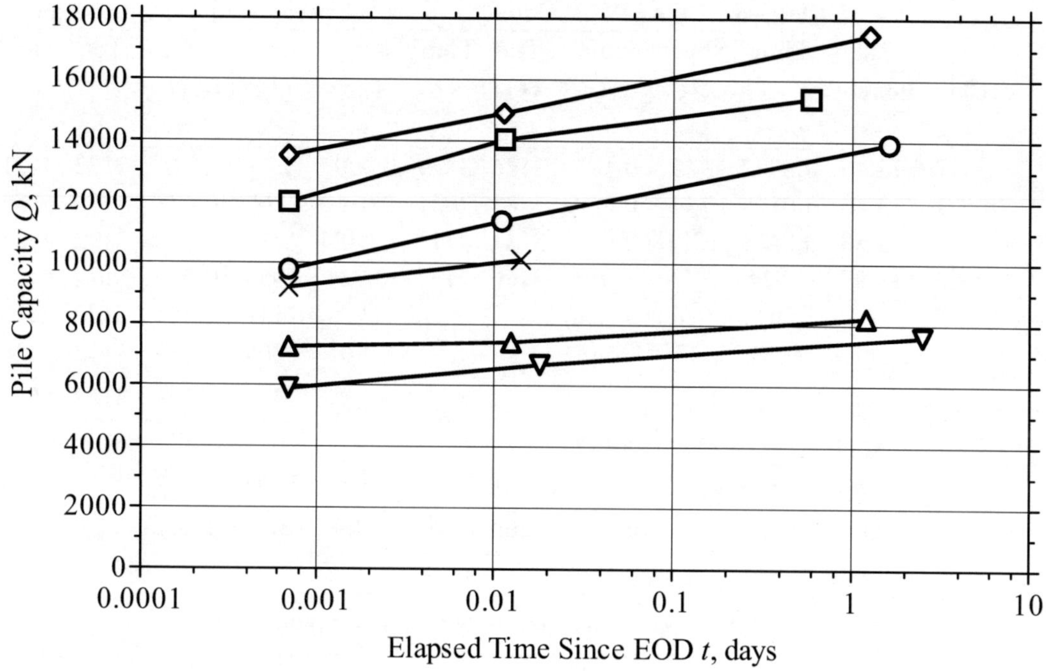

FIG. 7. CAPWAP Total Capacity from Restrikes of Six Red Sea Pipe Piles

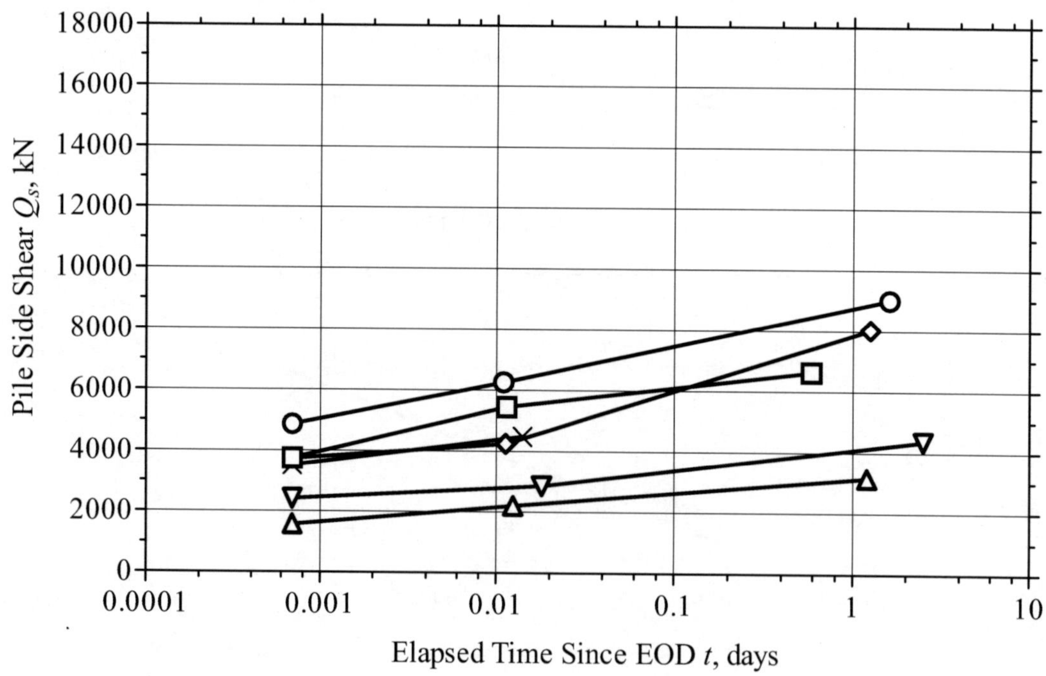

FIG. 8. CAPWAP Side Shear Capacity from Restrikes of Six Red Sea Pipe Piles

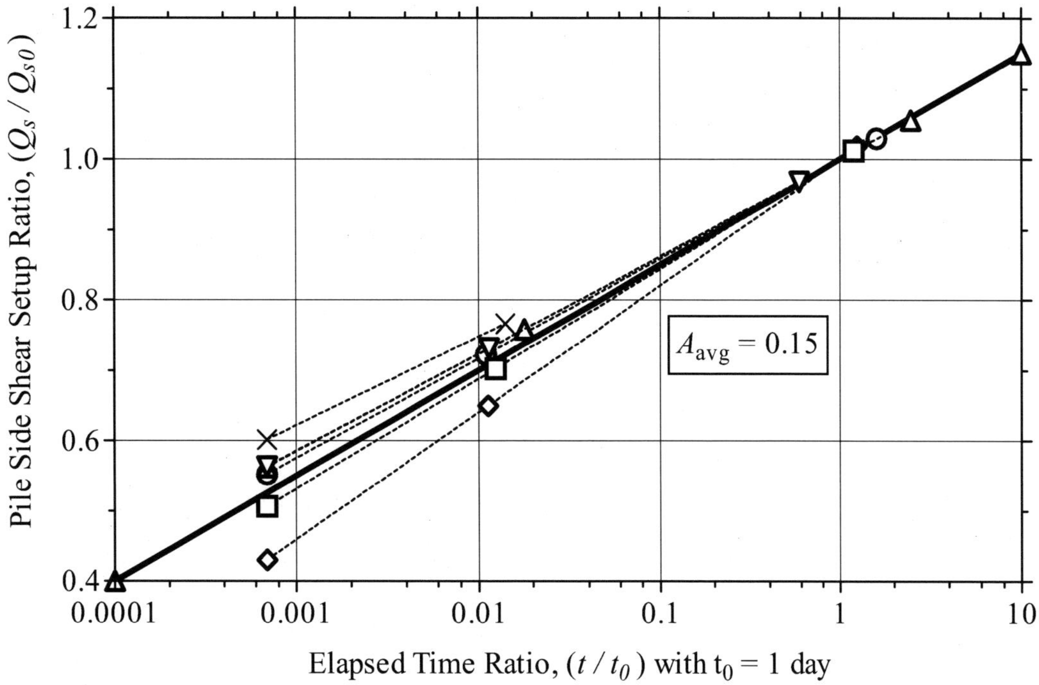

FIG. 9. Setup Factor from Restrikes of Six Red Sea Pipe Piles

APPLICATION OF SETUP FACTOR

After calculating a setup factor from the restrikes and analysis, what next? The list below provides a few possibilities:

- Estimate the static capacity of the pile at a future time (generally less than 90 days) to determine if it meets the design criteria. The engineer may specify additional restrikes to verify large extrapolations or estimates that barely comply. Foundations lacking redundancy and those with high capacity piles (especially offshore platforms) require greater conservatism, which may include testing of every pile, more tests per pile, or greater safety factors (lower resistance factors).
- Calculate an average setup factor from multiple test piles to validate the setup rate for the site. Apply this setup factor to production piles (after reduction for stage testing effects). Perform random restrikes during production driving to validate capacity.
- If the setup analysis of test piles indicates excess capacity, consider shorter, smaller, or fewer production piles. Some soil types provide a greater setup factor, and the designer should check the site stratigraphy to verify that a shorter pile does not eliminate too much of the setup. Also check that any design change does not result in undesirable settlement of the single pile or the pile group. Perform restrikes at the start of production driving to validate any changes in pile length or size.

- If the setup analysis of production piles indicates excess capacity, the engineer may reduce the blow count acceptance criteria and/or required penetration, possibly resulting in a foundation cost savings.
- Develop a database of setup factors for design use. Include location, geology, site stratigraphy, and pile type and size for future reference.

SUMMARY AND CONCLUSIONS

1. The published research shows that displacement piles exhibit side shear setup in most soil types, and that relaxation occurs with less frequency in relatively specific stratigraphy.
2. End bearing does not appear to contribute significantly to setup, and the engineer should avoid including it in setup calculations when possible.
3. Analysis of restrike capacity should generally result in constant end bearing so that the total and side shear capacity will have a similar slope versus the logarithm of elapsed time (unless stratigraphy indicates changing tip conditions).
4. Side shear setup follows a relatively continual process, with capacity increasing linearly versus the logarithm of time elapsed since the EOD, as a result first of pore pressure dissipation and consolidation, and then of mechanical aging. Pile capacity in clean sands may stabilize temporarily before the onset of aging.
5. Expect the setup factor A, describing the increase in side shear relative to the reference capacity at one day elapsed time since the EOD, in the range of 0.1 to 0.8.
6. Limited evidence of staged testing increasing the rate of setup for the tested pile suggests a 60% reduction of the setup factor before applying it to piles without staged tests.
7. Dynamic test measurements during restrikes provide a simple, cost-effective, means to establish a setup trend, with wide acceptance and reasonably accurate estimates of static capacity from signal-matching.
8. Use the estimated setup trend to validate design capacity, and possibly reduce the length, size, or number of production piles. (Provide static tests to verify dynamic estimates for unfamiliar soil conditions, large projects, or heavy loads.)
9. Early restrikes minimize project delay and provide adequate information to extrapolate capacity at later times. Except for piles driven in clean sands, the linear increase of capacity versus the logarithm of time results in rapid capacity gain, measurable and usable within 24 hours. Restrike times of 15 min, 60 min, and 1 day should normally prove adequate. Sands may require longer setup times to allow for mechanical aging effects.
10. Long-term restrikes provide verification of capacity estimates when extrapolating early restrikes to attain the desired capacity.

ACKNOWLEDGEMENTS

Dr. John Schmertmann's interest in, and contributions to, the investigation of the mechanical aging of soil began early in his career. The author gratefully acknowledges the guidance, support, and encouragement provided by Dr. Schmertmann and Dr. David Crapps during the author's research efforts in the topics described in this paper.

REFERENCES

American Society for Testing and Materials (2007). “Standard test method high-strain dynamic testing of piles.” D 4945-00, *Annual Book of ASTM Standards.* ASTM, Philadelphia, PA, 4:08.

Axelsson, G. (1998a). “Long term set-up of driven piles in non-cohesive.” *Licentiate Thesis.* Department of Civil and Environmental Engineering, Royal Institute of Technology, Stockholm, Sweden.

Axelsson, G. (1998b). “Long term set-up of driven piles in non-cohesive soils evaluated from dynamic tests on penetration rods.” *Proceedings of the First International Conference on Site Characterization.* P.K. Robertson and P.W. Mayne, eds., Balkema, Brookfield, VT, 2, 895-900.

Bullock, P.J. (1999). "Pile friction freeze: A field and laboratory study." Ph.D. Dissertation. Department of Civil Engineering. University of Florida. Gainesville, FL.

Bullock, P.J., Schmertmann, J.H., McVay, M.C., and Townsend, F.C. (2005). "Side Shear Setup. I: Test Piles Driven In Florida." *Journal of Geotechnical and Geoenvironmental Engineering.* ASCE, Reston, VA , 131(3), 292-300.

Bullock, P.J., Schmertmann, J.H., McVay, M.C., and Townsend, F.C. (2005). "Side Shear Setup. II: Results from Florida Test Piles." *Journal of Geotechnical and Geoenvironmental Engineering.* ASCE, Reston, VA , 131(3), 301-310.

Bullock, P.J. and Schmertmann, J.H. (2003). “Determining the effect of stage testing on the dimensionless pile side shear setup factor.” Research Report BC354, RPWO#27, Florida Department of Transportation.

Chow, F.C., Jardine, R.J., Brucy, F., and Nauroy, J.F. (1998). “Effects of time on capacity of pipe piles in dense marine sand.” *Journal of Geotechnical and Geoenvironmental Engineering.* ASCE, Reston, VA, 124(3), 254-264.

Fellenius, B.H., Riker, R.E., O’Brien, A.J.O., and Tracy, G.R. (1989). “Dynamic and static testing in soil exhibiting setup.” *Journal of Geotechnical Engineering.* ASCE, Reston, VA , 115(7), 984-1001.

Hannigan, P.J., Goble, G.G., Likins, G.E., and Rausche, F. (2006). “Design and construction of driven pile foundations, Volume 2.” Report No. FHWA-NHI-05-043, Federal Highway Administration, McLean, VA.

Karlsrud, K. and Haugen, T. (1985). “Axial static capacity of steel model piles in overconsolidated clays.” *Proceedings of the Eleventh International Conference on Soil Mechanics and Foundation Engineering.* Balkema, Brookfield, VT, 3, 1401-1406.

Kehoe, S.P. (1989). "An analysis of time effects on the bearing capacity of driven piles." *Master's Report.* Department of Civil Engineering, University of Florida, Gainesville, FL.

Likins, G. E. and Rausche, F. (2004). "Correlation of CAPWAP with static load tests." *Proceedings of the Seventh International Conference on the Application of Stresswave Theory to Piles 2004*, Balkema, Brookfield, VT, 153-165.

Likins, G., Rausche, F., and Goble, G. (2000). "High strain dynamic pile testing, equipment and practice." *Proceedings of the Sixth International Conference on the Application of Stresswave Theory to Piles 2000*, Balkema, Brookfield, VT, 327-333.

Lukas, R.G. and Bushell, T.D. (1989). "Contribution of pile freeze to pile capacity." *Proceedings of the Congress: Current Principles and Practice.* Kulhaway, F.H., editor, ASCE, Reston, VA, 2, 991-1001.

Miller, G.A. (1994). "Behavior of Displacement Piles in an Overconsolidated Clay." PhD Dissertation. Department of Civil and Environmental Engineering, University of Massachusetts Amherst, Amherst, Massachusetts.

Rausche, F., Thendean, G., Abou-matar, H., Likins, G., and Goble, G. (1996). "Determination of pile driveability and capacity from penetration tests." FHWA Contract No. DTFH61-91-C-00047, Federal Highway Administration, McLean, VA.

Schmertmann, J.H. (1991). "The mechanical aging of soils (the twenty-fifth Karl Terzaghi Lecture)." *Journal of Geotechnical Engineering.* ASCE, Reston, VA , 117(9), 1285-1330.

Seidel, J.P., Haustorfer, I.J., and Plesiotis, S. (1988). "Comparison of static and dynamic testing for piles founded into limestone." *Proceedings of the Third International Conference on the Application of Stress-Wave Theory to Piles.* Fellenius, B.G., editor, BiTech Publishers, Vancouver, BC, 717-723.

Skov, R., and Denver, H. (1988). "Time-dependence of bearing capacity of piles." *Proceedings of the Third International Conference on the Application of Stress-Wave Theory to Piles.* Fellenius, B.G., editor, BiTech Publishers, Vancouver, BC, 879-888.

Tavenas, F. and Audy, R. (1972). "Limitations of the driving formulas for predicting the bearing capacities of piles in sand." *Canadian Geotechnical Journal.* National Research Council of Canada, Ottawa, Ontario, 9(1), 47-62.

Downdrag in Pile Design: The Positive Aspects of Negative Skin Friction

Gary L. Kuhns[1], M. ASCE, P.E.

[1]Chief Engineer, Geotechnical and Environmental Consultants, Inc., 1230 East Hillcrest Street, Orlando, Florida 32803, glkuhns@g-e-c.com

ABSTRACT: Downdrag is pile settlement in response to settlement of the surrounding soil. The bearing capacity approach to downdrag leads to erroneous design calculations and the implementation of unnecessary construction mitigation techniques. Pile settlement calculations are required to properly evaluate downdrag, but geotechnical engineers have been reluctant to use settlement calculations in pile design. When viewed from a settlement perspective, it becomes clear that downdrag is a very common phenomenon that all pile foundations experience to some degree. In fact, downdrag can limit structure settlement by pre-compressing the pile and the soil at the pile tip prior to application of the structure load. Spreadsheet calculations using Florida Department of Transportation (FDOT) FB-Deep and FB-MultiPier computations for skin friction, end bearing and pile tip movement are presented to illustrate the effect of soil settlement on pile settlement, structural integrity and bearing capacity in numerical and graphical formats.

KEY WORDS: downdrag, negative skin friction, neutral plane, pile settlement, soil settlement

INTRODUCTION

Downdrag is pile settlement in response to settlement of the surrounding soil. A common case of pile downdrag is the soil settlement caused by the placement of fill at a bridge abutment after the piles have been driven. Since the soil is more compressible than the pile (3 orders of magnitude for concrete piles) the soil along the upper portion of the pile settles more than the pile and creates a downward force, known as

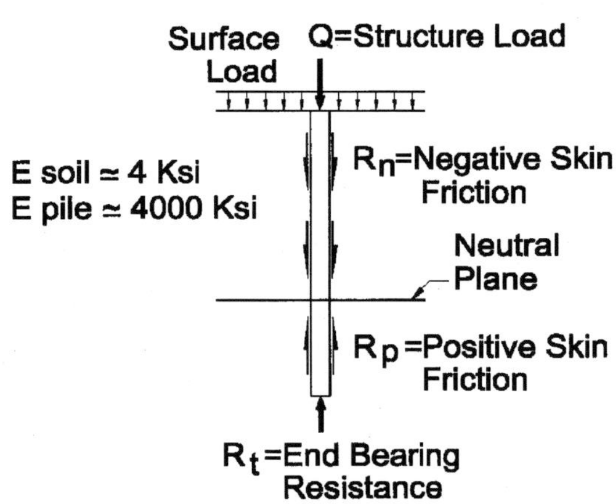

FIG. 1. Basic downdrag concept

negative skin friction (NSF), as shown in Figure 1. The point along the pile where the pile and soil settlement are equal is called the neutral plane. The settlement of the pile head is then equal to the settlement of the neutral plane plus the pile compression above the neutral plane. This is in comparison to a pile without downdrag where pile head settlement is equal to the pile tip settlement plus pile compression. Soil settlement can cause pile settlement, but as will be discussed in the following section, the bearing capacity approach is a poor model for evaluating the effects of downdrag.

English units are used throughout this paper because several key references (FDOT Structures Design Guidelines, 2007, FB-MultiPier, 2007 and FB-Deep, 2007) use English units as does the spreadsheet used to develop the example analysis at the end of the paper. English to SI unit conversions are provided in the **NOTATION** section.

CURRENT PILE DESIGN PRACTICE

Pile foundations resist structure loads through a combination of skin friction resistance and end resistance as shown in Figure 2. The allowable structure load (Q) is defined as,

$$Q = \frac{\left(R_{p\max} + R_{t\max}\right)}{FS} \qquad (1)$$

Where,

R_{pmax} = Ultimate skin friction resistance
R_{tmax} = Ultimate end bearing resistance
FS = Factor of safety

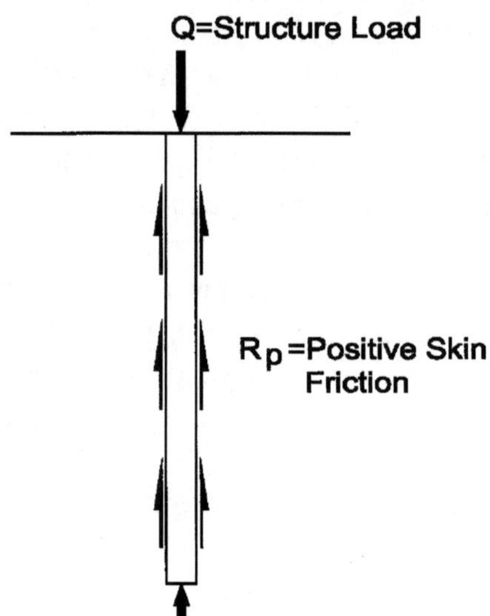

FIG. 2. Bearing capacity approach to pile design.

As a structure load is applied, the pile mobilizes skin friction and end resistance by moving downward, or settling. Very little pile movement is needed to mobilize the maximum skin friction resistance, typically approximately 0.15 inches (4 mm). Much more movement is needed to mobilize significant end bearing resistance in most soils.

A typical pile load-settlement curve is shown on Figure 3. The curve is plotted with the Davisson Offset Limit Criteria to determine maximum resistance. As shown, the Davisson maximum capacity is 250 tons (2,224 kN), and using a factor of safety of 2.5 the allowable service load would be 100 tons (890 kN). Since this allowable load is considerably less than the maximum skin friction resistance of about 200 tons (1779 kN), individual pile settlement is negligible (although pile group settlement should also be checked). This "bearing capacity approach" is commonly used in pile design and pile settlement calculations are often not performed. However, when the soil profile settles with respect to the pile, this methodology cannot be relied upon to evaluate the impact of downdrag on pile performance.

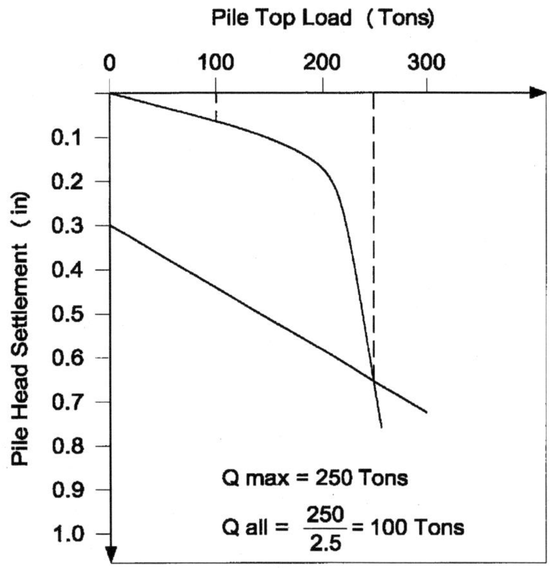

FIG. 3. Typical pile load settlement curve to determine allowable pile load

The Florida Department of Transportation (FDOT) *Structures Design Guidelines (SDG)* (2007) provides the following equation for calculating the required driving resistance (RDR) (which includes structure load and soil resistance safety factors) using load and resistance factor design (LRFD) methodology:

$$RDR = \frac{(FPDL + 2 \times NSF)}{0.65} \tag{2}$$

Where,

FPDL = Factored permanent design load (excludes transient loads)
NSF = Negative skin friction

In this approach, up to 2 times the skin friction above the neutral plane is added to the structure load to account for the loss of positive skin friction resistance. The SDG recommends a maximum RDR of 300 tons (2,669 kN) as a practical limit for pile driving equipment for an 18-inch (457 mm) wide square concrete pile. Using this 300-ton (2669-kN) limit, for a negative skin friction load of about 100 tons (890 kN), Equation 1 would indicate that the pile would not be able to support any structure load at all.

The bearing capacity approach proposed in the *National Cooperative Highway Research Program Report 393* (Briaud and Tucker, 1997) is similar, but incorporates a load factor of 1.7 for downdrag that results in the following equation (when rearranged in a manner similar to the FDOT formula):

$$RDR = \frac{(FPDL + 2.35 \times NSF)}{0.75} \tag{3}$$

Using the recommended FDOT 300-ton (2669-kN) limit for RDR in this relationship also reveals that for a pile subjected to about 100 tons (890 kN) of negative skin friction, the allowable structure dead load would be zero. A NSF load of 100 tons (890 kN) can occur when just 20 vertical feet (6.1 m) of silty sand, with an SPT blow count of 20, settles with respect to the pile.

The bearing capacity-based methodology presented in the USDOT FHWA Manual

Design and Construction of Driven Pile Foundations (Hannigan, P.J., et al) is essentially the same as the FDOT and NCHRP Report 393 procedures. However, it is worth noting that the *American Association of State Highway and Transportation Officials (AASHTO) LRFD Bridge Design Specifications 2004* state that downdrag loads are only a consideration for the structural capacity of the pile, not soil bearing capacity, and "downdrag may be regarded as a settlement issue."

It is evident that applying the bearing capacity concept to negative skin friction results in severe limitations for the use of piles in a settling soil. Since this is not a reasonable conclusion, it is revealed that the bearing capacity model is a poor analytical tool for evaluating negative skin friction effects. This traditional concept treats NSF like a structure load, *but a NSF load is not like a structure load* – its magnitude changes with pile settlement. A settlement-based approach is needed to correctly evaluate the impact of a settling soil on pile performance.

PILE LOAD-SETTLEMENT BEHAVIOR

A typical pile load transfer diagram is shown in Figure 4. The structure load is shown at the top of the pile, and the transfer curve indicates that the load in the pile is reduced with depth as load is transferred to the soil through skin friction. The load at the pile tip is resisted by the soil below the pile, or end bearing resistance. The maximum skin friction resistance is mobilized in most soils with a very small pile movement of approximately 0.15 inches (3.4 mm) as shown in the Figure 5 example.

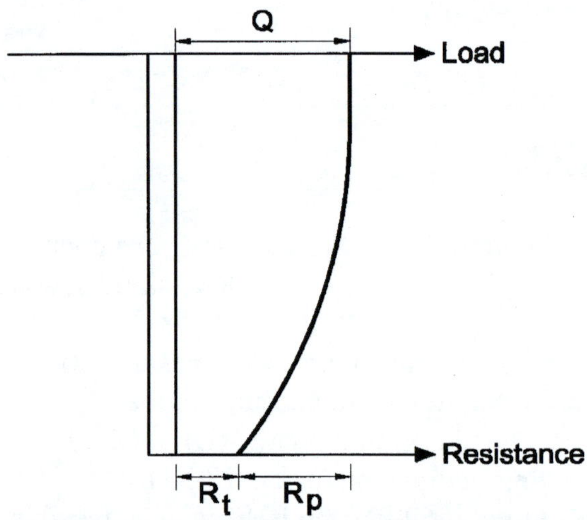

FIG. 4. Pile load transfer diagram

When the maximum skin friction resistance is achieved, skin resistance is typically considered constant for additional pile movement as shown by the solid vertical line in Figure 5. However, testing has shown that skin friction tends to decrease with continuing pile movement, or "strain softening" occurs as indicated by the dashed line in Figure 5 (Fellenius, 2004).

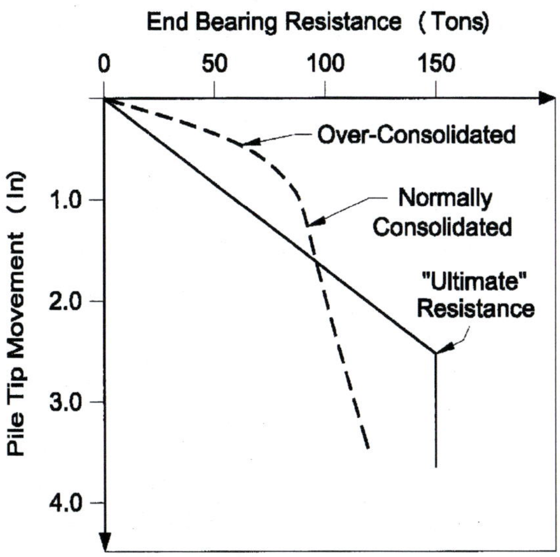

FIG. 5. **Pile movement versus friction resistance**

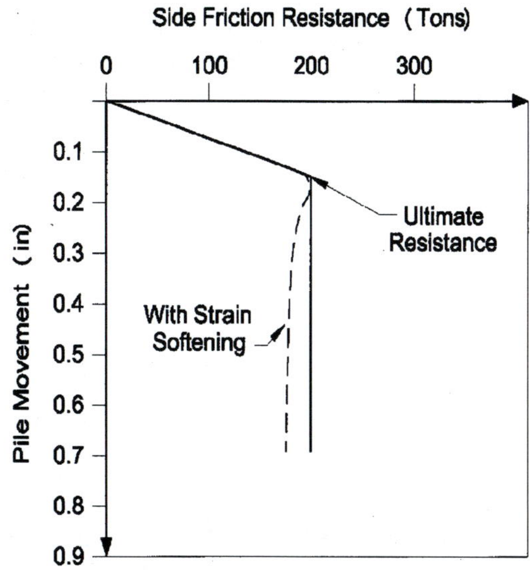

FIG. 6. Pile tip movement skin versus end bearing resistance

To mobilize end bearing resistance, more pile movement is usually required. It is widely believed that end bearing resistance increases at a roughly constant rate to an "ultimate" capacity, where a plunging failure occurs as shown by the solid line in Figure 6. However, field testing of foundations has shown that end bearing resistance often continues to increase with increasing pile movement, especially in non-cohesive soils. Also, end bearing resistance is found to consist of both over-consolidated and normally consolidated settlement behavior as illustrated by the dashed line in Figure 6.

Figure 7 illustrates the combined effects of strain softening and normally consolidated pile settlement behavior. Although end bearing resistance continues to increase with additional pile movement, the combination of strain softening and normally consolidated pile tip settlement gives the appearance of a much steeper load-resistance curve, suggesting something close to a plunging–type failure. However, even with these combined effects, pile load resistance still increases slightly with additional pile movement and a distinct "failure" point is not evident in many cases.

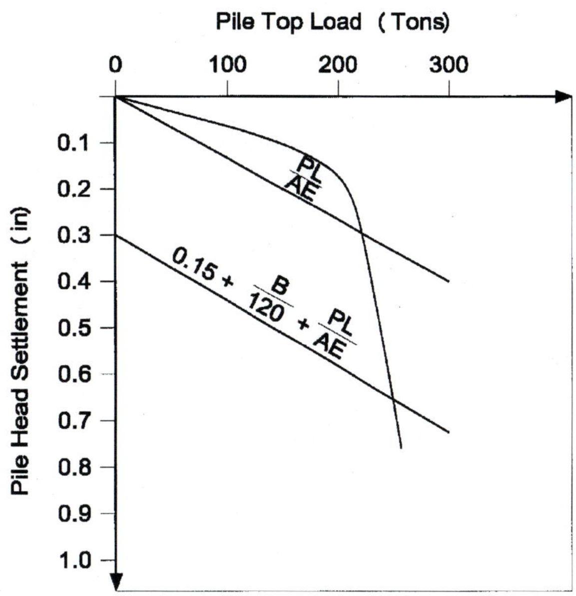

FIG 7. Pile load settlement curve with Davisson criteria

To establish an allowable structure load based on the load-settlement curve obtained from a pile load test, offset limit criteria have been developed. The most commonly used criterion in the United States is the Davisson Offset Limit Criteria, which is defined as follows:

$$\textit{Pile Head Settlement} = 0.15 + \frac{B}{120} + \frac{PL}{AE} \quad (4)$$

Where,

B = Pile width, inches (mm)
PL/AE = Pile compression, inches (mm)

The Davisson Offset Limit Criteria for a load test performed on an 18-inch (457 mm) square concrete pile is shown on Figure 7. For this pile, so-called failure would occur at a top load of 250 tons (2,224 kN) and a pile head settlement of 0.65 inches (17 mm). Pile tip movement at the Davisson failure point is a mere 0.3 inches (8 mm) and 0.35 inches (9 mm) of the pile head settlement is attributed to pile compression. This example illustrates that the Davisson offset limit criteria is very settlement restrictive and that pile "failure" occurs at a very small tip movement (in this case 0.3 inches (8 mm)). At a working load of 100 tons (890 kN), pile head settlement would be about .05 inches (1 mm) , a negligible amount.

Plunging failure of a pile at an "ultimate" pile load is a theoretical condition that does not reflect actual pile behavior, except possibly in the case of very weak soil at the pile tip, an unusual condition for driven piles. Pile end bearing resistance continues to increase with increasing pile load, well beyond the small amount of tip movement used to determine the "ultimate" pile load from a top-loaded pile load test.

This discussion is important since the bearing capacity approach implies that proper pile design is based on calculating an allowable load that achieves an adequate factor of safety against a plunging-type bearing failure. In fact, pile design methodology such as FDOT's FB-Deep computer program (see FDOT FB-Deep, 2007) has been developed, based on correlation with static load tests, to calculate end bearing resistances compatible with the small amount of tip movement allowed by the Davisson criteria, not plunging failure. This methodology assures acceptable pile settlement performance without the need to calculate pile settlement and is well-established in engineering practice. However, in the case of a pile subjected to soil settlement, application of the bearing capacity approach has led to the mistaken belief that negative skin friction forces should be included in bearing capacity-type calculations.

PILE SETTLEMENT WITH SOIL SETTLEMENT

Load transfer and settlement diagrams similar to those depicted in this section can be found in several published papers that address downdrag (e.g., Fellenius, 1989 and Briaud and Tucker, 1997). Consider the illustration shown in Figure 8, which depicts a pile that undergoes soil settlement due to embankment construction. The dashed line indicates the load transfer if soil settlement did not occur. The solid line indicates load transfer with soil settlement. The "neutral plane" is located at the depth where the soil and pile settlements are equal. Also, note the increase in end resistance (R_t) due to the NSF load.

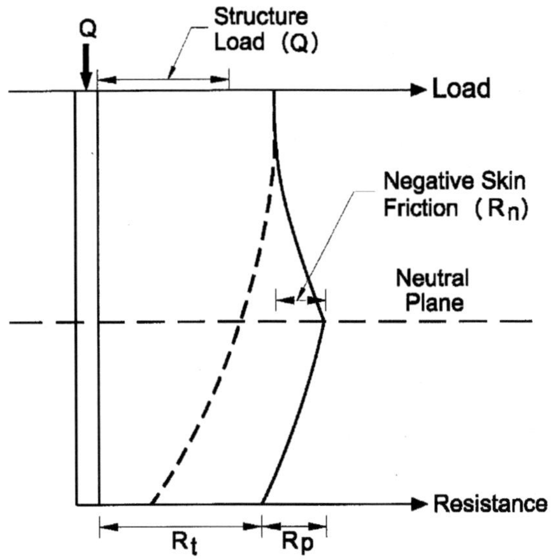

FIG. 8. Load transfer with downdrag

Figure 9 presents the pile settlement diagram including soil settlement. Note that the pile tip settlement is equal to the soil settlement at the pile tip prior to application of any structure or downdrag loads. Again, the neutral plane is located at the point where pile and soil settlements are equal after structure and downdrag loads have been applied. Pile head settlement has been increased by the additional pile compression and additional penetration of the pile tip. Pile head settlement is equal to the settlement of the neutral plane plus pile compression above the neutral plane. The load transfer and soil settlement relationships for this case are combined in Figure 10.

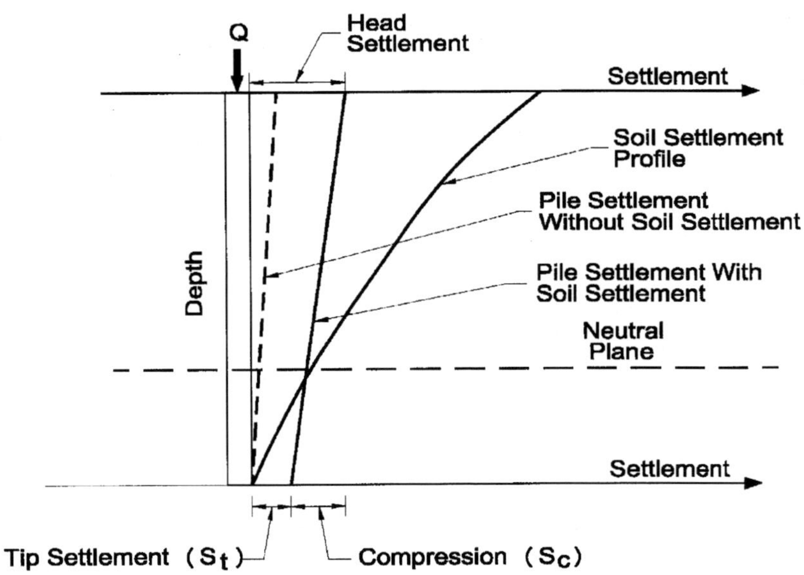

FIG. 9. Pile settlement diagram with downdrag

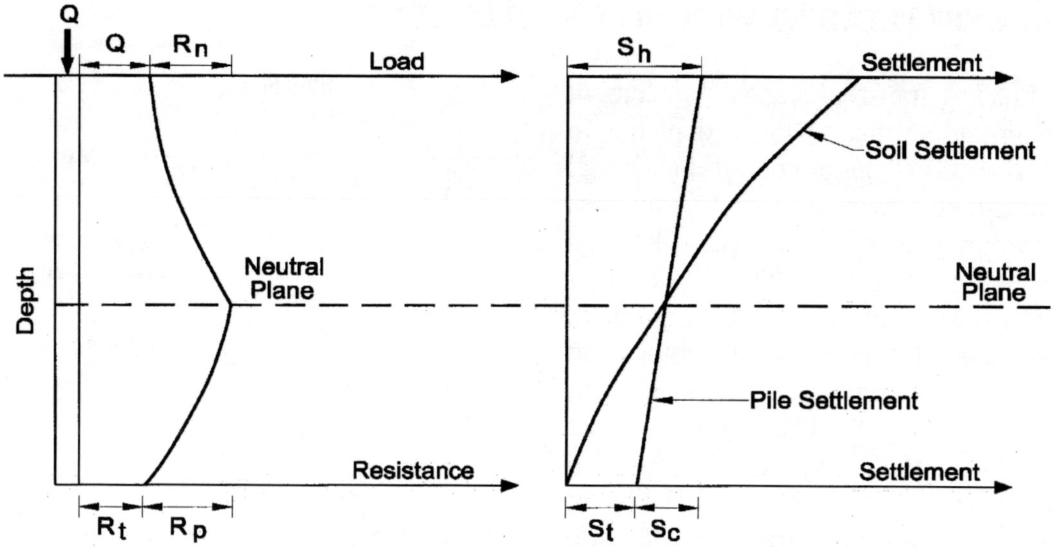

FIG. 10. Combined load transfer and settlement diagrams

In order to visualize the impact of soil settlement on piles, consider the idealized condition of a pile with no end bearing resistance and no structure load as shown in Figure 11. When the surrounding soil settles, the NSF load (R_n) is resisted only by positive skin friction (R_p). For the case of a uniform soil, with skin friction increasing from 0 at the ground surface, the neutral plane will occur at a depth of 2/3 of the pile length below ground surface, if there is no end bearing resistance or structure load. The pile head settlement will equal the soil settlement at the neutral plane plus the pile compression above the neutral plane. The importance of this idealized case is to demonstrate that even if end bearing resistance is zero, NSF cannot cause a bearing capacity failure, it only causes settlement of the pile with the soil at the neutral plane. Figure 12 demonstrates this situation with end bearing resistance, which results in a lowering of the neutral plane and less pile settlement. Figure 13 shows the addition of a structure load to this scenario that causes the neutral plane to occur at a higher elevation and pile settlement increases.

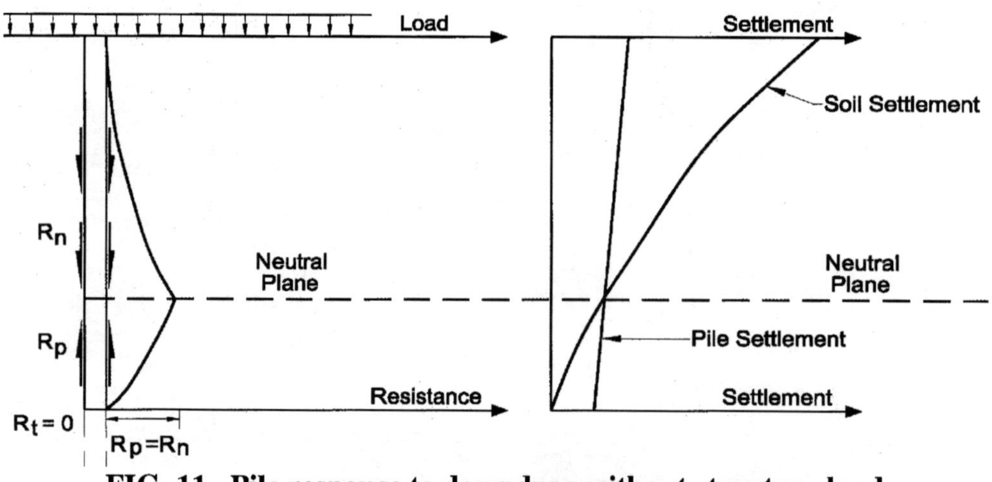

FIG. 11. Pile response to downdrag without structure load or end bearing resistance

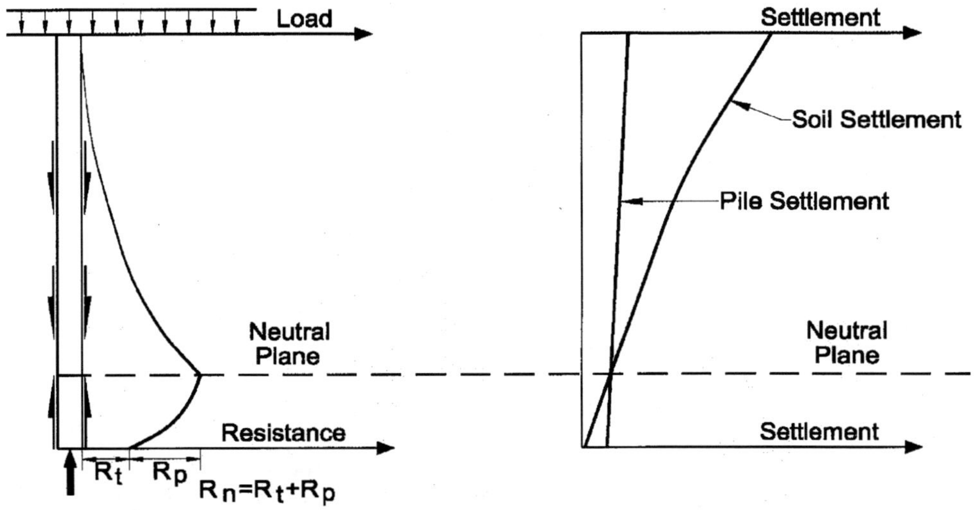

FIG. 12. Pile response to downdrag with end bearing resistance

Review of the pile behavior illustrated in Figures 11 through 13 indicates that the response of a pile to soil settlement is pile settlement, and the magnitude of pile settlement is determined by the depth of the neutral plane. End bearing resistance reduces pile settlement by lowering the neutral plane, but is not needed to resist NSF load. An NSF load cannot contribute to a so-called "plunging" failure of the pile since that can only occur under a structure load sufficient to cause the pile to move downward with respect to the soil along its entire length, which precludes the possibility of NSF forces.

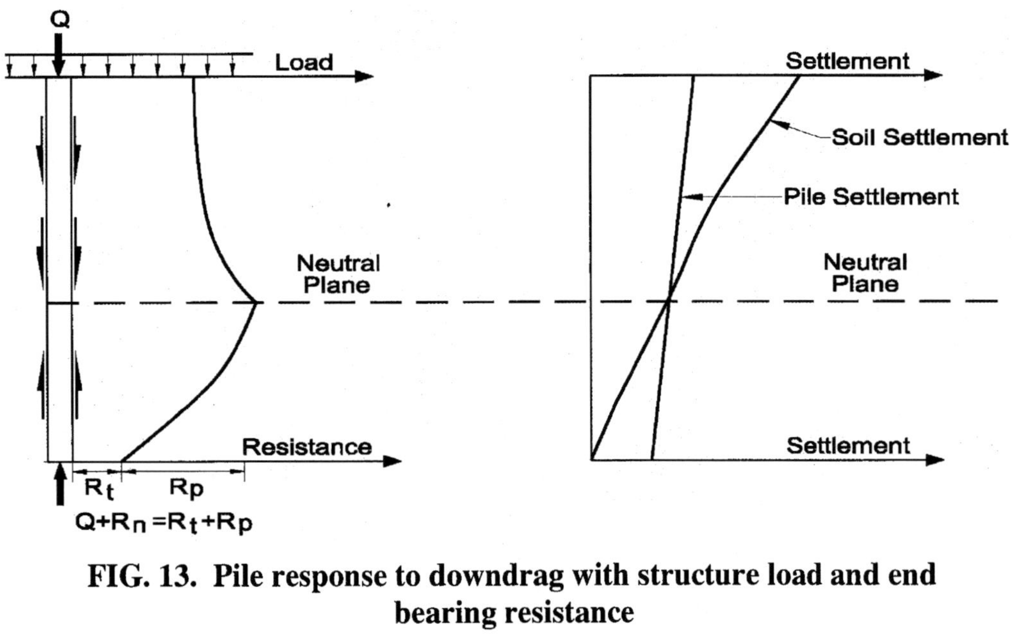

FIG. 13. Pile response to downdrag with structure load and end bearing resistance

Based on this review, pile downdrag must be evaluated from the standpoint of pile settlement, not bearing capacity. For a settling soil profile, a pile settlement analysis is needed to verify that pile head settlement remains within tolerable limits. Since downdrag increases the maximum load in the pile, pile structural integrity must also be considered. With respect to bearing capacity, the traditional bearing capacity model can be used to determine the allowable structure load applied at the pile head and NSF load should not be included in those calculations.

LIVE LOAD, RESIDUAL LOAD AND NSF LOAD

It is generally recognized that pile live load and NSF load do not occur simultaneously. Live load causes downward movement of the pile that relieves at least an equal amount of NSF load. That is why the FDOT and NCHRP Report 393 equations for RDR with downdrag do not include transient live load.

Residual load can remain in the pile after driving and would appear on a load-transfer diagram as shown in Figure 14a (Briaud and Tucker, 1997). Residual loads are due to compressive pile stresses induced during pile driving that are not released, as well as reconsolidation and settlement of the liquefied soil around the pile after pile driving stops. It is interesting that residual loads cause the same load-transfer characteristics as NSF loads, and have been observed in piles that were instrumented to evaluate downdrag forces.

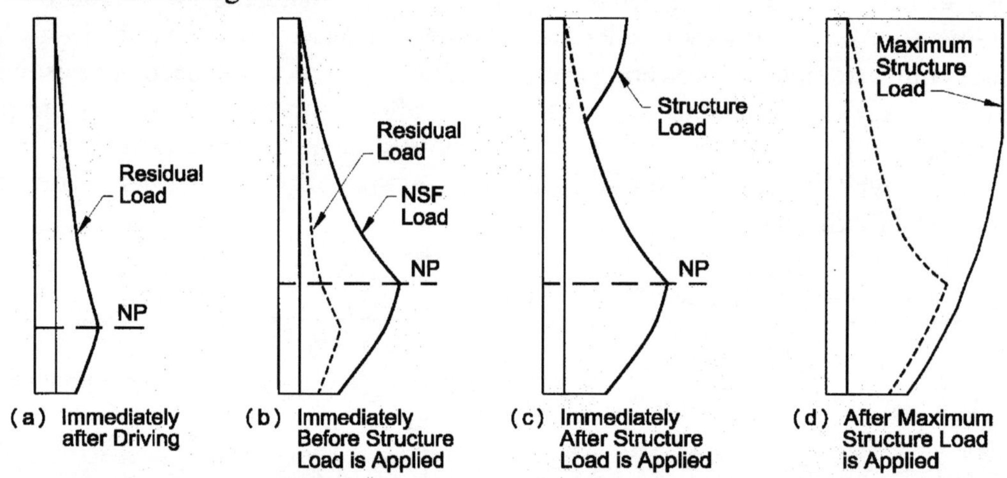

FIG. 14. Progression of load distribution in a pile when downdrag occurs before structure loading. Adapted from Briaud and Tucker (1997)

Figure 14 illustrates the progression of pile load distribution through four phases of pile loading when downdrag occurs before the structure load is applied: (a) residual load immediately after pile driving, (b) NSF load prior to application of structure load, (c) structure and NSF load immediately after a "small" structure load is applied, and (d) load transfer when a "large" structure load is applied. The settlement caused by application of load at the top of the pile (Figures 14c and 14d) will be limited since the pile and the soil below the pile tip have already been compressed by the NSF load. In effect, the pile has been "stiffened" against settlement by the downdrag load.

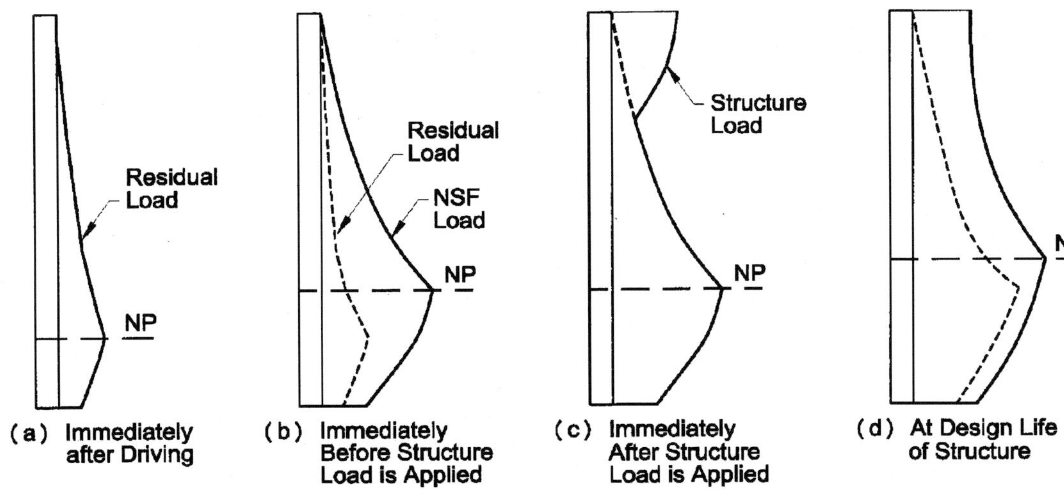

FIG. 15. Progression of load distribution in a pile when downdrag occurs before and after structure loading. From Briaud and Tucker (1997)

Figure 15 illustrates the progression of pile load distribution through four phases of pile loading when downdrag occurs before, *and after*, the structure load is applied: (a) residual load immediately after pile driving, (b) NSF load prior to application of structure load, (c) structure and NSF load immediately after structure load is applied, and (d) load transfer at the design life of the structure (Briaud and Tucker, 1997). In this case, the settlement caused by application of load at the top of the pile, and soil settlement that occurs after the load is applied (15c and 15d), will be limited since the pile and the soil below the pile tip have already been compressed by the NSF load. The final settlement of the pile will be determined by the load distribution in Figure 15d; however, the distribution shown in Figure 15b occurs prior to structure loading and will have no negative impact on the structure.

The most common case of pile downdrag is the construction of a bridge abutment fill after the piles are driven. In this case, NSF load is generated in the pile due to settlement of the soil profile before the pile cap is poured, since abutment fill construction can require several weeks or months. Therefore, all of the soil settlement that occurs prior to pouring the pile cap, will "stiffen" the pile response and limit the settlement that occurs after the structure load (i.e., bridge beams) is applied. Long-term soil settlement that occurs after the structure load is applied will cause additional pile settlement, or downdrag.

The pile head settlement that occurs after the structure load is applied can be determined by first calculating the pile compression and the soil compression at the pile tip that occurs due to downdrag before the structure load is applied. Then the pile head settlement caused by the structure load and the soil settlement that occurs after the structure load is applied is calculated. Finally the pile and soil compression that occurred prior to the structure load is subtracted to determine the pile head settlement that will impact the structure.

DOWNDRAG CALCULATIONS

The first step in analyzing NSF is to establish the depth of the neutral plane. The neutral plane depth is calculated by satisfying both force and settlement equilibrium. When the depth of the neutral plane is calculated, all pile loads and settlements can be determined. The following procedure is recommended to establish the neutral plane:

1. Determine pile type, dead load and length
2. Determine soil settlement profile
3. Assume a neutral plane depth
4. Calculate negative and positive skin resistances
5. Calculate toe resistance
6. Determine pile tip penetration
7. Determine pile settlement at assumed neutral plane
8. Iterate until pile settlement equals soil settlement at assumed neutral plane
9. Settlement of neutral plane plus pile compression above neutral plane equals pile head settlement

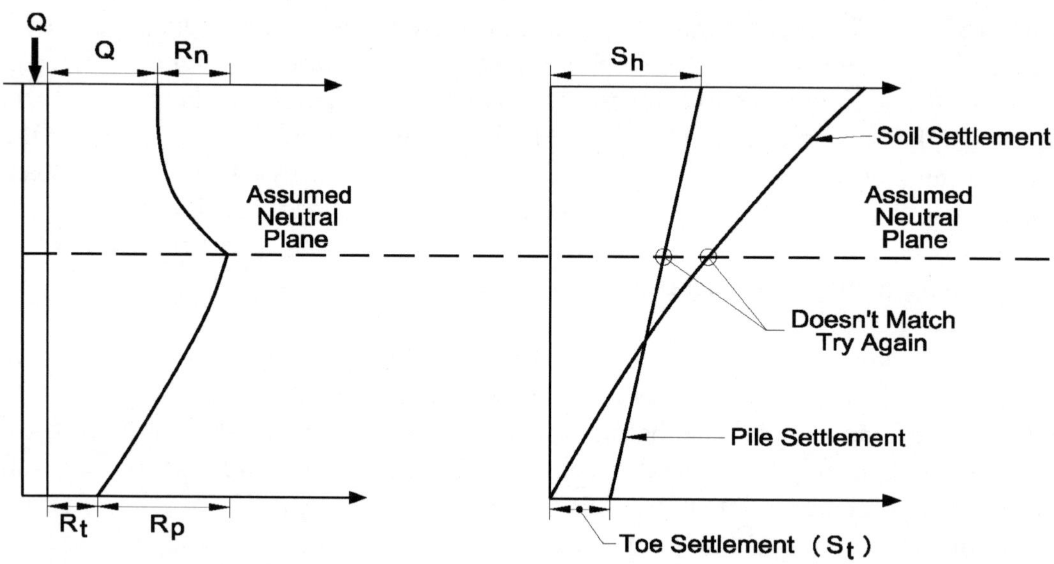

FIG 16. Downdrag Calculation Procedure

The above procedure is illustrated in Figure 16. An Excel "in-house" spreadsheet, SETTLE-DOWN, was developed by the author and Rachel Andre, P.E., to facilitate settlement calculations for concrete piles. The spreadsheet uses the same computations included in the FDOT FB-Deep (2007) and FB-MultiPier (2007) programs to calculate skin friction, end bearing and pile tip movement versus load behavior. The FB-MultiPier (2007) equation for pile tip movement is as follows:

$$S_t = \frac{Q_b(1-\nu)}{4rG\left(1-\frac{Q_b}{Q_f}\right)^2}$$

Q_b = End Bearing Resistance, tons, (kN)
Q_f = Ultimate Tip Resistance, tons, (kN)
r = Pile Radius or width, ft (m)
G = Shear Modulus, tsf (kPa)
< = Poisson's Ratio (Range: 0.3 to 0.45)
S_t = Tip Penetration, ft (m)

Figure 17 presents pile tip movement versus pile load for an 18-inch square concrete pile. The soil material at the pile tip is FDOT Soil Type 2 (silty sand) with SPT blow counts ranging from 20 to 100. The Davisson tip movement criterion of 0.3 inches (8 mm) is shown on the diagram, as well as the FB-Deep "ultimate" end bearing resistance for each SPT blow count. The FB-Deep program divides the calculated "ultimate" end bearing resistance by 3 to determine the Davisson end bearing capacity based on correlations with field load test data. It is interesting to note that the end bearing resistances shown in Figure 17 at 0.3 inches (8 mm) are roughly 1/3 of the ultimate resistances. In addition, the curves also approximate over-consolidated and normally consolidated settlement behavior, as previously illustrated in Figure 6.

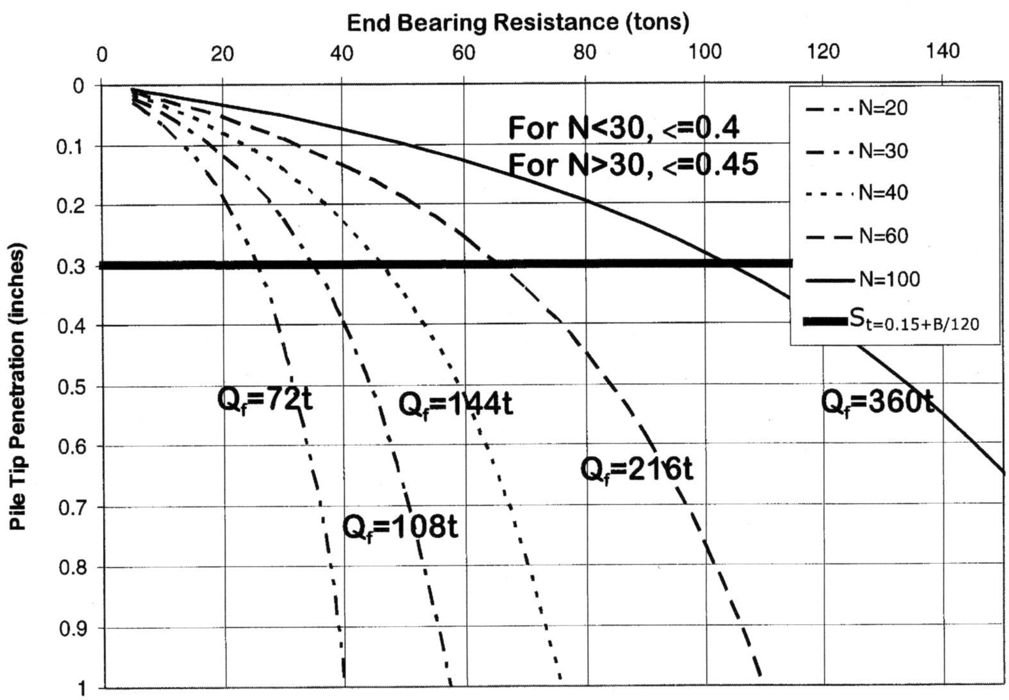

FIG. 17. Plot of tip penetration (in) vs. end bearing resistance (tons) for 18-in. PPC pile bearing in Soil Type 2 (Clay-Silt-Sand Mixture)

Case 1 –Pile Head Settlement before Structure Load is Applied

A pile foundation situation common to Central Florida is shown in Figure 18. A 25-foot (7.6-m) thick sand fill is constructed after piles are driven at a bridge abutment. The subsurface profile is comprised of a 40–foot (12.2-m) thick surface layer of medium dense sand, and a 10-foot (3.1-m) thick layer of soft sandy clay underlain by medium dense to dense sand. The pile length is 100 feet (30.5 m), or 75 feet (22.9 m) below natural ground surface, and the 120 ton (1068 kN) design load includes 100 tons (890 kN) of dead load and 20 tons (178 kN) of live load. The total short term soil settlement (before application of structure load) is 4 inches (102 mm) and is distributed between the soil layers as shown on Figure 18. It is assumed that the standard practice of wrapping the piles with plastic sheeting is effective in eliminating NSF load in the embankment sand.

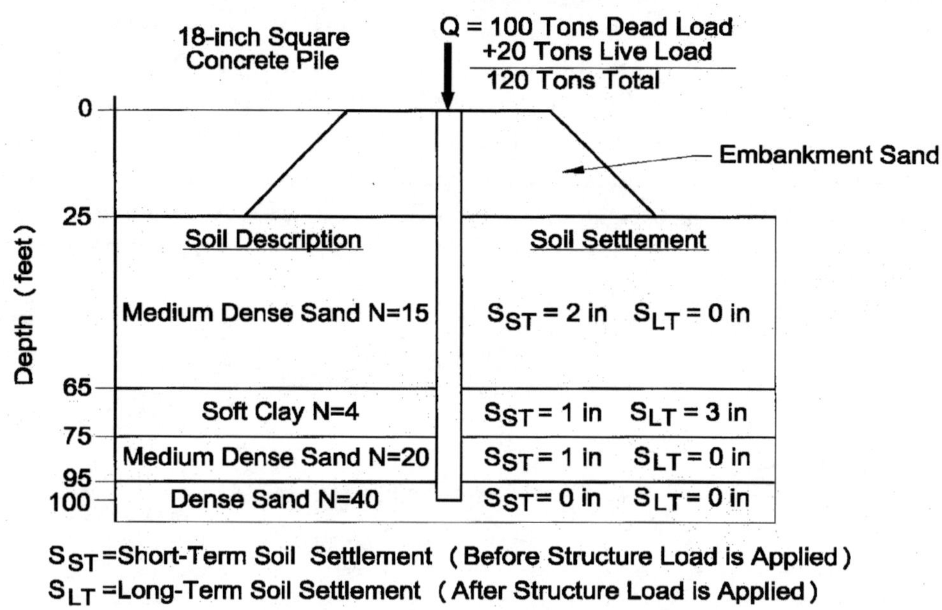

FIG 18. Soil and pile data for Case 1 and Case 2 example calculations

The SETTLE-DOWN graphical results are shown on Figure 19. The neutral plane is at 83 feet (25.3 m) deep and the negative skin friction load is 194 tons (1726 kN). Pile tip penetration is about 0.45 in (11 mm), which is greater than the 0.3 in (8 mm) Davisson criterion. Total pile head settlement is 0.63 inches (16 mm), but that would occur prior to pouring the pile cap, and would have no impact on the bridge structure.

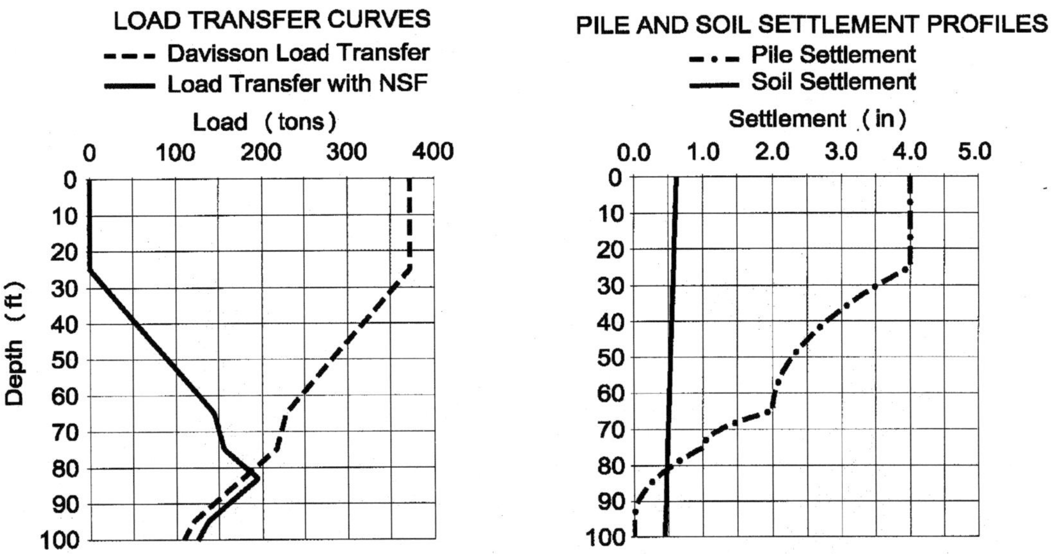

FIG 19. SETTLE-DOWN load transfer and settlement diagrams for Case 1

Case 2 – Pile Head Settlement after Structure Load is Applied

As shown on Figure 18, the long-term settlement of the clay layer is 3 in (76 mm), and no significant long-term settlement of the sand layers is expected. The SETTLE-DOWN graphic results for this case are shown on Figure 20. The neutral plane moves up to 71 feet (21.7 m), and the pile head settlement increases to 0.87 inches (21 mm).

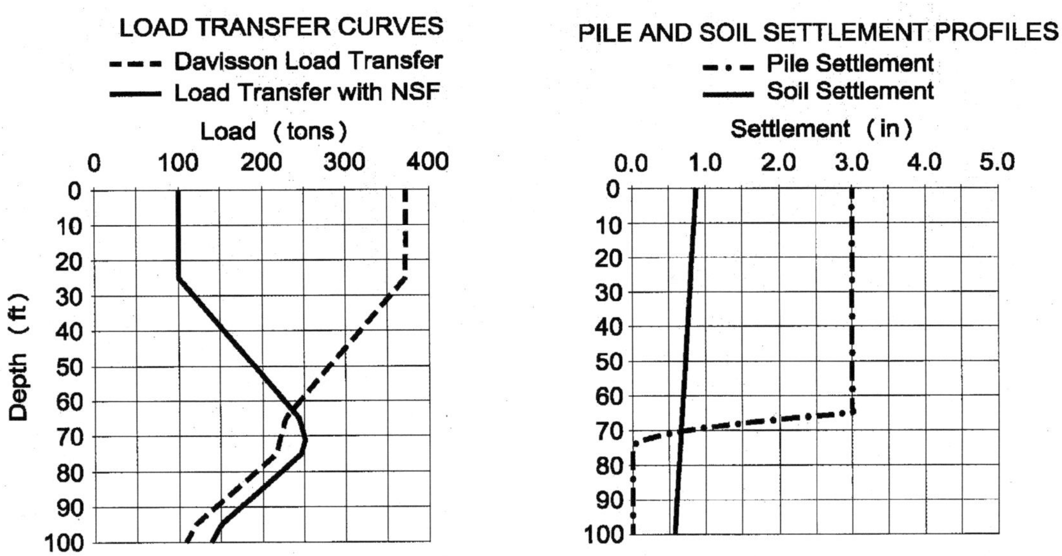

FIG 20. SETTLE-DOWN load transfer and settlement diagrams for Case 2

To determine the pile head settlement that will impact the structure, the previous pile compression and pile tip penetration from Case 1 are entered into the SETTLE-DOWN spreadsheet as shown on Figure 21. This reduces the pile head settlement after the load is applied to 0.24 inches (6 mm), a tolerable amount. Note that if the NSF load of 151 tons (1343 kN) were used in the FDOT equation to determine RDR, the pile would have to be driven to an RDR of 465 tons (1343 kN), ostensibly to provide an adequate factor of safety against bearing failure due to NSF load alone (no structure load included).

SETTLE-DOWN
Deep Foundation Settlement Analysis and Downdrag
For Square Precast Concrete Piles
By
Gary Kuhns, P.E. and Rachel Andre, P.E.

PROJECT INFORMATION

Project Name:	GMEC Presentation
Project Number:	14-Apr-05
Engineer:	glk
Run Description:	Pile Head Settlement After Structure Load

PILE DESIGN INFORMATION

Pile Length (ft) =	100	Pile Dead Load (tons) =	100
Pile Width (ft) =	1.5	Pile Live Load (tons) =	20
Pile Elastic Modulus (ksi) =	4415	Total Structure Load (tons) =	120

SOIL CONDITIONS AT PILE TIP

Soil Type =	4	Soil Shear Modulus (G, tsf) =	1440
Blow Count =	40	"Ultimate" Tip Resistance (tons) =	324
Poisson's Ratio (v) =	0.45		

SOIL LAYER INFORMATION

Layer Number	Layer Bottom Depth (ft)	Soil Type	Blow Count (N)	Soil Description	Soil Settlement (in)
1	25	2	0	Embankment Sand	0
2	65	2	15	Sand	0
3	75	1	4	Sandy Clay	3
4	95	2	20	Sand	0
5	120	4	40	Sand	0
				Total =	**3**

SETTLEMENT AND STRUCTURAL INTEGRITY

Neutral Plane (NP) Depth (ft) =	71	Pile Tip Resistance (tons) =	138
Negative Side Friction Load (tons) =	151	Pile Tip Penetration (in) =	0.570
Maximum Load in Pile (tons) =	251	Pile compression Below NP (in) =	0.095
Maximum Compressive Stress in Pile (psi) =	1551	Pile Settlement at NP (in)=	0.665
Positive Side Friction Resistance (tons)	113	Pile Compression Above NP (in)=	0.209
		Total Pile Compression (in)=	0.304
Pile Compression Before Load (in) =	0.18	Total Pile Head Settlement (in) =	**0.87**
Pile Tip Penetration Load (in) =	0.45	Pile Head Settlement After Load (in) =	**0.24**

BEARING CAPACITY

Total Structure Load (tons) =	120	Davisson Tip Resistance (tons)=	108
Ultimate Side Friction Resistance (tons) =	264	Davission Pile Resistance (tons)=	372
		Factor of Safety =	3.1

FIG. 21. SETTLE-DOWN calculations for pile head settlement after structure load is applied

CONCLUSIONS

The conclusions presented in the preceding sections can be summarized as follows:

1. Negative skin friction is a common phenomenon that occurs in all driven pile foundations to some degree
2. Negative skin friction load should not be included in pile bearing capacity calculations.
3. Pile settlement calculations are necessary to evaluate pile downdrag.
4. Soil settlement that occurs prior to application of the structure load will pre-compress the pile and bearing soil, and limit pile settlement after the structure load is applied.

NOTATION

Symbols

B	Pile width, ft (mm)
E_{pile}	Elastic Modulus of pile, ksi (MPa)
E_{soil}	Elastic Modulus of soil, ksi (MPa)
G	Soil shear modulus, tsf (kPa)
LRFD	Load and Resistance Factor Design
N	Standard Penetration Resistance, bpf
NSF	Negative skin friction, tons (kN)
PL/AE	Pile compression, in (mm)
Q	Structure load, tons (kN)
Q_b	End bearing resistance, tons (kN)
Q_f	Ultimate pile tip resistance, tons (kN)
r	pile radius or width, ft (mm)
RDR	Required driving resistance, tons (kN)
R_n	Negative skin friction, tons (kN)
R_p	Positive skin friction, tons (kN)
R_t	End bearing resistance, tons (kN)
S_c	Pile compression, in (mm)
S_h	Pile head settlement, in (mm)
S_t	Pile tip settlement, in (mm)
S_{LT}	Long-term soil settlement, in (mm)
S_{ST}	Short-term soil settlement, in (mm)
$<$	Poisson's ratio, dimensionless

Unit Conversions

inch = 25.4 mm
foot = 0.305 m
ksi = kips/in^2 = 6.895 MPa
ton = 2000 lbs = 8.896 kN
tsf = tons/ft^2 = 95.761 kPa

REFERENCES

Achari, G. and Joshi, R.C., (1994) "Effect of Surcharge Loading on Pile Capacity." Proceedings of International Conference on Design and Construction of Deep Foundations, December 1994, pp. 1485-1498.

American Association of State Highway and Transportation Officials (AASHTO) Load and Resistance Factor Design (LRFD) Bridge Design Specifications 2004.

Briaud, J.L., (1997). "Bitumen Selection for Reduction of Downdrag on Piles." American Society of Civil Engineers, *Journal of Geotechnical and Geoenvironmental Engineering*, December 1997, Volume 123, No. 12, pp. 1127-1134.

Briaud, J.L. and Tucker L., (1997) Design and Construction Guidelines for Downdrag on Uncoated and Bitumen-Coated Piles. Transportation Research Board. NCHRP Report 393.

Fellenius, B.H., (1989) "Unified Design of Piles and Pile Groups" Transportation Research Board, Washington, TRB Record 1169, pp. 75-82.

Fellenius, B.H., (2004) "Unified Design of Piled Foundations with Emphasis on Settlement Analysis" Proceedings of ASCE Conference Deep Foundations 2004, pp. 253-275.

Florida Department of Transportation (FDOT) Structures Design Guidelines. (2007).

Florida Department of Transportation (FDOT) FB-Deep (2007). Program to Estimate Static Axial Capacity of Drilled Shafts and Driven Piles. Available from University of Florida Bridge Software Institute.

Florida Department of Transportation (FDOT) FB-MultiPier (2007). Nonlinear Finite Element Analysis Program. Available from University of Florida Bridge Software Institute.

Hannigan, P.J., Goble G.G., Thendean, G., Likens, G.E., and Rausche, F. (1996) Design and Construction of Driven Pile Foundations – Volume 1. USDOT Federal Highway Administration.

Poulos, H.G. (1997) "Pile Subjected to Negative Friction: A Procedure for Design." *Journal of Southeast Asian Geotechnical Society.* June 1997, Volume 28, Number 1, PP 23-44.

Anatomy of Foundation Performance Involving Three Grain Silos Systematically Loaded to Impending Failure

James E. Laier, Ph.D., P.E., F. ASCE[1], Gary D.E. Cowles, P.E., M. ASCE[2] and Mark E. White, P.E., M. ASCE[3]

1. Chief Executive Officer and Principal Geotechnical Engineer, Southern Earth Sciences, Inc., Mobile, Alabama 36609 (jlaier@soearth.com)
2. Partner and Principal Structural Engineer, Cowles, Murphy, Glover and Associates, Mobile, Alabama 36602 (gcowles@cmg-a.com)
3. Materials Department Manager, Southern Earth Sciences, Inc., Mobile, Alabama, 36609 (mewhite@soearth.com)

ABSTRACT

In 2003, four grain silos, each 104 ft (32 m) in. diameter and 100 ft (31 m) tall, were constructed at a waterfront storage facility in Mobile, Alabama, to replace a cold storage facility that had been demolished. Foundation plans for the replacement silos included auger-cast piles to support the perimeter steel shell walls on a concrete ring, auger-cast piles to support the buried conveyor tunnel used to unload grain and a soil-supported interior floating floor slab (isolated from the perimeter ring and conveyor tunnel) to support grain loads on the order of 5,000 psf (240 kN/m^2). During the initial loading of Silo #1, as filling neared capacity, the structure suddenly shifted and began to lean toward the south. Large cracks appeared in the concrete foundation system and structural steel beams located at the top of the silo twisted. Because grain storage silos were in short supply in the aftermath of hurricane Katrina, the owners of the storage facility hired the authors to conduct a unique, full scale test loading program, by placing Silos #2, #3 and #4 into production. Using available historical data, along with supplemental data gathered specifically for settlement monitoring, the authors developed a full scale test loading protocol designed to help the owner temporarily store grain, while simultaneously evaluating foundation performance. This paper discusses geotechnical and structural assumptions used in design of the silo foundations. Then, using data collected from full scale loading of the silo foundations to impending failure, the authors describe lessons learned by comparing expectations of foundation performance to measured performance.

INTRODUCTION

In 1934, the Alabama State Docks Department constructed a cold storage facility along the waterfront in Mobile, Alabama. The concrete structure was supported on

14 in. (356 mm) and 18 in. (457 mm) square concrete piles bearing in the Pleistocene sand formation beginning about 45 ft (14 m) below existing ground surface at the project site.

In the mid-1970's, as part of a modernization and expansion program, the Alabama State Docks Department demolished the superstructure of the cold storage facility, leaving the remnant piles and pile caps in place. In 2003, a private venture received permission to build four grain silos at the waterfront site. Silo #4 was positioned closest to the Mobile River, over the heaviest concentration of remnant foundations. Silo #1 was located at the western end of the property, nearest the existing head house and storage complex. Construction plans indicate that each silo was approximately 100 ft (31 m) tall and 104 ft (32 m) in diameter. The maximum anticipated contact design pressure on each silo floor was approximately 5,000 psf (240 kN/m^2). Figure 1 shows a cross section of the typical silo elevation plan proposed by the original design team.

Construction of the new grain silos involved the installation of auger cast piles, each 16 in. (406 mm) in diameter and approximately 80 ft (25 m) long, to support the perimeter concrete ring wall foundation used to support the exterior steel shell wall of each silo. In addition, auger cast piles, also 16 in. (406 mm) in diameter and approximately 80 ft (25 m) long, were installed to support the rectangular concrete conveyor tunnel located beneath the floating concrete floor slab in each silo. The ring wall foundation piles were designed to support 100 tons (890 kN) of compressive loads. Each conveyor tunnel pile was also designed to support 100 tons (890 kN) of compressive loads. Plans called for sixteen tension piles, four per silo, with a tensile capacity of 40 tons (356 kN) each. The remnant foundations were left in place, primarily beneath Silos #3 and #4, as shown in Figure 2.

Prior to initiating the installation of production piles, the contractor installed seven reinforced auger cast test piles, each 16 in. (406 mm) in diameter and 80 feet (25 m) long at selected locations across the footprint of the project site. All of the piling were test loaded to 200 tons (1,423.4 kN) in compression in general accordance with ASTM Specification D-1143[1]. Construction records indicate that only test pile number 6, located in close proximity to Silo #2, successfully passed load testing. Despite the poor results recorded in the pile load test program, construction of the silo foundations commenced at the conclusion of the test load program.

After the pilings were installed, the ring walls were cast and the conveyor tunnels were constructed, approximately 12 ft (4 m) of compacted structural soil fill was placed within the confines of the ring walls to serve as foundation support for the 4 in. (102 mm) floating concrete floor slabs in each silo. The silo design team anticipated that, when fully loaded, the pile supported ring walls would experience minor, but tolerable settlements as the structural fill confined within the rings settled about 6 in. (153 mm) in Silos #1, #2 and #3. Due to the remnant foundations located beneath Silo #4, little floor slab settlement was anticipated. Based on these settlement assumptions, the aeration ducts and equipment platforms were installed in Silo #4.

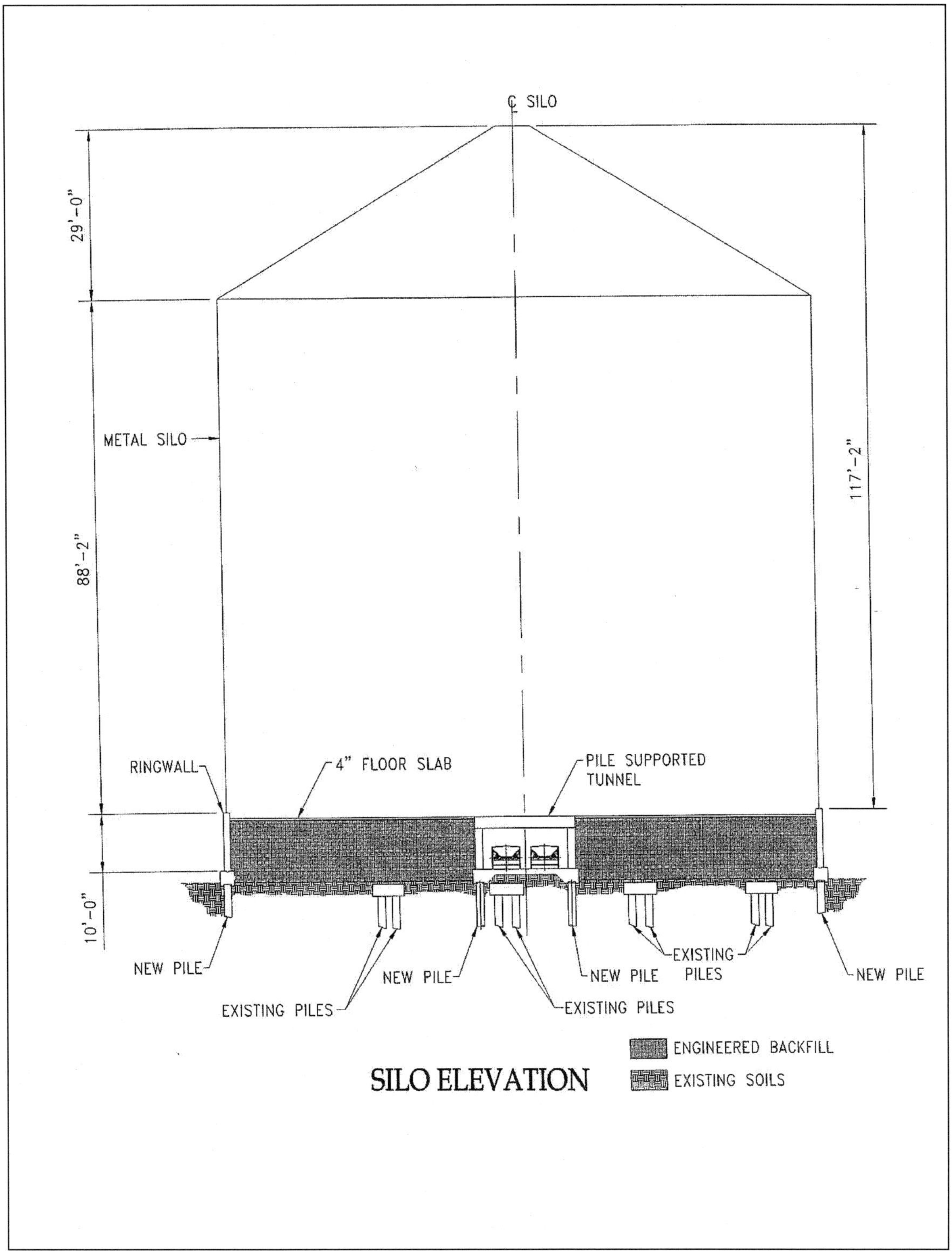

Figure 1 – Typical Silo Elevation Plan

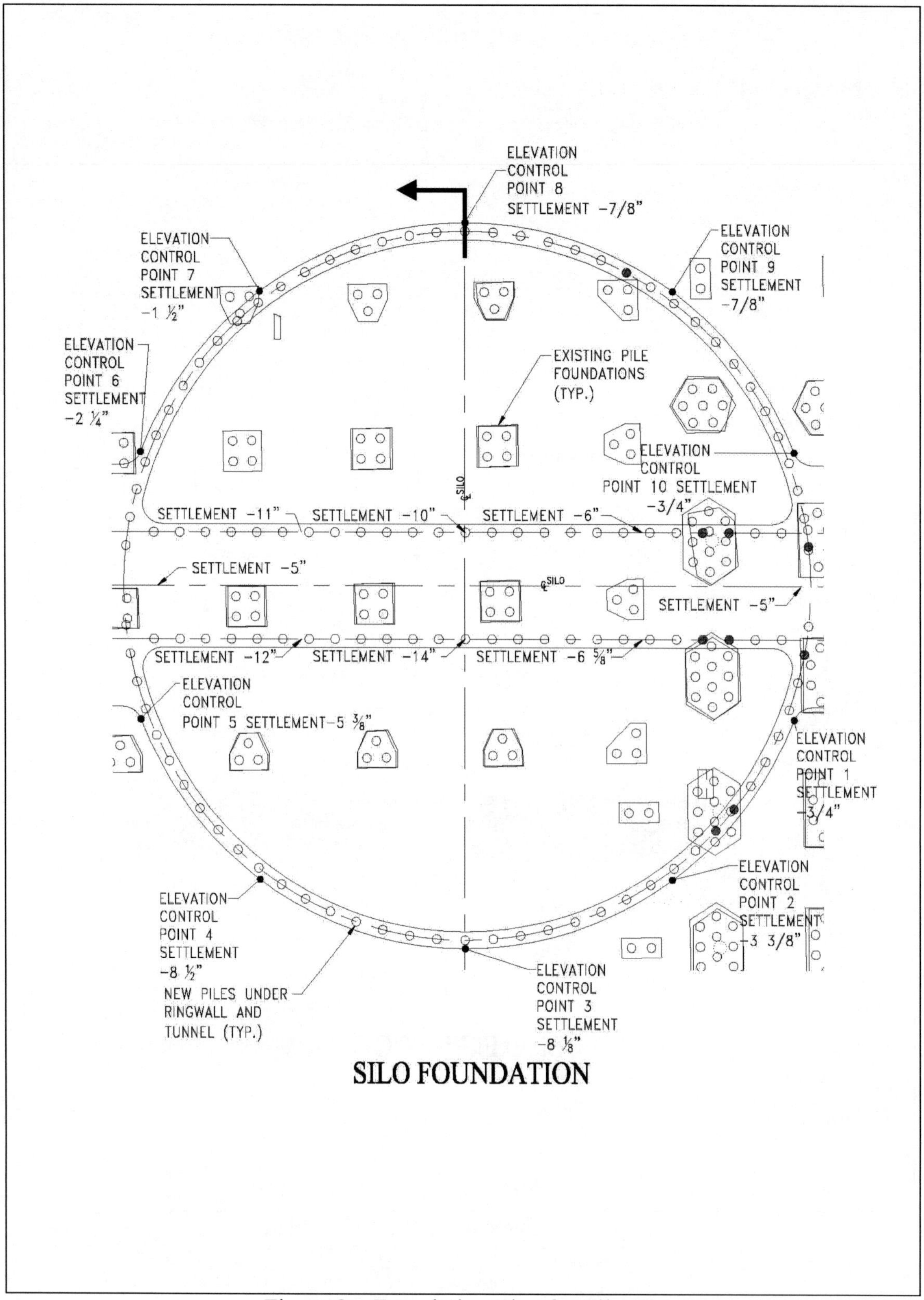

Figure 2 – Foundation Plan for Silo #3

After construction of the four silos was completed, Silo #4 was systematically loaded with grain weighing approximately 50 pcf (7.85 kN/m^3). Based upon the available storage volume when Silo #4 was fully loaded, the distributed compressive load across the floating floor slab was approximately 5,000 psf (192.0 kN/m^2). According to project records, in-place settlement monitoring devices recorded only minor vertical movements during test loading.

Following the successful loading of Silo #4, loading commenced at Silo #1. Grain weighing approximately 50 pcf (7.89 kN/m^3) was placed into Silo #1 at the same loading rate used to load Silo #4. When the in-place monitoring devices indicated noticeable settlements at the 50% loading threshold, the field crew notified the design team, but were directed to continue loading. When loading was approximately 90% complete, the field crew heard an explosive noise as Silo #1 simultaneously tilted toward the roadway on the south side of the site. Initial estimates reported a 6 in. (153 m) lateral displacement at the top of the silo, with visible distress in the overhead steel framework and the exterior concrete foundations as illustrated in Figure 3 and Figure 4. After Silo #1 was safely unloaded, work at the site stopped. Silos #2 and #3 were not load tested.

Figure 3 – Steel Framework

Figure 4 – Concrete Foundation

FORENSIC SURVEY

Following the foundation failure at Silo #1, the storage facility laid dormant for the next eighteen months while the owners and the design team studied the situation and contemplated a course of action.

In the aftermath of hurricane Katrina, which hit the Alabama Gulf Coast on August 29, 2005, the owners of the silos faced a shortage of suitable waterfront storage space for their grain. Consequently, the owner commissioned the authors to evaluate the as-built condition of the four silos to determine if Silos #2, #3 and #4 could temporarily be used for the safe storage of grain. After an extensive forensic review, the authors concluded that the three silos were structurally competent. The next step was to initiate a limited geotechnical analysis of foundation soils to determine what steps could be taken to safely load each of the silos.

The geotechnical survey focused on confirming the soil profile by drilling four soil test borings using conventional standard penetration testing(2) techniques. Each boring was drilled to a depth of 61.5 ft (18.7 m). Five undisturbed 3 in. (76 mm) diameter Shelby Tube samples (3) were extracted from the cohesive soils near ground surface for laboratory strength and consolidation testing. In addition, six seismic cone penetration soundings (CPTU(4)) ranging in depth from 34 ft (11 m) to 77 ft (24 m), were also performed across the site. Finally, we performed a series of four Parallel Seismic/Cone Penetrometer test (PS/CPT(5)) soundings to confirm in-place pile lengths and pile integrity at Silo #1.

Based upon our review of available geotechnical information, the authors concluded that the soil stratification at the site consisted of about 13 ft (4 m) of firm silty sand over approximately 25 ft (8 m) of soft clays interbedded with layers of very loose silty clayey sands, overlying a bearing stratum of firm to very dense sands and silty sands. Maximum unconsolidated undrained shear strength values measured in the cohesive soils were in the range of 450 to 500 psf (36 to 40 kN/m^2). Groundwater was encountered at about 5.5 feet (1.7 m) below existing ground surface.

Our review of project plans and specifications(6) showed that the cast-in-place piles used to support critical structural silo components were unreinforced. Further, plans showed that the weight of the grain within each silo was to be directly applied to (and supported by) a 4 in. (102 mm) thick concrete slab resting directly on top of the 10 ft (3 m) thick cylindrical soil plug encapsulated by the ring walls. Because the soil plugs and floor slabs were not structurally connected to the ring walls, the designers assumed that the predicted 6 in. (153 mm) of consolidation settlement would not adversely affect performance of the ring wall foundation systems or the pile supported conveyor tunnel. In fact, plans called for removing the relatively thin sacrificial concrete floor slabs after consolidation settlements stabilized, in order to replace them with a permanent, 6 in. (153 mm) thick, reinforced concrete floor.

Even though the design team clearly considered the issue of soil/pile interaction associated with vertical consolidation of in-place soft soils, one of the most significant design assumptions involved plans to fill the silos to full capacity without considering the impact associated with applying a 5,000 psf (239.4 kN/m^2) load to a 30 ft (9 m) thick cylindrical column of highly compressible, saturated soft soils with unconsolidated undrained shear strength values ranging from 250 psf (12.0 kN/m^2)

to 500 psf (24.0 kN/m^2). Furthermore, there were no provisions to minimize pore pressure effects during loading and no consideration was given to the effects of lateral soil displacement of the soft soils confined within the cylindrical matrix of unreinforced cast-in-place piling used to support the ring walls. In addition, the designers apparently ignored the potential effect of soil/pile interaction upon the structural integrity of the conveyor tunnel and the overhead loading system. Survey results following the failure of Silo #1 confirmed suspicions that the 10 ft (3 m) thick cylindrical plug of structural fill, (compacted to 100% Modified Proctor dry density) acted as a piston within the ring walls, exerting large down drag forces onto both the ring wall system and the piles that support the tunnel foundations and the overhead loading system. In addition, inclinometer and PS/CPT results confirmed significant lateral displacement of the near surface soft clays that squeezed through gaps between unreinforced concrete cast-in-place piling that supported the ring walls.

Load Testing of Silos #2 and #3

After reviewing the results of the forensic survey, the owners commissioned the authors to determine if Silos #2, #3 and #4 could be salvaged. In an attempt to take advantage of this unusual opportunity to collect and analyze settlement data by load testing full scale structures, the authors agreed to manage a full scale loading of the three silos.

The initial objective of our study was to assess the general structural condition of the three silos and to develop a field loading and settlement monitoring program that would virtually take the foundation systems of the silos to the point of impending failure. Based upon the forensic work that had been performed at the site following the failure of Silo #1, the authors developed an instrumentation[6] plan to monitor in-situ consolidation, pore pressures and lateral soil displacement during loading of the silos. The layout of our principal instrumentation is shown on Figure 6, and consists of eleven piezometers and four inclinometers. Based upon our reviewing of the foundation failure of Silo #1, we anticipated settlements on the order of 5 to 10 in. (125 to 250 mm) at both silo locations. Because our objective was to induce settlement without causing complete structural failure, we carefully monitored vertical and lateral movements, along with pore pressure, within the weak soil zone that extended to a depth of about 34 ft (11 m) below existing ground surface.

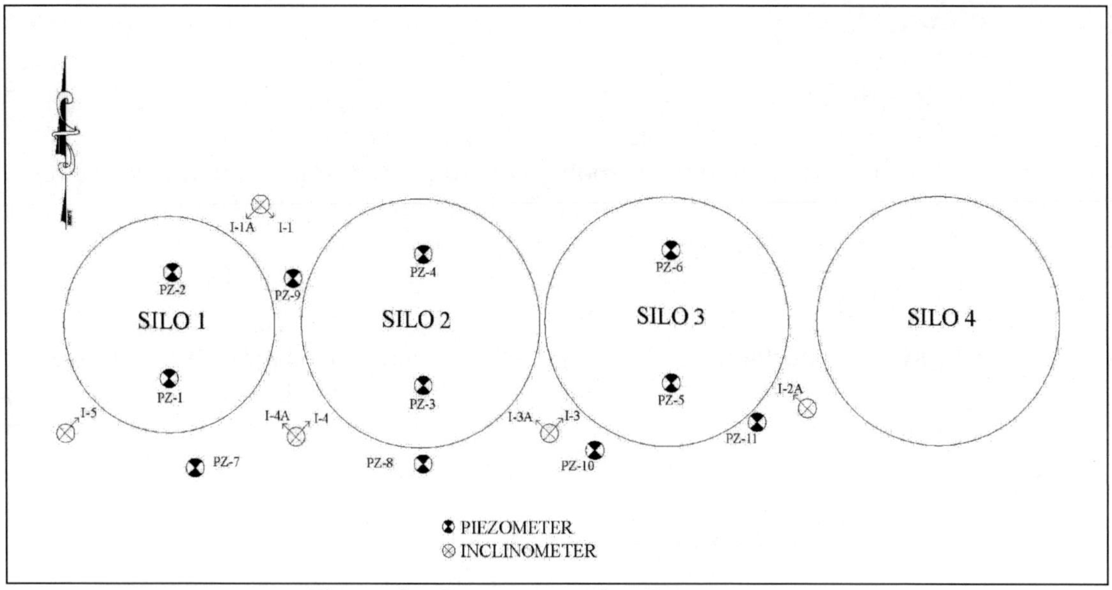

Figure 5 – Instrumentation Layout

Because it was necessary to close the silo doors when loading with grain, we were not able to measure vertical settlement of the floor slabs during the actual loading process. However, to help predict vertical soil movement when load was being applied, we relied upon inclinometer data collected from instruments placed adjacent to the outer ring walls as shown in Figure 2. We also installed two pneumatic piezometers in the floor of each silo to monitor pore pressures directly below the column of grain as load was applied to the soils.

The loading plan for Silos #2, #3 and #4 consisted of applying load to the silos (using grain), in roughly 5% increments, while monitoring the soils reaction to each load increment. Figures 4 and 5, along with Tables 1 and 2, illustrate the kinds of lateral soil deflection and pore pressure levels achieved during our loading program.

Typically, loads were applied after pore pressures had dissipated and the lateral movements had subsided. The loading sequence for Silos #2 and #3 was very slow and cyclic. Generally, load increments in the range of 200 psf (9.6 kN/m^2) were applied by placing grain into each silo. After each load increment was added, the instrumentation was monitored and the structure was visually observed.

During the early phases of loading the silos, we concluded that the weak soil zone between 18 and 35 ft (5 to 11 m) was highly stratified with interbedded clay and sand. Consequently, traditional methods for predicting consolidation and strain related settlement beneath the silos was erratic. Field monitoring showed that, in general, settlements following a given loading increment occurred simultaneously with the loading of grain. Because the loading of the silos generally took place at night, we often did not have access to the site until morning, which caused some difficulty in getting comprehensive pore pressure measurements. As a result, we relied heavily upon inclinometer displacements to determine when the soil response was complete. We typically recorded readings over a period of several days to ensure that all residual movements had stabilized. Depending on the client's needs, the silos were often unloaded for weeks.

OBSERVATIONS

The basic objective of providing the owners with a temporary grain storage facility was achieved. Five of the most significant observations and conclusions noted by the authors, while stage loading Silos #2, #3 and #4 are briefly outlined below:

1) <u>Settlements and Lateral Deformations</u>: Horizontal and vertical deflections measured in the range of 6.0 in. (152 mm) and 2.5 in. (64 mm), respectively, at Silos #2 and #3 were significant and greatly exceeded expectations for soft soil performance when subjected to design compressive loads. Measured foundation settlements associated with traditional consolidation and shear failure of near surface soft clay soils clearly demonstrated the negative effects of negative skin friction combines with lateral soil displacement on deep, unreinforced concrete pile foundations.

2) <u>Remnant Foundations</u>: Essentially, the buried (remnant) foundation elements consisted of pile caps located below original ground surface, directly beneath the structural footprints of Silos #2, #3 and #4. One of the most damaging assumptions made by the original silo design team involved the load transfer mechanism used to predict the settlement of individual structural components of each silo. The design team assumed that, when loads were applied to the floor slabs in each silo, the cylindrical column of soil confined by the ring walls would settle without affecting the structural integrity of any pile supported element, including the ring walls, the conveyor tunnel and other buried remnant foundations. Because the key foundation elements at the site were pile supported, they concluded that settlements associated with these individual foundation elements would be minimal. Actual test results collected by the authors showed that the combination of direct load application to the floor system, along with the arching effect of grain between the walls of the structures, resulted in significant load transfer to and between individual foundation elements.

3) <u>Pile Foundation Issues</u>: One of the most interesting observations encountered during full scale load testing of the silos involved the reaction of the pile supported conveyor tunnel foundations. During test loading, both Silos #2 and #3 tilted, leaning about 6.0 in. (153 mm) near the top, toward the southwest with little evidence of structural distress to visible silo components. The ring walls were essentially unaffected structurally except for small cracks at some contact points where the foundation piles made contact with the base of the ring wall. Conversely, the pile supported conveyor tunnel developed significant crack resulting from large differential movements that have, for all practical purposes, put the system out of commission. Survey results confirm that the foundation piles are unable to support design loads in compression and reinforce the need for reliable load testing programs that comply with

international and local building codes to assure that production piles have the same structural integrity and performance characteristics as the test piling.

4) Hybrid Foundation Issues: The hybrid foundation system proposed for this project, as it was initially conceived, was theoretically complex, impractical and complicated to build. Performance monitoring showed that soft soil consolidation occurred more quickly than expected due to sand seams in the soil matrix. Further, lateral displacement of the soft soils due to over stressing, compromised the integrity of the unreinforced ring wall piles and added to down drag forces on the ring wall and tunnel piles. Consequently, the combined effect of in-place (remnant) foundations, interacting with lateral (shear) displacements of soft clay soils, made it virtually impossible for the design team to accurately predict the reaction of the combination (hybrid) foundation system to compressive loading. In addition, some remnant foundations obstructed tunnel settlement, causing the tunnel floor, roof and walls to settle independently of one another, creating humps, heaving, in the tunnel floor and huge cracks in both the floor and walls, virtually rendering the reclaim conveyors unusable. The remnant foundations clearly distorted the expected settlement profile beneath the soils and contributed to unnecessary differential settlement distress to key structural elements of the silos.

5) Catastrophic Failure of Silo #1: The results of our forensic analysis following the catastrophic failure of Silo #1 during initial loading suggest that the unreinforced concrete piles used to support the ring wall foundations were subjected to significant lateral pressure in the soft soil zone. Because there were virtually no remnant foundations located beneath Silo #1, the possibility exists that the combination of down drag and/or structural pile failure may have resulted in global failure of the foundation system.

CONCLUSIONS

With the benefit of careful monitoring and cooperation between the authors and the owners, Silos #2, #3 and #4 were successfully placed into operation at approximately 60%, 90%, and 100% capacity, respectively, for eighteen months. Ultimately, the structural distress from settlements associated with loading Silos #2 and #3 to the point of impending failure will necessitate significant remedial repairs before the facility can be placed into long term operation. Never the less, the owners benefited significantly from having a temporary storage facility for their grain and the authors benefited from the unusual opportunity to load two full scale commercial structures to impending failure, while monitoring the influence of the settlement distress and foundation performance.

As a result of work performed and data collected during the cyclic loading of the three silos, the owners have decided to proceed with a program to restore and repair the foundations beneath Silos #1, #2 and #3 so that the silos can be used to store grain

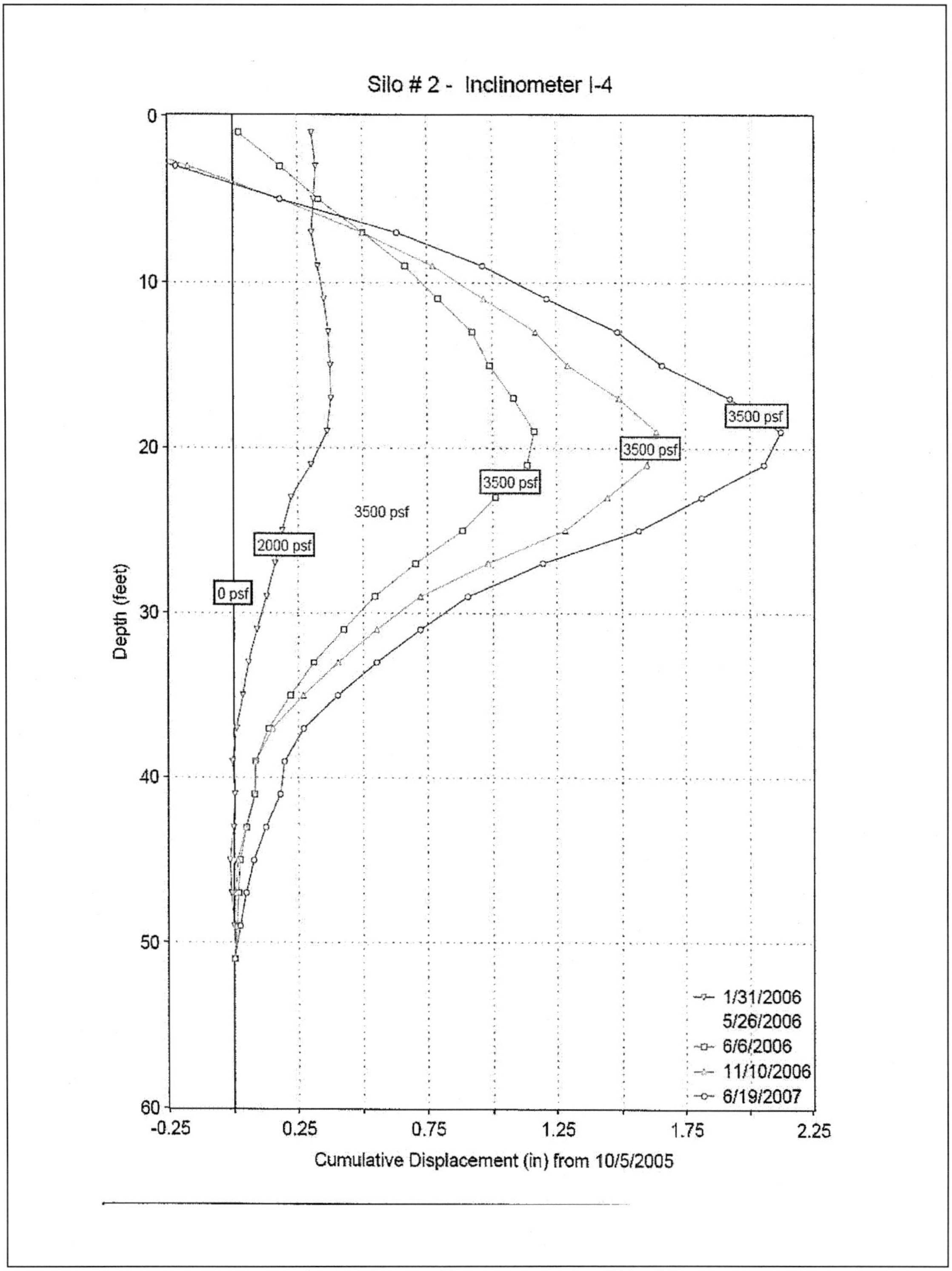

Figure 6 – Inclinometer I-4 – Silo #2

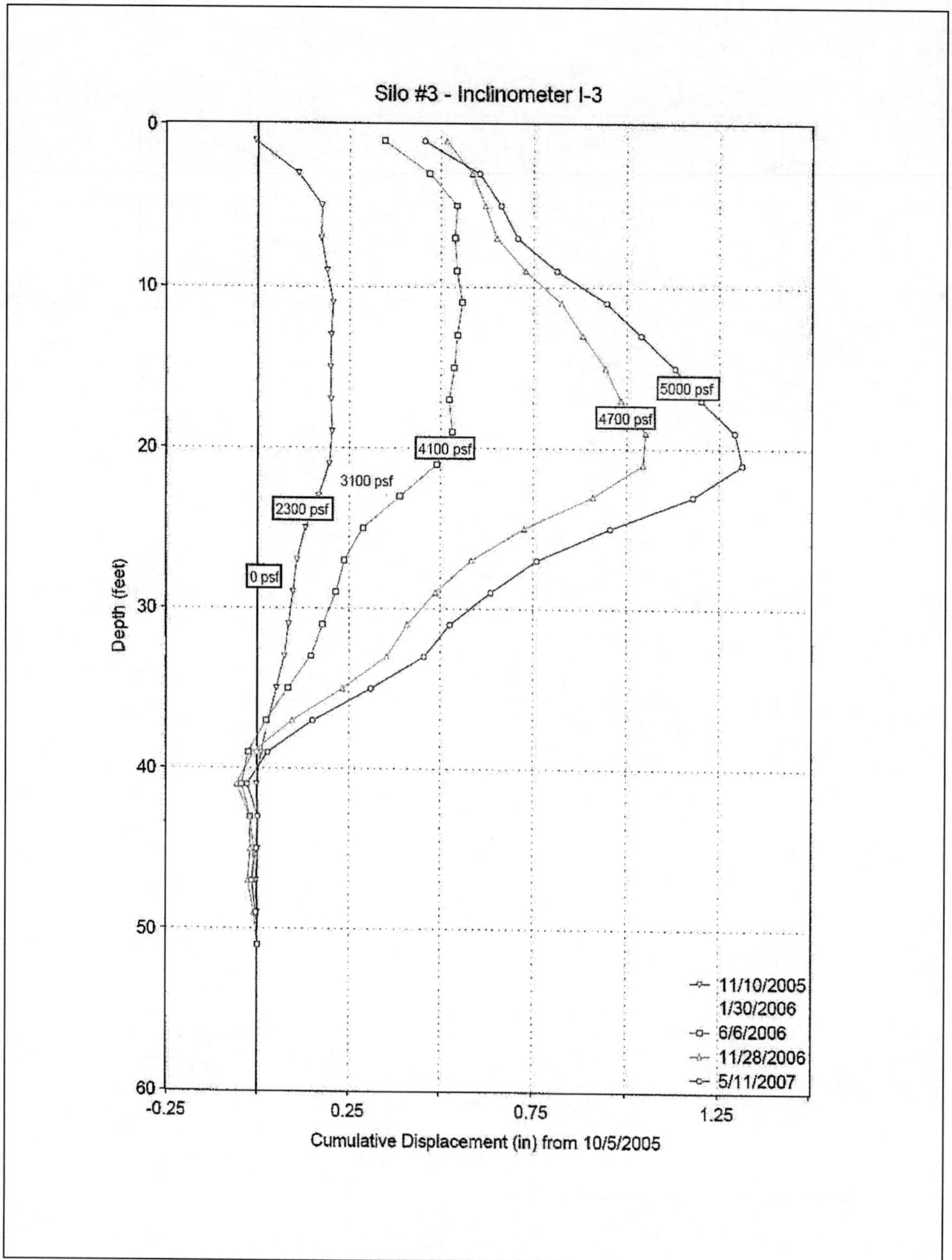

Figure 7 – Inclinometer I-3 – Silo #3

Silo #2 Peizometer Readings - Oct '05 - Feb '06					
Date	Approximate Floor Load (psf)	PZ-3 Pore Pressure (psi)	PZ-3 Variation from initial pore pressure	PZ-4 Pore Pressure (psi)	PZ-4 Variation from initial pore pressure
10/6/05	0.0	5.6	0.0	6.6	0.0
10/10/05	825	5.8	0.2	7.1	0.5
11/1/05	825	5.6	0.0	6.6	0.0
11/30/05	1205	6.0	0.4	6.8	0.2
12/9/05	570	5.6	0.0	6.6	0.0
1/6/06	2085	8.0	2.4	8.6	2.0
1/31/06	2085	7.6	2.0	8.8	2.2
2/16/06	2470	8.3	2.7	9.2	2.6

Silo #2 Peizometer Readings - May '06 - June '07					
Date	Approximate Floor Load (psf)	PZ-8 Pore Pressure (psi)	PZ-8 Variation from initial pore pressure	PZ-9 Pore Pressure (psi)	PZ-9 Variation from initial pore pressure
5/11/06	3000	7.6	0.0	7.0	0.0
5/24/06	3200	7.7	0.1	7.1	0.1
5/26/06	3500	8.3	0.7	7.7	0.7
5/30/06	3500	7.7	0.1	7.8	0.8
7/13/06	0	7.2	-0.4	6.9	-0.1
8/7/06	0	7.2	-0.4	6.6	-0.4
8/14/06	2500	7.3	-0.3	7.4	0.4
8/25/06	3500	8.0	0.4	7.4	0.4
11/28/07	2780	7.9	0.3	7.4	0.4
2/14/07	0	7.6	0.0	6.6	-0.4
5/31/07	3300	8.4	0.8	7.2	0.2
6/6/07	3300	8.0	0.4	7.0	0.0

*** Pore pressures listed above were measured using pneumatic piezometers placed approximately 20' below grade in a grouted borehole.**

TABLE 1 – Silo #2 Piezometer Readings

Silo #3 Peizometer Readings - Oct '06 - June '06					
Date	Approximate Applied Load (psf)	PZ-5 Pore Pressure (psi)	PZ-5 Variation from initial pore pressure	PZ-6 Pore Pressure (psi)	PZ-6 Variation from initial pore pressure
10/6/05	500	3.9	0.0	7.2	0.0
10/7/07	1400	5.7	1.8	8.6	1.4
10/11/05	1400	5.1	1.2	7.4	0.2
10/18/05	1900	3.3	-0.6	7.5	0.3
11/1/05	2300	4.8	0.9	7.6	0.4
11/10/05	2300	4.3	0.4	7.2	0.0
11/18/05	2700	5.2	1.3	8.1	0.9
1/17/06	3000	pinched line	N/A	7.2	0.0
5/11/06	3500	4.8	0.9	8.2	1.0
5/26/06	4000	6.2	2.3	8.5	1.3
6/9/06	4000	4.8	0.9	7.8	0.6
6/15/06	4400	5.2	1.3	8.0	0.8

Silo #3 Peizometer Readings - Aug '06 - June '07					
Date	Approximate Applied Load (psf)	PZ-10 Pore Pressure (psi)	PZ-10 Variation from initial pore pressure	PZ-11 Pore Pressure (psi)	PZ-11 Variation from initial pore pressure
08/07/06	4500	4.8	0.0	4.8	0.0
03/21/07	4000	4.6	-0.2	4.8	0.0
05/11/07	4700	5.0	0.2	5.4	0.6
05/18/07	5000	5.0	0.2	5.2	0.4
06/19/07	5000	4.8	0.0	4.8	0.0

*** Pore pressures listed above were measured using pneumatic piezometers placed approximately 20' below grade in a grouted borehole.**

Table 2 - Silo #3 Piezometer Readings

at their full capacity. Consequently, the authors are currently in the process of analyzing all of the loading and settlement data in order to assist the structural designers with plans and specifications for silo restoration. After the silo renovations are completed and the facility is back in full operation, we plan to write a supplemental paper comparing the performance characteristics of the upgraded foundation systems with the performance data collected during our forensic survey.

ACKNOWLEDGEMENTS

The authors gratefully acknowledge Mr. Roger Buzard FGDI, LLC, for his assistance in obtaining project design data, along with his continued support and encouragement during all phases of our forensic survey. Appreciation is also due to Mr. Scott H. Slaughter, P.E., Mr. Matt Coaker, E.I., and Mr. Danny Hines, E.I. of Southern Earth Sciences, Inc. for their technical support on the project. The authors are also grateful to Mr. David E. Murphy and Mr. John D. Glover of Cowles, Murphy, Glover & Associates for their support on this project.

REFERENCES:

1. American Society for Testing and Materials, ASTM (2006), Annual Book of Standards, ASTM D-1586-99, Standard Test Method for Penetration Test and Split-Barrel Sampling of Soil, Vol. 04.08

2. American Society for Testing and Materials, ASTM (2006), Annual Book of Standards, ASTM D-1587-99, Standard Practice for Thin-Walled Tube Sampling of Soil for Geotechnical Purposes, Vol. 04.08

3. American Society for Testing and Materials, ASTM (2006, Annual Book of Standards, ASTM D-1143-99, Standard Test Method for Piles Under Static Axial Compressive Load, Vol. 04/08

4. Campanella, R.G. and Robertson, P.K., Current Status of the Piezocone Test, Penetration Testing, 1988, ISOPT-1, DeRuiter (ed.)

5. Olson, L.D., Slaughter, S.H. and Sack, D.A., Combined Measurement of Unknown Foundation Depths and Soil Properties with NDE Methods, Presented at the TRB Conference, 2004

6. Buzard, Robert, "Unpublished Plans, Specifications and Notes" for design of the FGDI, LLC. Grain Storage Silos jn Mobile, Alabama, 2003.

7. Laier, James E. and Brenner, William H., Instrumentation Performance in Soft Clay Soils, Proceedings of a Symposium: Sedimentation Consolidation Models, ASCE/October, 1984

Seepage Through Rockfill Dams in Narrow Valleys

Ali Soleimanbeigi[1] and Fardin Jafarzadeh[2]

[1]Graduate Researcher, Department of Civil Engineering, Shiraz University, School of Engineering, Shiraz, Iran; ali.soleiman@gmail.com
[2]Assistant Professor, Department of Civil Engineering, Sharif University of Technology, Azadi St., Tehran, Iran; fardin@sharif.edu

ABSTRACT: Seepage analysis serves as one of the most significant stages in the design process of an embankment dam. In two-dimensional (2D) seepage analysis of embankment dams, little or no attention is paid to the seepage through side abutments. Moreover the role of grout curtain extensions into the side abutments and abutment material properties are inevitably neglected when performing a 2D seepage analysis. In this paper, two and three-dimensional (3D) models of a rockfill dam during operation state are generated and several unsteady and steady state seepage analyses are performed using finite element method (FEM). The results obtained from 2D and 3D seepage analyses were compared with measurements from the instrumentation system installed in the dam body and foundation during construction. It is concluded that for embankment dams to be constructed in narrow valleys, 2D seepage analysis results are far from reality and considering 3D modeling of such dam sites is vital. Several design graphs were developed to calculate 3D discharge rates and hydraulic gradients from those obtained from 2D seepage analysis.

KEY WORDS: 3D Seepage Analysis, Finite Element Method, Instrumentation, Rockfill Dam

INTRODUCTION

The study of 77 embankment dam failures occurred in the United States until early 1980s indicates that seepage piping accounts for approximately 50% of all dam failures, the majority of which progress within the dam embankment and foundation (Engemoen, 2003). When seepage rates or hydraulic gradients increases unexpectedly, undesirable conditions may prevail to progress the piping or internal erosion that leads to dam instability or ultimate failure. Despite several measures incorporated in the design process to control seepage and piping, inaccurate assessment of seepage rates, pressure heads and hydraulic gradients lead to improper design of seepage controlling elements. In this case study, two and three-dimensional model of a rock-fill dam were developed and several unsteady and steady state seepage analysis were performed for the operation state of the dam. Variation of pressure heads, discharge rates, hydraulic gradients and water levels in downstream parts were investigated and the data acquired from the instrumentation system installed during construction of the dam were utilized to verify the obtained results. In addition, a comparison was made to evaluate the quantitative difference between two and three-dimensional analysis results and actual data measured by different instruments installed in the dam body and foundation.

MATERIAL AND SITE CHARACTERISTICS

The dam considered in this study is an earth core rockfill dam (ECRD) called Masjed-E-Soleiman (MES) dam located in South-West of Iran. The dam is 177 m high with 490 m crest length. Figure 1 shows an aerial view of the dam at the end of first impounding and Figure 2 illustrates the typical cross section of the dam.

The foundation treatment method for seepage control consists of a grout curtain constructed along the dam axis in two-line grouting boreholes with 1.5 meter spacing. Figure 3 illustrates the longitudinal section of the grout curtain.

Monitoring of the dam behavior was based on the measured data from various instruments installed in the dam body and foundation. Some instruments, mainly those installed in the upstream parts of clay core, malfunctioned or were damaged during construction or first impounding (noted in Report on Review of Masjed-E-Soleiman Dam Behavior). Most seepage instrumentation systems are installed in four cross sections, A, B, C and D of the dam except Groundwater observation wells, GOs, and Casagrande piezometers, CPs, as illustrated in Figure 4. Table 1 summarizes the location of Stand-pipe piezometers, SPs, installed in the foundation.

MATERIAL PROPERTIES

Foundation

The permeability of a rock mass as a function of depth was investigated by carrying out 106 Lugeon tests, introduced by M. Lugeon (1993), in five boreholes. A total of 41 tests were performed at depths of less than 50 m, 31 tests at depths of less than 100 m and 34 tests at depths of more than 100 m.

FIG. 1. Masjed-E-Soleiman Rock-fill Dam During First Impounding

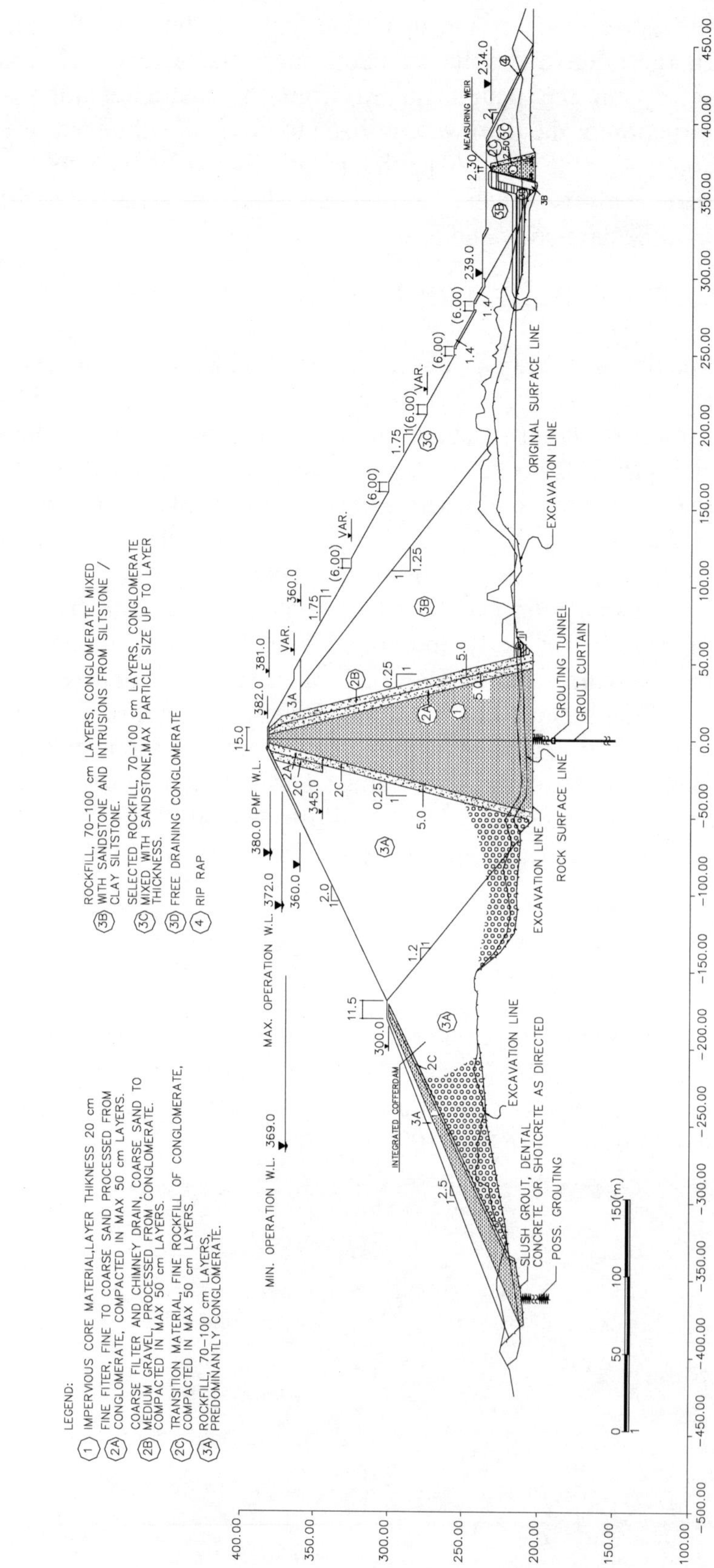

FIG. 2. Typical Cross Section of MES Dam.

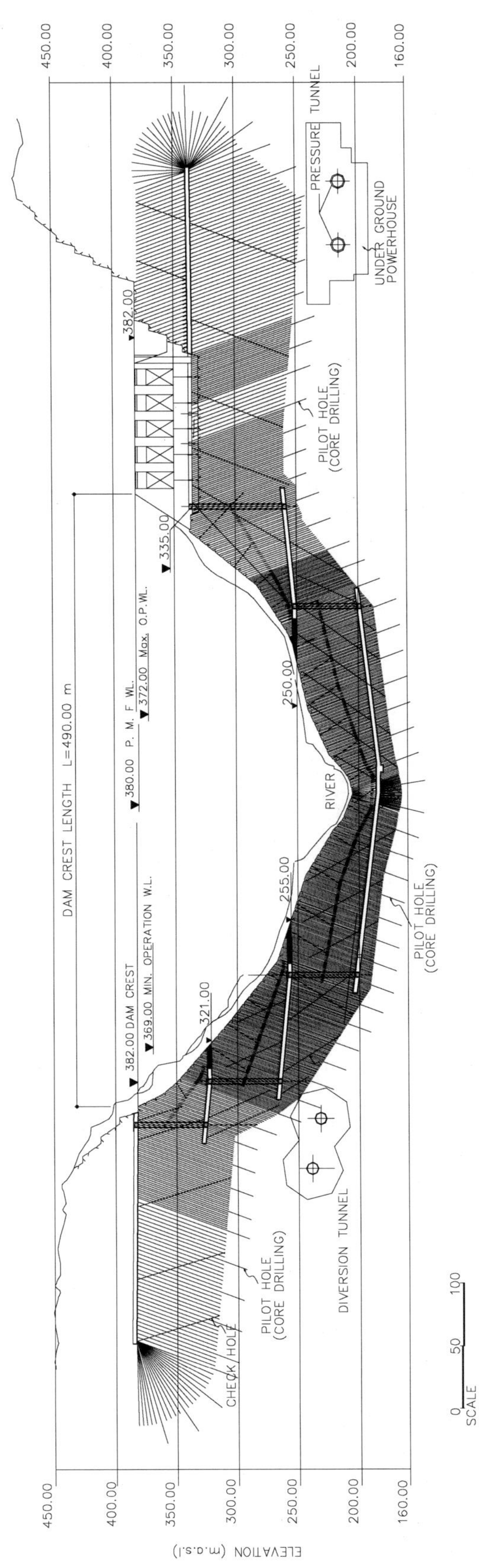

FIG. 3. Longitudinal Profile of Grouting Curtain for MES Dam

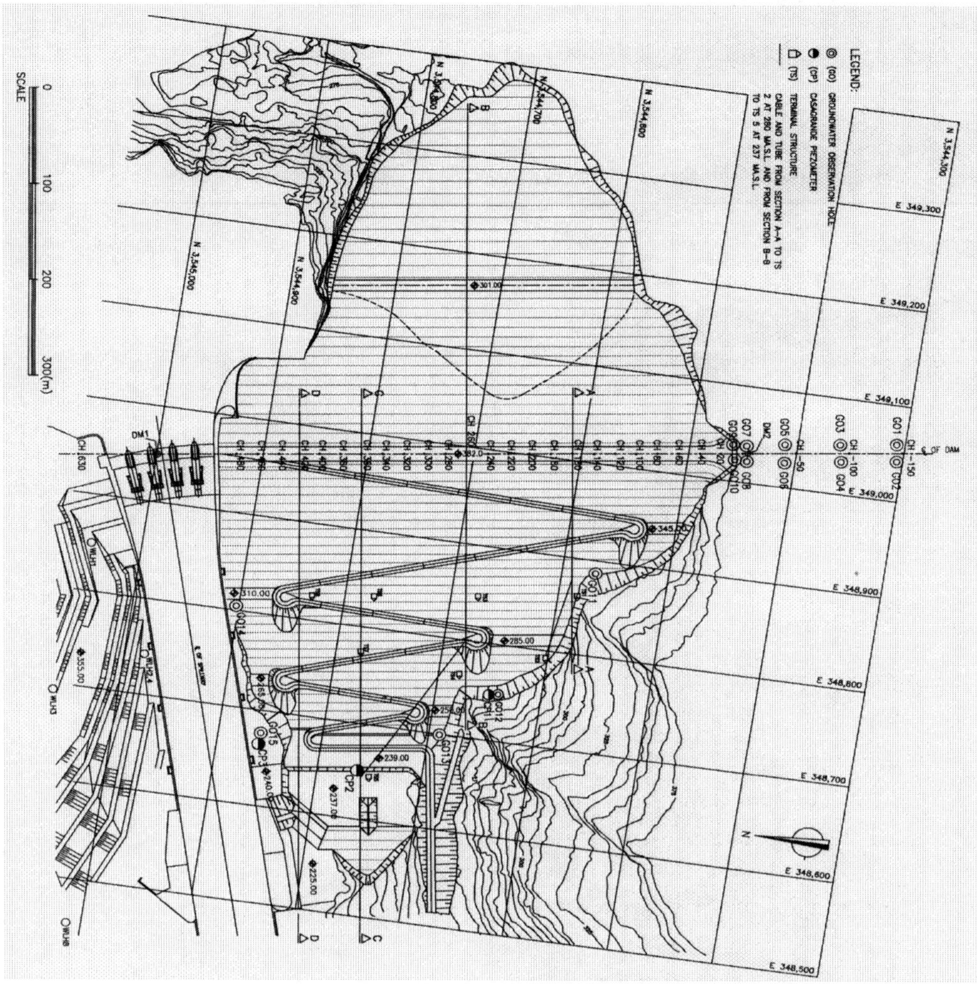

FIG. 4. Instrumentation Plan for Seepage Analysis.

Permeability coefficients with respect to the depths are presented in Table 2. The median results are more reliable than the average values since the effect of individual very large results is excluded. Laboratory tests were also performed on core and shell materials.

Core

Permeability tests have been done on core materials. The specimens were compacted in molds with 20.3 cm diameter and 30.2 cm height and were saturated with back pressure prior to the test. The test results are summarized in Table 3 which indicates that the range of coefficient of permeability is in the order of 10^{-9} to 10^{-8} m/s. Figure 5 shows the relationship between the results of compaction tests and permeability tests. The figure indicates that the permeability of the specimen having optimum water content is always lowest which agrees with the general tendency that the lowest permeability appears at slightly higher than the optimum moisture content. Hence the results of the permeability test seem reasonable. In order to obtain compressibility of core material, consolidation tests with 10 cm diameter and 4 cm height molds have been carried out on samples with optimum moisture content, OMC, OMC+3% and OMC-2%.

TABLE 1. Locations of Standpipe Piezometers Installed in the Dam Body and Foundation

Number (1)	Instrument (2)	Chainage (m) (3)	Offset from the dam axis (m) (4)	Elevation (m) (5)
1	SP-1*	105.1	25.0 U/S†	217.0
2	SP-2	105.2	17.0 D/S†	230.0
3	SP-3	123.4	23.0 D/S	250.3
4	SP-4	160.0	20.0 U/S	231.0
5	SP-5	160.0	20.0 U/S	205.0
6	SP-6	160.0	18.0 D/S	214.0
7	SP-7	160.0	21.0 D/S	232.0
8	SP-8*	210.0	15.0 U/S	200.9
9	SP-9	210.0	23.0 D/S	191.2
10	SP-10*	260.0	12.0 U/S	168.2
11	SP-11	260.0	6.0 U/S	178.2
12	SP-12*	260.0	26.0 U/S	192.5
13	SP-13	260.0	9.0 D/S	206.0
14	SP-14*	310.0	13.0 U/S	209.3
15	SP-15	310.0	20.0 D/S	220.0
16	SP-16*	359.8	17.0 U/S	225.1
17	SP-17	339.4	35.0 D/S	229.0
18	SP-18*	359.9	17.0 U/S	199.2
19	SP-19	359.9	24.0 D/S	218.1
20	SP-20*	419.9	19.0 U/S	224.6
21	SP-21	420.2	14.0 D/S	230.7
22	SP-22U	446.5	13.6 U/S	255.4
23	SP-22D	446.9	18.3 D/S	255.4
24	SP-23	478.7	28.3 U/S	338.1
25	SP-24r	473.0	16.5 D/S	338.0
26	SP-25	491.0	40.0 U/S	335.2
27	SP-26	492.3	32.6 D/S	335.3
28	SPF-27	530.0	25.1 U/S	295.6
29	SPF-28	530.0	1.5 D/S	311.2
30	SP-29	572.4	15.8 U/S	335.3
31	SP-30	572.7	17.2 D/S	335.2
32	SP-31	620.0	29.7 U/S	335.2
33	SP-32	620.2	18.7 D/S	335.2
34	SP-33	700.0	27.2 U/S	336.2
35	SP-34	700.1	16.8 D/S	336.1

*Unreliable standpipe piezometers

† U/S=Upstream, D/S=Downstream.

TABLE 2. Permeability Coefficient of Foundation Materials

Depth (m) (1)	Coefficient of Permeability (m/sec)	
	Average (2)	Median (3)
< 50	4.0×10^{-6}	2.5×10^{-6}
50-100	1.0×10^{-6}	5.0×10^{-7}
> 100	1.0×10^{-6}	1.0×10^{-7}

The value of m_v, coefficient of volume compressibility that represents the compressibility of soil is defined as follows:

$$m_v = \Delta\varepsilon / \Delta p \tag{1}$$

where $\Delta\varepsilon$ is compression strain and Δp is consolidation pressure. Table 4 summarizes the results of consolidation tests on clay core material. The range of m_v values corresponding to the load range between 100 kN/m^2 to 1000 kN/m^2 is 4×10^{-6} to 1×10^{-4} m^2/kN. The recommended value for the coefficient of volume compressibility is 2×10^{-6} m^2/kN.

TABLE 3. Permeability Test Results on Core Materials

Sample (1)	Water Content (%) (2)		Coefficient of Permeability (m/s) (3)
M1 (CL: 60%, GC: 40%)	OMC	13	5.3×10^{-9}
	OMC-2%	11	6.5×10^{-9}
	OMC+3%	15	6.3×10^{-8}
M1 (CL: 40%, GC: 60%)	OMC	10	1.1×10^{-9}
	OMC-2%	8	1.2×10^{-9}
	OMC+3%	12	1.9×10^{-8}
M1 (CL: 20%, GC: 80%)	OMC	9.3	3.2×10^{-9}
	OMC-2%	7.3	6.9×10^{-8}
	OMC+3%	11.3	1.9×10^{-8}

Zone 3B

A series of constant head permeability test was performed on zone 3B materials with the mold having a diameter of 35.5 cm and a height of 41.0 cm. the average permeability coefficient at dry density of 20.6 kN/m^3 and water content of 3.5% is 6.06×10^{-5} m/s.

Similar tests were carried out to obtain permeability coefficients of other embankment materials. The selected permeability coefficients of embankment materials are summarized in Table 5.

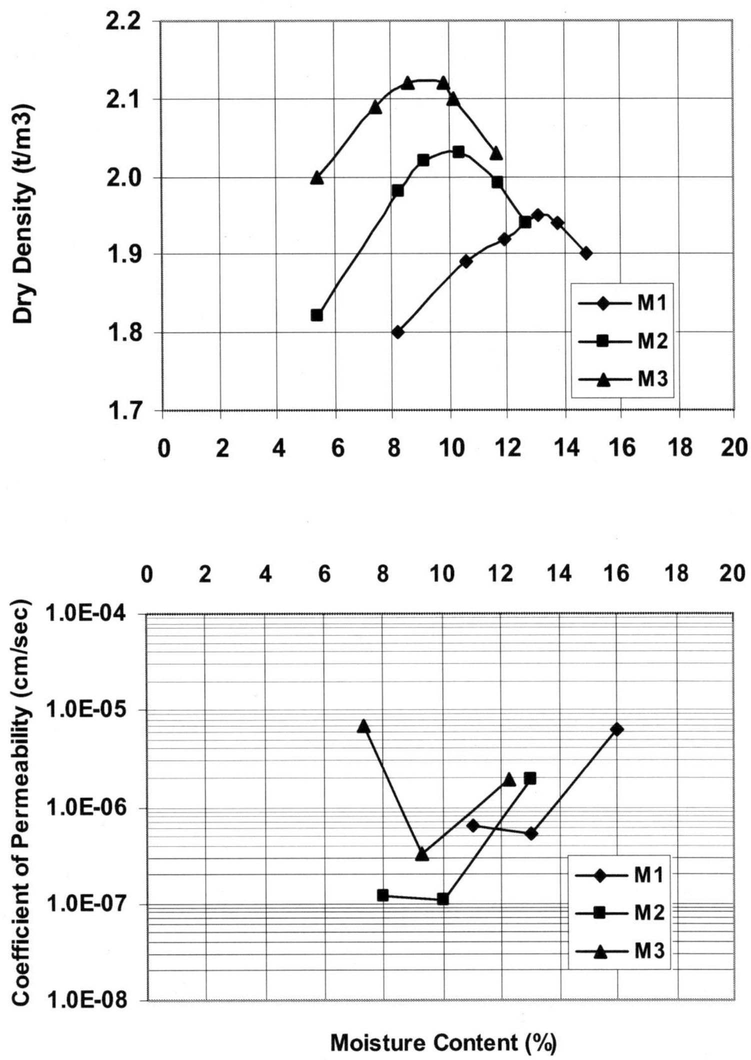

FIG. 5. Relationship between Moisture Content, Dry Density and Coefficient of Permeability

THEORY

Seepage condition underneath or through an earth dam is initially unsteady however, after elapse of enough time steady state seepage conditions is eventually established. The governing uncoupled formulation for unsteady state condition in 3D space is given by:

$$\frac{\partial}{\partial x}\left(k_{wx}\frac{\partial h_w}{\partial x}\right)+\frac{\partial}{\partial y}\left(k_{wy}\frac{\partial h_w}{\partial y}\right)+\frac{\partial}{\partial z}\left(k_{wz}\frac{\partial h_w}{\partial z}\right)+Q=\frac{\partial \theta}{\partial t} \tag{2}$$

where k_{wx}, k_{wy}, k_{wz} = hydraulic coefficients of permeability in x, y and z directions, respectively; h_w = total head; Q = external applied flux; θ = volumetric water content, the ratio of water retained within the soil structure to the total volume, and t = time.

TABLE 4. Summary of Consolidation Test Results on Core Materials

Sample	Water Content (%)		Density (kN/m^3)	Cc	m_v (m^2/kN)
(1)	(2)		(3)	(4)	(5)
	OMC	13	19.1	0.099	$1\times10^{-5} \sim 5\times10^{-5}$
M1	OMC-2%	11	18.6	0.122	$9\times10^{-6} \sim 4\times10^{-5}$
	OMC+3%	16	18.1	0.129	$8\times10^{-6} \sim 8\times10^{-5}$
	OMC	10	20	0.067	$7\times10^{-6} \sim 6\times10^{-5}$
M2	OMC-2%	8	19.3	0.119	$7\times10^{-6} \sim 2\times10^{-5}$
	OMC+3%	13	18.9	0.08	$8\times10^{-6} \sim 1\times10^{-4}$
	OMC	9.3	20.9	0.025	$4\times10^{-6} \sim 2\times10^{-5}$
M3	OMC-2%	7.3	20.3	0.078	$3\times10^{-6} \sim 1\times10^{-5}$
	OMC+3%	12.3	19.7	0.072	$5\times10^{-6} \sim 4\times10^{-5}$

TABLE 5. Permeability Coefficient of Embankment Materials

Material	**Permeability (m/s)**	$k_y / k_{x\,or\,z}$
(1)	(2)	(3)
3A/C	**> 1×10^{-3}**	**1**
3B	**>1×10^{-5}**	**1**
2C	**>1×10^{-5}**	**1**
2A/B	**>1×10^{-5}**	**1**
Core	**1×10^{-8}**	**0.1**
Grout Curtain	**< 5×10^{-7}**	**1**

Eq. 2 states that net rate of flow into a soil element plus the external applied flux is equal to the rate of change in the volumetric water content with time. Zienkiewicz et al. (1966) and Finn et al. (1967) applied the finite element method (FEM) to solve seepage problems in dams. The finite element formulation for unsteady state seepage in three dimensions has been derived using the Galerkin principle of weighted residuals applied to the governing differential equation (Papagiannakis and Fredlund, 1984):

$$\int_v \left([B]^T [C][B]\right) dv \{H\} + \int_v \left(\lambda \langle N \rangle^T \langle N \rangle dv \{H\},_t\right) = \int A \left(q \langle N \rangle^T \langle N \rangle\right) dA \tag{3}$$

where $[B]=$ gradient matrix; $[C]=$ element permeability matrix; $\{h\}=$ vector of nodal heads; $\lambda = m_v \gamma_w$; $m_v = \partial\theta/\partial u_w$; $u_w = \gamma_w (h_w - y)$; $y=$ is elevation; $\langle N\rangle^T \langle N\rangle = [M]=$ mass matrix; $\{h\},t = \partial h/\partial t=$ change in head with time; $q=$ unit flux across the faces of an element; $\langle N\rangle =$ vector of interpolating function. By solving Eq. 3, hydraulic heads at nodal points of finite elements are obtained.

The hydraulic head gradients in the *x, y* and *z* directions are computed for an element by taking the derivative of the element hydraulic heads with respect to *x, y* and *z*, respectively. Finally the element flow rates are obtained from the hydraulic head gradients.

MODEL DEVELOPMENT

The MES dam was modeled in two and three-dimensional space using Seep3D software version 1 (Geo-Slope ™, 2002). The software is able to model many classes of seepage problems that were impossible before. At first, different zones of the dam body were assembled according to their geometry. Since based on in-situ tests there are three successive layers of different permeability, abutment zones were added and extended to depths of 50 m, 100 m and beyond 100 m. The initial condition to perform an unsteady state analysis includes zero water head at all upstream faces and potential seepage conditions over all downstream surfaces. Boundary conditions applied to the model for steady state analysis comprised of 165 m water head over upstream faces of the dam and abutments. Potential seepage boundary conditions were assigned to the downstream shell and abutments. In order to perform the 3D finite element analysis the model was discritized into basic octahedral elements. The matrix contains 4850 elements and 5782 nodes. Figures 6a to 6c show the constructed 3D model of the dam, its finite elements and 2D model of the highest section.

Reservoir filling was started on January 2001 and after six months, in June; the water level reached to the current operating elevation of 370 m. The unsteady state FE analysis was performed at two-month interval time steps after the first impounding. The unsteady state seepage analysis results in this paper correspond to the time step number 21, representing May 2004, until when the instrumentation data were available. The steady state condition prevailed at time-step number 36, when the change in downstream discharge rates was negligible.

For the considered model with 4850 elements, a run for an unsaturated steady state and unsteady state (with 36 time-steps) seepage analysis would take nearly 1 and 30 hours; respectively. For a saturated steady state and unsteady state the run time reduces to nearly 20 minutes and 10 hours; respectively.

RESULTS AND DISCUSSIONS

Figs 7a and b show the result of the FE seepage analysis on the model dam site. Figure 7a shows a view of the water surface from downstream part that corresponds to the reservoir normal water level. It is illustrated from the figure that the side grouting curtains have lowered the water surface from upstream to downstream. This figure also shows seepage paths from the reservoir into the upstream abutments passing over the grout curtain and exiting to the downstream shell and riverbed. The concentration of seepage paths at the downstream toe accounts for the higher hydraulic gradient in comparison with that obtained from two-dimensional analysis, discussed later in this section. Figure 7b shows the water free surface and flow lines resulted from seepage analysis of the 2D model.

Figure 8 compares the measured and calculated water elevation in the locations of Standpipe piezometers installed in the foundation. In the majority of Standpipe piezometers, the water elevations obtained from the FE analysis agree well with those obtained from the measurements. Water elevation heads for piezometers SP-1, SP-8, SP-10, SP-12, SP-14, SP-16, SP-18 and SP-20 obtained from the FE analysis are much higher than the measured ones. This is due to the previously known malfunctioning of most upstream measuring devices in which these piezometers are installed. This malfunctioning was confirmed by the numerical investigation. The comparative results have been categorized to downstream water elevations, discharge rates and hydraulic gradients:

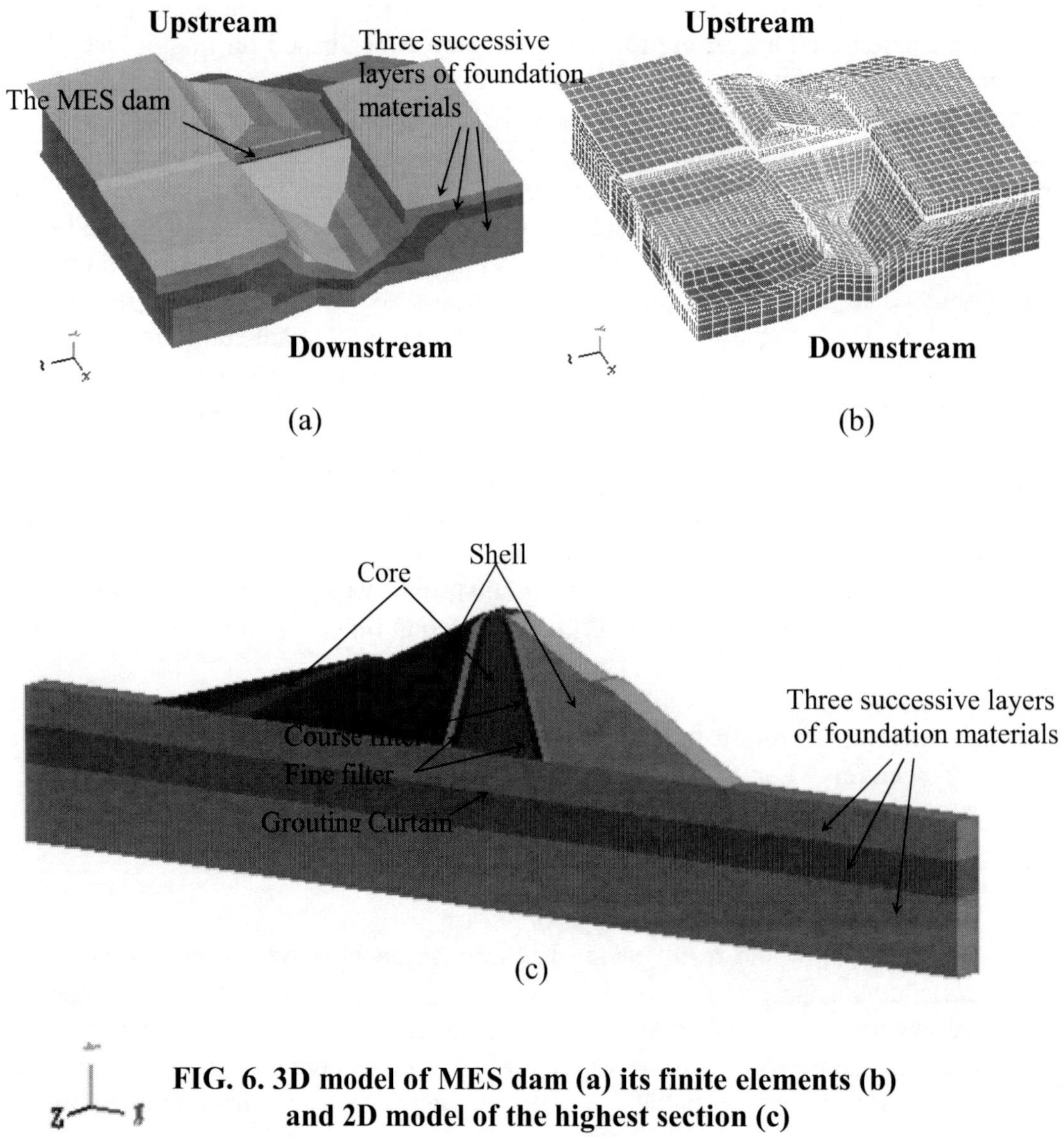

FIG. 6. 3D model of MES dam (a) its finite elements (b) and 2D model of the highest section (c)

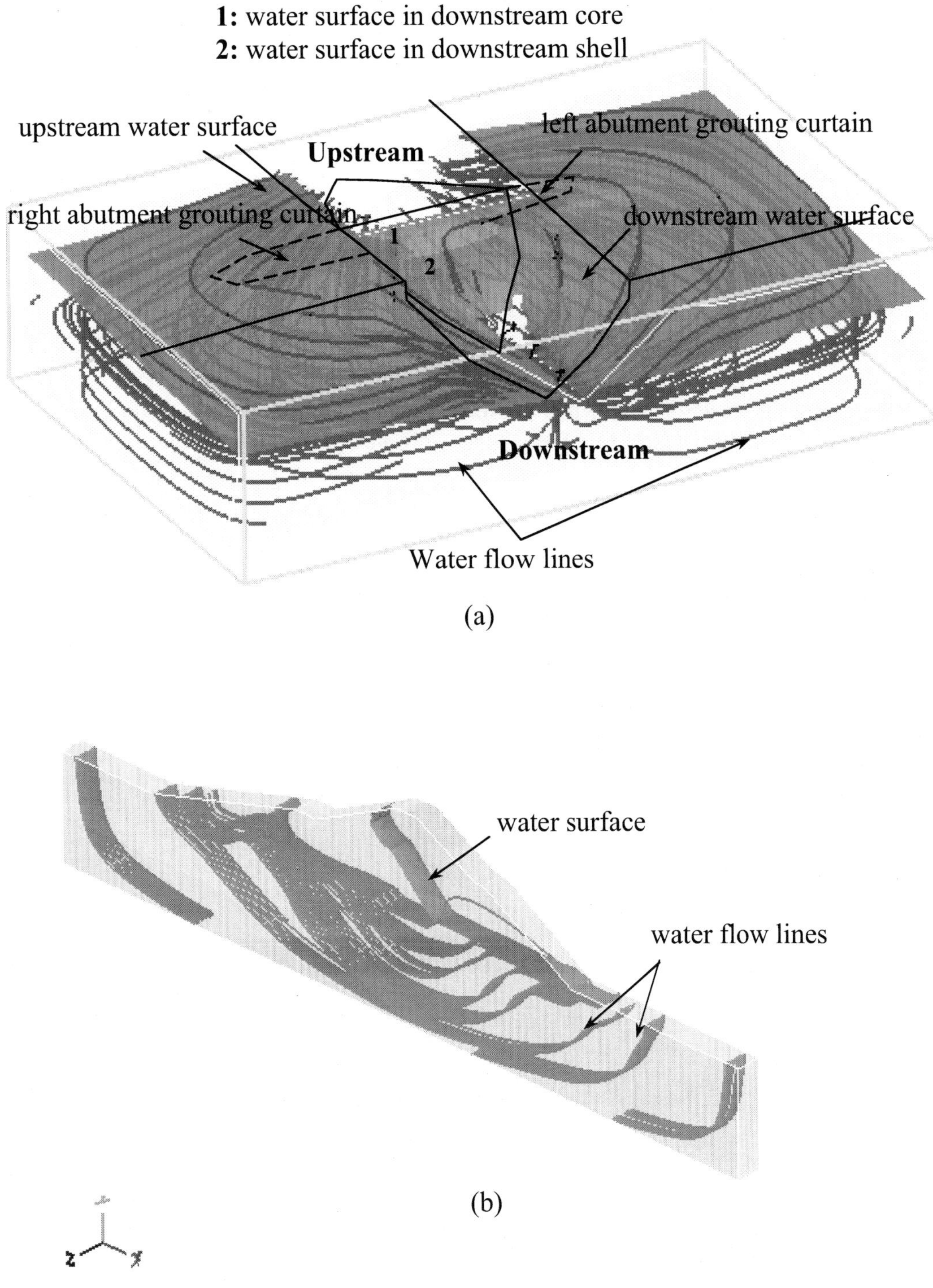

FIG. 7. Phreatic Surface around the Dam Structure and Distribution of Water Flow Lines from 3D Seepage Analysis (a) and those from 2D Seepage Analysis (b).

Water Elevation

Figure 9 shows measured water level at a downstream shell together with the water levels obtained from two and three-dimensional FE analysis for unsteady and steady state seepage conditions. At downstream distance of 100 m from the dam axis the observed water level stands at elevation of 235.6 m while the FE analysis for the unsteady and steady state seepage indicates that downstream water levels in that position are 237.4 m and 249.7 m for 3D model and 207.9 m and 214 m for 2D model seepage analysis; respectively. The results are summarized in Table 6. As indicated, the measured water level nearly coincides to the water level obtained from unsteady state seepage analysis. This predicts the fact that water flow has not yet reached to the steady state condition in the core and water discharge rate from the core is negligible. Therefore, the discharge rates from foundation and downstream side abutments account for the existing water level in the downstream shell.

Discharge Rate

The measuring weir (MW) installed at the downstream toe shows an average total discharge rate of 8.6 L/s from the core, part of foundation and side abutments. 3D finite element model analysis results in the total downstream discharge rate of 8.9 and 12.9 L/s for unsteady and steady state conditions; respectively. Total 2D downstream discharge rate obtained from seepage analysis of several cross sections of the dam are 5.12 and 6.88 L/s for unsteady and steady state conditions; respectively, indicating that the water discharge rate from the downstream side abutments to the downstream shell is 3.78 and 6.02 L/s for the two respective seepage conditions.

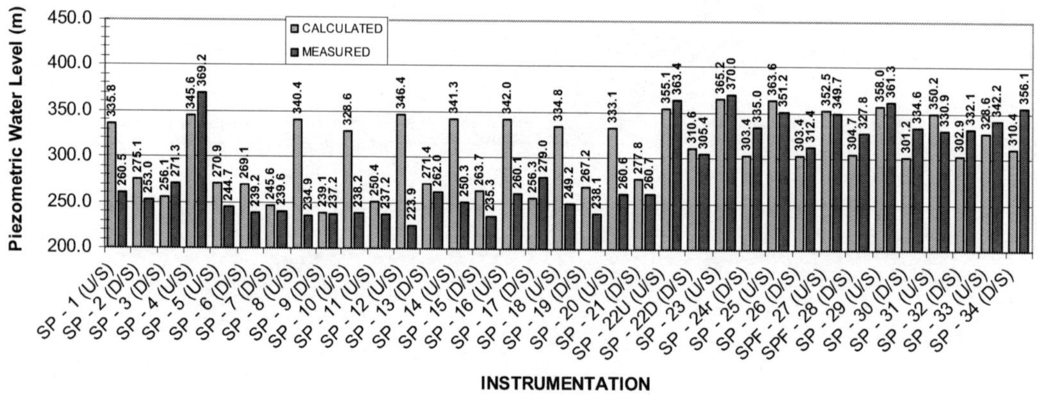

FIG. 8. Measured and Predicted Piezometric Water Elevation for Standpipe Piezometers.

Hydraulic Gradient

Hydraulic gradients at downstream toe obtained from three-dimensional seepage analysis are 0.422 and 0.495 for unsteady and steady state conditions; respectively while two-dimensional analysis renders hydraulic gradients of 0.102 and 0.110 for the respective seepage conditions; almost one-forth of 3D gradients. As demonstrated in Figure 7a, the concentration of flow lines from the side abutments to the river bed produces the higher gradients in the downstream toe of the dam. Figs 10 and 11 show changes in discharge rates and hydraulic gradients with variation in abutment and grouting curtain material permeability; respectively.

Figure 10a shows the variation of 3D downstream discharge rate and 2D downstream discharge rate per unit length of the highest section, with variation of abutment equivalent coefficient of permeability of 3.5×10^{-7} m/s for the steady state seepage condition. Figure 10b illustrates the variation of 2D and 3D hydraulic gradients at downstream toe with variation of abutment equivalent coefficient of permeability. Figures 11a and b illustrate the effect of side grouting curtain on downstream discharge rates and hydraulic gradients.

In order to facilitate the task of obtaining accurate results when performing two dimensional seepage analyses, two parameters are defined for this case study:

Flux Ratio (FR): The ratio of 3D discharge rate, with m^3/sec unit, from the downstream shell to the discharge rate per unit length, m^3/sec/m, obtained from the conventional 2D seepage analysis.

Gradient Ratio (GR): The ratio of 3D hydraulic gradient at downstream toe to the 2D hydraulic gradient.

Figures 12a and b represents the variation of FR and GR with respect to variation of abutment permeability. FR and GR increases when abutment coefficient of permeability increases to 3.5×10^{-8} m/s. Between permeability's of around 3.5×10^{-8} m/s and 3.5×10^{-6} m/s, FR remains constant while GR decreases indicating that 2D hydraulic gradient increases from a small value to a higher one while 3D gradient increases from a much higher value compared to the 2D result when the abutment permeability increases.

For the MES dam which is constructed in a narrow valley, based on the 2D seepage analysis of the highest section, the discharge rate per unit length from the downstream shell and hydraulic gradient at downstream toe is obtained. The two parameters are then multiplied by FR and GR; respectively, to give 3D downstream discharge rate and hydraulic gradient, much nearer to reality. Figures 13a and b shows the variation of FR and GR against the variation of grouting curtain permeability. FR and GR remain constant when the permeability of grouting curtain decreases to a very low value and remain constant again when the grouting curtain permeability approaches to the permeability of abutment material.

TABLE 6. Comparison Between Seepage Analysis Results Obtained from 2D and 3D Analyses.

Seepage Condition (1)	Observed (2)	2D		3D	
		Unsteady State (3)	Steady State (4)	Unsteady State (5)	Steady State (6)
Water Elevation at D/S Offset of 100 m	235.6	207.9	214	237.4	249.7
Percent Change	---	-90	-70	+6	+37
Hydraulic Gradient at D/S Toe	---	0.102	0.110	0.422	0.495
Discharge (L/s)	8.60	5.12	6.88	8.90	12.90

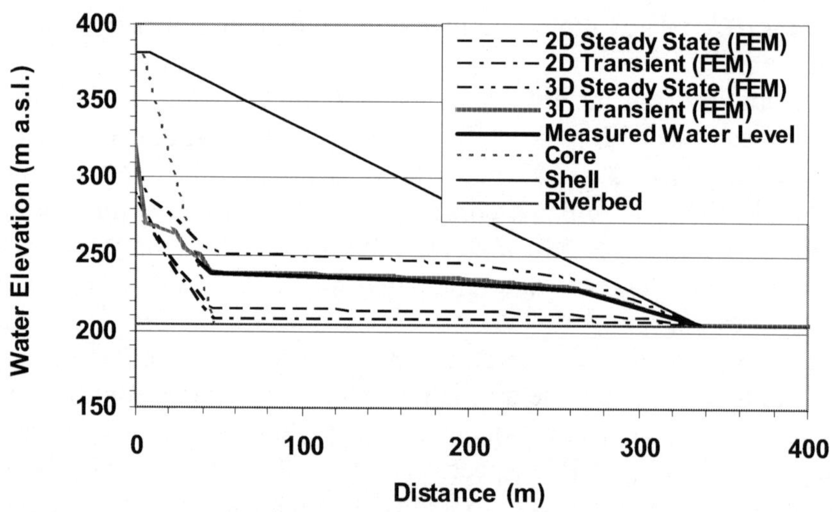

FIG. 9. Comparison between 2D and 3D water levels for unsteady and steady state seepage conditions.

(a) (b)

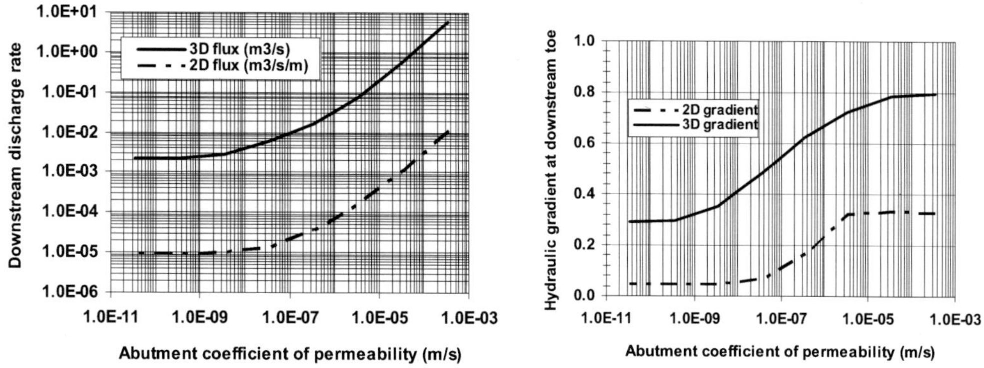

FIG. 10. Variation of 2D and 3D downstream discharge rate (a) and 2D and 3D hydraulic gradient at downstream toe (b) with abutment coefficient of permeability.

(a) (b)

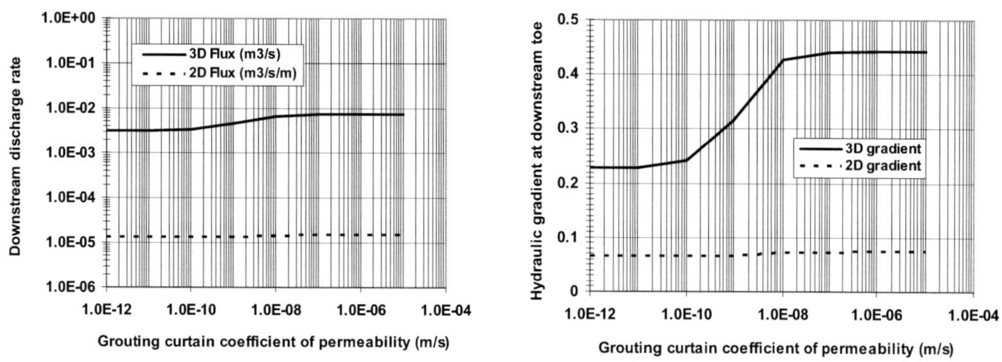

FIG. 11. Variation of 2D and 3D downstream discharge rate (a) and 2D and 3D hydraulic gradient at downstream toe (b) with grouting curtain coefficient of permeability.

(a) (b)

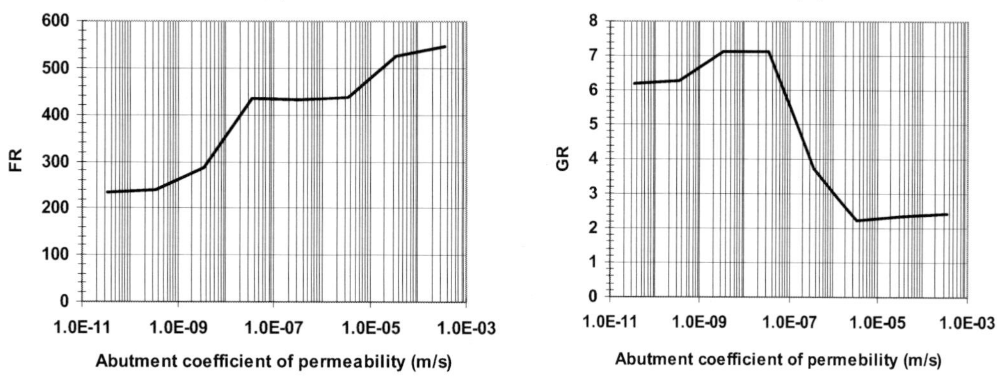

FIG. 12. Variation of FR (a) and GR (b) with abutment material permeability.

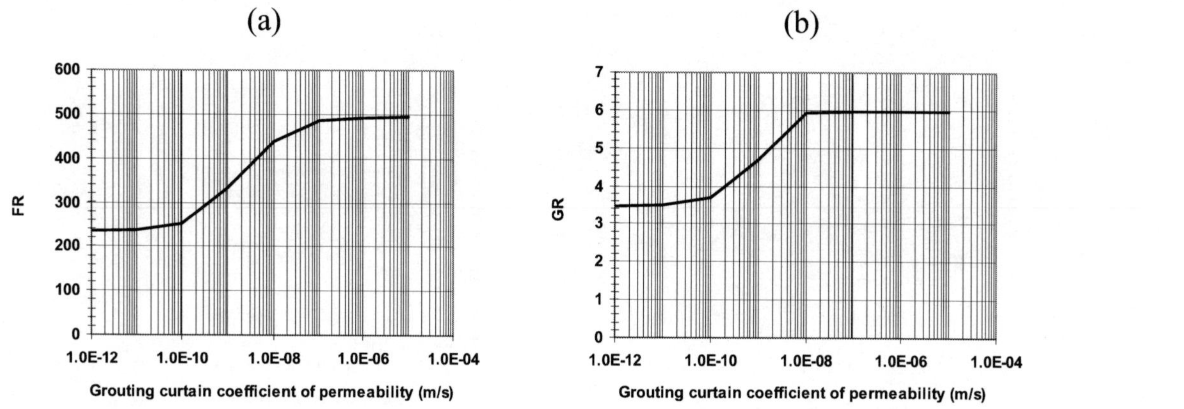

FIG. 13. Variation of FR (a) and GR (b) with grouting curtain coefficient of permeability.

CONCLUSIONS

In this paper several two and three-dimensional finite element analyses for unsteady and steady state seepage conditions were performed on an earth core rockfill dam. The results including pressure heads, discharge rates and hydraulic gradients were compared to those obtained from measurements of seepage instrumentation system installed in the foundation and dam body. Agreement between the calculated and the measured pressure heads for most of the instruments verified the suitability of the selected material properties and the finite element model for the seepage analysis. Due to the seepage through the side abutments into the downstream shell, the real water level, discharge rates and hydraulic gradients are much higher than those obtained from the two-dimensional seepage analysis. For the analysis results of the dam studied, the 3D steady state hydraulic gradient at downstream toe is 0.495, about 4 times higher than the corresponding 2D hydraulic gradient of 0.11. Total downstream discharge rate from 3D steady state seepage analysis is 12.9 L/s; nearly 2 times the discharge rate of 6.88 L/s obtained from sum of the 2D seepage analysis results of several cross sections. Therefore, in order to have more accurate seepage analysis results for an embankment dam located in a narrow valley, three-dimensional modeling of the dam site is vital. For this case study, several FEM based graphs were developed to calculate 3D discharge rates and hydraulic gradients from those obtained from 2D finite element seepage analysis.

ACKNOWLEDGMENTS

The authors are highly grateful to Manasazeh Corporation for providing Seep3D software and Moshanir Company for providing initial information of the case study to accomplish the investigation. Authors also would like to thank the anonymous reviewers of this paper for their suggestions and comments.

REFERENCES

Engemoen, W.O. (2003). "Assessing the risk of a seepage-related dam failure by means of failure mode identification risk analysis and monitoring practices." *Seminar on Safety Evaluation of Existing Dams*, U.S. Bureau of Reclamation.

Finn, W.D.L. (1967). "Finite element analysis of seepage through dams." *J. Soil Mech. and Found. Engrg.*, ASCE, 93(6): 41-48.

Geology and rock parameters for Godar-E-Landar hydroelectric power project. (1993). Iran Water and Power Resources Development Company.

Lugeon, M. 1933. *Barrages et Geologie,* Dunod, Paris.

Papagiannakis, A.T. & Fredlund, D.G. (1984). "A steady state model for flow in saturated-unsaturated soils." *Canadian Geotech. J.*, 21(13): 419-430.

Review of the Masjed-E-Soleiman dam behavior. (2004). Iran Water and Power Resources Development Company.

Seep3D User's guide. (2002). Calgary, Alberta: GEO-SLOPE International Ltd.

Segerlind, L.G. 1984. *Applied finite element analysis.* New York: Wiley.

Zienkiewics, O.C., Mayer, P. & Cheung, Y.K. (1966). "Solution of anisotropic seepage by finite elements." *J. Engrg. Mech.*, ASCE, 92(1): 111-120.

Zienkiewics, O.C. & Taylor, R.L. (1989). *The finite element method in engineering science.* London: McGraw Hill.

Site Characterization Philosophy and Liquefaction Evaluation of Aged Sands – A Savannah River Site and Bechtel Perspective

Michael R. Lewis[1], F. ASCE, P.E., Ignacio Arango[2], M. ASCE, P.E., and Michael D. McHood[3], M. ASCE

[1] Manager, Geotechnical Engineering, Bechtel Group, Savannah River Site, Aiken, SC 29808; *mike.lewis@srs.gov*
[2] Consultant and Former Manager Geotechnical Engineering, Bechtel Group, 22 Bowling Dr. Oakland, CA 94618; *iarango@bechtel.com*
[3] Senior Engineer Bechtel Savannah River, Inc., Savannah River Site, Aiken, SC 29808; *michael.mchood@srs.gov*

ABSTRACT: The authors describe two aspects of geotechnical engineering; site characterization utilizing the CPT and recognition of aging as a factor affecting soil properties. These were pioneered by Professor Schmertmann and are practiced by the Bechtel Corporation in general and at the Savannah River Site in South Carolina. This paper describes a general subsurface exploration approach that we have developed. It consists of "phasing" the investigation, employing the principles of the observational method suggested by Peck (1969) and others. We have found that the recommendations proposed by Sowers in terms of borehole spacing and exploration cost are reasonable for developing an investigation program, recognizing that through continuous review the final investigation program will evolve.

At the SRS a field investigation and laboratory testing program was devised to measure and account for aging as it relates to the cyclic resistance of the site soils. Recently, a panel of experts (Youd et al., 2001) has made recommendations regarding the liquefaction assessment of soils. This paper will address some of those recommendations in the context of re-assessing the liquefaction resistance of the soils at the SRS. It will be shown that, indeed, aging should be accounted for in liquefaction potential assessments for soils older than Holocene age.

KEY WORD: ageing, case history, CPT, earthquakes, liquefaction, sands, site characterization

INTRODUCTION

Professor Schmertmann's contributions to the Geotechnical Engineering profession, spanning over 50 years, have dealt with numerous aspects of soil mechanics important to practicing engineers. His papers, presentations, research reports, and technical discussions published in the ASCE Geotechnical Journals, at ASCE conferences, ASTM Special Technical Publications, at International Conferences, and Public Agency Research Reports cover many aspects of geotechnical engineering, but in particular they relate the application of laboratory and field testing to the strength and compressibility characterization of *in situ* soils. He published guidelines for the interpretation of Cone Penetration Tests (CPT) and Standard Penetration Tests (SPT) as early as 1970 (Schmertmann, 1970). His ideas about the potential of these two tests improved in the subsequent years through lessons learned from additional research and case histories. Although the SPT is giving way to many other *in situ* tests, the CPT and SPT still constitute two of the most important tools for geotechnical site characterization.

For the twenty-fifth Terzaghi Lecture, Professor Schmertmann chose the important topic of aging (Schmertmann, 1991) as it affects soil properties. In his lecture, he elaborated on the impact of aging on soil compressibility, stress-strain characteristics, static and cyclic strength, liquefaction resistance, and on other properties, based on numerous laboratory test results and observations compiled from well-documented case histories. The first part of this paper addresses exploration and the use of the CPT, while the second part of the paper addresses aging of soils and the role that it plays on the dynamic strength of soils.

BACKGROUND

The Savannah River Site (SRS) is located along the Savannah River in the upper portion of the Atlantic Coastal Plain of South Carolina, approximately 160 km upstream of Savannah, Georgia (Figure 1). The SRS occupies about 830 km^2 and is owned by the Department of Energy (DOE). Since its inception in the early 1950s, the SRS has been an integral part of our nation's defense. As a result, several critical facilities have been, and will continue to be, constructed and operated at the SRS. By their nature, these facilities demand the very best we can offer in design and construction and all of the trappings that follow nuclear and defense-related projects, all in an effort to ensure safety during construction, operation, and eventual decommissioning. From a geologic and geotechnical standpoint, the SRS presents a number of interesting challenges. We discuss two of those challenges in this paper; site characterization and how we utilize the CPT; and the effect that age plays in the cyclic strength of soil deposits.

SITE CHARACTERIZATION

For geotechnical engineers and geologists, site characterization is the most important aspect of our work, for without an accurate depiction of the subsurface conditions and the geology of a site, subsequent analyses are guesswork. In recent times, however, it appears this activity has been receiving less and less attention, or at least it may be taken for granted. We're not sure of the reasons, but we believe one aspect is the ever increasing reliance on modeling, parametric analyses, and statistical inference. While these activities are important and clearly play an integral role in site characterization, they are no substitute for carefully planned and executed subsurface exploration programs. In fact, they should go hand-in-hand.

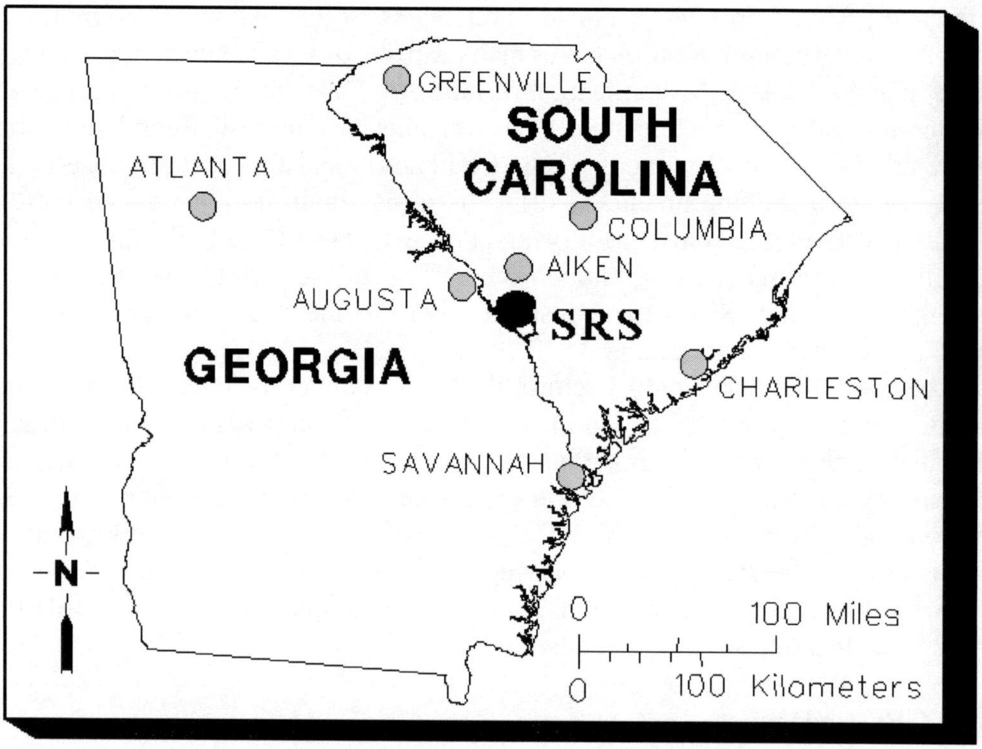

FIG. 1. Savannah River Site and Surrounding Region

What is site characterization? Gould (1985) described site characterization as "...to describe a site by a statement of its characteristics." Sowers (1979) described it as; "...a program of site investigation that will identify the significant underground conditions and define the variability as far as practical." More recently Baecher and Christian (2003) described site characterization as "a plan of action for obtaining information on site geology and for obtaining estimates of parameters to be used in modeling engineering performance." From our perspective, site characterization is the determination of subsurface conditions by;

- Understanding local/regional geology; (through site visits, geologic mapping and interpretation, and aerial photo interpretations),
- Performing appropriate geophysical surveys, borings, sampling, *in situ* testing (SPT, CPT, dilatometer, pressuremeter, field vane, etc.), and groundwater and piezometric observations,
- Completing appropriate laboratory testing, engineering analyses and modeling,
- Reviewing and interpreting the performance of nearby facilities, and
- Applying individual and collective experience, including site-specific (local) knowledge and general professional judgment.

Additionally, and probably more importantly, our experience is that an effective site exploration program must be flexible and continually reviewed and adjusted in real time as it proceeds. This can and does present challenges with regard to budget and schedule considerations.

The final objective is to predict the future performance of the proposed facility. As means to this end, we must understand; the geology, the groundwater conditions, the physical, mechanical, and dynamic properties of the affected strata, and the performance of existing facilities. In order to meet these objectives the characterization program needs to be well planned and communicated with the project team and the customer, and it must include two key components. First, the quality of the characterization data must be assessed continuously. Are the data adequate and accurate? Second, the data need to be interpreted and analyzed on a near real-time basis. What answers are suggested by the data? Do they make sense, and do we need additional information? In addition, all exploration programs have some uncertainty attached to the results; it's unavoidable. The question is how to keep uncertainty to a minimum given constraints such as budget and schedule.

Level of Effort

In developing an exploration program the scope is invariably reduced to cost. Historically our experience indicates that most non-geotechnical professionals will attempt to limit this expenditure, not because the cost isn't justified, but simply to "manage" the project. This so-called "low cost/high speed mentality" may be fine on some or even most projects, but it can be a recipe for disaster.

We clearly endorse project management principles and the need to manage the effort; as geotechnical professionals we also recognize the need for flexible investigation programs that take into account actual site-specific conditions and the inherent uncertainty we have in all of our site investigations. Therefore, it is incumbent upon the geoscience professional to ensure this philosophy is clearly understood by the decision makers on the project and the client; this is precisely where tools such as the CPT are invaluable.

At the SRS the cost of a routine boring to 50 m depth with split spoon sampling every 1.5 m is about three times the cost and takes four times longer than a seismic piezocone to the same depth (Note: It is the opinion of the authors that except for the most routine projects, if the CPT is to be utilized, it should be the seismic piezocone [$SCPT_u$] rather than the conventional CPT). Thus, at any stage of the investigation program, and for less time and money, much more stratigraphic detail can be obtained utilizing CPT technology as a first choice over conventional drilling and sampling methods. We do not advocate the abandonment of traditional borings as a technique for subsurface exploration. In fact, and as will be discussed in the next section, CPT technology should be combined with drilling and sampling (and other *in situ* testing) for a highly effective exploration program.

In our experience, on most major and critical projects the initial budget is normally at issue. Although heavily scrutinized, budgets generally are given to perform a scope of work, albeit ill-defined at the beginning of a project. Rather it is a combination of schedule, (having enough time to complete the initial program and evaluate the results), and/or revisions to the program based on actual conditions encountered (scope changes), that present the greatest challenge to investigation programs. In other words, changes to the original scope, even though they may be fully warranted, are difficult to get approved. Therefore, we are obligated to communicate risk and uncertainty, common to every program, to the decision makers. In this case risk can be in terms of money and time to complete a program, and/or technical risk if a program is not fully implemented or if it is cut short.

Risk and uncertainty can't be alleviated, but they can be managed. A discussion about risk and uncertainty is far beyond the scope of this paper; however based on experience, we can factor in actual results (cost and scope) for like projects and ensure that the project under consideration is not an outlier in terms of the proposed level of effort. For example, Figure 2 shows results of the number of borings/CPTs by facility area and hazard category for projects with which the authors have been involved, as well as results from familiar case histories. The hazard category is somewhat subjective and is not based on any hard and fast criteria; rather it is more qualitative based on the authors' experience. The results show considerable scatter in terms of hazard, but there is a distinct trend in terms of size; the larger the facility the more penetrations, but at a decreased density. These results are not unlike the suggestions of Sowers (1979) in the size range of about 1,000 to 100,000 m^2 for dams/dikes, multistory buildings, and manufacturing plants.

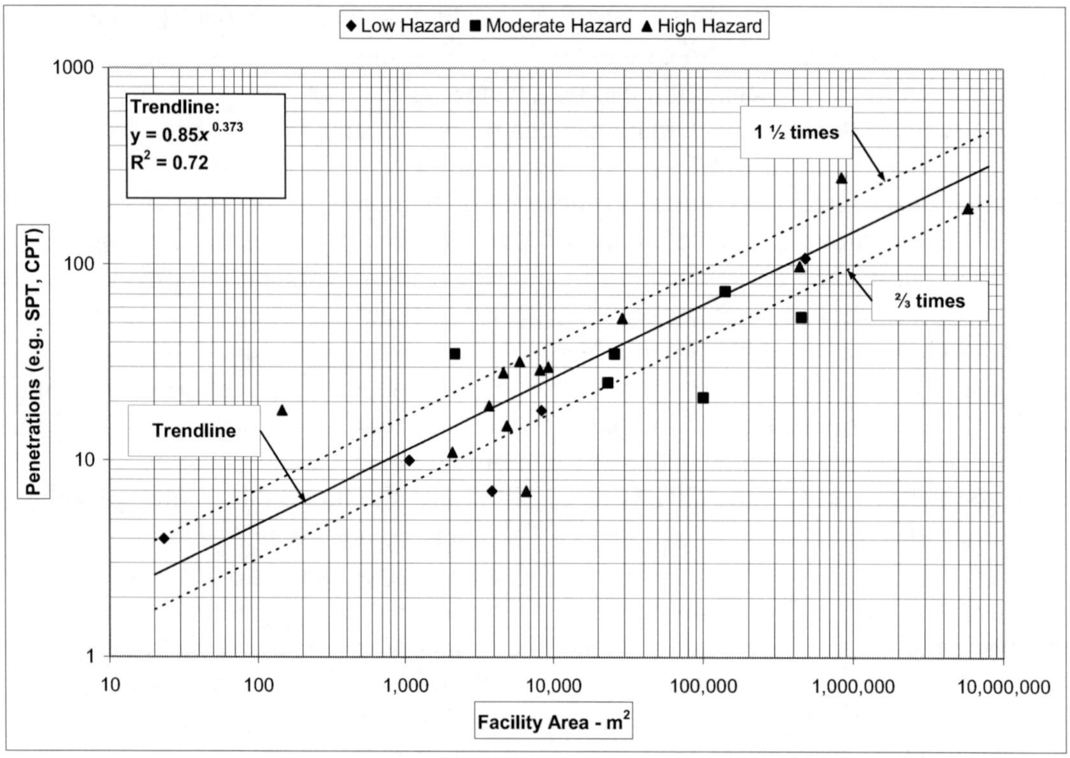

FIG. 2. Total Penetrations for various projects by hazard category

In the same way, Figure 3 depicts geotechnical cost in relation to the total estimated cost (TEC) of a particular project. The projects shown are those with which the authors have been involved or are projects found in the literature where reasonable cost information is available. The projects are categorized by focus on transportation, power, nuclear fuel handling, and liquefied natural gas (LNG). While the scatter is significant, trends are still obvious; the higher the estimated cost, the lower the geotechnical effort on a percent basis. The projects shown have an average geotechnical expenditure of approximately 0.6% for the range of TEC shown. The large differences result mostly from actual site conditions and the geologic variability associated with, particularly, the transportation projects, which traverse great distances and involve widely varying geologic and site conditions.

The results are not unlike other published cost information. For example, Sowers reports that for "an adequate investigation (including laboratory testing and geotechnical engineering)" the cost ranges from 0.05 to 0.2% of the TEC, but for critical facilities or facilities with unusual site or subsurface conditions the cost could increase to 0.5 to 1%. A range of site investigation costs as a function of TEC was reported by Sara (1994) for tunnels (0.3-2%), dams (0.3-1.6%), bridges (0.3-1.8%), roads (0.2-1.5%), and buildings (0.2-0.5%). Littlejohn, et al. (1994) report that for building projects in the United Kingdom the expenditure for site investigations ranges from 0.1 to 0.3% of TEC, however they also report that the perception of the respective clients was that the site investigation was, on average, five times higher. We're not sure how to interpret this discordance other than an apparent lack of communication coupled with geotechnical scope growth after award.

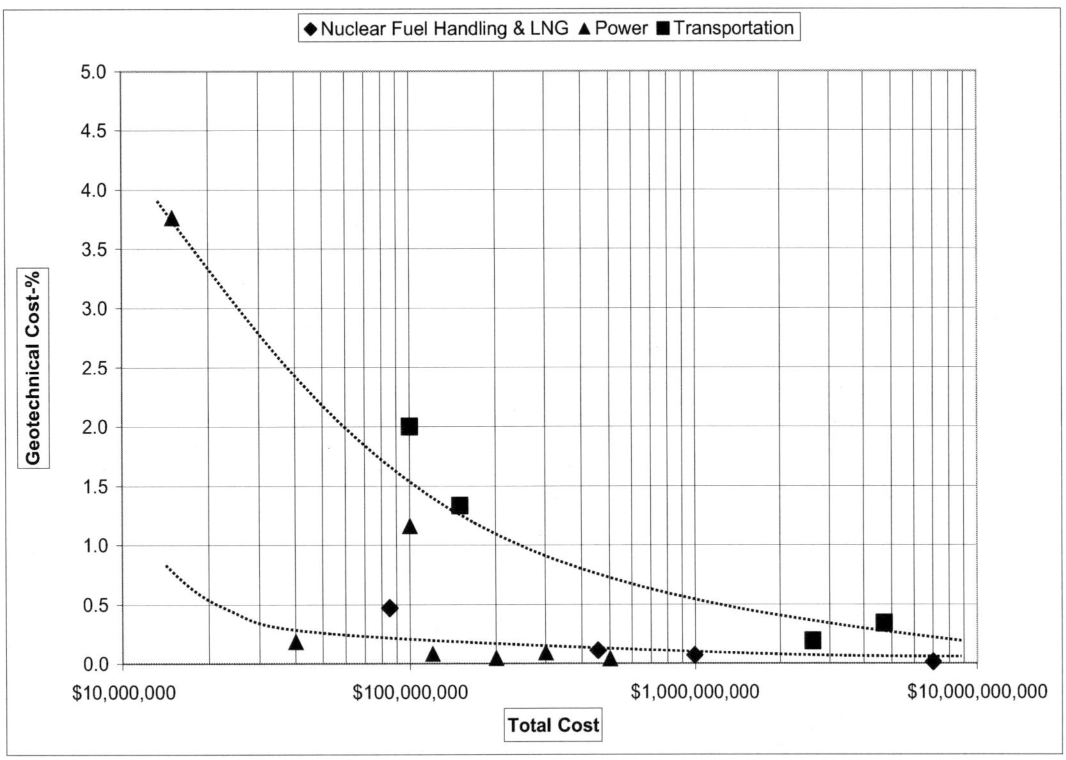

FIG. 3. Geotechnical cost as a function of total project estimated cost

Unfortunately, subsurface investigations are thought of as commodities that can be purchased off the shelf, based on a low cost/high speed mentality. Rather, each program is unique, designed to fit the project and the unique conditions that are inherent to every project with which we become involved. We suggest that information, such as that given in Figures 2 and 3 be developed and used in the planning stages of a project to "calibrate" the decision makers and clients on the level of effort required and to provide a sanity check on the baseline program established. In this way project managers and clients are included in the decision making process and are a part of the risk and uncertainty discussions.

These results are not meant as recommendations, as actual site conditions should dictate what is ultimately carried out. There are no building codes, regulatory documents or other hard and fast criteria that dictate the ultimate level of effort; there are only guidelines. There are however, particular attributes to every well planned and well executed characterization program above and beyond the level of effort discussed above. They are discussed subsequently.

Attributes of a Good Characterization Program

So, what constitutes a good characterization program? First, each site and facility is unique, thus all characterization programs are unique, or should be. Too often characterization programs (including the reporting) are re-cycled with a cut-and-paste mentality. Although this approach may suffice in some instances, it is a slippery slope that really should be avoided.

For a characterization program to be as successful as possible it must be tailored to the specific project under consideration and it must be sufficiently flexible to adapt to changing conditions as they are encountered. From our experience a successful program is done in five basic phases: 1) reconnaissance, 2) proposal or preliminary design, 3) detailed design, 4) construction, and 5) post-construction monitoring. Each phase has a specific purpose and can vary considerably given the specific project conditions.

The reconnaissance phase is generally done for planning purposes and feasibility studies. The effort generally entails researching the site and surrounding area by reviewing historical reports, topographic maps, geologic maps, soil surveys, aerial photographs, field visits, and performance surveys of existing structures.

The proposal or preliminary design phase may only include the reconnaissance phase, but it could also include a limited field exploration with widely spaced borings, CPTs, and geophysical tests. It can include some laboratory testing and simplified analyses for conceptual design, and/or cost estimating purposes.

The detailed design phase is where the bulk of the characterization program is performed. It includes detailed field exploration, such as sample borings (SPT and undisturbed sample borings), CPT soundings, dilatometer and pressuremeter soundings, field vane tests, and geophysics. Representative samples of the subsurface materials are taken and sent to a laboratory for testing. Testing generally includes index tests, and tests for static and dynamic strength and compressibility. Depending on the size of the project and the complexity of the subsurface, this phase may be subdivided into additional phases. For example, the initial phase might include CPT soundings to determine site stratigraphy. A second phase would then target specific horizons for undisturbed samples for laboratory testing, in addition to the more routine standard penetration test borings. In the experience of the authors, for critical projects (critical can be defined in terms of safety or monetary expenditure) a phased approach for the detailed design phase is highly recommended. It allows "pinpoint" sampling of specific horizons rather than sampling at pre-selected depths. This tends to focus the effort on those strata that have the greatest potential effect on the facility. It also adds needed flexibility to the program, which is required if the exploration program is to be successful. Without the flexibility to adjust locations, depths, sample types, and the type of exploration to meet the conditions encountered, the characterization program is doomed. Unfortunately, in many cases once the program has been agreed to and initiated, we tend to manage the cost and schedule at the expense of gathering needed data. Communication with the project team, and in particular the project manager, is critical to success. This also requires full time oversight and direction of the program by qualified geotechnical engineers and geologists dedicated to the effort and who will continue and follow through on the project as it moves from the investigation phase into the design and later construction phases.

Phases 1, 2, and 3 should be carried out on every project. Phases 4 and 5 will depend on the success of the initial program and on any scope changes or any unknown subsurface conditions encountered during construction. However, the level of effort required for each phase may vary considerably based on the type and size of the project and complexity of the subsurface. The key point is that whatever the program entails, it needs to be flexible and the engineer/geologist must be able and allowed to adapt the program to the conditions encountered. In the authors' experience, this doesn't necessarily mean the program will grow, however communication with the project team and/or owner is crucial. On any project, large or small, simple or complex, we are still working with a schedule and budget. And any deviation from either causes concern even though the deviations are valid given the subsurface conditions that dictated the change.

Use of the CPT

Following the trend observed in the industry in general, use of CPT technology at the SRS, and within Bechtel, has progressively increased since the late 1980s in an effort to meet the aforementioned objectives on a project by project basis. This evolution has resulted in a basic exploration philosophy for critical facilities. "Utilize the CPT early and often in a project following with borings and "pinpoint sampling" of targeted strata for further evaluation and laboratory testing."

Several particularly important advantages of CPT technology have been recognized and thus used to further enhance the quantity and quality of geotechnical exploration performed by the authors:

- More exploratory penetrations due to lower cost and less field time as compared to traditional drilled borings (CPTs at the SRS are about ½ to ⅓ the cost and take about $^1/_5$ the time of a SPT boring of equal depth),
- Higher vertical resolution due to nearly continuous measurements, allowing for superior stratigraphic interpretation, detection of layers of special interest, including very thin, loose, or compressible layers, which can be used to determine target intervals for further adjacent sampling and subsequent laboratory testing,

- Highly repeatable measurements within similar material types or layers because of standard and automatic testing and data acquisition methods, and
- Multiple measured parameters including tip stress, sleeve stress, friction ratio, pore pressure, and shear wave velocity for resolving material characteristics, including initial stiffness.

At the SRS, and within Bechtel, the CPT is used primarily to establish stratigraphy, identify any anomalous strata (soft or compressible soil), and acquire a preliminary estimation of specific engineering soil properties for design. However a word of caution, verification and calibration of site-specific correlations of engineering parameters determined with CPT parameters is highly recommended, as correlations shown in the literature do not fit all conditions.

AGED SOILS AT THE SRS

In the shallow subsurface beneath the SRS, Eocene and Miocene age (35 to 50 million years old) sediments of the Altamaha, Tobacco Road and Upper Dry Branch Formations are composed primarily of laminated, gap-graded, clayey sands deposited under alternating marginal marine and fluvial conditions. The clay, chiefly kaolinite and illite, binds the sand grains and appears to have been formed by *in situ* weathering. The CPT tip resistances of this material range from less than 1 MPa to about 15 MPa, and CPT friction ratios (sleeve resistance divided by tip resistance) range from less than 1% to over 10%. Corresponding Standard Penetration Test (SPT) N-values range from less than 10 to over 20 blows per 0.3 m. Because of the relatively low penetration values and the relatively high seismic exposure (proximity to Charleston, SC), studies of the liquefaction vulnerability utilizing the empirical liquefaction chart suggested by Seed et al. (1985) indicated that the site was potentially vulnerable to seismic liquefaction. It is well known that the empirical chart is based on observations of the performance of Holocene deposits. Since the soils at the SRS are geologically much older than Holocene-age, the question logically arose regarding whether the empirical chart was appropriate for the liquefaction evaluation at the SRS. To resolve the concern, two tasks were completed: an extensive field and laboratory geotechnical investigation at the site, and a review of available opinions and data in the technical literature on the liquefaction vulnerability of geologically old sand deposits.

Geotechnical Investigations at the SRS

For the formations of interest (Tobacco Road and Dry Branch Formations) there were no paleoliquefaction events (case histories) to draw upon at the SRS or in the vicinity. For this reason, the decision was made to perform a detailed geotechnical exploration program including field testing, undisturbed sampling, and dynamic testing of carefully sampled soil specimens in the laboratory. The program was developed and implemented by Bechtel Savannah River Incorporated (BSRI), and the dynamic laboratory testing was carried out at the University of California at Berkeley (UCB) laboratory (BSRI, 1993 and 1995).

Samples were obtained by a fixed-piston sampler using controlled techniques and sampling procedures well established at SRS. Measurements, including X-ray photography, were performed on each sample tube prior to packing and transporting and after being received at the UCB laboratory. The laboratory testing included index testing and the determination of dynamic strength, volumetric strain after liquefaction and an evaluation of the influence of confining pressure, leading to site-specific recommendations for K_σ, a factor that normalizes the cyclic resistance of a soil (CRR) to an overburden pressure equal to one atmosphere. All of the test results were correlated back to a sample specific $(N_l)_{60}$, or $(q_t)_l$, where $(N_l)_{60}$ is the SPT penetration resistance normalized to one atmosphere and 60% energy level, and $(q_t)_l$ is the CPT tip resistance normalized to one atmosphere.

Site-specific sampling and laboratory testing were completed for two facilities at the SRS. In the first, a series of 17 stress-controlled, isotropically consolidated undrained cyclic triaxial tests were performed for site A. The samples were obtained adjacent to (within 1.5 to 3 m) SPT boreholes at locations exhibiting low N-values. To aid in the evaluation, SPT energy measurements were obtained and used later to correct the field N-values to N_{60}. Though the fines content of the samples varied, a single "CRR design curve" for these soils was established. The overall assessment resulted in three data points relating $(N_1)_{60}$ to CRR. The samples tested to develop the SRS curve were classified as SC soils (unified classification system) and had plastic fines with contents ranging from about 9 to 29%, and an average of about 17%.

Secondly, to correlate CPT $(q_t)_1$ with the cyclic resistance values obtained in the laboratory, CPT soundings were pushed adjacent to the Site A borings described above. In the same way, 18 additional high quality fixed-piston samples were obtained at site B from boreholes adjacent to (1.5 to 3 m) 18 CPT soundings. These samples were also sent to the UCB laboratory for dynamic testing (stress-controlled, istropically consolidated, undrained cyclic triaxial tests). The laboratory results were evaluated in the same manner described above except that the CPT $(q_t)_1$ was utilized instead of $(N_1)_{60}$. The samples tested were SC soils and had plastic fines contents ranging from about 16 to 34%, with an average of about 24%. Table 1 summarizes the relevant data for the combined data set. Note that although 35 cyclic triaxial tests were performed, only 12 data points are shown. This is due to grouping of like material in terms of fines content and $(q_t)_1$.

Prior to the re-evaluation (discussed subsequently), a suite of curves based on plastic fines content was established. The lines representing various fines contents were constructed based on the laboratory test results described above and the trends of Seed et al.'s (1985) empirical chart, (i.e., the clean sand passing through the origin of coordinates), and engineering judgment, assuming the suite of curves was more or less parallel. That relationship (developed in 1994-1995) has recently undergone a re-evaluation to take into account newer information (since 1995), including results from Youd et al. (2001) and Idriss and Boulanger (2004).

The re-evaluation included a review of all the SRS data, but in particular, it centered on the shape of the clean (<5% fines) curve at low penetration resistances. For example, both Youd et al. (2001) and Idriss and Boulanger (2004) show the clean sand (≤5% fines) curve becoming flatter at low penetration resistances and intersecting the ordinate at a CRR value of 0.05.

However, for the re-evaluation of the SRS data we relied more on the Idriss and Boulanger relationship because; 1) the Idriss and Boulanger clean curve "fits" our site-specific data more closely than the Youd et al. curves, and 2) as expected, at higher penetration resistances the Idriss and Boulanger curve for Holocene soils results in a more conservative estimate of CRR. Thus, for the revised SRS clean curve we have adopted the Idriss and Boulanger (2004) relationship for the CPT to "construct" the revised SRS-specific relationships.

TABLE 1. SRS Data Summary

No.	Fac.	Geo. Form.	USCS	$(q_t)_1$ (MPa)	D_r (%)	$(V_s)_1$ (mps)	Fines (%)	LL (%)	PI (%)	γ_d (kN/m^3)	CRR_f
1	A	UTR	SC	0.9	<5	293	28.7	51	30	16.5	0.167
2	A	UTR	SC	2.0	15	268	18.5	49	26	15.4	0.138
3	A	LTR	SP-SM	2.3	20	255	9.6	NP	NP	15.6	0.095
4	A	LTR	SC, SM, SP-SC	3.3	32	257	15.7	41	17	15.1	0.135
5	A	LTR	SP-SM, SC	5.0	45	253	10.8	NP*	NP*	15.4	0.115
6	B	TR3	SC	1.7	10	247	34.0	47	28	16.4	0.152
7	B	TR3	SC	0.5	<5	223	33.6	48	32	17	0.173
8	B	TR3	SM-SC, SC	1.1	<5	267	26.3	50	29	16.2	0.165
9	B	TR3	SC	0.6	<5	201	20.5	48	29	16.4	0.149
10	B	TR3/TR1	SC	1.8	11	162	19.7	60	37	16.5	0.134
11	B	DB1/3	SC	1.9	14	269	16.1	39	18	16.1	0.117
12	B	DB1/3	SP-SC, SC	1.1	<5	206	18.6	80	62	15.1	0.139

UTR is Upper Tobacco Road; LTR is Lower Tobacco Road; TR1 is Tobacco Road 1; TR3 is Tobacco Road 3; DB1 is Dry Branch 1; DB3 is Dry Branch 3; USCS is the unified soil classification system; $(q_t)_1$ is the normalized cone tip resistance; $(V_s)_1$ is the normalized shear wave velocity (normalization per Andrus and Stokoe); LL is the liquid limit; PI is the plasticity index; γ_d is the dry density and CRR_f is the field-corrected cyclic resistance ratio. The estimate of relative density (D_r) is based on Tatsuoka et al., 1990 (given in Ishihara and Yoshimine, 1992). * Two of the three samples for point #5 were NP.

To develop the SRS "aged" clean curve, we utilized the low end of the strength gain factor range (1.3) proposed by Lewis et al. (1999) for clean sands (discussed below). Thus, for the revised SRS "aged" clean sand relationship, the y-intercept was 1.3 times 0.05, (the revised ordinate for the clean curve, Youd et al., 2001) or 0.065. In addition, the shape of the revised clean sand curve was assumed to be that of the Idriss and Boulanger (2004) clean sand curve, which is similar to the shape given in Youd et al. (2001) at low penetration resistances. Thus, for the clean "aged" curve, a factor of 1.3 was applied over all penetration resistances to the Idriss and Boulanger clean curve to derive the CRR corresponding to the "aged" clean SRS curve. Utilizing a constant factor to increase the curve across penetration resistances is consistent with the work of Polito (1999) and Polito and Martin (2001), for fines contents below about 40%. Using a constant factor for a given fines content independent of the penetration resistance and adopting the shape of the Idriss and Boulanger (2004) Holocene clean sand curve, the remainder of the SRS CRR curves were developed for various fines contents simply by the ratio of the site-specific data (CRR_f) to the adopted Holocene clean sand curve ($CRR_{I/B}$) of Idriss and Boulanger (2004), applied over all penetration resistances. For example, the 10% fines content (FC) curve utilized data points 3 and 5 from Table 1. Point 3 has a normalized tip stress ($[q_t]_1$) of 2.3 MPa and point 5 has a $(q_t)_1$ of 5.0 MPa. The ratios of the CRR at corresponding CRR from the Idriss and Boulanger clean curve is; 0.095/0.058 = 1.64 for point 3; and 0.115/0.079 = 1.45 for point 5; considering both points the ratio would be about 1.6. The resulting SRS 10% fines CRR curve would be 1.6 times higher than the Idriss and Boulanger clean curve. In the same way curves can be constructed for fines contents of 15, 20, 25 and 30%. Table 2 and Figure 4 summarize the evaluation results for each fines content curve (Figure 4 also shows the previous, 1995, relationship).

TABLE 2. Summary of CRR Design Curve Factors

FC Curve (%)	Data Point (Table 1)	Actual FC (%)	$(q_t)_1$ (MPa)	CRR_f	$CRR_{I/B}$	Ratio $CRR_f/CRR_{I/B}$	Ratio selected for FC group
10	3	9.6	2.3	0.095	0.058	1.64	1.6
10	5	10.8	5.0	0.115	0.079	1.45	1.6
15	4	15.7	3.3	0.135	0.064	2.10	2.1
15	11	16.1	1.9	0.117	0.056	2.11	2.1
20	2	18.5	2.0	0.138	0.056	2.46	2.6
20	9	20.5	0.6	0.149	0.051	2.93	2.6
20	10	19.7	1.8	0.134	0.055	2.44	2.6
20	12	18.6	1.1	0.139	0.052	2.67	2.6
25	1	28.7	0.9	0.167	0.052	3.24	3.1
25	8	26.3	1.1	0.165	0.052	3.17	3.1
25	9	20.5	0.6	0.149	0.051	2.93	3.1
30	1	28.7	0.9	0.167	0.052	3.24	3.4
30	7	33.6	0.5	0.173	0.050	3.43	3.4

Note: Point 6 from Table 1 is not in Table 2. This data point has not been incorporated in the evaluation because it is not consistent with the results of the entire data set. We believe this data point to be somewhat anomalous; $CRR_{I/B}$ refers to the CRR utilizing the Idriss & Boulanger (2004) clean curve for CPT.

The data show that compared to the Idriss and Boulager (2004) Holocene clean (less than 5% fines) curve for CPTs, the increase in dynamic strength ranges from 1.6 to 3.4 for fines contents from 10 to 30%.

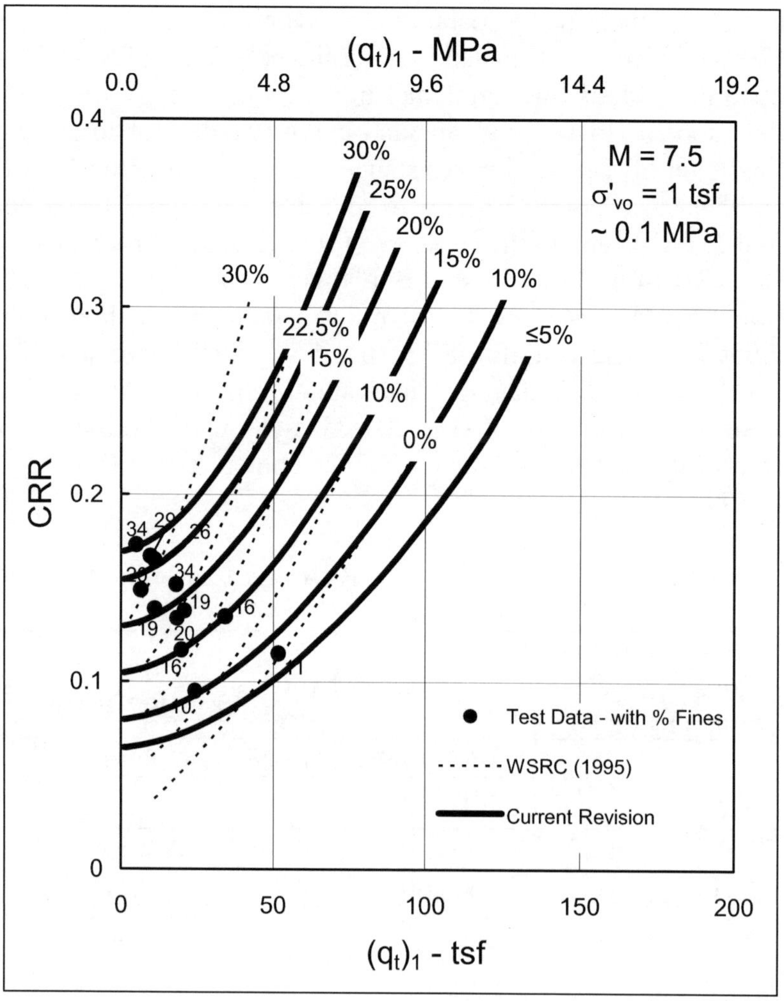

FIG. 4. CRR vs CPT $(q_t)_1$

Review of Data on the Performance of Aged Soil Deposits

Several investigators have addressed the issue of soil aging; among them are Youd and Hoose (1977), Seed (1979), Skempton (1986), Kulhawy and Mayne (1990), Martin and Clough (1990, 1994), Schmertmann (1991, 1993), BSRI (1993, 1995), Arango and Migues (1996), Lewis et al. (1999), and Leon et al. (2006). The results of some of the more significant findings are summarized below.

Seed (1979) considered the cyclic resistance of laboratory-prepared samples and of hydraulic fills of different ages (up to about 3000 years) and concluded that the data indicated "the possibility of increases in cyclic mobility resistance of the order of 75% over the stress ratios causing cyclic pore pressure ratios of 100% in freshly deposited laboratory samples, due to long periods of sustained pressure in older deposits."

Skempton (1986) discussed the evidence for increase in the deformation resistance of sand with the increased duration of sustained loading. He considered the increase in penetration resistance blow count (N-value) as a reflection of the increase in resistance to deformation. He found that the ratio between the normalized SPT,

$(N_1)_{60}$ blow count, and the square of the relative density, (D_r^2), varied with the period of sustained loading. Skempton reported strength gains (increases in $(N_1)_{60}/D_r^2$ relative to that predicted for samples in the laboratory) for normally consolidated sands of about 14% and 57% at 10 years and >100 years, respectively after deposition.

Kulhawy and Mayne (1990) compiled the values of the same parameter as Skempton, $(N_1)_{60}/D_r^2$, for several fine and fine to medium sand deposits of known geologic age, including some of the same data evaluated by Skempton (1986). They concluded that the parameter $(N_1)_{60}/Dr^2$, is influenced by particle size, overconsolidation, and aging. Although they acknowledge some of the data may be "imprecise", they developed a conservative relationship ($C_A = 1.2 + 0.05\log(t/100)$) to account for aging through the parameter $(N_1)_{60}/D_r^2$. We consider this relationship to be a lower bound of potential strength gain with time.

Lewis et al. (1999) reviewed published data compiled from the 1886 Charleston, S.C. earthquake. There are no quantitative data available regarding the magnitude of the event or of associated peak ground accelerations; however, the earthquake's moment magnitude has since been estimated to be between about 7 and 7.5. Independent studies carried out by several investigators estimated the epicentral acceleration somewhere between 0.3g and 1.0g. Lewis et al. (1999) concluded a reasonable range of acceleration was between 0.3-0.5g.

Relic liquefaction features have been investigated by many along the eastern seaboard (e.g., Talwani and Cox, 1985; Obermeier et al., 1986; Dickenson et al., 1988; Amick et al., 1990; and Martin and Clough, 1990 and 1994). Features were found primarily in the sands and silty sands of two ancient ridges dating back 130,000 to 230,000 years and located 10 miles inland. The beach processes led to sands and silty soils being concentrated in the highest portions of the beach ridges. For their study, Lewis et al. (1999) reviewed data collected by Martin and Clough (1990, 1994) and Dickenson, et al. (1988). These included borings, velocity profiles, piezocone probes and trenches. Grain size tests showed that the sands at the sites are non-plastic, clean, with fines content less than 5% (14 locations), and 10% (5 locations). The sands in the remaining sites were described as SP, SM material. For the evaluation of these data, Lewis et al. (1999) calculated the induced cyclic shear stresses at the depths of interest for each deposit utilizing peak ground surface acceleration of 0.5g. A lower boundary was established showing the minimum stress ratios required to cause liquefaction at those sites that experienced marginal liquefaction and liquefaction. The strength gain of the boundary, relative to the clean sands curve in the empirical chart by Seed et al. (1985), was found to be about 2.2. Similarly, an upper boundary was established that separates the maximum cyclic stress ratios tolerated by the soil with no liquefaction, from those sites that experienced limited to widespread liquefaction. The strength gain in this case was calculated to be about 3.0. In the same way, using a lower bound acceleration of 0.3g resulted in computed strength gains of 1.3 to 1.8. (Note: The 1.3 factor was applied above for the re-evaluated SRS "aged" clean curve).

Arango and Migues (1996) performed investigations after the occurrence of the January 17, 1994, Northridge earthquake. The area selected for the study was within the Gillibrand Quarry site in the Tapo Canyon, north of Los Angeles. Area acceleration levels exceeded 0.5g, resulting in the failure of a small water-retaining dam in the quarry. However, in nearby Simi Valley, an old deposit of sand showed no signs of liquefaction. This sand has been estimated to be approximately one million years old. Although this deposit is now exposed in outcrops, it was previously buried by as much as 460 m of overlying soil. It is relatively uniform, fine quartz sand (SP) with less than 5% non-plastic fines. In its current state, the sand is lightly cemented, such that it can support vertical faces when dry, but is weak enough to crush with the slightest pressure between one's fingers. Microscopic examination reveals a high degree of quartz grain overgrowth, evidence of age and burial.

The field exploration program utilized drilled and augered boreholes, CPT soundings, test pits, and undisturbed block sampling techniques. A total of 18 stress-controlled cyclic triaxial tests to classify and determine the static and dynamic strengths of the sand were carried out at the Geotechnical Laboratory at the University of California, Berkeley. The range of field cyclic resistance ratio (CRR), based on results of laboratory testing, was estimated to vary between 0.80 and 1.37. Based on these results, and adopting a predicted induced cyclic stress ratio equal to 0.50 for Holocene-age sands from Seed, et al., (1985), the increase in dynamic strength ranges from 1.6 to 2.7.

Leon, et al. (2006) investigated the effect of age at four sites in the South Carolina coastal plain. The parameters reviewed were SPT N-values, CPT $(q_c)_1$, and normalized shear wave velocity, $(V_s)_1$. The four sites ranged in age from 546 years to 450,000 years old. The fines content from samples at all of the sites ranged from 0-9%, averaging 4%. The results of their evaluation indicated that these coastal plain soils had increased resistance to liquefaction by a factor ranging from 1.3 to 2, with an average of 1.6 (compared to the Youd et al., 2001 relationships for Holocene soils), induced by a magnitude 7.5 earthquake. The specific factors and ages reported for each of the four sites were; Ten Mile Hill site A: 1.3, 3548 yrs., Ten Mile Hill site B: 2, 200,000 yrs., Sampit: 1.5, 546-450,000 yrs., and Gapway: 1.7, 3548-450,000 yrs.

Figure 5 compares the predicted strength gain from the SRS studies reported above, and the historical data reviewed. Note that for the SRS re-evaluated data, the strength gain is relative to the clean CPT curve from Idriss and Boulanger (2004). Utilizing the Seed et al. (1985) relationship would result in strength gains of approximately 10 to 20% less. In either case we note the consistency between the results of the SRS investigation and the field data from South Carolina, and Southern California. Furthermore, the results are also compatible with the extrapolated trends suggested by Seed (1979), Skempton (1986), and Kulhawy and Mayne (1990).

It is interesting to note that trends shown on Figure 5 relating strength gain with age utilizing the work based on SPT N-values (Skempton, 1986 and Kulhawy and Mayne, 1990) are at the low end of the data shown, particularly older than about 1,000,000 years. This may be an indication that the SPT N-value is a poor indicator of strength gain with time for very old deposits (Note: The trend of Kulhawy and Mayne is an acknowledged conservative trend, as they show other data with higher $(N_1)_{60}/D_r^2$ for ages up to 10^8 years). However, it appears the Kulhawy and Mayne relationship can be used as a lower bound for the data shown, and a similar relationship (utilizing the same functional form suggested by Kulhawy and Mayne) can be used for an upper bound trend ($C_A = 1.92 + 0.23\log(t/100)$).

CONCLUSIONS

The CPT has enhanced our capability to perform subsurface exploration within Bechtel and at the SRS. Utilizing the CPT early in a project adds flexibility to the program and allows greater site coverage with a given budget in a shorter period of time. Initial program development following the suggestions of Sowers provides a reasonable starting point. Properly verified site-specific correlations between CPT parameters and laboratory testing add a dimension that can be very powerful when assessing site conditions and performing design related activities.

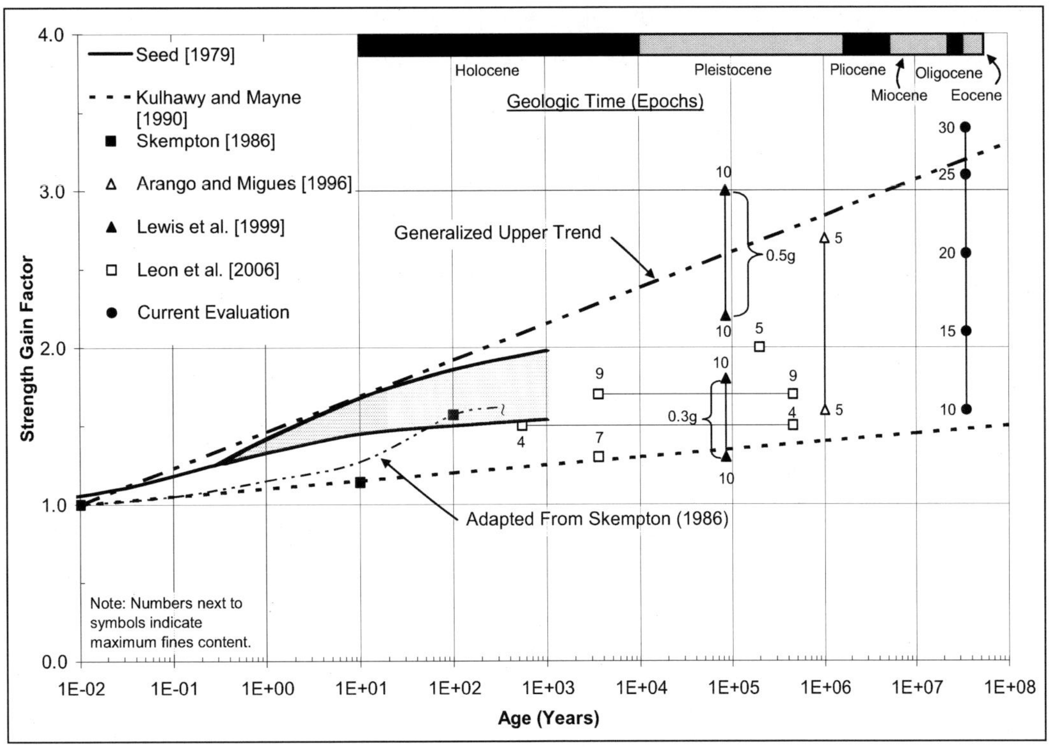

FIG. 5. Strength Gain with Age

The key component however in any exploration program is still effective communication with decision makers at all levels of the program. Full time geotechnical oversight will enhance and facilitate the needed communication and will allow quick and early decisions to be made during the program. With these attributes, utilizing the CPT will result in a program that allows maximum flexibility and affords superior stratigraphic definition through continuous or near continuous data, excellent repeatability and data reliability, and time and cost savings. The result is more high quality data, including "pinpoint" sampling and testing of targeted strata. While there will always be a need for soil borings and laboratory testing, the amount should decrease with increased use of the CPT. However, knowledge about the subsurface conditions will increase. This has been commonly recognized as far back as 1978 (Schmertmann, 1978): "Although engineers with much CPT experience in a local area sometimes conduct site investigations without actual sampling, in general one must obtain appropriate samples for the proper interpretation of CPT data. But, prior CPT data can greatly reduce sampling requirements."

In terms of aging, it is widely recognized by the technical community that the geotechnical properties of sand deposits are influenced by their age. Cyclic resistance data summarized in the widely accepted "empirical chart" by Seed et al. (1985) are limited to the relatively geological young soil of Holocene age. For use at the SRS the need arose to define the cyclic resistance of older sands. Lacking information about the performance of the sands at the site, it was necessary to carry out the field and laboratory test programs, and the literature review. The results, summarized in Figure 5, provide confidence in the validity of the investigations. The case histories reviewed in this paper confirm the observations of Professor Schmertmann, namely that age does play a major role in the strength of soil deposits and cannot be ignored and that strength does increase as time passes.

ACKNOWLEDGEMENTS

Through the years Professor Schmertmann has been involved with Bechtel projects including several projects at the SRS. It has been through this direct interaction that we have come to the methodologies and conclusions described herein. We owe a great deal to Professor Schmertmann for his wisdom, insight, and his fundamental knowledge of soil mechanics. We also acknowledge the contribution of Laura Bagwell and Rucker Williams for their patience and much needed assistance in preparing this manuscript.

REFERENCES

Amick, D., Gelinas, R., Maurath, G., Cannon, R., Moore, D., Billington, E., and Kemppinen, H. (1990). "Paleoliquefaction features along the Atlantic seaboard," *Report NUREG/CR-5613*, U.S. Nuclear Regulatory Commission, Washington, D.C.

Andrus, R.D., and Stokoe, K.H., II (2000), "Liquefaction resistance based on shear wave velocity," *Journal of Geotechnical and Geoenvironmental Engineering*, ASCE, 126(11), 1015-1025.

Arango, I., and Migues, R. E., (1996). "Investigation of the Seismic Liquefaction of Old Sand Deposits." *Report on Research, Bechtel Corporation,* National Science Foundation Grant No. CMS-94-16169, San Francisco.

Baecher, G. B. and Christian, J. T. (2003). *Reliability and Statistics in Geotechnical Engineering.* John Wiley & Sons, Chichester, West Sussex, England.

Bechtel Savannah River, Inc. (BSRI), (1993). "Savannah River Site, Replacement Tritium Facility, Geotechnical Investigation (U)," *Report No. WSRC-RP-93-606*, Aiken, SC.

Bechtel Savannah River, Inc. (BSRI), (1995). "In-Tank Precipitation Facility (ITP) and H-Tank Farm (HTF) Geotechnical Report (U), *Report No. WSRC-TR-95-0057*, Aiken, SC.

Dickenson, S. E., Clough, G. W., and Martin, J. R., (1988). "Evaluation of the engineering properties of sand deposits associated with liquefaction sites in the Charleston, SC area," *A report of first year findings*, U.S. Geological Survey, June.

Gould, J. P., (1985). "Problems of site characterization." *Seminar on Site Characterization*, ASCE National Capitol Section, Geotechnical Engineering Committee: 4-5.

Idriss I. M., and Boulanger R. W., (2004), "Semi-empirical procedures for evaluating liquefaction potential during earthquakes," *Proceedings Joint 11th International Conference on Soil Dynamics and Earthquake Engineering and the 3rd International Conference on Earthquake Geotechnical Engineering*, Berkeley, CA 32-56.

Ishihara K. and Yoshimine M., (1992). "Evaluation of settlement in sand deposits following liquefaction during earthquakes," *Soils and Foundations*, 32(1), 173-188.

Kulhawy. F. H., and Mayne, P. W., (1990). "Manual on Estimating Soil Properties for Foundation Design." *Final Report 1493-6. EL-6800,* Electric Power Research Institute, Palo Alto, California.

Leon, E., Gassman, S. L., and Talwani, P., (2006). Accounting for soil aging when assessing liquefaction potential." Journal of Geotechnical and Geoenvironmental Engineering, 132(3), 363-377.

Lewis, M. R., Arango, I., Kimball, J. K. and Ross, T. E., (1999). "Liquefaction Resistance of Old Sand Deposits," *Proceedings 11th Pan-American Conference on Soil Mechanics and Geotechnical Engineering*, Foz do Iguasu, Brazil, 821-829.

Littlejohn, G. S., Cole, K. W., and Mellors, T. W., (1994). "Without site investigation ground is a hazard." *Proceedings Institute of Civil Engineers.* Civil Engineering, Vol. 102, May: 72-78.

Martin, J. R. and Clough, G. W., (1990). "Geotechnical setting for liquefaction events in the Charleston, South Carolina vicinity," *H. Bolton Seed Memorial Symposium*, Vol. 2, 313.

Martin, J. R. and Clough, G. W., (1994). "Seismic parameters from liquefaction evidence," *Journal of Geotechnical Engineering*, ASCE, Vol. 120(8).

Obermeier, S. F., Jacobson, R. B., Powars, D. S., Weems, F. E., Hallbick, D. C., Gohn, G. S., and Markewick, H. W., (1986). "Holocene and Late Pleistocene (?) earthquake-induced sand blows in coastal South Carolina," *Proc. Of the Third U.S. national conference on Earthquake Engineering*, Charleston, SC, Vol. 1, p. 197.

Peck, R. B., (1969). "Advantages and limitations of the observational method in applied soil mechanics." Ninth Rankine Lecture, *Geotechnique* 19(2): 171-187.

Polito C.P., (1999), "The effects of nonplastic and plastic fines on the liquefaction of sandy soils." *PhD thesis*, Viginia Polytechnique Institute and State University, Blacksburg, VA.

Polito CP and Martin JR, (2001), "Effects of non-plastic fines on the liquefaction resistance of sands," *Journal of Geotechnical and Geoenvironmental Engineering*, ASCE, 127(5), 408-415.

Sara, M. N. (1994). "*Standard Handbook for Solid and Hazardous Waste Facility Assessments*." Lewis Publishers, Boca Raton, FL.

Schmertmann, J. H., (1970). "Static cone to compute static settlement over sand." *J. of the Soil Mechanics and Foundation Division,* ASCE, 96(SM3): 1011-1043.

Schmertmann, J. H., (1978). *"Guidelines for Cone Penetration Test Performance and Design."* US Department of Transportation, Federal Highway Administration, Offices of Research and Development, Washington, DC, Report No. FHWA –TS-78-209.

Schmertmann, J. H., (1991). "The mechanical aging of soils." *Journal of Geotechnical Engineering*, ASCE, 117(9): 1288-1330.

Schmertmann, J. H., (1993). "Update on the mechanical aging of soils," for the symposium "*Sobre Envejecimiento de Suelos,*" The Mexican Society of Soil Mechanics, Mexico City.

Seed, H. B., (1979). "Soil liquefaction and cyclic mobility of evaluation for level ground during earthquakes," *Journal of Geotechnical Engineering Division*, ASCE, 105(2), 201-255.

Seed, H. B., Tokimatsu, K., Harder, L. F., and Chung, R. M., (1985). "The influence of SPT procedures in soil liquefaction resistance evaluations," *Journal of Geotechnical Engineering*, ASCE, 111(12), 1425-1445.

Skempton, A. W., (1986). "Standard penetration test procedures and the effects in sands of overburden pressure, relative density, particle size, aging and overconsolidation," *Geotechnique*, 36(3), 425-447.

Sowers, G. F., (1979). "*Introductory Soil Mechanics and Foundations: Geotechnical Engineering.*" Macmillan Publishing Co., Inc., New York, 4th ed.

Talwani, P., and Cox, J. (1985). "Paleoseismic evidence for recurrence of earthquakes near Charleston, South Carolina." *Sci.*, 228(4711), 378-381.

Tatsuoka F., Zhou S., Sato T., and Shibuya S., (1990), "Method of evaluating liquefaction potential and its evaluation," *Seismic Hazards in the Soil Deposits in Urban Areas*, 75-109 (in Japanese).

Youd, T. L., and Hoose, S. N., (1977). "Liquefaction susceptibility and geologic setting in dynamics of soil and soil structures," *Proceedings, Sixth World Conference on Earthquake Engineering*, Vol. 3, Prentice-Hall, Englewood Cliffs, NJ, 2189-2194.

Youd, T. L., Idriss, I. M., Andrus, R. D., Arango, I., Castro, G., Christian, J. T., Dobry, R., Finn, W. D. L., Harder, L. F., Jr., Hynes, M. E., Ishihara, K., Koester, J. P., Liao, S. S. C., Marcuson, W. F., III, Martin, G. R., Mithcell, J. K., Moriwaki, Y., Power, M. S., Robertson, P. K., Seed, R. B., and Stokoe, K. H., II (2001). "Liquefaction resistance of soils: Summary report from the 1996 NCEER and 1998 NCEER/NSF workshops on evaluation of liquefaction resistance of soils." *J. of Geotechnical and Geoenvironmental Engineering*, ASCE 127(10): 817-833.

Design and Construction of Earthen Reservoirs in Florida (1967 – 2007)

John E. Garlanger[1], P.E., Member, ASCE, and Thomas J. Leto[2], P.E.

[1]Senior Vice President & Chief Engineer, Ardaman & Associates, Inc., P.O. Box 593003, Orlando, Florida, 32859-3003; jgarlanger@ardaman.com,
[2]Senior Vice President, Ardaman & Associates, Inc., P.O. Box 812, Bartow, Florida, 33831-0812; tleto@ardaman.com

ABSTRACT: The Florida Phosphate Industry has been constructing earthen reservoirs for water storage and clay settling/disposal for more than 70 years. During the first 35 years of operation, the perimeter dams for the clay settling areas were simply overburden spoil (primarily silty sand) cast up in windrows and shaped to provide a crest road. The spoil windrow was excavated down to natural ground at one or two locations to install an outlet pipe for the decant structure(s). Little or no engineering was performed. Between 1936 and 1969, there were 32 reservoir failures. The first rule regulating the design and construction of earthen dams used in phosphate mining and processing operations was adopted by the state of Florida in 1969 after a reservoir failure near Brandon, Florida turned the Alafia River a brownish-white 3.2 kilometers into the Gulf of Mexico. The rule was subsequently modified in 1972 after a reservoir failure occurred near Fort Meade and again in 1999 after a reservoir failure occurred near Hopewell. All of the reservoirs constructed after 1969 were engineered.

The authors' firm has investigated, designed, monitored construction, and inspected more than 120 earthen reservoirs constructed in Florida after 1967. The paper describes the advances that have been made during the past 40 years in the investigation, analysis, design, construction and inspection of the embankment dams surrounding these reservoirs. The lessons learned in the phosphate industry have and should be applied to other earthen reservoirs constructed in Florida and elsewhere.

KEY WORDS: construction, dams, design, failures, Florida, flownets, history, piping, seepage, tailings

INTRODUCTION

Phosphorous is one of the essential elements necessary for plant growth. Fertilizer containing phosphorous has been used since the early 1800's. The first fertilizers containing phosphorous were derived from animal bones, which are mineralogically similar to the carbonate-apatite mineral dahlite (Roseberry et al.1930). In 1881, Captain J. F. Le Baron, a civil engineer with the US Army Corps of Engineers, discovered phosphate pebbles in the Peace River near the town of Arcadia, Florida (Frisbee, 1974). These phosphate pebbles, containing the mineral fluorapatite, were part of an outcrop of the Bone Valley formation, an extensive deposit of phosphatic sand and gravel underlying approximately 5,000 km^2 of land in west-central Florida. The price of land in Florida boomed during the late 1880s from under $5 per hectare to near $740 per hectare when more than 200 companies opened to mine the Florida 'Black Gold' (Frisbee, 1974).

Phosphate mining along the Peace River began both in the river itself and on the adjacent land between 1888 and 1890. Because of higher costs, river pebble mining could not compete with land pebble mining and by 1908 river mining had ceased. Land pebble mining was initially conducted using pick, shovel and wheelbarrow, or water jetting and pumping. Steam shovels, with bucket capacities up to one cubic meter, came into use in the early 1900's and electric and diesel powered draglines, with bucket capacities up to 4 cubic meters, followed in the 1920's and 1930's. Today, modern electric draglines remove the overburden and phosphate matrix from depths in excess of 20 meters using dragline buckets with capacities up to 40 cubic meters.

Digging the phosphate out of the ground is only the first step in the mineral recovery process. Phosphate comes out of the ground as part of a matrix of sand- and gravel-sized particles of fluorapatite, sand-sized particles of quartz and calcite, and silt- and clay-sized particles of smectite, polygorskite, apatite, quartz, dolomite and wavellite (Wissa et al. 1982).

The phosphate pebble is separated from the matrix by washing and screening. The silt- and clay-sized particles are separated from the sand-sized particles using a series of hydrocyclones. The overflow is pumped to clay settling ponds. The underflow, i.e., sand-sized fluorapatite, is separated from the sand-sized quartz and calcite particles in a two-stage flotation process. Prior to 1927 and the advent of flotation, only the pebble product was recovered. The remainder of the matrix, including the fine-grained sediment, was pumped back to the mine pits or unmined lowlands from which the wash water was discharged to local streams and rivers.

The Industry did not begin constructing earthen reservoirs for water storage and clay settling until 1936 after the Florida legislature passed a law (FS 533) prohibiting off-site discharge of mining waste. The perimeter dams for these reservoirs were typically overburden spoil (primarily silty sand) cast up in windrows and shaped to provide a crest road. The spoil windrow was excavated down to natural ground at one or two locations to install an outlet pipe for the decant structure(s). Little or no engineering was performed and many of these dams failed. During the period between 1936 and 1972, 33 of approximately 100 clay settling areas released water or dilute clay slurry to the environment. One notable release occurred near Fort Meade in 1967 when approximately 2 million cubic meters of dilute clay slurry flowed into the Peace River.

The first rule regulating the design and construction of earthen dams used in phosphate mining and processing operations was adopted in 1969 (Florida Department of Pollution Control) after a reservoir failure near Brandon, Florida turned the Alafia River "brownish-white 3.2 kilometers into the Gulf of Mexico". The rule was substantially modified in 1972 after another reservoir failure occurred near Ft Meade, Florida in December 1971 and dilute clay slurry was again released into the Peace River.

One of the major deficiencies of the 1969 version of the rule was that it did not require any type of seepage control within the dam. Nor did the 1969 rule require a roadway at the toe of the dam for inspection and maintenance. Both of these requirements were included in the 1972 revision. The revised rule also required that all dams be designed and inspected during construction by a professional engineer experienced in the design, construction and maintenance of earthen dams, that all dams be inspected weekly during operation by a trained employee of the Owner, and that the dams be inspected annually by a professional engineer experienced in dam construction and maintenance.

Approximately 175 clay settling areas have been constructed in the Florida phosphate mining district since 1972. As discussed below, only one of the dams associated with these settling areas has released water or dilute clay slurry to the environment as a result of an embankment failure.

SUBSURFACE CONDITIONS

Although quite variable, the subsurface conditions in the central Florida mining district generally consist of a 4.5- to 6-meter thick layer of silty sand (10 % to 20 % minus 200 sieve size), locally referred to as overburden overlying a 4.5- to 6-meter thick layer of clayey phosphatic sand (15 % to 30% minus 200 sieve size), locally referred to as matrix. The overburden is of Holocene and Pleistocene age and consists of both windblown and waterborne beach and near-shore terrace deposits containing surficial deposits of muck or peat and deeper deposits of organically-cemented and iron-stained sand locally referred to as hardpan. The average horizontal hydraulic conductivity of the overburden is typically in the range of $3x10^{-3}$ to $4x10^{-3}$ cm/sec, but can occasionally be as high as $1.5x10^{-2}$ or as low as $1x10^{-4}$ cm/sec. The hardpan does not contribute to the transmissivity of the overburden. The vertical hydraulic conductivity of the hardpan is on the order of 10^{-5} cm/sec.

The matrix is of Pliocene age and consists of shallow-water marine and estuarine deposits of phosphatic sand and gravel with variable percentages of silt and clay. Layers of cemented shell are occasionally present within the matrix. The hydraulic conductivity of the matrix is typically less than 10^{-4} cm/sec but can be significantly higher depending on the extent and thickness of occasional layers of pebble phosphate or cemented shell.

The matrix is typically underlain by a very stiff to hard calcareous sandy clay of Miocene age that is referred to locally as bedclay. Where the bedclay is absent, the underlying deposits are typically calcareous mudstones or siltstones containing phosphatic sand and gravel. These deeper layers, although rich in phosphate, are not mined. The bedclay or bedrock formation is relatively impervious and is the upper part of the confining layer separating the surficial aquifer from the underlying limestones of the Floridan aquifer. The average vertical hydraulic conductivity of the confining layer, which varies in thickness from less than 15 meters to greater than 60 meters, is approximately 10^{-7} cm/sec. Occasionally, a layer of weathered limestone is encountered at the base of the matrix. This layer can be highly transmissive (Wissa et al. 1975).

CONSTRUCTION PRACTICES PRIOR TO 1972

Most of the earthen reservoirs used for clay settling and disposal during the period between 1936 and 1972 were constructed without the benefit of geotechnical engineering. Engineering typically consisted of determining the location of the area to be used for waste clay disposal and instructing the dragline operator where to cast the overburden. The overburden (spoil) pile was then dozer-shaped to widen the crest of the dam and sometimes to flatten the outside slope. Outside slopes as steep as 2H:1V were not uncommon. Sometimes low 'starter dams' were constructed followed by incremental raises by the upstream method using the more granular portion of the waste stream. The maximum water level behind a dam was typically limited to 5 to 6 meters above natural ground surface and the width of the dam was selected to prevent the hydraulic gradient through the dam from exceeding a 'rule of thumb' slope of 5H:1V. There were no other provisions for positive seepage control and typically no perimeter roads to inspect the critical areas at the downstream toe of the embankment. Adequate vegetation was not established to protect the embankment surfaces against wind and water erosion.

A hydraulic ditch was excavated along portions of the perimeter embankment to return clarified water back to the plant site or to a nearby stream for off-site discharge. Water was decanted from the settling area using a half-round corrugated riser structure connected to a corrugated metal discharge pipe. The discharge pipe was installed by excavating a notch through the embankment into natural ground, installing the pipe, and then backfilling around the pipe using the excavated material. The riser structure was typically anchored to a concrete foundation if it was taller than 3 meters. The height of the embankment at the location of the decant spillway was typically significantly lower than the embankment on either side of the structure because the adjacent spoil piles were typically cast higher than needed for the design height of the dam.

Figure 1 is a cross section through a typical earthen dam used to form a reservoir for waste clay settling/disposal in the Florida phosphate industry prior to 1972 (See Fig. 2 for description of soil layers). There are a number of potential failure mechanisms for this type of construction: 1) piping along the outlet conduit; 2) concentrated seepage resulting in piping into the hydraulic ditch; 3) seepage-induced slope failure followed by piping at the toe of the outside slope; and 4) concentrated surface erosion leading to overtopping at the location of the decant structure.

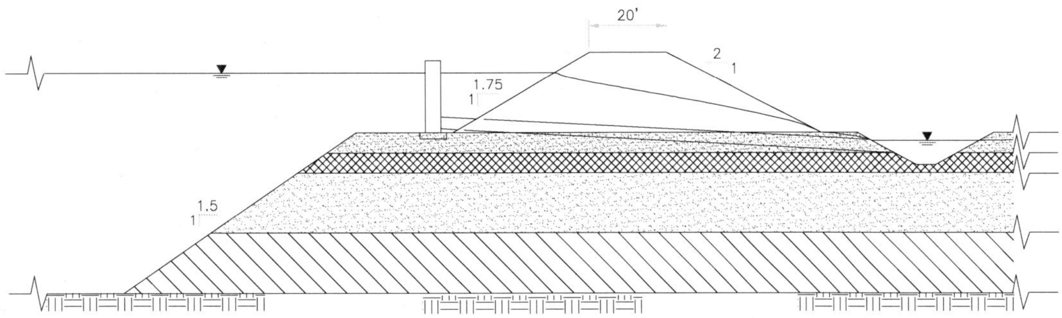

FIG. 1. Cross Section of Typical Pre-Rule Dam (Note: 1 ft = 0.305m)

Piping along the Outlet Conduit

This is the most probable failure mechanism for dams constructed prior to 1972. It is essentially impossible to place and compact soil along the underside of a circular conduit passing through an earthen embankment. Attempts to compact the soil immediately adjacent to the pipe below the spring line will almost certainly lift the pipe off its foundation. No special bedding was specified for pre-1972 dams. The gap created between the pipe and the sandy foundation during placement and compaction of backfill can easily erode leading to an erosional piping failure during initial filling of the reservoir. The Cities Service dam failure near Fort Meade, Florida in 1971 is believed to have failed by piping along the spillway conduit.

Concentrated Seepage Resulting in Piping into the Return Ditch

A hardpan layer or other relatively impermeable layer in the overburden that is only partially penetrated during excavation of the hydraulic ditch at the toe of the embankment can lead to heave of the ditch bottom, concentrated seepage, and eventually piping into the ditch bottom. Sand boils large enough to require remediation have been observed in the hydraulic ditch at a number of settling areas constructed both before and after 1972. However, prior to 1972, the hydraulic ditch was not routinely inspected and the sand boils could have gone unnoticed or been washed away by the return flow.

Seepage-Induced Slope Failure Resulting in Piping at the Toe of the Outside Slope

Few, if any, of the embankments constructed prior to 1972 incorporated positive seepage control features. Consequently, even if the embankments were wide enough to maintain an average hydraulic gradient through the embankment of 5H:1V, the

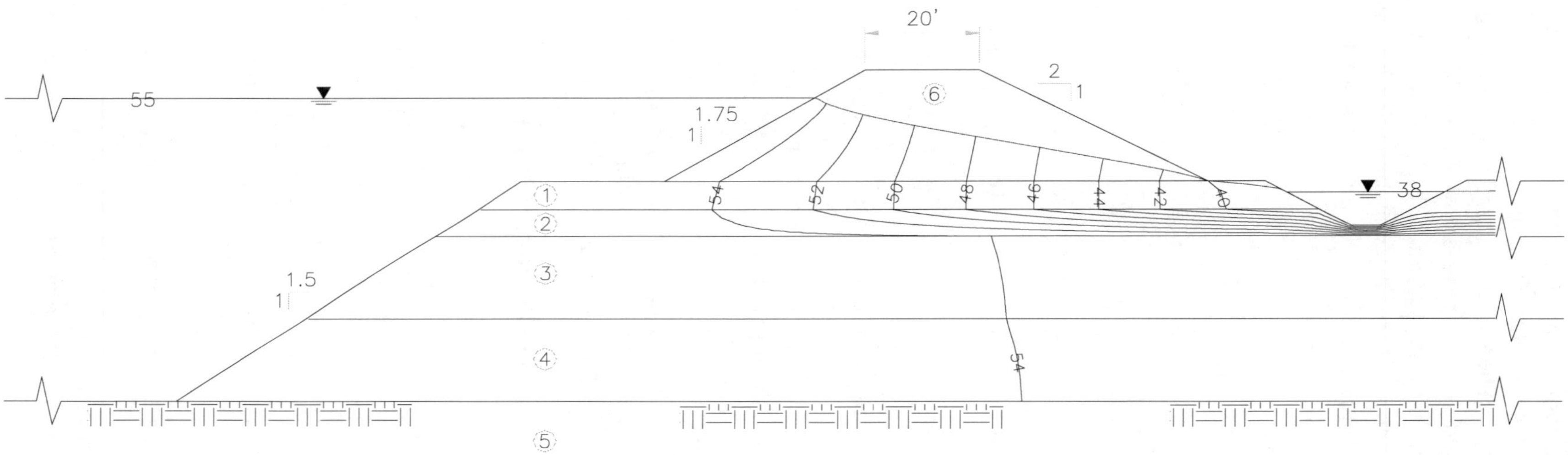

SOIL LAYER	1	2	3	4	5	6
SOIL TYPE	SILTY SAND	CEMENTED SAND	SILTY SAND	CLAYEY SAND	CONSOLIDATED CLAY	CAST OVERBURDEN
HYDRAULIC CONDUCTIVITY K_V (CM/SEC)	2×10^{-3}	1×10^{-5}	5×10^{-3}	5×10^{-4}	1×10^{-7}	5×10^{-4}
ANISOTROPY K_H/K_V	4	4	4	4	1	1

FIG. 2. Piezometric Levels for Typical Dam Constructed Prior to 1972 (Note: 1 ft = 0.305m)

gradient at the toe of slope could have been significantly higher. The flow net in Figure 2 indicates a spring line several feet above the toe of slope. The cast overburden, although loose (dry density < 14 kN/m^3), typically has an angle of internal friction of 32°. The computed factor of safety against seepage instability for a cast embankment with a 2H:1V slope is less than 1.0.

Piping at the location of seepage-induced sloughing is almost inevitable because of the high gradient and has been observed by both authors (See Fig. 3).

FIG. 3. Piping Failure at Toe of Dam

Concentrated Surface Erosion Leading to Overtopping at the Location of the Decant Structure

Inspection of many of the pre-1972 embankments almost always disclosed severe erosion gullies on the outside slope of the embankment at the location of the decant spillway. Water running along the crest road from the higher cast embankment on either side of the spillway conduit concentrated at the low point and ran down the outside slope causing significant erosion. Although no evidence exists that a clay settling area overtopped as a result of this erosion, it is certainly a possible failure mechanism that required serious attention.

CONSTRUCTION PRACTICES AFTER 1972

Five obvious changes to the design and construction of settling area embankments were required and implemented after promulgation of the 1972 rule revisions: 1) seepage control along the spillway conduit; 2) positive seepage control within the embankment; 3) positive seepage control within the foundation; 4) surface runoff control on crest road; and 5) incorporation of an all-weather trafficable inspection road along the outside toe of the embankment.

Seepage Control along the Spillway Conduit

The only way to eliminate the gap between a circular conduit and its foundation is to use flowable cementitious fill as the bedding material. The replacement for the spillway conduit for the Cities Service embankment that was rebuilt after the 1971 failure was the first to use a lean concrete cradle beneath the spillway conduit. Figure 4 shows the typical design currently used by the authors' firm for the concrete cradle. In earlier designs, the cradle extended up the side of the conduit to only 1/6th the diameter of the pipe to minimize the amount of lean concrete used in the cradle. Increases in the cost of labor to backfill around the lower ½ of the pipe led to the current design, which is also structurally superior to the earlier design.

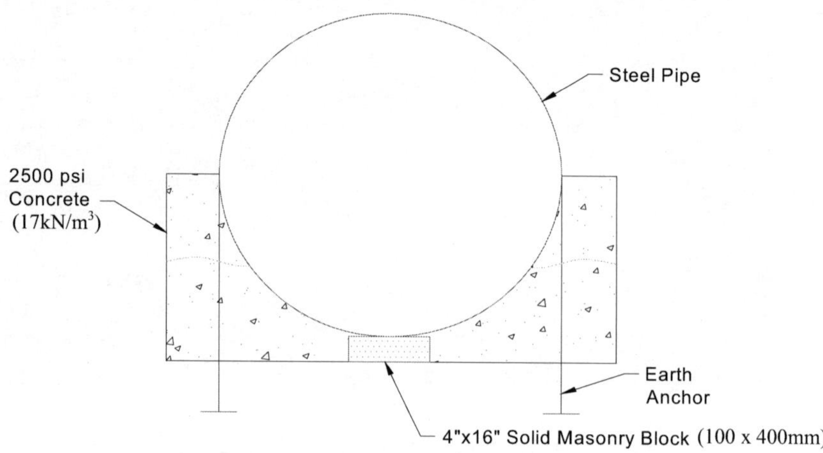

FIG. 4. Concrete Cradle for Spillway Conduit

Approximately 110 of the 145 clay settling areas constructed between 1972 and 1994, including all of those designed by the authors' firm, used a lean concrete cradle under the spillway conduit to avoid any potential gap beneath the pipe. The remaining 35 settling areas relied on special design and construction procedures, including shaped bedding and multiple seep shields (required by the 1972 rule), to avoid piping failure along the pipe.

On November 19, 1994, during initial filling, a settling area located near Hopewell, Florida released approximately 2 million m^3 of water into nearby wetlands and the Alafia River. Forensic investigation of the failure, which occurred at one of the two spillways, identified piping at the interface between the sand bedding and the conduit as the cause of failure. The gap was visually observed in the unused sister spillway conduit (Wissa, 1994). After the Hopewell failure, Chapter 62-672 was revised to require a concrete cradle beneath the lower third of the spillway conduit passing through the embankment of all settling areas.

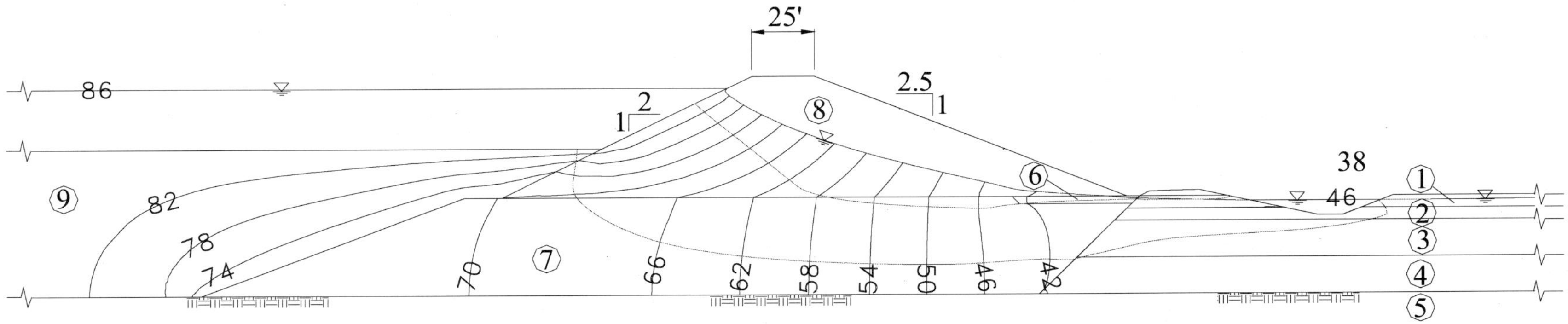

SOIL LAYER	1	2	3	4	5	6	7	8	9
SOIL TYPE	SILTY SAND	CEMENTED SAND	SILTY SAND	CLAYEY SAND	CONSOLIDATED CLAY	SAND TAILINGS	CAST OVERBURDEN	COMPACTED OVERBURDEN	WASTE CLAY
HYDRAULIC CONDUCTIVITY K_V (CM/SEC)	2×10^{-3}	1×10^{-5}	5×10^{-3}	5×10^{-4}	1×10^{-7}	1×10^{-2}	5×10^{-4}	5×10^{-5}	1×10^{-5}
ANISOTROPY K_H/K_V	4	4	4	4	1	1	1	2	1

FIG. 5. Piezometric Levels in Typical Dam Constructed after 1972 (Note: 1 ft = 0.305m)

None of the approximately 140 clay settling areas constructed in Florida after 1972 that included a concrete cradle have experienced any problems that have resulted in the release of waste clay slurry to the environment.

Positive Seepage Control within the Embankment

Positive seepage control in settling area embankments have typically consisted of both an upstream core of lower permeability silty or clayey sand and a blanket drain of much higher permeability tailings sand from the flotation plant. Not only does selection of the lower permeability soil for the upstream blanket reduce seepage, but also prevents a more permeable soil layer from extending across the entire width of the embankment. Figure 5 is a flow net through a typical embankment constructed entirely within the mine cut. The embankment below natural grade typically consists of cast or dozer shaped overburden, which is somewhat more permeable than the compacted soil in the embankment. Piezometers and observation wells installed within the embankment and foundation of numerous settling area embankments have documented the efficacy of this typical design section.

Where sand tailings are not available or have been used to construct the embankment, the design has incorporated geotextile-wrapped gravel drains beneath the outside slope for positive seepage control. Observed performance of these designs is also good. It should be noted that the clay slurry after settling and consolidation provides additional seepage control on the inside slope of the embankment, although the quantity of seepage is only moderately reduced as long as there is several feet of clear water above the consolidated clay.

Positive Seepage Control within the Foundation

When it is necessary to construct the embankment on natural ground, e.g., for an initial settling area or when the phosphate reserves within a proposed settling area are not mined, additional seepage control within the foundation is required. The concern is development of elevated hydrostatic pressures in a relatively permeable layer that outcrops in the mine pit (e.g., Layer 3 in Figs. 2, 6 and 7) and extends beyond the downstream toe of the embankment beneath a relatively impermeable layer (e.g., Layer 2 in Figs. 2, 6 and 7). This condition can lead to soil heave and possibly piping failure, particularly when there is a return water ditch at the toe of the embankment. Return water ditches typically carry a normal flow in excess of 6 m^3/sec and normally have a water depth in excess of 2 meters. Not only does the depth of the water make visual inspection nearly impossible, but the high flow removes any evidence that heave or piping has occurred.

An adequate soil exploration program is critical in uncovering this condition, which requires special design, operational and monitoring considerations. If the impermeable layer is not too deep, the seepage collection system within the embankment can be extended down into the permeable layer as shown in Figure 6. Excavating the return water ditch through the impermeable layer and backfilling across the bottom and up the inside slope with sand tailings is another positive seepage control measure that has been used to address this concern.

Because the waste clay in the impoundment acts as a liner on the inside face of the mine cut (and the embankment), the effectiveness of the seepage control measures described above is enhanced by limiting the elevation of the water in the reservoir until the sedimented clay surface rises above the permeable stratum. A comparison of the head in the sand stratum below the return water ditch in the flow nets presented in Figures 2 and 7 shows a reduction in head of more than 90 percent after the waste clay rises above the natural ground surface.

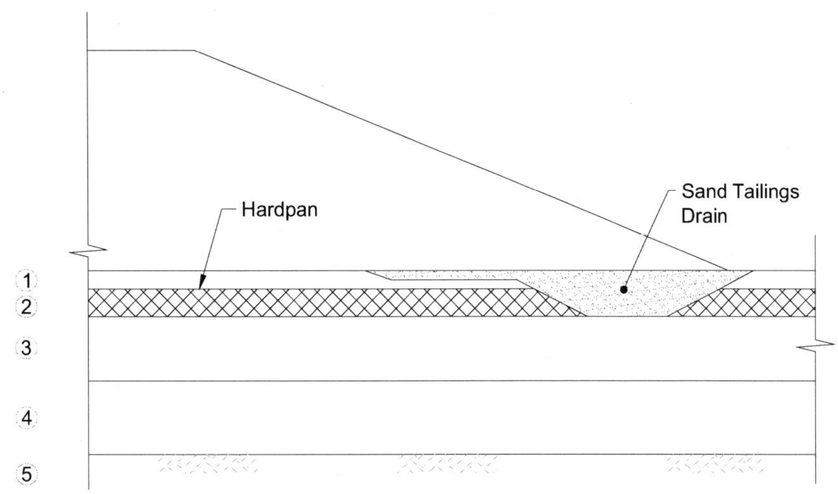

FIG. 6. Method used to Relieve Excess Toe Pressures

To monitor the effectiveness of seepage control measures implemented for embankments constructed on natural ground, piezometers are installed within the permeable foundation layers at the toe of embankment. When excess pore pressures have been detected, gravity relief wells have been installed to correct the situation.

Surface Runoff Control on Crest Road

Concentrated erosion of the downstream slope of a settling area embankment during storm events is prevented by keeping the centerline of the crest road level along the dike alignment and by grading the top of the embankment so that the outside edge is 15 cm higher than the inside edge. Rainfall runoff is either allowed to sheet flow down the inside slope or is diverted by a soil curb to polyethylene drain pipes spaced approximately 60 meters apart that carry the runoff down the inside slope into the reservoir. General wind and water erosion is minimized by grassing the crest road and both the inside and outside slopes of the embankment.

The inside slope of settling area embankments is not rip-rapped. Wave action is controlled by minimizing the depth of water above the clay, minimizing fetch using

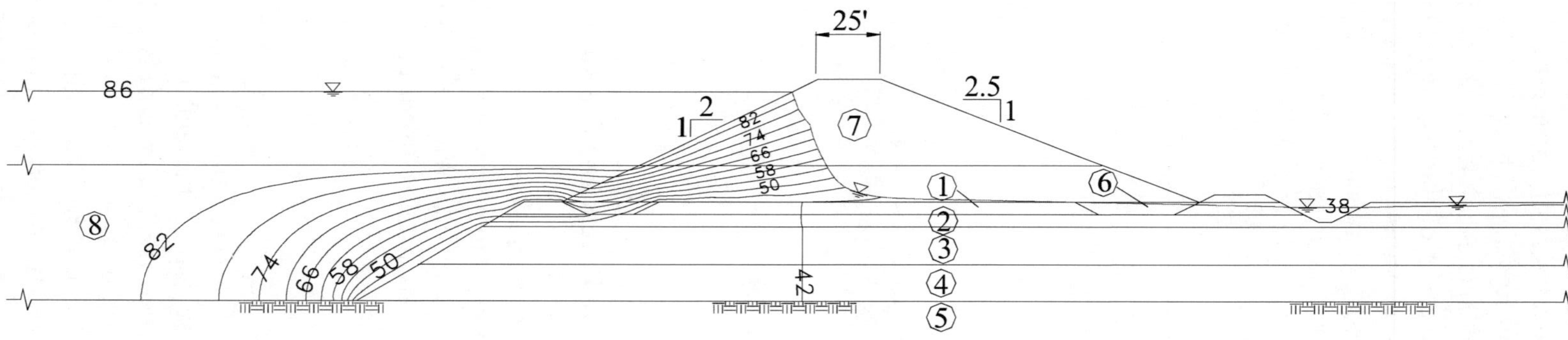

SOIL LAYER	1	2	3	4	5	6	7	8
SOIL TYPE	SILTY SAND	CEMENTED SAND	SILTY SAND	CLAYEY SAND	CONSOLIDATED CLAY	SAND TAILINGS	COMPACTED OVERBURDEN	WASTE CLAY
HYDRAULIC CONDUCTIVITY K_V (CM/SEC)	2×10^{-3}	1×10^{-5}	5×10^{-3}	5×10^{-4}	1×10^{-7}	1×10^{-2}	5×10^{-5}	1×10^{-5}
ANISOTROPY K_H/K_V	4	4	4	4	1	1	2	1

FIG. 7. Effect of Sedimented Clay on Toe Pressures (Note: 1 ft = 0.305m)

interior partitions, or, occasionally, breaking the waves using floating breakwaters. Because settling areas are typically filled in a few years, minor wave erosion at each new water level is buried as the level of the sedimented clay increases. On rare occasions, wave action results in erosion severe enough to require repair of significant portions of the slope by placing additional fill material on the eroded area. The additional fill material is typically oversized material from the washer known as mudball debris. In the summer of 2004, the paths of three major hurricanes passed over a number of active settling areas in central Florida causing only minor wave erosion.

Incorporation of Inspection Road at Toe of Slope

All active settling areas are inspected weekly by a qualified engineer or technician employed by the owner and trained by the design engineer. An all-weather toe inspection road is included in the design of the settling area to facilitate these inspections. Inspectors are instructed to look for any of the following items and, if observed, to immediately bring them to the attention of the design engineer: (a) seepage on the outside slope or in the area within 15 meters downstream from the toe of slope; (b) silt accumulations, boils, deltas, or cones in the area within 15 meters downstream from the toe of slope; (c) seepage, damp areas, or boils in the vicinity of any conduit through the dam; or (d) cracking or bulging on the outside slope of the embankment. The inspector also makes note of overgrowth patches of vegetation, wet or soggy areas, or any surface erosion on the inside or outside slope of the embankment. The inspectors also record the water level in all piezometers and observation wells installed within the embankment.

The toe inspection road has allowed several critical conditions at and beyond the toe of the slope to be identified and corrected before they developed into much more serious problems or resulted in failure of the embankment. The authors have investigated and provided remedial designs to collect seepage, relieve excess hydrostatic pressure, and stop potential piping at the toe of slope for a number of settling areas and believe that regular inspections, particularly at the toe of an embankment, has significantly reduced the risk of failure for settling areas constructed since 1972.

CONCLUSIONS AND RECOMMENDATIONS

1. The high rate of failure for settling areas constructed prior to 1969 (one per year) was a result of the absence of engineering. Seepage control was non-existent in the early dike construction technique and most failures were likely the result of piping along the spillway conduit, sloughing and subsequent piping at the downstream toe of the embankment, or soil heave and subsequent piping into the return water ditch.

2. Use of a concrete cradle beneath the outlet pipe combined with positive seepage control at the downstream end of the pipe has substantially reduced

the potential for piping along the spillway conduits for modern settling areas. None of the more than 140 settling areas constructed with a concrete cradle since 1972 has experienced any problems that could have resulted in the release of waste clay slurry to the environment.

3. Positive seepage control consisting of less permeable soils within the upstream portion of the embankment and more permeable soils in the downstream portion of the embankment combined with a properly designed sand (or gravel) drain for collecting seepage beneath the downstream slope has substantially reduced the potential for sloughing at the downstream toe of the embankment.

4. Identifying the stratification that can lead to excess hydrostatic pressure in the foundation soils beneath a settling area embankment constructed on unmined natural ground and designing a seepage collection system to relieve the potentially harmful pressure has substantially reduced the potential for soil heave and piping near the downstream toe of the embankment.

5. The clay deposited within a settling area acts as a relatively impervious blanket reducing seepage through the foundation and embankment of the settling area. The presence of the sedimented clay further reduces the potential for seepage related failure of the settling area.

6. Frequent inspection, particularly at the toe of the embankment, combined with routine piezometric monitoring further reduce the potential for failure by identifying critical conditions that can be fixed before they could worsen and lead to failure.

7. All of the lessons learned from the design, construction, and operation of settling area embankments can be applied to earthen reservoirs designed for water storage. The benefits of the sedimented clay in reducing seepage through sandy foundation soils can be achieved by installing a soil-bentonite cutoff wall through the foundation soils using the slurry trench technique. A soil-bentonite cutoff wall can also be used to minimize seepage through the embankment. Earthen reservoirs used for water storage will require upstream slope protection, i.e., rip rap, to prevent erosion from wave action.

ACKNOWLEDGMENTS

The authors would like to acknowledge our colleagues, Dr. Rafeek Nashed and Dr. Reinaldo Rolo, for performing the seepage analyses and preparing figures for the paper.

REFERENCES

Florida Department of Pollution Control. (1969). *Minimum Requirements for Earthen Dams, Phosphate Mining and Processing Operations*. Chapter 17-9 Florida Administrative Code (Now Chapter 62-672).

Frisbee, L. (1974). *Peace River Pioneers*, E. A. Seemann Publishing, Inc., Miami, Florida.

Roseberry, H. H., A. B. Hastings & J. K. Morse (1930). "X-Ray Analysis of Bone and Teeth". *The Journal of Biological Chemistry*, Vol. XC, No. 2.

Wissa, A. E. Z. (1994) Personal Communication.

Wissa, A. E. Z., N. F. Fuleihan, & T.S. Ingra. (1982). "Evaluation of Phosphatic Clay Disposal and Reclamation Methods, Volume 2: Mineralogy of Phosphatic Clays". Florida Institute of Phosphate Research.

Wissa, A. E. Z., R. T. Martin, & J. E. Garlanger. (1975). "The Piezometer Probe". *Proceedings*, *ASCE Specialty Conference on In-Situ Measurement of Soil Properties*, *Vol. 1*, pp. 536-545.

FAILURE OF THE NEW ORLEANS 17TH STREET CANAL LEVEE & FLOODWALL DURING HURRICANE KATRINA

Robert G. Bea, Ph.D, P.E., F. & Life Member, ASCE[1]

ABSTRACT: Hurricane Katrina resulted in the single most catastrophic failure of a civil engineered system in the history of the United States – failure of the flood defense system for the greater New Orleans area. This paper summarizes results from several forensic studies that have examined the causes for failure of one of the most important components of the flood protection system - failure of the levee and floodwall on the 17th street canal. This failure has been publicly cited as an 'engineering failure' (Walsh 2006) that involved 'unforeseen and unforeseeable' (Marshall 2006) conditions. This paper illustrates why the engineering failure was firmly rooted in a failure to translate research to practice. Geotechnical engineering aspects of the levee and floodwall failure are developed including description of the soil and geologic conditions, analyses of the loading conditions, and analyses of the soil-structure-loading performance characteristics. In addition, the human and organizational aspects that played key roles in development of this failure are detailed. It is concluded that this was a predictable failure whose causes were embedded in a dysfunctional Technology Delivery System.

INTRODUCTION

Hurricane Katrina was a very tough teacher. This severe natural event interacted with a pervasively flawed flood defense system for the Greater New Orleans area and developed into a disaster. This disaster was compounded into a catastrophe by similarly flawed evacuation and recovery systems. The result was approximately 2,000 deaths (immediate, delayed, on-site, off-site) and total costs estimated to approach U.S. $500 billion (direct, indirect, immediate, delayed, on-site, off-site). Currently, there are more than $500 billion in Katrina flood damage claim lawsuits in New Orleans Federal District Court.

---

[1] Department of Civil & Environmental Engineering, University of California Berkeley, Berkeley, CA 94720: bea@ce.berkeley.edu

FAILURE AT 17TH STREET CANAL

During Hurricane Katrina, a large segment of the levee and floodwall (Fig. 1) lining the 17th Street canal in New Orleans (Fig. 2) failed catastrophically before the design water elevations (+4.4 m above mean sea level, reference for all vertical elevations) were realized. Eyewitness reports and recorded water levels (Fig. 3) at the drainage pump station located at the head of the 17th Street canal indicated that the failure initiated about 6:00 a.m. on August 29, 2005 when the water level was at about +2 m (USACE IPET 2007).

Prior to arrival of the storm surge, strong winds from the northwest had blown over large oak trees growing at the toe of the levee thereby opening paths for accelerated water seepage (Fig. 4). Water was first seen coming through the 'water stops' (vertical rubber joints) that separated the concrete floodwall panels. Differential movements of the supporting soil levee were large enough to open these joints.

About 8:30 a.m. an eyewitness reported that the water had reached the top of the levee on the west side (+2 m) and a whirlpool had developed on the east side at the south end of the developing breach. Due to the off-centerline dredged profile of the canal, the top of the levee on the east outboard (canal) side was approximately 2 m lower than on the west side (normally covered with water, Fig. 5). Photographs taken at this time showed the vegetation which had grown at the bottom of the concrete wall on the west canal side (Fig. 4). At this point (about 9:30 a.m. Fig. 3), a deep breach opened due to large horizontal displacements between two adjacent sections of the floodwall indicating that there had been a failure of the sheet pile interlocks at that section. In the end, the levee, floodwall, and supporting sheet pile were displaced laterally toward nearby homes by more than 15 m (Fig. 6).

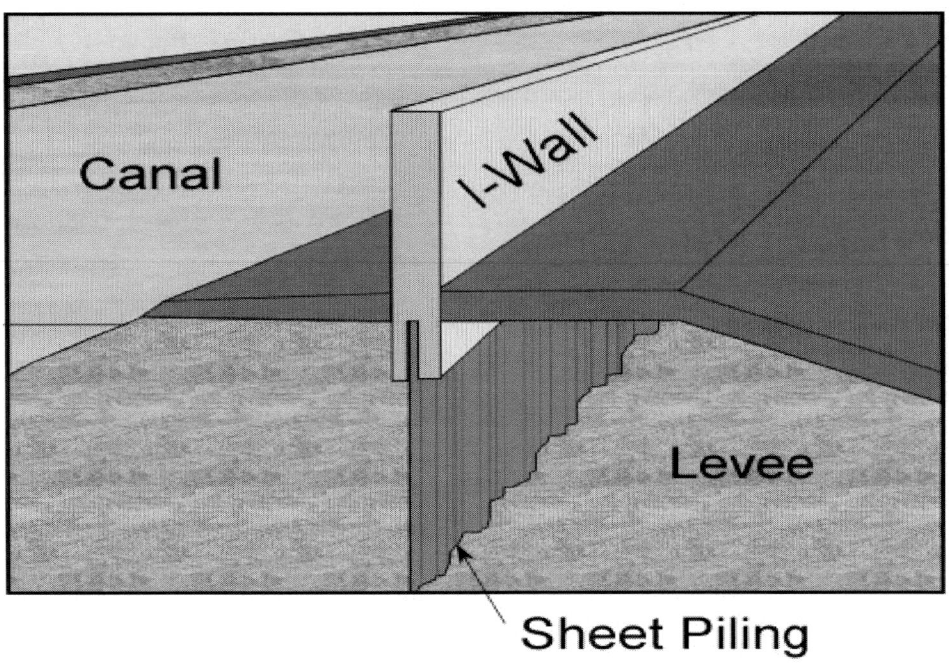

FIG. 1: 17th Street canal levee, floodwall, and sheet piling (USACE IPET 2007)

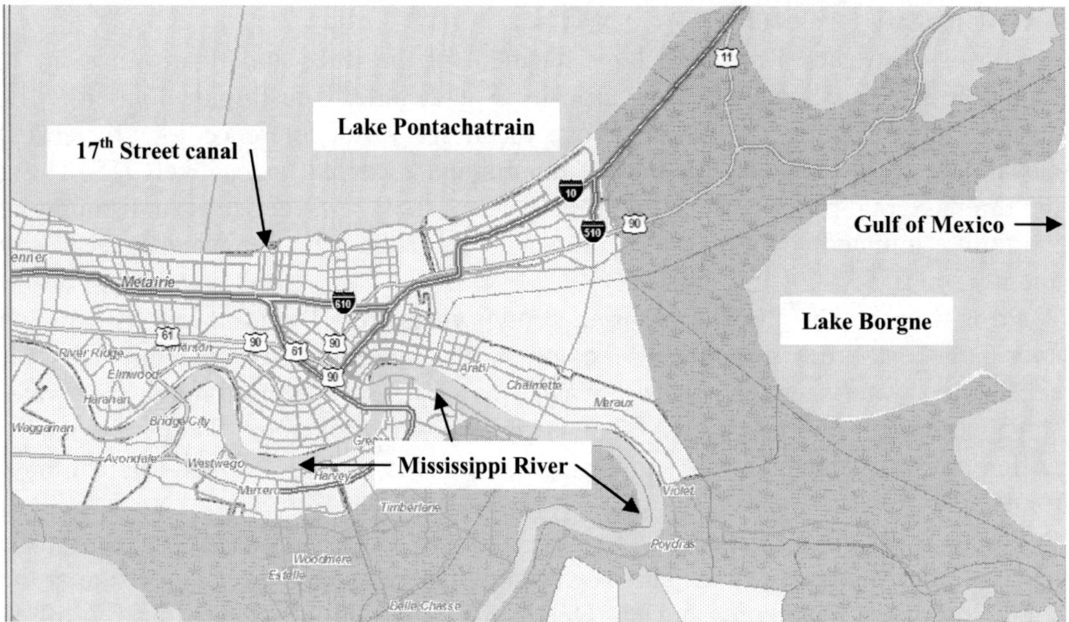

FIG. 2: Location of 17th Street canal breach

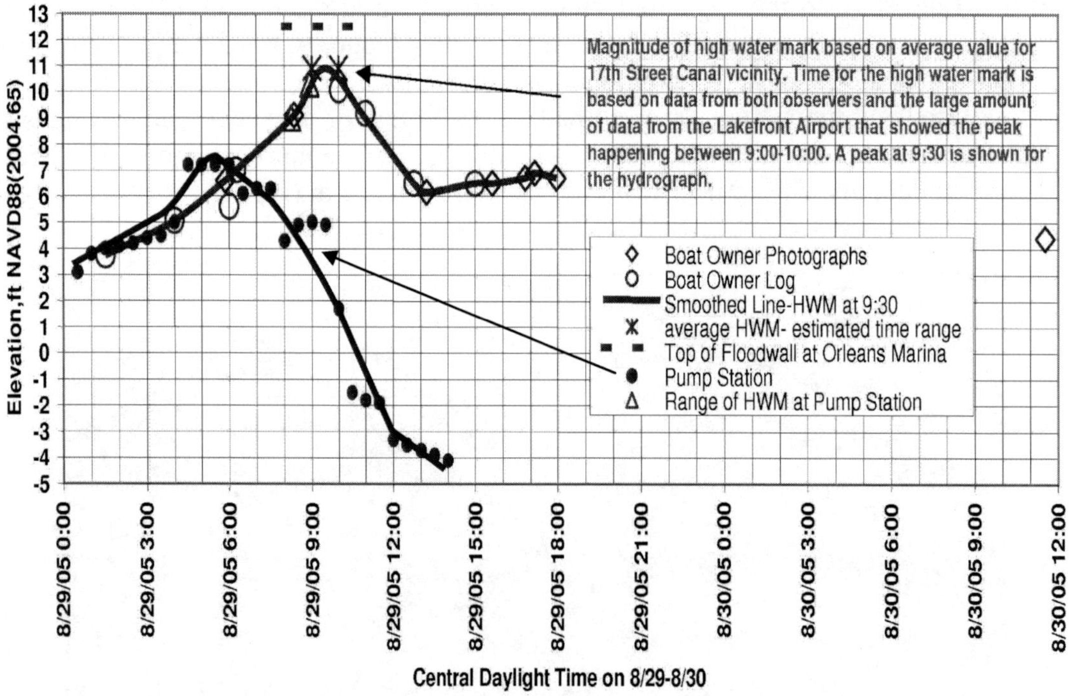

FIG. 3: Hurricane surge hydrograph showing time history of water levels in the 17th Street Canal (USACE IPET 2007)

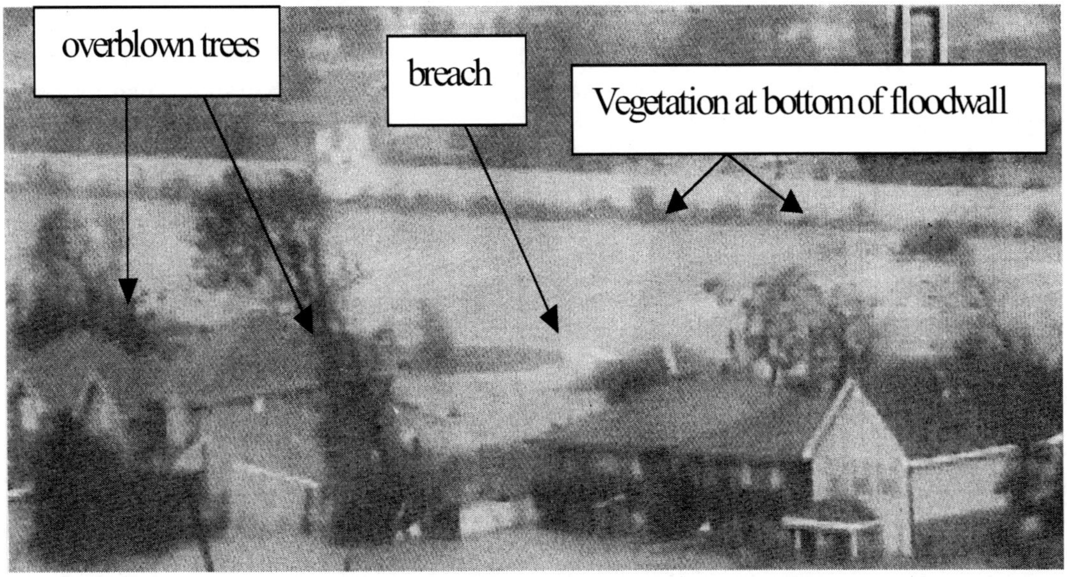

FIG. 4: Initiation of breach on 17th Street canal (USACE IPET 2007)

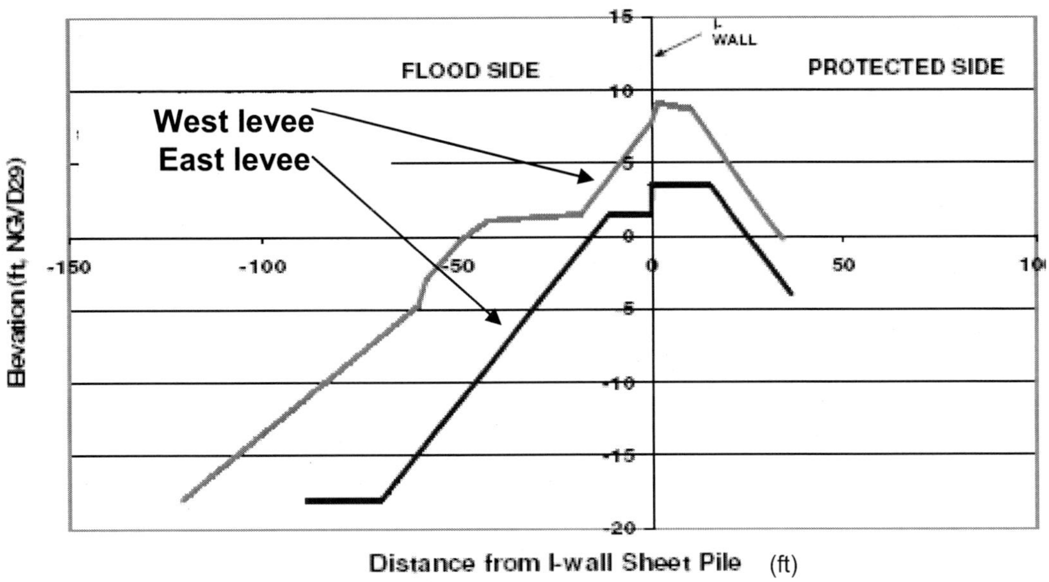

FIG. 5: Comparison of cross-sections of east levee that failed with west levee that did not fail (1 ft = 0.3 m, Team Louisiana 2007)

Shown in Fig. 7 is a cross-section through the breach – at the completion of the failure. Analyses performed by the Independent Levee Investigation Team (ILIT 2006) indicated a lateral translational failure with the inboard side of the levee being pushed laterally by the elevated canal water. These lateral pressures opened a gap on the canal side between the sheet piles supporting the concrete floodwall and the soils, allowing water intrusion and further increasing the lateral force acting against the section (Fig. 8). This action effectively cut the levee in half longitudinally resulting in dramatic increases in horizontal forces and decreases in the lateral soil resistance (USACE IPET 2007, ILIT 2006, Team Louisiana 2007).

The upper portion of the levee (Fig. 9) consisted of brown silty organic clays. This was underlain by older sections comprised largely of locally available gray marsh and lacustrine (lake deposits) clays. The levee was underlain by marsh (swamp, peat) deposits varying from a few feet to as much as 2.5 m in thickness. These were underlain by a transition zone of interbedded marsh and clay deposits with a similar thickness. The transition zone was underlain by a layer of soft gray clay. Sands were present at the base of the profile.

FIG. 6: Conclusion of east side breach (Environmental Protection Agency).

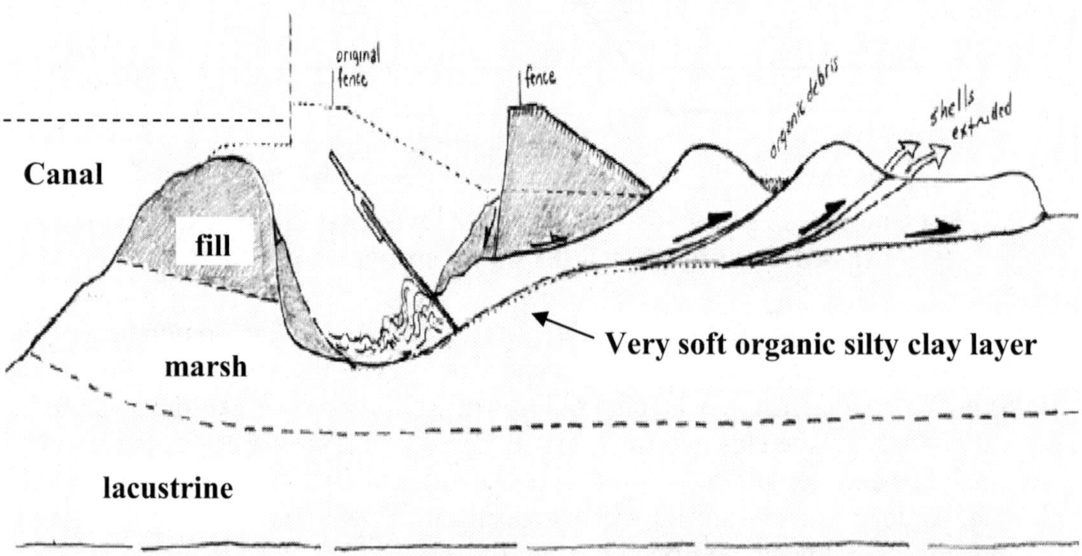

FIG. 7: Cross-section through 17th street canal breach at conclusion of failure (looking north, top of floodwall at elevation +3.8 m, bottom of canal at –4.9 m, top of levee inboard at +1.8 m, and land at inboard at –1.8 m, Rogers 2006)

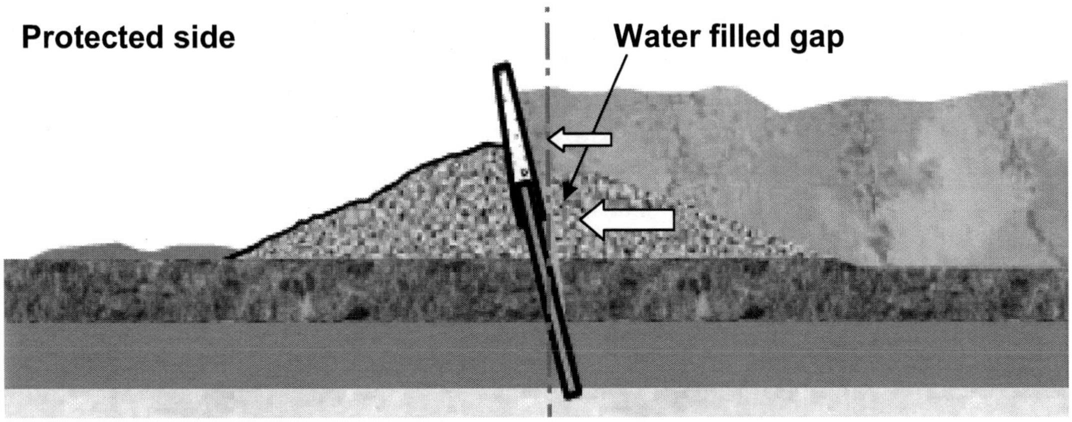

FIG. 8: Water filled gap - cut the levee in half failure mode (USACE IPET 2007)

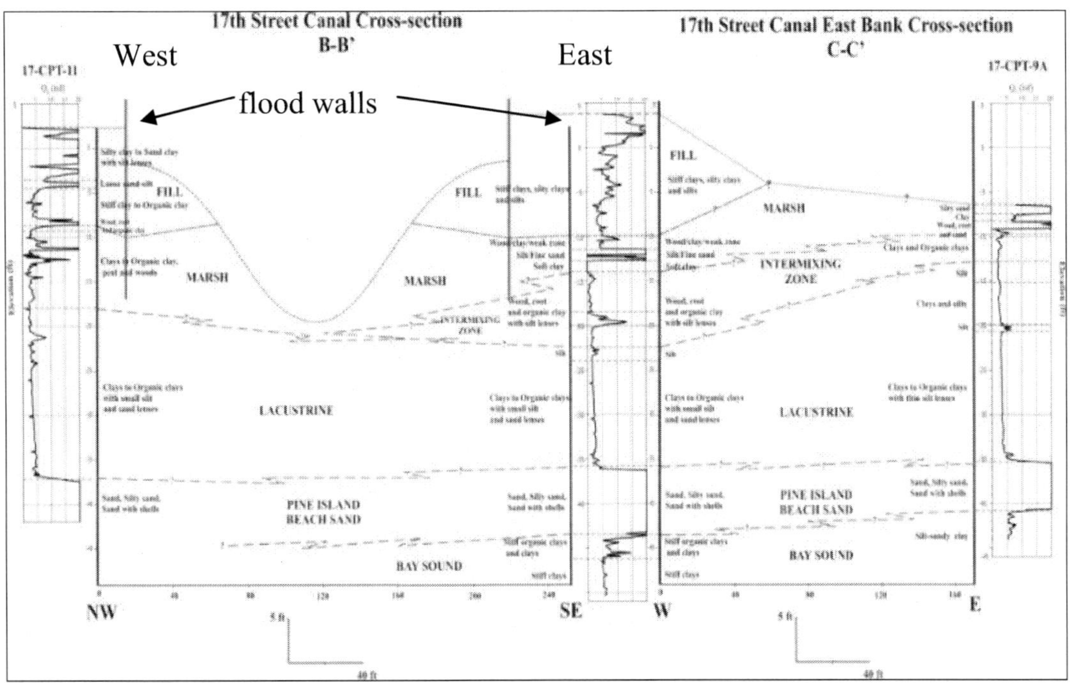

FIG. 9: Stratigraphic cross-sections through 17th street canal breach (looking north). Underlying marsh (peat) layer thins northward as does the underlying Pine Island Beach sand. The lacustrine organic silty clays thicken southward (Rogers 2006)

A large number of borings and cone penetration test probes were performed at this location by the U.S. Army Corps of Engineers (USACE) Interagency Performance Evaluation Taskforce (IPET 2007), ILIT (2006), and Team Louisiana (2007) investigation teams. In addition, many borings had been performed earlier as part of the initial design studies (USACE 1990). The soft gray clay stratum and the overlying marsh and intermixing zones exhibited three distinct over-consolidation profiles representing the effects of desiccation at three times during the history of these deposits (Fig. 10, results at inboard side toe). Geologic studies indicated there were no erosional periods during the deposition of these layers. The desiccation profiles were less pronounced beneath the levee crest due to the increased overburden pressures.

The profile of shear strengths within the marsh layer beneath the crest and inboard toe of the levee showed pronounced differences (Fig. 11). This layer is a complex and inter-layered zone of organic marsh depositions including peat, marsh clays, and organic silts. The critical layer (sensitive zone) proved to be a thin layer of dark organic clayey silt near the base of the marsh layer. This layer had the appearance, consistency and strength of black toothpaste. Geologic studies indicated this stratum was the result of a previous major hurricane that changed the local depositional regime (Rogers 2006, ILIT 2006). The storm churned up organics and silts and mixed them with the locally available clays. A layer was formed of heavily flocculated (highly sensitive) organic silty clay due to the increased salinity caused by the storm. This layer was then covered by settled sticks, leaves, and twigs. The layer was thickest in the vicinity of bayou sloughs shown on an 1872 map of the area (Fig. 12). The map showed that these sloughs crossed the location of the 17th street canal in the immediate vicinity of the breach.

Two-dimensional nonlinear finite element analyses (PLAXIS) were performed by the ILIT team for two sets of conditions – with and without the tension gap between the sheet piles and the outboard soils (Fig. 13). These results were validated with limit equilibrium methods (e.g. Spencer's Method). A water-filled tension crack begins to open at the outboard side of the sheet piles at a canal water elevation of about +1.8 m – at this point, the horizontal displacements of the sheet piles exceed that of the levee soils. The water-filled tension crack results in applying additional lateral water pressures against the sheet piles and the inboard and outboard side of the levee. Under these conditions, the lateral translational failure mechanism has a significantly lower factor of safety than the deeper rotational failure mechanism (Fig 14). The factor of safety becomes less than unity at a water elevation of about + 2.4 m.

The ILIT investigation disclosed a section on the opposite (west) bank of the 17th street canal that was displaced laterally several inches hurricane Katrina. Analyses showed that this very similar section also developed a very low factor of safety, but substantially larger than the east side (Fig. 15). The soil characteristics and properties at the east and west sides were very similar – the geologic conditions were the same (ILIT 2006).

At the time that the east wall failed, the factor of safety for the west wall was approximately 1.6 (Fig. 15). Failure would not have developed until the water reached an elevation of approximately +3.4 m (top of floodwall at elevation of approximately +3.8 m). The difference in performance of the floodwalls and levees on the two different sides of the 17th street canal was due fundamentally to the differences between the east and west bank levee cross sections. The investigation performed by Team Louisiana (2007) confirmed these findings. The differences in the levee cross-sections developed when the canal was dredged off centerline (Fig. 5). The differences in the crest elevations of the supporting levees were evident before the failure. Photographs show that the east side was normally submerged and the west side was normally above water and therefore could support vegetation.

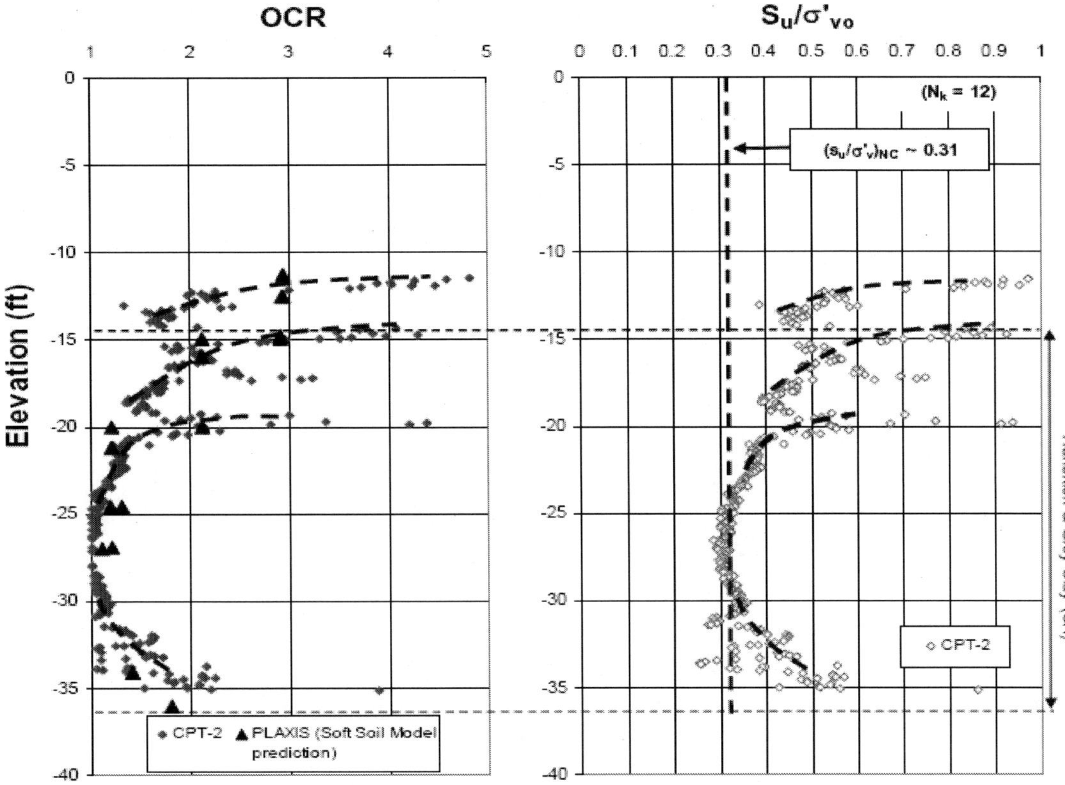

FIG. 10: Over Consolidation Ratio (OCR) versus elevation and ratio of Undrained Shear Strength (Su) to effective vertical pressure (σ'_{vo}) (1 ft = 0.3 m, ILIT 2006).

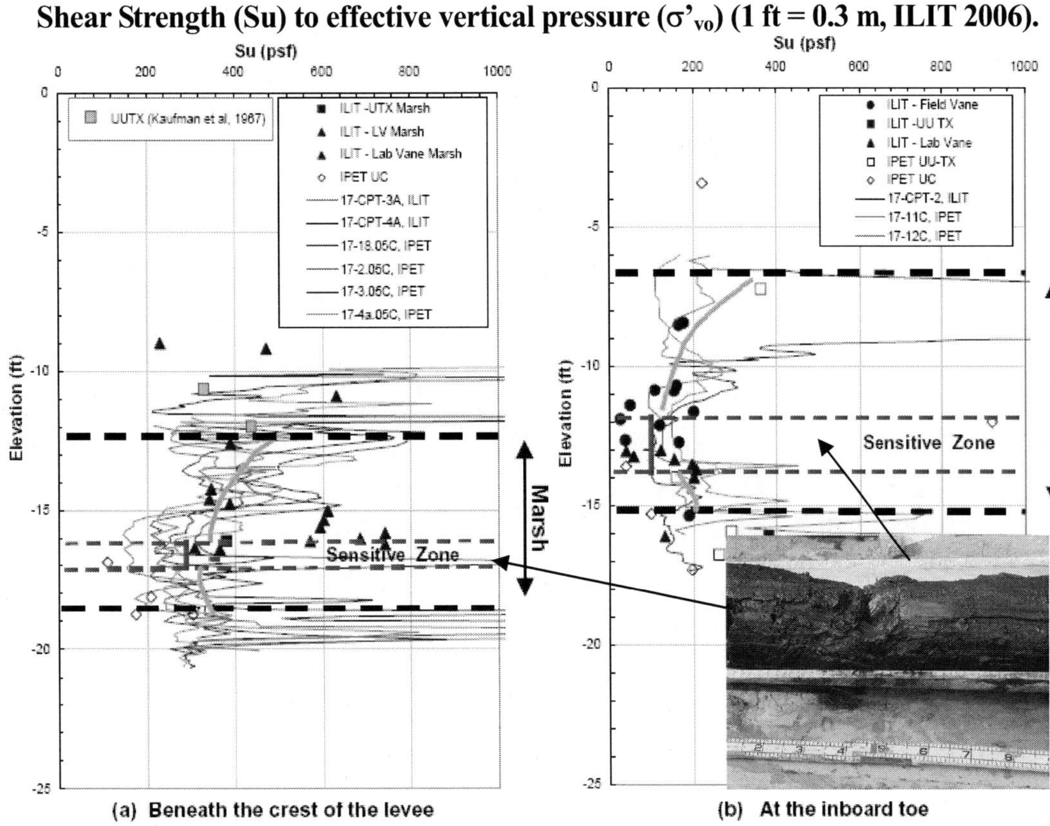

FIG. 11: Shear strengths of the marsh layer (1 ft = 0.30 m, 1 psf = 48 Pa, ILIT 2006)

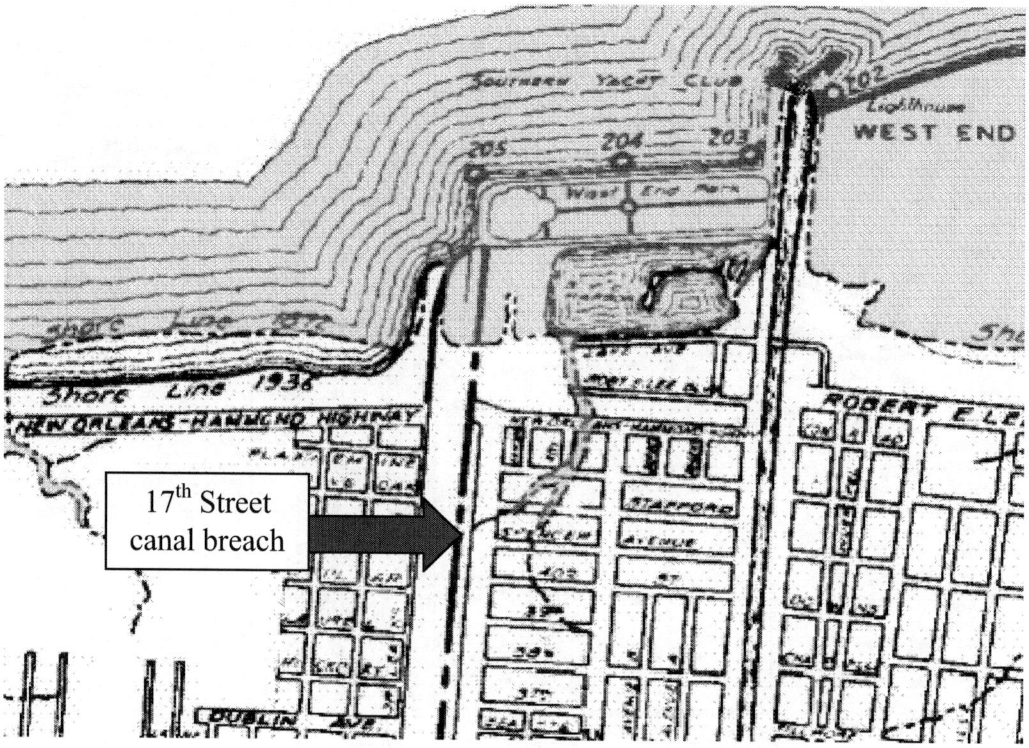

FIG. 12: 1872 Map overlaid with 1936 street map showing intersections of bayou sloughs with 17th Street canal alignment in vicinity of the 2005 breach (Rogers 2006)

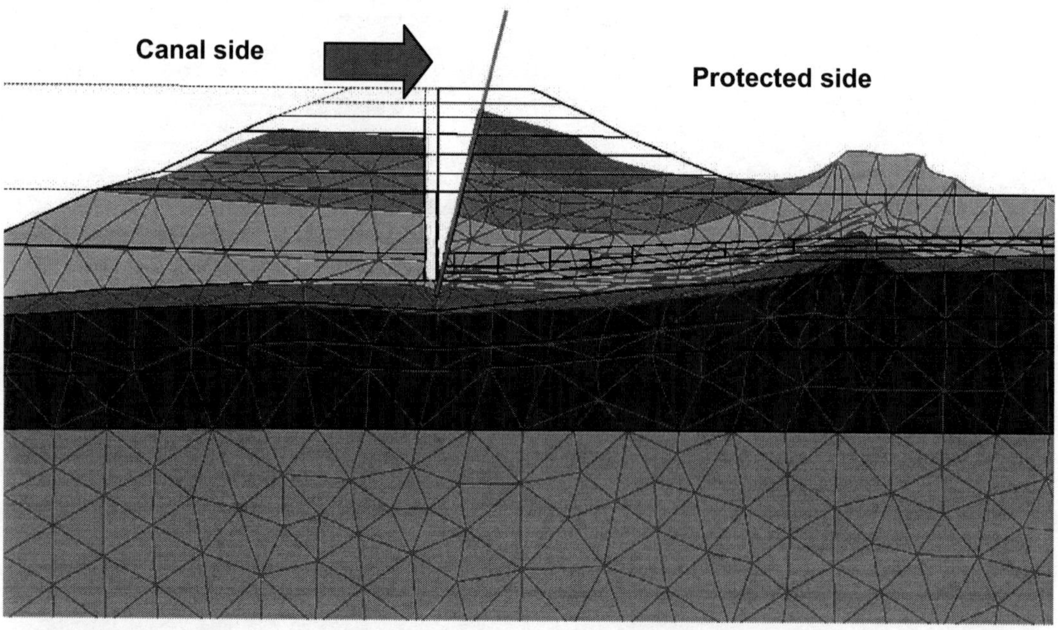

FIG. 13: Deformed mesh plot at failure from PLAXIS two-dimensional finite element analyses (ILIT 2006)

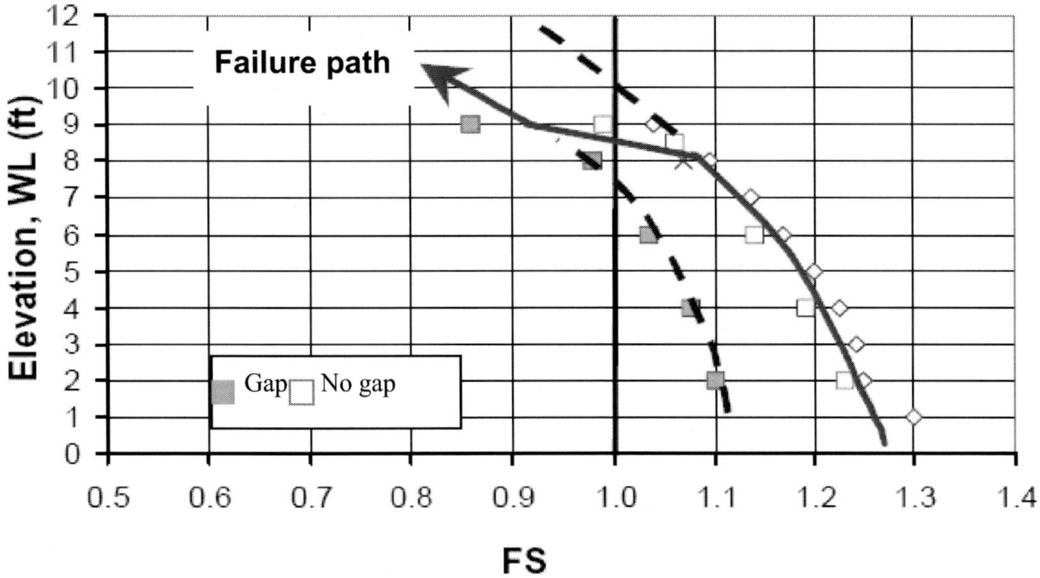

FIG 14: Factor of Safety (FS) for east breach site as function of water level (1 ft = 0.3 m, ILIT 2006).

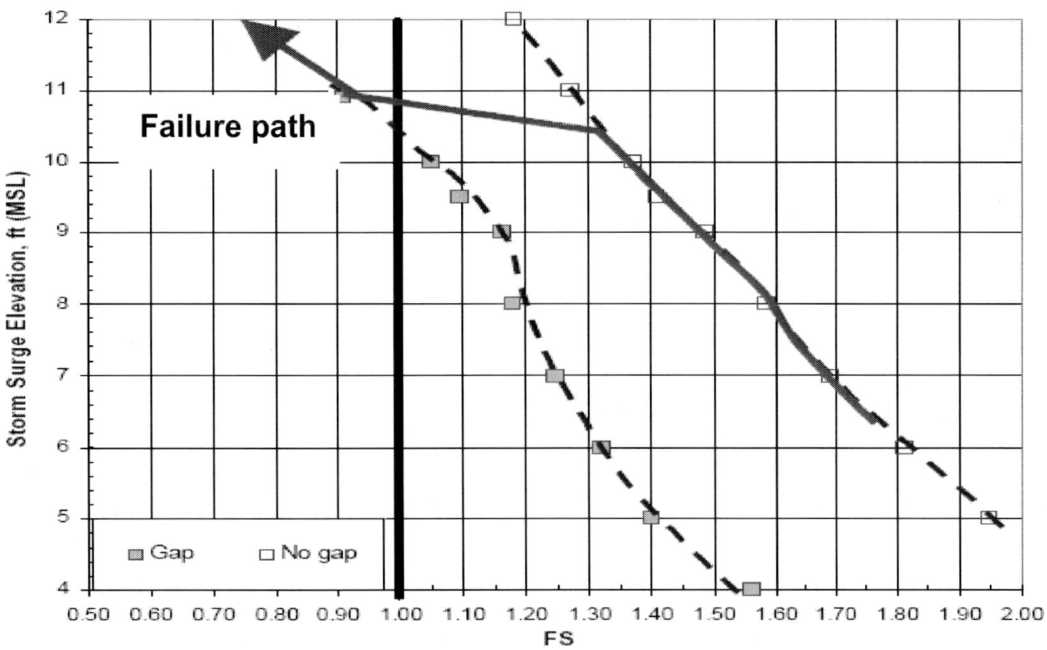

FIG 15: Factor of Safety (FS) for west breach site as function of water level (1 ft = 0.3 m, ILIT 2006).

Three detailed forensic studies were conducted to determine the engineering mechanics that were responsible for the failure of the east side of the 17th street canal levee and floodwall. These studies arrived at the same general conclusions: the levee and floodwall failed prematurely. The water filled gap cut-the-levee in half behavior was critical in this development. This behavior had not been anticipated in the design analyses. The analyses performed by the USACE IPET team indicated a failure mode involving the deeper lascustrine clays. The ILIT team indicated a failure mode involving the shallower marsh intermixed zone that contained the very weak and sensitive organic silty clay layer that was missed in the IPET soil characterizations. The analyses performed by Team Louisiana also confirmed these findings.

One of the potentially important issues involved in development of the breach was the subterranean water flow and pressure effects associated with the underlying marsh and transition zone layers overlying the lacustrine clay and beach sand layers found at this site. The USACE IPET studies associated with the centrifuge model testing (Appendix 5 of IPET 2007 report) and the final report (Appendix 17) did specifically address such conditions. These studies were based on soil characterizations in which the tips of the sheet piling penetrated into the lacustrine clay layer effectively cutting off water flow and pressure effects.

In contrast, the ILIT studies were based on soil characterizations in which the tips of the sheet piling did not penetrate into the lacustrine clay layer, but rather penetrated into an overlying intermixing zone that could allow water migration and pressure effects. Water pressure and migration effects were specifically included in the ILIT analyses (Figs. 13,14,15).

This important difference was based on the very detailed soil characterization field work, geologic studies, and historic mapping studies conducted during the ILIT study (e.g. Figs. 9, 10 and 11, ILIT 2006). Observations made in the field during the ILIT soil borings performed in this area showed that the buried marsh and transition zone sediments (swamp deposits) had very high horizontal water conductivity (permeability). The field observations indicated that the horizontal conductivity was much higher than the vertical conductivity due to the horizontally layered structure of the organic – marsh materials.

During the ILIT studies, it was not possible to perform in situ tests to determine the hydraulic conductivity characteristics of the buried marsh – transition layers. In the initial set of results, a horizontal hydraulic conductivity (permeability coefficient) of 10^{-2} cm/sec was used. A value 10 times less was used for the vertical permeability. These results were based on published results for intrinsic – insitu horizontal permeabilities for soils and on results from recent high quality laboratory and field measurements performed on peat – marsh deposits similar to those found at the 17th Street canal location (Beckwith et al 2002, Mesri and Ajlouni 2007). It is important to note that there are potentially very important differences between permeability coefficients determined from conventional laboratory tests and those that are actually present in the field. Permeability coefficients based on results from consolidation tests fail to capture important effects of sampling and anisotropic permeability caused by the structure of the organic materials (Mesri and Ajlouni 2007, Mitchell 1976).

Recently, this same issue developed in evaluations of levee vulnerability in the Sacramento Delta area. This area has geologic depositional and geotechnical characteristics that are very similar to those of the Mississippi River Delta area. In this very extensive study of about 100 failures of levees during flood stages it was concluded: "*The Vulnerability team believes 80 percent of the past failures (levees) can be* attributed *to seepage-induced failures. The team also believes that both through and under seepage-induced failures occurred in equal numbers. The remaining 20% of past failures can be attributed to overtopping*" (URS 2007). In their analytical studies of levee vulnerabilities based on results from laboratory testing, field testing, and analyses of previous breaches, horizontal permeabilities in the range of 10^{-3} to 10^{-4} cm/sec were used for the organic peat and marsh layers.

Based on all available information on the geologic and geotechnical characteristics of the east side breach area, transient flow analyses were performed (Fig. 16) using GEOSTUDIO SEEP/W (Cobos-Roa and Bea 2007).

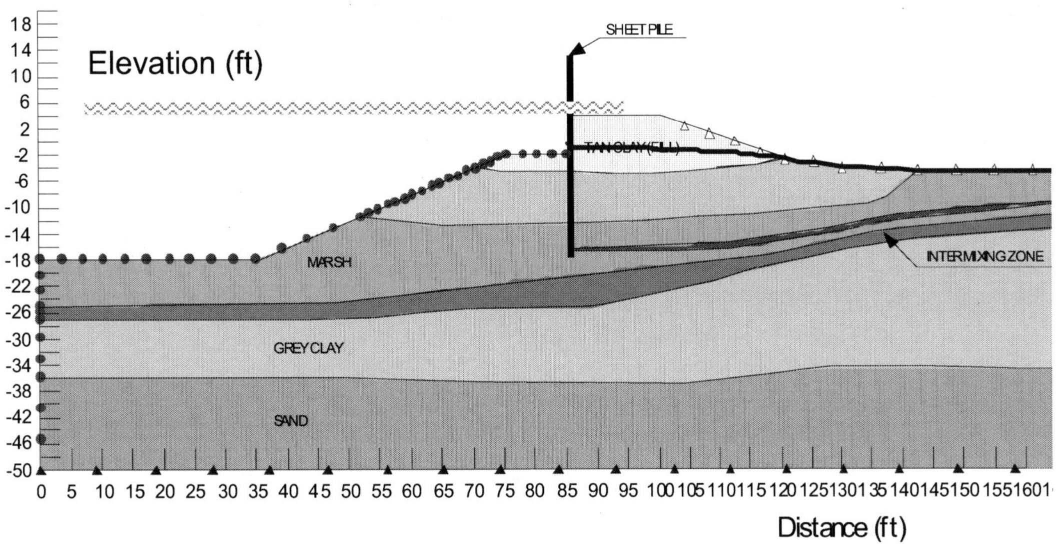

Fig. 16: Transient flow analyses model geometry (1 foot = 0.30 m)

The water elevation versus time characteristics provided in Fig. 3 were utilized in these analyses. These analyses incorporated a very wide range in horizontal permeabilities (10^{-2} to 10^{-6} cm/sec) of the marsh – intermixing zone soils that connect the bottom of the 17th Street canal with the protected landside (note that the bottom of the canal is at an elevation lower than the bottoms of the sheet piling). The transient flow analyses indicated that even for the very lowest horizontal permeability, there was almost direct water pressure communication under and through the levee trough the marsh – intermixing zone soils (Fig. 17). As the water level in the canal increased there were rapid increases in the water pressures in the marsh –intermixing zone soils. The transient pore pressure analyses indicated that in excess of 95% of the steady-state pressures for a water level of +2 m were generated by 6:00 a.m. on August 29, 2005.

Fig. 18 shows the variation of pore pressure along the bottom of the marsh – intermixing zone layers that underlie the east breach site for a water level of +1.8 m. The pore pressures along the bottom of the marsh – intermixing zone layers vary less than 5% when the in situ horizontal hydraulic conductivity (k) decreases from 10^{-2} to 10^{-6} cm/sec. The insensitivity of the pore pressures to the hydraulic conductivity is due to the saturated conditions of the layers – the layers are able to quickly communicate the pressures from the rising water in the canal. There is virtually no obstacle in pressure development underneath the levee toward the landside toe. The sheet piling were not driven deep enough to form an effective barrier to the seepage – water pressure development. Most important, the results indicate that the hydraulic pressures are sufficient at the toe of the levee to exceed the overburden pressures from the soils and hence indicate a very high probability of levee toe 'blow-out'. It is clearly apparent for a very wide range in plausible in situ horizontal permeabilities that the marsh – intermixing zone water conductivity is a key player in the hydraulic behavior and stability at this location.

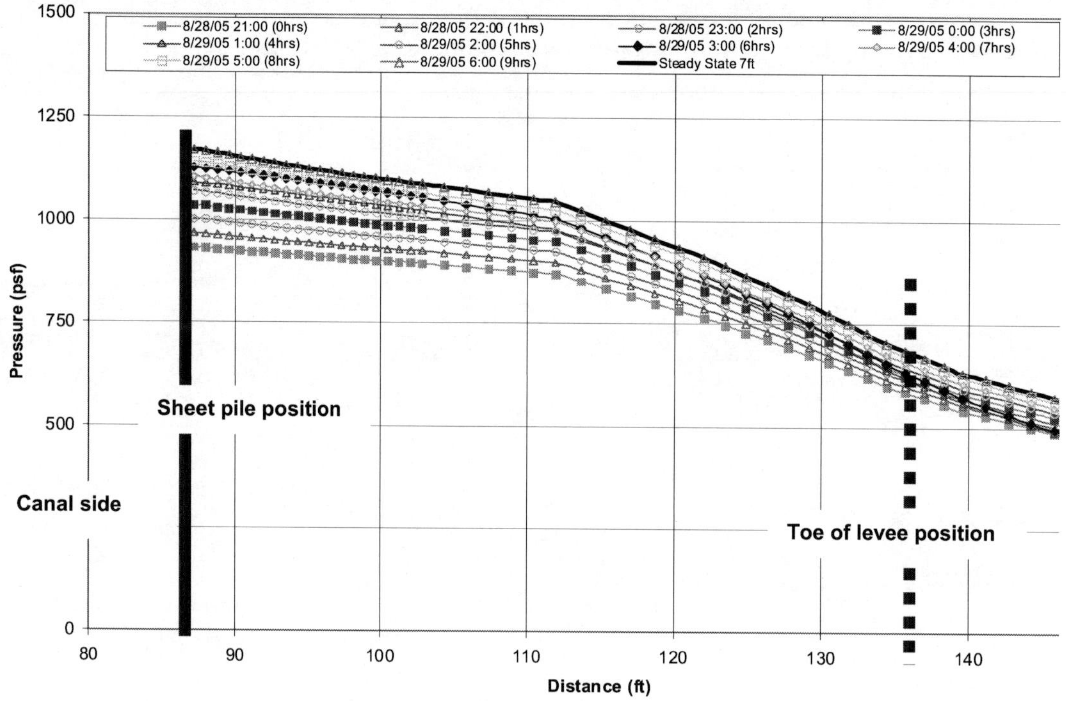

FIG. 17: Pore water pressure development along bottom of marsh deposit (horizontal permeability of 10^{-6} cm/sec).

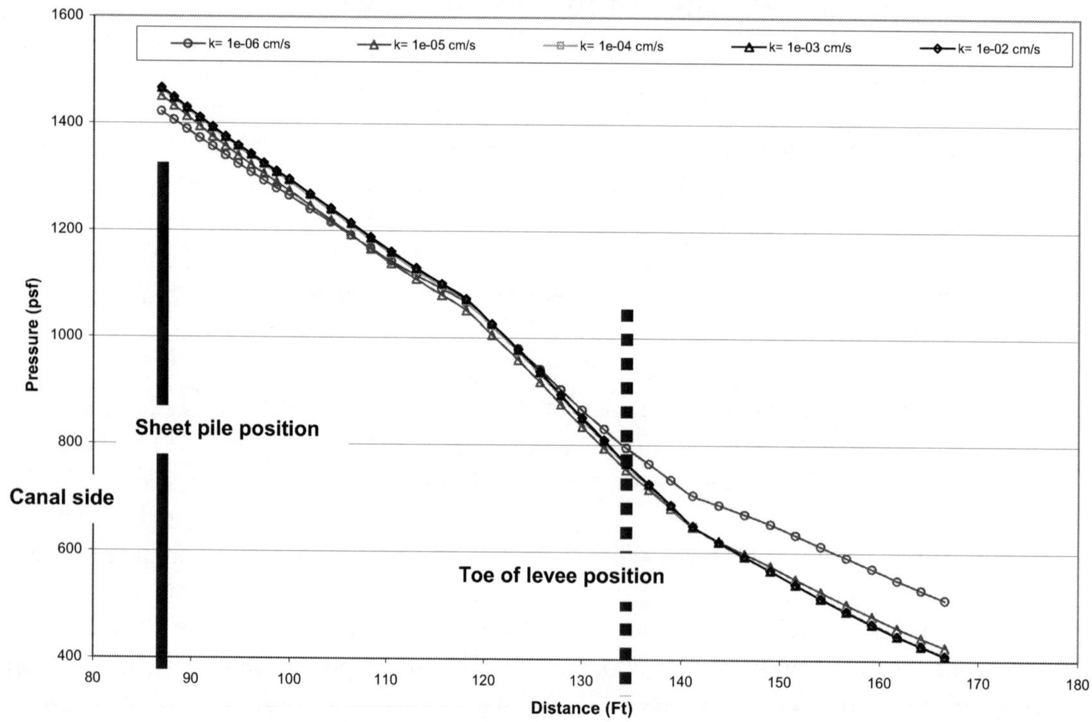

FIG. 18: Pore water pressures along bottom of marsh – transition zone layers for SSC water level elevation of + 1.8 m

Transient and steady-state pore pressure development analyses also were performed for the 'blown-over tree root-ball' at the toe of the levee condition (4.6 m diameter, 1.5 m thick). To model this condition, elements were removed in three stages starting at 5:00 a.m. (to simulate the tree being blown over) from the finite elements and new boundary conditions assigned to the appropriate nodes. Results for the entire range of permeabilities showed increasing hydraulic gradients as the hole got deeper and wider and as the storm surge increased in elevation. For a surge elevation of +6 feet, vertical hydraulic gradients larger than unity were found in the marsh deposit and in the underlying intermixing deposit in the vicinity of the toe of the levee and root-ball removal zone (Fig. 19). Such vertical hydraulic gradients would have led to additional softening and/or erosion of the marsh – intermixing zone soils leading to a 'blow-out' at this location. These analyses corroborated the observations made in the field during and after the failure concerning the high likelihood that the overblown trees played instrumental roles in triggering the failure and in defining the location of initiation of the failure on the east side of the canal.

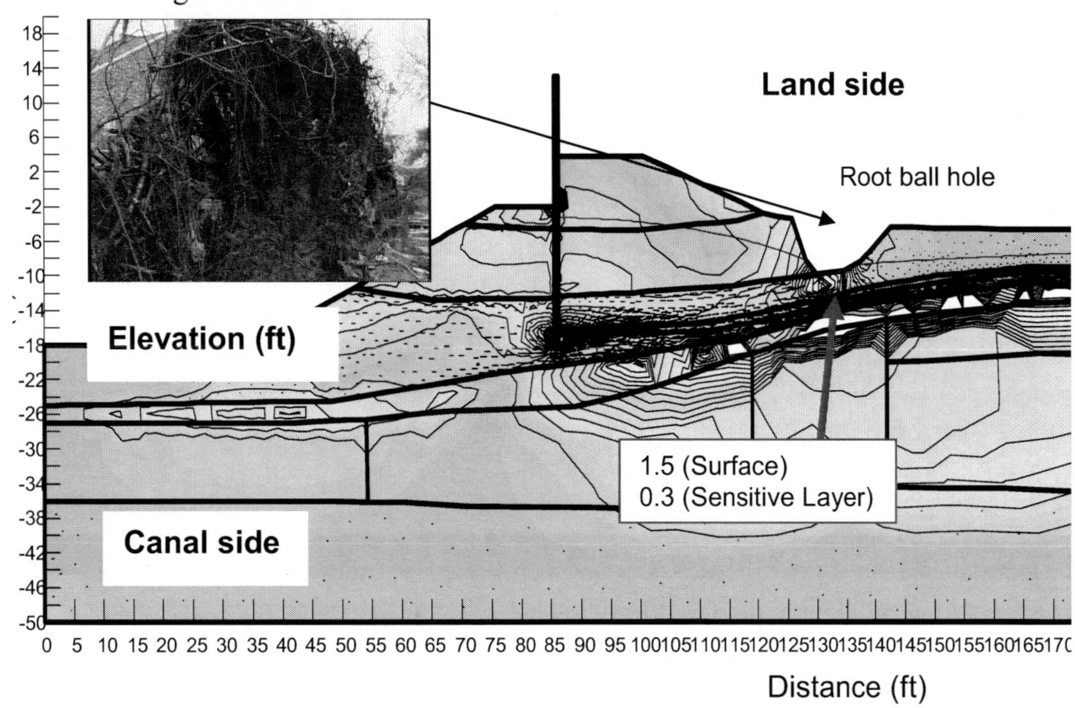

FIG: 19: Tree removal simulation (final stage) vertical hydraulic gradients for surge at +1.8 m (1 foot = 0.3 m)

The finite element analytical models showed vertical hydraulic gradients that exceed unity in the immediate vicinity of the toe of the protected side levee. Interestingly, they are the highest for the in situ horizontal permeabilities for the marsh and intermixing zone in the range of 10^{-6} cm/sec. The marsh and intermixing zone become an effective blanket that acts to develop very high pore water pressure gradients leading to uplift and vertical gradients that cause reductions in levee – flood wall – sheet piling lateral resistance and/or blowout failures. The combination of the geometric, hydro-geologic, geotechnical and mechanical aspects of this location combine to generate a very important weak zone that very likely led to the initiation of the failure that developed into the catastrophic east side breach.

FROM RESEARCH TO PRACTICE

During the year that followed hurricane Katrina, it was contended publicly that the failure at the 17th street canal was "unforeseen and unforeseeable" (Marshall 2006) and later that it was a "design failure" (Walsh 2006). Certainly, many of the critical elements had been unforeseen and there were design failures. However, there were many elements that were foreseen and foreseeable, and in some cases, officially permitted (e.g. dredging the canal off center line resulting in significant reduction in the levee elevation and cross section). The information and technology existed – it was not accessed nor properly used.

But, these failures were not the causes (Bea 2006, 2007a, 2007b). Extensive forensic investigations have shown that the causes had much deeper roots founded firmly in the organizational and institutional elements responsible for creation of the flood protection system that evolved before and was put in place following the flooding and levee breaches that accompanied hurricane Betsy (1965) (ILIT 2006; Team Louisiana 2007; Woolley and Shabman 2007; American Society of Civil Engineers 2007; National Academy of Engineering / National Research Council 2006).

For example, insufficient boring samples and other tests were performed to detect the variations in soil properties and characteristics. Borings were widely spaced (more than 150 m). Samples that could not be retrieved failed to attract proper attention (soils too weak to sample using conventional procedures). Soil shear strengths were averaged through different soil layers (horizontally and vertically) from different depositional environments (over horizontal distances exceeding 300 m) – thus, failing to recognize the weak soil layers that could control the failure mechanics (USACE IPET 2007). During the design review process, concerns were raised over how the analyses had been performed and these concerns were negated. Concerns were also raised about the depths of the sheet piling and their potential inability to arrest seepage from the canal during high water. But again, these concerns were negated.

The levees – floodwalls were designed and built to a minimum factor of safety (1.3 for lateral stability at temporary high water conditions), based on experience with rural – agricultural land rather than a major metropolitan area. This factor of safety was far too small given the extremely high human and economic costs of failures protecting a major metropolitan area like New Orleans and the almost complete lack of robustness and resilience in the 'series-like' flood protection system. Effective life-cycle quality assurance and control processes were not in place to assure that the minimum factor of safety was actually achieved and maintained.

The tops of the floodwalls were almost two feet lower along the 17th street canal than intended because of subsidence. This was the result of a deliberate decision made in the mid-1980's to 'freeze the reference elevation' to allow the work to be completed to authorized elevations. Even when surveys in 2004 clearly showed that the floodwalls and levees were not high enough, no corrective action was taken (Woolley and Shabman 2007).

Field tests performed in 1965 should have provided warning that the levee itself was having important effects on underlying soil strengths and that conventional laboratory tests combined with conventional stability analysis methods would tend to over-predict the factors of safety. Borings taken at the crests of the levees in New Orleans were taken to be representative of the soil strength at and beyond the toes of the levees. The method of planes was applied in a prescribed manner that did not allow analyses of the critical modes of failure.

In 1964 - 1965 the USACE ran a full scale levee test in the Atachafalaya basin in which advanced studies were conducted regarding characterizations of the soil strengths and performance and stability characteristics of the levee (USACE 1968; Kaufman and Weaver 1967). The levee test sections were thoroughly instrumented and their performance monitored during and after construction. Various analytical methods were used to evaluate the usefulness and reliability of the various methods. These developments clearly indicated the need to understand the geologic soil depositional processes and the associated variations in soil strengths (horizontal and vertical) in order to understand the performance and stability characteristics of levees. The importance of local soil conditions to the performance of the levee was clearly pointed out. Additional reports and professional papers were published that resulted in significant advances to the engineering knowledge (Duncan 1970, Ladd et al. 1972; Edgers et al. 1973; Foott and Ladd 1973, 1977).

Even earlier, geologic studies performed by the USACE indicated the treacherous nature of the soil conditions and the need to be on high alert for zones of unusually low soil strength and very high permeability. In-depth background on the geologic and depositional environment of vital importance to understanding the characteristics of the Mississippi Basin soils were developed in the 1950s and 1960s (Kolb and Van Lopek 1958; Krinitzsky and Smith 1969). The USACE lead in development of this background. Of particular importance was recognition that the marsh and swamp deposits were "treacherous" and highly variable. It was repeatedly pointed out that "careful and detailed characterization of the soil properties was required." Further the studies that the method based on traditional USACE soil characterization and stability analyses gave factors of safety that were too large (Foott and Ladd 1977). As in the first instance, these developments were not reflected in the design guidelines and practices that were used.

Full scale floodwall loading tests performed in 1985 should have warned the engineers that the sheet pile – soil interactions could result in the 'cut the levee in half' failure mode. The failure mode involved lateral deflection of the concrete floodwall and the sheet piles that supported that floodwall. This deflection resulted in separation between the stiff supporting sheet piling and the soft soil of the levee on the water side of the wall. Water was then able to enter the gap and exert additional lateral forces on the remaining half of the levee-floodwall. Now, the levee only had about half of its width able to transmit the lateral forces to the underlying soils. This combination resulted in lowering the lateral resistance with a commensurate lowering of the factor of safety.

In 1985, the New Orleans District of the USACE conducted a full scale instrumented lateral load test of a 200-foot long sheet pile flood wall in the Atachafalaya basin (USACE 1988a). This particular location (south of Morgan City, Louisiana) was chosen because of the close correlation of the soil conditions in the New Orleans area with those at the test location. "The foundation soils are relatively poor, consisting of soft, highly plastic clays, and would be representative of near worst case conditions in the NOD (New Orleans District)."

Test data (from slope indicators) from the highly instrumented sheet pile wall and adjacent supporting soils indicated a gap behavior (separation of the sheet piles from the soils). The test was designed to take a 2.4 m height of water (above the supporting ground level) with a factor of safety of 1.25. But the wall was already in a failure condition (rapidly increasing lateral displacements with no increase in loading) when the water level reached 2.4 m instead of the calculated 3.1 m. Strain gage readings on the sheet piles indicated that they were well below the steel yield point, thus the yielding had to have been developing in the supporting soils. Two very important pieces of information developed by the E-99 sheet pile tests were that there was potential soil separation from the sheet piles (allowing water to penetrate below the ground surface between the piles and the soils) and that the calculated safety factor was not reached (it was over-estimated due to unanticipated deformations in the soils).

Additional reports and professional papers further developed the experimental information. Advanced analytical models were developed that could be used to help capture the observed behavior (USACE Waterways Experiment Station 1989). Later developments in this work were reported by Oner, Dawkins and Mosher (1997): "*As the water level rises, the increased loading may produce separation of the soil from the pile on the flooded side (i.e., a "tension crack" develops behind the wall). Intrusion of free water into the tension crack produces additional hydrostatic pressures on the wall side of the crack and equal and opposite pressures on the soil side of the crack. Thus part of the loading is a function of system deformations.*"

These developments in technology were *not* reflected in the design guidelines used (USACE 1988b, 1989, 1990). A traditional method of active and passive pressures acting along the length of the sheet piles embedded in the earth levee was used to determine stresses induced in the concrete wall - sheet pile joint and in the sheet piles. It was assumed that the soils in the supporting levee were not deforming under the varying water levels. This traditional method did not incorporate the information developed from the E99 floodwall test. The traditional design guideline-based method used to design and engineer the floodwall system did not possess the required attributes of validity and reliability (Kardon, Bea, Williamson 2006).

The long-term deficiencies in translating research to practice were combined with flaws incorporated during the operating and maintenance phases. The decision to permit the off-centerline dredging that resulted in the significant decrease in the outboard side levee elevation and cross section on the east side (rising water could initiate the prying water intrusion mode much earlier) provided critical flaws. The decisions to allow large trees to grow at the toes of the levees proved to provide additional important flaws and defects. Although the records reflect that inspections were conducted and the levees received acceptable ratings, the records appear to be incomplete or inaccurate. In the end, given this combination of critical flaws and defects, when the system was tested, it failed at a water level far below that intended.

CONCLUSIONS

Failure of the levee and floodwall at the 17th street canal provides an excellent example of how research was not effectively translated to practice in geotechnical engineering. Geotechnical engineering technology is very able to demonstrate why this catastrophic failure developed. This technology existed at the time this failure was incubated and developed. As one investigator insightfully observed: "we failed to connect the dots." The geotechnical engineering contributions to this failure were not causes – they were results of multiple breakdowns in the Technology Delivery System that developed over a 40-year period (Bea 2007a, 2007b; ILIT 2006). The challenge of the future is to learn the painful lessons from this disaster so that they will not be repeated.

ACKNOWLEDGEMENTS

John Schmertmann provided the fundamental knowledge of geotechnical engineering that was critically important to development of my understanding of the engineering mechanics of this failure. I owe a large measure of the success of my career to the gifts John gave me.

This paper could not have been written without the leadership, knowledge, and professionalism of Drs. Raymond Seed and David Rogers. Ray and Dave provided the important insights into the engineering mechanics of this failure. University of California Berkeley geotechnical engineering graduate students Diego Cabos-Roa, Xavier Vera-Grunauer and Adda Athanasopoulos provided vital field, laboratory and analytical support for the work summarized in this paper. The GEO Institute manuscript reviewers are also due recognition for their constructive detailed suggestions; these suggestions had major effects on the contents of this paper. Because of these people and the other members of the ILIT team, failure to deeply understand the lessons of this disaster – to see through the fog of this failure - was not an option.

REFERENCES

ASCE External Review Panel (2007). *The New Orleans Hurricane Protection System: What Went Wrong and Why.* ASCE, Reston, VA. www.pubs.asce.org.

Bea, R.G. (2006). "Reliability and Human Factors in Geotechnical Engineering." *ASCE J. Geotechnical and Geoenvironmental Engineering*, 132 (5), 631-643.

Bea, R.G. (2007a). "Reliability Assessment & Management Lessons from Hurricane Katrina," *Proceedings 6th International Conference on Offshore Mechanics and Arctic Engineering*, OMAE 2007-29650, 1-12, ASME, New York.

Bea, R.G. (2007b). "Lessons from Failure of the Flood Protection System for the Greater New Orleans Area During Hurricane Katrina," *Proceedings 6th International Conference on Offshore Mechanics and Arctic Engineering*, OMAE 2007-29649, 1-10, ASME, New York.

Beckwith, C.W., Baird, A. J., and Heathwaite, A. L. (2002). "Anistotropy and depth-related heterogeneity of hydraulic conductivity in a bog peat. I: laboratory measurements," *Hydrological Processes*, 17, 89-101.

Cobos-Roa, D. and Bea, R. G. (2007). *Parametric Analyses for the Breach on the 17th Street Canal During Hurricane Katrina, New Orleans, LA*. Report to Risk Assessment and Management Services, Moraga, CA, September 30.

Duncan, J.M. (1970). *Strength and stress-strain behavior of Atchafalaya foundation soils*. Research Report TE70-1, Department of Civil Engineering, University of California, Berkeley.

Edgers, L. et al (1973). *Undrained creep of Atchafalaya levee foundation clays*. Research Report R73-16, Soils Publication No. 319, Dept. of Civil engineering, Massachusetts Institute of Technology, Cambridge MA.

Foott, R., and Ladd, C. C. (1973). *The behavior of Atchafalaya test embankments during construction*. Research report R73-27, Dept. of Civil Engineering, Massachusetts Institute of Technology, Cambridge MA.

Foott, R., and Ladd, C. C. (1977). "Behaviour of Atchafalaya levees during construction." Geotechnique, 27(2), 137-160.

ILIT (2006). *Investigation of the Performance of the New Orleans Levee Systems in Hurricane Katrina on August 29, 2005*, National Science Foundation and Center for Information Technology Research in the Interest of Society Report, Berkeley, CA, www.risk.berkeley.edu.

Kardon, J.B., Bea, R.G., Williamson, R.B. (2006). "Validity and Reliability of Forensic Engineering Methods and Processes." *Proceedings 4th Forensic Congress*, ASCE, 1-11.

Kaufman, R. I., and Weaver, F. J. (1967). "Stability of Atchafalaya levees." ASCE J. Soil Mechanics and Foundations Division, 93(4), 157-176.

Kolb, C.R., and Van Lopik, J.R. (1958). *Geology of the Mississippi River Deltaic Plain, Southern Louisiana*. Technical Report No. 3-483, U.S. Army Engineer Waterways Experiment Station, Vicksburg, MS.

Krinitzsky, E. L., and Smith, F. L (1969). *Geology of backswamp deposits in the Atchafalaya basin*, Louisiana. Technical Report S-69-8, U.S. Army Engineer Waterways Experiment Station, Vicksburg, MS.

Ladd, C. C. et al. (1972). *Engineering properties of soft foundation clays at two south Louisiana levee sites*. Research Report R72-26, Dept. of Civil Engineering, Massachusetts Institute of Technology, Cambridge MA.

Marshall, B. (2006). "Report: Levee failure could not be foreseen," *The Times Picayune*, quotation attributed to Dr. Ed Link, March 11.

Merri, G. and Ajlouni, M. (2007). "Engineering Properties of Fibrous Peats," *ASCE J. Geotechnical and Geoenvironmental Engineering*, 177, 7, 850-866.

Mitchell, J.K. (1976). *Fundamentals of Soil Behavior*, John Wiley & Sons, New York.

National Academy of Engineering / National Research Council (2006). *First, Second, and Third Report of the Committee on New Orleans Regional Hurricane Protection Projects*, The National Academies Press, Washington DC.

Oner, et al (1997). "Soil-Structure Interaction Effects in Floodwalls." Electronic Journal of Geotechnical Engineering, <http://www.ejge.com/1997> (Jan,1, 2006).

Rogers, J.D. (2006). "Engineering Geologic Characterization of Levee Failures in New Orleans During Hurricane Katrina," Proceedings 2006 Annual Meeting Association of Envioronmental and Engineering Geologists, Boston, MA.

Team Louisiana (2007). *The Failure of the New Orleans Levee System During Hurricane Katrina and the Pathway Forwards*, Louisiana State University Hurricane Center, Baton Rouge LA.

URS Corporation / Jack R. Benjamin & Associates, Inc. (2007). *Levee Vulnerability*, Technical Memorandum, Delta Risk Management Strategy (DRMS) Phase 1, Report to Department of Water Resources, Sacramento Ca.

USACE (1988a). *E-99 Sheet Pile Wall Field Load Test Report*. Technical Report No. 1, U.S. Army Engineer Division, Lower Mississippi Valley, Vicksburg, MS.

USACE (1989). *Development of Finite-Element-Based Design Procedure for Sheet-Pile Wall*. USACE Waterways Experiment Station, Tech. Report GL-89-14, Vicksburg, MS.

USACE IPET (2007). *Performance Evaluation of the New Orleans and Southeast Louisiana Hurricane Protection System*, Final Report, Washington, DC.

USACE, New Orleans District (1968). *Field tests of levee construction, test sections I, II, and III, Atchafalaya Basin Floodway, Louisiana*. Interim Rpt,. New Orleans, LA.

Walsh, B (2006). "Corps chief admits to 'design failure'", *The Times-Picayune*, quotation attributed to Lt. General Carl Strock, April 6.

Woolley, D. and Shabman, L. (2007). *Decision-Making Chronology for the Lake Pontchartrain & Vicinity Hurricane Protection Project*. Draft Final Report for the Headquarters, USACE, Institute for Water Resources of the USACE, June.

Geotechnical Issues for Closure of Ammonia Soda Ash Waste Ponds

Gary R. Schmertmann[1], M. ASCE, P.E., CPEng., Robert C. Bachus[2], M. ASCE, P.E. and Keith D. Roberts[3]

[1]Golder Associates, PO Box 1302, Crows Nest, NSW 1585, Australia; gschmertmann@golder.com.au
[2]Geosyntec Consultants, 1255 Roberts Blvd NW Ste 200, Kennesaw, GA 30144; rbachus@geosyntec.com
[3]Olin Corporation, PO Box 248, Charleston, TN 37310; kdroberts@olin.com

ABSTRACT: The paper describes the application of geotechnical site characterization, material testing, test fill construction and settlement analysis to the closure design for two settling ponds at a former chemical plant in southwestern Virginia, USA. The ponds contained by-product material from ammonia soda ash manufacturing that had been slurry-deposited to thickness up to 20m over a total area of approximately 480,000m^2. The material is a man-made particulate exhibiting a unique structure with high moisture content, high compressibility, and unusual geotechnical properties. Closure construction was successfully completed in 2001. The paper demonstrates that although the by-product material exhibits unusual characteristics, the application of geotechnical engineering principles allowed the successful execution of pond closure.

KEY WORDS: calcium carbonate; industrial waste; soil structure; settlement

INTRODUCTION AND SITE DESCRIPTION

This paper presents geotechnical investigations and modelling performed between 1998 and 2000 for capping design of two large waste by-product settling ponds in southwestern Virginia, USA. The ponds principally contain the by-product from the manufacture of soda ash. The by-product material is particulate and exhibits unusual geotechnical properties, which presented several challenges to the capping design.

The site is located near Saltville, Virginia in a narrow floodplain of the Holston River, North Fork. The two settling ponds, referred to as Pond 5 and Pond 6, were constructed by forming starter dikes along the riverbank and depositing the by-product material as a slurry. The filling of the ponds proceeded through a series of dike raisings and slurry depositions between 1925 and 1971, resulting in final dike heights of up to 30 m and deposited material thicknesses greater than 20m.

Pond 5 and Pond 6 occupy areas of approximately 300,000m^2 and 180,000m^2, respectively. A plan view of the ponds is provided in Figure 1 with a conceptual cross section provided in Figure 2. The maximum depths of settled solids in 1998 in Pond 5 and Pond 6 were 20m and 10m, respectively.

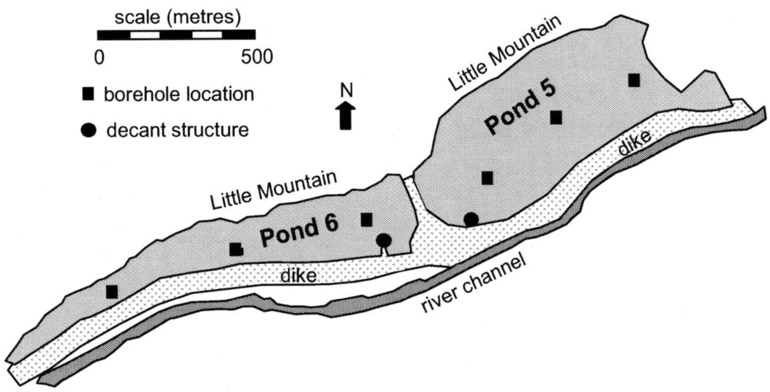

FIG. 1 – Site Plan

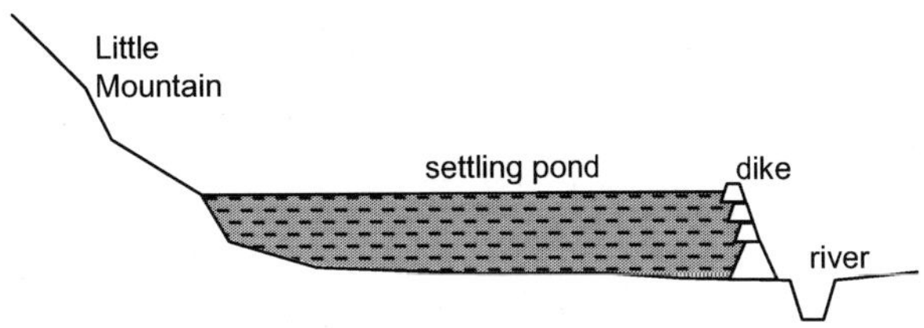

FIG. 2 – Conceptual Cross Section through Pond

Notable features of Pond 5 construction and filling are as follows:

- Pond 5 was operated from 1925 until 1971. It was constructed adjacent to a previous similar pond that experienced a dike failure in 1924.
- Historical aerial photos and anecdotal descriptions indicate that slurry discharge points were in the eastern end of the pond and that the slurry was directed along the inside of the dike through a series of channels.
- The wastes contain trace amounts of mercury from a manufacturing effluent that was co-disposed with the soda ash wastes from 1951 to 1971.
- A decant structure was constructed in the southwestern corner of the pond along the upstream dike face to drain water from the settled solids. The decant structure contained stoplogs that could be removed or replaced to adjust the internal drainage elevation for pond water.

Pond 6 is the smaller of the two ponds and was operated from 1964 until 1971. The by-product material has lower mercury concentration than the materials in Pond 5. A decant structure was located in its southeastern corner.

Ponds 5 and 6 were designated collectively as a Superfund site in 1983, with a remedial investigation and feasibility study (RI/FS) completed by Golder Associates in 1995. The RI/FS concluded that the ponds should to be capped to minimize water infiltration. Due the size of the ponds, capping might require placing a significant thickness of fill over the ponds to provide a suitable final landform. It was recognized that capping design would therefore need to address stability and settlement concerns due to the physical properties of the by-product material, as described subsequently. Cap instability and excessive total and/or differential settlements could lead to ineffective cap performance due to poor surface drainage and excessive infiltration.

The work described in this paper is from capping design studies performed by Geosyntec Consultants, on behalf of the Olin Corporation in collaboration with the former Law/Gibb Group.

AMMONIA SODA ASH WASTE MATERIAL (ASAW)

The material in the ponds is a white-colored, fine-grained particulate by-product of soda ash (i.e., sodium carbonate) manufacturing at the site by the Solvay process. The Solvay process consumes ammonia and the resulting solid by-product is referred to as ammonia soda ash waste, or ASAW. These by-products may be present at other production facilities that use the Solvay process in the USA and worldwide (e.g., Kulhawy et al. 1977). The composition of the ASAW is described subsequently.

SITE CONDITIONS

The following observations were noted during a site walkover in 1998.

- The settled solids surface of the ponds was dry and could generally be walked on without difficulty, although local areas were encountered with extremely loose surface exposures of the ASAW materials. Conventional rubber-tired vehicles could traffic on established access roads. Anecdotal information indicated that tracked equipment would become bogged after a few passes on the virgin ASAW surface. Vegetation was present in most areas of the ponds, ranging from sparse grass to dense reeds, and occasional small trees.

- Indications of significant settlement of the pond surface included: (i) the dike crest was as much as 6m to 9m above pond surface levels; (ii) pond water monitoring wells installed in the late 1980s had approximately 1m of excess pipe stick-up; and (iii) invert levels of a surface water channel along the northern edge of Pond 5 were irregular despite previous repairs (Houlihan and Rogers, 1988).

- The surface of the ponds was relatively flat on a macro-scale. For example, the Pond 5 surface generally sloped at less than one percent toward a low point in the west-central portion of the pond. In the vicinity of the low point, however, the surface was often irregular with numerous small hummocks, one to two meters in height. Similar hummocks were present in Pond 6. Occasional surface “wormholes” up to 200mm diameter were observed in the hummocky areas. These wormholes appeared to be dissolution features offering preferential surface water infiltration pathways into the ASAW.

- Aerial photos of the site showed an apparent large-scale cracking pattern in the west-central portion of Pond 5, coinciding with the hummocky areas. Although the pattern was not discernable from ground level observations, the aerial photos revealed an approximate hexagonal arrangement of “crack” tracings with spacing of 20m to 50m Anecdotal information indicated that large crevasse-like cracks had been present on the pond surface in the period after stoplog removal in 1978. Stoplog removal resulted in lowering of pond water levels. The authors later conjectured that deep surface cracking occurred due to pond dewatering and that the cracks were then enlarged by erosion and dissolution and subsequently infilled by side wall collapse, leaving a hummocky surface.

- The ASAW exhibited significant sensitivity, or strength loss upon remolding. Strength loss, along with release of pore water, was clearly apparent upon reworking a handful of the material.

Information available from periodic pond surface surveys indicate that between 1981 and 1999, or approximately 10 to 28 years after slurry deposition ceased, the surface of Pond 5 settled as much as 2.4m. This magnitude of settlement is considerably greater than the authors would have expected for this time interval, based on their experience with settlements in ponds containing dredged spoil materials. This relatively large settlement is believed to have been initiated primarily by drawdown due to the events listed below. As will be indicated subsequently, the ASAW exhibits a very low buoyant unit weight. Consequently, drawdown-induced stress increases can be significant.

- Stoplogs were removed from the Pond 5 decant structure in 1978 to lower the drain elevation and enhance dewatering of the settled solids. No data regarding actual pond water levels are available for this period.

- Improvements were made to surface water and groundwater controls at the site in 1989 to reduce the amount of water entering the pond. Data from a monitoring well within the western portion of Pond 5 measured a gradual reduction in annual minimum pond water level of nearly 4m between 1989 and 1994.

Data from monitoring wells within Pond 5 provide water levels within the ASAW from 1981. Since 1994, when levels appear to have stabilized following the 1989 surface water and groundwater improvements, data indicated the following:

- pond water levels exhibit seasonal fluctuation, being highest in the first half of the year and lowest in the second half;

- near the center of the pond, annual high pond water levels are 2m to 3m below the pond surface and annual low pond water levels range from 5m to 9m below the pond surface; and

- seasonal fluctuation of groundwater levels is significant, exceeding 6m in the western area of Pond 5, and being 2m to 5m at other well locations in Pond 5.

SITE INVESTIGATION

Although the ASAW was clearly not a conventional soil material, a site investigation and laboratory characterization study was initiated to assess ASAW behaviour in the context of a geotechnical material. Limited test results reported by Kulhawy et al. (1977) appeared to support this strategy. Near-surface and deep subsurface investigation programs were developed.

The near-surface investigation was conducted at 214 locations on the two ponds. The near-surface investigation included miniature insitu vane shear testing using hand equipment and ASAW sample collection to a depth of 1m. The investigation relied on equipment that could be easily transported to all areas of the ponds using a small, low ground pressure, off-road vehicle, thereby precluding the need to establish access roads. The primary purpose of the near-surface investigation was to screen for material sensitivity and to establish material index properties to identify any areal trends significant for capping design. It was thought that trends might be present, perhaps reflecting the ASAW slurry deposition or dewatering sequence.

Geotechnical boreholes were advanced at six locations, three on each pond (Figure 1). The purpose of the boreholes was to assess pond depth and ASAW stratigraphy, while allowing samples of ASAW to be collected for laboratory analysis and testing. At each location the boreholes extended through the entire thickness of ASAW and into underlying native materials. The thickness of ASAW encountered in Pond 5 ranged from 11m to 19m, and in Pond 6 from 3m to 9m.

The boreholes were successfully advanced using hollow stem augers and a truck-mounted drill rig operating on a 300mm-thick fill pad on the pond surface. Thin-wall tube sampling (75mm diameter) and Standard Penetration Test (SPT) driven sampling was performed. Conventional push sampling with thin-wall tubes was successful in obtaining intact ASAW samples exhibiting minimal apparent disturbance. Borehole sampling was performed at frequent intervals, often with continuous push tubes. Sampling depths for each borehole were selected to recover

ASAW from each of three depth intervals: (i) upper interval- above the maximum pond water level; (ii) intermediate interval- annual groundwater fluctuation range; and (iii) lower interval- below the minimum pond water level.

NEAR-SURFACE SCREENING TESTS

Vane shear testing was conducted at 214 locations at depths of up to 1m. Results were interpreted for undrained shear strength and sensitivity; the latter defined as the ratio of peak shear strength to residual shear strength measured with the miniature insitu vane device. The measured sensitivity of the near-surface ASAW was generally slight to moderate, with values as high as ten occasionally recorded. These results confirmed that the near-surface ASAW could behave as a relatively sensitive material and affect capping construction activities. The results across the ponds were generally consistent and did not reflect any areal trends that were considered significant for capping design.

Geotechnical index tests were conducted on bulk samples of near-surface ASAW from a number of locations on each pond. The tests included moisture content, grain size distribution and Atterberg limits. These test results indicated that the behaviour of the near-surface ASAW would lead to soil classification as a medium to high plasticity clayey silt with moisture content exceeding the liquid limit. As with the vane shear test results, no significant areal trends were identified.

LABORATORY TESTING OF BOREHOLE SAMPLES

Moisture Content and Dry Unit Weight

Ninety-five (95) intact samples of ASAW from the three identified ASAW depth intervals were tested to assess moisture content and dry unit weight using conventional test procedures. Measured moisture contents ranged from 90% to 360% with the majority of measurements indicating greater than 200%. Measured total unit weights ranged from 11 to 13 kN/m^3 and calculated dry unit weights ranged from 2.5 to 7.5 kN/m^3 (note: 10 $kN/m^3 \approx$ 64 lb/ft^3). Moisture contents and dry unit weights were generally consistent with a saturated material exhibiting a specific gravity of solids between 2.0 and 3.0 (Figure 3). A weak trend of increasing dry unit weight with depth is discernable for the ASAW, suggesting that the ASAW deposit has undergone some degree of self-weight consolidation.

There was a slight trend toward higher values of dry unit weight in the eastern portion of Pond 5 and lower values in the western portion. This trend may be consistent with slurry discharge points located in the eastern portion of the pond and the consequent settling of coarser ASAW in this area. This trend may also be consistent with the possible presence in eastern portion of the pond of remnant ASAW material from the original (pre-1924) pond. Such remnant material would likely be well consolidated and relatively dense. In any case, however, the magnitude of the observed trend was not considered significant for capping design. No lateral trends were observed in Pond 6.

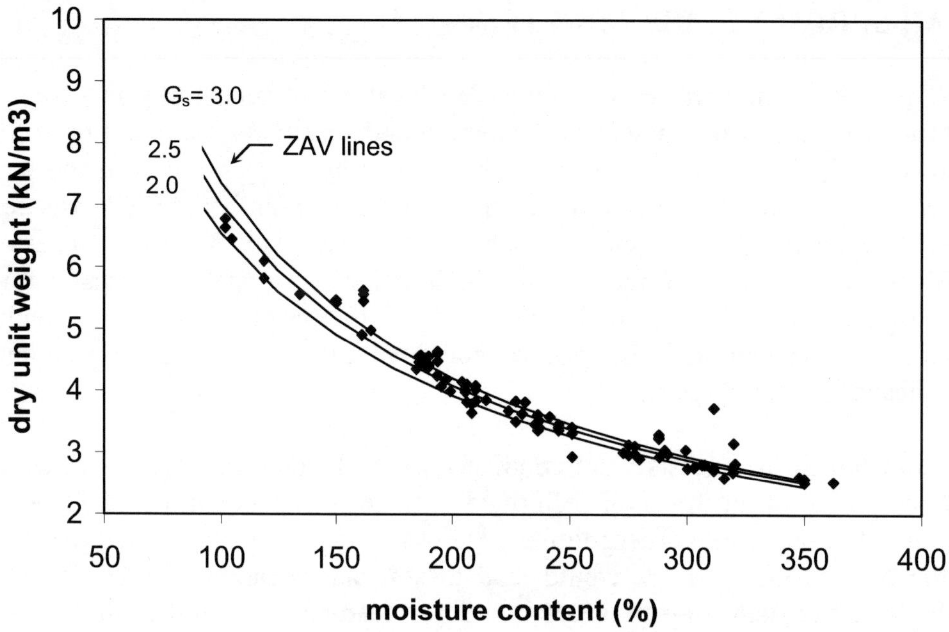

FIG. 3 – Measured ASAW Dry Unit Weight versus Moisture Content

As noted above, measured values of dry unit weight are significantly lower than for typical soil deposits, and also lower than for typical slurry-deposited natural mineral tailings deposits of similar thickness. The measured moisture contents are relatively high. Assuming a specific gravity of solids of 2.7, the ASAW has an unusually high calculated total porosity in the range of 70% to 90%, equivalent to a void ratio of 2.3 to 9.0. The fact that the material could support light to moderate equipment loads as long as repetitive loading was avoided indicated considerable undisturbed shear strength despite the high porosity. These observations led to a preliminary assessment that there may be significant effects of soil structure or grain interlocking within the ASAW.

The measured values of total unit weight indicate that buoyant unit weight of the ASAW would be in the range of 1 to 3 kN/m^3. One implication of such small buoyant unit weight, given the relatively large annual pond water level variation, is that effective vertical stresses in the ASAW can vary significantly on an annual basis. At many depths, the effective vertical stress during the dry season would be twice that during the wet season. Similarly, these low buoyant unit weights would support the observation that drawdown effects within the ponds would result in significant settlement.

Atterberg Limits

Atterberg Limits tests were conducted on 16 SPT samples of ASAW from the three identified ASAW depth intervals. Measured liquid limits, LL, ranged from 36 to 242, with a majority of values being greater than 100. The average plasticity index, PI, for the majority of samples is 70. Considering measured moisture contents from adjacent intact samples, the results indicate that at 15 of 16 sample locations the actual moisture content exceeded the liquid limit. This is a common characteristic of sensitive soils, and when expressed by the liquidity index (i.e., LI= [w-PL] / PI), the ASAW measurements ranged between LI of 1.5 and 5 for the majority of samples. No lateral or depth trends were apparent in the Atterberg limits testing results.

Shear Strength

Consolidated-undrained triaxial shear tests with pore pressure measurements (CU-bar tests) were conducted on 16 intact samples of ASAW from the three identified ASAW depth intervals. Individual tests at different effective consolidation pressures (isotropic) were performed on three specimens from each sample. Effective consolidation pressures ranged from 19 kPa to 235 kPa and were generally selected with respect to the estimated current maximum (i.e. dry season) effective vertical stress on the samples as follows: (i) the lowest consolidation pressure was 0.5 to 1.0 times the estimated current maximum vertical effective stress (σ_{vo}'); (ii) the intermediate consolidation pressure was 1.0 to 2.0 times σ_{vo}'; and (iii) the highest consolidation pressure was 1.5 to 3.0 times σ_{vo}'. It was therefore expected that the test samples would be generally normally consolidated, with the samples at the lowest consolidation pressure possibly lightly overconsolidated.

The ASAW materials generated positive pore pressures in all tests and generally exhibited a moderate strain softening response at the intermediate and highest consolidation pressures. The observed strain softening is indicative of a sensitive material, although the degree of sensitivity, or brittle behavior, is significantly lower than would be expected in fine-grained soils with similar liquidity index values of 1.5 to 5.0 (eg. Holtz and Kovacs, 1981, p. 587). No lateral or depth trends were apparent in the triaxial shear testing results.

Measured peak undrained shear strengths ranged from 20 to 135 kPa. The ratio of these measured values to the effective consolidation pressure for each test ranges from 0.4 to 1.0. For typical natural clay soils, these values would be indicative of light to moderate overconsolidation. In addition, the value of the Skempton $\bar{A}$ parameter at failure, $\bar{A}_f$, obtained from each test varied from 0.2 up to 1.0 at the highest consolidation pressure, also consistent with lightly to moderately overconsolidated behavior. These results indicated a more overconsolidated response than expected.

The drained angle of internal friction values determined from the triaxial tests, assuming zero effective cohesion, ranged from 44 to 68 degrees. These values are high compared to typical fine-grained soils, particularly clays with plasticity index as high as the ASAW typical value of 70 percent (eg. Kulhawy and Mayne, 1990, pp. 4-22).

Composition

X-ray diffraction (XRD) testing was carried out by K/T GeoServices on 18 intact samples of ASAW from the three identified ASAW depth intervals. The XRD procedure was able to identify up to two-thirds of the solid materials. The identified materials comprised calcite ($CaCO_3$) at 40% to 60% of total dry sample mass, with minor amounts of other calcium containing minerals including aragonite, bassanite, and gypsum. The specific gravity of Calcite is 2.71. Conventional soil minerals comprising quartz, kaolinite, illite and mica were detected at a combined mass percentage of 5% or less. Halite (NaCl) was detected in some samples at a mass percentage up to 5%. Between 30% and 60% of the solid mass of the samples was not able to be identified through XRD, although the unidentified materials were judged to be crystalline.

Scanning electron microscope (SEM) imaging was conducted by K/T GeoServices on specimens from the same samples tested by XRD. The SEM images provide information on grain size and shape as well as overall material structure. In addition, energy dispersive spectroscopy was performed in conjunction with the SEM imaging to provide an indication of the elemental composition of specific mineral grains within each specimen and to support the mineral identifications made by XRD.

The SEM images indicate that the ASAW comprises mineral particles with a wide range of shapes and sizes. The particles are predominantly crystalline with rhombic, spar, acicular (needle), tabular and star configurations as shown in Figure 4. Particle sizes typically range from 1μm to 5μm, in the fine silt to clay size range. Although there were a wide range of particle shapes, the occurrence of spar and needle-shaped particles within a highly structured mass was evident throughout. The SEM images indicate that the ASAW is highly porous. Evidence is seen of in-place crystallization of calcite in some samples, meaning some crystals likely formed in place after sedimentation rather than prior to sedimentation. It is not possible to identify from the SEM images whether particle contacts are cemented.

FIG. 4 – Crystal Shapes in SEM Images of ASAW

Elemental analyses for metals were conducted on 16 samples of ASAW. Results indicated that calcium comprised 20 to 38% of the solid mass, generally consistent with the XRD results for calcite. Other metals consistently detected were aluminum, iron, and sodium, each at 1% or less of solid mass.

Compressibility

Conventional one-dimensional (1-D) consolidation tests were conducted on 16 intact samples of ASAW from the three identified ASAW depth intervals. Consolidation pressures ranged from 25 kPa to 750 kPa for the majority of the tests. A load increment ratio of one was employed and an unload-reload cycle was performed in the virgin compression range in each test. Deionized water was used as the submerging fluid, although one comparison test was performed using simulated high pH pond leachate as the submerging fluid. No significant effect of the submerging fluid was detected.

The ASAW samples were highly compressible with the majority of samples undergoing total vertical strains of 40 to 50% during the 1-D testing. An example strain versus log pressure curve is given as Figure 5. Values of modified compression index from the tests, being the slope of the strain-log pressure curve, ranged from

$C_{c\varepsilon}$= 0.2 to 0.64, with the majority greater than 0.4. The ratio of modified recompression index to modified compression index ranged from 0.03 to 0.08.

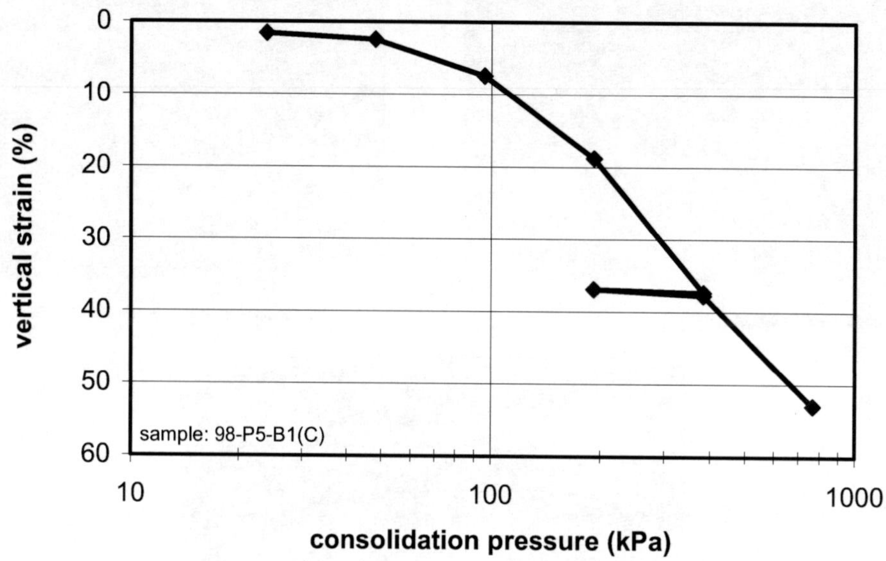

FIG. 5 – Example ASAW Consolidation Test Result

It is recognized that the high compressibility and internal friction of the ASAW may have resulted in significant load transfer by side friction to the consolidometer walls. It was not possible, however, to assess this effect with available data. This effect, along with sampling disturbance, may account for the lack of a well defined transition point from reloading to virgin loading (i.e., preconsolidation pressure) as expected from the test fill results described subsequently. Final compressibility parameters used for cap design were not based on the 1-D test results alone but were calibrated with test fill settlement measurements and historical pond surface settlements.

The measured values of modified compression index are considered consistent with very high porosity soils, but are in the order of two times greater than typical values for fine-grained soils with liquid limits similar to that of the ASAW. For example, the well-known correlation C_c= 0.009 (LL-10) gives a compression index of C_c= 1.3 for a typical ASAW liquid limit of 150%. For comparison, a modified compression index of $C_{c\varepsilon}$= 0.4, typical for ASAW, corresponds to a standard compression index of C_c= 2.4, assuming a typical ASAW void ratio of five (porosity 83%) and using $C_{c\varepsilon}$= C_c / (1+e).

Conventional Casagrande construction techniques were used to evaluate preconsolidation pressure. The resulting values of preconsolidation pressure ranged from 60 to 200 kPa as indicted in Figure 6. These values were compared to the current maximum vertical stress on the sample, which was estimated using the measured ASAW unit weight, and assuming that pond water level was at typical annual low levels. The ratio of preconsolidation pressure to current maximum vertical stress is calculated as the apparent overconsolidation ratio (OCR). The apparent OCR varies from 0.7 to 10.6, with all values greater than 2.5 occurring in the upper 3m and probably related to surface effects. The deeper values of OCR are in the order of two and are not considered explainable based on known loading history. The predicted overconsolidated conditions are consistent with the triaxial results discussed above.

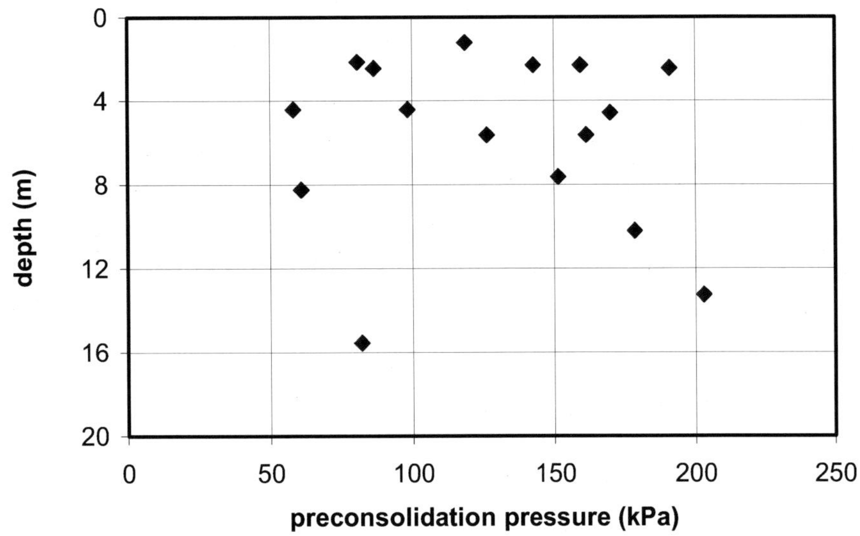

FIG. 6 – ASAW Preconsolidation Pressure versus Depth

Values of the coefficient of consolidation, c_v, calculated from the tests ranged from 0.005 to 0.020 cm^2/sec for consolidation pressures between 100 and 500 kPa. These values reflect rapid consolidation in the order of that expected for silty sand, rather than much lower values typically expected for soil with the plasticity and grain-size characteristics of ASAW.

Values of the modified secondary compression index, $C_{\alpha\varepsilon}$, were computed for each loading step of the tests. The majority of the values for the loading steps at pressures above the preconsolidation pressure range from 0.01 to 0.03, indicative of a high rate of secondary compression in virgin loading and similar to that for high water content soils.

The effects of 1-D consolidation on the ASAW structure was considered by taking two subsamples from one sample tube, sending one directly for SEM imaging, and subjecting the other to a 1-D consolidation test, using a maximum applied stress of 3000 kPa and recording 57% total vertical strain, prior to SEM imaging. Although the consolidated specimen did not exhibit a highly porous texture as seen in the unconsolidated specimen, image resolution was not sufficient to assess potential changes in structure or grain interlocking.

Dispersion Potential

The potential for internal erosion of ASAW material along any preferential flow paths within the ponds was considered of interest for capping performance and was therefore investigated by performing pinhole dispersion tests on intact samples using both upgradient site groundwater and site surface water. The test results indicated that the potential was minor and not critical for purposes of the capping design.

Dissolution Potential

Dissolution and other geochemical activity within calcite-rich soil deposits, such as calcareous marine sediments, is known to potentially alter geotechnical characteristics (Lee, 1982). The presence of surface "wormhole" features and the known solubility of calcite suggested that the potential for ongoing dissolution of the ASAW was of interest with respect to capping performance. Dissolution potential tests were thereby conducted on five intact samples of ASAW from depths of 6 to 16m. The samples were permeated with upgradient site groundwater in a triaxial permeability device and physical changes and effluent chemistry were measured. The tests were conducted using hydraulic gradients ranging from 5 to 30 and test durations ranging from 14 to 65 days. Three to six pore volumes of groundwater were passed through each sample.

Total dissolved solids (TDS) in the test effluent was measured periodically during the course of each test. Measured values ranged from 3300 to 123,000 mg/l, with all values greatly exceeding the influent value of 110 mg/l. Effluent TDS levels progressively decreased in each test. Effluent was analysed for metals in some cases with the measurements indicating calcium being the only metal detected at significant concentrations (up to 9370 mg/l). Effluent pH was high, ranging from 12.0 to 12.8.

Physical measurements of the sample dimensions, weight, and moisture content before and after testing were used to assess solid mass loss. Significant mass loss was measured in three of the five tests, ranging from 13% to 46% loss. The other two tests indicated negligible mass loss, although the effluent data indicate that some dissolution was occurring. The physical measurements and effluent data were consistent in that larger physical changes were associated with higher effluent TDS concentrations.

The test results are consistent with dissolution of calcite, known to occur in natural geological environments. Since the tests were conducted on materials from the pond rather than "virgin" by-product material, the role of dissolution in explaining the current structure and behaviour of the ASAW is not known. On the basis of these results, however, the potential impact of dissolution within ASAW was judged to be significant for design of a capping system for Ponds 5 and 6.

Laboratory Testing Summary

The engineering properties of ASAW were evaluated using conventional geotechnical testing approaches and the results yielded some interesting observations. The testing indicated that ASAW is saturated, fine-grained particulate material with a very high porosity and liquidity index. The following characteristics of the ASAW material measured in the laboratory testing were considered most significant for capping design.

- The material is not a natural soil. The predominant solid mineral is calcite ($CaCO_3$) and grain shapes are unusual compared to typical natural soils. There is evidence of insitu calcite crystal formation and growth. The solid material grains are susceptible to significant dissolution when leached with upgradient site groundwater.

- The material is moderately sensitive in terms of undrained shear strength, but to a significantly lesser degree than would be expected for a natural clay soil with similar liquidity index.

- The material has an unusually high effective angle of internal friction relative to natural clay soils with similar plasticity index and liquidity index.

- The material exhibits overconsolidated response in shear and 1-D compression even at stresses well above estimated maximum past values.

- The material is highly compressible in virgin loading, with primary compression index more than two times larger than would be expected for a natural clay soil with similar plasticity index. The rate of primary consolidation is very high compared to natural clay soils.

- The secondary compression index after virgin loading is also high, but similar to that expected for natural clay soils with very high water content.

In natural soils, sensitivity and overconsolidated response can develop as a result of soil aging. This phenomenon has been described by John Schmertmann, the first author's father, in his 1989 Terzaghi lecture (Schmertmann, 1991). The authors believe that the explanation for the sensitive nature and apparent overconsolidation of the ASAW material is the inherent structure of the mass. This may be related to high particle interlocking between unusually shaped crystals, and/or to light cementation between crystals and other grains, resulting in highly porous and slightly brittle solids matrix. The unusually high moisture content of the ASAW is similarly due to the water that is effectively held within the crystalline structure. Although the ASAW is explicitly not a conventional soil material, its macroscopic behaviour appears consistent with lessons learned from studying the behaviour of soils at a microscopic/particulate level.

TEST FILL

To assess the potential performance of a soil cap placed on the pond, a surcharge test pad was constructed by others in 1993 in the far western portion of Pond 5 using materials excavated from the pond perimeter dike. Limited data from this test pad were available, but these data indicate that the pad was 30m by 45m in plan dimension and 3m thick. Measured settlements at two points near the center of the pad were 0.15m and 0.50m shortly after completion of test pad construction; settlement at the same two points were 0.59m and 1.02m after eight months. These limited data demonstrated that short-term total and differential settlements under capping fill loads could be large and that settlements would increase with time.

An additional test fill was conducted under direction of the authors in 1999 in the west central portion of Pond 5. The test fill was instrumented and monitored as described below. The primary purposes of the test fill were to improve understanding of the likely performance of a soil cap during and after construction, and to assess the settlement mechanism and likely time-dependent settlements. A secondary purpose was to assess fill constructability on the ASAW surface. The test fill response was used, along with laboratory measurements, to calibrate settlement analysis parameters for ASAW as described subsequently. The test fill also incorporated trial capping construction procedures using geomembranes and cushion geotextiles; however, findings regarding these materials are not discussed in this paper.

The 1999 test fill was constructed using two distinct types of fill. One section used imported conventional soil materials, while a second portion was constructed using former perimeter dike materials excavated from the 1993 test pad. It was anticipated that the former dike materials would compact to a significantly lower unit weight than the soils.

The 1999 test fill was 50m by 70m in plan dimension and 2.5m to 3.1m thick. A cluster of three vibrating wire (VW) piezometers and a standpipe piezometer were installed in the center of the test fill, prior to construction, at depths in the ASAW below the pond water level. Five settlement plates were installed at the pond surface. The VW piezometers and settlement plates incorporated corrosion resistant features due to the high pH water conditions. The installed instrumentation is shown in Figure 7.

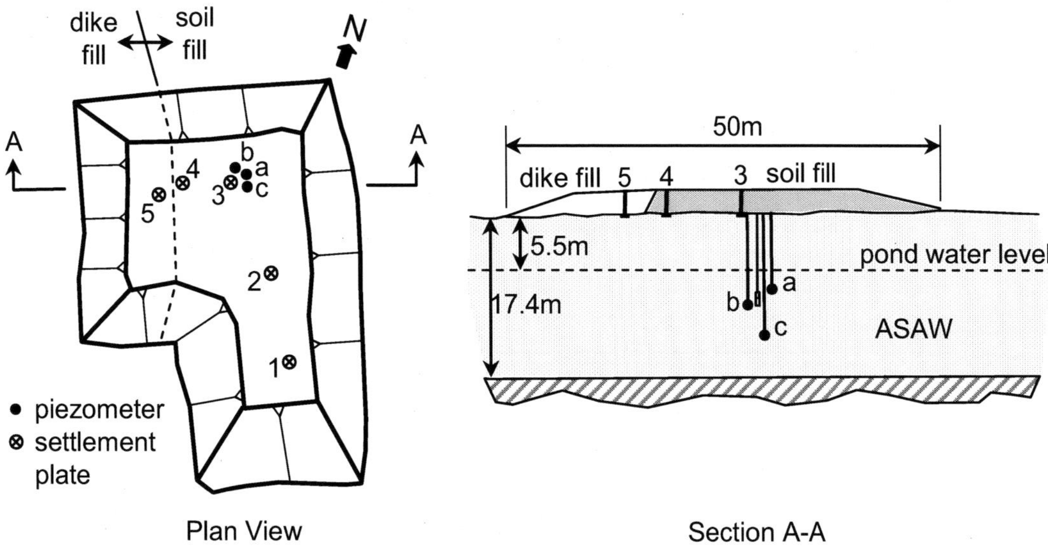

FIG. 7 – Plan View and Cross Section of 1999 Test Fill

Existing vegetation in the vicinity of the test pad consisted of sparse long grass that was left in place. Fill was successfully placed on the pond surface without any major construction difficulties. Several construction lessons were noted, as indicated below.

- There was no apparent difference in ease of initial fill lift placement in areas with and without a geotextile separator placed on the pond surface. The pond surface with sparse vegetation adequately supported the construction equipment.
- An initial construction lift was placed using low ground pressure equipment. Rutting was minor and local instability (i.e., mud waving) was not detected.
- Loaded 30-ton capacity trucks could traffic the test fill surface without rutting when an initial 450-mm thick construction lift was placed.
- Field measurements of the placed and compacted fills indicated that the average total unit weight of the soil fill was 19.3 kN/m^3 and that of the dike fill was 15.7 kN/m^3. These results indicate that adequate compaction could be achieved over the relatively soft and compressible ASAW foundation

The test fill was constructed after a period of low rainfall and the dry conditions may have contributed to the relative ease of fill construction on the pond surface.

Measured settlement data from the test fill (Figure 8) are summarized below.

- Rapid settlement initiated in the vicinity of the piezometer cluster sometime before the soil fill thickness at this location reached 1.4m. This suggests that the ASAW may have a sharp transition from reloading to virgin compression, or a "trigger" point, and that the loading increment to reach the trigger may be small.

- Settlements during the two-day fill construction period were approximately 180mm for the four plates in the soil fill area and 80mm for the plate in the dike fill area. The lighter weight dike fill apparently caused smaller settlements.

- Total settlements for the first 44 days after fill placement ranged from 300mm to 395mm for the four plates in the soil fill area; corresponding settlement for the plate in the dike fill area was 150mm. The time of 44 days appears to coincide with the end of the period of relatively rapid settlement.

- Total settlements for the first 92 days after fill placement ranged from 350mm to 460mm for the four plates in the soil fill area; corresponding settlement for the plate in the dike fill area was 180mm. Settlement rates during the time interval between 44 and 92 days after construction appear to be relatively constant at 1.0 to 1.3 mm/day in the soil fill area and 0.7 mm/day in the dike fill area.

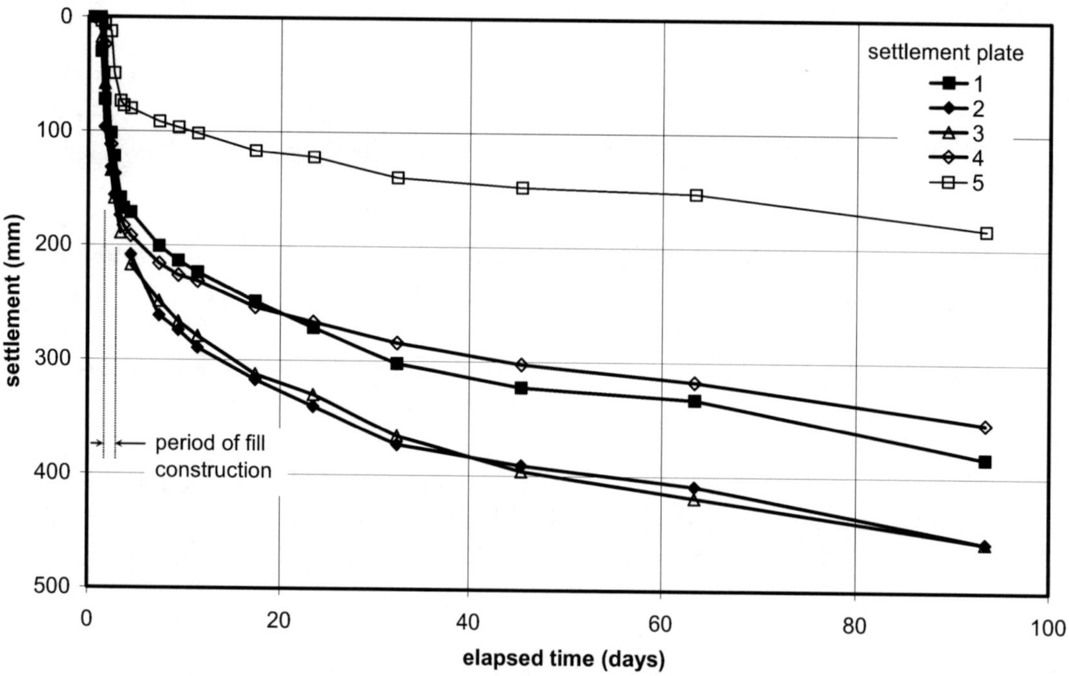

FIG. 8 – Settlement of 1999 Test Fill

Measured piezometer data from the test fill (Figure 9) are summarized below.

- Pore pressure response to fill placement was immediate. During the day when a 2.4m thick layer of fill was placed, the upper and middle VW piezometers responded with pore pressure increases equal to 70% of the estimated total stress increase from the fill. The lower VW piezometer had a 55% response.

- The standpipe piezometer had a 45% response on the primary day of fill placement, but continued to record further increases subsequent to placement (i.e., a lag in response). However the frequency of monitoring of the standpipe was not sufficient to specifically relate the subsequent increases to specific fill placement periods. The standpipe piezometer became blocked shortly after fill construction was completed and no further measurements could be made.

- The VW piezometers indicated that rapid pore pressure dissipation began as soon as fill placement ceased. These piezometers also recorded slight pressure increases during the subsequent placement of thin fill levelling and capping layers. The pore pressure increases from fill placement had dissipated by about 50% within the first ten days of construction. For example, Figure 9 indicates that the excess pore pressure at the middle VW piezometer decreased from a peak of 39 kPa to 21 kPa in this period.

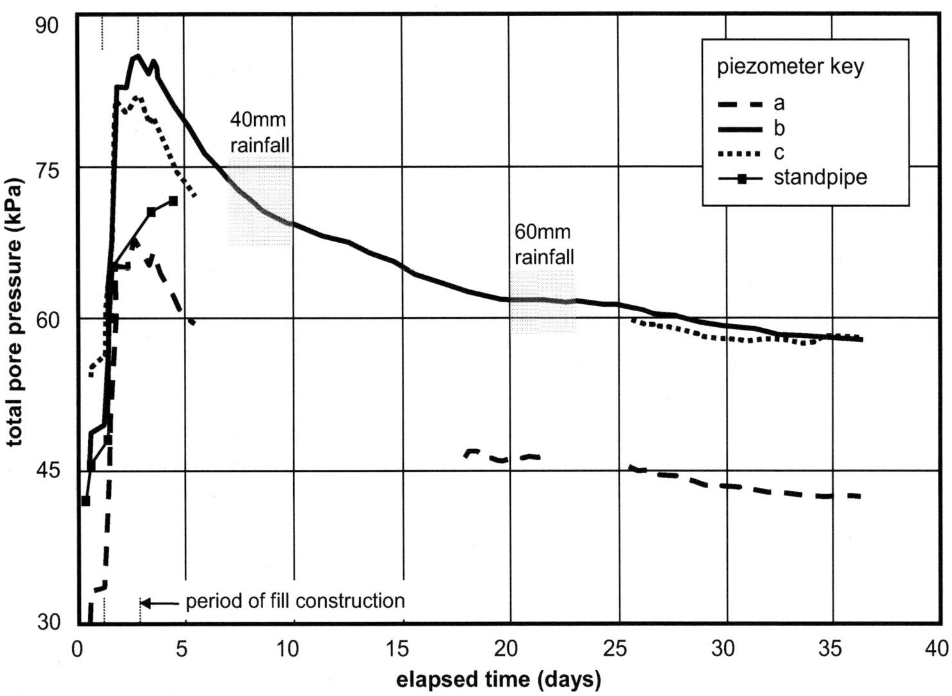

FIG. 9 – Foundation Pore Pressure for 1999 Test Fill

- A storm event with a 3-day total rainfall of 40mm occurred immediately after fill construction was completed. A second storm event with a 3-day total rainfall of 60mm occurred 12 days later. These rainfall events caused a detectable response in the measured pore pressures. Specifically, pore pressure dissipation slowed when the rainfall started and then increased after the end of the rainfall. This effect suggests rapid infiltration of rainfall and overall rise in pond water level.

The 1999 test fill demonstrated that fill placement and compaction are feasible on the ASAW at the pond surface provided that reasonable methods and equipment are employed. The measured settlements and pore pressure response confirm that the ASAW responded to the test fill loading in a manor consistent with a saturated fine-grained soil, although with relatively high compressibility and rapid dissipation of excess pore pressure. The test fill provides data to allow "calibration" of the settlement parameters, as described in the next section of this paper.

SETTLEMENT ANALYSIS

A settlement model was developed for the ASAW and used to guide the design of the cap for the two ponds. The model considered mechanical compression and used traditional consolidation calculations for saturated, fine-grained soils. This calculation approach was selected on the basis that measured ASAW characteristics (e.g., Figure 3) and the 1999 test fill response were consistent with traditional consolidation behavior.

The model parameters were initially selected based on the laboratory measurements, and then successfully calibrated using observed settlements. For example, in the west central portion of Pond 5 the predicted settlement between 1981 and 1991 was 0.95m compared to the observed value of 0.92m. This time interval comprised a period of secondary settlement in response to the 1978 removal of stoplogs followed by a period of primary settlement in response to the 1989 improvements in site water controls. Predicted settlement at the same location extending to 1999 was 1.54m compared to the observed value of 1.50m and also included periods of both primary and secondary settlement. Predicted settlements of the soil fill portion of the 1999 test fill, at 44 days after fill placement, ranged from 0.3 to 0.4m and were within ten percent of measured values. The test fill involved overconsolidated and normally consolidated response.

The calibration efforts for the 1999 test fill highlighted the importance of considering both the overconsolidated and normally consolidated phases of the settlement response. The transition between these phases for ASAW was considered to be abrupt, possibly due to the disruption of the light cementation bonds and the relatively sensitive structure observed in the laboratory strength testing. Although a transition of the settlement response is anticipated for natural soils when moving from an overconsolidated to a normally consolidated condition, the magnitude of the "post-triggered" settlement and the relatively small increase in loading required to reach a trigger load in the ASAW was significant for cap design (as discussed subsequently).

The authors recognized that ASAW dissolution may also contribute to the long-term response on the ASAW in the pond, in addition to the mechanical compression effects discussed above. However, by incorporating the historical pond settlement response in the calibration, the calibrated model may indirectly account for a component of dissolution-related settlements. Overall, the lack of specific data regarding past and future factors affecting dissolution was considered to preclude the direct development of a dissolution component within the settlement model.

The calibrated settlement model was used to predict the settlement response of potential caps due to cap self-weight and the gradual lowering of pond water level that would be anticipated due to the effectiveness of the cap. The model indicated that upon the placement of fill as a capping layer, relatively rapid and large primary settlements would be possible once the trigger load was exceeded, followed by relatively large secondary settlements. For example, placement of 2.0m of fill was estimated to cause 1.0m to 1.3m of long-term total settlement at a typical pond location.

IMPLICATIONS FOR CAPPING DESIGN

Results from the calibrated settlement model clearly indicated significant advantages in placing the smallest feasible amount of capping fill in order to maintain a factor of safety regarding the trigger load for settlement. This would minimize the potential for initiating large total settlement and associated differential settlement across the cap. The cap design considered a number of grading concepts for the cap (Table 1).

Ultimately, the 'landform' grading concept was adopted for the cap. This concept has minimal and relatively uniform fill placement, as well as negligible cut. The limited thickness of the cap offered many significant advantages, primarily in minimizing the potential for triggering large total and differential settlements which could lead to surface grade reversals. It was also considered that minimal fill loading would be least likely to perturb the existing pseudo equilibrium condition of the current pond surface, and thus reduce the need to anticipate unknowns, such as possible irregularities in pond depth and location of preferential internal drainage paths, that might be significant if large settlements were to occur. Finally, because the 'landform' cap used the least quantity of fill, the construction period was shortened, minimizing the potential of activating a sensitive response of the ASAW to reworking and construction traffic.

TABLE 1 – Grading Concepts for Pond Capping

Designation	Maximum Capping Fill Thickness for Pond 5 (m)	Surface Water Drainage Configuration
sawtooth	2.4	two parallel main drainage channels with numerous minor side channels
washboard	1.5	central drainage channel with multiple side channels
grid	1.8	grid network with two main channels and numerous side channels
landform	0.6	matches subtle drainage on existing pond surface

CAP CONSTRUCTION

Utilizing the results of the studies conducted for this project, a cap was designed and constructed in 2001 using the landform grading concept. The construction contractor was Sevenson Environmental Services (Niagara Falls, NY). The construction of the cap advanced satisfactorily and the limited regrading required by the landform grading concept resulted in very few instances of activating the sensitive response of the ASAW. Although the cap was constructed utilizing relatively flat grades over a relatively compressible foundation, the contractor was able to establish the design grades. To date, performance of the capping systems has been satisfactory and monitoring is ongoing.

SUMMARY

This paper describes the results of geotechnical investigation of ASAW (ammonia soda ash waste) settling ponds at the Saltville, Virginia project site. The settling ponds were to be capped to reduce water infiltration and the goal of the investigation was to develop material parameters for capping design. The ASAW material is a man-made particulate that exhibits unusual geotechnical behavior.

Conventional geotechnical sampling and laboratory testing provided primary information for ASAW characterization, yet there were some inconsistencies in behaviour compared to natural soils. For example, the material occurs at high moisture contents above the liquid limit, yet its shear strength is relatively high. Additionally, the material is believed to be generally normally consolidated, yet its consolidation response is more consistent with an overconsolidated material. These inconsistencies were resolved when the conventional testing was supplemented with non-standard techniques, in particular XRD and SEM, which gave insight into material composition and structure and allowed geochemical factors to be considered along with mechanical response.

A full-scale test fill on the pond surface proved useful in establishing viable construction practices and measuring insitu response to capping loads. The test fill response confirmed that the ASAW was responding to loading in a manner that could be represented with conventional geotechnical analysis and with material parameters consistent with those measured in the laboratory testing. A capping settlement analysis model was developed which was calibrated using laboratory test results and test fill response, as well as historical pond surface settlement data.

Application of the analysis models lead to an innovative capping design that minimized the use of fill by preserving the existing pond surface drainage pattern. This approach did not achieve slopes as steep as in standard cap designs, but reduced the potential for triggering large total and differential settlements across the constructed cap. The authors consider that the thoroughness of the ASAW characterization made it possible to demonstrate that the cap design was best suited to site conditions. The cap was constructed in 2001 and has performed as anticipated.

ACKNOWLEDGEMENTS

The authors acknowledge the cooperation of the Olin Corporation in allowing the publication of this paper. The authors would like to express appreciation to Mr. John Burns, formerly of Olin Corporation, Mr. Steve Youngs, currently with Mactec Engineering and Consulting, and Mr. Neil Davies of Geosyntec Consultants, for their support during the execution of the work. The first author thanks his father for introducing him to the geotechnical world.

REFERENCES

Holtz, R.D. and Kovacs, W.D. (1981), *An Introduction to Geotechnical Engineering*, Prentice-Hall, pp. 733.

Houlihan, M.F. and Rogers, W.K. (1988), "Design and Construction of Geosynthetic-Lined Diversion Ditch", *Geotechnical Fabrics Report*, March/April, pp. 8-10.

Kulhawy, F.H. and Mayne, P.W. (1990), Manual on Estimating Soil properties for Foundation Design, EPRI report EL -6800.

Kulhawy, F.H., Sangrey, D.A., and Grove, C.S. Jr. (1977), "Geotechnical Behavior of Solvay Processes Wastes," *Proceedings of the Conference on Geotechnical Practice for Disposal of Solid Waste Materials*, June 1977, pp. 118-135.

Lee, H.J. (1982), "Bulk Density and Shear Strength of Several Deep Sea Calcareous Sediments," ASTM STP 777, *Geotechnical Properties, Behavior, and Performance of Calcareous Soils*, Demars, K.R. & Chancy, R.C. eds., pp. 54-78.

Schmertmann, J.H. (1991), "The Mechanical Aging of Soils," ASCE *Journal of Geotechnical Engineering*, v. 117, no. 9. pp. 1285-1330.

Slope Failure at a Major Cut for a Motorway in Austria – its Cause and its Stabilisation

Martin Fuchsberger[1]

[1]em.o.Univ.-Professor, Institute for Soil Mechanics and Foundation Engineering, Graz University of Technology, Austria, martin@fuchsberger.name

ABSTRACT: In this case study of a cut for a motorway in the Alpine region of Austria it could be demonstrated that an inadequate pre-construction site investigation can lead to unexpected problems connected with a slope instability. In this instance a dormant fossil slide, discovered by a later geotechnical survey, was in part re-activated by the slope cutting process.

A geological model of the slide was made which gave sufficient insight into the event to establish a multi-stage remedial-work scheme. The motorway axis was shifted, a toe load placed and bored piles were constructed within the slope, and in addition, an efficient drainage system was provided. Inclinometers were installed to monitor the deformations of the slope and piles. In case the structural deformation of the piles would have exceeded a set limit, further piles were foreseen as a second stage. The remedial measures taken proved to be successful as demonstrated by plotting the frequent deformation measurements on a time scale.

KEY WORDS: case history, drainage, failure, piles, remedial, slope stability, stabilization

INTRODUCTION

General

The A-9 motorway "Pyhrn" connects the motorway network of Germany and Austria along the river Danube with the South-East of Europe. The route crosses the eastern Alps via several tunnels and runs for lengthy stretches along steep narrow valleys and over minor passes which represent local watersheds. This requires in many instances deep cuts into the flanks of a valley thereby disturbing the existing equilibrium of a stable slope. Such a situation arose where the motorway route crosses the 841 m high Schober-Pass with surrounding mountain tops up to 1500 m above sea level. This section being particular narrow, there was always a bottleneck for traffic due to an existing railway line, a principal road and a local village with its own road system. The new motorway section comprises two tunnels, totalling 2500 m in length, a cut and cover section on the western end, two major cuts and minor bridges and embankments. One of the major cuts described here is about 500 m long with a max. vertical height of 30 m and was designed as an unsupported slope with a 2V to 3H inclination (Fig.1).

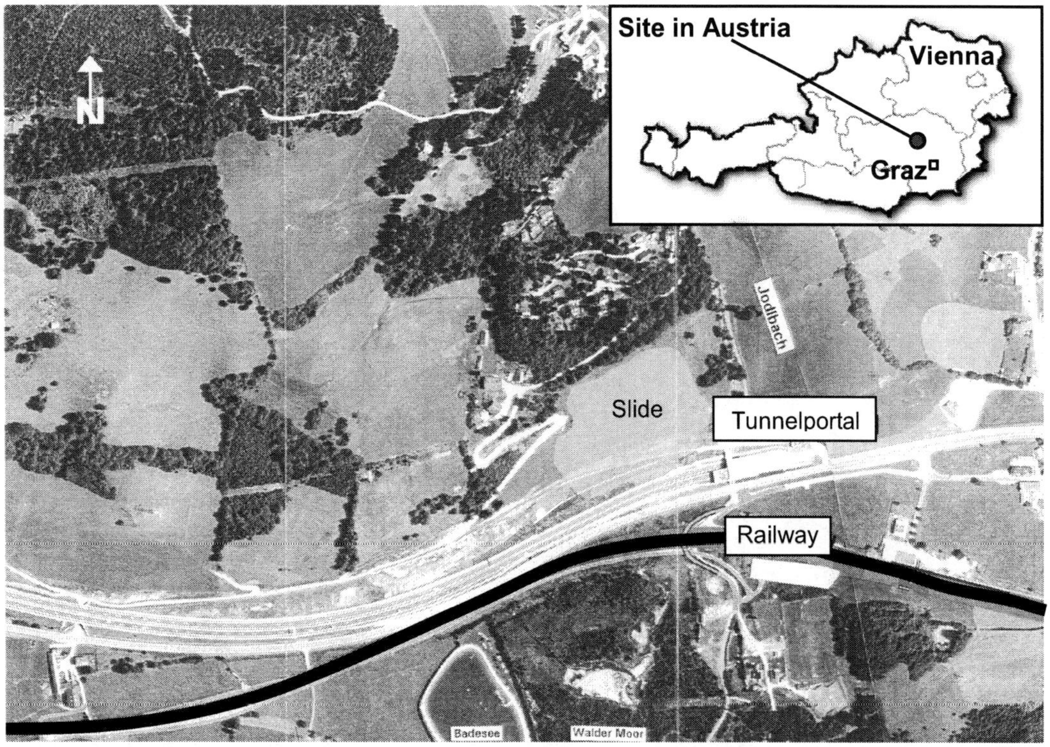

FIG.1. Location and aerial view of slide area.

Due to the pressure of time between selecting the final route for the motorway, the pre-construction investigation and the start of the construction itself, combined with limited funds available at this stage, the initial site investigation turned out to be insufficient under the circumstances. When geotechnical problems arose one had to revert largely to on-site decisions for a final design solution.

Geotechnical aspects

According to the post-failure investigation by the geologists *Brosch & Riedmüller (1994)* "…the project is situated almost in the middle of the tectonic contact between two major systems of the Eastern Alps "… shaley graphite phyllites of low strength with few intercalated thin layers and lenses of dark banded fine-grained marble… "encountered along the cut. There "…the rock mass is faulted – folded, sheared, squeezed, partially altered to graphitic black, soft clayey gouge…". This material was found in the 300 m long part of the slope which eventually failed. It was not clearly revealed as such in the two boreholes sunk to a depth of 4 to 5 m below sub-grade level at the original site investigation. Shear-box tests on severely disturbed samples taken from these borings indicated an angle of internal friction of φ=32° to 39°, with negligible cohesion. These values provided enough confidence for the designers to choose initially a 2V to 3H slope for the cut. (Fig.2)

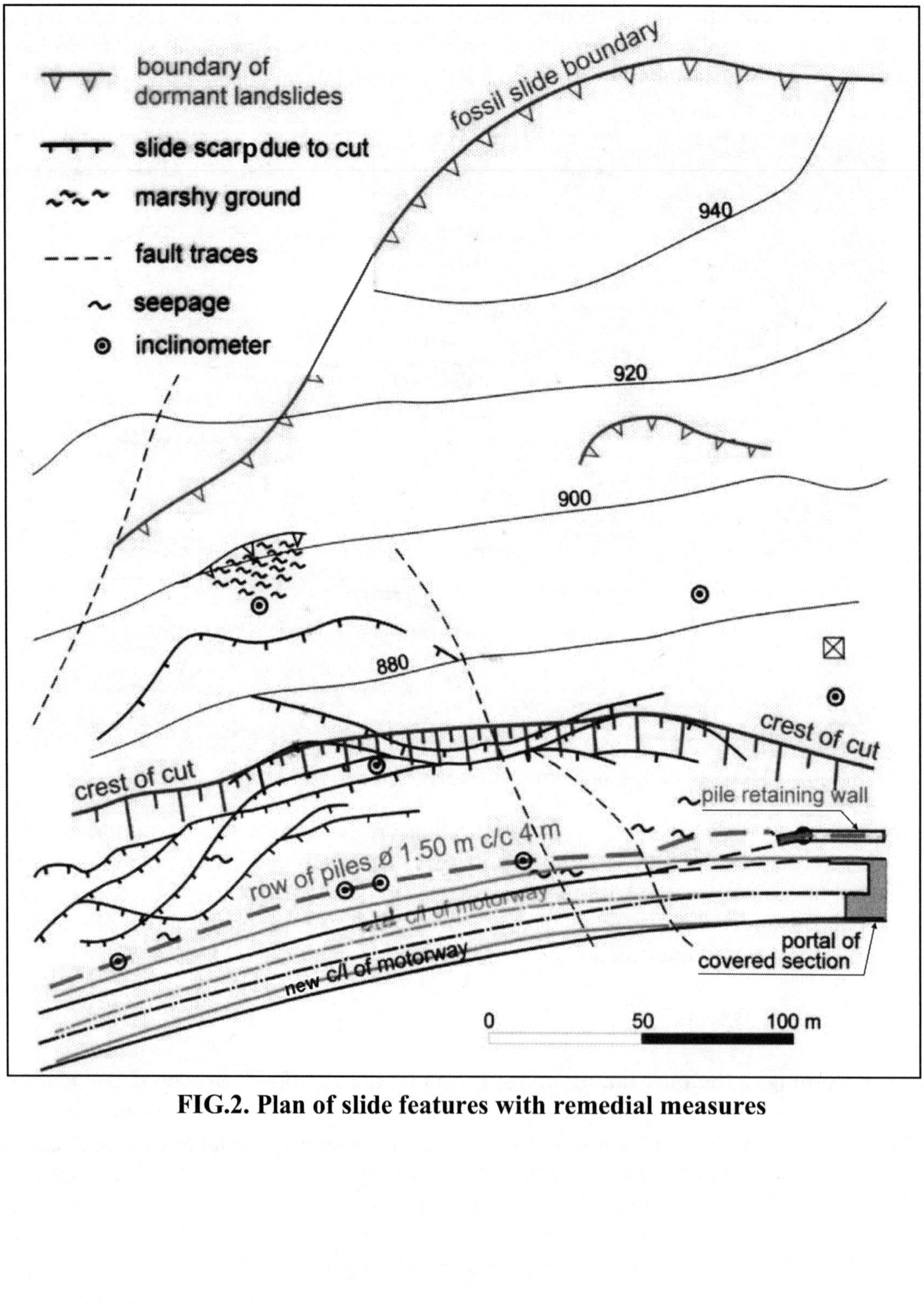

FIG.2. Plan of slide features with remedial measures

THE CONSTRUCION PERIOD

Bulk excavation and early signs of failure

The cut into the existing slope for the road construction commenced from the western end of the cut-and-cover section progressing westward in several layers and rising gradually to ground level some 350m away from the portal of the covered section. During this work a number of superficial shallow fissures developed in the slope containing the soft graphite phyllite indicating instability of this material. At the same time significant water seepage occurred, apparently emanating from the debris within the slope. Initial remedial measures consisted in collecting the seepage discharge by trench drains and replacing the soft fissured phyllite with rockfill in 2 to 3m wide trenches. A geodetic survey of the cut was instigated to monitor any movement at the surface.

FIG. 3. Scarps on the slope.

After the excavation for the motorway had progressed about 2.5m above formation level, further large fissures and small scarps in the upper part of the cut developed over a weekend break. The monitoring of the surface movement showed distortions and bulging of the whole slope including the rockfill, and downhill deformations of several centimetres per day were observed. Clearly, a mass movement had been triggered (Fig.3). Only then geotechnical advice was sought from the above mentioned geologist and from the author as geotechnical consultant.

Further bulk excavation was stopped immediately and some of the most recent excavated material was replaced. The monitoring of the slope deformation was intensified by installing five inclinometers into further exploration boreholes to a depth of 15 to 25m below formation level. A fan of drainage pipes was installed from two levels of the cut, within slightly rising 50m deep boreholes drilled into the slope, and the seepage water was collected in a longitudinally falling drainage line connecting up to the drainage system of the tunnel at the lower level.A simple back calculation using a circular failure surface, led to the construction of a rockfill wedge at the toe of the cut and to the replacement of the whole soft material within the slope by a rockfill, stepped at its base. This work had to be carried out in short sections (about 15 m long) over the whole length of the cut.

However, after extensive rainfall the drainage system became only partly effective because of clogging, and more fissures and scarps appeared further uphill, indicating that new failure surfaces had developed. Moreover, the downhill deformation of the slope increased again at a significant rate. (Fig.3) At this point in time it was decided to place a substantial counterweight of fill material at the toe of the slope, a measure which eventually brought about a decrease in the rate of movement and finally the deformation of the slope came to a halt. (Fig.5)

Geological model

At this stage a detailed geological survey was instigated in order to provide a geological model of the slide. This model has been described in detail by *Brosch & Riedmüller (1994)* and is in principle explained below.

From the overall geology of the site, the available site investigations and the sequence of events leading to the slide, and further from more site surveys, a model was developed which explained the progressive failure that had occurred. In essence the model infers the presence of a deep-seated fossil principal sliding-surface which is believed to continue below the motorway subgrade level and is believed to correspond with the rock surface in the valley floor. It also infers that a mass movement had been triggered by the depletion of resisting masses at the toe of the natural slope and thereby had activated parts of the dormant fossil landslide. (Fig.4)

When initially some counterweight was placed at the toe, new sliding planes became active higher up involving about the same slide mechanism with subunits of the rock mass. Only after placing a massive counterweight in front of the toe of the cut an apparent equilibrium condition was achieved.

Stabilisation measures

From the geological model and from the fact that the massive counterweight at the toe had, at least for the time being, stopped the slope deformation, it became clear that any remedial measures would require meeting the following conditions:

- minimum disturbance of the counterweight placed in front of the toe
- intercepting the deep seated sliding surface by structural means
- drainage of the water efflux from the slope

It was accepted that a pure soil/rock anchor design would not be workable on account of the high deformability and the low strength of the phyllitic rock.

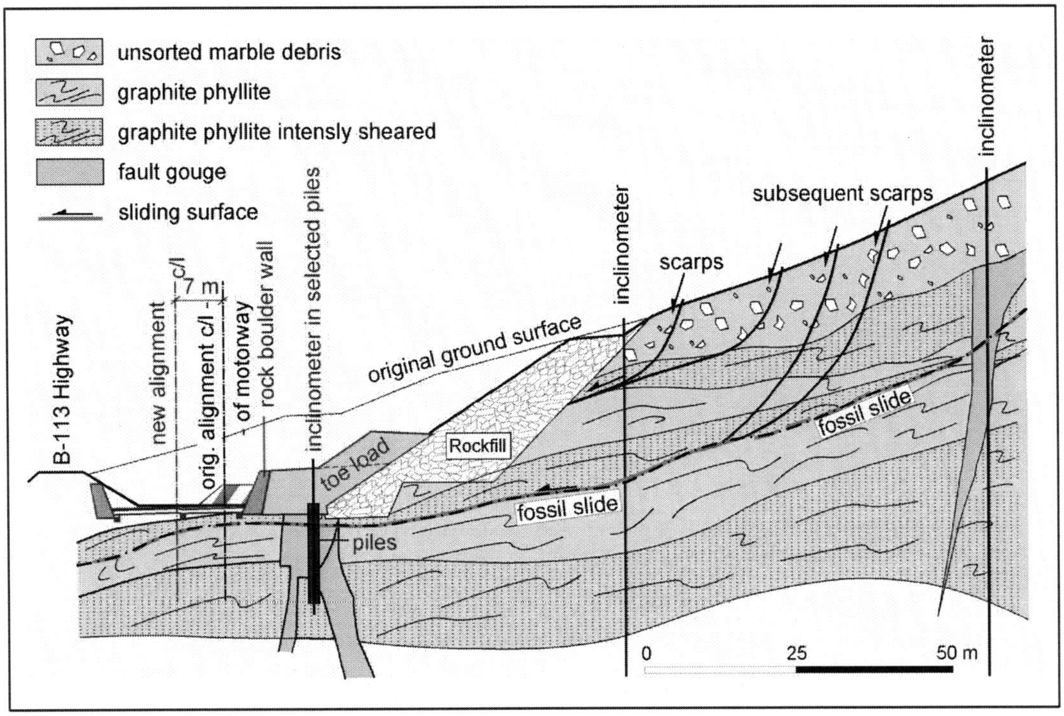

FIG. 4. Typical cross-section with geological features.

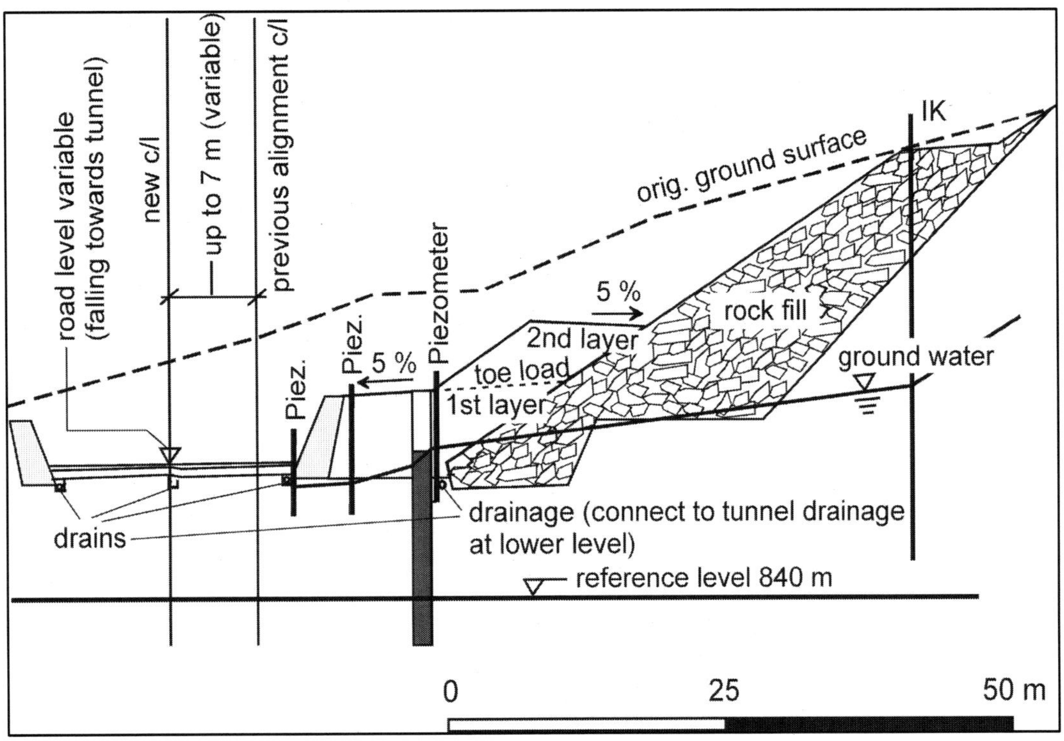

FIG. 5. Typical section with measures instigated.

A number of alternative solutions were proposed, such as extending the tunnel by a cut-and-cover section, or, constructing large diameter deep piers, or, slope-parallel diaphragm wall sections, or, dowelling the slope with a grid of small diameter piles. Any of the proposals also had to meet the requirements of the road alignment and of the overall very restricted construction programme, as well as clearly defined financial limitations.

Eventually the following measures were proposed and accepted (Fig.4) & (Fig.5):

(i) Re-alignment of the motorway over the length of the slide by shifting its centre line up to 7m away from the toe of the slope. In this way it was not necessary to remove the toe load. Only a small part of it had to be substituted by a boulder wall.

(ii) Construction of a row of reinforced bored piles, 1.50m diameter, at 4m centres, entirely within the first layer of the toe load, to a depth of 13m below formation level, thereby clearly intercepting the slip surface with sufficient embedment underneath it. Only near the portal of the covered-section a contiguous bored pile wall with a capping beam was required on account of the given road geometry, this cantilevered some 4 m above ground.

(iii)Installation of inclinometers pipes within selected piles to permit the monitoring of the deformation of the piles at any time.

(iv)Construction of an efficient drainage system at the toe of the slope together with piezometers to monitor its effectiveness.

This was considered as the first stage of the slope stabilisation. A second stage would be required if the structural design limit for the deformation of the piles was reached. In this case a second row of piles would be installed further uphill through the upper toe-load to increase further the factor of safety of the slope.

Design considerations

The 1.50m diameter reinforced concrete piles within the slope are acting like dowels across the sliding surface thereby hindering the free movement of the sliding mass. A single pile is considered to behave structurally like a reinforced concrete member fully embedded in the ground above and below the sliding boundary. A linear-elastic bedding was assumed and the sliding mass was assumed to move in a block-like manner. (Fig.6)

A limit state concept was used in the structural design of the pile. The maximum bending moment and the maximum shear-force at failure for the at optimum symmetrically reinforced cross-section was established and from this the maximum deformation of the sliding mass at pile failure was derived using design charts by *Schwarz (1987)*, as there was insufficient time available for the required detailed laboratory testing to find the time-deformation behaviour of the actual soil.

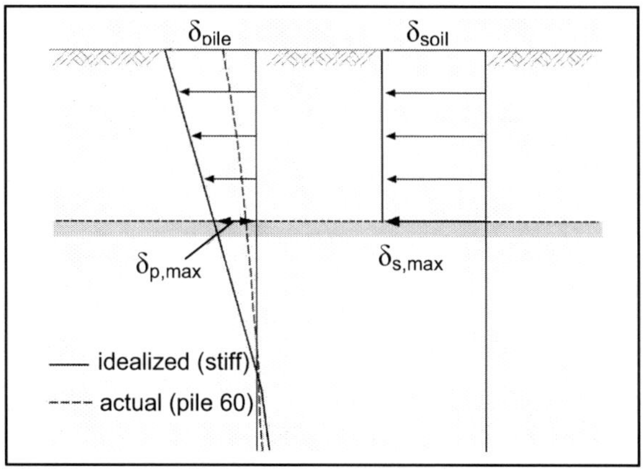

FIG.6. Simplified design principle for pile and soil deformation

Taking into account uncertainties in the choice of the coefficient of subgrade reaction (derived from in-situ pressiometer tests) and to allow for a certain factor of safety, the maximum allowable deformation of the pile shaft of δ_{pmax}=60mm was obtained. This corresponds to a maximum deformation of the pile shaft of δ_{smax}=25mm at slip surface level. A revised slope stability analysis was made, guided by the geological model and by including the toe-load into the cross-section. The soil parameters used in this calculation (φ=27°) were gained from lab tests on high quality soil samples extracted from new set of site investigation borings(using double-barrel core samplers under the guidance of an engineering-geologist on site).
The new analysis established in the overall shear-force required at the slip surface to arrest the block deformation of the sliding mass incorporating a factor of safety of 1.15. By equating this shear-force with the available resisting shear force from the piles at slip surface level a centre/centre distance of 4m for the fully embedded piles was thus derived.

POST-CONSTRUCTION PERFORMANCE

The installation of the piles and the construction of the re-aligned motorway together with the boulder retaining wall on either side of the road and the drainage system, was completed in time before the official opening of the road. Since then the 5 inclinometers installed in the slope and the 4 inclinometers within selected piles are being monitored at frequent intervals, especially after periods of high atmospheric precipitations. The initial time dependent deformation has been plotted in Fig.7 and has not changed significantly up to date.

This diagram clearly shows that even after the installation of the piles the slide continued to move initially, though at a receding rate, until finally the piles established their effectiveness and no further movement took place. The maximum deformation of the pile at the slip surface at this stage was about 7mm, well below the set limit of 25mm.

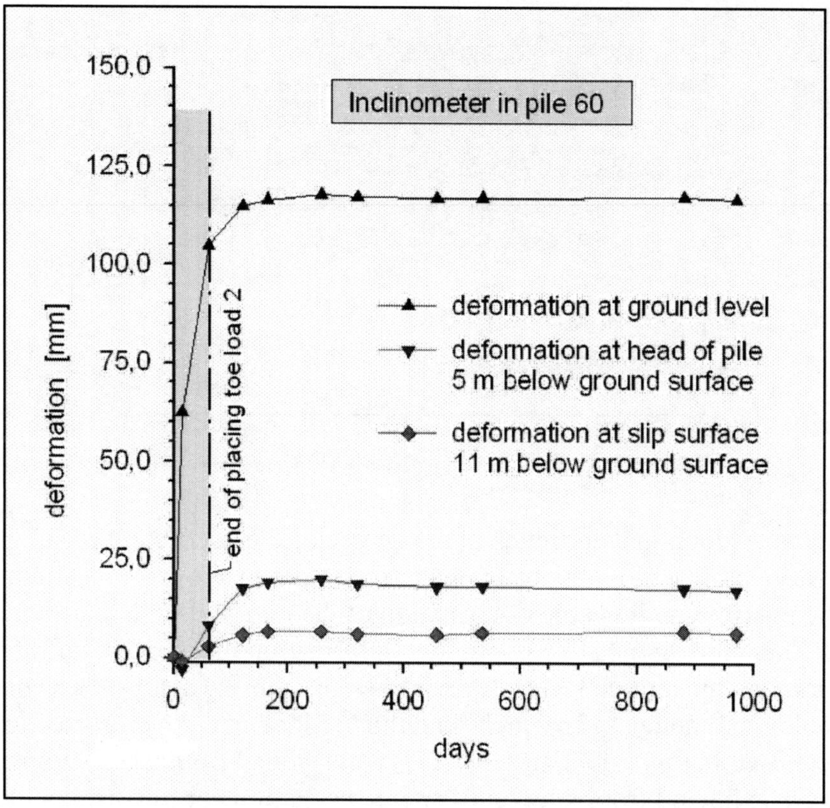

FIG.7. Bending deformation of pile No. 60 with time.

As foreseen, the repeated monitoring of the pile deformation is an essential part of the remedial work scheme, an acceptable price to be paid for the very economical measures taken to achieve a stable slope.

SUMMARY AND CONCLUSIONS

During the construction of the motorway A9, in Styria/Austria a large cut had to be made with a 2V to 3H slope. The stability of this cut in metastable material of shaley graphite phyllites was overestimated on account of insufficient site investigation. The excavations for the cut triggered a mass movement of the slope whereby clearly visible fissures and slide scarps developed progressively uphill to a distance of up to 50 m. Deformation measurements at the surface of the slope indicated movements of several centimetres per day. Several inclinometers were installed in the slope and a detailed geological survey was instigated which proved the presence of a deep-seated fossil slip surface extending well below the planned formation level of the road.

In order to counteract the slope movement further bulk excavation was stopped and a counterweight of rock- and soil was placed at the toe of the slope. This brought the deformations eventually to a halt.

Based on a detailed geological model of the slide resulting from an additional site investigation and a provisional failure analysis, various remedial measures were considered for a permanent stabilisation of the cut. Structural aspects and economic considerations led to a remedial work proposal which was to be executed in various stages:

As a first stage the alignment and the cross-section of the motorway was altered in order to keep the stabilizing counterweight of the toe-fill largely in place. Then, a row of large diameter bored piles was installed within the lower part of the slope, penetrating deep enough to act as dowels across the slip surface. The bored piles were designed to withstand only a limited deformation and by means of inclinometers installed in the piles their lateral deformation was monitored at frequent intervals. A second row piles was contemplated further uphill if the deformations reached the calculated limit.

The deformation measurements so far have demonstrated the effectiveness of the remedial works and, for practical purposes, the slope can be considered as being stable.

REFERENCES

Brosch, F.J & G. Riedmüller (1994). "Dormant landslide activated by excavation – A case study from the Austrian Alps." *7th Intl. IAEG Congress,* Lissabon 1994.: Balkeema, Rotterdamm; 1653-1659

Schwarz, W. (1987). "Verdübelung Toniger Böden." *Veröffentlichung des Inst. f. Bodenmechanik und Felsmechanik der Universität Fridericiana,* Karlsruhe; Heft 105

Benefits of In-House Collaboration of Geotechnical and Structural Engineering Services

David L. Freed, M.ASCE, P.E.

Geotechnical Associate, Gibble Norden Champion Brown Consulting Engineers, Inc., 130 Elm Street, Old Saybrook, Connecticut 06475: freed@gncbengineers.com

ABSTRACT: Traditionally, separate geotechnical and structural engineering firms contribute to design and construction of building projects. This paper discusses the unique opportunity that Gibble Norden Champion Brown Consulting Engineers, Inc., (GNCB) a structural engineering firm in Connecticut, took when it combined expertise from both services into one firm. This innovative, in-house collaborative practice adds value to its menu of client services and demonstrates the benefits of incorporating professional and experienced geotechnical engineering into a project, from concept to completion. The geotechnical engineering staff enhanced its presence on the design team with direct participation in early design and construction decisions, where knowledge of subsurface soil conditions are used to benefit the important early stages of project planning. The benefits, illustrated by three case studies, have been far reaching and include enhanced contract drawings, expansion into new services, reduced company risk, greater structural engineer's awareness of geotechnical services, and increased recognition of geotechnical engineering as a significant factor in project success.

KEY WORDS: examples, experience, geotechnical, collaboration, practice, risk reduction, structural

INTRODUCTION

This paper describes the writer's experiences at a traditional structural engineering firm, Gibble Norden Champion Brown Consulting Engineers, Inc. (herein identified as "my firm") of Old Saybrook, Connecticut, that innovates its services by including geotechnical engineering. To the best of my knowledge, this logical collaboration of services has not occurred to any degree within the United States. Some structural engineering firms profess to provide geotechnical engineering services, usually at a rudimental level and by personnel with limited experience and no formal training. The model presented herein represents a true collaboration of the two services, led by registered professional engineers having appropriate education, training, and experience in each discipline. Roots for this collaborative model date back to the beginnings of the geotechnical engineering discipline, as developed by a renowned structural engineer, Karl Terzahgi.

My firm's experiences will hopefully encourage growth and advancement of collaborative geotechnical/structural engineering services into the building design and construction process.

WHERE ARE WE NOW?

Traditionally, two separate firms complete geotechnical and structural engineering services for building projects. Architects and their team of consultants, including structural, mechanical, electrical, cost estimators, and perhaps a specialty service pertinent to the building type, normally lead the owner's design team. The architect typically assembles the design team; a team that generally has some previous working relationship, and confidence in each other's skills. The services provided by the architect's team all relate to the *new* building, to be constructed on a particular site.

Owners have traditionally, and occasionally independently, been responsible for identifying site conditions and foundation requirements for *their* building. The rationale for this approach centers on the owner's responsibility to provide information on his/her property. These "geotechnical engineering" services, which may include drilling and materials testing, are best provided by a firm specializing—by education, training and experience—in this work. This philosophy, which has prevailed for years, reinforces the separation of building design vs. site design services. In fact, the standard architect's AIA contract identifies geotechnical engineering as an additional service, one that lies outside the limits of architectural building design services, and one to be provided by the owner. The original concept for this separation intended to avoid the higher liability of geotechnical services; however, across the country today, insurance premiums for geotechnical services are at lower rates than for structural services.

However, are services of a geotechnical engineer really outside the limits of new building design? Certainly not! While it may seem like a logical separation, the input that a geotechnical engineer provides to the construction process is quite integrated with the building system. There are advantages, which will be discussed herein, to having the geotechnical engineer as part of the architect's design team. Structural engineering is the one discipline that relies on information developed in a geotechnical study of a building site; the foundations and structural framing of the building are integral. Logically, collaboration of geotechnical and structural engineering services can add to the success of a project.

In 1995, this traditional structural engineering firm took advantage of the opportunity to add geotechnical engineering services to its other work disciplines. This paper identifies five significant benefits of in-house geotechnical/structural marriage of services.

This eighteen-person engineering firm is located in Old Saybrook, Connecticut. Today, the professional and technical staff totals 15, the majority of the professional staff are registered engineers covering 15 states. One of the registered professional engineers has a post-graduate degree in geotechnical engineering. The successful collaboration of services makes it difficult to identify the geotechnical staff, since many of the traditional structural engineers are also involved with, and make decisions, regarding the geotechnical side of the practice.

With this groundwork, the remainder of the paper discusses in greater detail the experiences and benefits of in-house integration of geotechnical and structural engineering services.

GEOTECHNICAL INVOLVEMENT THROUGHOUT THE PROJECT

When GNCB serves as both geotechnical and structural engineer on a building project, the geotechnical involvement, from conceptual design to completion of construction, benefits the design team. Too often, owners that contract directly for geotechnical engineering services require only limited participation by the geotechnical firm. At completion of the "report," the geotechnical engineer often takes on reduced project involvement; further participation frequently occurs on an "as needed" (i.e. I'll call you) basis. This somewhat limited participation leaves the design team without geotechnical engineering support during the critical period of final design, where changes in the project may occur, and construction, where unanticipated conditions may be encountered.

Michael Fortuna AIA of TLB Architecture, Chester, Connecticut, comments on the geotechnical involvement throughout the project *"Having geotechnical expertise on the Design Team during the Construction Document phase of the project allows the Design Team to fully coordinate the myriad subsurface requirements, including foundation design, excavation requirements, waterproofing and/or dampproofing, backfill and compaction requirements and groundwater management. Effectively working with the Architect, Structural Engineer and Civil Engineer, the Geotechnical Engineer can provide real time feedback, leading to well coordinated and detailed documents...... Often during construction, soil conditions over the breadth of an excavation vary from the conditions taken at individual boring points. Having the Geotechnical Engineer and Structural Engineer working hand-in-hand allows rapid decision making in the context of the overall project goals."*

On some projects, if it were not for a structural engineer requesting geotechnical design information, a project might even proceed without a professional geotechnical engineer. Alternatively, structural engineers who do not necessarily have post-graduate training may attempt to fill the role of a qualified geotechnical engineer, depending on their comfort level. We all know of projects where the structural engineer has arranged for test borings and, as long as the driller reports dense soils, the structural engineer proceeds with a conservative foundation design, often times at increased cost to the project. It is not uncommon, however, for the "dense" soils to be misidentified, and while they may be dense, they may not be a suitable bearing material, but rather fill. The design proceeds without taking steps necessary to deal with the fill; only after the contractor starts the work and uncovers a problem is it dealt with properly. The owner is ultimately the one who loses out, in lost time and cost.

My firm's early involvement on a project has assisted the architect in locating a building on a site, to take advantage of optimal soil conditions. Techniques of "reading the landscape," use of surficial geology maps, and preliminary test borings (Freed 2003), are relatively low cost services that can help reduce project costs early in the planning stages, when such changes have little or no impact on the design.

As the design progresses, the geotechnical engineer within a structural firm has opportunities to view and comment on the structural foundation drawings, make value engineering design changes, and confirm that the design intent is followed. Too often, limited services by geotechnical engineering firms do not follow up their reports with a check of the structural engineer's foundation drawings to confirm that the report recommendations have been properly interpreted. My firm's proposals and contracts for geotechnical engineering services typically include a full scope of work, from design through construction.

ENHANCED CONTRACT DRAWINGS

Contract drawings, along with specifications (i.e. contract documents), normally convey design requirements to contractors. Enhancement of these drawings, by including geotechnical engineer requirements, provides a benefit to all participants on a project: owner, design team, and contractor. Projects, regardless of size or complexity, have a list of important foundation design criteria and requirements for implementation. Documents usually rely only on geotechnical engineering reports to convey this information and requirements. Contractors can be at a loss with respect to interpreting geotechnical engineering reports, and ultimately, understanding the impact of these conditions on the construction process. Enhancement of the contract drawings, whether it is by detail on the structural drawings or preparation of a separate geotechnical engineering contract sheet, conveys necessary foundation design requirements to the contractor. Too often contractors state they either did not have, or understand, the geotechnical engineering report. However, information, if shown on the contract drawings, is clear and less likely to be misunderstood.

On my firm's projects, when serving as both the geotechnical and structural engineer-of-record, the decision has been made to prepare separate contract drawing sheet(s) that clarify the project geotechnical requirements. A few examples of what is included on the geotechnical contract document drawing are:

- locations of test borings, with pertinent information summarized, such as ground surface elevation, elevation top of suitable bearing soil, groundwater levels, or top of bedrock.
- approximate limits and depths of overexcavation of unsuitable soils
- limits of compacted structural fill placed below footings and/or slabs
- locations and installation details of foundation drains
- limits of concrete mudmats or crushed stone placed below footings to preserve bearing surfaces.
- underpinning details
- limits and details of earth surcharges
- limits and details of lateral support systems

As with any information inserted on contract drawings, the appropriate loss prevention clauses, typically included in the geotechnical engineering report, also appear. Contractor payment items, estimate of quantities, and pricing can be easily referenced on the geotechnical drawing sheet.

If my firm only serves as structural engineer on the project, much of the above items can be alternately shown on the structural foundation drawings. GNCB has made an internal decision to show geotechnical information in this manner, when not the geotechnical engineer-of-record; however, the information is coordinated with the design professional.

EXPANSION INTO NEW SERVICES

My firm's integration of geotechnical and structural engineering services enabled an expansion of in-house services. These additional services relate to areas not typically completed by a traditional or specialty service firm. Furthermore, the services have several components that require both geotechnical and structural engineering disciplines. Such areas of work include:

- demolition specifications and contract drawings
- design of building foundation perimeter and underslab drainage systems.
- waterfront structures
- retaining wall design, particularly segmented walls

Design of foundation drainage systems requires integration with structural foundation systems. While many coordination meetings and reviews may eventually end up with a drainage system with proper grades set for footings and pipe inverts, one firm handling the entire design simplifies the process.

Retaining wall and waterfront structures, particularly permanent sheeting and lateral support designs, similarly require design input of both geotechnical and structural engineers. At my firm, the geotechnical staff has brought a number of newer technologies, including high strength "HZ" sheeting, slurry walls, and segmented retaining walls, to the realm of design solutions. We routinely use computer solutions for design of lateral support systems.

The geotechnical services group also works on projects on which my firm is not the structural engineer-of-record. For these projects, new clients and additional work have been brought to the company. On a number of occasions, a geotechnical engineering project has expanded to include structural peer review and special inspection services into the firm's work load.

A direct bottom line benefit to GNCB of the expanded new services, has been an increase in company revenues. Currently, approximately 15 percent of our services relate directly to geotechnical engineering studies and the new services mentioned above.

REDUCED COMPANY RISK

The geotechnical presence on my firm's structural projects reduces the overall risk of conducting business. Every member of the design-construction team has also benefited. A few examples include:

- Renovation projects, where existing foundations may receive increased load. Geotechnical input with respect to most probable subsurface conditions, allowable bearing capacity and anticipated settlements are integral to the structural analysis.
- Construction monitoring and third party review projects have assigned geotechnical staff to handle geotechnical-related items, which were otherwise handled by structural engineers.
- Residential work, which does not routinely have a geotechnical engineer, can be structurally designed with less risk of a problem.

Risk management and loss prevention education at my firm has been expanded to include tools and programs offered by the national trade association ASFE: The Best People on Earth, of which we are a member. Proper contracts, limitation of liability, handling a claim, etc. have been the subject of in-house training courses and seminars.

GREATER (STRUCTURAL) STAFF AWARENESS

Since integration of geotechnical and structural engineering services, the structural engineers have all gained a valued and beneficial understanding of geotechnical issues. Now, GNCB structural engineers routinely request preliminary information on subsurface conditions, even during the proposal stage; this enables an early incorporation of geotechnical information into the design process. During the design, there is an on-going dialogue between the in-house geotechnical and structural disciplines, to ensure that all avenues of the design are covered. A recent museum project included issues of lateral support and underpinning, over-excavation of unsuitable soils, and hydrostatic design of a basement foundation mat. These intertwined issues, and the in-house communications between disciplines enabled a cost effective design.

With geotechnical issues on their minds, my firm's structural engineers have also been more proactive with clients and owners, to ensure that a geotechnical engineering study is completed for all projects.

EXAMPLES OF BENEFITS TO COLLABORATION OF SERVICES

Case I – Design of Foundations for a Residential House

- Scope of Service: Structural foundation design for two residences

- Available Subsurface Information: Eight shallow test pits excavated by the site civil engineer for septic design. The pits show a surface 0.9 m thick layer of man-placed fill underlain by at least 0.6 m of sand.
- Structural Design Before Collaboration: Both houses designed for a conventional shallow spread footing foundation and slab on-grade first floor
- Structural Design With Collaboration: Since the site is located next to wetland areas, in-house geotechnical input questioned the possibility of buried organic soil below the naturally-deposited sand. Further test borings and test pits confirmed the presence of organic soils below the sand. At one residence, the organic soils were removed and replaced with compacted structural fill, to a depth of 3.7 m below ground floor slab grade. At the other residence, a sufficient thickness of sand underlain by a relatively thin layer of organic soil, permitted the residence to be constructed, while leaving the organic soil in-place, and with only minimal additional cost for upgrading the near surface footing subgrade conditions.
- Benefits: GNCB firm avoided a design that may have resulted in an unacceptable foundation performance, and possible litigation.

Case II – Additions to a 100-Year Old Structure

- Scope of Service: Full structural engineering services for two additions to a one-story building with partial below grade basement.
- Available Subsurface Information: None, except the existing heavy concrete structure has performed satisfactorily
- Structural Design Before Collaboration: Design team proceeds without a geotechnical engineering study, in view of satisfactory performance of existing building. Foundation design and drawings prepared for a shallow spread footing foundation system.
- Structural Design With Collaboration: My firm recommended a geotechnical engineering study for which test borings were completed with continuous sampling within the upper 3.7 m of soil, and installation of a groundwater observation well. Test borings revealed a surface man-placed fill, topsoil, and subsoil layers considered unsuitable for support of the building frame but suitable to support the basement slab, underlain by a deep deposit of sand that grades finer with depth; groundwater is near the surface of the natural sand. While the geotechnical engineering study did not change the overall building foundation system, the presence of unsuitable materials to 0.9 m below the proposed building foundation level, and presence of groundwater, were revealed, and appropriate details for handling the work were shown on the contract drawings.
- Benefits: The near surface unsuitable soil conditions revealed by the exploration program were appropriately addressed on the construction documents, resulting in no extra foundation costs. The owner had a

realistic contractor fee and there was no lost time due to unforeseen conditions.

Case III – Helical Pile Foundations Save the Day

- Scope of Service: Structural foundation design for a one-story structure with a basement
- Available Subsurface Information: Following basement excavation to footing grade, man-placed fill remained from the former gas station site. Test pit excavations revealed the fill extended several feet below groundwater; the excavation was in close proximity to an existing one-story structure with exterior masonry walls. The natural inorganic soils below the fill consisted of sand.
- Structural Design Before Collaboration: Design proceeded without a geotechnical study; the specified spread footing and slab-on-grade basement floor system were not appropriate for the actual conditions encountered.
- Structural Design With Collaboration: In-house geotechnical staff, which was familiar with subsurface conditions in the area of the site, excavated a few test pits and recommended installation of drilled helical piles (i.e. screw anchors) to support the building frame, with an earth supported basement slab (Freed 2006). The helical piles eliminated the need to expand the size of the excavation needed for over-excavation of man-placed fill, which would otherwise require a lateral support system, as well as minimized vibrations from installation of conventional driven piles.
- Benefits: My firm presented an innovative design in a timely manner (i.e. in-house services vs. engaging a separate geotechnical firm) that adequately addressed the presence of fill, expedited the construction solution, saved cost (vs. driven piles and need for lateral support), and eliminated vibrations that would have likely damaged the nearby masonry wall construction.

INTEGRATING GEOTECHNICAL REVIEW AT GNCB

GNCB has implemented a policy for geotechnical engineering review of all projects on which the firm serves as structural engineer-of-record (Freed 2007), regardless of whether we also serve as the geotechnical engineer-of-record. The review, which identifies the pertinent engineering recommendations for the project, ensures that the structural drawings appropriately address these issues. If no geotechnical engineering evaluation has been completed, for whatever reason, our structural drawings will so state. The geotechnical review policy at the firm follows:

- Geotechnical engineering staff reviews applicable reports or documentation; if appropriate, a call is made to the geotechnical engineer-of-record if there is no written documentation. Staff prepares a brief summary of the recommendations, to be reviewed by the in-house structural (project manager) engineer.

- A geotechnical contract drawing is developed, or structural drawings are prepared, with the pertinent geotechnical engineering information.
- If my firm is not the geotechnical engineer-of-record, the geotechnical engineer-of-record reviews the geotechnical information shown on the structural drawings for accuracy.
- As needed, any bid items are coordinated with the project specifications.
- My firm reviews all foundation related information for consistency with the design intent.

BENEFITS TO THE PROFESSION

Lastly, the geotechnical engineering profession realizes a benefit as a result of the collaborative effort. Specifically, the geotechnical engineer contributes to all aspects of the design and construction process, and is recognized as a significant participant in the entire project. Too often, the geotechnical engineer may not be involved with observations of the construction process. It is during this observational phase of work that the foundation design recommendations are finalized, and the geotechnical engineer's experience level is increased. The geotechnical engineer, as part of a collaborative effort with the structural engineer, is contracted at the beginning of the project to be involved through all of its phases.

CONCLUSIONS

The collaboration of geotechnical and structural engineering services in one firm provides benefits for the entire design and construction team. The significant benefits include:

- Geotechnical Involvement Throughout the Project
- Enhanced Contract Drawings
- Expansion of Services
- Reduced Company Risk
- Greater (Structural) Staff Awareness of Geotechnical Issues

At my firm, the collaboration supports and elevates the knowledge base of both disciplines. The better understood soil-structure interaction reduces conservative design, which would otherwise add unnecessary cost to a project. But most of all, the enhancement of contract drawings identifies and highlights the geotechnical design issues to both the design team members and contractor.

GNCB's in-house policy for geotechnical engineering oversight of structural engineering designs, ensures that the geotechnical issues of a project are brought to everyone's awareness.

ACKNOWLEDGEMENTS

The writer gratefully acknowledges the fundamental teachings and engineering discipline provided by John Schmertmann during the writer's studies at University of Florida in the early 1970s. Dr. Schmertmann's attention to detail, thought process for dealing with geotechnical engineering problems, and professional attitude provided a sound basis which the writer has used throughout his career.

In addition, the writer acknowledges the entire staff at GNCB for their acceptance of the geotechnical engineering practice into their once traditional structural engineering work. A debt of gratitude goes to the current firm principals, Kenneth Gibble, James Norden, Laura Champion, and Charles Brown, for their encouragement and acceptance during this process. A special thanks to Ken, who had the foresight to start the collaborative process, and who read and assisted on this paper; and to Grace Waldvogel who edited and prepared the final manuscript.

My wife Arlene has also provided insight and assistance in this paper and throughout my career as a geotechnical engineer.

REFERENCES

Freed, D.L. (2003). "Reading the Landscape: Preliminary Foundation Design", *GNCB Technical Note 3*

Freed, D.L. (2006). "Helical Piers To the Rescue", *GNCB Technical Note 6*

Freed, D.L. (2007). *GNCB In-House Memo to Project Managers*

THE QUEST FOR QUALITY IN DEEP FOUNDATIONS

John. A. (Jack) Hayes[1], D.I.C., Member ASCE

[1]President, LOADTEST Inc., Gainesville, Florida, hayes@loadtest.com

ABSTRACT: The natural desire to improve the way we design and build deep foundations provides a fundamental motive for achieving excellence and high quality. Unfortunately, perceived high cost and the lack of rewards or incentives to make the effort that quality work demands, often leads us to complacency and mediocrity. This complacency can result in substantial reductions in bored pile capacity that dwarf the improvements derived from good quality engineering design. Examples from case histories illustrate how seemingly minor factors can cause capacity reductions of 80% or more. Faced with the daunting risks and uncertainties associated with deep foundations many engineers fall back on code-value engineering and from there to unnecessarily expensive foundations. A review of the impact of shaft diameter on measured side shear resistances leads to cautionary advice about how relying too much on intuition can affect the quality of our analytical work. The engineer and the contractor do have several options for improving and maintaining quality including but not limited to - better assessment of the high cost of risk and uncertainty, changing the low-bid approach to awarding foundation contracts, including a special bid item to encompass all of the QA/QC work on a project and generally making a greater effort to focus on the value that high quality brings to any deep foundation project.

KEY WORDS: quality control, assurance, deep foundations, risk, uncertainty, assessment, code values, design complacency, intuition, second low bid, value, misconceptions, side shear, bored piles, drilled shafts, cost reduction.

INTRODUCTION

The history of the human race provides clear evidence that striving to do better, to improve the quality of life, exists as a natural human instinct. When we do something well, we feel good (I have yet to see anyone who feels good after doing anything poorly). A key element of most successful endeavors whether in business, academia, the arts, or science involves improving and focusing on the quality of what they do. Certainly John Schmertmann exemplifies this very human trait in everything that he does and this volume provides a great reason to say something about quality and the pursuit of excellence. As we have progressed in the deep foundations testing business, Dr. Schmertmann has provided us at Loadtest with a great standard for quality in our work. "Will your report pass muster if reviewed by Schmertmann?" has become the mantra and incentive for our engineers when completing a report for a client. The Schmertmann review process has resulted in the continuing improvement of our analyses and reports. In his Key Speech Ove Arup (1970) emphasized that *"for our own sake we need the stimulation produced by excellence."* ...... *"We must therefore strive for quality in what we do, and never be satisfied with the second-rate."*

But before we get too far on the subject of quality we need to have some acceptable definitions. In reality, the perception of "quality" varies a lot. (For example in the world of high fashion clothing some see quality in the "grunge" look where others see only rags) Many of us see quality, subjectively, in things that appear unique or better than anything else commonly available (a hand-painted scarf for example). In engineering, we usually see quality, more objectively, as something that significantly exceeds a minimum acceptable standard. We believe that we can measure quality and therefore that we can use "Quality Control" (QC) to maintain a desired level or standard. Or we test things to satisfy ourselves and others that a desirable level of quality exists via Quality Assurance (QA) programs. In the deep foundation industry we would like the contractor to provide the QC while we engineers do the QA testing. But having a definition of and paying lip-service to QA and QC does not mean that you will get any; that depends largely on the attitude toward quality.

In this paper I refer to the deep foundation industry as that concerned with the design and construction of bored piles (also known as drilled shafts).

QUALITY AND ENGINEERING ATTITUDES

Although we might believe that good engineering characteristically evolves from the innate desire to do things better, it does not follow that engineers will make the effort that quality requires. Clearly, however, making any change for the better requires effort and when we make it, things generally do get better. Often, we then have a subsequent period of Contentment or Complacency (C^2). Too many engineers fall into this cycle of Mediocrity, Complacency and Contentment - - the MC^2 trap which has absolutely no connection to Einstein's formulation for energy.

Those that fall into this trap then begin to design by code, use standard specs, make no effort to investigate new processes for design or construction, make no effort to evaluate the benefits of higher quality, make no effort to put a value on higher quality and become satisfied with the mediocre. (If it ain't broke, don't fix it; let sleeping dogs lie etc.) Indeed, some engineers begin to resist any non-standard, non-code process or idea and ultimately get caught in the MC^2-Code trap - - - MC^3.

In his address to the Ohio River Valley Soil Seminar Professor Jorj Osterberg (1999) had this to say about the lack of value engineering in foundation design practice – "*The reasons for the present situation are many: the tendency of designers ....to "play it safe" by being overly conservative; the reluctance to deviate from "accepted practice" for fear of later being sued; the antiquated and inflexible building codes; the attitude that "it's not my money, so why would I care"; do it the way it was always done; the ignorance or indifference of the owner about potential savings.*"

As evidence of the global nature of the issue Professor Heinz Brandl (2004) in his state of the profession paper offered these observations- "*Young engineers and especially professionals of a low qualification frequently stick slavishly to codes, standards and guidelines. If a theory finds its way into a textbook, it is considered by many readers a law. Moreover, too many regulations, standards or codes, which are too confining hinder innovation in geotechnical engineering. They act like a brake, slowing down new development. ....Engineers are increasingly afraid to design outside of standards or codes because they fear legal problems in case of a failure. This also has been dramatically reducing the willingness to take responsibility. Fear for liability or litigation is stifling innovation in civil engineering, especially in geotechnics, and pushing engineers towards over-reliance on standards. But over-reliance on standards or codes hampers also engineering judgment and kills "engineering intuition". The tendency to do things as we've always done is another hindrance to innovation and development in construction.*" Codes by their nature tend toward the lowest common denominator and these minimum standards often become maximum standards.

Looking outside our own profession we generally find important quality assurance and quality control procedures well-established in most manufacturing and service industries. Most successful companies have very good internal quality control systems which have contributed measurably to their success in the same way that good safety programs have contributed to corporate profitability. Why then, does quality control in the deep foundation (bored pile) industry seem so difficult to establish? I have suggested that most companies involved in the bored pile industry do not think of themselves as manufacturers (Hayes, 2002). Because of that we often find that poor attitudes toward quality assurance (QA) and quality control (QC) lead to making it someone else's problem. ("We have a job to do – don't slow us down with testing.")

The consequence of that attitude during construction along with the concern for structural integrity leads to a situation where designers attempt to improve quality by means of detailed specifications. This process may work if the design engineer has a great deal of expertise in the construction and/or manufacturing aspects of bored piles (unfortunately not a common occurrence). A detailed and rigid specification, written by a designer who does not understand the manufacturing process for a bored pile foundation, becomes an open invitation to either high costs or poor workmanship (or perhaps both depending on the nature of the contractor/manufacturer doing the work).

QUALITY, RISK AND UNCERTAINTY

Those of us with even a little experience in the deep foundation industry know of its perils and inherent risks. Not only must we deal with all the vagaries and variability of Mother Nature but also with the impact of construction technique on foundation capacity. There are many ways to deal with this but again as Brandl (2004) points out- *"A courageous engineer is willing to take a calculated geotechnical risk, which involves detailed ground investigation, geotechnical prognoses, "active" design (i.e. adaptable), proper risk assessment, and monitoring with back analyses. The more [the] prerequisites of such a serious geotechnical risk are missing, the more it approaches a "Geopoker" (or "Geo-Gambling"). Geopoker is favored by brutal competition, time pressure, economy and lack of knowledge (thus underestimating risk and danger). Geopoker has led to the paradox that the number of geotechnical failures has continuously increased during the past years (statistically confirmed), despite the advanced geotechnical education at the universities. The opposite of "Geopoker" is costly overdesigning, which is definitively not an engineering art."*

In the author's opinion the unfortunate adoption, and continued use, of Factor of Safety (FS) as a design concept in geotechnics lies at the root of our apparent inability to handle the risk and uncertainty associated with deep foundations in a rational way, mainly because the many uncertainties in deep foundation geotechnics do not permit accuracy in "factor of safety" estimates. We should have admitted in the beginning to a *F*actor of *Unc*ertainty (FU_{nc}). This would allow a much more rational discussion and review (at the design stage) of the inherent risks or uncertainties that exist when dealing with deep foundations. As it now stands there becomes a psychological barrier when an engineer actually has the temerity to suggest spending money on testing, or QA/QC, to *lower* the factor of safety. Why should an owner get excited about lowering the "safety" of his foundation, especially when he will have to spend money to do it? If, on the other hand, we had started by considering the high FU_{nc} (and therefore higher risk) the rationale for funding a program to mitigate risk and lower the FU_{nc} becomes much more palatable to both the owner and his insurance company. See the subsequent discussion relating to Figures 6 to 8 for a variation of this approach, which tries to lower the cost of testing by using small-diameter, and therefore potentially unrepresentative shafts. In such cases the testing process may actually add to the uncertainty.

Geotechnical engineers often do not show much interest in the risk management approach to design. Usually we handle the uncertainty of resistance capacity by "cheating" on ultimate resistance estimates. (This would qualify as a Geopoker bluff). We carry out the bluff with a straight face by underestimating the ultimate capacity, using the lowest measured resistance values or by falling back on code values (usually, and necessarily, very conservative). When we don't have to "show our cards" (no prototype testing) no one knows that we have just saddled the owner with the big "E" (a very expensive foundation). If the specifications do call for testing, the bluff can still succeed by ensuring that the test load does not exceed twice the design load.

All of this neatly follows from the "Theory of Elusive Design Deception" (or "Theory of Elucivity" for short) expressed as an acronym in Equation 1.

$$\Delta E = MC^3 \tag{1}$$

where ΔE = Extra Expense, M = Mediocrity and C = Code values, Contentment and Complacency.

An Example of MC^3 Engineering

From our load testing experience with different agencies around the world we often encounter the MC^3 phenomenon when trying to carry out an O-cell test in the most effective way. Because the upward and downward movements at a given loading generally differ, the preferred test procedure requires small regular load intervals carried out to the ultimate resistance load with no intermediate unloading and reloading. This generally conforms to the "Quick Test" under the ASTM Standard D 1143.

The basic ASTM Standard D 1143 load test for conventional top loading describes procedures for carrying out maintained load tests with loads held for particular time intervals (usually, but not necessarily, equal time intervals) and then requires unloading and reloading at one or more intermediate load levels (usually at multiples of the design load). Most project test specifications include this basic Standard.

When originally written, these standards appear to have given little consideration (probably because of a lack of related knowledge) to the impact of residual stresses on the measured performance of the test piles. Fellenius (2002) has described the important effect that residual stresses can have on the interpretation of test data. Unload-reload sequences, although now widely recognized as providing little if any value, (in fact they tend to complicate the analysis of pile performance) still get included in most static pile test specifications. The superior method, we now know, has constant interval maintained loads up to the maximum test load. Even maintaining loads of irregular intervals without unloading complicates the analysis of the test data. We rarely succeed however, in convincing the C^3 engineer that slavishly following such outdated standard procedures actually reduces the quality and value of the testing.

The Widening Gulf between Analysis and Construction

Ralph B. Peck (1991) predicted a gradual widening of the gap between theory and practice: *"Researchers will take refuge in increasingly esoteric investigations; practitioners will pay little attention to the research results. Reading learned journals will become less interesting and profitable to practitioners, scientific oriented workers will find themselves more or less writing to each other."*

Engineers often get absorbed in analytical techniques, fuss and fret over factors that might change things by 5% while ignoring factors that probably affect capacity by 50% or more. We know from personal experience that some engineers will pay more attention to an out-of-date gage calibration certificate than to an obvious lack of proper base cleaning of a bored pile.

The evidence suggests that many engineers fail to appreciate the importance of construction technique on bored pile capacity. Some engineers, and most contractors, sincerely believe that "the dry hole method" provides the best way to install a bored pile. The reason usually given: we can examine a dry hole by eye or with a camera and such inspection ensures good end bearing conditions. (i.e. visual quality control) The evidence from our testing, however, suggests that the worst cases of poor end bearing capacity occur in dry holes (Schmertmann and Hayes, 1997, Schmertmann et al, 1998).

We know several reasons for this, namely:

- Excessive stress relief in a dry hole can disturb a large zone below and around the base of the pile
- The inability to use air lifting or pumping procedures and the ineffectiveness of cleanout buckets in dry holes generally make them more difficult to clean.
- Apparent minor seepage at the base of a "dry hole" can dramatically loosen or soften the base

Mounting evidence from static load testing indicates that, unless the founding material consists of reasonably competent rock or hardpan, we should construct every bored pile for important structures using a stabilizing fluid that suits the conditions.

We have also found that both designers and contractors often do not appreciate the impact of some construction techniques on side shear. We know from our testing data that several construction processes can dramatically affect side shear namely:

- Poor concrete/rebar design
- Poor temporary casing techniques
- Improper use of slurry
- Shape, roughness, and vertical alignment

Poor Concrete/Rebar Design

An example of the impact of mismatched rebar design and concrete, reported by Hayes and Simpson (2003) illustrates how incompatibility of design and constructability impacts foundation capacity. Test results demonstrated how tightly spaced spirals on a heavily reinforced cage, along with concrete having no retarder, reduced side shear by a factor of 10. Table 1 summarizes the test data from this case and one other to illustrate the dramatic reductions that can occur in these situations.

Table 1- Concrete Placement and Rebar Effects on Side Shear

Case Reference	Proper Technique Side Shear	Improper Technique Side Shear	Shear Reduction Factor
A – 723-1, 2	1867 kPa	187 kPa	10
B – 627	1915 kPa *	105 kPa	18
* Estimated unit side shear for shaft in 10,000 psi rock.			

Table 2-Casing Effects on Side Shear

Case Reference	Correct Use of Temp. Casing	Improper Use of Temp. Casing	Side Shear Reduction Factor
C – 932	1341 kPa	67 kPa	20
D – 748	>400 kPa	72 kPa	>5.5
E – 582	192 kPa *	10 kPa	20
* Estimated unit side shear for an uncased shaft in glacial till soils.			

Poor Temporary Casing Technique:

Seemingly insignificant aspects of using temporary casing can also have a major detrimental impact on side shear capacity of bored piles. The results of comparison tests of casing technique (Table 2) further illustrate the important effects that construction details can have on side shear.

The construction sequence in Case D (Figures 2 to 5) provides an interesting example of the detrimental effect of inserting temporary casing in a cored slot, trapping sludge behind the casing. The trapped sludge remained at the limestone concrete interface during the concreting and subsequent removal of the temporary casing. An O-cell® test revealed a side shear in the limestone of only 72 kPa. Changing the construction sequence, in a subsequent test pile, by placing the temporary inner casing after drilling through the limestone resulted in a measured unit side shear exceeding 400 kPa. (The test ended well before reaching ultimate shear resistance). The "before and after" load-movement curves (Figure1) establish dramatically how seemingly innocuous changes in construction sequence can have a profoundly deleterious impact.

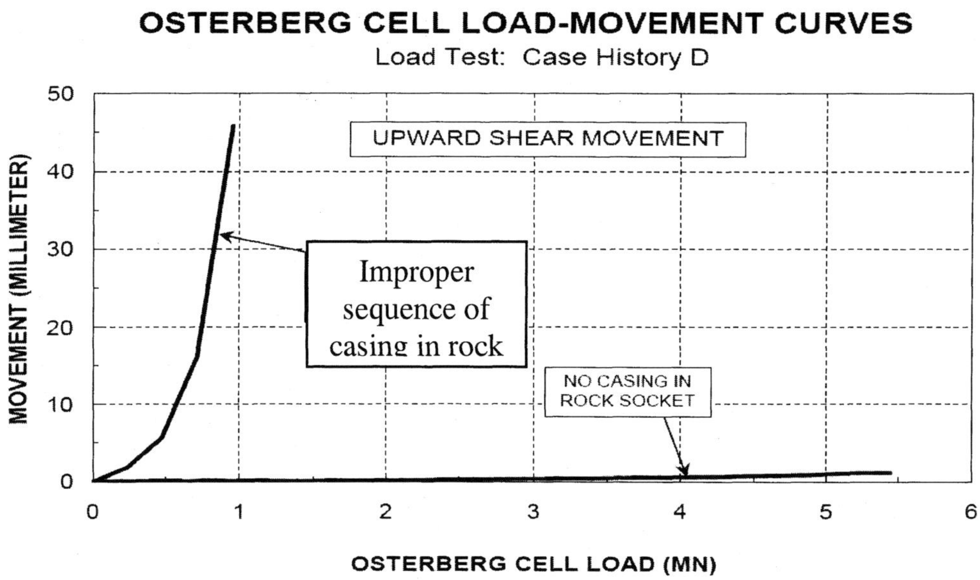

Figure 1 – The Impact of Temporary Casing on Side Shear

Construction Sequence – Case "D"

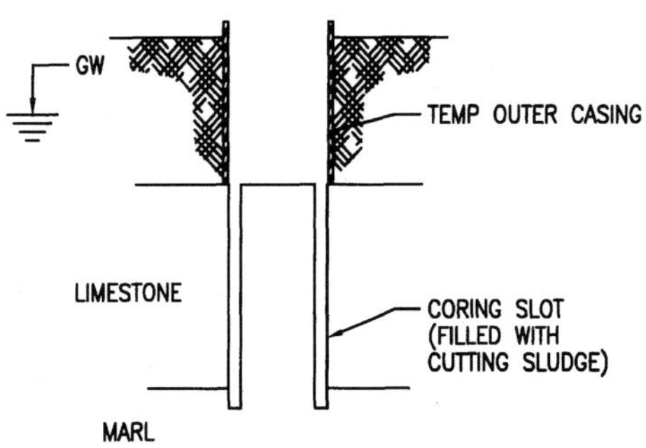

Figure 2- Coring through limestone

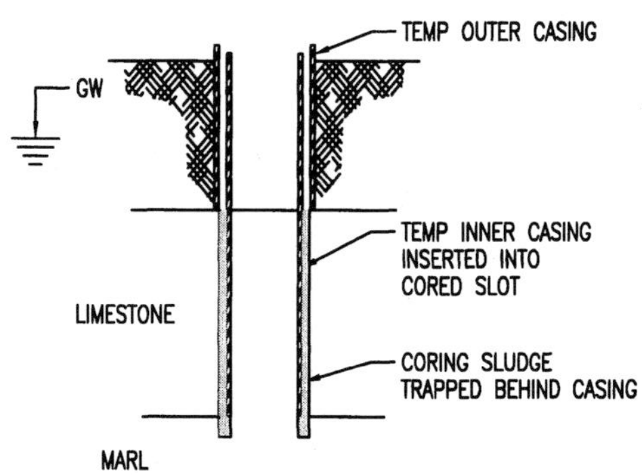

Figure 3- Temp. casing inserted into slot

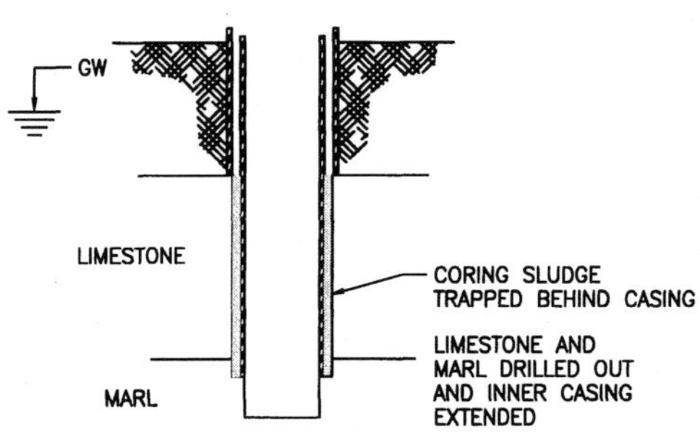

Figure 4- Limestone drilled out

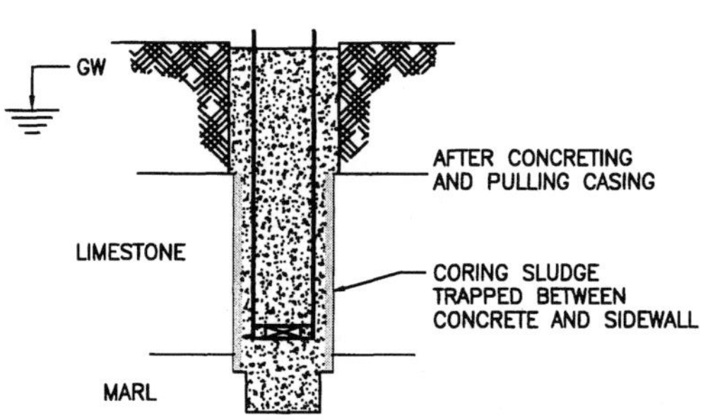

Figure 5- O-cell® assembly inserted, concrete placed and temporary casing removed.

Engineering Misconceptions and Intuition

The foregoing examples of the impact of construction technique should make us very aware of the crucial importance of quality assurance testing of the foundation "product" in exposing design and construction techniques that may reduce capacity by 50% to 90% or more. They also indicate the counterintuitive nature of some engineering perceptions. Intuition suggests, for example that a "dry hole" should produce better end-bearing or that temporary casing would not have a significant effect on side shear capacity.

Engineers setting up a testing program for quality assurance of bored piles should take careful note of the sometimes misleading reliance on intuition. For many years, as a consulting engineer, the author intuitively assumed that he could extrapolate the side shear test data from a smaller bored pile to a larger diameter bored pile, all else remaining equal. Intuitively one would assume that such measured unit side shears would reasonably represent a site characteristic for all bored piles. Well, as it turns out, maybe not.

The important work of Seidel and Collingwood (2001) and remarks by M.W. O'Neill (1998) related to the influence of side wall roughness on all side shear capacity of bored piles in rock sockets, indicated to the author that a significant inverse relationship might exist between unit side shear and pile diameter. Seidel and Collingwood formulated the shaft resistance coefficient (SRC) as follows:

$$q_s \approx SRC = \eta_c \cdot \frac{E_m}{(1+\nu)q_u} \cdot \frac{\Delta r}{d_s} \tag{2}$$

wherein Δr denotes mean roughness height; d_s denotes the socket diameter; η_c denotes a construction method reduction factor, ν denotes Poisson's Ratio, E_m denotes rock mass modulus and q_u denotes the unconfined compressive strength of the rock. (E_m/q_n commonly defines the modular ratio).

Earlier work by Horvath et al (1983) also produced an empirical equation for shaft resistance as follows:

$$q_s = 0.8\left(\frac{\Delta r_h}{0.5 d_s}\frac{L_t}{L_s}\right)^{0.45} .\sigma_{cw} \tag{3}$$

where q_s denotes unit shaft resistance, σ_{cw} denotes the smaller unconfined compressive strength of the rock or the concrete shaft, L_s denotes socket length, d_s denotes socket diameter, r_h denotes mean roughness height and L_t denotes traversed length along the rough surface of the socket. Both of these formulations suggest an inverse relationship between unit shear resistance, q_s and shaft diameter, d_s. Out of curiosity the author compiled unit shear data from O-cell® tests carried out in rock sockets in shale at various USA locations as shown in Table 3.

Table 3- Data from Loadtest Files Used for Side Shear Study

Project	Material	Shaft Dia.	Max. Test Load	Unit End Bearing (EB)		Unit Side Shear (SS)		Ratio EB/SS
				Max. Meas'd	At 12.5 mm	Max Meas'd	At 5 mm	
		m	MN	kPa	kPa	kPa	kPa	
8785	Clay Shale	1.98	43.9	5267	5267	306	306	17.2
8998	Clay Shale	1.22	21.4	8427	8427	418	418	20.2
8547	Shale	1.37	15.3	4745	3064	517	517	9.2
8932	Shale	2.44	16.1	6876	958	718	479	10 est'd
8879	Shale	1.83	14.7	4668	4668	747	747	6.3
8932	Shale	2.44	160.6	19152	7661	958	383	20.0
8626	Shale	1.07	19.6	9260	9260	982	982	9.4
8817	Blue Shale	1.52	49.4	6416	6416	1025	1025	6.3
8929	Brittle Shale	1.37	34.0	8523	7182	1087	1087	7.8
8626	Shale	1.07	19.6	7613	9911	1187	814	6.4
8958	Massive Shale	1.22	41.8	17093	6033	1197	862	14.3
8739	Claystone	1.07	25.0	11874	11874	1211	1211	9.8
8640	Shale	1.07	16.1	5138	5138	1336	896	3.8
8958	Massive Shale	1.22	30.7	12353	12353	1344	838	9.2
8817	Blue Shale	1.52	65.4	13502	13502	1570	1197	8.6
8516	Black Shale	1.17	34.8	9624	9624	1880	1880	5.1
8640	Shale	1.07	32.5	11539	11539	2062	862	5.6
8621	Shale	1.02	40.7	27723	27723	2385	2394	11.6
8528	Shale	1.22	43.2	18051	18051	3064	1939	5.9
8528	Shale	1.22	49.4	18769	23940	3069	2107	6.1
8854	Clay Shale	1.07	38.0	18099	8379	3800	503	4.8

This preliminary compilation of the data, plotted on Figures 6 to 8 with a linear regression trend line, appears to support the proposition that an inverse relationship between unit side shear and socket diameter does exist, at least in shale rock. Apparently the effect can reduce unit shear capacity by as much as 40% to 50% in soft to medium shales and even more in harder shales. Further analysis and research, currently underway, will examine this phenomenon in more detail.

The only case history known to the author, that provided evidence of the pile diameter (or roughness ratio) effect, evolved from the foundation study for a high rise structure in Miami. In their paper describing the study Frizzi and Meyer (2004) concluded that - "The use of other correlations in the literature based on lighter loaded, smaller diameter shafts (i.e. smaller scale tests) yield un-conservative side shear predictions."

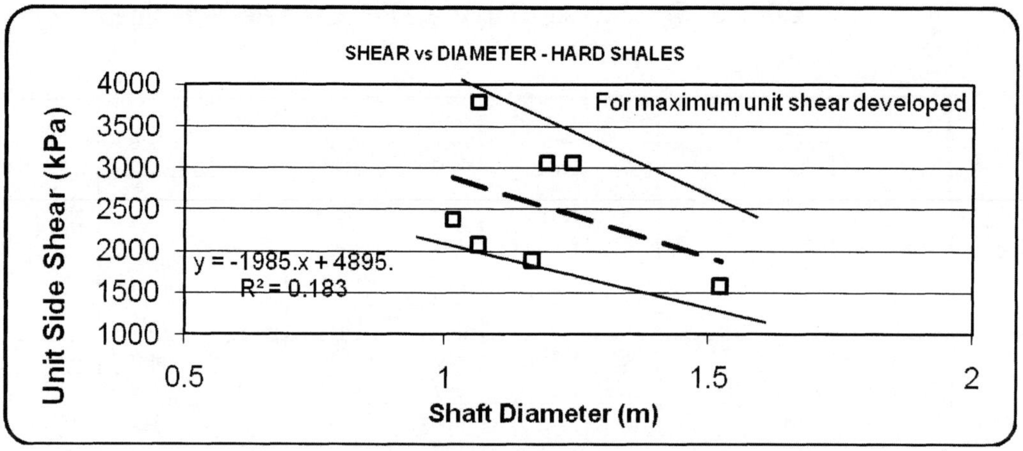

Figure 6- Diameter effect on shear in hard shale

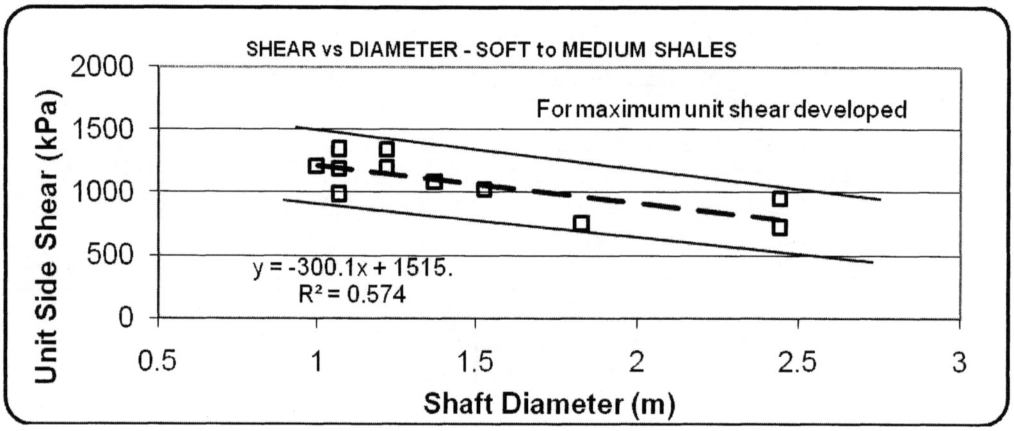

Figure 7- Diameter effect on shear in medium shale

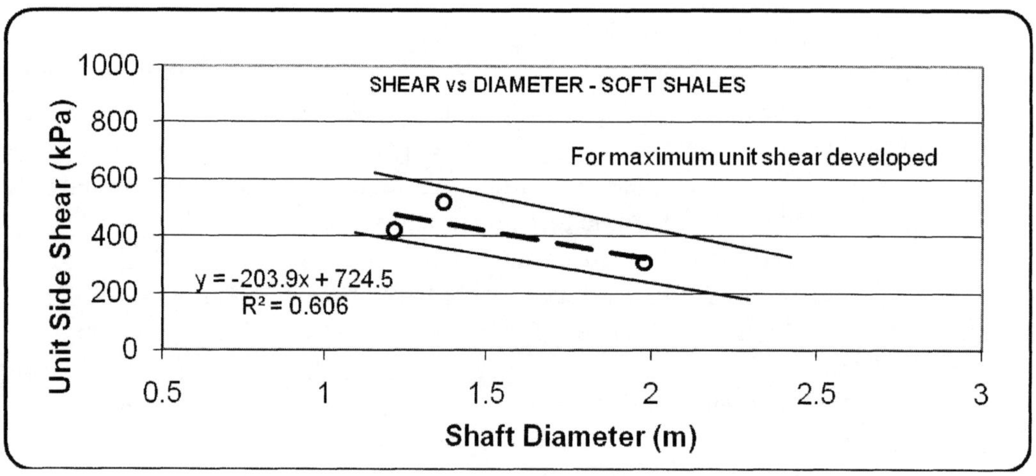

Figure 8- Diameter effect on shear in soft shale

The importance of this "diameter effect" often comes into play when designers of quality assurance test programs consider whether or not to use a reduced-diameter (or model) test shaft instead of a full size prototype for a load test. Engineers usually reason that testing a smaller pile would reduce costs or that testing a full size pile to three or four times design load exceeds normal testing capabilities (the required test loads being too high). Clearly, however, using the data from testing a smaller pile without making a currently uncertain correction for roughness may lead to an unsafe extrapolation of the measured unit shear to larger diameter piles.

The advent of the O-cell® method (Osterberg 2001) for pile load testing makes it possible to apply (formerly inconceivable) high loads to large diameter bored piles. We no longer have to test "model" piles. Beginning in 1993 the magnitude of applied static test loads has increased steadily from 58 MN to the current world record of 284 MN achieved on the New Incheon Bridge in March, 2005. The chart in Figure 9 summarizes the trend.

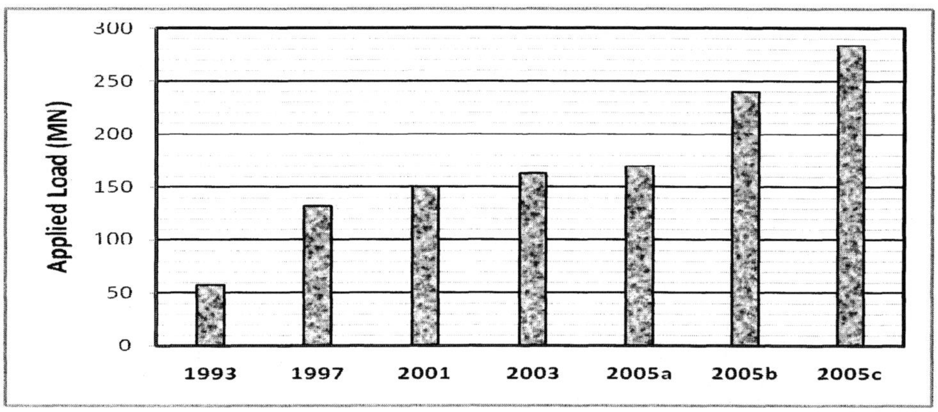

Figure 9- Historical Trend in World Record Test Loads

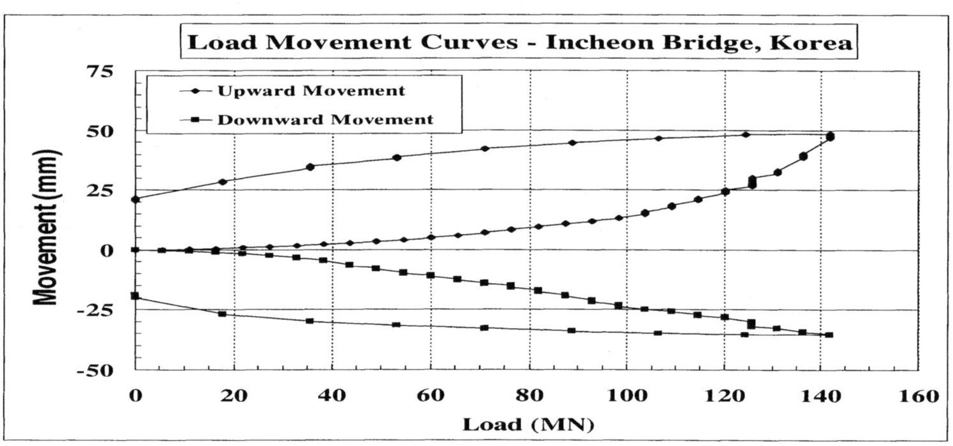

Figure 10 - Load-movement curves, Incheon Bridge, South Korea

These important tests provide convincing evidence that we can achieve quality assurance, not only of design procedures but also of construction technique, even for very high capacity bored piles.

PURSUING THE VALUE IN QUALITY

With respect to deep foundations how should we define value and how should we tie it in to quality? Usually we think of anything with a high monetary value as "valuable" and we generally associate high quality with high value. Almost by definition we consider anything cheap as not valuable. Only rarely do we perceive cheaper as a better value and certainly the cheapest often does not qualify as the "Best Buy." Do we want the cheapest foundation or the most effective (valuable) foundation? What kind of message do we send when we award foundation drilling contracts to the lowest (cheapest) bidder? Surely we should always have the goal of getting the best quality, or the least risk, for our money. The low bid process does not convey that message.

A Possible Guideline to Value and Quality

Consider the development guideline promoted by NASA..... "Faster, better, cheaper". This mantra gave incentive to improve quality and reduce costs (mainly because of the "faster' part of it – i.e. time means money). On construction projects such a guideline tends to "morph" into faster, better, cheaper (with less emphasis on *better* and more on the faster, cheaper part). This inevitably leads to - cheaper regardless of how fast. For our industry we should say - "First better, then faster, then lower cost". Reduced cost has more value than cheaper and relates better to improved reliability and reduced risk.

Looking for Value

Our experience suggests that 95% of bored pile foundations, because of overly-conservative designs, (meaning more expensive than necessary) provide a potential source of hidden value. We have already discussed at length how uncertainty about strength parameters, variable conditions, construction methods and lack of quality control lead to understandable conservativism and mediocrity in design. We know that more effective deep foundation designs would result if we could somehow reduce the uncertainty and mitigate the related risks. We know that testing programs and QA/QC processes can provide more certainty and fewer risks. But how do we justify the expense? "There's the rub."

A proper risk assessment, including an estimate of the costs associated with it, would go a long way to giving the geotechnical engineer the financial clout to institute a valuable testing program. It seems to me that making use of the LRFD approach to deep foundation design makes it easy to put a dollar amount against the mitigating of risk and uncertainty.

By way of example let us make a few assumptions about a generic drilled shaft foundation design.

- Assume that the engineer carries out a routine geotechnical investigation.
- The C^3 engineer estimates ultimate capacity of the shaft using code values.
- Based on a resistance factor of 0.4 (maximum allowed under these conditions) and the structural loads, the engineer calculates a foundation cost of $20 million.
- Assume a new resistance factor of 0.5 on the basis of a load testing program that tests 2% of the shafts and calculate new foundation cost of $16 million. If the test program will cost less than $4 million, get it approved to reduce the total project cost.
- Assume another resistance factor of 0.8 and calculate a foundation cost of say $12 million. The engineer has just calculated a risk and uncertainty cost of at least $8 million. This amount becomes the carrot for convincing the owner to accept only the best contractor using the best QA/QC techniques including an extensive load testing program. Even if this QA program cost $6 million, the owner has acquired a high quality, risk-mitigated foundation for $18 million which cuts $2 million from the one first envisioned by the C^3 engineer.

Those who have paid attention will note that these numbers probably represent the "tip of the foundation cost iceberg". A good engineer will move to the next phase based on real ultimate resistance values instead of code values. He or she will discover many situations where testing every bored pile will result in high quality, lower cost, non-redundant, deep foundations with virtually no risk of poor performance.

Undoubtedly expanding the quality control to the extent of testing every pile would sometimes reveal unexpected problems. Isolating such problems at that time would increase construction costs. These costs, however, pale in comparison to those of remediating unrevealed problems later on.

Putting Quality into the Bidding Process

In my opinion we as engineers need to take on the task of changing, modifying or discarding the low bid process as it applies to deep foundation. We should explore, and promote the use of, the "second-low-bid" concept advocated by Walter Vickrey (1997), Nobel Prize winner in Economics. More realistic pricing for construction work would result from a process whereby all bidders know in advance that contract award always goes to the second lowest bid.

In addition we could improve construction quality, and the attitude toward it, if we provided a special QC bid item in the project specifications and used it to assess best buy. This would allow the contractor to include all the quality control work into one bid item, the price of which would get deducted from his total bid price. Table 4 illustrates how the contractor putting the most effort into quality wins the contract as the second low bidder.

Table 4 – Illustrating the Use of a Special QA/QC Bid Item

Best Value Assessment			
Contractor	A	B	C
Total Bid	110	115	122
QA/QC bid item	5	12	18
"Net Bid"	105	103	104

The QA/QC bid item would include:

- Material testing for concrete, steel (slump test etc)
- Monitoring of shaft shape and alignment (caliper)
- Slurry control (viscosity, sand content etc)
- Base cleanliness assessment (Shaft Inspection Device)
- Base cleaning, side wall roughening, reaming etc.
- Proof testing for capacity (load testing)

Payment for this bid item would require completion all of the work described in it to the satisfaction of the engineer-of-record.

SUMMARY and CONCLUSIONS

High quality, a desirable goal in all foundation works, becomes especially so for deep foundation projects. Good quality requires the proper state of mind and attitude. Trying to achieve high quality by "force" through rigid specifications often leads to trouble and high cost. The best results occur when the pile constructor (manufacturer) takes responsibility for the construction quality, leaving the engineer-designer free to concentrate on the geotechnical aspects. Quality in design engineering happens when we avoid Complacency, Contentment and Code reliance. Remember $\Delta E = MC^3$.

Both design engineers and pile constructors need to pay attention to the devilish but important details that produce high quality and high capacity in bored piles. Documented examples show that improper casing technique can cause 80 to 95% reductions in shear capacity. Similarly mismatching rebar design and concrete mix can result in measured side shear reductions of 90 to 95%. Having a great attitude about QA/QC provides a basis for eliminating such risks and uncertainties during the construction process.

Design engineers should always make sure to question their intuition about geotechnical issues, especially when it might affect the quality of their work. Available test data and project experience indicates, for example, that using smaller diameter test piles (scaled-down) to get unit shear parameters may produce unit values much higher (40% or more) than those achievable on the actual larger production pile sizes. Resisting the temptation to do things cheaply requires effort. Recognize that the sweetness of cheapness soon wears off leaving the bitterness of bad work.

Engineers should always make the effort to quantify the cost of the inherent risk and uncertainty associated with designing and constructing a deep foundation. We can use the LRFD approach to get a sense of the magnitude of such costs. The engineer then has a much better idea of how much QA/QC expense the owner should commit to the task of reducing risk and uncertainty.

Few in the deep foundation industry question the fact that the low-bid award process has a deadening impact on getting high quality results. The design geotechnical engineering profession should take the lead in changing the way we award contracts for deep foundation work. The "second-low-bid" concept advocated by Walter Vickrey, along with a special bid item for QA/QC work, deserves our attention. As John Schmertmann has often said when suggesting a new approach - - "*try it, you might like it*".

References

Arup, Ove. (1970). "The Key Speech" delivered to assembled leaders of Arup at Winchester.

Brandl, Heinz. (2004). "The Civil and Geotechnical Engineer in Society – Ethical and Philosophical Thoughts; Challenges and Recommendations" DFI, Vancouver

Fellenius, B. H. (2002). Determining the resistance distribution in piles. Part 2: Method for Determining the Residual Load. Geotechnical News Magazine 20(3) 25 - 29.

Frizzi, R.P., Meyer, M.E. and Zhou, L. (2004). "Full Scale Field Performance of Drilled Shafts Constructed Utilizing Bentonite and Polymer Slurries'. GeoSupport 2004: Deep Foundations, Soil Mixing, Ground Improvement (Geotechnical Special Publication No. 124), ASCE. Page 584

Hayes, J.A. (2002) "ADSC – An Association of Drillers or Manufacturers?"; ADSC, Foundation Drilling Magazine, May 2002.

Hayes, J.A., & Simmonds, T. (2002). "Interpreting Strain Measurements from Load Tests in Bored Piles" Proceedings of the Ninth International DFI Conference on Piling and Deep Foundations, Nice, France.

Horvath, R.G., Kenney, T.C., and Kozicki, P. (1983). "Methods for Improving the Performance of Drilled Piers in Weak Rock. Canadian Geotechnical Journal. **20**: 758-772

Osterberg, J.O., and Hayes, J.A. (2001). "The Osterberg Load Cell as a Research Tool". Proceedings of the 15th International Conference on Soil Mechanics and Geotechnical Engineering, Istanbul Turkey.

O'Neill, M. W. (2001). "Side Resistance in Piles and Bored piles", 34th Karl Terzaghi Lecture, Vol. 127, No. 1, Journal of Geotechnical and Geoenvironmental Engineering, Jan., pp. 3 -16.

Peck, R. B., Powderham, A. (1999). Talking point - The Observational Method. *Ground Engineering,* Vol. 32, No. 2.

Schmertmann, J. and Hayes, J. (1997) "Observations from Osterberg Cell Tests of Bored Piles", FULCRUM, the Newsletter of the Deep Foundations Institute, Winter '96:97, pp 11-14.

Schmertmann, J. & Hayes, J. & Molnit,T. & Osterberg, J.O. (1998). "O-cell® Testing Case Histories Demonstrate the Importance of Bored Pile (Drilled Shaft) Construction Technique". Proceedings of the Fourth International Conference on Case Histories, University of Missouri, Rolla, Missouri.

Seidel J.P. & Collingwood B. (2001). "A new socket roughness factor for prediction of rock socket shaft resistance" – Canadian Geotechnical Journal **38**: 138-153

Vickrey, W. (1997). "Public Economics". Cambridge University Press, 1997. ISBN10: 0521597633

Some Simple Methods I Used in Research and Practice

John H. Schmertmann[1], F. ASCE, P.E., NAE

[1]Professor Emeritus, Dept. of Civil and Coastal Engineering, Univ. of Florida, Gainesville, FL 32611, schmert@ufl.edu

ABSTRACT: Making research and practice as simple as you can has many advantages, and the writer uses this approach. This paper gives ten personal examples. They vary from demonstrations of engineering behavior, direct insitu test methods, making field measurements, the installation and testing of piles, to a method to improve technical writing.

KEY WORDS: Compressibility, field measurements, movements, permeability, piles, report writing, shear strength

INTRODUCTION

Einstein once said: "Make everything as simple as possible, but not simpler." Simplicity has the possible advantages of focusing effort, making teaching and client-convincing easier and saving time and resources. Making something *too* simple runs the risk of significant error. The trick involves striking a balance suitable to the problem. I advocate "If in doubt – simplify!" and herein give some examples from my own research and practice.

1. Lateral Load Effect Demonstration

Sometimes one needs a simple, easily understood demonstration of a concept to get others to understand and possibly use it. Anyone can imagine something and explain in words why it should work. Engineers will want a demonstration. I faced this situation when recently describing something I had experienced in practice, namely the significant increase of a pile's axial load capacity in sandy soils when simultaneously subjected to a lateral load. I decided on a field model demonstration – the simpler and more visual the better.

Using a tripod, some wire cable, two spring scales and some turnbuckles to control horizontal loading, an engineer's scale with an optical level for movement, a 1.95m (6 ft) length of galvanized pipe for the pile, plus a posthole digger and several bags of children's play sand for the surrounding soil, an assistant and I prepared a demonstration on the ground immediately adjacent to our office building. It required a time investment of approximately two days to design, assemble, perform and analyze the demonstration test, all at a cost to me of under $500.

Figure 1 shows photos and a schematic of this demonstration test setup and the results obtained. It seemed to provide convincing evidence that the simultaneous lateral loading of a pile in sand can dramatically increase axial pile capacity. Also see Schmertmann (2006) wherein I discuss a simple prediction theory for the lateral pressures on a rigid pile. Multiply these by the sand/pile friction coefficient to get axial capacity. Using a coefficient of 0.25 and this theory gave the good linear prediction shown in *Figure 1d.*

(a) Setting up Tripod Lift, Level for Measuring Movements, Placing Clean Sand at 3 Densities around Pipe Pile in Posthole

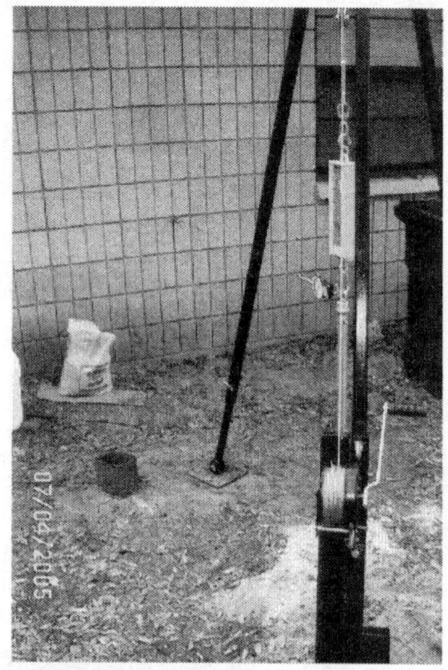

(b) Mechanical Scales Inline with Vertical & Horizontal Loading Wires

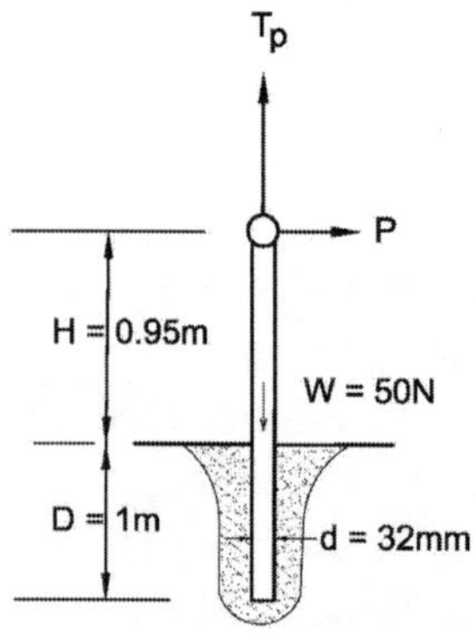

(c) Schematic

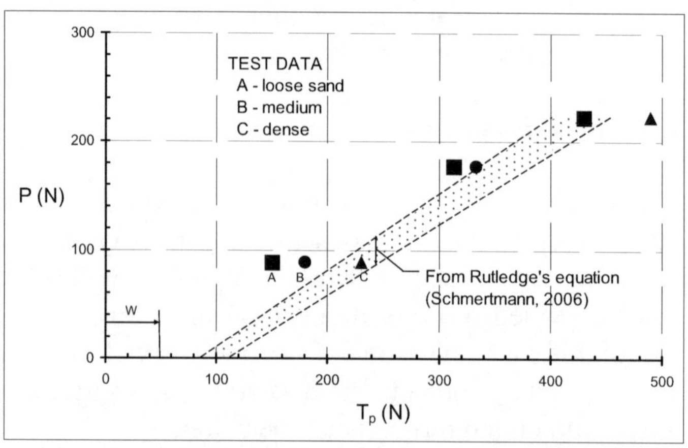

(d) Test Results from the Demonstration

FIG. 1 - Demonstration that Lateral Loading of a Pile in Sand Increases its Axial Capacity

2. Handcone Penetration Test

For over 30 years I have found it convenient to carry a handcone penetrometer with me for possible use when investigating new sites. It can provide quick, personally obtained answers to near-surface stratigraphy and other problems by providing quantitative bearing capacity and qualitative soil type information to depths as great as 12 meters (40 feet). The writer has used such a hand penetrometer on numerous occasions and for a variety of purposes. For example: to locate muck pockets trapped under highway embankments under construction, to check soil compaction around buried pipes, to search for the depth to shrink-swell active clays, to determine the depth or thickness of soft clays or organic deposits, to determine the depth of scour where subsequent soil infill has occurred, and to check underlying soils to help evaluate the behavior of concrete-clad slopes along beaches, dam faces, and at the space shuttle launch pads.

Figure 2 shows the report insert used by Schmertmann and Crapps, Inc., to explain the handcone test method to the reader of our reports. It shows a photo of the equipment, describes the method for estimating soil type, and illustrates a sounding profile as presented in one of our reports from testing in the Gainesville area. It shows a loose surface sand with near-surface compaction, and an underlying shrink-swell clay. When it seemed advisable, we confirmed the soil types by sampling with 2 inch (50 mm) diameter hand auger equipment, which can also provide samples for disturbed-sample classification tests such as water content and grain size. Try it – you might like it!

revised Apr 83

SCHMERTMANN & CRAPPS, INC.
— Consulting Geotechnical Engineers —

EXPLANATION OF THE STATIC HAND CONE PENETRATION TEST (HC)

This test represents a form of quasi-static cone penetration test, similar to the Dutch cone penetration test (CPT) described in ASTM Standard D-3441. However, it uses smaller diameter equipment suitable for testing near-surface soils by hand methods. The cone bearing pressure in sand from this test equals about 1x to 2x q_c from the CPT.

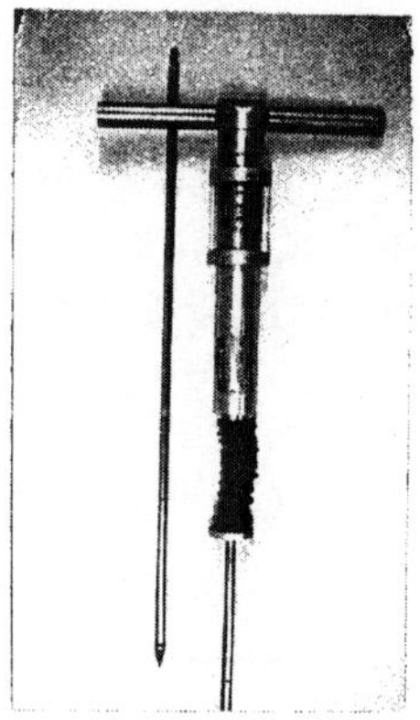

The photo at right shows the test equipment. It includes a load cell graduated in 10 pound divisions to 300 lbs, with a slide that retains the maximum force. The operator pushes vertically on 0.44 inch diam. rods, screwed together flush-jointed in 2 ft lengths, and tipped with a 30° cone point 0.50 inches in diam. He enters the soil from the ground surface in typical 6 inch increments, reads and records the maximum force, resets the slide to zero, pulls the rods up this 6 inches and pushes down again to get a maximum force estimate of soil/rod friction only, pushes to penetrate the next six inches, etc. The equipment has no special provision for eliminating soil/rod parasite friction, other than the reduced rod diameter. But, the operator obtains a measure of the friction during the above push-down-again and subtracts this from the previous point+friction force reading to obtain the net point bearing force.

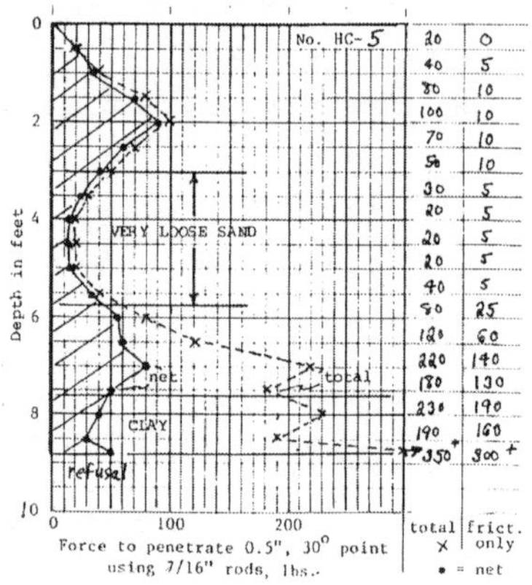

The operator records both the total (bearing+friction) and friction-only forces. The typical HC log at the left shows the total (x) and the net cone bearing (•) forces at the bottom of each 6" test depth increment, in lbs. The connected pts. show the vertical variation of the net bearing force with depth. Divide by 2.73 to get bearing pressure in tsf. The log may also show soil layer descriptions as estimated from the HC log only, or by visual examination from a parallel soil boring.

FIG. 2 – Report Handout for Handcone Penetration Test

3. Improvised Tape Extensometer

I use an optical, self-leveling level to simply measure vertical movements. Horizontal movements present more of a challenge. To meet this challenge I improvised a light-weight tape extensometer which has a typical precision for horizontal measurements of 1.5 mm and a repeatability and estimated accuracy of 3 mm over a horizontal measuring distance of 3-15m (10-50ft). My improvised extensometer requires a dedicated nylon survey tape calibrated for tension and temperature, a spring scale to apply a fixed tension, grip clamps to hold the tape and the scale, embedded screw heads to measure between, and a thermometer to measure ambient temperature. *Figure 3a* shows a photo of this simple, but reliable equipment.

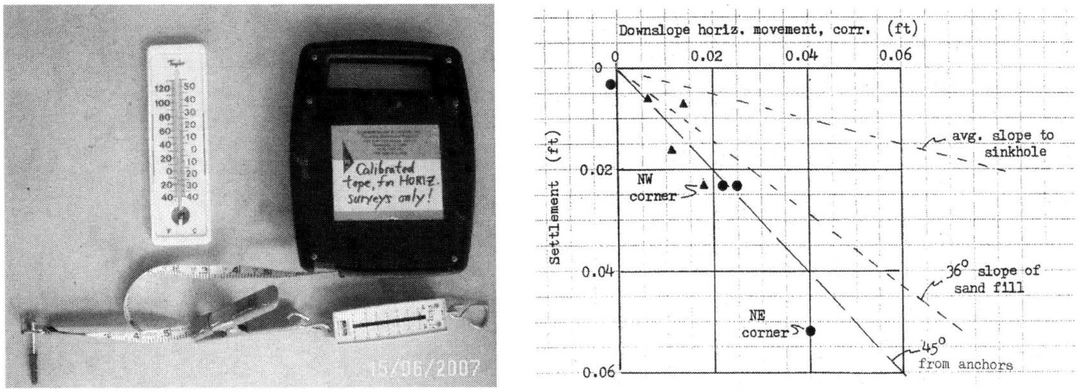

(a) Equipment Needed **(b) Downslope Corner House Movements over 526 Days (Note: 0.02ft. = 6.1mm)**

FIG. 3 – Improvised Tape Extensometer

Figure 3b shows a set of measurements made with a level and this extensometer to determine the vertical and horizontal movement of the downslope corners of a house over a time of 526 days. The data clearly showed horizontal movements. But, did they follow the 10° slope of an underlying clay layer and the pre-construction ground surface?, the 36° backfill slope of the sand under the downslope part of the house?, or a 45° incline forced by the 45° backward (into slope) slope of the Chance helical anchor underpinning used previously in an attempt to stabilize the house movements? The data suggest that the anchors controlled the movement.

4. Slope Test for Shear Strength

Sometimes I have had the problem of estimating the shear strength parameters, c and Φ, of a surface layer. The simple solution: Dig into this layer to create a high enough and steep enough slope until it fails and backfigure reasonable combinations of c and Φ to match that failure geometry. Shallow slopes and trenches need only hand digging methods, deeper ones require any handy construction equipment that one can use for excavation. It helps to have a convenient graph along to back-calculate the shear parameters. *Figure 4* presents an example of one I used. See below for an example of its use:

Starting with a horizontal surface, dig a trench H deep and undercut it until it fails at an overhang angle β. Say at failure H = 1.5 meters (5 feet) and β = +10 degrees. Assuming $\gamma = 16$ kN/m^3 (100 pcf), and $\Phi = 30°$, then $c/\gamma H = 0.19$ and c = 4.5 kPa (95 psf). In this example the estimated critical failure plane with no significant tension crack would occur at approximately $\alpha = 65°$ from the horizontal.

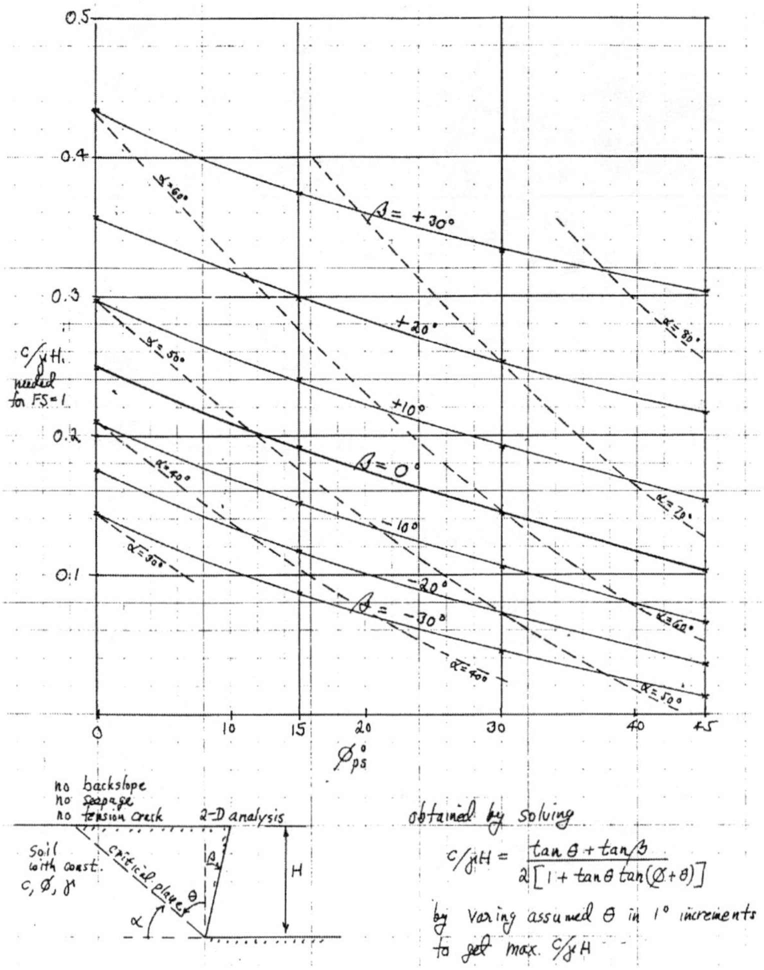

FIG. 4 – Chart for Estimating Near-Surface Shear Strength

A simple method such as that described above can quickly, and insitu, provide an estimate of a soil's friction coefficient or the often difficult-to-determine value of the appropriate cohesion intercept of possible importance, for example in bearing capacity, slope stability, and arching problems.

5. The Conical Test Load (CTL)

Perhaps like yourself, I have occasionally had the problem of evaluating or demonstrating the consolidation settlement behavior of a near-surface soil layer. Assuming suitable site conditions, why not just pile up some loose (to minimize arching) soil on top of a plate and measure the behavior directly in the field at the location of interest? I and others have done just that, as explained in my Conical Test Load (CTL) technical note (1993) and expanded upon in a Penn DOT (1994) lecture. Not only does a CTL test have the advantage of simplicity, but also the advantages of speed, economy, safety and impressiveness.

Figure 5a shows a photo of its on-site use prior to the 1999 move of the Cape Hatteras Lighthouse (LH). *Figure 5b* shows a schematic with the ratio of (CTL/LH) vertical stress increase from the 4.6m (15 ft) CTL height. At the 4.6m (15 ft) height the CTL weighed 6.7MN (1500 kips). The 1-hour, axial settlements at each successive height stage = 5, 8 and 12 mm. The LH settled a maximum of about 50 mm during its move.

(a) Conical Test Load at Cape Hatteras Lighthouse

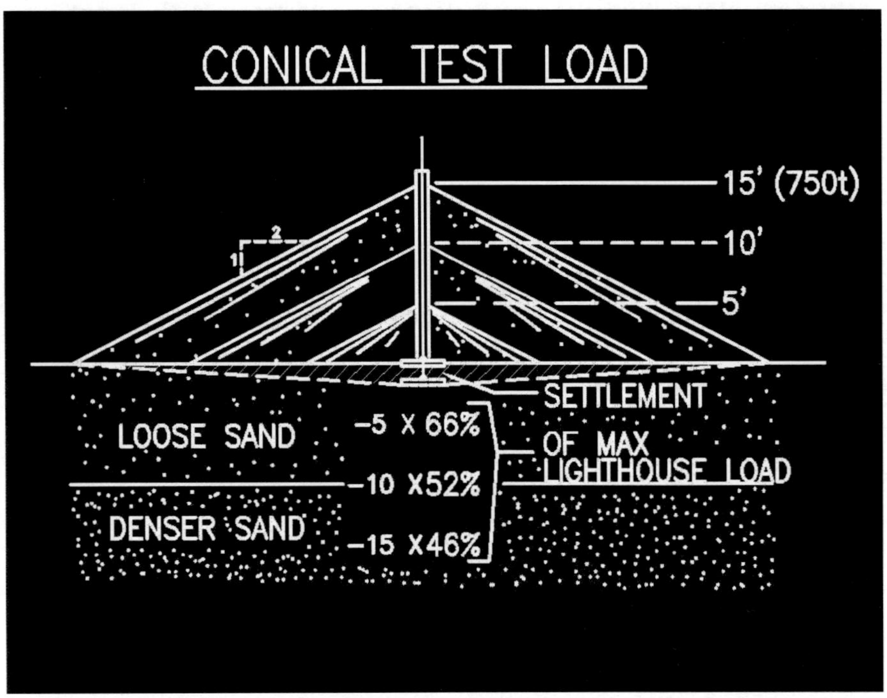

(b) Schematic of 3-Stage CTL with Vertical Stress Applied at Maximum Stage at 5, 10 and 15 ft. Depths

FIG. 5 – Conical Load Test (CTL) Example

I chose the conical shape because of its ease of construction and volume efficiency in getting significant surcharge pressures to significant depths, and have used the CTL for such problems as:

1. Calibrating a settlement prediction method, as in the *Figure 5* example.
2. Evaluating the effectiveness of:
   a. Dynamic compaction on a garbage landfill.
   b. A roller-compacted blanket over a construction debris fill.
   c. Wick drains under a bridge approach embankment before construction.
   d. Compaction grouting vs. sinkhole subsidence under the liner of a new landfill.
   e. The site preparation for a prestressed mat foundation under large storage tanks.

Try it - you might like it!

6. Parallel Trench Permeability Testing

Following the breaching failure of a low dam in Florida, probably caused by piping, it became important to evaluate the permeability of a particular very shelly sand layer exposed near the bottom of the breach. The situation called for a quick and accurate evaluation and I decided to use the direct, simple and convincing approach of measuring this permeability insitu and on a large scale. This involved the digging of a trench, 14m long, 1.5m deep, parallel to and 8m from a longer existing exploratory trench, each one intersecting the shelly layer in question. I then put water in the trenches and used pumps to maintain a constant head difference between them. After allowing time for equilibrium, I introduced an upstream dye marker and measured a 51 minute time of seepage transit between the two trenches. The rate of pumpage into the upstream trench (0.18 m^3/min) gave the equilibrium rate of flow between trenches.

Figure 6 illustrates the plan and section of this test, with sketched flownets, and it includes a photograph. After considering the effects at the trench boundaries and making an estimate of the ratio of horizontal to vertical permeability (k_H/k_V) of between 4 and 25, I obtained a maximum $k_H \approx 50$ mm/s along the top of the layer and an average $k_H \approx 10$ mm/s for the total layer thickness of *c.* 0.75 to 1.5m (2.5 to 5 ft).

(a) Test in Progress. 2 = Ditch A, 3 = Ditch B

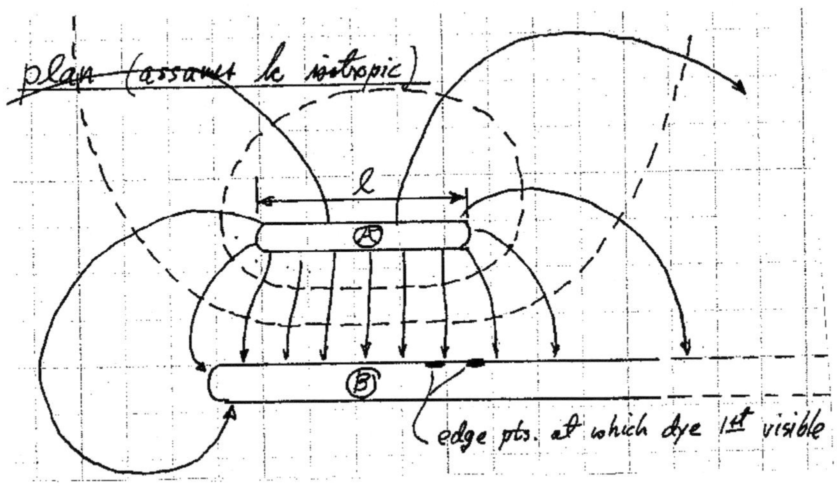

(b) Assumed Horizontal Flow Pattern Between Ditches

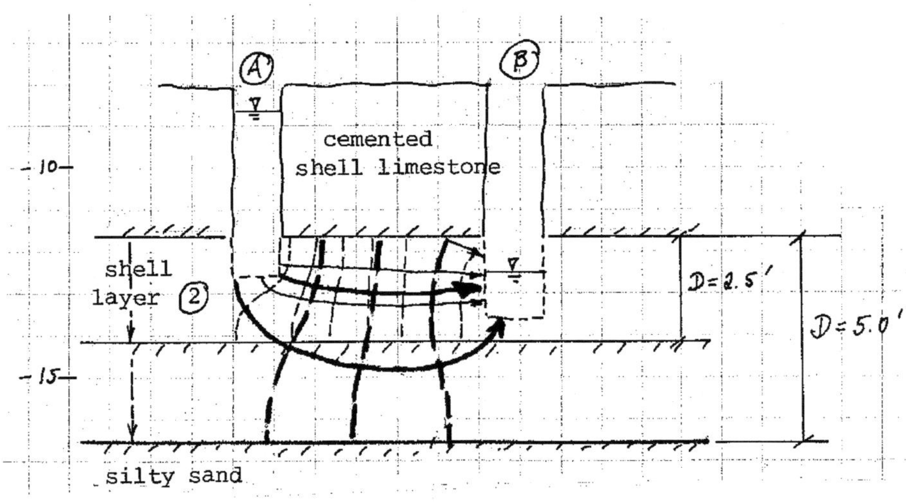

(c) Assumed Vertical Flow Pattern for $k_h/k_v = 25$

FIG. 6 – Improvised Field Permeability Test

This test quickly and convincingly proved and quantified the very high permeability of the shelly layer relative to the surrounding shelly limestone and fine sand layers. It also showed that the limestone-shell layer contact provided a possible path for the piping. Note that a later nearby, much more elaborate, expensive and time consuming well pumping test that included the same shelly layer yielded a similar avg. $k_H \approx 12$ mm/s for this layer. But it could not isolate the high-k of the contact zone.

7. Hand Driven Rebar Pinpiles

Small diameter displacement piles can carry a surprisingly large compression or tension loading. When faced with the problem of providing foundation support for light structures, such as private homes or small commercial buildings, for either the initial construction or remedial work, I have taken economical advantage of this knowledge by using steel reinforcing bars as piles and driving them by hand using electric or air hammers.

The typical situation in my experience involves trying to bypass or control active clay, loose sand, or poorly constructed manmade fill in the upper 10-15 feet below foundation level. Such a rebar pile has a typical length of 6 meters (20 feet), in 2 meter lengths with compression/tension connectors, using #10 rebars driven plumb with a 0.4kN (90lb) air hammer. You can then easily test such a driven pile in tension with a center-hole jack and in compression by jacking against a suitably heavy truck, bulldozer, etc. The writer has found typical allowable loading in the 18 to 45kN range (4-10 kip), depending on rebar size, pile length and soil conditions. Such rebar piles may need appropriate corrosion protection over all or part of their lengths, but exhuming old, unprotected piles has so far indicated negligible corrosion effects in the Gainesville, FL area.

The installation of such pinpile foundations has varied from small groups of five piles to support outside fire stairs added to a structure, to *c.* 300 piles used to completely underpin a large house against clay swelling uplift. First we jacked up and releveled all its wall footings on *c.* 100 jacks. *Figures 7a and 7b* show photos of driving 20 ft (6m) long, #7 end-threaded rebars in coupled 1.5m (5ft) and 3m (10ft) lengths, inside and outside this house. Prior load tests showed an allowable tension capacity of 30 kN (7 kips) to resist clay swelling (see the *Figure 7c and 7d* test example). When in production, a crew of four laborers could hand drive 30-40 such pinpiles per day. Again, try it - you might like it!

(a) Driving 6m Piles inside House with Air Hammer

(b) Driving Outside House

(c) A Pile Tension Load Test

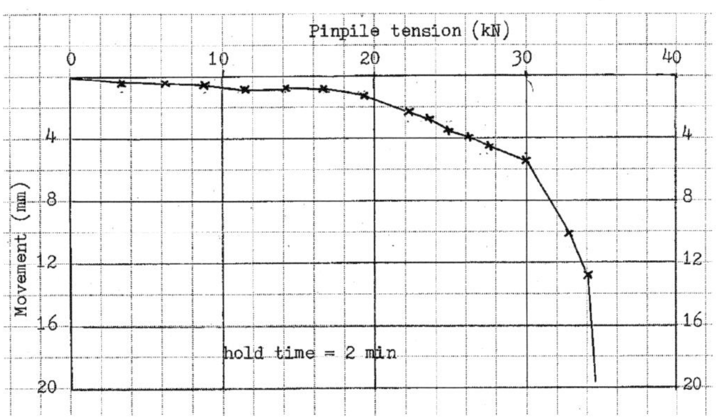

(d) 6m Pile Test Results

FIG. 7 – Rebar Pinpiles to Resist Clay Swelling

8. Crane Tension Test

Consider this problem in an excavation for a power plant addition in Florida: A base slab, required to dewater a sheetpile-enclosed *c.* 37x46m excavation, needed the addition of tension piles to resist water pressure uplift. This meant I also needed tension tests inside and along the perimeter of a flooded excavation. This called for a 'sky hook' and fortunately the contractor had one available. I suggested installing the piles and then using a crane sitting alongside the excavation to reach over and pull the piles, using a load cell to measure the tension force of the pull and an optical level sighting on an engineer's scale to measure movement. It took the contractor only a few days to mobilize for the tests. Altogether we performed 28 tests on 14 sheet, H soldier and drilled shaft piles to a maximum safe tension load of 1.34MN (300 kips).

Figure 8 shows photos of the 9-part line tackle block used, the 0.22MN (50 kip) load cell in the final line, the 0.14MN concrete blocks used to calibrate the cell (the factor = 9.07), and a test in progress. *Figure 8(c)* shows the results, using 5 minute, approximate constant load increments. This simple and rapid load testing procedure allowed the quick testing of many piles, which proved important in this highly variable site. Try it – you might like it! But, of course, stay within the crane's load safety limits.

(a) Nine Part Tackle Block with Load Cell and Calibration Weights

(b) Ready to Test a Soldier Pile

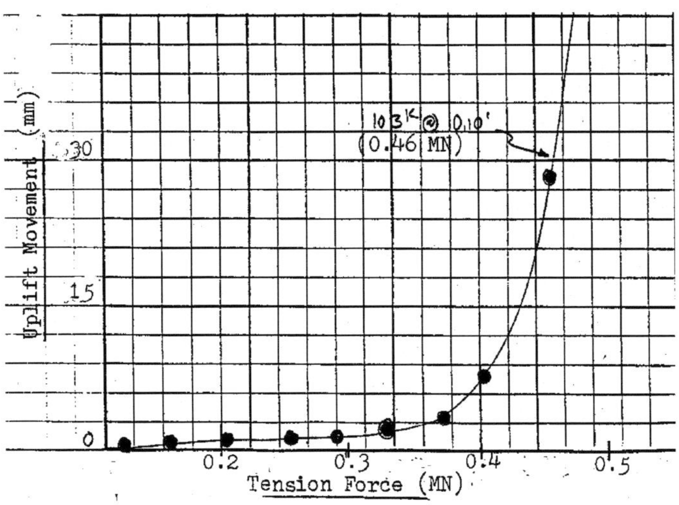

(c) Test Results

FIG. 8 – Improvising A Crane for Pile Tension Load Tests

9. Improvised Wellpoints for Pressure Relief

During a difficult dewatering problem in sand over porous Florida limestone it became necessary to relieve the uplift water pressure on an unexpected, low permeability silty sand layer under the bottom of the planned excavation. The contractor, who had previously failed to dewater adequately when pumping over 400 wellpoints, pulled them out and had them readily available. I decided to use some of these wellpoints to increase the permeability of the silty layer by placing them just below the silty sand and letting them self-pump into an overlying gravel blanket. The contractor and I successfully used 120 wellpoints with 0.75m (2 ½ ft) screens and 1.5m (5 ft) of riser pipe to bypass the offending silt when dewatering the excavation. *Figure 9* illustrates the installation and Schmertmann (1974) describes in detail the dewatering problem and this improvised use of sacrificial wellpoints.

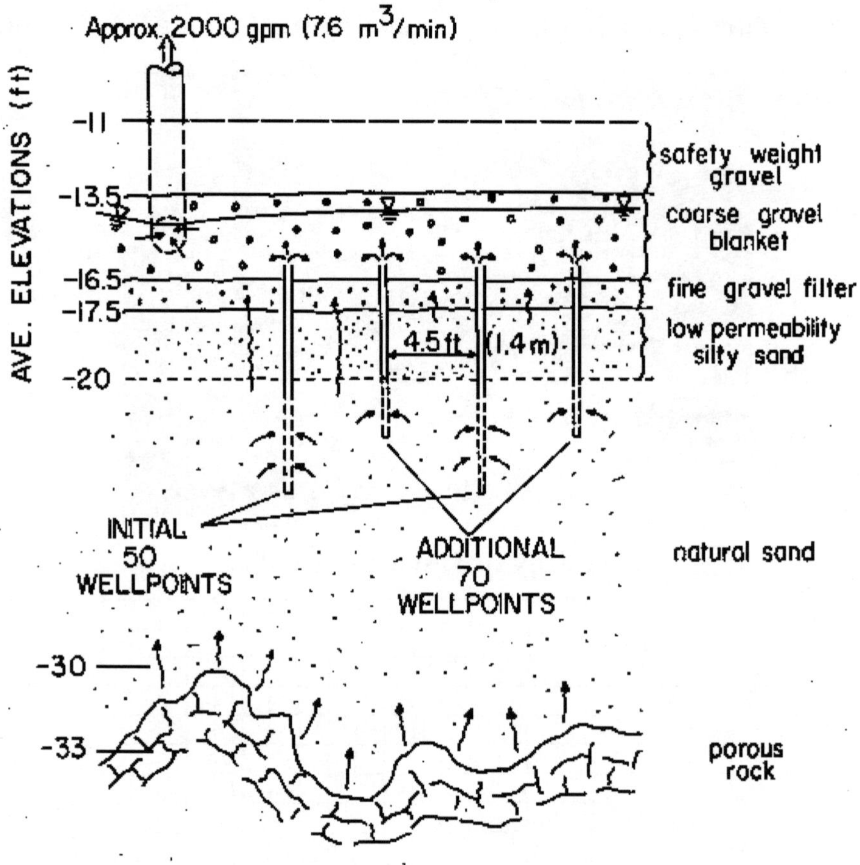

FIG. 9 – Self-Pumping Wellpoints for Pressure Relief (from Schmertmann, 1974)

10. E-Prime English

In 1969 I happened to read a short article in a popular magazine (Time, 1969) on the subject of a new philosophy for improving clarity in English writing, developed by semanticist Alfred Korzybski and disciple D. D. Bourland, Jr. Bourland named it "E-prime English". His bottom line seemed elegant, simple and very useful for technical writing. I immediately applied it in part to a paper I had in process and that paper won the ASCE's 1971 Norman Medal Prize. This very positive feedback convinced me, and I have used E-prime English ever since – including in this paper. Also note the paper by J. Hayes.

In practice E-prime means not using the verb "to be". This forces the active voice and thus improves readability, promotes precision by forcing the writer to assign a subject to the verb, encourages brevity, and avoids the implication of permanence sometimes associated with the verb "to be". Initially you may find it awkward, but try it - you might like it!

CONCLUSIONS

I have often found simple, but not too simple, methods useful in my geotechnical engineering research and practice. This paper demonstrates, with examples, methods to measure near-surface stratigraphy, movement, shear strength, compressibility and permeability; and how to improvise other economical methods and tests under difficult conditions. It concludes with a suggestion for improving your technical writing. When you can use simple but appropriate and dependable methods, do so! Some of them also provide the satisfaction, assurance, and enjoyment of using personal initiative to solve a problem.

REFERENCES

Schmertmann, J. H., (1974). "A Dewatering Case History in Florida", ASCE, *Journal Construction Division*, Vol. 100, C03, Sep: 377-393.

Schmertmann, J. H., (1993). "Conical Test Load Measuring Compressibility", Technical Note, *ASCE Journal of the Geotechnical Engineering Division*, Vol. 119 (5): 965-971.

Schmertmann, J. H., (1994). "A Conical Test Load (CTL) Load Test", *Proc. Soil-Structure Interaction and Environmental Issues*, Hershey, PA, 12-14 Sep, Central PA ASCE Section, Penn DOT, 19 pp.

Schmertmann, J. H. (2006). Discussion of "Ultimate Lateral Resistance to Piles in Cohesionless Soils" by L. Zhang, F. Silva, and R. Grismaia, *ASCE J. Geotechnical & Geoenv. Engrg.*, Vol. 132 (8): 1108-1109.

Time Magazine (1969), "The Un-Isness of Is", May 23, p.69

Subject Index

Page number refers to first page of paper

Aging, 540
Alabama, 507
Ammonia, 594
Austria, 616

Bearing capacity, 49
Bored piles, 177
Boreholes, 82
Boundaries, 230
Bridge decks, 407
Bridges, 170, 417

Canals, 574
Case reports, 95
Clays, 1, 45, 112, 193, 230, 384
Compaction, 149
Comparative studies, 259
Compression, 163, 653
Cone penetrations tests, 92
Construction, 559
Cylinders, 366

Dams, earth, 183
Dams, embankment, 181, 522
Deep foundations, 312, 636
Density measurement, 133
Design, 489, 559
Dewatering, 95
Dilatometers, 130, 292
Drilled shafts, 177
Dynamic loads, 312
Dynamic tests, 471

Erosion, 181
Experimentation, 277

Failures, 230, 407, 616
Fills, 133
Finite element method, 522
Florida, 82, 95, 112, 170, 417, 559
Footings, 112
Foundation design, 351
Foundation settlement, 49, 206
Foundations, 95, 407, 507
Fuzzy sets, 259

Geogrids, 277
Georgia, 540
Geotechnical engineering, 626
Grouting, 149

Hurricanes, 407, 574
Hydraulics, 251

In situ tests, 92, 170, 292
Industrial wastes, 594
Infiltration, 183

Laboratory tests, 1, 170
Lateral loads, 351
Lateral Stress, 92
Levees, 574
Liquefaction, 251, 292, 433, 540
Load tests, 163, 177, 312, 452

Measurement, 34, 163, 177, 653

Netherlands, 92

Penetration, 82
Penetrometers, 82
Permeability, 183, 653
Physical properties, 206
Piers, 251
Pile driving, 331, 471
Piles, 206, 366, 384, 489, 636, 653
Pipes, 181
Ponds, 594
Predictions, 331, 452
Probability, 259

Quality control, 636

Raft foundations, 206, 351
Rainfall, 183
Research, 653
Reservoirs, 559
Risk management, 626
Rivers, 540
Rock fills, 522

Sampling, 1
Sand, 49, 92, 277, 433, 540
Saturation, 133
Seepage, 522
Seismic effects, 292
Settlement, 130, 489, 594
Shafts, 384, 452, 636
Shear resistance, 34
Shear Strain, 45
Shear strength, 653
Shear tests, 230
Shear waves, 417
Shrinkage, 112, 193
Silts, 82, 433
Sinkholes, 149
Skin friction, 489
Slope stability, 183, 616
Slopes, 112
Soil classification, 259
Soil compaction, 251
Soil consolidation, 1
Soil structure, 594
Soil-structure interaction, 351
Specifications, 331
Stiffness, 452
Storm surges, 407
Strain, 34
Stress analysis, 384
Structural engineering, 626
Structural failures, 507, 574
Subsurface investigations, 170
Swelling, 45, 112, 193

Time factors, 471
Turbines, 351

United States, 522

Velocity, 292, 417
Vibration, 417
Voids, 34, 133

Walls, 277, 574
Waterfront structures, 507

Author Index

Page number refers to first page of paper

Abu-Farsakh, Murad Y., 259
Arango, Ignacio, 540

Bachus, Robert C., 594
Badri, Dhuruva, 366
Bea, Robert G., 574
Bloomquist, David, 366
Briaud, Jean-Louis, 407
Bullock, Paul J., 471

Chen, Hamn-Ching, 407
Cowles, Gary D. E., 507
Crapps, David K., 112, 331,417

Décourt, Luciano, 452

Fellenius, Bengt H., 384
Fioravante, Vincenzo, 206
Freed, David L., 626
Fuchsberger, Martin, 616

Garlanger, John E., 559
Giretti, Daniela, 206

Handy, Richard L., 251
Hayes, John A. (Jack), 636
Hussein, Mohamad H., 312

Jafarzadeh, Fardin, 522
Jamiolkowski, Michele, 206

Kuhns, Gary L., 489

Lai, Peter, 366
Laier, James E., 507
Landva, Arvid, 230
Leto, Thomas J., 559
Lewis, Michael R., 540
Likins, Garland, 312
Lutenegger, Alan J., 193

Marchetti, Diego, 292
Marchetti, Silvano, 292
McHood, Michael D., 540
McVay, Michael, 366
Mitchell, James K., 433
Monaco, Paola, 292

Navarrete, Fernando, 277

Rausche, Frank, 312
Reddy, D. V., 277
Reese, Lymon C., 351
Roberts, Keith D., 594

Schmertmann, Gary R., 594
Schmertmann, John H., 1, 34, 45, 49, 82, 92, 95, 112, 130, 133, 149, 163, 170, 177, 181, 183, 653
Soleimanbeigi, Ali, 522

Totani, Gianfranco, 292
Tumay, Mehmet T., 259

Wang, Shin-Tower, 351
White, Mark E., 507
Woods, Richard D., 417

Zhang, Zhongjie, 259